BIOTECHNOLOGY IN AGRICULTURE, INDUSTRY AND MEDICINE

HANDBOOK OF CHITOSAN RESEARCH AND APPLICATIONS

BIOTECHNOLOGY IN AGRICULTURE, INDUSTRY AND MEDICINE

Additional books in this series can be found on Nova's website under the Series tab.

Additional E-books in this series can be found on Nova's website under the E-books tab.

BIOTECHNOLOGY IN AGRICULTURE, INDUSTRY AND MEDICINE

HANDBOOK OF CHITOSAN RESEARCH AND APPLICATIONS

RICHARD G. MACKAY
AND
JENNIFER M. TAIT
EDITORS

Nova Science Publishers, Inc.
New York

For permission to use material from this book please contact us:
Telephone 631-231-7269; Fax 631-231-8175
Web Site: http://www.novapublishers.com

Additional color graphics may be available in the e-book version of this book.

LIBRARY OF CONGRESS CATALOGING-IN-PUBLICATION DATA

Handbook of chitosan research and applications / editors, Richard G. Mackay and Jennifer M. Tait.
p. ; cm.
Includes bibliographical references and index.
ISBN 978-1-61324-455-5 (hardcover)
1. Chitosan. I. Mackay, Richard G. II. Tait, Jennifer M.
[DNLM: 1. Chitosan--pharmacology. 2. Chitosan--therapeutic use. QU 83]
TP248.65.C55H36 2011
573.7'74--dc22

2011010933

Published by Nova Science Publishers, Inc. † New York

Contents

Preface		**ix**
Chapter 1	Electrodialytic Optimization and Rationalisation of Chitosan Transformation in Bioactive Oligomers *Laurent Bazinet*	**1**
Chapter 2	Chitosan and Chitosan Derivatives: A Spotlight on Cellular Interactions *Noha M. Zaki*	**27**
Chapter 3	Chitosan and Derivatives–Behavior in Dispersed State and Applications *A. Krim Azzouz*	**51**
Chapter 4	Chitosan as a Binding Agent for Modification of Endotoxin Biological Activity *V. N. Davydova, T. F. Solov'eva, G. A. Naberezhnykh, V. I. Gorbach and I. M. Yermak*	**85**
Chapter 5	Radiation Crosslinked Chitosan Grafted with Polyaniline *A. T. Ramaprasad and Vijayalakshmi Rao*	**103**
Chapter 6	Bio-applications of Chitosan *Magdy Elnashar*	**139**
Chapter 7	Chitosan-Induced Modulation of Signal Pathways for Value-Addition to Horticultural Commodities *G. Manjunatha, V. Lokesh, Nandini P. Shetty and N. Bhagyalakshmi*	**163**
Chapter 8	Chitosan Edible Films and Coatings – A Review *Henriette M.C. Azeredo, Douglas de Britto and Odílio B. G. Assis*	**179**
Chapter 9	Encapsulation of Functional Food Ingredients Using Chitosan-Based Materials *Lu-E Shi and Zhen-Xing Tang*	**195**

Chapter 10 Chitosan Nanoparticles and Nanofibers: Preparation and Application for Enzyme Immobilization **211**
Zhen-Xing Tang, Lu-E Shi and Qing-Bin Guo

Chapter 11 Research Progress on the Preparation and Application of Amphoteric Chitosan **227**
Hu Yang, Yaobo Lu and Rongshi Cheng

Chapter 12 Chitosan Reduces Dry Rot of Potato Tuber: Fungistatic Effects and Induce Resistance **241**
Yongcai Li, Yang Bi, Yonghong Ge, Yi Wang and Di Wang

Chapter 13 The Use of Chitosan in Surgery: Prevention of Adhesions **259**
Neusa Margarida Paulo and Ângela Maria Moraes

Chapter 14 Solid-Phase Production and Modification of Chitosan under Conditions of Shear Deformation **271**
S. Z. Rogovina

Chapter 15 The Use of Chitosan in Basic Studies and Clinical Application in Urinary Bladder Epithelium **289**
Andreja Erman and Peter Veranic

Chapter 16 Functionalization and Applications of Chitosan **301**
Congming Xiao

Chapter 17 The Unusual Bell-Like Dependence of the Activity of Chitosan against Penicillium Vermoesenii on Chitosan Molecular Weight **315**
Vladimir Tikhonov, Evgeniya Stepnova, Sergey Lopatin, Valery Varlamov, Alla Ilyina and Igor Yamskov

Chapter 18 Chitosan-Silicate Hybrids via Sol-Gel Method for Scaffold Applications **327**
Yuki Shirosaki, Akiyoshi Osaka, Satoshi Hayakawa and Kanji Tsuru

Chapter 19 Application of Chitosan in Environmental Engineering **337**
Ashura A. Dulazi and Hui Liu

Chapter 20 Lipase Immobilization in Chitosan for Use in Non-Aqueous Media Catalysis **345**
Carlos E. Orrego and Natalia Salgado

Chapter 21 Effect of Chitosan on the Cells of Pea Leaf Epidermis **379**
Vitaly D. Samuilov, Dmitry B. Kiselevsky, Elena V. Dzyubinskaya, Julia E. Kuznetsova and Lev A. Vasil'ev

Chapter 22 Chitosan as Elicitor of Health Beneficial Secondary Metabolites in In Vitro Plant Cell Cultures **389**
Maura Ferri and Annalisa Tassoni

Chapter 23 Effect of Chitosan on the Production of Secondary Metabolites in Embryogenic Cultures of Nutmeg, *Myristica Fragrans* Houtt **415**
R. Indira Iyer, G. Jayaraman and A. Ramesh

Chapter 24 Whole-Cell Biocatalyst for Utilization of Chitosan by Yeast Cell Surface Engineering of Chitosanase **425**
Danya Isogawa, Kouichi Kuroda and Mitsuyoshi Ueda

Chapter 25 Amide Bond Formation During Oxidative Destruction of Chitosan **435**
Yu. I. Murinov, A. R. Kuramshina, R. A. Hisamutdinov and N. N. Kabal'nova

Chapter 26 Chitosan Microspheres **453**
Sonia Touriño, Tzanko Tzanov, N. Skirtenko, U. Angel and Aharon Gedanken

Index **471**

PREFACE

Chitosan is a partially deacetylated derivative of chitin, a natural polysaccharide extracted from crustaceans, insects and certain fungi. Owing to its unique properties such as biodegradability, biocompatability, biological activity and capacity of forming polyelectrolyte complex with anionic polyelectrolytes, chitosan has been widely applied in the food and cosmetics industry, as well as the biomedical field in relation to tissue engineering, and the pharmaceutical industry relating to drug delivery. This handbook gathers current research from around the globe in the study of chitosan and its applications.

Chapter 1 - Chitosan is a partially deacetylated derivative of chitin, a natural polysaccharide extracted from crustaceans, insects and certain fungi. Since chitosan is biodegradable, non-toxic and biocompatibile in animal tissues, it is widely used in the medical industry in products ranging from burn dressings to drug delivery capsules. Much attention has been paid to convert chitosan to functional chito-oligomers, since these low molecular weight saccharides exhibit bioactivities. Chitosan oligomers are oligosaccharides composed of β-(1-4) linked glucosamine residues. They exhibit pharmaceuticals and biological properties such as antimicrobial activity, antitumor and immunostimulating activity, and a protecting effect against phytopathogens. Several methods were proposed for preparing chitosan oligomers, including chemical hydrolysis with inorganic acid and enzymatic hydrolysis with chitosanolytic enzymes, particularly chitosanase. However, addition of salts occurs with chitosan acidification because of the use of chemical acid, resulting in a decrease of purity of the oligomers. Furthermore, it has also been demonstrated that the main physiological activities and nutraceutical properties of chitosan oligomers depend clearly on their molecular weight and chain length. However, in the two main methods usually used in the industry for chitosan oligomer production, the final product is a mixture of molecules of different molecular weights and contains minerals. In this chapter, the main methods for preparing chito-oligomers and their biological activities will be first described. After that, recent results on electrodialysis with bipolar membrane (EDBM) and with filtration membrane (EDFM) processes for the preparation of chito-oligomers from chitosan will be presented. EDBM is an innovative technology for acid and base generation using protons and hydroxyl ions produced by water dissociation at the interface of bipolar membranes. This electromembrane technology coupled to EDFM process would complete the production of purified oligomers. Indeed, EDFM is a hybrid process, combining conventional electrodialysis and ultrafiltration membranes. This technology was demonstrated to allow the separation of oligomers according to their molecular weights from an oligomer mixture.

Chapter 2 - Chitosan is one of the most sought-after polymers for biomedical and drug delivery applications (>1100 articles and > 700 patent applications published in 2009). It is, therefore, fundamental to understand and control its interactions with cells, focusing on e.g. cytotoxicity and cellular access issues. Chitosan based nanosysetems are capable of delivering their active payloads, such as low MW drugs, genetic material or other complex biopharmaceutical agents intracellularly. The surface properties of the polymer/nano-carriers play a crucial role in determining their interactions with cells and, as a result, the possibility and the mechanism of intracellular transport, thereby significantly affecting the efficacy of encapsulated drugs. Physical contact with the cell membrane or cell surface interactions, initiate a number of different internalisation pathways. Endocytosis is the formation of endocytic vesicles that have engulfed the polymer and the subsequent trafficking of these endosomes through the cell. Once inside the cell, the intracellular fate of the endosomal contents is an important determinant of successful delivery of the cargo. Moreover, cytotoxicity will rely on either membrane damage with leakage of intracellular components or a subtle decrease in metabolic activity of cells. In this chapter, methods giving essential information about the interactions of nanoparticles with cells grown in culture are compiled with a focus on the cytocompatibility issue commonly tested by the formazan assay and the cellular access. One limitation is the inability to test high concentrations of polymer/nanoparticles. We have developed a modification that allows testing elevated amounts of nanocarriers with particular awareness for physiological pH and osmolarity. Concerning cellular access, this can be observed and quantified employing fluorescently labeled chitosan, and trypan blue as an effective quencher of any extracellular fluorescence.

Chapter 3 - The chemical behavior of chitosan in dispersed state determines most of its applications, and the mere presence of chitosan in the cuttlefish bone structure and its ability to combine with limestone in the crustaceans' shells accounts for most of its physicochemical features. This natural compound exhibits a strong but specific tendency to coagulation-flocculation in aqueous media, making it to be suitable for water treatment and other clotting processes. Unlike conventional aluminum salts, such as alum or other synthetic polyelectrolytes, used so far for this purpose, chitosan offers more economical and ecological advantages, being a natural, biodegradable and nontoxic product. In addition, chitosan does not generate harmful or toxic by-products, and produces less sludge. The latter are fully recyclable, because they do not contain aluminum, and, thereby, do not present risks to human health.

Chapter 4 - The ability of chitosan to form specific complexes with polyanions opens up broad opportunities for its application as a drug and gene delivery vector as well as a constituent component of biospecific adsorbents and composites. By virtue of its polycationic origin, chitosan can bind to endotoxin (LPS) and can be selected as a specific endotoxin ligand.

Date about the interaction of chitosan and its derivatives with enterobacterial LPS obtained by using of a broad range of advanced methods and techniques are considered. The process of forming the stable complexes chitosan with LPS, their stoichiometry and morphology are described. The influence of a number of factors (temperature, ionic strength, concentration) as well as the structure of LPS, chitosan and its derivatives on the binding process is discussed. Along with the binding of chitosan with the isolated LPS is considered the interaction of chitosan with the endotoxin in the bacterial cell. The three-dimensional structure models of complexes of LPS and chitosans and their derivatives are presented.

Modification of biological activities of endotoxin by chitosan in experiments *in vitro* and *in vivo* is considered. The protective effect of chitosan on the damaging effect of endotoxin in mice with bacterial intoxication is shown. The mechanisms of modulation of biological activity of LPS by chitosan are discussed.

Chapter 5 - The chapter deals with the crosslinking of crosslinked chitosan using 8MeV electron beam and γ-rays and grafting of polyaniline on to the crosslinked chitosan. The crosslinked chitosan is characterized by dissolution, Fourier transform infrared spectroscopy (FTIR), X-ray diffraction (XRD), scanning electron microscopy (SEM) and differential scanning colorimetry (DSC). Mechanical properties such as elastic modulus and hardness increase with increase in the irradiation dose due to the increase in degree of crosslinking. E-beam crosslinked chitosan shows better mechanical properties compared to gamma irradiated samples.

Conducting polymer polyaniline has been grafted onto crosslinked chitosan to improve the processsibility and mechanical properties of polyaniline. Chitosan grafted with polyaniline has got potential applications such as biosensors, actuators, polymer battery electrode etc. But for these applications grafted polyaniline should be insoluble in the whole range of pH. Hence the chemical grafting of polyaniline onto the radiation crosslinked chitosan is carried out. Grafted polymer is characterized by dissolution, swelling, UV–vis–NIR spectroscopy, FTIR, XRD, SEM, TGA, DC conductivity and nanoindentation studies. From TGA it is observed that grafted polymer is thermally stable. DC conductivity of the grafted polymer is improved after doping with 1M HCl. The change in volume conductivity is from 10^{-11} to 10^{-5} S/cm and surface conductivity from 10^{-10} to 10^{-2} S/cm

The conductivity, mechanical properties and stability of grafted polymer under ambient condition are sufficient to make it an interesting material for diverse applications. Yang et al. (1989) has reported polyaniline grafted chitosan with highest surface conductivity of 10^{-2} S/cm. But they were able to get flexible film of graft polymer having surface conductivity of maximum10^{-4} S/cm. In the present study flexible film of polyaniline grafted radiation crosslinked chitosan with surface conductivity of 10^{-2} S/cm is obtained. Present graft polymer can be used for applications, such as sensors and electrodes of polymeric batteries, since it is insoluble in whole range of pH.

Chapter 6 - Efficient commercial carriers suitable for the immobilization of enzymes are economically expensive. Contrarily, the immobilization techniques would enable the reusability of enzymes for tens of times, thus significantly reducing both enzyme and product costs.

To prepare a novel carrier for enzyme immobilization it is, accordingly advantageous, that the utilizable starting materials are among those already, permitted for pharmaceutical or food industrial use. Carrageenans and chitosans are two commercially available polysaccharide families, belonging to this category having in addition, diverse features and reasonable costs. Unfortunately, chitosan is not individually manipulated, as it has low physical/mechanical stability, while carrageenan, in addition to its low thermal stability, is lacking the active functionalities required to covalently bind enzymes.

In their laboratory, the authors have succeeded in assembling a combination of the two biopolymers, in such a way to gain the benefits of both, such as the abundance of active functional amino (NH2) groups of chitosans, and being shapeable, while having good thermal stability for carrageenan gel.

Carrageenan gels were treated with protonated polyamines "chitosan" to form a polyelectrolyte complex, which was then followed by glutaraldehyde treatment. The newly developed carrier revealed an outstanding gel's thermal stability as it was augmented from 35 to 95 °C. The novel gel incorporating the aldehydic chemical functionality has been efficaciously manipulated to covalently immobilize enzymes.

FTIR techniques, as well as Schiff's base color development were used to elucidate the structure of the newly grafted carrier biopolymer. FTIR, equally confirmed the incorporation of the aldehydic carbonyl group to the carrageenan coated chitosan at via tracing the band 1720 cm^{-1}. Interestingly, the operational stability retained 97% of the enzyme activity, even after 15 time uses. In brief, the newly developed immobilization methodology is simple and the carriers are economically favored vis à vis the commercially available Eupergit C® or Agaroses®, yet effective and utilizable for the immobilization of other enzymes.

Chapter 7 - Once known as a mere surfactant, chitosan is now recognized as a marvelous signal-inducer capable of eliciting a wide array of secondary and primary metabolic networks in plants, imparting tremendous improvements to field crops and harvested flowers, fruit and vegetables. The presence of amino group in C_2 position of chitosan provides major functionality allowing various modifications to meet the needs of horticultural, biotechnological and food applications. Jasmonic acid (JA), calcium (Ca^{++}), ethylene (Et), abscisic acid (ABA), nitric oxide (NO) etc., in plants act as chief signal molecules modulating the networks involved in regulation and turn-over of shikimates and terpenoids implicated in defense, and polyamine pools needed for supporting most of such biochemical pathways. Such chitosan-induced regulations point towards developing better synergies for desirably eliciting specific type of biochemical changes. Chitosan-induced JA alteration also modulates genes of ethylene-responsive networks that affect senescence and other biochemical, sensorial and keeping quality of fruits and vegetables. Chitosan induced enzymatic alterations accomplished through different signal transducers alter shikimic acid pathways modulating the transcription activation and translational regulations of phenylalanine ammonia-lyase (PAL), tyrosine ammonia-lyase (TAL), peroxidase (POD), polyphenol oxidase (PPO), catalase (CAT) etc., resulting in conferring beneficial changes to fruit and vegetables such as conversions of phenolic compounds to more desirable ones, increase in soluble solids, ascorbic acid accumulation, pigment biosynthesis, while simultaneously eliciting phytoalexins, defense compounds and anti-oxidants imparting resistance against decay of fresh harvests. The present review highlights progress made recently in chitosan-induced elicitation of signal networks for accomplishing desirable changes in horticultural commodities.

Chapter 8 - The use of conventional food packaging materials is usually effective in terms of barrier. On the other hand, their non-biodegradability creates serious environmental problems, motivating researches on edible biopolymer films and coatings to at least partially replace synthetic polymers as food packaging materials. Chitosan is a biopolymer obtained by N-deacetylation of chitin, which is the second most abundant polysaccharide on nature after cellulose. Chitosan forms clean, tough and flexible films with good oxygen barrier, which may be employed as packaging, particularly as edible films or coatings, enhancing shelf life of a diversity of food products. One of the main applications of the film forming properties of chitosan is to develop coatings to decrease respiration rates and moisture loss in fresh fruits and vegetables. Chitosan has some advantages over other biomaterials because of its antimicrobial activity against a wide variety of microorganisms. The main purpose of this review is to summarize the literature on development and applications of chitosan films and coatings, and their properties/performance as affected by intrinsic factors (molecular weight and degree of deacetylation), chemical/physical treatments, and incorporation of nanoreinforcements.

Chapter 9 - Consumers prefer food products that are tasty, healthy and convenient. Encapsulation, a process to entrap active agents into particles, is an important way to protect them during processing does not affect their availability in the body. For example, this technology may allow taste and aroma differentiation, mask bad tasting or bad smelling components, stabilize food ingredients and/or increase their bioavailability. Chitosan is a natural polysaccharide prepared by the N-deacetylation of chitin. It has been widely used in food industrial and biomedical aspects, including in encapsulating active food ingredients, in enzyme immobilization, and as a carrier for controlled drug delivery, due to its significant biological and chemical properties such as biodegradable, biocompatible, bioactive, and polycationic. In this chapter chitosan-based carriers used to encapsulate for food active ingredients are reviewed.

Chapter 10 - Enzyme immobilization has attracted continuous attention in the fields of fine chemical engineering and bio-chemical engineering. The performance of immobilized enzyme largely depends on the supports. Compared with traditional enzyme immobilized carriers, nano-structured carriers show many advantages including surface area, mass transfer resistance, and effective enzyme loading. Nano-structured carriers are believed to be able to retain the catalytic activity as well as ensure the immobilization efficiency of enzyme to a high extent. Various nanomaterials, such as nanoparticles, nanofibers and nanoporous matrices, have shown potential for revolutionizing the preparation and application of enzyme immobilized carriers. Chitosan is a natural polysaccharide prepared by the N-deacetylation of chitin. It has been widely used in many industrial and biomedical aspects, including in wastewater treatment, in enzyme immobilization, and as a carrier for controlled drug delivery, due to its significant biological and chemical properties such as biodegradable, biocompatible, bioactive, and polycationic. This chapter mainly discussed the recent advances in using chitosan nanoparticles and nanofibers supports as hosts for enzyme immobilization. Preparation of chitosan nanoparticles and nanofibers, the work the authors achieved in their group on this topic was introduced. Some problems encountered with chitosan nanoparticles and nanofibers carriers for enzyme immobilization were discussed together with the future prospects of such systems.

Chapter 11 - Chitosan is one of the high-performance polysaccharide materials, which has been already applied widely in various fields such as biotechnology, biomedicine, wastewater treatment and food process. However, some disadvantages, such as inactive chemical properties and bad solubility, always limit chitosan in practical application. In order to improve the performances of chitosan, amphoteric chitosan, containing both cation and anion on the chain backbone, have been prepared by chemical modifications, such as etherification, esterification and graft copolymerization. The amphoteric chitosan derivatives present some interesting features and distinguishing performances. Therefore, based on the ionic strength of groups on the amphoteric chitosan, the preparation and applications of amphoteric chitosan have been reviewed in this paper.

Chapter 12 - Potato dry rot caused by *Fusarium sulphureum* is one of the most important postharvest diseases restricting potato production and incidence of disease is up to 30% during storage in Gansu, China. Fungistatic characteristic and induced disease resistance of chitosan on dry rot of potato tuber (cv. Atlantic) were studied. *In vitro* tests, spore germination and mycelial growth of *F. sulphureum* were inhibited by chitosan treatment and the inhibitory effect was highly correlated with chitosan concentration. Morphological changes such as intertwisting hyphal, distortion, and swelling with excessive branching were observed by scanning electron microscopy (SEM) observation. Transmission electron microscopy (TEM) observation further indicated the ultrastructural alterations of hyphae. These changes included abnormal distribution of cytoplasma, non-membraneous inclusion bodies assembling in cytoplasm, considerable thickening of the hyphal cellular walls, and very frequent septation with malformed septa. Application of chitosan at higher concentration caused serious damage to fungal hyphae, including cellular membrane disorganisation, cell wall disruption, and breaking of inner cytoplast. New hyphae (daughter hyphae) inside the collapsed hyphal cells was often detected in the cytoplasm of chitosan-treated hyphae. Chitosan at 0.25% increased the activities of peroxidase (POD) and polyphenoloxidase (PPO), and the contents of flavonoid compounds and lignin in tissues. Increased activities of β-1, 3-glucanase (GLU) and phenylalanine ammonialyase (PAL) were observed, but there were no significant differences between the treated and the control. This suggests chitosan is promising as fungicide to reduce potato tuber disease and to partially substitute for the utilization of synthetic fungicides in fruit and vegetables.

Chapter 13 - Adhesions consist on fibrotic union among tissues and organs, as a consequence of trauma and/or tissue ischemia, infection, and foreign body reactions, among other causes. Even though the term adhesion has been classically employed to designate peritoneal fibrosis formation, it also refers to fibrosis reaction due to ophthalmic, orthopedic, cardiovascular (pericardium), neurological (dura mater or dura) and gynecological surgical procedures. The experimental research on adhesion formation has often been associated to the implantation of synthetic biomaterials as polypropylene meshes for the treatment of patients with hernia. It is known that when a macroporous biomaterial as a polypropylene mesh is put into direct contact with visceral peritoneum there is a risk of triggering complications as adhesions over the mesh, what may lead to intestine obstruction processes or fistulae development. Chitosan is a highly biocompatible polysaccharide that has drawn attention lately regarding adhesion prevention. This biopolymer and some of its derivatives can be employed as gels or films for this purpose, and some of the interesting results attained with their use will be discussed in this chapter.

Chapter 14 - The development of new methods for production of chitosan and its derivatives is among the most important problems of the modern polysaccharide chemistry. The solid-phase processes carried out under conditions of joint action of high pressure and shear deformation is the promising methods of chitin and chitosan modification. The advantage of this method is a significant reduction of the process time, reagent consumption, and wastewater amounts that leads to the improvement of the economical and ecological parameters of the process. The reactions were carried out in a two-screw extruder, where the joint action of high pressure and shear deformation on material is realized. The properties and fractional composition of chitosan prepared by solid-phase deacetylation of chitin, as well as chitosan interaction with stearic, oxalic, malonic, and succinic acids and with phthalic, succinic and maleic anhydrides under conditions of shear deformation were investigated. Also, as a result of chitin interaction with monochloroacetic acid in the presence of sodium hydroxide under these conditions, carboxymethyl ether of chitosan was obtained. Based on the data of elemental analysis, IR spectroscopy, viscometry, potentiometry, and other experimental techniques, the formation of the corresponding chitosan derivatives is corroborated and the possible mechanisms of the reactions of modification are discussed.

Chapter 15 - Chitosan is a cationic polysaccharide mainly used as an absorption enhancer. Its known low toxicity for most tissues gave it a wide applicability in the local delivery of pharmaceuticals to selected tissues. The positive charge of the chitosan enables it to attach to the mainly negatively charged proteoglycans that cover the apical surface of most epithelial cells and allow a gradual release of the drugs trapped in this polysaccharide cage. The intravesical application of chitosan to urinary bladder recently revealed an unpredicted deteriorating effect on superficial epithelial cells. This discovery broadens the application of chitosan to basic research of urinary bladder regeneration and gave it a possible application in treating chronic bacterial cystitis and detachment of superficial urothelial tumors. The presented commentary examines our current understanding of the effects of chitosan on urinary bladder epithelial cells in normal rodent bladder, with its concentration and time dependent effects including reversible dysfunction of tight junctions at mildest conditions, over selective removal of superficial epithelial cells and at the severest conditions the complete eradication of the urinary bladder epithelium down to the basal lamina. Its local and very controllable effects make it a perfect model system for the studies of urothelial tissue regeneration without involving inflammatory side effects. The stimulated exfoliation of the bladder epithelium provides a new and very intriguing model for the removal of intracellular pools of uropathogenic E.coli responsible for the majority of recurrent uroinfections and possibly enables a noninvasive removal of superficial bladder tumors predictably with fewer side-effects in comparison to classical treatment.

Chapter 16 - Chitosan, the second abundant naturally occurring polysaccharide, is an N-deacetylated derivate of chitin. Owing to its unique properties such as biodegradability, biocompatibility, biological activity, and capacity of forming ployelectrolyte complex with anionic polyelectrolytes, chitosan has been widely applied in food, cosmetics, biomedical, pharmaceutical and the other fields. On the other hand, chitosan has its inherent limitations. Thus, many attempts to render more functions to chitosan via chemical or physical modifications have been made and reported. Herein, we describe several means to functionalize chitosan in order to improve its water-solubility or provide it some useful properties. The applications of these chitosan derivates are also involved.

Chapter 17 - Chitosan has been reported to be a non-toxic, biocompatible antibacterial, antifungal and insecticidal agent. In many publications it has been shown that the biocidal activity of chitosan can grow or reduce with the reduction of chitosan molecular weight (MW). In some experiments it has been shown that the antimicrobial activity of chitosan does not depend on MW or even controversially differs for Gram-positive and Gram-negative bacteria. In this work the fungistatic efficacy of a set of low polydispersity chitosans varying in molecular weight from 1.2 to 90.0 kDa towards the filamentous palm parasitic fungus *Penicillium vermoesenii* has been studied *in vitro*. It has been shown for the first time that the activity of chitosan against the fungus has unusual bell-like dependence on chitosan molecular weight so that chitosans having the molecular weight between 5÷10 kDa possess the highest fungistatic activity while chitosans with both lower and higher molecular weights are significantly less active.

Chapter 18 - Novel strategies for regenerating or reconstructing damaged bone tissues are of urgent necessity, because of limitations in conventional therapies for trauma, congenital defects, cancer, and other bone diseases. The tissue engineering approach to repair and regeneration is founded upon the use of biodegradable polymer scaffolds, which may manipulate bone cell functions, encourage the migration of bone cells from border areas to the defect site, and provide a source of inductive factors to support bone cell differentiation. During the past decades, scaffolds from natural biodegradable polymers such as collagen, gelatin, fibrin, and alginates, or synthetic biodegradable polymers such as polyglycolide, polylactides, and copolymers of glycolide with lactides have been extensively explored. On the other hand, it has been confirmed that the bonelike apatite layer, deposited spontaneously on the biomaterial surfaces, can enhance osteoconductivity. The presence of such bone-like apatite layers is also believed to be a prerequisite to conduction of osteogenic cells into various porous scaffolds or onto the surface of bioactive glasses. The formation of such bone-like apatite is favored by the cooperative behavior of a hydrated silica or titania gel surface with many Si-OH or Ti-OH groups, and calcium ions to be released into the body fluid when implanted. Thus, hybrid materials derived from the integration of biodegradable polymers with bioactive inorganic species may construct a new group of scaffolds appropriate for tissue engineering. Moreover, tissue engineering approach depends on the use of porous scaffolds that serve to support and reinforce the regenerating tissue. Controlled porous structures of these scaffolds allow cell attachment and migration, tissue generation, or vascularization. We have been studying the synthesis of novel chitosan-silicate hybrids derived from the integration of chitosan and γ-glycidoxypropyltrimethoxysilane (GPTMS). We introduce some of the results on this hybrid for biomedical application.

Chapter 19 - Chitosan, the deacetylated form of chitin, has become a material of high potential in different fields such as environment, health, agriculture and medicine. The present review provides an insight of different applications of this biopolymer in the environment engineering especially in wastewater treatment, support for enzyme and microorganisms immobilization. Past and recent investigations are discussed in this paper.

Chapter 20 - The most recent efforts in the development of cleaner sustainable chemistry are being driven by the production of raw materials from biofeedstocks using white biotechnology or industrial biotechnology instead of the conventional petrochemical technology.

Lipases are versatile biocatalysts and find applications in organic chemical, agrochemical, food, fine chemicals and pharmaceutical processing. Chitosan is a biopolymer prepared by deacetylation of chitin— a renewable resource obtainable from the wastes of the seafood industry or fungal sources— that has been used as a matrix for immobilization of enzymes for use in aqueous and non aqueous media because it has suitable properties such as insolubility, mechanical stability and rigidity in organic solvents, high affinity to water and proteins, availability of reactive functional groups, biodegradability, regenerability, and ease of preparation in different geometrical configurations. In this chapter, a comprehensive survey is made of different aspects of lipase activation and immobilization in chitosan hydrogels for catalysis in non-aqueous media. Topics on water activity, polarity of solvents, engineering support characteristics by physical-chemical modification of chitosan and, with special emphasis, surface characterization of chitosan supports and chitosan-lipase systems with differents techniques, are also discussed.

Chapter 21 - The effect of chitosan on epidermal and guard cells in the epidermis isolated from pea leaves was investigated. Chitosan induced the condensation and marginalization of chromatin and the subsequent destruction of nuclei in epidermal, but not in guard cells. The addition of H_2O_2 produced a similar effect on the cells. Nitroblue tetrazolium, pyruvate and mannitol, possessing antioxidant properties, inhibited the destruction of nuclei caused by chitosan or H_2O_2. The chitosan-induced destruction of epidermal cell nuclei was prevented under anaerobic conditions. Diphenylene iodonium and quinacrine, inhibitors of the NADPH oxidase of the plasma membrane, suppressed the destruction of nuclei induced by chitosan. Chitosan caused the generation of reactive oxygen species and, similar to H_2O_2, impaired the permeability barrier of the plasma membrane in guard cells. The data obtained demonstrate that chitosan and H_2O_2 produce similar effects on plant cells. Chitosan initiates the production of reactive oxygen species which induce programmed cell death, presumably by activating the NADPH oxidase of the plasma membrane.

Chapter 22 - Chitosan is an important structural component of the cell wall of several plant pathogen fungi. In plants, it can be applied *in vivo* as antimicrobial agent against fungi, bacteria and viruses, and *in vitro* can be used as elicitor of defense mechanisms. Its degrees of polymerization and of acetylation strongly influence the elicitation activity. Chitosan seems to be particularly effective in increasing the content of a large spectrum of plant polyphenols, such as stilbenes (i.e. resveratrol) and flavonoids (i.e. anthocyanins), very important antioxidants for plants, animals and humans. All these compounds, can be ingested by humans through daily diet and are absorbed in the small intestine, providing health benefits, such as increase of the antioxidant capacity of blood and potential prevention of cancer and cardiovascular diseases.

In this chapter the authors present a comprehensive overview of the applications and effects of chitosan in plant cell cultures. Although the exact mechanism of chitosan action in plants is still unknown, different hypotheses have been proposed. Plant cell cultures have been viewed as a promising alternative to whole plant extraction for obtaining valuable metabolites, with the elicitation of the interesting compounds and the scale-up process as key points. In particular, in grape cell suspensions, the synthesis and release of several metabolites, were up-regulated by chitosan treatment. Concomitantly, the expression levels of some enzymes of the phenylpropanoid pathway as well as those of pathogen related proteins, together with the *de-novo* synthesis and/or accumulation of different proteins, were promoted. With the aim to further ameliorate the polyphenol yield, batch and fed batch grape cell fermentations were optimized also with the addition of chitosan, which proved to be highly effective in improving metabolite production in bioreactors.

Chapter 23 - Nutmeg (*Myristica fragrans* Houtt.) is a tropical evergreen dioecious tree which yields the spices, nutmeg (endosperm) and mace and has a long history of medicinal usage. It has a rich diversity of phytochemicals of commercial and therapeutic value. However it is of restricted distribution. Though nutmeg tissues are extremely recalcitrant to tissue culture technology owing to the high content of phenolics, embryogenic cultures of nutmeg were established from zygotic embryos by using media with activated charcoal. The present study investigated the effect of chitosan, a biomolecular tool with tremendous potential in biotechnology, on the production of phytochemicals by the embryogenic cultures of nutmeg using GCMS analysis. The induction of the biosynthesis of several novel metabolites by chitosan including methyl salicylate, farnesol and anthraquinones as compared with unelicited cultures is being reported . The results indicate the enhancement of the biosynthetic potential of embryogenic cultures of nutmeg by chitosan and point to its application in strategies for large scale *in vitro* production of valuable secondary metabolites from nutmeg.

Chapter 24 - Chitosan is mainly degraded by chemical methods and the degree of degradation cannot be easily controlled. On the other hand, the degree of degradation of chitosan can be mildly controlled with enzymatic degradation properties of chitosanase. Yeast cell surface engineering is an attractive strategy for molecular preparation of novel whole-cell biocatalysts that can degrade chitosan polymers. Using this technique, chitosanase from *Paenibacillus fukuinensis* could be displayed on the yeast cell surface with the retention of the enzymatic activity and its application has been widely developed. Interestingly, chitosanase from *P. fukuinensis* is a unique enzyme that belongs to glucanase family 8 and has dual hydrolysis activities of chitosanase and □-1, 4 glucanase. A mutant library of chitosanase in which the putative catalytic proton acceptors were comprehensively substituted by the other amino acids was constructed using yeast cell surface engineering. As the results of this research, we demonstrated that the amino acid residues of Glu302 and Asn312 were catalytic proton acceptors and essential for having both chitosanase and □-1, 4 glucanase activities. Utilization of chitosanase for the production of useful chitosan oligosaccharides has been made further attractive. Promotion of efficient utilization of chitosan by enzymatic degradation may contribute to the reduction of the amount of waste crustacean shell and to the remaking of functional materials with bioactivities for the construction of a sustainable society with recycling of natural resources.

Chapter 25 - The interest to investigation of the properties of nature polysaccharide chitosan is caused by this wide prevalence in the nature, low toxicology and high biological activity. The unique physicochemical properties of chitosan, its derivatives and water-soluble oligomers allow to use them as supports, prolongating the action of medicine forms. The authors found that in the reactions of the destruction and oxidation of chitosan by the different reagents (ozone, hydrogen peroxide, NaClO, $NaClO_2$, ClO_2) can be occur amide bond formation. Formation of the amide bond was shown by the methods of IR and ^{13}C NMR, potentiometric titration. The inverse gas chromatography (IGC) is a useful method to observe the interaction between polymer and solutes. In the present work the given method is applied for study of adsorption properties of chitosan and its modified analogue in relation to some *n*-alkanes and *n*-alcohols. It is shown, that ability of adsorbents to dispersive interactions with test probes is higher on the modified chitosan. The amide bond formation were observed when the chitosan oxidation occur which rupture of glycoside bond and formation of the carboxyl groups at C(6).

Chapter 26 - Over the last few decades, researchers have tried to develop systems capable of delivering active compounds to specific target sites. Microparticulate carriers, e.g. microspheres, are considered good potential oral delivery systems to provide a constant therapeutic level of the drug and, in this way, to avoid frequent administration of the solid oral dosage form. Microspheres are generally defined as small spheres made of any material and sized from one micron (one one-thousandth of a millimetre) to a few mm. Similarly, smaller spheres, sized from 10 to 500 nm are called nanospheres.

In: Handbook of Chitosan Research and Applications
Editors: Richard G. Mackay and Jennifer M. Tait
ISBN 978-1-61324-455-5

Chapter 1

ELECTRODIALYTIC OPTIMIZATION AND RATIONALISATION OF CHITOSAN TRANSFORMATION IN BIOACTIVE OLIGOMERS

Laurent Bazinet*

Institute of Nutraceuticals and Functional Foods (INAF),
Université Laval, Department of Food Sciences and Nutrition,
Pavillon Paul Comtois, Sainte-Foy (QC), Canada

ABSTRACT

Chitosan is a partially deacetylated derivative of chitin, a natural polysaccharide extracted from crustaceans, insects and certain fungi. Since chitosan is biodegradable, non-toxic and biocompatibile in animal tissues, it is widely used in the medical industry in products ranging from burn dressings to drug delivery capsules. Much attention has been paid to convert chitosan to functional chito-oligomers, since these low molecular weight saccharides exhibit bioactivities. Chitosan oligomers are oligosaccharides composed of β-(1-4) linked glucosamine residues. They exhibit pharmaceuticals and biological properties such as antimicrobial activity, antitumor and immunostimulating activity, and a protecting effect against phytopathogens. Several methods were proposed for preparing chitosan oligomers, including chemical hydrolysis with inorganic acid and enzymatic hydrolysis with chitosanolytic enzymes, particularly chitosanase. However, addition of salts occurs with chitosan acidification because of the use of chemical acid, resulting in a decrease of purity of the oligomers. Furthermore, it has also been demonstrated that the main physiological activities and nutraceutical properties of chitosan oligomers depend clearly on their molecular weight and chain length. However, in the two main methods usually used in the industry for chitosan oligomer production, the final product is a mixture of molecules of different molecular weights and contains minerals. In this chapter, the main methods for preparing chito-oligomers and their

* E-mail: Laurent.Bazinet@fsaa.ulaval.ca.

biological activities will be first described. After that, recent results on electrodialysis with bipolar membrane (EDBM) and with filtration membrane (EDFM) processes for the preparation of chito-oligomers from chitosan will be presented. EDBM is an innovative technology for acid and base generation using protons and hydroxyl ions produced by water dissociation at the interface of bipolar membranes. This electromembrane technology coupled to EDFM process would complete the production of purified oligomers. Indeed, EDFM is a hybrid process, combining conventional electrodialysis and ultrafiltration membranes. This technology was demonstrated to allow the separation of oligomers according to their molecular weights from an oligomer mixture.

I. INTRODUCTION

Chitosan is derived from chitin, which is the second more abundant biopolymer after cellulose (Shahidi *et al.*, 1999). Chitin, the raw material used for the production of chitosan, is a white polysaccharide, harsch and with an inelastic structure, presents in the cellular walls of fungi and in the exoskeletons of crustaceans (Kim and Rajapakse, 2005). Chitosan is obtained by N-deacetylation of chitin and is mainly formed of 2-amino-2-deoxy-D-glucopyranose (GlcN) units bonded in $\beta(1\rightarrow4)$: its typical degree of acetylation is lower than 35% (Ravi Kumar, 2000). Chitosan is a voluminous molecule (1000 à 2000 kDa, 5000 residues) insoluble in water (Kumar *et al.*, 2004). The chitosan solubilization, by protonation of its amine groups into the NH_3^+ form, can be carried-out by addition of mineral or organic acids. Chitosan is mainly known and used for its properties of cationic polyelectrolyte. However, its high viscosity and its low solubility at neutral pH make difficult its use in the food and pharmaceutical fields. Hence, during the last decade, new classes of compounds with different physico-chemical and biological properties had been developped : oligomers coming from the hydrolysis of chitosan.

Chitosan oligomers are oligosaccharides composed of β-(1-4) linked glucosamine residues. They exhibit pharmaceutical and biological properties such as antimicrobial activity, antitumor and immunostimulating activity, and a protecting effect against phytopathogens. Several methods were proposed for preparing chitosan oligomers, including chemical hydrolysis with inorganic acid and enzymatic hydrolysis with chitosanolytic enzymes, particularly chitosanase. However, addition of salts occurs with chitosan acidification because of the use of chemical acid, resulting in a decrease of purity of the oligomers. Furthermore, it has also been demonstrated that the main physiological activities and nutraceutical properties of chitosan oligomers depend clearly on their molecular weight and chain length. However, in the two main methods usually used in the industry for chitosan oligomer production, the final product is a mixture of molecules of different molecular weights and contains minerals.

In this chapter, the main methods for preparing chitosan oligomers as well as oligomer biological activities will be first described. After that, electrodialysis principle, the different types of membranes used in electrodialytic processes and their intrinsic characteristics will be presented. At the end, recent results on electrodialysis with bipolar membrane (EDBM) and electrodialysis with filtration membrane (EDFM) processes for the preparation of oligomers from chitosan will be highlighted and discussed.

II. CHITOSAN OLIGOMERS

Chitosan oligomers have numerous advantages in comparison with non-depolymerised chitosan. Their lower molecular weights make them soluble in water and oligomer solutions have a viscosity lower than the one of chitosan solutions. In addition, on the contrary of chitosan which acts as a fiber, oligomers are easily absorbed through the intestine, and pass quickly in the blood (Kim and Rajapakse, 2005).

2.1. Production of Oligomers

As all polysaccharides, chitosan can be depolymerised by hydrolytic agents by disrupting the glycosidic links. Chitosan degradation leads to the production of oligomers having different polymerisation degrees (PD) and specific biological properties. Amongst the chitosan oligomer preparation methods, one can mentioned : mechanical depolymerisation (Kasaai *et al.*, 2003), chemical hydrolysis (Horowitz *et al.*, 1957 ; Domard and Cartier, 1989) and enzymatic hydrolysis (Izume and Ohtakara, 1987).

2.1.1. Mechanical Depolymerisation

The mechanical depolymerisation of chitosan can be obtained by microfluidization (Kasaai *et al.*, 2003). This process implies the transfer of mechanical energy to the fluid particles under the application of a high pressure. The solution is pumped and separated in two micro-currents which knocks against one another in an interaction chamber. The chains of the chitosan polymer are then submitted to a high shearing, breaking themselves and the degree of polymerisation is consequently decreased. Microfluidization can be used for a partial depolymerisation of chitosan in order to reduce the viscosity of the final chitosan solutions. However, this method is not appropriated for the production of chitosan with low molecular masses.

2.1.2. Chemical Hydrolysis

The chemical hydrolysis of chitosan consists in heating the polymer dissolved in concentrated inorganic acid aqueous solutions (Horowitz *et al.*, 1957 ; Domard and Cartier, 1989 ; Nagasawa *et al.*, 1971 ; Defaye *et al.*, 1989 ; Allan and Peyron, 1995). Chitosan oligomers with a polymerization degree of 2 to 5 had been prepared by hydrolysis of chitosan with concentrated hydrochloric acid (35% HCl). Concentrated nitric acid has also been used to hydrolyzed chitosan leading to a mixture of oligomers with a variety of PD. An other method employing heated phosphoric acid was reported ; Yields of 10-20% of oligomers with PD from 6 to 8 were obtained (Hasegawa *et al.*, 1993). Generally speaking, the chemical hydrolysis of chitosan produces a large quantity of D-glucosamine monomers (Horowitz *et al.*, 1957), resulting in a low oligomer yield. At last, oligomers produced by chemical hydrolysis could be toxic due to the chemical residuals which are present after the reaction. The chemical hydrolysis is then not adapted to the production of biologically active oligomers.

2.1.3. Enzymatic Hydrolysis

Due to the limitations of all the chemical methods, very few industrial applications of chemical hydrolysis appeared (Bosso *et al.*, 1986) and consequently, it is necessary to develop an alternative method of chitosan hydrolysis, using more soft conditions and allowing a better monitoring of the reaction : the enzymatic hydrolysis. During enzymatic hydrolysis chitosan can be degraded by different specific or non-specific enzymes.

2.1.3.1. Specific Hydrolysis

The specific hydrolysis of chitosan is performed using the chitosanase enzyme. This last one was found in abundancy in certain varieties of micro-organims including bacteria, moulds and plants. Since the majority of these chitosanases hydrolyzes the chitosan on an endolytic way, these enzymes produce oligomers from the dimers to the octamers (PD 2-8). The process of enzymatic hydrolysis is generally speaking carried-out in batch mode, but some authors also proposed continuous systems. Amongst others, Jeon and Kim (2000) developed an ultrafiltration system coupled with an enzymatic reactor allowing the separation of chitosanase and its re-use in a continuous way. The reaction was stopped by boiling the solution during 10 minutes. The oligomers produced ranged mainly from the trimers to the hexamers (PD 3 to 6), and a low quantity of monomers and dimers was found. However, in this application, the ultrafiltration step was limited by the increase in transmembrane pressure due to the chitosan fouling of the membrane related to its high viscosity.

2.1.3.2. Non-Specific Hydrolysis

The non-specific hydrolysis can be obtained using enzymes such as chitinase or the lysozyme which degrades partially acetylated chitosan and produce hetero-oligosaccharides (Aiba, 1994a and 1994b). Many other non-specific enzymes have been studied, amongst others we find papaïn, lipase (Muzzarelli *et al.*, 1999), a mix of cellulase, amylase, proteinase (Zhang *et al.*, 1999), celloviridin (Il'ina *et al.*, 2002) and pectinase (Kittur *et al.*, 2003). The non-specific enzymes are interesting because there are less expansive than the chitosanase ; On the other hand, they must be added at relatively high concentrations while, a specific enzyme such as chitosanase demontrates a substantial activity at low concentration.

2.2.Oligomer Biological Properties

Following their decreases in polymerisation degree, numerous studies have shown that oligomers from chitosan possess specific biological properties (Table 1) : antimicrobial activity, antiviral activity, anitumoral activity, antioxidant activity ...

2.2.1. Antimicrobial Activity

The antimicrobial activity (antibacterial and antifungi) is one of the more important applications of the chitosan oligomers. The antimicrobial activity of chitosan oligomers depends mainly on their polymerisation degree (Park *et al.*, 2004a,b; Park *et al.*, 2002; Yun *et al.*, 1999), their deacetylation degree (Chung *et al.*, 2004; Tsai *et al.*, 2002) and the type of microorganisms (Gerasimenko *et al.*, 2004; Park *et al.*, 2004a,b; Uchida *et al.*, 1989).

Table 1. Some biological activities of chitosan oligomers

Activity	Applications or conditions of applications	References
Antimicrobial	Antibacterial effects : – Inhibition of *Escherichia coli* by 12 kDa oligomers – Inhibition of *Bacillus cereus* by 11kda oligomers – Reduced *Escherichia coli* by 2.47-2.84 log cycle at 1000 ppm Antifungal effects : – Inhibition of *Fusarium oxyporum* by oligomers with PD of 2-8. – Inhibition of *Phytophthora capsici*	Gerasimenko *et al.*, 2004 No *et al.*, 2002 Yalpani *et al.*, 1992 Hirano et Nagano, 1989 Xu *et al.*, 2007
Antiviral	Decrease in necroses on leaves due to different viruses. Increase immunity against influenza A and B in mouse. Inhibition of bacteriophages	Pospieszny *et al.*, 1991 Bacon *et al.*, 2000 Kochkina and Chirkov, 2000a,b
Antitumoral/ anticancer	Inhibition of Sarcoma 180 tumoral growth by oligomers with PD of 4-5 in mouse. Inhibition of tumoral cell growth by oligomers with molecular weights ranging from 1.5-5.5 kDa Chemopreventive effect of chitosan oligosaccharides against colon carcinogenesis	Qin *et al.*, 2002 Suzuki *et al.*, 1986 Nams *et al.*, 2007
Antioxydant	Neutralization of free radical (hydroxyls and superoxides) by oligomers of 1-3 kDa. High antioxidant activity of oligomers with molecular weights of 1 to 3 kDa Protective effect of oligomers against hydrogen peroxide (H_2O_2)-induced oxidative damage on human umbilical vein endothelial cells High scavenging activities on 1,1-diphenyl-2-picrylhydrazyl (DPPH), hydroxyl, superoxide and carbon-centered radicals	Park *et al.*, 2003a Kim and Rajapkse, 2005 Liu *et al.*, 2009 Je *et al.*, 2004; Chen *et al.*, 2003; Huang *et al.*, 2005
Hypo-cholesterolemic	Reduction of blood cholesterol, and increase in HDL cholesterol	Sugano *et al.*, 1988
Hypotensive	Inhibition of ACE for the control of blood pressure	Park *et al.*, 2003b
Anticoagulant	Anticoagulant effect of oligosaccharide with deacetylation degree of 90%	Park *et al.*, 2004c;
Intestinal absorption	Improvement of intestinal absorption of insulin in rats and hypoglycemic effect with hexamer	Gao *et al.*, 2008
Prebiotic	Effect on *Bifidobacterium bifidum* and *Lactobacillus sp* of oligomers with PD of 2-8	Lee *et al.*, 2002

The antimicrobial activity of chitosan oligomers would be linked to the presence of amine groups on their structure (Chen *et al.*, 2002). Three explanations had been proposed for the antimicrobial and antifungi activities : the changes in the permselective properties of the bacterial or fungi cells (Choi *et al.*, 2001; Sudharshan *et al.*, 1992), the chelation of metal and/or the blocking of the ARN transcription following the adsorption of oligomers on the bacterial ADN (Kim *et al.*, 2003).

2.2.2. Antiviral Activity

The oligomers were also tested for suppressing viral infections in plants (Pospieszny *et al.*, 1991) and mices (Bacon *et al.*, 2000) as well as for the inhibition of bacteriophages (Kochkina & Chirkov, 2000a,b). The reaction mechanism for antiviral activity is not clearly identified and contradictory observations had been reported.

2.2.3. Antitumoral Activity

The chitosan oligomers would also have an effect on the inhibition of tumoral cell growth (Qin *et al.*, 2002). The results reported in the litterature on the inhibition of tumoral cell growth suggest that this activity would not be due to the direct elimination of these tumoral cells but to the higher production of lymphokins and consequently the reinforcement of the immunitary system (Tokoro *et al.*, 1998; Nishimura, *et al.*, 1986). As previously mentioned for the antimicrobial activity, the antitumoral activity of oligomers depends on their structural characteristics, molecular weights and deacetylation degrees (Jeon & Kim, 2002). It was observed that oligomers having molecular weights from 1.5 to 5.5 kDa could avoid the growth of tumoral cells. Furthermore, inhibition of tumoral cells from 55 to 100% on mices was observed with chitohexaose (hexamers) (Suzuki *et al.*, 1986).

2.2.4. Antioxidant Activity

Numerous studies were published on the antioxidant activity of chitosan oligomers (Chiang *et al.*, 2000; Park *et al.*, 2003a,b; Yin *et al.*, 2002; Xie *et al.*, 2001; Huang *et al.*, 2005; Guzman *et al.*, 2003). Chitosan oligomers are natural antioxidants allowing the neutralization of free radicals (Park *et al.*, 2003a) or the chelation of Fe^{2+} ions (catalyst of oxidation reaction) (Huang *et al.*, 2005). The polymerisation and deacetylation degrees of the chitosan oligomers are the main factors of their antioxidant activity. A high antioxidant activity was correlated with molecular weights of 1 to 3 kDa and a deacetylation degree of 90% (Kim and Rajapakse, 2005).

2.2.5. Other Biological Activities

The oligomers have a hypocholesterolemic effect, mainly due to their cationic nature responsible for the fixation of negatively charged lipids (Sugano *et al.*, 1988). The chitosan oligomers have also been used to inhibit the enzyme responsible for the conversion of angiotensine I which has an effect on the regulation of the blood pressure in mammals (Park *et al.*, 2003b). Okamoto *et al.* (2003) and Park *et al.* (2003a) reported that this activity would be also linked to the deacetylation degree of the oligomers. An anti-coagulant effect of oligomers was also reported in the literature (Park *et al.*, 2004c; Klokkevoid *et al.*, 1999; Hong *et al.*, 1998; Rao & Sharma, 1997).

III. Electrodialytic Technologies and Their Applications to Chitosan Transformation in Bioactive Oligomers

3.1. Electrodialysis and Hybrid Technologies

Electrodialysis (ED) is an electrochemical process in which the driving force or the mass transfer force is an electrical potential. This electric field is created by the application of a potential difference between two electrodes immersed in an aqueous solution enriched in mineral or organic ionic species (Leitz, 1986). Under the influence of an electric field electrically-charged ionic species migrate through perm-selective membranes. Electrical energy supplied is then only used for the transfer of charged species. During this process, the chemical nature of species in solution remains unchanged.

Conventional electrodialysis or electrodialysis with ion-exchange membranes consists of a series of cation- (CEM) and anion- (AEM) exchange membranes arranged in an alternating pattern between an anode and a cathode to form individual cells (Figure 1). A frame spacer equipped with a grill serving both to support the membrane and to promote turbulence, is placed between the AEM and the CEM to allow circulation of a solution between them (Figure 1). Cations migrate toward the cathode, leave the dilution compartment by crossing the CEM and are then retained in the concentration compartment since they cannot pass the anionic membrane. Simultaneously, the anions migrate towards the anode. They leave the dilution compartment, pass through the AEM and are retained in the concentration compartment, as they cannot go through the CEM (Bazinet and Firdaous, 2009).

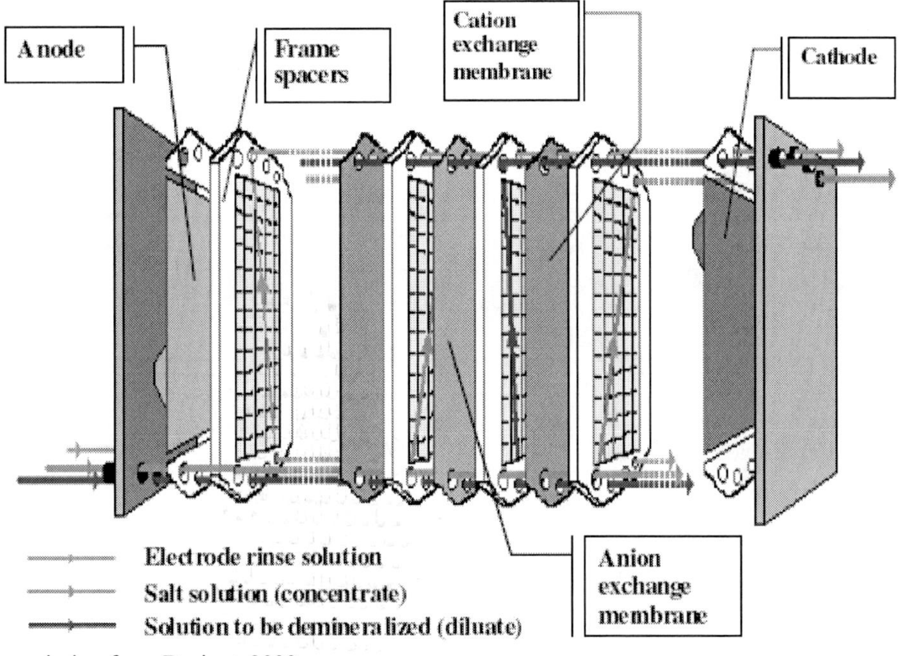

With permission from Bazinet, 2000.

Figure 1. Scheme of a filter-press ED stack.

In hybrid electrodialytic technologies, some of the ion-exchange membranes in the stacking and configuration can be changed for membranes with different specificities or characteristics such as bipolar membrane for water dissociation (electrodialysis with bipolar membrane) or porous membrane for electrofiltration (electrodialysis with filtration membrane).

3.2. Electrodialysis with Bipolar Membranes

3.2.1. Bipolar Membrane

Bipolar membrane (BPM), appeared commercially at the end of the 1980's. Bipolar membranes have the property to dissociate water molecules under an electric field (Mani, 1991) and to generate a flow of protons H^+ at the cationic interface and a flow of hydroxyl ions OH^- at the anion-exchange interface (Pourcelly and Bazinet, 2008) (Figure 2). Generally speaking, these membranes are composed of three parts: an anion-exchange layer, a cation-exchange layer, and a hydrophilic transition layer at their junction (Mani, 1991). In electrodialysis with bipolar membranes (EDBM), cation- and anion-exchange membranes are stacked together with bipolar membranes in alternating series in an electrodialysis cell. This technology combines the demineralization property of ion-exchange membranes and the unique particularity to produce acid and base in-situ without salt addition.

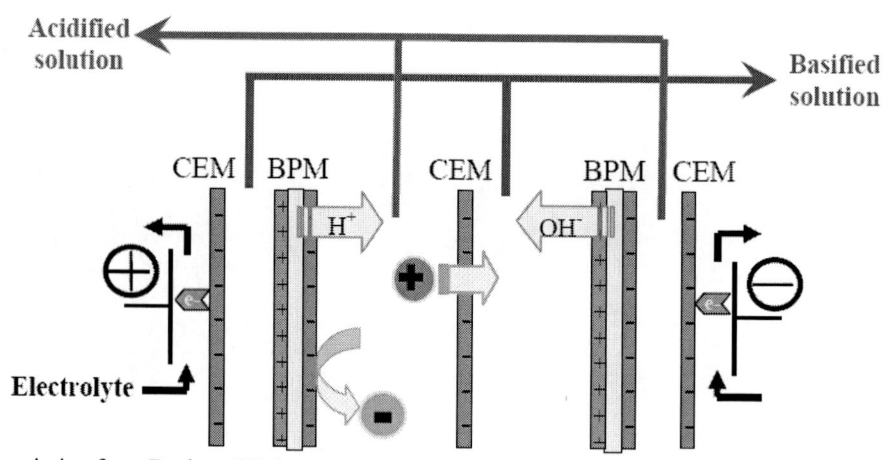

With permission from Bazinet, 2005.

Figure 2. Principle of electrodialysis with bipolar membrane (EDBM). BPM, bipolar membrane and CEM, cation-exchange membrane.

3.2.2. Application to the Transformation of Chitosan in Oligomers

In 2007, Bazinet and Lin Teng Shee proposed an integrated process for chitosan transformation into its oligomers using a three-compartment EDBM configuration (bipolar / anionic/ cationic / bipolar). Each compartment was used to carry-out a specific operation (chitosan solubilization, chitosanase inactivation and chito-oligomer demineralization) and that simultaneously (Lin Teng Shee *et al.*, 2008a). The solubilisation of chitosan was performed by its acidification in compartment number 1, the chitosanase inactivation after hydrolysis of chitosan was in compartment n°3 and the purification of the chitosan oligomers by demineralization was conducted in compartment n°2 (Figure 3). Before being used at full capacity and optimum energy efficiency, the technology needs three consecutive steps.

3.2.2.1. The Three Steps of the EDBM Integrated Process

In the first step, the acidification compartment (n°1) is used to solubilize the chitosan powder. After pH 2.0 is reached, the acidified chitosan solution can be taken out of the EDBM system and the solution is adjusted at the pH of incubation of chitosanase by adding chitosan powder (Figure 4). The addition of a base would increase the ionic strength and the mineral content of the final oligomers. Only one compartment is really operational and in both other compartment a salt solution is circulated.

Bipolar membrane electroacidification step allows the complete solubilization of chitosan up to 5 g/L with a final pH of 2.0. The choice of current intensity, chitosan feeding mode and the type of monopolar membrane were important factors to consider for the optimisation of the solubilization step of chitosan (Lin Teng Shee, 2007; Lin Teng Shee *et al.*, 2006; 2007). On another hand, the precipitation phenomenon of chitosan in the spacers and at the ionic membrane interfaces was a main obstacle to be overcome ; the chitosan was revealed very sensitive to pH variation during electroacidification. One of the solutions proposed by Lin Teng Shee *et al.*, (2007) to reduce and prevent this chitosan fouling by precipitation was to work at a current density under the limiting current density to avoid water molecule dissocation at the interface of the monopolar ionic membranes.

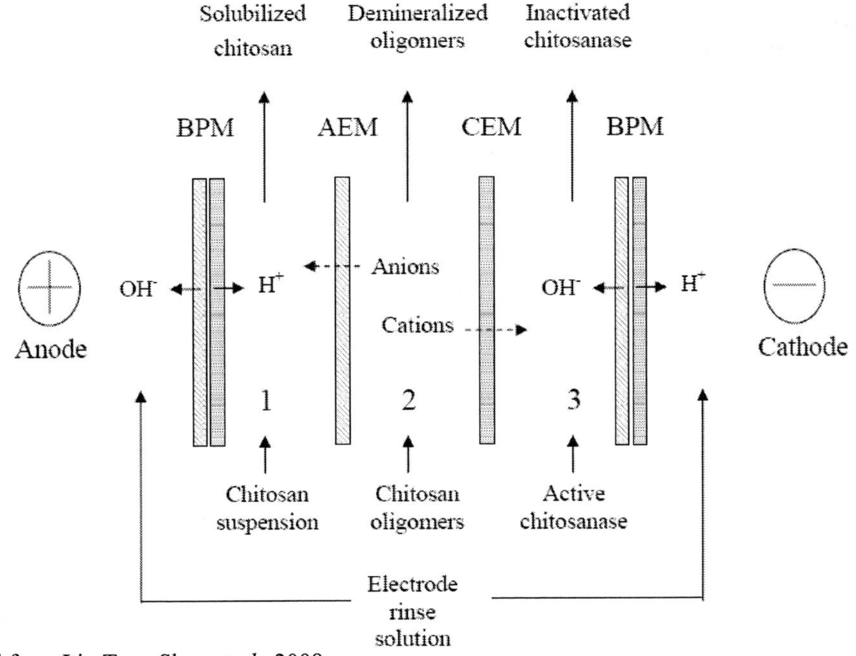

Adapted from Lin Teng Shee *et al.*, 2008a.

Figure 3. Elementary unit used in the «3-in-1» integrated bipolar membrane electrodialysis process for production of chito-oligomers. (1) Electroacidified compartment; (2) electrobasified compartment; (3) demineralized compartment. BPM, bipolar membrane; AEM, anion-exchange membrane; CEM, cation-exchange membrane.

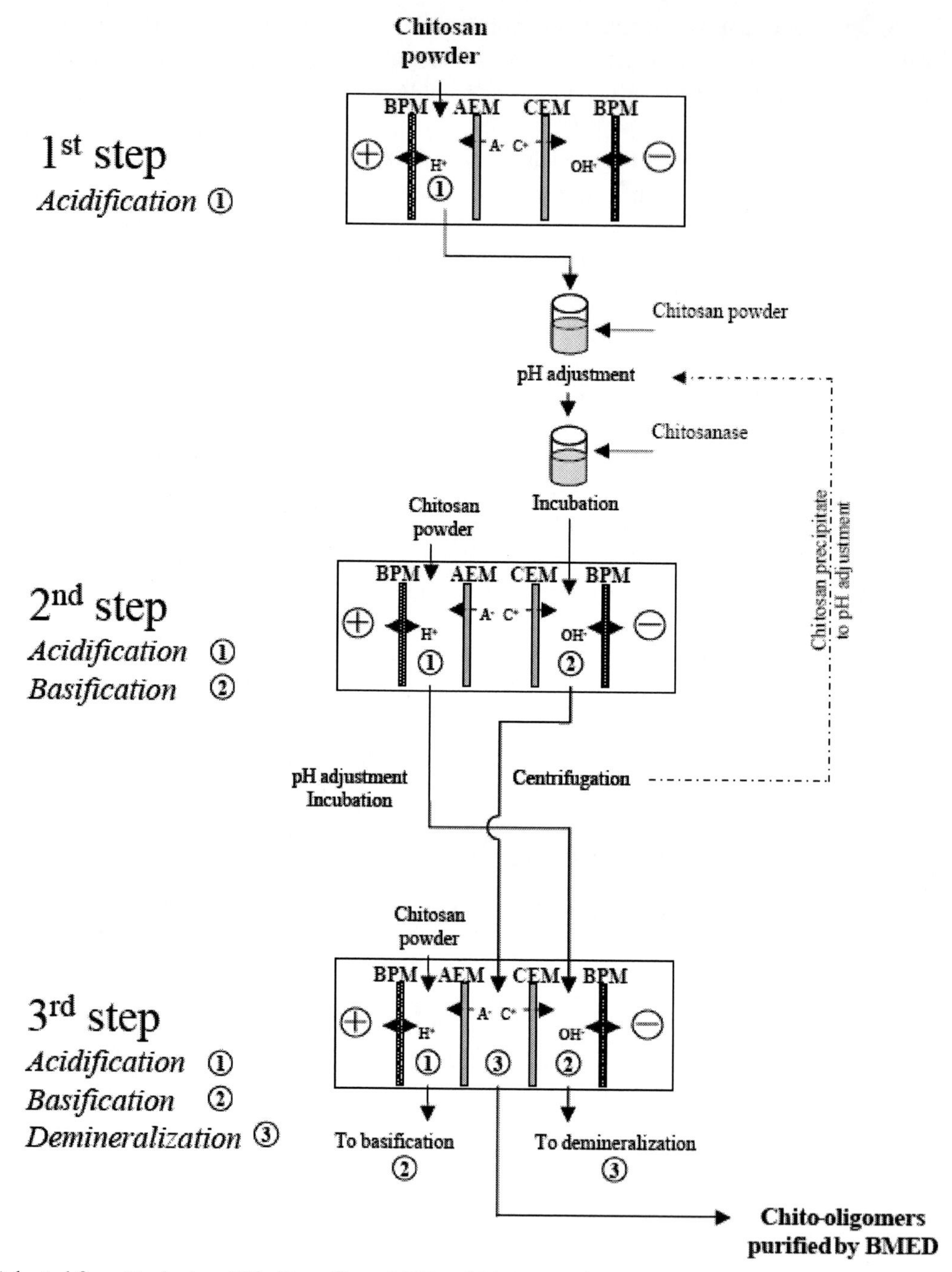

Adapted from Bazinet and Lin Teng Shee, 2007 and Lin Teng Shee *et al.*, 2008a.

Figure 1. Integrated process flow for chitosan hydrolysis into oligomers using bipolar membrane electrodialysis.

Before the second step was performed, the acidified solution from compartment n°1 is transferred in compartment n°3. In parallel, a new batch or solution of chitosan to be solubilized is added in compartment n°1. Compartment n°3, the basification compartment is then used to terminate the chitosan hydrolysis by elevating the pH of the solution (pH > 7) during EDBM. At this step, two compartments of the EDBM cell are used and operational (Figure 4), since in compartment n°2 a salt solution is circulated. The resulting acidified solution is treated as described before, by adjusting the pH and incubating with chitosanase.

Lin Teng Shee *et al.* (2008b) demonstrated that EDBM could be used for terminating the enzymatic reaction of chitosan with chitosanase by electrobasification in a bipolar / cationic membrane configuration. Enzyme inhibition was obtained by denaturation at highly acidic (pH < 3) or basic pH (pH > 8). Indeed, chitosanase was sensitive to basic or acid extreme pH values. Complete inhibition of the enzymatic reaction was ensured by the extreme basic pH of solution (pH > 11). As shown by Lin Teng Shee *et al.* (2008b), the catalytic activity of chitosanase during chitosan hydrolysis decreased after pH adjustment by electrobasification. The reaction rate decreased by 50 % from pH 5.5 to pH 6, whereas the reaction was completely inhibited at pH > 7.0. Apart from denaturation of chitosanase, Li *et al.* (2005) showed that the alkaline pH values have also an effect on chitosan, by precipitating the high molecular weight chitosan, while the low molecular weight material remains soluble. The decrease of reaction rate at alkaline pH values was consequently due to the simultaneous chitosan insolubilization and chitosanase denaturation (Lin Teng Shee *et al.*, 2008b). PH adjustement to a value of 7.0 could allow the inhibition of the enzymatic reaction and the precipitation of the high molecular weight chitosan, but keeping in solution the oligomers of low molecular weights and the chitosanase in a non denaturated form. A centrifugation step at the outlet of the basification compartment would allow the separation of the compounds of interest and the reuse of chitosan and/or chitosanase for the hydrolysis process (Lin Teng Shee, 2007) (Figure 4). Otherwise, if chitosanase has to be completely inactive in the end product containing chitosan oligomers, adjusting the pH at a value higher than 9 will be necessary for destruction of the enzyme structure.

In the third Step, the new acidified solution is transferred in compartment n°3 and the basified oligomer solution is transferred in compartment n°2 to be demineralized. Once again, a new solution of chitosan to be solubilized is added in compartment n°1 (Figure 4). At this step, the three compartments are running simultaneously and the energy efficiency of the process is maximum. The acidification and basification compartments are used for chitosan solubilization and chitosanase inactivation respectively. Meanwhile, the diluate compartment situated between the acidification and the basification compartments is used for demineralization of the basified solution to produce purified chitosan oligomers.

The addition of a central demineralization compartment completes the integrated EDBM system and constitutes a major improvement of the traditional process. Indeed, the demineralization step allows a 53% decrease and more of the mineral content of the oligomers. In the «3-in-1» integrated process the ionic flux is controlled by the place of the bipolar and monopolar membranes in the cell configuration. Hence, minerals which are withdrawn during the demineralization play a major role in the adjacent compartments of acidification and basification, since they allow the conduction of the current and increase the conductivity of these solutions (Figure 5) (Lin Teng Shee and Bazinet, 2009). These balances in conductivity and in ionic fluxes between the compartments are consequently guarantees of good performance and interesting current efficiency of the integrated system (Figure 5).

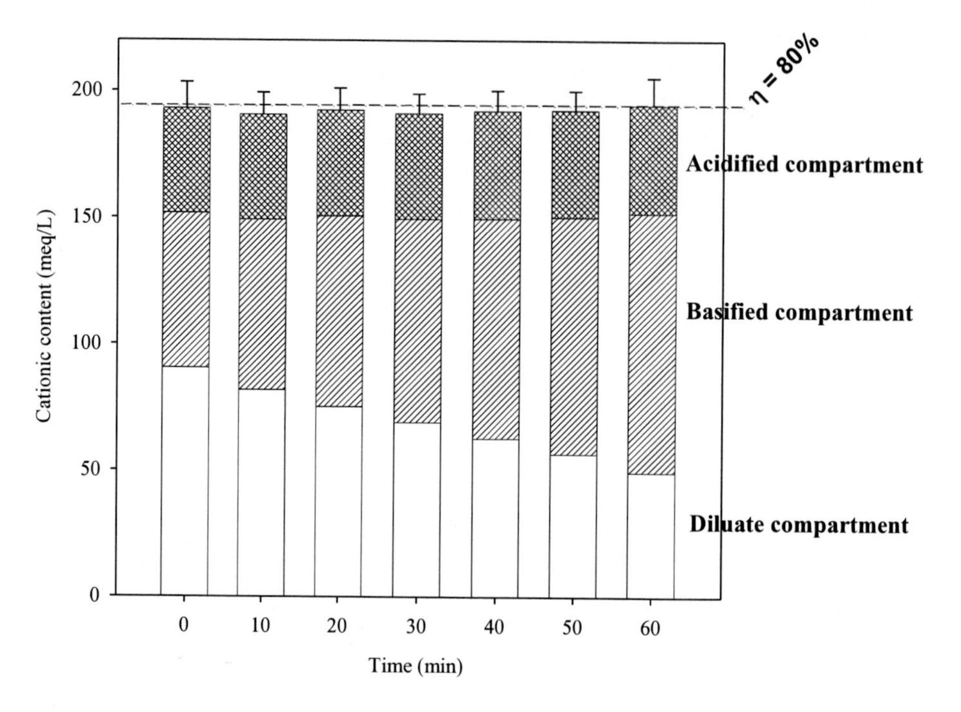

Figure 5. Evolution of total cationic content (expressed as meq/L) and current efficiency (η) of the integrated three-compartment bipolar membrane electrodialysis system during the preparation of chitosan oligomers (Adapted from Lin Teng Shee & Bazinet, 2009).

3.2.2.2. Advantages and Limitations of the «3-in-1» Integrated Process

The configuration shown on figure 3 was identified as the best configuration in terms of membrane nature and position in the stack. It was demonstrated that the electrochemical solubilization of chitosan by EDBM depends on the type of monopolar ionic membranes since the stacking of an anion-exchange membrane facing the acidified compartment (anionic configuration) was less energy-consuming in comparison with a cationic configurations in which a cation-exchange membrane was in contact with the acidified compartment (Lin Teng Shee *et al.*, 2006).

The EDBM process was compared to a traditional process using chemical acid and base. One of the advantages of EDBM over the chemical method is the possibility to remove the salts previously added during acidification and basification, so that at the end of a one hour treatment, the ash content returns to its initial value (about 0.26 %). Moreover, no storage and handling of chemical acids and bases are required since the protons and hydroxyls ions are produced *in situ* in the EDBM system, and thus creating a safe working environment. No effluents are generated since all streams in the 3 compartments are used to produce the purified oligomers. This innovative "3-in-1" process for chitosan transformation into oligomers has great potential for industrial application, since it is convenient, ecological and produces oligomers with a lower mineral content in comparison with the conventionnal method. The relative energy consumption was estimated to 0.02 kWh/L of solution (Lin Teng Shee *et al.*, 2008a). In addition, the energy was used for running three treatments simultaneously, allowing efficient use of the electrical current. Considering all the observations and conclusions done by Lin Teng Shee and collaborators during their different studies, the integrated EDBM process is an innovative process for the simultaneaous transformation of chitosan in oligomers of high purity. This process integrates within one only and even unit : acid solubilization of chitosan, the control of the enzymatic reaction and the purification of the final bioactive oligomers.

The main limitation of this EDBM integrated process applied to the production of chitosan oligomers is the relatively high cost of the initial investment of the ED system as well as the renewing of the bipolar and monopolar membranes. However, during the last decade, the advances in bipolar membranes researches allowed the increase in the operation life of these particular membranes as well as the performances of these membranes. Nowadays, the duration of bipolar membranes during acid/base production reach operation life over 20 000 hours. This explains why an increasing number of applications using the bipolar membrane technology appeared recently, particularly for its environmental aspect of sustainability, incenting equipment suppliers to develop this technology for industrial applications such as wine deacidification.

3.3. Electrodialysis with Filtration Membranes

Recently, in fundamental studies on chitosan D-glucosamine and oligomers electrophoretic mobility, it was demonstrated that these molecules migrated under the effect of an external electric field, according to the solution conditions, mainly pH (Aider *et al.*, 2006a, b). Under pH conditions of 4, these molecules are positively charged because of the protonation of the amine group. Furthermore, the electrophoretic mobility of the dimer was higher than the one of trimer and tetramer due to their differences in molecular weight and electric charge. It was proposed that the exploitation of the electric properties of these molecules could offer a solution to separate mixtures of chitosan oligomer into pure or enriched fractions by electrodialysis with filtration membranes.

3.3.1. Filtration Membrane

Filtration membrane are selective porous barriers that allow the transmission of certain components present in a solution while retaining other components by means of an applied driving force generally pressure. The mechanism behind the selectivity is generally the size of the component. Particles larger than the membrane pores are retained while the smaller ones pass through. The filtration spectrum of membrane covers a broad range of pore size (Ps) ; microfiltration membrane (Ps >0.1 μm), ultrafiltration membrane (1 nm< Ps< 500 nm), nanofiltration (0.1 nm< Ps< 1 nm) and reverse osmosis membrane (Ps <0.1 nm).

A new alternative method to separate bio-molecules into more homogeneous fractions and to demineralise them was recently developed (Bazinet *et al.*, 2005a). This new method is named electrodialysis with filtration membrane (EDFM). It is a hybrid process which combines conventional electrodialysis cell and filtration membranes (Figure 6). The filtration membrane is stacked as a molecular barrier in an electrodialysis cell and since no pressure is applied in the electrodialysis cell, the electric field is the only driving force of this technology

(Bazinet and Firdaous, 2009). Consequently, only the charged molecules migrate under the

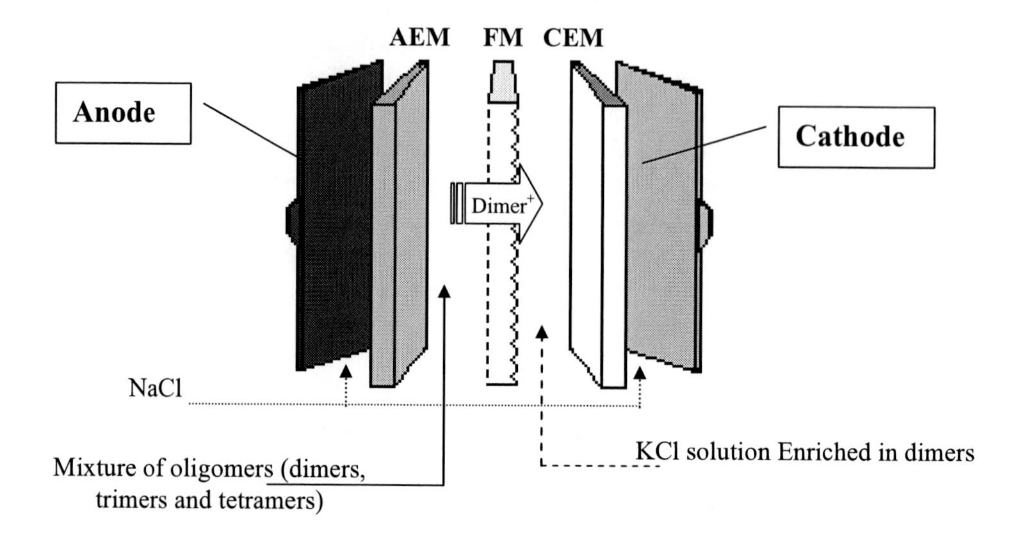

Figure 6. Electrodialysis with filtration membrane cell configuration for the separation of chitosan oligomers. FM, filtration membrane; AEM, anion-exchange membrane; CEM, cation-exchange membrane.

effect of the electric field and the neutral molecules theoretically stay in the primary solution and do not reach or pass the filtration membrane. This process allows the separation of molecules according to their molecular size and their electrical charge. This new separation method was successfully used for tobacco polyphenol separation (Bazinet *et al.*, 2005b,c), green tea catechin separation (Labbé *et al.*, 2005) and for bio-active peptide purification (Poulin *et al.*, 2006, 2007, 2008; Poulin, 2007; Firdaous *et al.*, 2009, 2010).

3.3.2. Application to the Separation of Oligomers

Considering that potential applications of chitosan oligomers require pure or highly enriched fractions of defined molecular weight, it appears important to develop more effective means for the production on a large scale of these bioactive molecules. In this context, EDFM was applied to the purification of chitosan oligomers, and parameters for the optimisation of the process were studied : membrane molecular weight cut-off (MWCO), pH, electrical field strength, processing time.

3.3.2.1. Impact of Membrane Molecular Weight Cut-Off

MWCO of the UF-membrane was the primary factor affecting the electromigration rate of the chitosan oligomers. Aider *et al.* (2008a) tested five cellulose ester ultrafiltration membranes of different molecular weight cut-offs (500, 1000, 5000, 10000 and 20000 Da) in the EDFM configuration. A constant voltage of 5 V was used during the EDFM process. When the 500 Da UF-membrane was used, no electromigration was observed throughout the treatments and that whatever the oligomer, while for UF-membrane with higher MWCO, the electro-migration of the chitosan oligomers was effective but different according to the MWCO, the chain length and also the duration of the process. 1000 Da MWCO UF-membrane was permeable to the dimer and trimer during the first 3 h of treatment while the tetramer was found in the adjacent migration compartment only after 4 hours of electro-separation process. 5000 and 10000 Da MWCO UF-membranes were not permeable to the tetramer only during the first 2 h of the EDFM treatment. The 20000 Da MWCO UF-membrane presented no selectivity for all three oligomers. The dimer showed a linear electro-migration kinetic in all cases. The electro-migration behavior of the trimer was linear when 1000 and 5000 Da MWCO UF-membranes were used and showed an exponential electro-migration behavior when the EDFM cell was stacked with 10000 and 20000 Da MWCO UF-membranes. The 10000 Da MWCO UF-membrane appeared to be the best membrane to carried-out separation of the tetramer.

3.3.2.2. Impact of pH

Electroseparation of chitosan oligomers treatments with EDFM system were carried-out on 200 ml of chitosan oligomer solution in a batch process at 6 pH values (4, 5, 6, 7, 8 and 9). These pH values were selected following previous work on the effect of the pH value on chitosan oligomer electrophoretic mobility (Aider *et al.*, 2006a,b). A constant voltage of 5 V was used as previously by Aider *et al.* (2008a).

For all pH values, the dimer showed the highest electromigration rates compared to the others chitosan oligomers (Figure 7). Moreover, by increasing pH of the medium, electromigration rate of the oligomers decreased significantly. The effect of the pH on chitosan oligomers electromigration through 10000 Da MWCO UF-membrane is directly related to the protonation and deprotonation phenomenon of the chitosan oligomers amine group, which is pH dependent. Chitosan oligomer electrophoretic mobility depends on the degree of protonation of the amine group. At pH values of 4 and 5, the oligomers showed the highest electromigration rates because of the high protonation level of the amine functional group. By increasing pH of the medium up to 7, electromigration rate of all chitosan oligomers decreased significantly compared with data recorded at pH between 4 and 6. This was caused by high deprotonation level (Juang and Hao, 2002) of the amine function, which had as consequence to decrease the electrophoretic mobility of the oligomers. At pH 8 and 9 no oligomer had migrated through the 10000 Da MWCO UF-membrane because of the complete deprotonation of the amine function. All these results recorded at different pH values show that the migration of chitosan oligomers through 10000 Da MWCO UF-membrane was caused by the effect of the electric field on charged amine groups and not by the diffusion. If diffusion acted as driving force, chitosan oligomer migration at pH 8 and 9 would have been possible.

3.3.2.3. Effect of Electrical Field Strength

It was demonstrated by Aider *et al.* (2007c) that the electric field strength showed a significant effect on each chitosan oligomer electromigration rate while the effect of solution flow velocity was not significant. Electroseparation by EDFM treatments were carried-out on 200 mL of chitosan oligomer solution in a batch process at three applied voltages (5, 10 and 20 V) which corresponded to three average electric field strengths of 2.5, 5 and 10 V/cm.

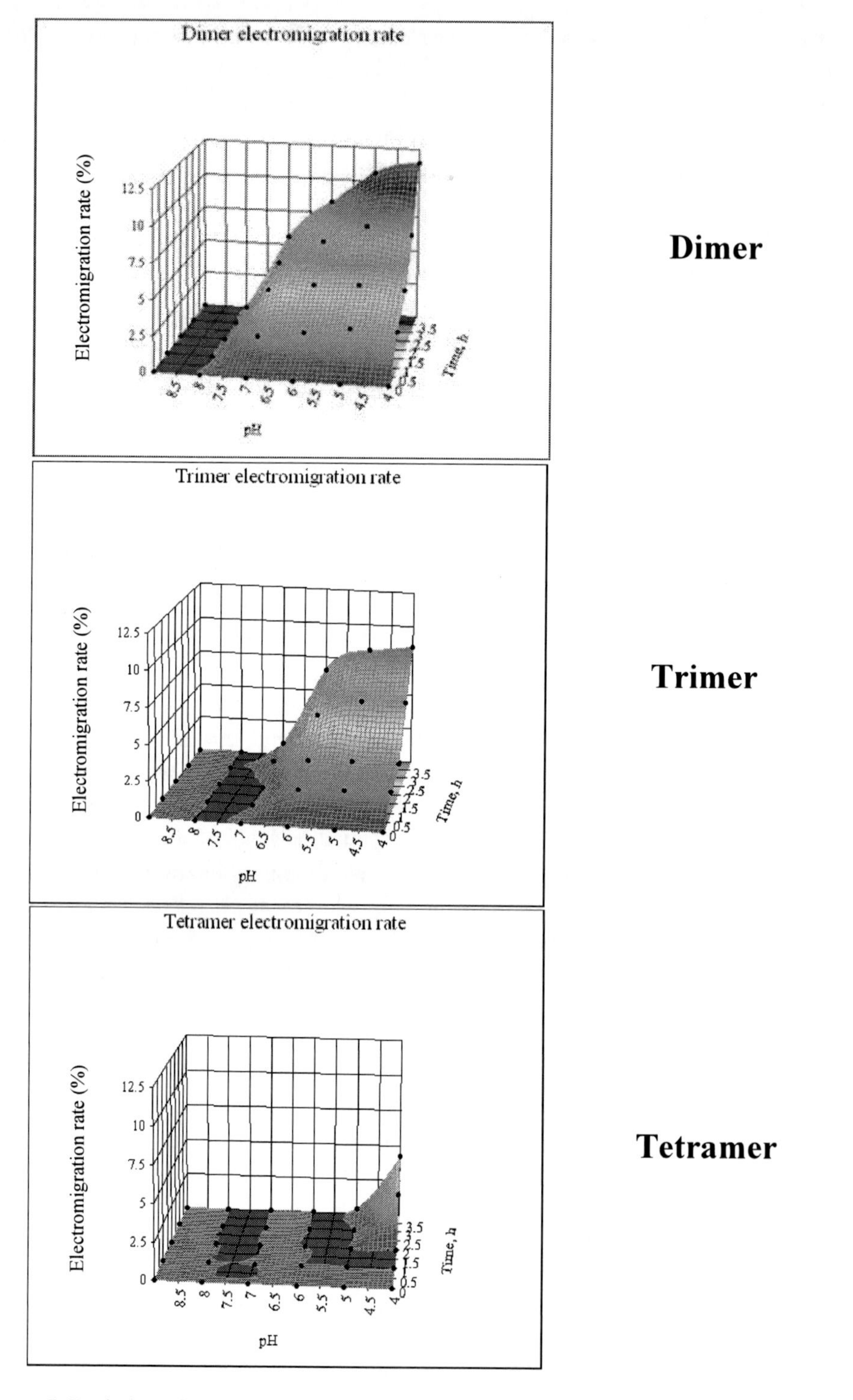

Figure 7. Evolution of each oligomer electromigration rates as a function of time and pH value during EDFM process with a 10 kDa MWCO membrane (Adapted from Aider *et al.*, 2008b).

The dimer showed a linear electromigration kinetic whatever the applied average electric field (Figure 8). This behavior indicates that the dimer migrated quite freely through the 10000 Da MWCO ultrafiltration membrane used, under the effect of the applied external electric field as a driving force. The trimer electromigration behavior followed a curvilinear type equation and showed an inflexion point after 2h of treatment, when an electric field strength of 2.5 V/cm was applied. But in the cases when electric field strengths of 5 and 10 V/cm were used, its electromigration behavior followed a linear kinetic. Indeed, by increasing the applied voltage, this molecule was subjected to a higher driving force that makes it possible to significantly minimize the friction exerted by the membrane. As result, the trimer migrated quite freely at the applied electric fields of 5 and 10 V/cm. The given explanation about the trimer is also applicable to the tetramer. Concerning the separation possibility of the chitosan oligomers, when electric field strength of 2.5 V/cm was applied, a solution composed only by the dimer and trimer was obtained until 2h of electroseparation treatment. At the pH used in the present study, no separation was possible at any time when electric field strengths of 5 and 10 V/cm were applied to the electroseparation system with the 10000 Da MWCO cellulose ester ultrafiltration membrane stacked in the EDFM cell; the applied electric field strength was sufficient for the tetramer to migrate through the membrane together with the dimer and trimer since the first hour of the treatment.

The effect of the electric field strength was also significant on the separation possibility of the studied oligomers. It was shown that by using electric field strength of 2.5 V/cm, it was possible to obtain a solution composed only of the dimer and trimer until an operating time of 2 h. By increasing the electric field up to 5 and 10 V/cm, it was no more possible to separate the chitosan oligomers. Chitosan oligomer transport numbers were measured. It was found that they contribute to about 7% of total electric current carrying, while about 93% of the total electric current would be carried by the electrolytes.

3.3.2.4. Effect of Processing Time

Processing time has an effect on both chitosan oligomer electromigration rates and possibility of their electro-separation via electrophoretic mobility of each oligomer which itself is function of time (Figure 8). The highest electromigration rate was obtained in the pH range of 4 and 7. Chitosan oligomer with highest electrophoretic mobility will migrate more quickly through the 10000 Da MWCO UF-membrane. Consequently, at a given time of the EDUF operation, the most mobile chitosan oligomer will show the highest electro- migration rate than the other molecules.

All these data can be used to determine the optimal conditions for the production of pure or enriched chitosan oligomer fractions taking time as variable of the process : Hence, it was possible to obtain pure solution of dimer at pH 7 after 2h because it was the only chitosan oligomer which migrated through the 10000 Da MWCO UF-membrane, or to produce fractions composed only by the dimer and trimer without tetramer with a process duration of 2h at pH 4, 3h at pH 5 and 4h at pH 6. The performance of the electroseparation process, in term of quantity, of the studied chitosan oligomers can be optimised by increasing the effective ultrafiltration membrane area under applied electric field strength of 2.5 V/cm. This technique can be successfully used in the basic research laboratory as well as for commercially important scales for chitosan oligomers separation.

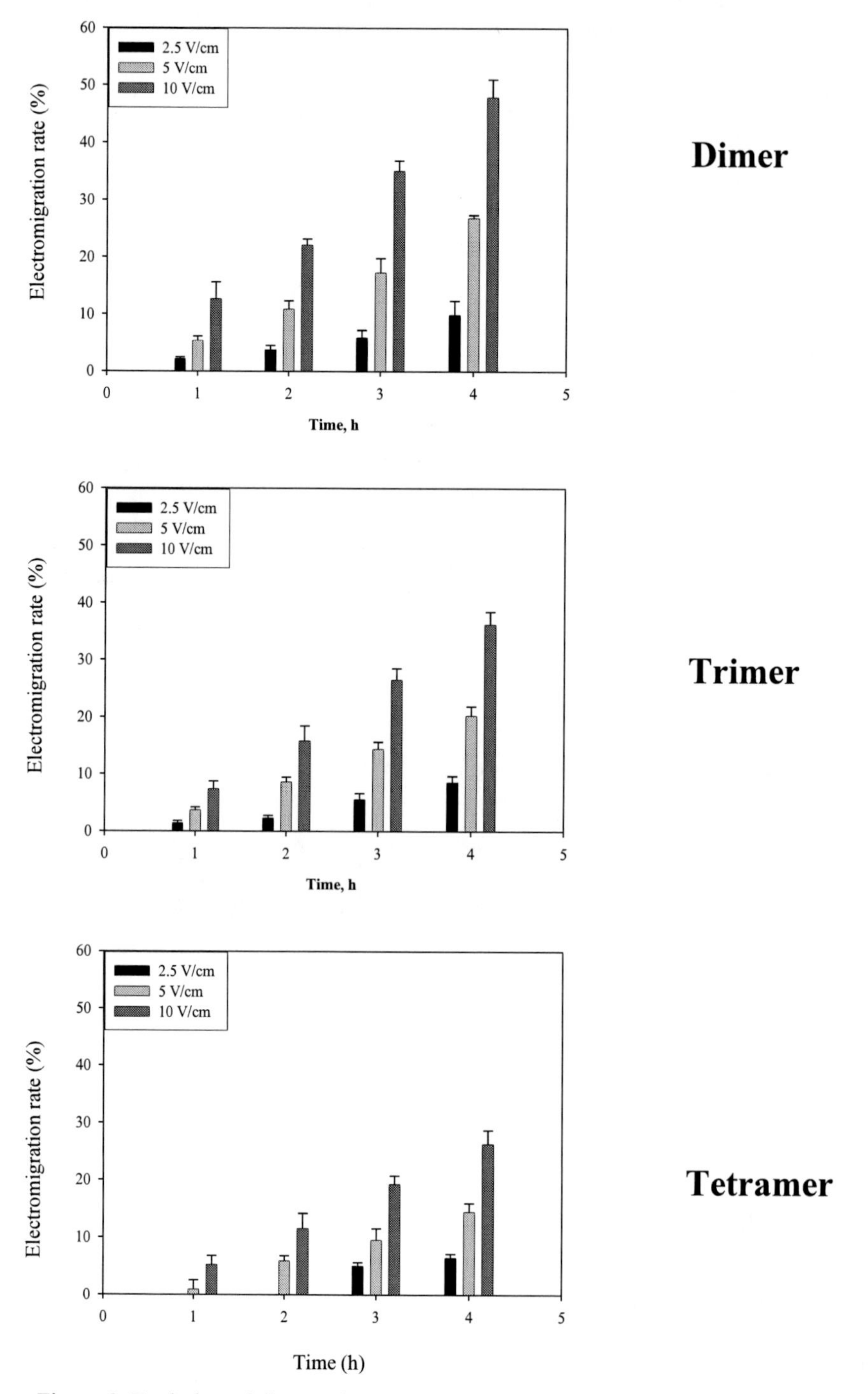

Figure 8. Evolution of dimer, trimer and tetramer electromigration rates as a function of time and the applied electric field strength during EDFM process with a 10 kDa MWCO membrane (Adapted from Aider *et al.*, 2009).

IV. CONCLUSION

Chitosan have many applications in food, pharmaceutical, and medicinal areas, and its oligomer biological and physiological properties made them compounds of great interest since their fields of application will continue to increase in the future. Hence, methods and technologies for their separation will be needed. In addition, in a context of sustainability, innovative and more ecological technologies will have to be developed to limit their impacts on the environment (e.g production of chemicals during chitosan chemical hydrolysis). In this context, electrodialytic technologies and processes are promising technologies for the removal, transformation, separation and purification of bioactive charged molecules from complex solution such as chitosan and its oligomers.

EDBM can be in the future a new way for the production of oligomers from cationic polysaccharides such as chitosan but also from anionic polysaccharides. In fact, to applied the EDBM technology developed by Bazinet and Lin Teng Shee (2007) to anionic polysaccharides, the process would only have to be inversed by solubilising the anionic polysaccharides in the basification compartment and stopping the enzymatic reaction in the acidification compartment, the demineralization compartment always been there to purify the final product. In the case of EDFM, this method could be a powerful process for the separation of bioactive chitosan oligomers of interest from complex feed solution under mild conditions or for the production of polysaccharides fractions of specified molecular weight ranges. Furthermore, as demonstrated in this chapter, based on the physicochemical and electrokinetic properties of solutes in the feed solution, 1) chitosan could be solubilised and hydrolysed in an environmental friendly process, 2) fouling of chitosan on membranes could be prevented to ensure the best efficiency of the process, 3) minerals could be removed from the feed solution in order to obtain pure or enriched fractions and 4) chitosan oligomers according to their polymerization degree can be purified very selectively from a mixture.

Both processes were studied separately, but in a context of optimisation and rationalisation of chitosan transformation in bioactive oligomers, EDBM and EDFM could be coupled in one production process. Hence, EDBM could be used first to produce oligomers from chitosan and then EDFM to purify the oligomers or to produce fractions of oligomers of specific polymerisation degree. This global process will be an environment friendly process completely integrated.

REFERENCES

Aiba, S. (1994a). Preparation of N-acetylchitooligosaccharides by lysozymic hydrolysates of partially N-acetylated chitosans. *Carbohydrate Research*, 261, 297-306.

Aiba, S. (1994b). Preparation of N-acetylchitooligosaccharides by hydrolysis of chitosan with chitinase followed by N-acetylation. *Carbohydrate Research*, 256, 323-328.

Aider, A., Arul, J., Mateescu, M.A., Brunet, S., Bazinet, L. (2006a). Electromigration of Chitosan D-Glucosamine and chitosan oligomers in Dilute Aqueous media. *Journal of Agricultural and Food Chemistry*, 54, 6352-6357.

Aider, M., Arul, J., Mateescu, M.A., Brunet, S., Bazinet. L. (2006b). Electromigration behaviour of a mixture of chitosan oligomers at different concentrations. *Journal of Agricultural and Food Chemistry, 54*, 10170-10176.

Aider, M., Brunet, S., Bazinet, L. (2008a). Electro-separation of Chitosan Oligomers by Electrodialysis with Ultrafiltration Membrane (EDUF) and Impact on Electrodialytic Parameters. *Journal of membrane Science*, 309, 222-232.

Aider, M., Brunet, S., Bazinet, L. (2008b). Effect of pH and Cell Configuration on the Selective and specific electrodialytic separation of Chitosan oligomers. *Separation and Purification Technology,* 63, 612-619.

Aider, M., Brunet, S., Bazinet, L. (2009). Effect of Solution Flow Velocity and Electric Field Strength on Chitosan Oligomer Electromigration Kinetics and their Separation in an Electrodialysis with Ultrafiltration Membrane (EDUF) system. Separation and Purification technology, 69, 63–70.

Allan, G.G., Peyron, M. (1995). Molecular weight manipulation of chitosan. I: kinetics of depolymerization by nitrous acid. *Carbohydrate Research,* 277, 257-272.

Bacon, A., Makin, J., Sizer, P.J., Jabbal-Gill, I., Hinchcliffe, M., Illum, L. (2000). Carbohydrate biopolymers enhance antibody responses to Mucosally delivered vaccine antigens. *Infection and Immunity*, 68, 5764-5770.

Bazinet, L. (2000). Séparation des caséines du lait bovin par électro-acidification avec membranes bipolaires et caractérisation des isolats. Ph. D. thesis n°18470, Université Laval, Québec, Canada.

Bazinet, L. (2005). L'électrodialyse et ses applications en bio-industrie. *BioTechno* CQVB, Vol.4, n°2 , Mai 2005

Bazinet, L., Amiot, J., Poulin, J.-F, Tremblay, A., Labbé, D. (2005a). Process and system for separation of organic charged compounds. *Patent Pending* PCT/CA2005/000337.

Bazinet, L., DeGrandpré, Y., Porter, A. (2005b). Electromigration of tobacco polyphenols. *Separation and Purification Technology*, 41, 101-107.

Bazinet, L., DeGrandpré, Y., Porter, A. (2005c). Enhanced tobacco polyphenol electromigration and impact on membrane integrity. *Journal of Membrane Science*, 254, 111-118.

Bazinet, L., Firdaous, L. (2009). Membrane processes and devices for separation of bioactive peptides. *Recent Patents on Biotechnology*, 3, 61-72.

Bazinet, L., Lin Teng Shee, F. (2007) Method for transforming polysaccharides in oligosaccharides with bipolar membrane electrodialysis. *Patent Pending* US 60/975,681.

Bosso, **C.,** Defaye, J., Domard, A., Gadelle, A., Pedersen, C. (1986). The behavior of chitin towards anhydrous hydrogen fluoride. Preparation of β-(1→4)-linked 2-acetamido-2-deoxy-d-glucopyranosyl oligosaccharides. *Carbohydrate Research*, 156, 57-68.

Chen, A.S., Taguchi, T., Sakai, K., Kikuchi, K.,Wang, M.W., Miwa, I. (2003). Antioxidant activities of chitobiose and chitotriose. *Biological and Pharmaceutical Bulletin*, 26, 1326–1330.

Chen, Y.M., Chung, Y.C., Wang, L.W., Chen, K.T., Li, S.Y. (2002). Antibacterial properties of chitosan in water born pathogen. *Journal of Environmental Science and Health* A, 37, 1379-1390.

Chiang, M.T., Yao, H.T., Chen, H.C. (2000). Effect of dietary chitosans with different viscosity on plasma lipids and lipid peroxidation in rats fed on a diet enriched with cholesterol. *Bioscience, Biotechnology and Biochemistry*, 5, 965-971.

Choi, B.K., Kim, K.Y., Yoo, Y.J., Oh, S.J., Choi, J.H., Kim, C.Y. (2001). In vitro antimicrobial activity of chitooligosaccharide mixture against *Actinobacillus actinomycetemcomitans* and *Streptococcus mutans*. *International Journal of Antimicrobial Agents,* 18, 553-557.

Chung, Y.C., Su, Y.P., Chen, C.C., Jia, G., Wang, H.L., Wu, J.C. (2004). Relationship between antibacterial activity of chitosan and surface characteristics of cell wall. *Acta Pharmacologica Sinica*, 25, 932-936.

Defaye, J., Gadelle, A., Pedersen, C. (1989). Chitin and chitosan oligosaccharides. *In* Chitin and chitosan, Skjak-Braek, G.; Anthonsen, T.; Sandford, P. (eds.), Elsevier applied Science, Barking, England, pp.415-429.

Domard, A., Cartier, N. (1989). Preparation, separation and characterisation of the D-glucosamine oligomer series. *In* Chitin and Chitosan, Skjak-Braek, G.; Anthonsen, T.; Sandford, P. (eds), Elsevier, London, pp.383-387.

Firdaous, L.; Dhulster, P.; Amiot, J.; Doyen, A.; Lutin, F.; Vézina, L.-P.; Bazinet, L. (2010). Investigation of the large scale bioseparation of an antihypertensive peptide from alfalfa white protein hydrolysate by an electromembrane process. *Journal of Membrane Science*, 355, 175-181.

Firdaous, L.; Dhulster, P.; Amiot, J.; Gaudreau, A.; lecouturier, D., Kapel, R., Lutin, F.; Vézina, L.-P.; Bazinet, L. (2009). Concentration and selective separation of bioactive peptides from an alfalfa white protein hydrolysate by electrodialysis with ultrafiltration membranes. *Journal of Membrane Science*, 329, 60-67.

Gao, Y., He, L., Katsumi, H., Sakane, T., Fujita, T., Yamamoto, A. (2008). Improvement of intestinal absorption of insulin. *International Journal of Pharmaceutics*, 9, 70-78.

Gerasimenko, D.V., Avdienko, I.D., Bannikova, G.E., Zueva, O.Y., Varlamov,V.P. (2004). Antibacterial effects of water-soluble low- molecular-weight chitosans on different microorganisms. *Applied Biochemistry and Microbiology*, 40, 253-257.

Guzman, J., Saucedo, I., Revilla, J., Navarro, R., Guibal, E. (2003). Copper sorption by chitosan in the presence of citrate ions:Influence of metal speciation on sorption mechanism and uptake capacities. *International Journal of Biological Macromolecules*, 33, 57-65.

Hasegawa, M., Isogai, A., Onabe, F. (1993). Preparation of low-molecular-weight chitosan using phosphoric acid. *Carbohydrate Polymers*, 20, 279-283.

Hirano, S., Nagano, N. (1989). Effects of chitosan, pectic acid, lysozyme and chitinase on the growth of several phytopathogens. *Agricultural and Biological Chemistry*, 53, 3065-3066.

Hong, S.P., Kim, M.H., Oh, S.W., Han, C.H., Kim, Y.H. (1998). ACE inhibitory and antihypertensive effect of chitosanoligosaccharides in SHR. *Korean Journal of Food Science and Technology*, 30, 1476-1479.

Horowitz, S.T., Roseman, S., Blumenthal, H.J. (1957). The Preparation of Glucosamine Oligosaccharides. I. Separation. *Journal of the American Chemical Society*, 79, 5046–5049.

Huang, R., Mendis, E., Kim, S.K. (2005). Factors affecting the free radical scavenging behavior of chitosan-sulfate. *International Journal of Biological Macromolecules*, 36, 120-127.

Il'ina, A.V., Tkacheva, I.V., Varlamov, V.P. (2002). Depolymerization of High-Molecular-Weight Chitosan by the Enzyme Preparation Celloviridine G20x. *Applied Biochemistry and Microbiology,* 38, 112–115.

Izume, M., Ohtakara, A. (1987). Preparation of d-glucosamine oligosaccharides by the enzymatic hydrolysis of chitosan. *Agricultural and Biological Chemistry*, 51, 1189–1191.

Je, J.Y., Park, P.J., Kim, S.K. (2004). Free radical scavenging properties of hetero-chitooligosaccharides using an ESR spectroscopy. *Food Chemical Toxicology*, 42, 381–7.

Jeon, Y.J., Kim, S.K. (2000). Production of chitooligosaccharides using an ultrafiltration membrane reactor and their antibacterial activity. *Carbohydrate Polymers*, 41, 133–141.

Jeon,Y.J., Kim, S.K. (2002).Antitumor activity of chisan oligosaccharides produced in an ultrafiltration membrane reactor system. *Journal of Microbiology and Biotechnology*, 12, 503-507.

Juang, R.S., Hao. D.J. (2002). Effect of pH on Competitive Adsorption of Cu(II), Ni(II), and Zn(II) from Water onto Chitosan Beads. *Adsorption*, 8, 71-78.

Kasaai, M.R., Charlet, G., Paquin, P., Arul, J. (2003). Fragmentation of chitosan by microfluidization process. *Innovative Food Science and Emerging Technologies*, 4, 403–413.

Kim, J.Y., Lee, J.K., Lee, T.S., Park, W.H. (2003). Synthesis of chitooligosaccharide derivative with quaternary ammonium group and its antimicrobial activity against *Streptococcus mutans*. *International Journal of Biological Macromolecules*, 32, 23-27.

Kim, S.K, Rajapakse. N. (2005). Enzymatic production and biological activities of chitosan oligosaccharides (COS): A review. *Carbohydrate Polymers*, 62, 357-368.

Kittur, F. S., Vishu Kumar, A.B., Tharanathan, R. N. (2003). Low molecular weight chitosans-preparation by depolymerization with Aspergillus niger pectinase, and characterization. *Carbohydrate Research,* 338, 1283–1290.

Klokkevoid, P.R., Fukuyama, H., Sung, E.C., Bertolami, C.N. (1999). The effects of chitosan (Poly-N-acetylglucosamine) on lingual Hemostasis in heparinized rabbits. *International Journal of Oral and Maxillofacial Surgery,* 57, 49-52.

Kochkina, Z.M., Chirkov, S.N. (2000a). Effect of chitosan derivatives on the development of phage infection in cultured *Bacillus thuringiensis*. *Mikrobiologiia*, 69, 266-269.

Kochkina, Z.M., Chirkov, S.N. (2000b). Effect of chitosan derivatives on the reproduction of coliphages T2 and T7. *Mikrobiologiia,* 69, 257-260.

Kumar, M.N.V., Muzzarelli, R.A.A., Muzzarelli, C., Sashiwa, H., Domb, A.J. (2004). Chitosan Chemistry and Pharmaceutical Perspectives. *Chemical Review*, 104, 6017-6084.

Labbé, D., Araya-Farias, M., Tremblay, A., Bazinet, L. (2005). Electromigration feasibility of green tea catechins. *Journal of Membrane Science*, 254, 101-109.

Lee, H.-W., Park, Y.-S., Jung, J.-S., Shin, W.-S. (2002). Chitosan oligosaccharides, dp 2–8, have prebiotic effect on the *Bifidobacterium bifidium* and *Lactobacillus* sp. *Anaerobe,* 8, 319-324.

Leitz, F.B. (1986). Measurements and control in electrodialysis. *Desalination,* 59, 381-401.

Li, J., Du, Y., Yang, J., Feng, T., Li, A., Ping, C. (2005). Preparation and characterization of low molecular weight chitosan and chito-oligomers by a commercial enzyme. *Polymer Degradation and Stability,* 87, 441-448.

Lin Teng Shee, F. (2007). Optimisation et rationalisation de la transformation du chitosane en oligomères par électrodialyse avec membranes bipolaires. Ph. D Thesis n°24826, Université Laval, Québec (QC), Canada.

Lin Teng Shee, F., Arul, J., Brunet, S., Bazinet, L. (2007). Chitosan solubilization by bipolar membrane electroacidification : Reduction of membrane fouling. *Journal of Membrane Science,* 290, 29-35.

Lin Teng Shee, F., Arul, J., Brunet, S., Bazinet, L. (2008a). Performing a three-step process for conversion of chitosan to its oligomers using an unique bipolar membrane electrodialysis system. *Journal of Agricultural and Food Chemistry,* 56, 10019-10026.

Lin Teng Shee, F., Arul, J., Brunet, S., Bazinet, L. (2008b). Effect of Bipolar Membrane Electrobasification on Chitosanase Activity during Chitosan Hydrolysis. *Journal of Biotechnology.* 134, 305-311.

Lin Teng Shee, F., Arul, J., Brunet, S., Mateescu, A.M., Bazinet, L. (2006). Solubilization of Chitosan by Bipolar Membrane Electroacidification. *Journal of Agricultural and Food Chemistry,* 54, 6760-6764.

Lin Teng Shee, F.; Bazinet, L. 2009. Cationic balance and current efficiency of a three-compartment bipolar membrane electrodialysis system during the preparation of chitosan oligomers. *Journal of membrane Science*, 341, 46-50.

Liu, H.-T., Li, W.-M., Xu, G., Li, X.-Y., Bai, X.-F., Wei, P., Yu, C., Du, Y.-G. (2009). Chitosan oligosaccharides attenuate hydrogen peroxide-induced stress injury in human umbilical vein endothelial cells. *Pharmacological Research*, 59, 167-175.

Mani, K.N. (1991). Electrodialysis water splitting technology. *Journal of Membrane Science*, 58, 117-138.

Muzzarelli, R.A.A., Stanic, V., Ramos, V. (1999). Enzymatic depolymerization of chitins and chitosans. *In* Carbohydrate biotechnology protocols, Bucke, C. (ed.), Humana Press, Totowa, New Jersey, pp.197-209.

Nagasawa, K., Tohira, Y., Inoue, Y., Tanoura, N. (1971). Reaction between carbohydrates and sulfuric acid : part I. Depolymerization and sulfation of polysaccharides by sulfuric acid. *Carbohydrate Research*, 18, 95-102.

Nam, K.S., Kim, M.K,, Shon, Y.H. (2007). Chemopreventive effect of chitosan oligosaccharide against colon carcinogenesis. *Journal of Microbiology and Biotechnology*, 17, 1546–1549.

Nishimura, S., Nishi, N., Tokura, S., Nishimura, K., Azuma, I. (1986). Bioactive chitin derivatives. Activation of mouse-peritoneal macro- phagesby O-(carboxymethyl) chitins. *Carbohydrate Research*, 146, 251-258.

No, H.K., Park, N.Y., Lee, S.H., Meyers, S.P. (2002). Antibacterial activity of chitosans and chitosan oligomers with different molecular weights *International Journal of Food Microbiology,* 74, 65-72

Okamoto, Y., Yano, R., Miyatake, K., Tomohiro, I., Shigemasa, Y., Minami, S. (2003). Effect of chitin and chitosan on blood coagulation. *Carbohydrate Polymers*, 53, 337-342.

Park, P.J., Je, J.Y., Byun, H.G., Moon, S.H., Kim, S.K. (2004a). Antimicrobial activity of hetero-chitosans and their oligosaccharides with different molecular weights. *Journal of Microbiology and Biotechnology*, 14, 317-323.

Park, P.J., Je, J.Y., Jung, W.K., Kim, S.K. (2004c). Anticoagulant activity of heterochitosans and their oligosaccharide sulfates. *European Food Research and Technology*, 219, 529-533.

Park, P.J., Je, J.Y., Kim, S.K. (2003a). Angiotensin I converting (ACE) inhibitory activity of hetero-chitooligosacchatides prepared from partially different deacetylated chitosans. *Journal of Agricultural and Food Chemistry*, 51, 4930-4934.

Park, P.J., Je, J.Y., Kim, S.K. (2003b). Free radical scavenging activity of chitooligosaccharides by electron spin resonance spectrometry. *Journal of Agricultural and Food Chemistry,* 51, 4624-4627.

Park, P.J., Kim, S.K., Lee, H.K. (2002). Antimicrobial activity of chitooligosaccharides on *Vibrioparahaemolyticus*. *Journal of Chitin and Chitosan*, 7, 225-230.

Park, P.J., Lee, H.K., Kim, S.K. (2004b). Preparation of hetero-chitooligosaccharides and their antimicrobial activity on *Vibrioparahaemolyticus*. *Journal of Microbiology and Biotechnology,* 14, 41-47.

Pospieszny, H., Chirkov, S., Atabekov, J. (1991). Induction of antiviral resistance in plants by chitosan. *Plant Science*, 79, 63-68.

Poulin, J.F. (2007). Étude du fractionnement d'un hydrolysat trypsique de β-lactoglobuline par électrodialyse avec membrane d'ultrafiltration. Master thesis, Université Laval, Québec, Canada..

Poulin, J.F., Amiot, J., Bazinet, L. (2006). Simultaneous separation of acid and basic bioactive peptides by electrodialysis with ultrafiltration membrane. *Journal of Biotechnology*, 123, 314-328.

Poulin, J.-F., Amiot, J., Bazinet, L. (2007). Improved peptide fractionation by electrodialysis with ultrafiltration membrane: influence of ultrafiltration membrane stacking and electrical field strength. *Journal of Membrane Science*, 299, 83-90.

Poulin, J.-F., Amiot, J., Bazinet, L. (2008). Impact of feed solution flow rate on peptide fractionation by electrodialysis with ultrafiltration membrane. *Journal of Agricultural and Food Chemistry,* 56, 2007-2011.

Pourcelly, G.; Bazinet, L. (2008). Developments of Bipolar Technology in Food and Bio-industries. *In* Handbook of Membrane Separations: Chemical, Pharmaceutical and Biotechnological Applications. Pabby, A.K.; Rizvi S.S.H. and Sastre, A.M. (Editors), CRC Press, taylor and Francis Group, Boca Raton, Floride. pp.581-633.

Qin,C., Du, Y., Xiao, L., Li, Z., Gao, X. (2002) .Enzymic preparation of water-soluble chitosan and their antitumor activity. *International Journal of Biological Macromolecules*, 31, 111-117.

Rao, S.B., Sharma, C.P. (1997). Use of chitosan as biomaterial: Studies on its safety and hemostatic potential. *Journal of Biomedical Material Research*, 34, 21-28.

Ravi Kumar, M.N.V. (2000). A review of chitin and chitosan applications. *Reaction and Functionality of Polymers,* 46, 1-27.

Shahidi, F., Arachchi, J.K.V., Jeon, Y.J. (1999). Food applications of chitin and chitosans, *Trends in Food Science and Technology*, 10, 37-51.

Sudharshan, N.R., Hoover, D.G., Knorr, D. (1992). Antibacterial action of chitosan. *Food Biotechnology*, 6, 257-272.

Sugano, M., Watanabe, S., Kishi, A., Izume, M., Ohtakara, A. (1988). Hypocholesterolemic action of chitosans with different viscosity in rats. *Lipids*, 23, 187-191.

Suzuki, K., Mikami, T., Okawa, Y., Tokoro, A., Suzuki, S., Suzuki, M. (1986). Antitumor effect of hexa-N-acetylchitohexaose and chitohexaose. *Carbohydrate Research*, 151, 403-408.

Tokoro, A., Takiwaki, N., Suzuki, K., Mikami, T., Suzuki, S. (1988). Growth inhibitory effect of hexa-*N*-acetylchitohexaose and chitohexaose against Meth-A solid tumor. *Chemical and Pharmaceutical Bulletin*, 36, 784–790.

Tsai, G.J., Su, W.H., Chen, H.C., Pan, C.L. (2002). Antimicrobial activity of shrimp chitin and chitosan from different treatments and applications of fish preservation. *Fisheries Science,* 68, 170-177.

Uchida, Y., Izume, M., Ohtakara, A. (1989). Preparation of chitosan oligomers with purified chitosanase and its application. In G.Braek, T. Anthonsen, and P. Sandford (Eds.), chitin and chitosan: Sources, chemistry, biochemistry, physical properties and applications. Barking, UK: Elsevier Applied Science, London, pp.373–382.

Xie, W., Xu, P., Qing, L. (2001). Antioxidant activity of water-soluble chitosan derivatives. *Bioorganic and Medicinal Chemistry Letters*, 11, 1699-1701.

Xu, J.G., Xhao, X.M., Han, X.W., Du, Y.G. (2007). Antifungal activity of oligochitosan against Phytophthora capsici and other plant pathogenic fungi in vitro. *Pest Biochemistry and Physiology*, 87, 220–228.

Yalpani, M., Johnson, F., Robinson, L.E. (1992). Antimicrobial activity of some chitosan derivatives. In: Brine, C.J., Sandford, P.A., Zikakis, J.P. (Eds.), Advances in Chitin and Chitosan. Elsevier, London, pp. 543– 548

Yin, X.Q., Lin, Q., Zhang, Q., Yang, L.C. (2002). O_2^- Activity of chitosan and its metal complexes. *Chinese Journal of Applied Chemistry*, 19, 325-328.

Yun, Y.S., Kim, S.K., Lee, Y.N. (1999). Antibacterial and antifungal effect of chitosan. *Journal of Chitin and Chitosan*, 4, 8-14.

Zhang, H., Du, Y., Yu, X., Mitsutomi, M., Aiba, S. (1999). Preparation of chitooligosaccharides from chitosan by a complex enzyme. *Carbohydrate Research*, 320, 257-260.

In: Handbook of Chitosan Research and Applications
Editors: Richard G. Mackay and Jennifer M. Tait
ISBN 978-1-61324-455-5

Chapter 2

CHITOSAN AND CHITOSAN DERIVATIVES: A SPOTLIGHT ON CELLULAR INTERACTIONS

Noha M. Zaki[*]
Department of Pharmaceutics. Faculty of Pharmacy,
Ain Shams University, El Abbassia,
Cairo, Egypt

ABSTRACT

Chitosan is one of the most sought-after polymers for biomedical and drug delivery applications (>1100 articles and > 700 patent applications published in 2009). It is, therefore, fundamental to understand and control its interactions with cells, focusing on e.g. cytotoxicity and cellular access issues. Chitosan based nanosysetems are capable of delivering their active payloads, such as low MW drugs, genetic material or other complex biopharmaceutical agents intracellularly. The surface properties of the polymer/nano-carriers play a crucial role in determining their interactions with cells and, as a result, the possibility and the mechanism of intracellular transport, thereby significantly affecting the efficacy of encapsulated drugs. Physical contact with the cell membrane or cell surface interactions, initiate a number of different internalisation pathways. Endocytosis is the formation of endocytic vesicles that have engulfed the polymer and the subsequent trafficking of these endosomes through the cell. Once inside the cell, the intracellular fate of the endosomal contents is an important determinant of successful delivery of the cargo. Moreover, cytotoxicity will rely on either membrane damage with leakage of intracellular components or a subtle decrease in metabolic activity of cells. In this chapter, methods giving essential information about the interactions of nanoparticles with cells grown in culture are compiled with a focus on the cytocompatibility issue commonly tested by the formazan assay and the cellular access. One limitation is the inability to test high concentrations of polymer/nanoparticles. We have developed a modification that allows testing elevated amounts of nanocarriers with

* Monazamet El Wehda El Afrikia St., El Abbassia, Cairo, Egypt. Tel: 002 0126144687, Fax: 002 24032059, Email: noha.zaki@pharm.asu.edu.eg. Current position is assistant Professor at Department of Pharmaceutics, Taif University, Saudi Arabia

particular awareness for physiological pH and osmolarity. Concerning cellular access, this can be observed and quantified employing fluorescently labeled chitosan, and trypan blue as an effective quencher of any extracellular fluorescence.

Keywords: *chitosan , nanoparticles; cytotoxicity; endocytosis; internalization; trypan blue*

INTRODUCTION

Water-soluble polycations have been described for various biomedical applications. Most commonly investigated is the polycationic poly(ethylenimine) (PEI) as a non-viral vector systems for the transfer of DNA and RNA into cells and tissues. These polymers need to be biocompatible, nontoxic, non-immunogenic, biodegradable, with a high drug-carrying capacity and controlled release of drugs at the target site. The term ''biocompatibility'' encompasses many different properties of the materials, however, two important aspects of the biomaterial screening refers to their in vitro cytotoxicity and blood-compatibility behaviour [1, 2].

Chitosan (essentially a highly deacetylated derivative of chitin, poly(N-acetyl-1,4-β-D-glucosamine) is currently one of most investigated polycation polymers in the field of pharmaceutics and biomaterials.

There are indeed a number of advantageous points in the use of chitosan: it is sourced from natural products [3] |It is abundant in nature being obtained from chitin present in crustacean shells., it is enzymatically degradable [4-6], it is an easily functionalizable and/or complexable [5], and above all it has a relatively low toxicity, in comparison to most polycations. Although depending on its molecular weight, on the nature of the counterions of its protonated amines and on the cell model, chitosan's IC50 values often exceed 1 mg/mL[6]. These values compare very favourably with those of other polycations: 0.1 - 1 mg/mL for poly(viny pyridinium bromide [7, 8], 0.01-0.1 mg/mL for poly(L-lysine) and poly (ethyleneimine) [9, 10] or diethylaminoethyl dextrane[10], <0.01 mg/mL for poly(2-dimethylamino ethyl methacrylate)[11] and even lower values for permanently quaternarized derivatives such as poly(diallyl dimethyl ammonium chloride) (polyDADMAC)[9], or also trimethylcthitosan[11], although the cytotoxicity of the latter decreases at high degrees of acetylation[12].

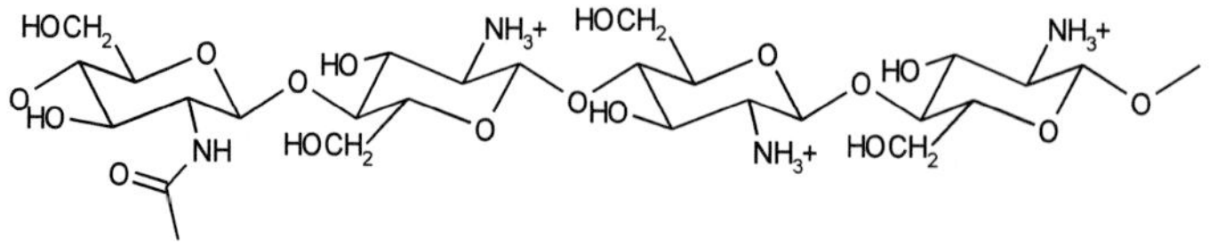

Figure 1. Structure of chitosan.

Its relatively low toxicity makes chitosan comparable to considerably more expensive synthetic polycations, such as starbust dendrimers[9, 13] or polyamidoamines[14]. Low toxicity is also reported in animals: the *LD*50 of chitosan in laboratory mice is 16 g/kg body weight, which is close to that of salt or sugar. Chitosan is safe in rats up to 10% in the diet.

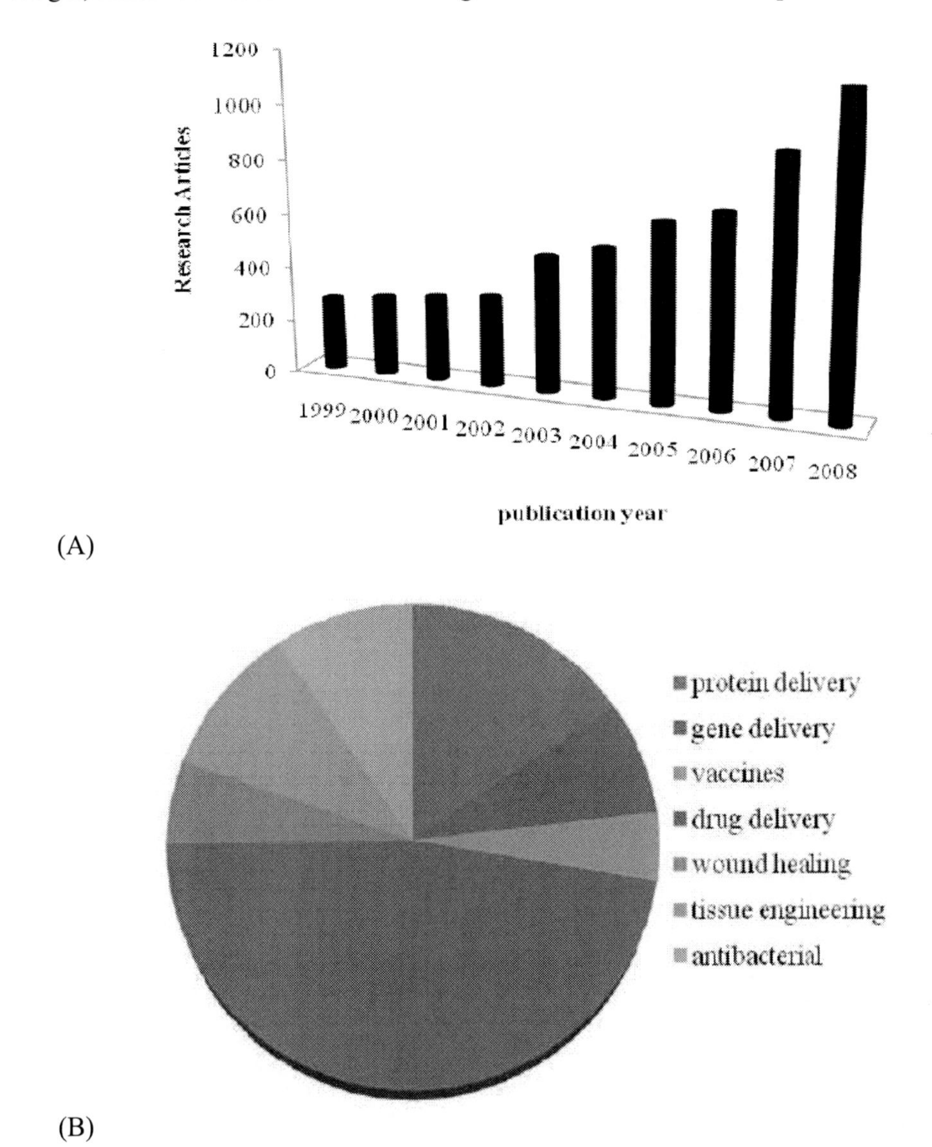

Figure 2. A PubMed search of the biomedical applications of chitosan. (A) Research articles published from 1999 to 2009. (B) Main applications of chitosan as a biomedical material. This survey did not include review articles.

The above properties render chitosan an appealing building block, in particular when electrostatics is at the basis of the functional and structural behaviour of the biomaterial; a number of reviews provide detailed information about these uses of chitosan, ranging from tissue engineering[15], a variety of pharmaceutica/nutraceutical applications[16, 17] and also clinical uses, due its hemostatic action[18, 19]. Estimation of the number of papers published on the use of chitosan in drug delivery over the past 10 years suggests that this molecule is attracting increasing interest in the drug delivery community. On closer examination of the published articles on the biomedical applications of chitosan in the PubMed database, it is evident that more than half are related to drug delivery, and a significant fraction of these are related to the delivery of macromolecules such as peptide and protein drugs, vaccines or genes. Another major area of chitosan research is related to tissue engineering (Figure 2)

Biocompatibility of Chitosans

In general, polycations are considered to be cytotoxic [9], with the mechanism of cytotoxicity being mediated by interactions of polycations with cell membranes [10]. As a polycation, the biocompatibility of chitosan would be a major concern. Biocompatibility is influenced by different properties of chitosans such as MW, charge density and type of the cationic functionalities, structure and sequence (block, linear, branched), and conformational flexibility [7].

Illum et al. [8] was the first to report in 1994 that high molecular weight chitosan solutions significantly increased the transport of insulin across the nasal mucosa in rats and sheep. Numerous human studies, that followed, have confirmed the potential of chitosan to improve the mucosal absorption of peptides [9–11]. However, the poor aqueous solubility and loss of penetration-enhancing activity above pH 6 are major drawbacks of chitosan at physiological conditions [20]. Derivatization of chitosan was attempted to solve these problems (Table 1) for example partial quaternization of chitosan's primary amine groups has been used to obtain a chitosan derivative that is soluble at physiological conditions namely trimethylchitosan (TMC) .

Table 1. Rationale for synthesis of chitosan derivatives

Derivatives of chitosan for biomaterial applications

- Low molecular weight (Oligomers)
  - improve aqueous solubility
  - decrease viscosity
  - provide better control over bioactive binding
- High degree of deacetylation (DD)
  - provide positive charges for drug delivery, gene delivery and antimicrobial activity
  - provide reactivity groups for synthesis of chitosan derivatives
- Quaternization (e.g. TMC)
  - improve aqueous solubility at physiological pH
  - Higher positive charge
  - Less cytotoxicity

On the other hand, chitosan as non viral gene delivery was first described in 1995 [15]. As a polycation, this polymer can form neutral or positively charged complexes with DNA due to electrostatic interactions [3]. Polycation/DNA complexes were usually found to be less cytotoxic than uncomplexed polycations [6]. Most importantly an appropriate balance between complex stability and ease of complex dissociation is needed to provide the highest in vivo transfection efficiency. This has triggered the synthesis of chitosan oligomers [21] and subsequent studies demonstrated improved chitosan-mediated gene delivery based on easily dissociated chitosan polyplexes of highly defined chitosan oligomers [22-24]. Other chitosan derivatives such as quaternized chitosan oligomers have also been shown to improve the physicochemical properties of chitosan and to enhance chitosan-mediated gene delivery [25, 26]. In the endeavour to obtain chitosans with better physicochemical and biological properties for drug, protein and gene delivery applications, testing the biocompatibility/ cytotoxicity becomes an essential prerequisite.

The methods for studying the biocompatibility of chitosan polymers and nanosystems are described herein:

Methods for Studying the Biocompatibility of Chitosans

1. Hemolysis Test

Erythrocyte hemolysis is commonly used as a model to investigate membrane interactions, because erythrocytes are readily available and their lysis is easily measured. The interaction of different polymers/nanomaterials with erythrocyte membranes can be compared with this *in vitro* method [27]. The blood compatibility of the polycations was quantified by spectrophotometric measurement of hemoglobin release from erythrocytes after polymer treatment.

2. MTT Assay

The MTT assay is an early indication of cellular damage where a reduction in metabolic activity of mitochondria in cells is detectable. The yellow tetrazolium salt [3-(4,5-Dimethylthiazole-2-yl)-2,5-diphenyltetrazolium-bromide] is metabolized by mitochondrial succinic dehydrogenase enzyme of proliferating cells to yield a purple formazan reaction product which is largely impermeable to cell membranes, thus resulting in its accumulation within healthy cells (Figure 3). Solubilization of the cells using dimethylsulfoxide (DMSO) results in the liberation of the product which could then be detected by a simple colorimetric assay. The ability of cells to reduce MTT provides an indication of the mitochondrial integrity and activity, which, in turn, may be interpreted as a measure of viability and/or cell number [28, 29]. A mouse connective tissue fibroblast cell line, L929 is recommended by USP 26 to evaluate cytotoxicity as a direct contact test. The test was previously described by [8]. Briefly, L929 was cultured in DMEM supplemented with 10% fetal calf serum and 2 mM glutamine without antibiotics. The cells were cultivated in an incubator at 37°C, 95% relative humidity and 10% CO_2. Chitosan was first dissolved in 0.5% acetic acid solution, and then diluted with equal volume of double DMEM medium. The pH of this stock solution was adjusted to 6.5 for all the chitosans tested. Thereafter, DMEM (pH 6.5) was used to prepare serial dilutions of the polymer. L929 cells were seeded into 96-well microtiter plates at a density of 8000 cells/well. 24 hour later, culture medium was replaced with 100 μl serial dilutions of chitosan (0.6-10 mg/ml) and cells were incubated for 24 h. Subsequently, 20 μl MTT (5 mg/ml) was added to each well. After 2 h, unreacted dye was aspirated and the formazan crystals were dissolved in 200 μl/well DMSO giving a purple color. Absorbance [A] at 580 nm was measured using a plate reader.

The relative cell viability compared to control cells containing cell culture medium (pH 6.5) without polymer was calculated as:

$$\text{Cell viability (\%)} = \frac{[A]test}{[A]control} \text{ X } 100 \qquad (1)$$

The IC50, the drug concentration at which inhibition of 50% cell growth occurs in comparison with that of the control sample, is calculated by the curve fitting of the cell viability data.

We have previously reported a modification of this test for testing high concentrations of nanomaterials/polymers [29, 30]. For cell culture studies, unpurified (synthesized) polymers or engineered nanoparticles should not be used owing to the non-physiological medium in which the polymer/modified polymer are dispersed/prepared in and the possible residual presence of reactants, catalysts, initiators or other deleterious reagents which could be cytotoxic. Hence purification is crucial, the most common method is by ultrafilteration followed by freeze drying in the presence of cryo-protectant. The freeze dried product must be dispersed in the least amount of water to serve as a stock to be further diluted with 5X concentrated culture medium. This enables testing of high concentrations of polymers or nanoparticles and avoids the dilution resulting from addition of the ready-made culture medium. The test and control solutions are obtained by mixing one part 5x culture medium with 4 parts of nanoparticles dispersion and sterile MilliQ water respectively.

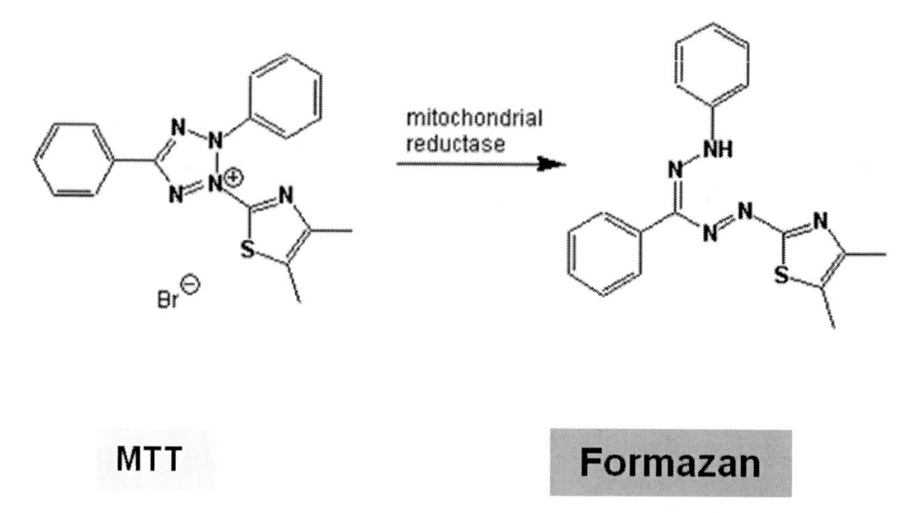

Figure 3. Reduction of yellow tetrazolium salt by mitochondrial succinic dehydrogenase enzyme of proliferating cells to yield a purple formazan product.

The MTS cell viability reagent is 3-(4,5-dimethylthiazol- 2-yl)-5-(3-carboxymethoxyphenyl)-2-(4-sulfophenyl)-2H-tetrazolium works with the same principle as MTT reagent except that the formazan crystals do not need to be solubilised by DMSO [31].

3. Trypan Blue Exclusion Test

Trypan blue does not penetrate unperturbed cell membranes and it is therefore easy to observe dye exclusion using light microscopy. Trypan blue dye is effluxed by vital cells. The test is performed by incubating the cells with 0.1% trypan blue solution in PBS (0.01 M) at 37 °C and then examined for dye exclusion by light microscopy. The intact monolayers would show no inclusion of dye, whereas the damaged cells would stain blue [32]

4. Lactate Dehydrogenase (LDH) Release Test

LDH reflects the damage /leakage of plasma membranes. LDH is a stable cytosolic enzyme present in the cytosol that is released upon cell lysis. This assay permits the investigation of chemicals that may induce alternations in cell integrity and to measure the membrane-damaging effects via the quantity of LDH in the lavage fluid/culture media at different time points [27, 33],[34]. The LDH content in these samples was assayed spectrophotometrically utilizing a commercial kit to determine the amount of reduced nicotinamide adenine dinucleotide (NAD) at 492 nm in the presence of lactate and LDH. Control experiments were performed with 0.1% (w/v) Triton X-100 and set as 100% cytotoxicity. LDH release was calculated by the following equation:

$$\text{LDH (\%)} = \frac{[A]sample - [A]medium}{[A]100\% - [A]medium} \text{ X } 100 \qquad (2)$$

where [A]sample, [A]medium, [A]100% denote the absorbance of the sample, medium control and Triton X-100 control, respectively.

5. Live/Dead Assay Kit

Double Staining Kit is utilized for simultaneous fluorescence staining of viable and dead cells. This kit contains Calcein-AM and Propidium Iodide (PI) solutions, which stain viable and dead cells, respectively. Calcein-AM, acetoxymethyl ester of calcein, is highly lipophilic and cell membrane permeable. Though Calcein-AM itself is not a fluorescent molecule, the calcein generated from Calcein-AM by esterase in a viable cell emits strong green fluorescence (excitation: 490 nm, emission: 515 nm). Therefore, Calcein- AM only stains viable cells. On the other hand, PI, a nuclei staining dye, cannot pass through a viable cell membrane. It reaches the nucleus by passing through disordered areas of dead cell membrane, and intercalates with the DNA double helix of the cell to emit red fluorescence (excitation: 535 nm, emmision: 617 nm). Since both calcein and PI-DNA can be excited with 490 nm, simultaneous monitoring of viable and dead cells is possible with a fluorescence microscope (Figure 4). With 545 nm excitation, only dead cells can be observed [29].

Others

Neutral red is a vital dye that is endocytosed by viable cells and internalized within lysosomes and determined colorimetrically at 540 nm using a plate reader. The combination of this test with the MTT assay has been carried out to add reliability to the final evaluation of cytotoxicity for the chitosan materials [35].

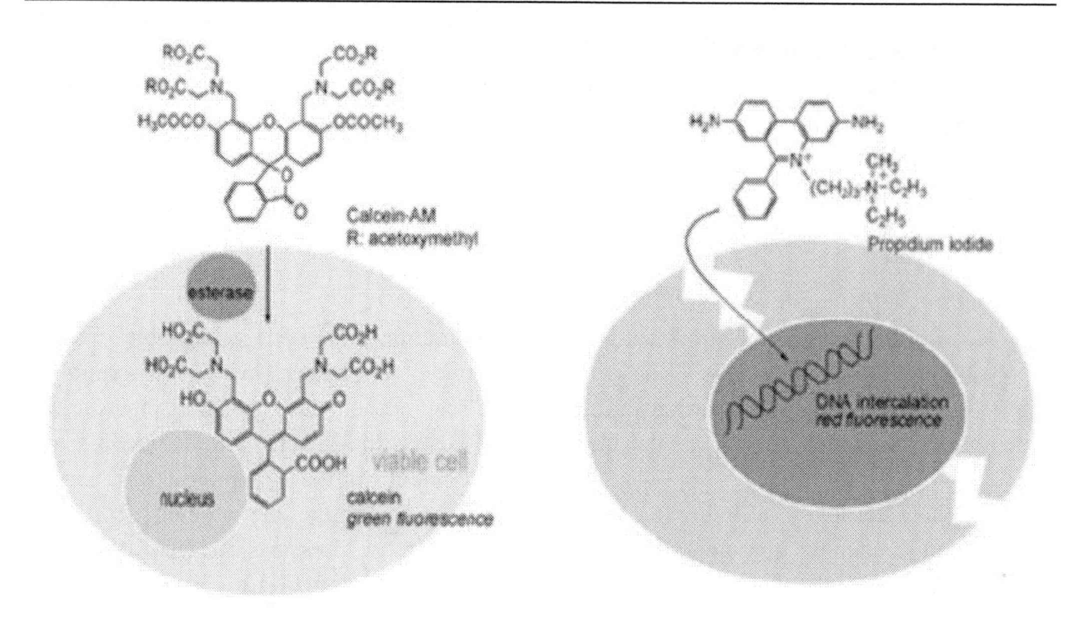

490 nm **545 nm**

Figure 4. Principle of live/dead assay. The Kit contains Calcein-AM and Propidium Iodide (PI) solutions, which stain viable and dead cells respectively.

The BrdU-based enzyme-linked immunosorbent assay is used to investigate the ability of cells to proliferate [36]

Chitosan Derivatives

Trimethyl chitosan(TMC). TMC is synthesized based on the method first published by Domard and coworkers [37]. They showed the alkylation of primary amines of chitosan by reaction of this polymer in strong alkaline conditions with an excess of iodomethane using N-methyl-2-pyrrolidone (NMP) as solvent. N,N,N-Trimethylated chitosan (TMC) has been shown to have mucoadhesive properties and is able to open tight junctions above a certain degree of quaternization (DQ) [38]. Several studies have determined that the optimal DQ for transepithelial delivery of low molecular weight drug molecules and/or proteins, was 40–50% [39-41]. In addition, TMC has been used to complex and condense DNA to yield polyplexes for gene delivery purposes [42]. The biocompatibility of TMC was studied and results revealed an increase in cytotoxicity as a function of MW. For example trimethylchitosan (TMC) 400 kDa was found to display the highest cytotoxicity, whilst TMC 25 kDa and TMC 5 kDa were almost nontoxic [43]. Whilst it has been found that chitosan was nontoxic irrespective of the MW, the cytotoxicity of TMC can probably attributed to the positive charge carried by the polymer even in neutral environment, which can subsequently interact with the negatively charged cell membrane, resulting in membrane damage. Similar results were reported for other polycations, such as DEAE-dextran [8]and polyethylenimine (PEI) [44]. PEGylation decreased the cytotoxicity of TMC considerably, the extent of which was substitution degree-, TMC MW- and PEG MW- dependent. The effect of PEG was attributed to steric effects, which acts to shield a proportion of the positive charges present on TMC since parts of the primary amino groups in TMCs were substituted by PEG, therby decreasing the overall positive charge density [43].

In a recent study, Verheul et al. [45] reported a two-step method to synthesize TMC with tailorable DQ avoiding O-methylation of the C-3 and C-6 hydroxyl groups of chitosan. The synthesized TMC with different DQs were studied for their physicochemical properties namely solubility. Their cytotoxicity (MTT assay and LDH release) and the ability to open tight junctions (transepithelial electrical resistance, TEER), were measured and compared with TMC synthesized via the traditional method. Results showed that the DQ of the TMC could be tailored by varying the reaction time. O-Methyl free TMCs with a DQ>33% were readily soluble in aqueous solutions at pH 7. The O-methyl free TMCs demonstrated a larger decrease in TEER of Caco-2 cells than O-methylated TMCs pointing at their higher penetration enhancing effect. Also, with increasing DQ, an increase in cytotoxicity (MTT) and membrane permeability (LDH) was observed. The new well-characterized O-methyl free TMCs demonstrated more favourable biological characteristics over the traditional TMC and allowed to effectively study the influence of the DQ of TMC in various delivery systems.

TMC has higher positive charges than chitosan and hence should be superior to chitosan in forming polyelectrolyte complex with negatively charged DNA. The use of TMC oligomers (~20 monomers) with luciferase plasmid in COS-1 and Caco-2 cell lines has shown that as gene transfecting agents they outperform native chitosan. The transfection efficiency and cytotoxicity was investigated with respect to DQ in chitosans using pGL3 luciferase reporter gene in COS-7 and MCF-7 cell lines. DQ of 44% on the chitosan backbone gave an optimum derivative for transfection. The cytotoxicity increased with increasing DQ but all the derivatives were less toxic than cationic linear PEI. At a similar DQ, higher toxicity was seen in polymeric chitosan derivatives than oligomeric chitosan ones. It appears that quaternization of 44% on the chitosan backbone gave an optimum derivative for transfection. TMC has been employed as a stabilizer for preparation of nanoparticles of DNA with poly-(e-caprolactone) by emulsion-diffusion evaporation using a blend of poly-(vinyl alcohol) and chitosan derivatives as stabilizers [46]. The cytotoxicity was found to be moderate and virtually independent of the stabilizer's concentration (TMC DQ 4%, 10%, 18%, 66%) with the exception of the highly quaternized TMC (DQ 66%) being significantly more toxic. In another study, the cytotoxicity and ciliotoxicity of TMC polymers with different DQ were studied using intestinal Caco-2 cell monolayers and ciliated chicken embryo trachea [47]. No substantial cell membrane damage could be detected on the Caco-2 cells, while the effect on the ciliary beat frequency of chicken in vitro was found to be marginal at a concentration of 1.0% (w/v). No acute toxicity was found with TMC and TMC oligomers by means of the ciliary beat frequency (CBF) tests.

Aryl Chitosan. Sajomsang et al. [48] reported the synthesis of methylated N-aryl chitosan derivatives by a single treatment with iodomethane in the presence of N-methyl pyrrolidone (NMP) and sodium hydroxide and studied their mucoadhesive property and biocompatibility. The physicochemical properties of chitosan derivatives were determined by ATR-FTIR, NMR, X-ray diffraction (XRD) and thermogravimetric (TG) techniques. The XRD analysis showed that the crystallinity and thermal stability of methylated chitosan derivatives were lower than those of chitosan. The effects of degree of quaternization (DQ), polymer structure and positive charge location on mucoadhesive property and cytotoxicity were investigated by using a mucin particle method and MTT assay. The cytotoxicity of methylated chitosan derivatives increased with increasing DQ. It was noted that the methylated N-(4-N,N-dimethylaminocinnamyl) chitosan chloride showed the highest cytotoxicity compared to other methylated chitosan derivatives at similar DQ level. The effect of the DQ and polymer structure on the mucoadhesive property was notable when DQ was higher than 65%. It was demonstrated that the DQ and the polymer structure played an important role on mucoadhesive property while the cytotoxicity correlated with DQ, the polymer structure, positive charge location and molecular weight after methylation [48].

Thiolated chitosan. Sulfhydryl bearing agents can be covalently attached to the primary amino group via the formation of amide bonds. In the case of the formation of amide bonds the carboxylic acid group of the ligands cysteine and thioglycolic acid reacts with the primary amino group of chitosan mediated by a water soluble carbodiimide to produce chitosan-cysteine conjugates and chitosan-thioglycolic acid conjugates respectively [49].

Recently a novel chitosan derivative has been obtained by the introduction of a 4-thiobutylamidine (TBA) substructure leading to a chitosan-TBA conjugate. This conjugate is a more significant exponent of the thiolated polymers or so-called thiomer familywhere the immobilization of sulfhydryl groups on the polymer backbone leads to strongly enhanced mucoadhesive as well as permeation enhancing features [50]. The conjunction of these properties renders this chitosan derivative a promising excipient in particular for the non-invasive administration of peptide drugs. the influence of the covalent linkage of TBA substructures to different d-glucosamine-derivatives on their cytotoxicity was investigated using MTT assay and results obtained confirmed the safety of these conjugates [51].

Chemical modification of d-Glucosamine and chitosan was attempted by the immobilization of thiol groups utilizing 2-iminothiolane [52]. The toxicity profile of the resulting d-glucosamine-TBA(4-thiobutylamidine) conjugate, of chitosan-TBA conjugate and of the corresponding unmodified controls was evaluated by haemolysis experiments in vitro. The erythrocyte membrane contains anionic glycoproteins able to bind to the protonated amino group of glucosamine and of their derivatives resulting in membrane curvature, rupture and release of haemoglobin [52, 53]. Compounds bearing primary amines demonstrated a significant toxic effect on red blood cells causing them to agglutinate [54] in contrast to the lower haemolytic effect of the d-glucosamine-TBA conjugate. This could be attributed to the fact that the thiolated compound had their primary amines derivatized to secondary functionalities. Compounds with secondary amino groups have demonstrated lower toxicity with than those with primary amino residues [55]. Moreover, the lower membrane-damaging effect of thiolated chitosan in comparison to the corresponding unmodified polymer might be explained by the formation of intra- as well as inter- molecular disulfide bonds, which impart rigidness to the molecules with more difficulties to attach to the cellular membrane than flexible molecules [8, 56].

A new in situ gelling system based on thiolated chitosan was reported [57]. The system comprises chitosan-TGA and different oxidizing agents to enhance the dynamic viscosity and consequently to reduce the time of sol –gel phase transition. Oxidizing agents chosen were hydrogen peroxide, sodium periodate, ammonium persulfate and sodium hypochlorite for their oxidizing potential and safety. The toxicity of the thiomer/oxidizing agent systems on Caco-2 cells was evaluated via membrane integrity (LDH assay) and cellular viability (MTT assay). Results revealed that neither the gel alone nor the gels with oxidizing agents induced severe cytotoxicity. Quantitative assessment of viable cells by MTT assay after 3 h of contact with the conjugate alone (control) showed 90% mitochondrial activity indicating a likely non-toxic nature of chitosan-TGA. In cases where the oxidizing agents were incorporated in the gel, viability of the cells was about 70–85% after 3 h and was not dramatically altered after 24h incubation. More than 50% of the cells were viable after 24 h of the experiment. The LDH release test revealed that hydrogen peroxide was the most harmless oxidizing agent in comparison to all others.

N-succinyl-chitosan (Suc-Chi). Suc-Chi was initially developed as wound dressing materials[58], it is currently also applied as cosmetic materials (Moistfine liquids)® [59]. In addition, water-soluble chitin and chitosan including Suc-Chi were applied for a patent as a treatment of arthritis in Japan [60]. In contrast to ordinary chitosan which is dissolved only in acidic water, highly succinylated Suc-Chi (degree of succinylation:>0.65) can dissolve in alkaline medium. As a drug carrier, Suc-Chi has favourable properties such as biocompatibility, low toxicity and long-term retention in the body. Water-insoluble and water-soluble conjugates with mitomycin C were prepared using a water-soluble carbodiimide[61]. In vitro release characteristics of these conjugates showed a pH-dependent manner, except that water-insoluble conjugates showed a slightly slower release of MMC than water-soluble ones. The conjugates of mitomycin C with Suc-Chi showed good antitumour activities against various tumours [61-63]

Chitosan-Based Nanoparticles

Chitosan nanoparticles have been made by chemical cross-linking with glutaraldehyde, glyoxal, and ethylene glycol diglycidyl ether[64]. Although these are very good cross-linkers, they are not preferred owing to their physiological toxicity. Chitosan is polycationic in acidic media (pKa 6.5) and can interact with negatively charged species such as TPP, sodium sulfate, dextran sulphate, polyacrylic acid, sodium alginate. This characteristic can be employed to prepare cross-linked chitosan nanoparticles. The interaction of chitosan with TPP leads to formation of biocompatible cross-linked chitosan nanoparticles, which can be efficiently employed in protein and vaccine delivery [65]. The cross-linking density, crystallinity, and hydrophilicity of cross-linked chitosan can allow modulation of drug release and extend its range of potential applications in drug delivery.

Huang et al. [35] found no difference in cytotoxicity between CS polymer and nanoparticles (NPs) toward the A549 cells. Both chitosan entities exhibited significant cytotoxicity only at concentrations higher than 0.741 mg/ml, the cytotoxicity not being significantly reduced by a lowering of the polymer Mw to 10 kDa. On the other hand, decreasing the DD of the polymer from 88% to 61% was found to attenuate the cytotoxicity of CS and NPs to comparable degrees. The disparity could be attributed to the smaller effect of Mw on the potential of CS and NP compared with changes in DD. The reported cytotoxicity of cationic polymers, such as poly-L-lysine, poly-L-arginine, and protamine, corroborated these results. Although the number of primary amino groups was important, the charge density resulting from the number of groups and the three-dimensional arrangement of the cationic residues were also important contributors of cytotoxicity for a material. Chitosans with high degrees of deacetylation have extended conformation because of charge repulsion, which might allow them to bind more readily to cell membranes than coiled chitosans of lower degrees of deacetylation.

A novel formulation of nanoparticles was developed comprised of two natural marine-derived polymers, namely chitosan and carrageenan, and evaluated for their potential for the association and controlled release of macromolecules [66]. Nanoparticles were obtained in a hydrophilic environment, under very mild conditions, avoiding the use of organic solvents or other aggressive technologies for their preparation. The developed nanocarriers presented sizes within 350–650 nm and positive zeta potentials of 50–60 mV. Polymeric interactions between nanoparticles' components were evaluated by Fourier transform infrared spectroscopy. Using ovalbumin as model protein, nanoparticles evidenced loading capacity varying from 4% to 17% and demonstrated excellent capacity to provide a controlled release for up to 3 weeks. Furthermore, nanoparticles have demonstrated to exhibit a noncytotoxic behavior in the MTS assay performed using L929 fibroblasts.

Recently, two publications demonstrated that hyaluronic acid (HA) markedly reduced the cytotoxicity of chitosan nanoparticles either by incorporation of [67] or coating with HA [29]. The original idea of incorporating HA into the nanoparticles was conceived to improve their cellular targeting capacity. Indeed, apart from its biocompatibility, biodegradability, and mucoadhesive character, HA is known for its implication in several processes, such as the regeneration of corneal and conjunctival epithelial cells, through an interaction with the CD44 receptor[68]. CD44 is expressed in the human cornea and conjunctiva and participates in a wide variety of cellular functions, including receptor-mediated internalization and degradation of hyaluronan[67]. For example, this receptor-mediated process has been reported to be the mechanistic explanation for the cellular uptake of HA-targeted liposomes to tumor cells that overexpress the CD44 receptor [69].

In our recent publication [29], we have demonstrated that the uncoated, positively charged CS nanoparticles have higher cytotoxicity than the hyaluronic acid-coated ones using the MTT assay and live/dead assay in two cell lines namely: fibroblasts L929 and macrophages J774.2. The combination of two different methods viz MTT assay and live/dead assay gives a clue about the *modus operandi* of NPs toxicity whether intracellularly and/or membrane-related. The cytotoxicity by the MTT assay and also the live/dead stain revealed the higher cytotoxicity of uncoated positively charged CSNPs compared with HA-coated ones. The cytotoxicity was nanoparticles dose-dependent. This was consistent with previously reported results [35, 70]. The cytotoxicity was higher in case of macrophages 744.2 as compared to the non phagocytic cell line fibroblasts L929 at all concentrations of NPs indicating that the phagocytic capacity of the macrophages is involved in the uptake and cytotoxicity of CSNPs. This was further confirmed when experiments were done at 4 ^{o}C (a condition when phagocytosis is inhibited), and lower cytotoxicity values were obtained as compared to the values found at 37 ^{o}C in macrophages J774.2 cell line.

These results are consistent with previous reports by Lamerchand et al. [71] and Illum et al. [72]. At non cytotoxic concentrations, the live/dead assay indicated that the membrane integrity was disrupted in case of uncoated but not HA-coated chitosan nanoparticles possible due to electrostatic interactions between positively charged nanoparticles and negatively charged cell membrane (Figure 5). Consistent results were reported by Zwiorek et al. [73] for cationic gelatin nanoparticles and Harush-Frenkel et al.[74] working on positively and negatively charged polylactide NPs

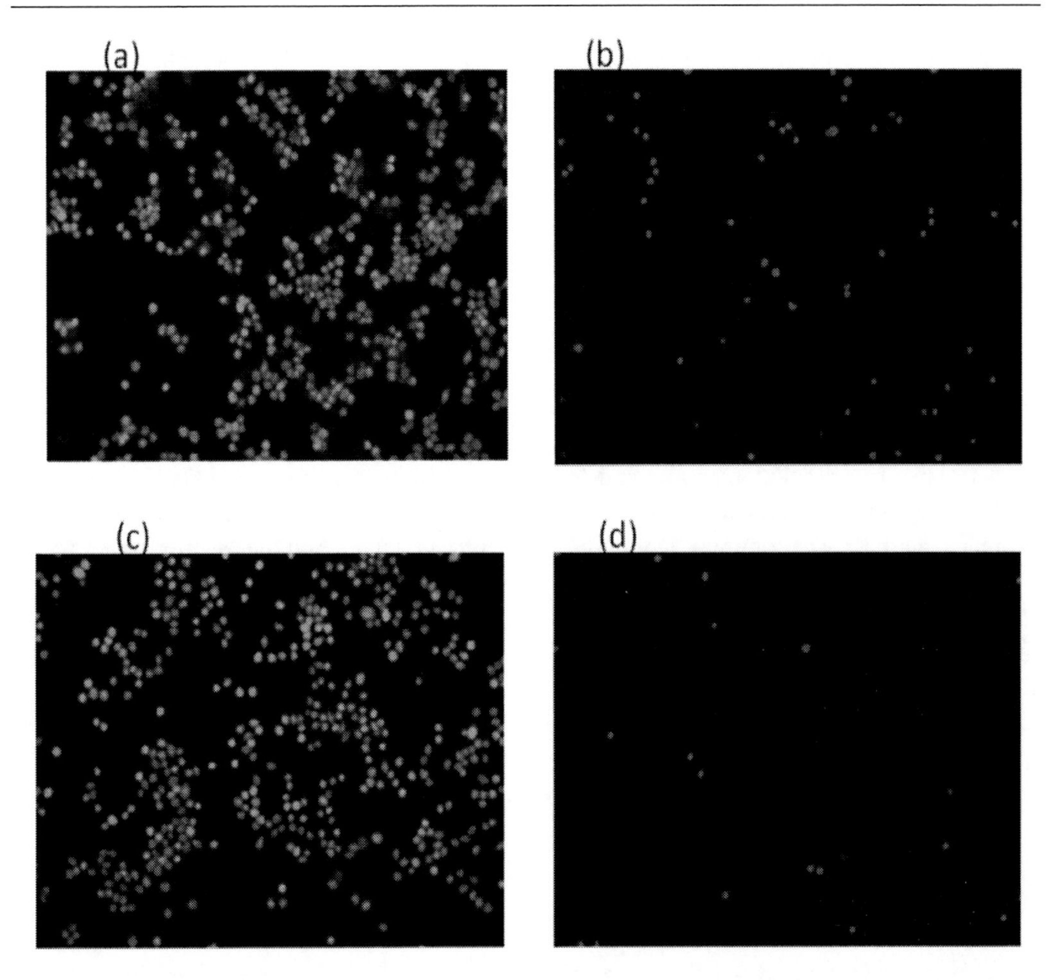

Figure 5. Representative fluorescent images of macrophages J774.2 cells showing the proportion of viable (stained green with Calcein AM) and dead (stained red with propidium iodide) cells 24 h-post treatment with uncoated CSNPs (a and b) and HA-coated CSNPs (c and d).

Chitosan Polyplexes

Another polyplex based on different molecular-weight chitosan-EDTA conjugates was developed as a carrier matrix for nanoparticulate gene delivery systems [75]. Covalent binding of EDTA to more than one chitosan chain provides a cross-linked stabilized particles. pDNA/chitosan-EDTA particles, generated via coazervation, were characterized in size and zeta potential by electrophoretic light scattering and electron microscopy. Stability was investigated at different pHvalues by enzymatic degradation and subsequent gel retardation assay. Lactate dehydrogenase assay was performed to determine toxicity. Furthermore, transfection efficiency into Caco-2 cells was assessed using a beta-galactosidase reporter gene. Chitosan-EDTA produced from low-viscous chitosan with 68% amino groups being modified by the covalent attachment of EDTA showed the highest complexing efficacy resulting in nanoparticles of 43 nm mean size and exhibiting a zeta potential of +6.3 mV. These particles were more stable at pH 8 than chitosan control particles. The cytotoxicity of chitosan-EDTA particles was below 1% over a time period of 4 hours. These new nanoplexes showed 35% improved in vitro transfection efficiency compared with unmodified chitosan nanoparticles[75].

The high toxicity of PEI limits its application in gene therapy. Therefore, the combination of chitosan and PEI might be a promising approach for enhancing transfection efficiency while decreasing the cytotoxicity at the same time. Neither chitosan/DNA complex nor PEI/chitosan/DNA complex showed cytotoxicity in Hela cells. In contrast PEI/DNA complex demonstrated significant cytotoxicity to the same cells even though the same amount of PEI was used in both experiments. The mechanism for such reduced toxicity was suggested to be interaction between CS nanoparticles and PEI [76].

Chitosan Oligomers. The polyplexes formed by chitosan with pDNA exhibited low cytotoxicity and have been used for delivery of genes to mucosal tissues in vivo [77]. Bearing in mind the positive results reported for sugar-substituted poly-L-lysine, PEI and chitosan in terms of improved cellular uptake and transfection efficacy, a series of chitosan oligomers were substituted with the trisaccharide GlcNAc-GlcNAc-2,5 anhydroMan branch at their primary amines. The optimal length of the chitosan chain and degree of trisaccharide substitution were selected from structure–activity studies in cell cultures. Using linear (unsubstituted) chitosan oligomers and PEI as controls, the trisaccharide-substituted chitosan oligomers (TCO) demonstrated higher efficacy in targeting cell lines expressing cell-surface lectins capable of binding to the GlcNAc residue at the free end of TCO. The trisaccharide substitution enhanced the in vitro colloidal stability and the buffering capacity of the polyplexes. The in vivo transfection efficacy of the TCO was superior to linear chitosan after intratracheal administration to the mouse lung and analysis of luciferase gene expression [78].

Cellular Access of Chitosans

Table 1. Fluorescent markers for visualization of chitosans interactions with cells

Marker	Application	Excitation λmax (nm)	Emission λmax (nm)	Reference
DiI red	Intracellular probe	577	590	[84]
Lysotracker red-DND99	Late endosomes and lysosomes	577	590	[80]
DAPI /Hoechst	Nuclear stain	358	461	[89]
Phalloidin, Texas red labeled	Actin filaments	595	615	[90]
Concanavalin A, Alexa Fluor® 594	Cell surface glycoprotein	590	617	[91]
FITC	Fluorescent label for chitosan polymer or nanoparticles	490	518	[79, 84]
Trypan blue	Extracellular probe (quench non- internalized FITC labeled nanoparticles fluorescent signals)			[79]

Chitosan polymer or nanoparticles are taken up by mammalian cells grown in culture primarily through non-specific endocytosis [79], at which point they are localized to endosomes. The proton sponge effect of chitosan [78] or incorporation of other polymers [80] provides a mechanism for endosomal escape of chitosan/polyplexes otherwise they will ultimately be trafficked to lysosomes. Confocal microscopy is used for estimation of cellular access and intracellular trafficking of chitosans [81-83] as illustrated in Figure 6. For quantification of cellular uptake, flow cytometry [84-86] or microfluoremetric techniques [81, 87, 88] are commonly used. Table 1 shows the fluorescent dyes commonly used in confocal microscopic studies.

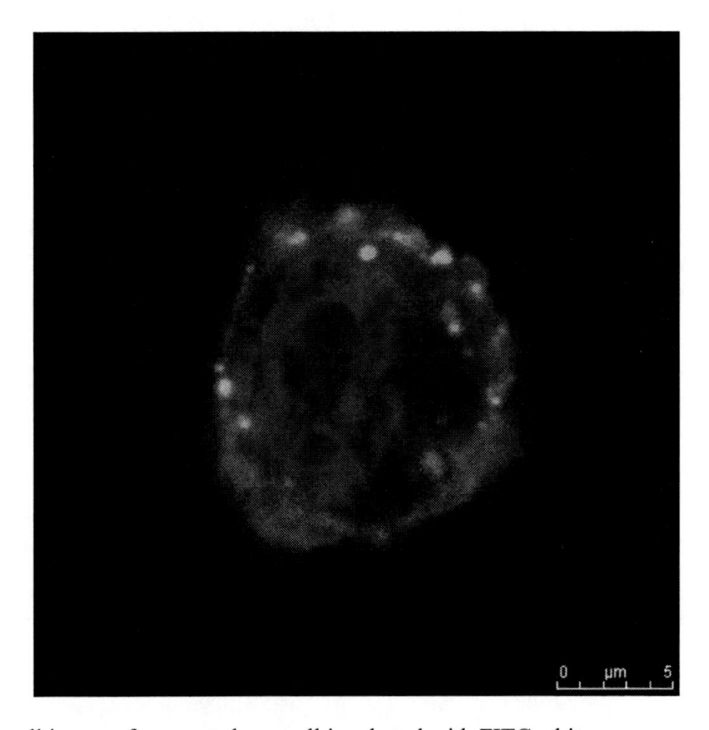

Figure 6. Single cell image of a macrophage cell incubated with FITC–chitosan nanoparticles showing cellular access of chitosan nanoparticles and localization in cytosol (green spots are the FITC labelled chitosan nanoparticles, blue is the nucleus stained with DAPI and red is the actin filaments stained with texas-red phalloidin).

In vitro transfection studies confirmed that the incorporation of poly(propyl acrylic acid) (PPAA) into the chitosan-DNA nanoparticles enhanced gene expression in both HEK293 and HeLa cells compared to chitosan nanoparticles alone. The dose and time at which PPAA was incorporated during the complex formation affected the release of DNA and transfection efficiency. The author hypothesized that PPAA would reduce the stable association between chitosan and DNA to facilitate DNA release. PAA has the ability to disrupt lipid bilayer membranes of endosomes within a sharply defined pH range. This was demonstrated by a lack of colocalization between PPAA and Lysotracker [80]

Apart from reduction of the cytotoxicity, PEGylation led to improved colloidal stability of TMC polyplexes and significantly increased cellular uptake compared to unmodified TMC.

Zaki and Tirelli [92] have recently reviewed the endocytic mechanisms with the identification of appropriate targets. In this updated overview, we have also discussed the cell surface receptors most commonly used in drug delivery based on their reported internalization mechanisms. In addition we have presented examples, taken from the most recent scientific literature, of their use as nanocarrier targets. The modification of TMC or chitosan with sugars provides means of achieving cellular recognition ability. TMC DQ 80%, bearing antennary galactose residues were synthesized and TMC–DNA complexes have demonstrated enhanced transfection of HepG2 cells which carries the galactose receptor. The transfection efficiency was significantly inhibited in the presence of an inhibitor indicating that the conjugates were specifically internalized via the galactose receptor present on the cellular surface of HepG2 cells[93].

It was demonstrated that clathrin-mediated endocytosis plays a key role in the internalization of chitosan nanoparticles in the Y79 [94] and A549[79] cell line. In vitro experiments [95] with endocytic inhibitors suggested that several distinct uptake pathways (e.g., clathrin-mediated endocytosis, caveolae-mediated endocytosis, and macropinocytosis) are involved in the internalization of hydrophobically modified glycol chitosan (HGC). Some HGC nanoparticles were found entrapped in the lysosomes upon entry, as determined by TEM and colocalization studies. Given such favorable properties including low toxicity, biocompatibility, and fast uptake by several nondestructive endocytic pathways, HGC nanoparticles may serve as a versatile carrier for the intracellular delivery of therapeutic agents. However for successful gene delivery application, the low specificity and low transfection efficiency of chitosan need to be solved prior to clinical application. To achieve this goal, different ligands were attached to chitosan to achieve receptor- mediated rather than adsorptive endocytosis. Targeting ligands include antibodies, growth factors, peptides, transferring and folate have been conjugated to several polymers[96]. Transfection efficiency of the folate-TMC/pDNA complex in KB cells and SKOV3 cells (folate receptor over-expressing cell lines) was enhanced up to 1.6-fold and 1.4-fold compared with those of the TMC/pDNA complexes [97]. In another approach, galactose or mannose ligand modification of chitosan was attempted for enhancement of cell specificity and transfection efficiency via receptor-mediated endocytosis in vitro and in vivo[93].

Interactions with Microbial Cells

In recent years, chitosan and its derivatives have attracted much attention as antimicrobial agents against fungi, bacteria, and viruses and as elicitors of plant defense mechanisms [98]. In fact, a number of commercial applications of chitosan benefit from its antimicrobial activity, including its use in food preservation, manufacture of wound dressings and antimicrobial-finished textiles It is generally assumed that the cationic nature of chitosan (pKa = 6.3), conveyed by the positively-charged NH_3^+groups of glucosamine, might be a fundamental factor contributing to its interaction with the negatively charged microbial cell surface. Binding of chitosan to cell wall polymers would trigger secondary cellular effects: destabilization and subsequent disruption of bacterial membrane function occurs ultimately resulting in impairment of vital bacterial activities [99]. On the other hand, a recent publication investigated some of chitosan's characteristics, such as its water-binding capacity as well as its ability to chelate trace metals and to interact with DNA, in relation to its antimicrobial mode of action [100].

Quaternization of chitosan (triethylation or trimethylation) offered enhanced antibacterial activity than unmodified chitosan [101]. This effect increased with acidic condition (e.g. by acetic acid) as well as molecular weight. The water soluble chitosan derivative, N-(2-hydroxyl) propyl-3-trimethyl ammonium chitosan chloride (HTCC), synthesized by reaction of chitosan with glycidyl-trimethyl- ammonium chloride fostered antimicrobial properties too [101].

REFERENCES

[1] Morgan DM, Clover J, Pearson JD. Effects of synthetic polycations on leucine incorporation, lactate dehydrogenase release, and morphology of human umbilical vein endothelial cells. J Cell Sci. 1988 Oct;91 (Pt 2):231-8.

[2] Morgan DM, Larvin VL, Pearson JD. Biochemical characterisation of polycation-induced cytotoxicity to human vascular endothelial cells. J Cell Sci. 1989 Nov;94 (Pt 3):553-9.

[3] Kumar M. A review of chitin and chitosan applications. Reactive & Functional Polymers. 2000 Nov;46(1):1-27.

[4] Nordtveit RJ, Varum KM, Smidsrod O. Degradation of Fully Water-Soluble, Partially N-Acetylated Chitosans with Lysozyme. Carbohydrate Polymers. 1994;23(4):253-60.

[5] Rinaudo M. Chitin and chitosan: Properties and applications. Progress in Polymer Science. 2006 Jul;31(7):603-32.

[6] CarrenoGomez B, Duncan R. Evaluation of the biological properties of soluble chitosan and chitosan microspheres. International Journal of Pharmaceutics. 1997 Mar;148(2):231-40.

[7] Sgouras D, Duncan R. Methods for the Evaluation of Biocompatibility of Soluble Synthetic-Polymers Which Have Potential for Bio-Medical Use .1. Use of the Tetrazolium-Based Colorimetric Assay (Mtt) as a Preliminary Screen for Evaluation of Invitro Cytotoxicity. Journal of Materials Science-Materials in Medicine. 1990 Jul;1(2):61-8.

[8] Fischer D, Li Y, Ahlemeyer B, Krieglstein J, Kissel T. In vitro cytotoxicity testing of polycations: influence of polymer structure on cell viability and hemolysis. Biomaterials. 2003;24(7):1121-31.

[9] Fischer D, Li YX, Ahlemeyer B, Krieglstein J, Kissel T. In vitro cytotoxicity testing of polycations: influence of polymer structure on cell viability and hemolysis. Biomaterials. 2003 Mar;24(7):1121-31.

[10] vandeWetering P, Cherng JY, Talsma H, Hennink WE. Relation between transfection efficiency and cytotoxicity of poly(2-(dimethylamino)ethyl methacrylate)/plasmid complexes. Journal of Controlled Release. 1997 Nov;49(1):59-69.

[11] Kean T, Roth S, Thanou M. Trimethylated chitosans as non-viral gene delivery vectors: Cytotoxicity and transfection efficiency. Journal of Controlled Release. 2005 Apr;103(3):643-53.

[12] Verheul RJ, Amidi M, van Steenbergen MJ, van Riet E, Jiskoot W, Hennink WE. Influence of the degree of acetylation on the enzymatic degradation and in vitro

biological properties of trimethylated chitosans. Biomaterials. 2009 Jun;30(18):3129-35.

[13] Malik N, Wiwattanapatapee R, Klopsch R, Lorenz K, Frey H, Weener JW, et al. Dendrimers: Relationship between structure and biocompatibility in vitro, and preliminary studies on the biodistribution of I-125-labelled polyamidoamine dendrimers in vivo. 9th International Symposium on Recent Advances in Drug Delivery Systems; 1999 Feb 22-25; Salt Lake City, Utah: Elsevier Science Bv; 1999. p. 133-48.

[14] Richardson S, Ferruti P, Duncan R. Poly(amidoamine)s as potential endosomolytic polymers: Evaluation in vitro and body distribution in normal and tumour-bearing animals. Journal of Drug Targeting. 1999;6(6):391-404.

[15] Kim IY, Seo SJ, Moon HS, Yoo MK, Park IY, Kim BC, et al. Chitosan and its derivatives for tissue engineering applications. Biotechnology Advances. 2008 Jan-Feb;26(1):1-21.

[16] Kumar M, Muzzarelli RAA, Muzzarelli C, Sashiwa H, Domb AJ. Chitosan chemistry and pharmaceutical perspectives. Chemical Reviews. 2004 Dec;104(12):6017-84.

[17] Illum L. Chitosan and its use as a pharmaceutical excipient. Pharmaceutical Research. 1998 Sep;15(9):1326-31.

[18] Rao SB, Sharma CP. Use of chitosan as a biomaterial: Studies on its safety and hemostatic potential. Journal of Biomedical Materials Research. 1997 Jan;34(1):21-8.

[19] Wedmore I, McManus JG, Pusateri AE, Holcomb JB. A special report on the chitosan-based hemostatic dressing: Experience in current combat operations. Journal of Trauma-Injury Infection and Critical Care. 2006 Mar;60(3):655-8.

[20] Kotze AF LH, de Boer AG, Verhoef JC, Junginger HE. Chitosan for enhanced intestinal permeability: prospects for derivatives soluble in neutral and basic environments. Eur J Pharm Sci 1999;. 1999;7:145–51.

[21] MacLaughlin FC, Mumper RJ, Wang J, Tagliaferri JM, Gill I, Hinchcliffe M, et al. Chitosan and depolymerized chitosan oligomers as condensing carriers for in vivo plasmid delivery. J Control Release. 1998 Dec 4;56(1-3):259-72.

[22] Issa MM, Koping-Hoggard M, Tommeraas K, Varum KM, Christensen BE, Strand SP, et al. Targeted gene delivery with trisaccharide-substituted chitosan oligomers in vitro and after lung administration in vivo. J Control Release. 2006 Sep 28;115(1):103-12.

[23] Koping-Hoggard M, Mel'nikova YS, Varum KM, Lindman B, Artursson P. Relationship between the physical shape and the efficiency of oligomeric chitosan as a gene delivery system in vitro and in vivo. J Gene Med. 2003 Feb;5(2):130-41.

[24] Koping-Hoggard M, Varum KM, Issa M, Danielsen S, Christensen BE, Stokke BT, et al. Improved chitosan-mediated gene delivery based on easily dissociated chitosan polyplexes of highly defined chitosan oligomers. Gene Ther. 2004 Oct;11(19):1441-52.

[25] Strand SP, Danielsen S, Christensen BE, Varum KM. Influence of chitosan structure on the formation and stability of DNA-chitosan polyelectrolyte complexes. Biomacromolecules. 2005 Nov-Dec;6(6):3357-66.

[26] Thanou M, Florea BI, Geldof M, Junginger HE, Borchard G. Quaternized chitosan oligomers as novel gene delivery vectors in epithelial cell lines. Biomaterials. 2002 Jan;23(1):153-9.

[27] Zaki NM, Mortada ND, Awad GA, Abd ElHady SS. Rapid-onset intranasal delivery of metoclopramide hydrochloride Part II: Safety of various absorption enhancers and pharmacokinetic evaluation. Int J Pharm. 2006 Dec 11;327(1-2):97-103.

[28] Mosmann T. Rapid colorimetric assay for cellular growth and survival: application to proliferation and cytotoxicity assays. J Immunol Methods. 1983 Dec 16;65(1-2):55-63.

[29] Nasti A, Zaki NM, de Leonardis P, Ungphaiboon S, Sansongsak P, Rimoli MG, et al. Chitosan/TPP and chitosan/TPP-hyaluronic acid nanoparticles: systematic optimisation of the preparative process and preliminary biological evaluation. Pharm Res. 2009 Aug;26(8):1918-30.

[30] Zaki NM, Tirelli N. Assessment of Nanomaterials Cytotoxicity and Internalization. In Series: Methods in Molecular Biology-3D Cell Culture. 2010;Haycock J (eds), Human Press (publisher).

[31] Desai A, Vyas T, Amiji M. Cytotoxicity and apoptosis enhancement in brain tumor cells upon coadministration of paclitaxel and ceramide in nanoemulsion formulations. J Pharm Sci. 2008 Jul;97(7):2745-56.

[32] Wong HL, Rauth AM, Bendayan R, Manias JL, Ramaswamy M, Liu Z, et al. A new polymer-lipid hybrid nanoparticle system increases cytotoxicity of doxorubicin against multidrug-resistant human breast cancer cells. Pharm Res. 2006 Jul;23(7):1574-85.

[33] Nafee N, Schneider M, Schaefer UF, Lehr CM. Relevance of the colloidal stability of chitosan/PLGA nanoparticles on their cytotoxicity profile. Int J Pharm. 2009 Nov 3;381(2):130-9.

[34] Zhang WF, Zhou HY, Chen XG, Tang SH, Zhang JJ. Biocompatibility study of theophylline/chitosan/beta-cyclodextrin microspheres as pulmonary delivery carriers. J Mater Sci Mater Med. 2009 Jun;20(6):1321-30.

[35] Huang M, Khor E, Lim LY. Uptake and cytotoxicity of chitosan molecules and nanoparticles: effects of molecular weight and degree of deacetylation. Pharm Res. 2004 Feb;21(2):344-53.

[36] Guggi D, Langoth N, Hoffer MH, Wirth M, Bernkop-Schnürch A. Comparative evaluation of cytotoxicity of a glucosamine-TBA conjugate and a chitosan-TBA conjugate. International Journal of Pharmaceutics. 2004;278(2):353-60.

[37] Domard A RM, Terrassin C. New method for the quaternization of chitosan. Int J Biol Macromol. 1986;8(105-107).

[38] Thanou M, Verhoef JC, Junginger HE. Chitosan and its derivatives as intestinal absorption enhancers. Advanced Drug Delivery Reviews. 2001;50(Supplement 1):S91-S101.

[39] Di Colo G BS, Zambito Y, Monti D, Chetoni P. . Effects of different N-trimethyl chitosans on in vitro/in vivo ofloxacin transcorneal permeation. J Pharm Sci. 2004;93:2851–62.

[40] Hamman JH SM, Kotze AF. Effect of the degree of quaternisation of N-trimethyl chitosan chloride on absorption enhancement: in vivo evaluation in rat nasal epithelia. Int J Pharm 2002;232:235–42.

[41] Boonyo W JH, Waranuch N, Polnok A, Pitaksuteepong T. . Chitosan and trimethyl chitosan chloride (TMC) as adjuvants for inducing immune responses to ovalbumin in mice following nasal administration. J Controlled Release 2007. 2007;121:168–75.

[42] Thanou M, Florea BI, Geldof M, Junginger HE, Borchard G. Quaternized chitosan oligomers as novel gene delivery vectors in epithelial cell lines. Biomaterials. 2002;23(1):153-9.

[43] Mao S, Shuai X, Unger F, Wittmar M, Xie X, Kissel T. Synthesis, characterization and cytotoxicity of poly(ethylene glycol)-graft-trimethyl chitosan block copolymers. Biomaterials. 2005;26(32):6343-56.

[44] Fischer D BT, Li Y, Elsaesser HP, Kissel T. . A novel non-viral vector for DNA delivery based on low molecular weight, branched polyethylenimine: effect of molecular weight on transfection efficiency and cytotoxicity. Pharm Res. 1999;19:1273-9.

[45] Verheul RJ, Amidi M, van der Wal S, van Riet E, Jiskoot W, Hennink WE. Synthesis, characterization and in vitro biological properties of O-methyl free N,N,N-trimethylated chitosan. Biomaterials. 2008 Sep;29(27):3642-9.

[46] Haas J, Ravi Kumar MN, Borchard G, Bakowsky U, Lehr CM. Preparation and characterization of chitosan and trimethyl-chitosan-modified poly-(epsilon-caprolactone) nanoparticles as DNA carriers. AAPS PharmSciTech. 2005;6(1):E22-30.

[47] Thanou MM, Verhoef JC, Romeijn SG, Nagelkerke JF, Merkus FW, Junginger HE. Effects of N-trimethyl chitosan chloride, a novel absorption enhancer, on caco-2 intestinal epithelia and the ciliary beat frequency of chicken embryo trachea. Int J Pharm. 1999 Aug 5;185(1):73-82.

[48] Sajomsang W, Rungsardthong Ruktanonchai U, Gonil P, Nuchuchua O. Mucoadhesive property and biocompatibility of methylated N-aryl chitosan derivatives. Carbohydrate Polymers. 2009;78(4):945-52.

[49] Bernkop-Schnurch A, Weithaler A, Albrecht K, Greimel A. Thiomers: preparation and in vitro evaluation of a mucoadhesive nanoparticulate drug delivery system. Int J Pharm. 2006 Jul 6;317(1):76-81.

[50] Werle M, Bernkop-Schnurch A. Thiolated chitosans: useful excipients for oral drug delivery. J Pharm Pharmacol. 2008 Mar;60(3):273-81.

[51] Martien R, Loretz B, Thaler M, Majzoob S, Bernkop-Schnurch A. Chitosan-thioglycolic acid conjugate: an alternative carrier for oral nonviral gene delivery? J Biomed Mater Res A. 2007 Jul;82(1):1-9.

[52] Guggi D, Langoth N, Hoffer MH, Wirth M, Bernkop-Schnurch A. Comparative evaluation of cytotoxicity of a glucosamine-TBA conjugate and a chitosan-TBA conjugate. Int J Pharm. 2004 Jul 8;278(2):353-60.

[53] Carreno-Gomez B, Duncan, R., 1997. . Evaluation of the biological properties of soluble chitosan and chitosan microspheres. IntJ Pharm 1997;148:231–40.

[54] Dekie L, Toncheva, V., Dubruel, P., Schacht, E.H., Barrett, L., Seymour, L.W. Poly l-glutamic acid derivatives as vectors for gene therapy. J Cont Rel 2000;65: 187–202.

[55] Ferruti P, Knobloch, S., Ranucci, E., Gianasi, E., Duncan, R. A novel chemical modification of poly-l-lysine reducing toxicity while preserving cationic properties. . In: Proc Int Symp Control Rel Bioact Mater. 1997.

[56] Singh AK, Kasinath, B.S., Lewis, E.J. Interactions of polycations with cell-surface negative charges of epithelial cells. Biochim Biophys Acta 1992;1120:337–42.

[57] Sakloetsakun D, Hombach JM, Bernkop-Schnurch A. In situ gelling properties of chitosan-thioglycolic acid conjugate in the presence of oxidizing agents. Biomaterials. 2009 Oct;30(31):6151-7.

[58] Kuroyanagi Y SA, Shirasaki Y, Nakakita N, Yasutomi Y, Takano Y, Shioya N. . Development of a new wound dressing with antimicrobial delivery capability. Wound Repair Regen. 1994;2.:122-9.

[59] Izume M. The application of chitin and chitosan to cosmetics. Chitin Chitosan Res. 1998;4(12-17).

[60] Nakagawa A, Myata S, Shimozono J, Soejima Y, Saida M. Water-soluble chitosan or chitin for treatment of arthritis. JP Patent No 06107551. 1994.

[61] Kato Y, Onishi H, Machida Y. A Novel water-soluble N-succinylchitosan– mitomycin C conjugate prepared by direct carbodiimide coupling:physicoch emical properties, antitumor characteristics and systemic retention. STP Pharm Sci. 2000;10:133-42.

[62] Kato Y, Onishi H, Machida Y. . Tumor cell uptake of lactosaminated and intact N-succinyl-chitosans and antitumour effects of conjugates with mitomycin C. Anticancer Res. 2002;22:2771-6.

[63] Sato M, Onishi H, Takahara J, Machida Y, Nagai T. . In vivo drug release and antitumor characteristics of water-soluble conjugates of mitomycin C with glycol–chitosan and N-succinyl-chitosan. Biol Pharm Bull. 1996;19:1170-7.

[64] Bhumkar DR, Pokharkar VB. Studies on effect of pH on cross-linking of chitosan with sodium tripolyphosphate: a technical note. AAPS PharmSciTech. 2006;7(2):E50.

[65] Calvo P, Remunan-Lopez C, Vila-Jato JL, Alonso MJ. Chitosan and chitosan/ethylene oxide-propylene oxide block copolymer nanoparticles as novel carriers for proteins and vaccines. Pharm Res. 1997 Oct;14(10):1431-6.

[66] Grenha A, Gomes ME, Rodrigues M, Santo VE, Mano JF, Neves NM, et al. Development of new chitosan/carrageenan nanoparticles for drug delivery applications. J Biomed Mater Res A. 2009 Mar 25.

[67] de la Fuente M, Seijo B, Alonso MJ. Novel hyaluronic acid-chitosan nanoparticles for ocular gene therapy. Invest Ophthalmol Vis Sci. 2008 May;49(5):2016-24.

[68] de la Fuente M, Seijo B, Alonso MJ. Bioadhesive hyaluronan-chitosan nanoparticles can transport genes across the ocular mucosa and transfect ocular tissue. Gene Ther. 2008 May;15(9):668-76.

[69] Eliaz RE SFJ. Liposome-encapsulated doxorubicin targeted to CD44: a strategy to kill CD44-overexpressing tumor cells. Cancer Res. 2001;61(6):2592–601.

[70] Ma Z, Lim LY. Uptake of chitosan and associated insulin in Caco-2 cell monolayers: a comparison between chitosan molecules and chitosan nanoparticles. Pharm Res. 2003 Nov;20(11):1812-9.

[71] Lemarchand C, Gref R, Passirani C, Garcion E, Petri B, Muller R, et al. Influence of polysaccharide coating on the interactions of nanoparticles with biological systems. Biomaterials. 2006 Jan;27(1):108-18.

[72] Illum L, Jacobsen LO, Muller RH, Mak E, Davis SS. Surface characteristics and the interaction of colloidal particles with mouse peritoneal macrophages. Biomaterials. 1987 Mar;8(2):113-7.

[73] Zwiorek K, Bourquin C, Battiany J, Winter G, Endres S, Hartmann G, et al. Delivery by cationic gelatin nanoparticles strongly increases the immunostimulatory effects of CpG oligonucleotides. Pharm Res. 2008 Mar;25(3):551-62.

[74] Harush-Frenkel O, Debotton N, Benita S, Altschuler Y. Targeting of nanoparticles to the clathrin-mediated endocytic pathway. Biochem Biophys Res Commun. 2007 Feb 2;353(1):26-32.

[75] Loretz B, Bernkop-Schnurch A. In vitro evaluation of chitosan-EDTA conjugate polyplexes as a nanoparticulate gene delivery system. Aaps J. 2006;8(4):E756-64.

[76] Zhao QQ, Chen JL, Han M, Liang WQ, Tabata Y, Gao JQ. Combination of poly(ethylenimine) and chitosan induces high gene transfection efficiency and low cytotoxicity. J Biosci Bioeng. 2008 Jan;105(1):65-8.

[77] Lee M, Nah JW, Kwon Y, Koh JJ, Ko KS, Kim SW. Water-soluble and low molecular weight chitosan-based plasmid DNA delivery. Pharm Res. 2001 Apr;18(4):427-31.

[78] Issa MM, Köping-Höggård M, Tømmeraas K, Vårum KM, Christensen BE, Strand SP, et al. Targeted gene delivery with trisaccharide-substituted chitosan oligomers in vitro and after lung administration in vivo. Journal of Controlled Release. 2006;115(1):103-12.

[79] Huang M, Ma Z, Khor E, Lim LY. Uptake of FITC-chitosan nanoparticles by A549 cells. Pharm Res. 2002 Oct;19(10):1488-94.

[80] Kiang T, Bright C, Cheung CY, Stayton PS, Hoffman AS, Leong KW. Formulation of chitosan-DNA nanoparticles with poly(propyl acrylic acid) enhances gene expression. J Biomater Sci Polym Ed. 2004;15(11):1405-21.

[81] Panyam J, Sahoo SK, Prabha S, Bargar T, Labhasetwar V. Fluorescence and electron microscopy probes for cellular and tissue uptake of poly(,-lactide-co-glycolide) nanoparticles. International Journal of Pharmaceutics. 2003;262(1-2):1-11.

[82] Davda J, Labhasetwar V. Characterization of nanoparticle uptake by endothelial cells. International Journal of Pharmaceutics. 2002;233(1-2):51-9.

[83] Al-Jamal KT, Ruenraroengsak P, Hartell N, Florence AT. An intrinsically fluorescent dendrimer as a nanoprobe of cell transport. J Drug Target. 2006 Jul;14(6):405-12.

[84] Park JS, Han TH, Lee KY, Han SS, Hwang JJ, Moon DH, et al. N-acetyl histidine-conjugated glycol chitosan self-assembled nanoparticles for intracytoplasmic delivery of drugs: endocytosis, exocytosis and drug release. J Control Release. 2006 Sep 28;115(1):37-45.

[85] Lee D, Mohapatra SS. Chitosan nanoparticle-mediated gene transfer. Methods Mol Biol. 2008;433:127-40.

[86] Douglas KL, Piccirillo CA, Tabrizian M. Cell line-dependent internalization pathways and intracellular trafficking determine transfection efficiency of nanoparticle vectors. Eur J Pharm Biopharm. 2008 Mar;68(3):676-87.

[87] Panyam J, Labhasetwar V. Sustained cytoplasmic delivery of drugs with intracellular receptors using biodegradable nanoparticles. Mol Pharm. 2004 Jan 12;1(1):77-84.

[88] Panyam J, Zhou WZ, Prabha S, Sahoo SK, Labhasetwar V. Rapid endo-lysosomal escape of poly(DL-lactide-co-glycolide) nanoparticles: implications for drug and gene delivery. Faseb J. 2002 Aug;16(10):1217-26.

[89] Howard KA, Rahbek UL, Liu X, Damgaard CK, Glud SZ, Andersen MO, et al. RNA interference in vitro and in vivo using a novel chitosan/siRNA nanoparticle system. Mol Ther. 2006 Oct;14(4):476-84.

[90] des Rieux A, Ragnarsson EG, Gullberg E, Preat V, Schneider YJ, Artursson P. Transport of nanoparticles across an in vitro model of the human intestinal follicle associated epithelium. Eur J Pharm Sci. 2005 Jul-Aug;25(4-5):455-65.

[91] Borges O, Cordeiro-da-Silva A, Romeijn SG, Amidi M, de Sousa A, Borchard G, et al. Uptake studies in rat Peyer's patches, cytotoxicity and release studies of alginate coated chitosan nanoparticles for mucosal vaccination. J Control Release. 2006 Sep 12;114(3):348-58.

[92] Zaki NM, Tirelli N. Gateways for the intracellular access of nanocarriers: a review of receptor-mediated endocytosis mechanisms and of strategies in receptor targeting. Expert Opin Drug Deliv. 2010 Aug;7(8):895-913.

[93] Kim TH, Park IK, Nah JW, Choi YJ, Cho CS. Galactosylated chitosan/DNA nanoparticles prepared using water-soluble chitosan as a gene carrier. Biomaterials. 2004 Aug;25(17):3783-92.

[94] Parveen S, Mitra M, Krishnakumar S, Sahoo SK. Enhanced Antiproliferative Activity of Carboplatin loaded Chitosan-Alginate Nanoparticles in Retinoblastoma Cell Line. Acta Biomater. Feb 8.

[95] Nam HY, Kwon SM, Chung H, Lee SY, Kwon SH, Jeon H, et al. Cellular uptake mechanism and intracellular fate of hydrophobically modified glycol chitosan nanoparticles. J Control Release. 2009 May 5;135(3):259-67.

[96] Ghaghada KB, Saul, J., Natarajan, J.V., Bellamkonda, R.V., Annapragada, A.V. Folate targeting of drug carriers: a mathematical model. . J Control Release. 2005;104:113-28.

[97] Zheng Y, Cai Z, Song X, Yu B, Bi Y, Chen Q, et al. Receptor mediated gene delivery by folate conjugated N-trimethyl chitosan in vitro. International Journal of Pharmaceutics. 2009;382(1-2):262-9.

[98] Xing K, Chen XG, Liu CS, Cha DS, Park HJ. Oleoyl-chitosan nanoparticles inhibits Escherichia coli and Staphylococcus aureus by damaging the cell membrane and putative binding to extracellular or intracellular targets. Int J Food Microbiol. 2009 Jun 30;132(2-3):127-33.

[99] Muzzarelli R, Tarsi R, Filippini O, Giovanetti E, Biagini G, Varaldo PE. Antimicrobial properties of N-carboxybutyl chitosan. Antimicrob Agents Chemother. 1990 Oct;34(10):2019-23.

[100] Raafat D, von Bargen K, Haas A, Sahl HG. Insights into the mode of action of chitosan as an antibacterial compound. Appl Environ Microbiol. 2008 Jun;74(12):3764-73.

[101] Mourya VK, Inamdar NN. Trimethyl chitosan and its applications in drug delivery. J Mater Sci Mater Med. 2009 May;20(5):1057-79.

In: Handbook of Chitosan Research and Applications
Editors: Richard G. Mackay and Jennifer M. Tait
ISBN 978-1-61324-455-5

Chapter 3

CHITOSAN AND DERIVATIVES–BEHAVIOR IN DISPERSED STATE AND APPLICATIONS

A. Krim Azzouz
Université du Québec à Montréal, Montréal (Québec) Canada

1. INTRODUCTION

The chemical behavior of chitosan in dispersed state determines most of its applications, and the mere presence of chitosan in the cuttlefish bone structure and its ability to combine with limestone in the crustaceans' shells accounts for most of its physicochemical features. This natural compound exhibits a strong but specific tendency to coagulation-flocculation in aqueous media [1,2], making it to be suitable for water treatment and other clotting processes. Unlike conventional aluminum salts, such as alum or other synthetic polyelectrolytes, used so far for this purpose, chitosan offers more economical and ecological advantages, being a natural, biodegradable and nontoxic product. In addition, chitosan does not generate harmful or toxic by-products, and produces less sludge. The latter are fully recyclable, because they do not contain aluminum, and, thereby, do not present risks to human health.

Nevertheless, other applications may also be envisaged, inasmuch as isolated chitosan fibers exert interactions upon dispersed solids to form various types of matrices. Being a deacetylated derivative of chitin [3], chitosan has a linear structure comprising a random sequence of D-glucosamine and N-acetyl-D-glucosamine monomers, which bear [$-NH_2$] and [$-NH-CO-CH_3$] groups, respectively. Thus, chitosan macromolecules bear certain numbers of free amino sites and electrical charges depending on the degree of acetylation (DA) or of deacetylation (DDA). Interactions of different kinds and strengths towards various dissolved or dispersed particles may occur in both liquid media and solid matrices. Organic and inorganic compounds, salts, metal cations, clay minerals, zeolites, mixed oxides or any chemical species having the ability to reduce the electrical charges on chitosan [1,4-9] or its derivatives [10,11] can spark coagulation and/or flocculation. Judicious chemical modifications of amino groups on chitosan can improve the coagulation capacity, offering new prospects for diversified applications. A primary type of modification of chitosan or chitin consists precisely in increasing the DA value, far beyond the 50-70% level, and accurate determination of this parameter [12,13] can ease the control of clotting properties and solubility of chitosan in aqueous media.

2. Chitosan Behavior In Aqueous Media

In aqueous environments, chitosan displays specific behavior, more particularly a variable solubility feature, due to its very structure of polysaccharide co-polymer bearing reactive [-NH_2] and [-NH-CO-CH_3] groups. The relative proportions of these chemical functions will determine not only the dispersion grade of chitosan in aqueous media, but also the spatial configuration of the stacks of chitosan chains in the dehydrated state. Regular and ordered arrangements of macromolecules should generate a pseudo-crystallinity that corresponds to a completely homogeneous structure similar to that of pure and insoluble chitin. Partial deacetylation will introduce imperfections in this ordered structure, and will affect its cohesiveness, enhancing, thereby the tendency to dispersion behavior.

The solubility or the dispersion grade are probably the key-factors that control the way chitosan combines to other chemical species dispersed in aqueous media. The solubility of chitosan involves hydrophobic interactions between chitosan macromolecules within the very polymer structure, and the formation of hydrogen bridges with water molecules. The competition between the attractive and repulsive interactions that favor either aggregation or dissolution is controlled by chemical equilibrium. In dilute acid, protonation of amino groups produces ruptures in the crystal structure, along with swelling and dissolution processes, as well illustrated by Figure 1.

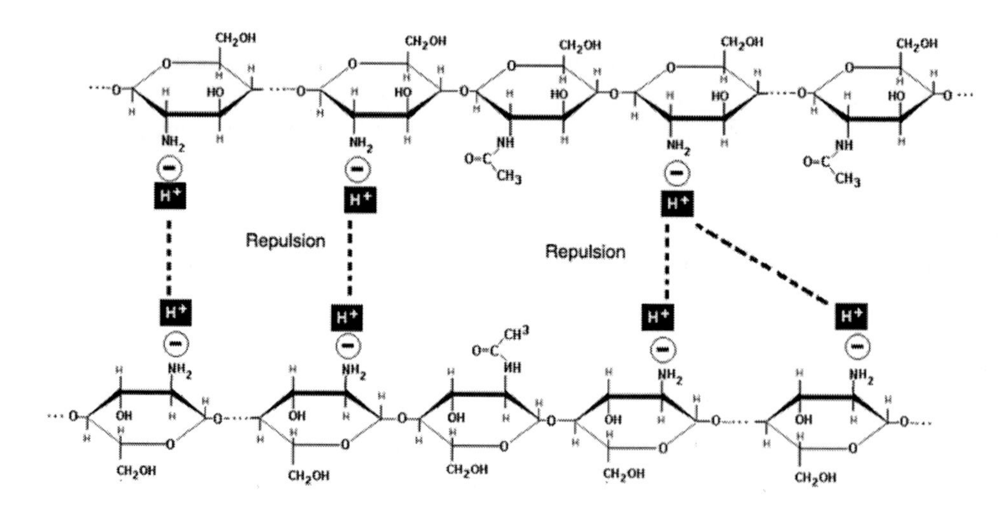

Figure 1. Repulsion between two polymer chains of chitosan in acidic media.

An increase in the solution pH may enhance deprotonation of amino groups, making them to be available for hydrogen bridges. At a critical pH, which depends on the degree of deacetylation and molecular weight, the chitosan molecules develop enough hydrogen bridges to form a gel [14]. Nevertheless, the number of these bridges is limited by the number of available amino groups on the chitosan macromolecules. This number varies according to the origin and structure of chitosan. For instance, chitosan extracted from squid feathers [15] is colorless and has a lower coagulation power than that originating from crustacean shells [16-18].

2.1. Colloidal State of Chitosan in Aqueous Media

In a first step, coagulation induces a destabilization of the colloidal suspension of chitosan, increasing, thereby, the attraction interactions between the dispersed particles. Here, one must mention that a colloidal state of chitosan may contains both isolated polymer chains and clusters of polymer fibers. The relative proportion of both chitosan colloidal forms is strongly depending on the degree of deacetylation, the pH of the medium and other factors.

In a second step, these particles will aggregates, generating small flakes under the action of Brownian motion. Flocculation involves the formation of bulkier and heavier flakes upon moderate stirring [19], followed by sedimentation when the flake density attains a certain level. In most cases, coagulation and flocculation take place simultaneously and cannot be distinguished, more particularly for pronounced coagulation and prolonged stirring times.

The colloidal suspensions are maintained stable as long as the chitosan particles still remain charged, and repel each other by electrostatic forces induced by same sign charges. In aqueous media, repulsion forces prevail at low distance between particles displaying similar electrical charges. In the meantime, attraction forces also exert, enhancing cohesion between clustered particles [20]. A stable colloidal suspension results from a balance between these forces of attraction and repulsion, which exert at a certain distance between two dispersed colloidal particles.

This balance between the forces of repulsion and attraction is very unstable and precarious. A mere change in pH or an increase in acid-base interactions due to foreign particles can cause destabilization of the suspension, enhancing particles collisions. At each contact during a collision, the surface charge is diminished, leading to particle aggregation into flakes more or less stable, which will behave like whole particles.

2.2. Chitosan Behavior towards Dissolved and Dispersed Species

The behavior of chitosan in aqueous media is strongly depends on the nature of the chemical species dissolved or dispersed in water. Destabilization of the colloidal state needs an energy equal to that required by the repulsion between colloidal particles. This can be achieved by reducing the zeta potential, and a judicious route should consist in the addition of minerals, multivalent cations or protons (by acidification) as coagulating impurities. The latter can adsorb on the surface of colloidal particles, and can neutralize the electric charges, thereby reducing the zeta potential, causing coagulation. Today's theories in force can distinguish four basic mechanisms to explain the destabilization of colloidal particles [20,21], namely the so-called electrostatic coagulation that involves the compression of the double layer, the route via the charge adsorption and neutralization, the particle capture in a precipitate and the particle adsorption on a chitosan fiber followed by bridging or cross-linking processes.

An electrostatic attraction may occur between a colloidal particle of chitosan and ions of opposite charge present in a polar environment. The same phenomenon is supposed to occur on both isolated fiber and clusters formed many chitosan fibers. According to the double layer theory [22], the electric charges carried by the surrounding ions are distributed in two layers in the vicinity of the particle surface (Fig. 2).

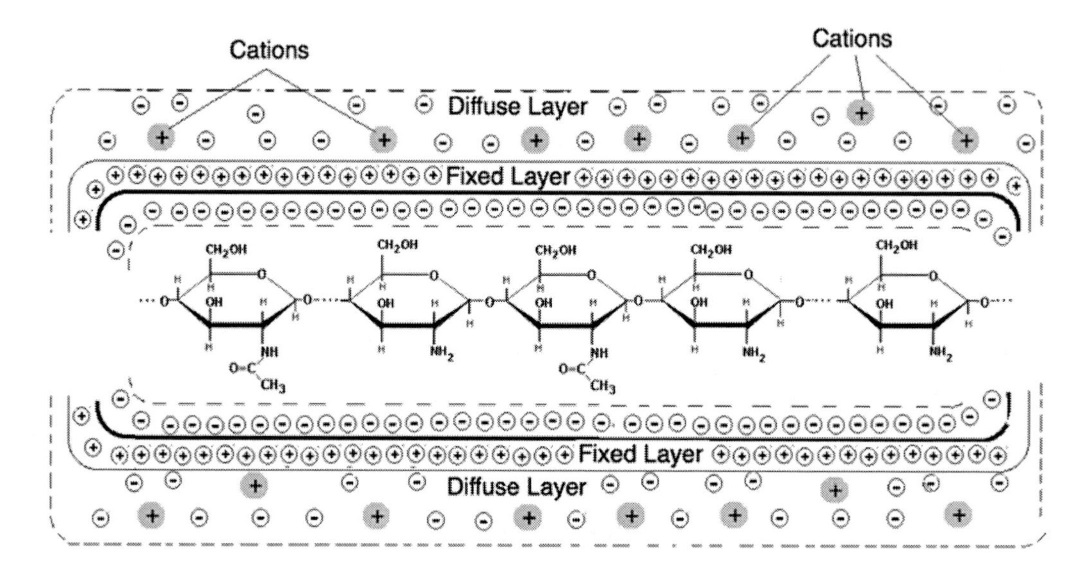

Figure 2. Double layer around an isolated chitosan fiber in aqueous solution of cations.

In aqueous environment, the lone pair of electrons borne by the amino group of chitosan induces negative charge that exerts attraction upon oppositely charged particles, such as cations and hydrogen atoms of water molecules. This results in the formation of a fixed layer of positive charges (Stern's layer), which, in turn, attracts by means the so-called "STERN potential" [23] counter-ions into a diffused cloud (Gouy's layer) with progressively decreasing concentration.

Addition of a polar substrate or an electrolyte to a colloidal dispersion induces an increase of the charge density in the diffuse layer, promoting thereby its compression. This modifies the distribution of the repulsive forces of the double layer around the colloidal particles, resulting in a reduction of surface potential. Nonetheless, the amount of electrolyte required for coagulation via compression of the double layer is not stoichiometric and does not depend on the concentration of colloid. However, an increase of electrolyte concentration can enhance the strength of van der Waal's attraction forces, favoring the particle aggregation. At a high concentration of electrolyte, the repulsive force diminishes, inducing a more pronounced agglomeration of particles. making coagulation to become irreversible. Thus, re-stabilization of a dispersion of particles is not possible anymore (or is no longer possible) after addition of an excess of coagulating agent [21].

Chemical species can adsorb on the surface of colloidal particles, especially when opposite charges are involved. This causes a decrease in surface potential and the destabilization of colloidal particles. The destabilization pathway for adsorption differs from that involving the compression of the double layer. Adsorption is presumably the major cause of destabilization even at low amounts of adsorbed particles. The latter are supposed to be capable of destabilizing the colloidal state of chitosan to a much smaller dosage than non adsorbable species. However, destabilization must involve stoichiometric adsorption. It is likely that an excess of adsorbable species causes a re-dispersion by reversing the colloidal particle charge. The adsorption of ions beyond the neutralization point, till charge reversal can be explained by the fact that the chemical interactions are predominant as compared to the electrostatic repulsion forces in some cases [21].

2.3. Flocculation Tendency of Chitosan

The clotting properties is a common feature of many other natural polymers like starch, cellulose, diverse polysaccharides, proteins,...) and a wide variety of synthetic polymers bearing partial electrical charges [21]. In many cases, flocculation tendency may also arise owing to these electrical charges. This why chitosan and most of natural polymers can also behave as flocculating agents. Colloidal particles can be destabilized by both cationic and anionic polymers.

After destabilization, the colloidal particles show a strong tendency to agglomerate into flakes. During flocculation, dispersed particles are trapped in a net comprised of tangled polymer fibers. This trapping phenomenon does not involve stoichiometry [21], and does not seems to be influenced by the charge of the coagulating agent [24]. Here, as already reported [25,26], chemical bridges must form between colloidal particles by means of macromolecules of chitosan, and conversely, as illustrated in Figure 3.

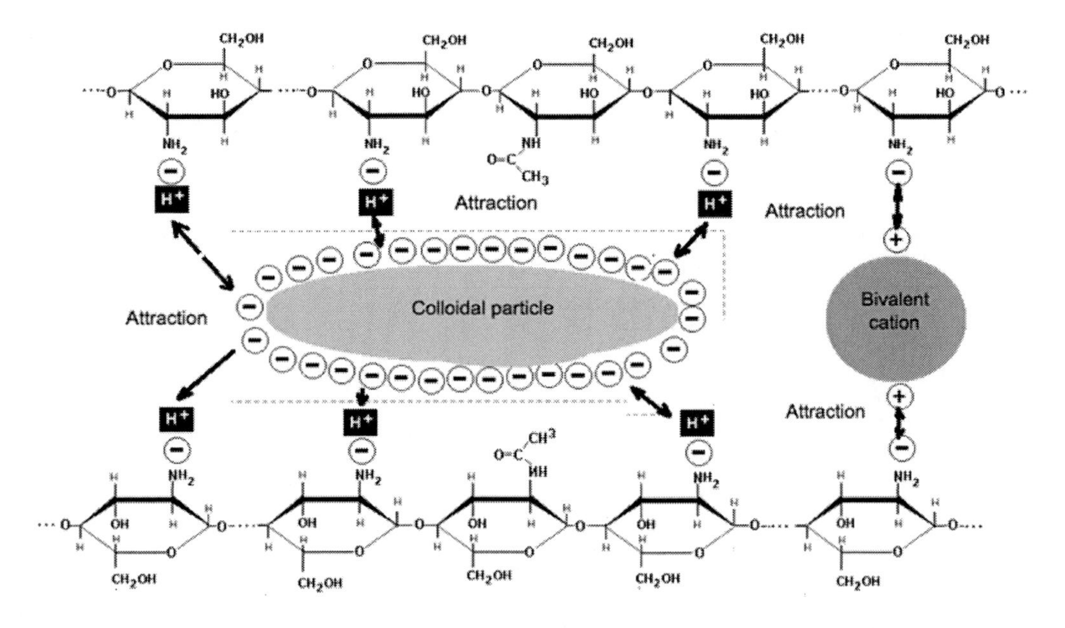

Figure 3. Bridging scheme of chitosan macromolecules in the presence of a colloidal particle.

Re-stabilization of the suspension may occur, and then the free sites of several polymeric chains cling to a colloidal particle. The retention of these chains by the particle may result from coulombic attraction, hydrogen bonds, van der Waals interactions, or ion exchange [27,28].

Flocculation and coagulation are often difficult to distinguish, because parts of both processes take place almost simultaneously. So far, many schemes have been proposed to describe and distinguish coagulation and flocculation [29]. However, it is well established that flocculation takes place after coagulation, although there is no clear boundary between both processes. As a general behavior, when the stability of the colloidal dispersion is altered, coagulation occurs spontaneously, while flocculation is triggered once coagulation have produced sufficiently destabilized colloidal particles.

3. Stability Concept and Destabilization Mechanisms

In aqueous media, a chitosan particle is stable as long as its density is similar to that of water, and the destabilization of a chitosan particle is an essential requirement of the clotting process. To be stable a particle must be coated by a water layer sufficiently thick, so as it can merge into the bulk liquid (Figure 4).

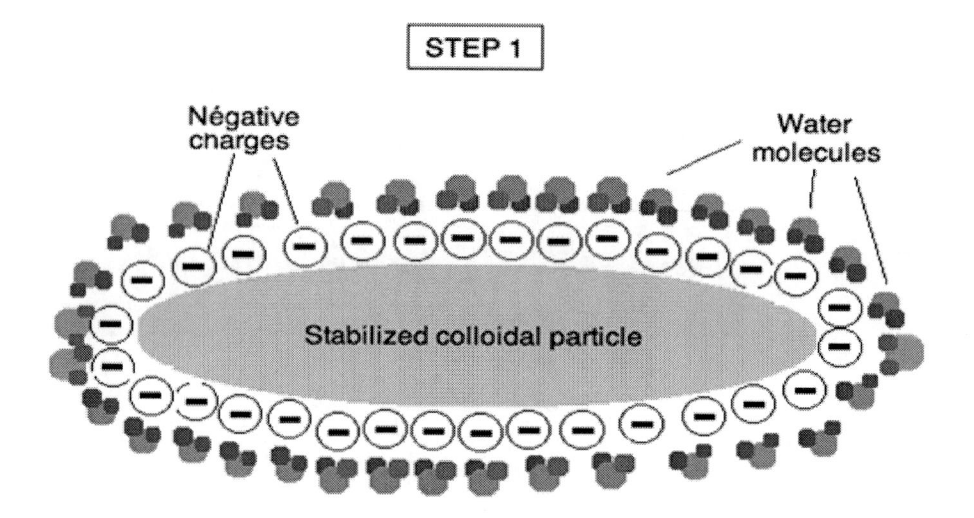

Figure 4. Scheme of stabilization of colloidal particle dispersion in aqueous media.

Stability seems to be proportional to the thickness of the layer of electrostatically bound water. In this condition, gravity does not exert any effect upon the particle, which remains suspended in water. Such a stability scheme is based on the double layer theory, and is often used to explain fat emulsions in milk, proteins dispersions in aqueous media, or the gel formation.Nevertheless, a stable colloidal dispersion is a precarious equilibrium state, which can be disrupted by thinning or even removal of the coating water layer. This generally can take place by mere pH change.

Because protons and water molecules are weakly retained on a chitosan particle, this effect may be accentuated by upon mechanical stirring, which may provoke collisions with another chitosan particle (Figure 5) or with another chemical species (Figure 6). However, destabilization via mechanical stirring without pH increase is not sufficient and has no effect upon the auto-coagulation of chitosan. Consequently, deprotonation of amino groups turns out to be necessary.

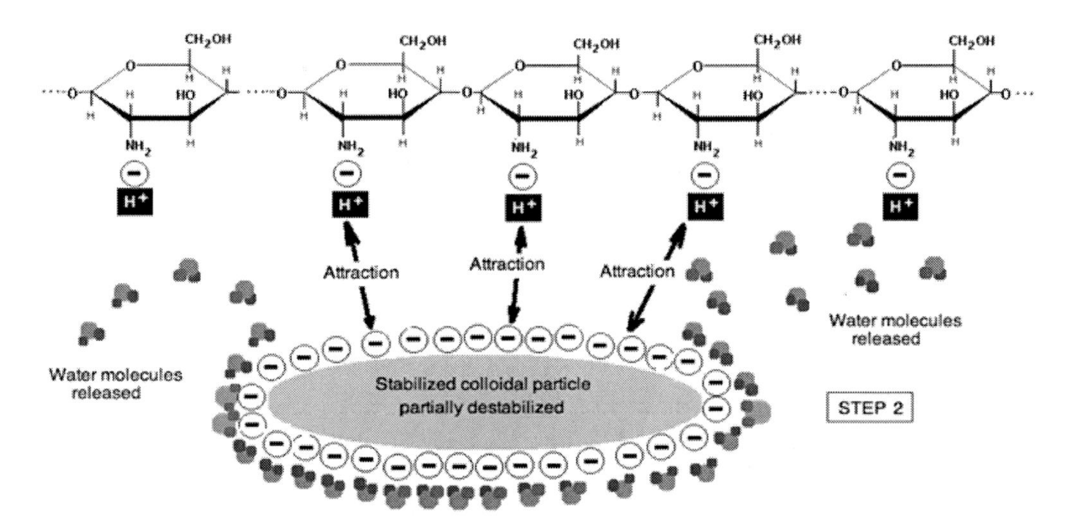

Figure 5. Destabilization in acidic media of a colloidal dispersion in the vicinity of a chitosan macromolecule.

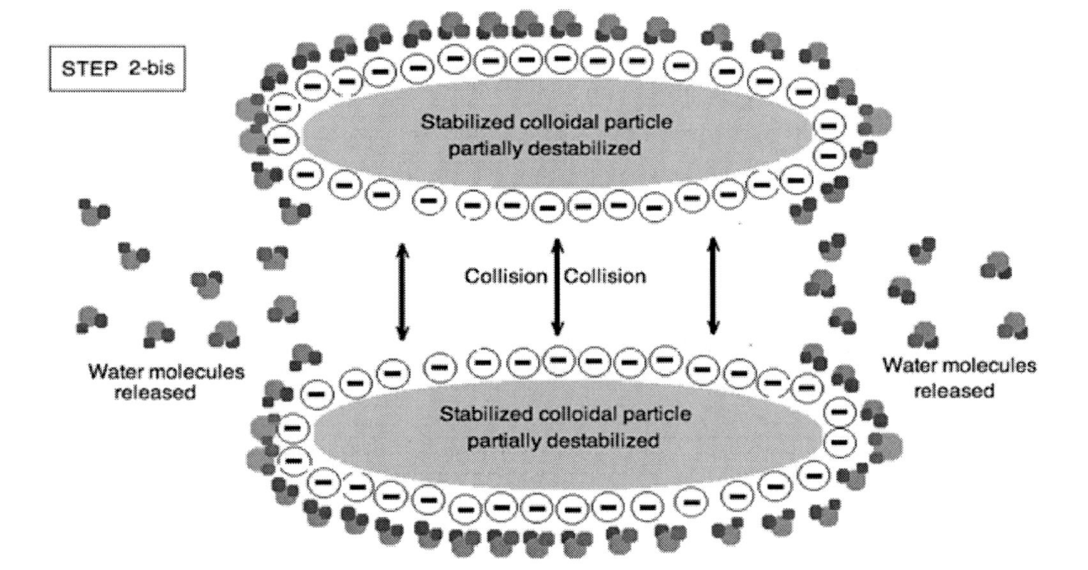

Figure 6. Mechanical destabilization of colloidal dispersion upon stirring.

3.1. Interactions between Destabilized Particles

Here, one assumes that the destabilization of the colloidal suspension occurs during coagulation, while flocculation takes place starting from the first irreversible collisions to form the first flakes. The theory according to which coagulation implies only destabilization of the colloidal suspension is prone to controversy. Many scientists still assert that there is no clear boundary line between coagulation and flocculation, inasmuch as destabilization of the colloidal dispersion still continues during flocculation.

The chemical bridges theory supposes that the polymeric molecule is linked to the colloidal particle by one or several sites [27-29]. This can occur via different pathways: coulombic attraction if the polymer and the particle have opposite charges, exchange of ions, hydrogen bonds, or van der Waal's forces if they have similar [27].

A chitosan macromolecule already linked to a destabilized particles can bind to another colloidal particle, and then to other macromolecules of chitosan through chemical bridges as shown in Figure 7. This bridging process generates progressively thicker and denser flakes till precipitation. Re-stabilization of free non-linked chitosan macromolecules is also possible. Such an entangled structure is not rigid because of the weak strength of the chemical bridges. Mere mechanical stirring can destroy the chemical bridges already formed. This may result in re-dispersion of the as-obtained flakes, leading to a re-stabilization of the colloidal dispersion [21].

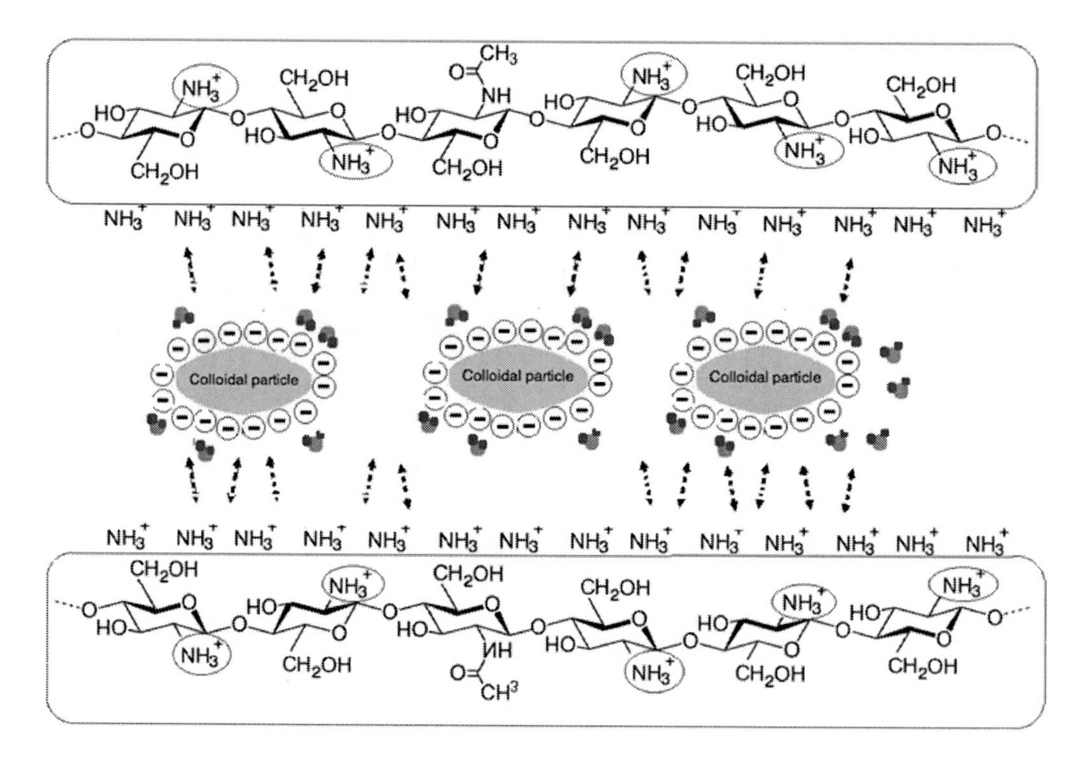

Figure 7. Scheme of flocculation through bridging process.

Addition of an excess of chitosan can lead rather to its own auto-coagulation, reducing dramatically the number of available sites supposed to form chemical bridges with other particles. This makes coagulation-flocculation to be completely ineffective towards any colloidal suspension.

According to the degree of acetylation (DA) and the pH of the liquid medium, chitosan can be partially or completely dispersed in the form of isolated macromolecules. In the case of a thorough dispersion of chitosan when all the polymeric chains are separated, the capacity of coagulation-flocculation should be maximum, and depends only on the strength of interaction with the substrate particles to be precipitated.

For low dispersion grade, chitosan will behave as solid dispersion in water, and its porosity will depend on the entanglement density of its macromolecules. The available surface of chitosan clusters may exert interactions upon various particles dispersed or chemical species dissolved in water. Whatever the nature of these interactions, adsorption on a solid surface is needed. According to Weber's model [28], these interactions involve a migration of the substrate particle towards chitosan (External diffusion), followed by its diffusion through the entanglement of chitosan fibers (Internal diffusion) and its adsorption on the surface of a chitosan macromolecule. Usually, adsorption is the fastest step. Both external and internal diffusion can be significantly improved upon sufficient mechanical stirring of the liquid media, taking into account obviously that this may also affect flocculation.

3.2. Interactions with Metal Cations

To date, several mechanisms of interactions between metal cations and chitosan have been proposed. Theoretically, a macromolecule chitosan displays chemical interactions towards cations dispersed in aqueous media, owing to the cation acidity and the base character of the chitosan amino group. According to some researches [30], cations should bind to chitosan via coordination bonds not only with the NH_2-groups of chitosan, but also through its hydroxyl groups. Other works [31,32] assume that only one monomer may be involved in the coordination. Nevertheless, they do not exclude the possibility that two different monomers of a same polymeric chain or of two different chains are implied in the formation of a stable complex with the metal cation. A deep analysis of some literature data [31-33] allow to establish interactions pathways involved between macromolecules of chitosan and cations present in the liquid media, as shown in Figures 8 and 9.

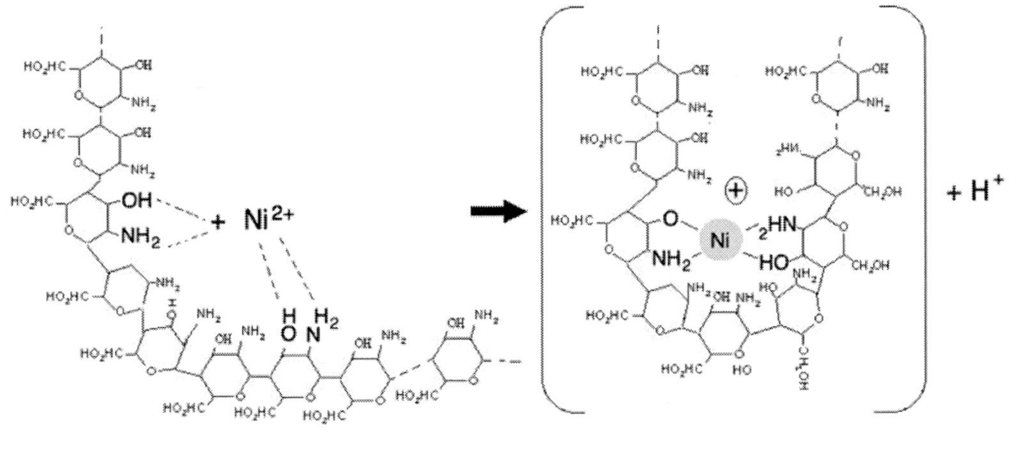

Figure 8. Ni^{2+} chelation by two chitosan monomers belonging to the same chain.

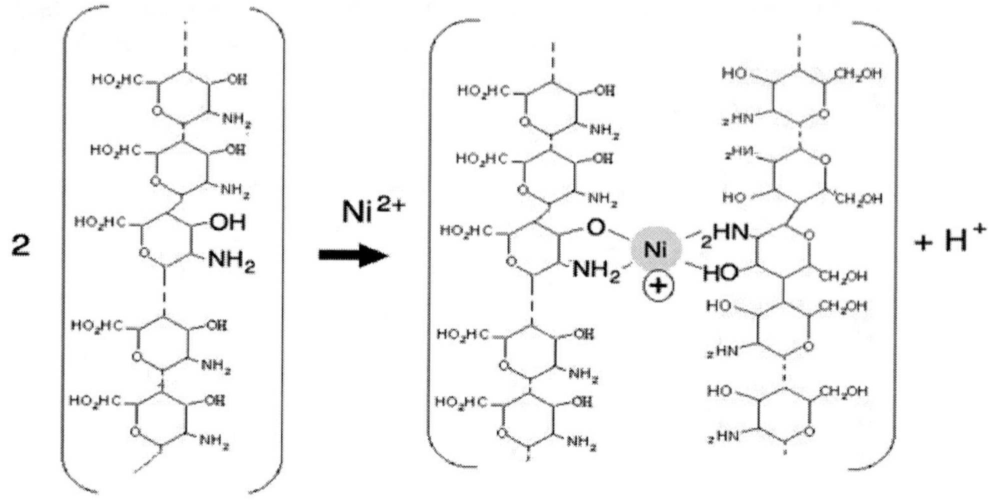

Figure 9. Ni^{2+} chelation by two chitosan monomers belonging to two different chains.

According to this scheme, a divalent cation binds to two amino groups and two OH groups. One OH group coordinated should release a proton, and a chemical equilibrium is expected to takes place between the released protons and the OH groups of chitosan. In this case, pH should play a key-role in the formation of such a cation-chitosan [31-33], and pH changes are expected to shift the chemical equilibrium, inducing the release of the cation retained by chitosan. Here, cations are supposed to interact with non-protonated amino groups, which are capable to retain a proton within a certain pH range. In this regard, Inoue et al. [34,35] suggest that cation binds to chitosan in aqueous nitric acid via previous protonation of a $-NH_2$ group, according to the scheme presented in Figure 10.

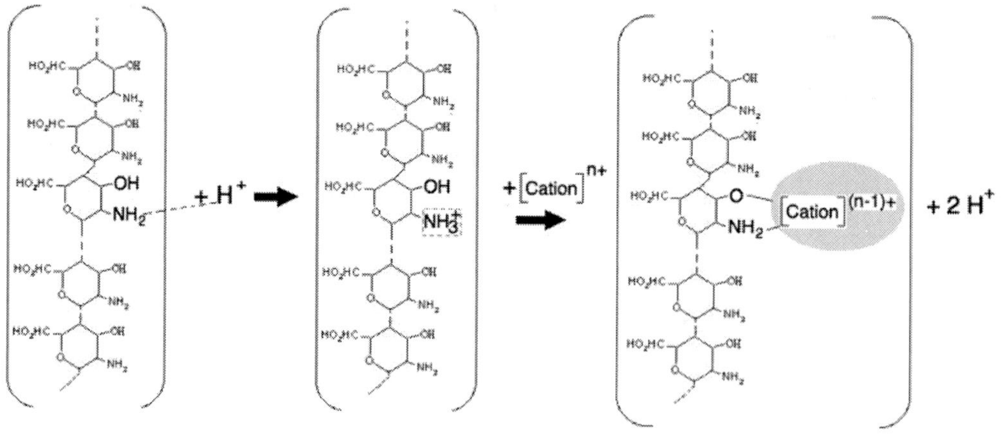

Figure 10. Chelation of a multivalent cation with protonated chitosan in acidic media.

For pH values close to the pKa of chitosan, the non protonated -NH2 form prevails, and hydrolyzed metal cations may appear. At this pH level, cations will combine with many hydroxyls via chelation. At lower pH, the ammonium groups resulting from protonation of -

NH2 groups will presumably act as exclusive ion exchange sites, by replacing a proton by a single charge of the cation.

The predominant complex form with these cations is assumed to have two OH groups and a -NH2 group as the ligands. The fourth site is occupied either by a water molecule or the OH group of the C3 carbon atom of the monomer of chitosan. When two amino groups are sufficiently close each other, the hydroxylated metal cation can simultaneously bind to these two groups.

Cations heavier than alkaline and alkaline earths are well known to behave as acidic species. The heavier the cation, the more acidic character is. Thus, one must expect that the base character of the amino groups of chitosan promote interactions with very acidic cations, especially of heavy metal cations and those of the lanthanide or actinide series. This is presumably why heavy metal cations like Hg^{2+}, Cr^{6+}, and Cd^{2+} can efficiently be removed from contaminated water by means of chitosan [36]. The uptake capacity of chitosan decreases in the following sequence: Hg^{2+}, Cr^{6+} and Cd^{2+}.

Another example in this regard is that of the retention by chitosan of ReO_4^+ cations, known as being hazardous radioactive pollutants [37]. ReO_4^+ cations are also acidic species, and exert very strong interactions upon the amino groups. These interactions seem to be strongly influenced by the ionic strength and pH of the aqueous media, and must obey the diffuse electrical double layer model. This model, combined with the model of Langmuir-Freundlich, has been already applied to chitosan before.

The theory of acid-base interactions between chitosan and cations must also be applicable for other polymers bearing base character groups. Clear evidence of this theory is provided by the lack if interaction between ReO_4 cations exhibit and acidic polymers, like poly-galacturonic acid (PGA) or poly-styrene sulfonate (PSS), which bear acidic carboxyl (-COOH) and sulfonate ($-SO_3H$) groups.

Chitosan is expected to display strong interactions with uranyl cations. In water, these interactions depend greatly on the mobility of polymer macromolecules [38]. The pseudo-crystallinity of dry chitosan, the particle size, along with the ability of solvatation and swelling of the polymer in water are key-parameters. The latter should strongly influence both the kinetics of cation diffusion in the entanglement of chitosan chains and adsorption of on these chains.

At neutral pH, metal cations adsorb by chelation on the amino groups of chitosan. In acidic media, adsorption of metal anions involves electrostatic attraction on protonated amino groups. The metal cation uptake mechanism can change chelation into the electrostatic interaction or vice-versa according to the conditions and composition of the liquid mixture. Thank to its special features, chitosan can interact with all kinds of metal cations, including precious metals (Pd, Pt), oxo-anions (Mo, V) and heavy metals (Cu, Ag) [39]. This is why chitosan can be employed for water treatments, waste-waters decontamination, valuable metals recovery, or merely reused as metal-loaded chitosan for others purposes (chitosan-supported catalysts, novel materials for electronics, medicine, chemistry, pharmacy, technologies, agriculture and others).

4. INTERACTIONS WITH INORGANIC PARTICLES

In aqueous environment, chitosan displays interactions with any solid particle that possesses surface charges. Most solids, particularly those of inorganic nature like clays, zeolites and other silicates, can remain suspended in water, provided that the contact surface with water is very large (very fine particle size and/or high porosity), and the surface of the particle displays sufficient affinity towards water.

Among the solids whose surfaces meet these conditions, clays occupy a privileged position. Chitosan interactions with clay minerals have already been thoroughly examined in a previous work [40]. It was shown that clay dispersion in water produces relatively persistent turbidity, but the addition of chitosan can overcome this drawback [10]. Nonetheless, it appears that the mere presence of an initial turbidity in water due the presence of traces of clay minerals can produce a synergistic effect that improves coagulation and flocculation by chitosan [41]. Significant decreases in the residual turbidity and in the sedimentation time are key-criteria for evaluating the improvement of the chitosan performances.

This is why chitosan can also be used as a co-coagulating agent to improve the performance of conventional coagulating agents like aluminum salts and to reduce their consumption [12,40]. Indeed, aluminum salts-chitosan mixtures seem to display synergy by promoting higher performance in terms of settling time and flakes size, with lesser amounts of clotting agents, more particularly at acidic pH [41].

4.1. Chitosan Behavior in the Presence of Clay Dispersions

Clay layered structures, more particularly those having electric charges on both sides of the lamella, such as smectites, including bentonite [43-46], can be of great interest for coagulation and flocculation. The main component of bentonite is montmorillonite, a layered structure where certain aluminum atoms (trivalent) of the central octahedral layer are substituted by divalent ones. Such an isomorphic substitution generates structural defects with an excess of negative charges. The latter are always counterbalanced by cations, creating, thereby, a cation-exchange capacity (CEC), which extends its influence even to the two outer layers of tetrahedral silicon oxide. Besides, in these peripherical silicon oxide layers, the oxygen atoms alternate with those of silicon, and generates additional partial charges that should undoubtedly contributes to the colloidal properties of clay minerals.

Indeed, many clays are known for their colloidal properties [40,47] thanks to their strong interactions with water [48]. Under certain conditions, they can undergo both self-coagulation and coagulation when contacted with others substances. It was already reported [49] that chitosan behaves as coagulating agent for bentonite suspensions in water, and the effectiveness of coagulation seems to be strongly dependent on the concentrations of both chitosan and clay mineral. A linear correlation between the optimum chitosan concentration and water turbidity due to the bentonite suspension has even been noticed.

The chitosan interactions depends greatly on the nature of the clay mineral, inasmuch as more effective clotting was obtained with bentonite that with kaolinite [49]. The same authors also reported that it was even possible to use small quantities of bentonite to enhance the coagulation of kaolinite by chitosan. Unlike many other works, they assert that pH has weak effect on the efficiency of the chitosan coagulation, and they conclude that the neutralization of surface charge is not the predominant step that controls the formation of chitosan flakes.

Chitosan also contributes to the coagulation and flocculation of bentonite suspensions in acetic or hydrochloric acids [50]. In acidic pH, protonation of amine groups induces pronounced coagulation even with small amount of chitosan. In the vicinity of the neutral pH, rather large amounts of chitosan are required for effective coagulation. Here, the degree of deacetylation (DDA) and the molecular size of chitosan appear to play a significant role.

When dispersed in water, montmorillonite lamellae behave as colloidal particles. According to the hydration grade, the exchangeable cations will rearrange in a double electric layer at a certain distance from the lamellae surface, giving rise to a colloidal micelle. These cations can also be replaced by others present in water. Such a process must depend on a series of parameters like the valence, hydration grade, and ionic radius of the ions substituted [51].

The interactions of chitosan with clays vary and involve different mechanisms depending on the DDA value and pH. In the presence of acetic acid, acetylated chitosan adsorbs differently on montmorillonite exchanged with sodium ions [40,52]. Adsorption may imply the formation of a single or double charge layer, depending on the exchange capacity of the clay mineral. Electrical interaction occurs between negatively charged clay sheets and dissolved chitosan acetate. The first charge layer forms via exchange of Na^+ by [chitosan-$NH_3]^+$ cations. while the acetate anion remains in the solution. This first charge layer is strongly retained on the particle surface, generating composite materials called organo-clays with significant rigidity for various applications.

This first layer exerts electrostatic dipole-dipole attraction upon chitosan-acetate dispersed in water, resulting in a purely physical adsorption. In this case, the whole chitosan acetate salt adsorbs at the upper layers. The latter will present peripheral sites [chitosan-acetate-NH_3^+] that can act as anion exchangers.

Interactions of chitosan and derivatives can induce changes in the clay mineral properties. For instance, carboxymethylated chitosan alters the microstructure, rheology, and water holding capacity of the clay mineral [53]. This result demonstrates the possibility to obtain chitosan-clay composite with various hydrophillic or hydrophobic characters depending on their composition for diverse purposes. A possible application in this regard would consists in producing coagulation-flocculation sludge displaying pronounced hydrophobic character and quick settling which, do not require drying. Similar researches [54] have shown that the interaction strength between chitosan and clay mineral have direct influence on the degree of compaction or expansion of the coagulated flakes.

The cohesiveness of the flake microstructure seems to be proportional to the dispersion level between chitosan and clay minerals, and increases with increasing the homogeneity of the blend of clay sheets and chitosan fibers. For montmorillonite, high homogeneity levels can be attained via previous full exfoliation in a potassium persulfate de potassium solution [55]. In this case, chitosan adsorbs in parallel positions on the clay lamellae, yielding a rigid and chemically stable composite.

Montmorillonite can be used as an adjuvant to accelerate the coagulation-flocculation of chitosan or vice versa. Each of the two agents may help improve the consistency and density of the precipitate flakes. This is probably why the coagulating behavior of chitosan is more pronounced in waters that already contain certain turbidity [40,41,56].

Waters with low turbidity are difficult to treat by coagulation even at optimum operating conditions. In such cases, addition of clay, more particularly bentonite, is often advantageous [19,41,56]. This opens news prospects in using both chitosan and clay traces for waters treatments, especially in removing heavy metals from polluted waters.

4.2. Binary and Ternary Interactions in Chitosan - Clay - Cation Mixtures

Both chitosan and clays, particularly montmorillonite, modified or not, can behave as very powerful agents in the removal of heavy metal cations by coagulation-flocculation [57,58]. In such a process, the size, charge and hydration degree of the exchangeable cation are regarded as being the key parameters [44,59-62]. Nonetheless, the effects of cations concentration and of the cation to clay mineral amount ratio must not be neglected [40,41].

So far, clay minerals have shown appreciable efficiency in the removal of metal cations from aqueous media [40,63-68]. This property is closely dependent on the nature of the exchangeable cation that compensates the permanent negative charge of the clay sheets. In the presence of montmorillonite exchanged with potassium (K-Mt) [41,56], it appears that the clay affinity of clay in retaining the metal cations varies in the following sequence: Cu^{2+}> Ni^{2+}> Co^{2+}, i.e. in the same order as the divalent cations radius. The balance between the Me^{2+} and $Me(OH)^{+}$ forms of a given bivalent cation can be more or less shifted by pH changes. This should greatly influence the mechanism of the cations capture by montmorillonite. In addition, pH adjustment with HCl, for instance, may introduce chloride ions into the solution, favoring the formation of species such as $CoCl^{+}$, $NiCl^{+}$ and $CuCl^{+}$ that can even more modify the interaction pathway.

As a general feature, cation removal by montmorillonite may be controlled by two different routes [62,69]. The first way would involve a clay ion-exchange independent of pH, via electrostatic interactions between cations and the permanent charge of montmorillonite. The second route would involve an adsorption via complexation that depends on pH, similar to those occurring with oxides.

Whatever the pH may be, montmorillonite can retain cations. At low pH, cations are retained even after clay auto-transformation upon acid attack. In this regard, the permanent negative charge of montmorillonite plays a major role, and removal by ion-exchange predominates on the clay surface [70]. Increasing pH slightly accentuates retention and removal of cations. This is probably due to the ionization of silanol and aluminol terminal groups located on the edges of the clay mineral structure [71].

In aqueous mixtures containing chitosan, clay mineral and metal cations, the latter are expected to exert binary interactions separately with each of the two substances involved, beside those occurring between chitosan and clay mineral (Figure 11).

Here, ion-exchange on the clay surface must not only take place unavoidably, but must also prevail in certain pH range. Nevertheless, this does not explain the relatively high amount of cations removed from the solution [40,41,56]. This is why synergy phenomena via ternary interactions of the three components of the aqueous mixture must also be taken into account. Such interactions become more complex when the role of water is also involved.

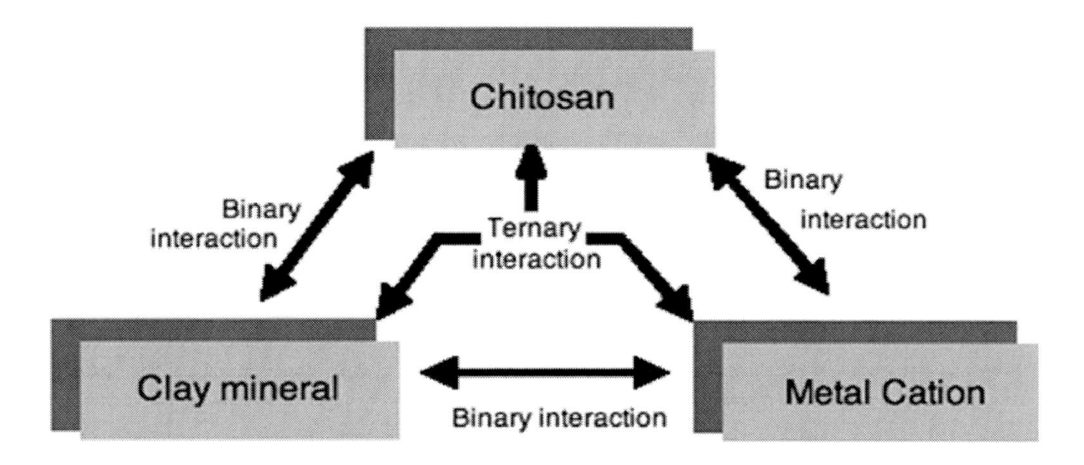

Figure 11. Cation-chitosan-clay interaction in aqueous media.

The interactions between montmorillonite, cations, chitosan and water molecules can take place via several mechanisms. The latter are influenced by various factors such as pH, nature and concentration of cations dispersed in solution, ion-exchange, clay mineral concentration, the concentration and molecular weight of chitosan, etc. A single lamella of montmorillonite dispersed in aqueous media will act as any colloidal particle that is stabilized by the formation of a double layer, as shown in figure 12.

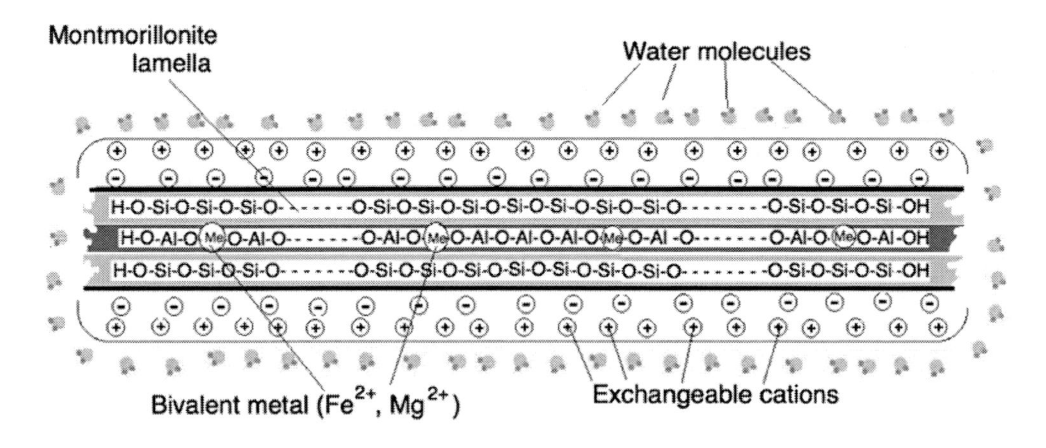

Figure 12. Stabilization of a single montmorillonite sheet dispersed in aqueous media.

When adding an electrolyte to a colloidal dispersion, unlike the surface charge arising from structural defects that remains unchanged, the surface potential decreases in the diffuse layer, and the latter will shrink [72].

At a critical electrolyte concentration, the van der Waals forces are stronger than those of repulsion and the particles coagulate [44,72]. Clotting may also occur with an excess of chitosan, which is expected to quickly saturate the particle surface, reducing, thereby, the number of bridging sites available. A re-peptization by steric stabilization and re-stabilization of the particles are still possible, especially for prolonged stirring time. These undesirable phenomena can affect the coagulation-flocculation effectiveness [21]. Thus, once coagulation started, any chitosan excess has negative effect , producing residual turbidity.

In diluted aqueous media containing traces of cations, auto-repulsion prevails and the clay dispersion is maximal. Addition of cations and/or polymers bearing electric charge shifts this equilibrium towards destabilization of the dispersed clay particles by ion exchange, altering the double layer. This may induce opposite charges in opposite faces of two dispersed clay sheets, producing coagulation (Figure 13).

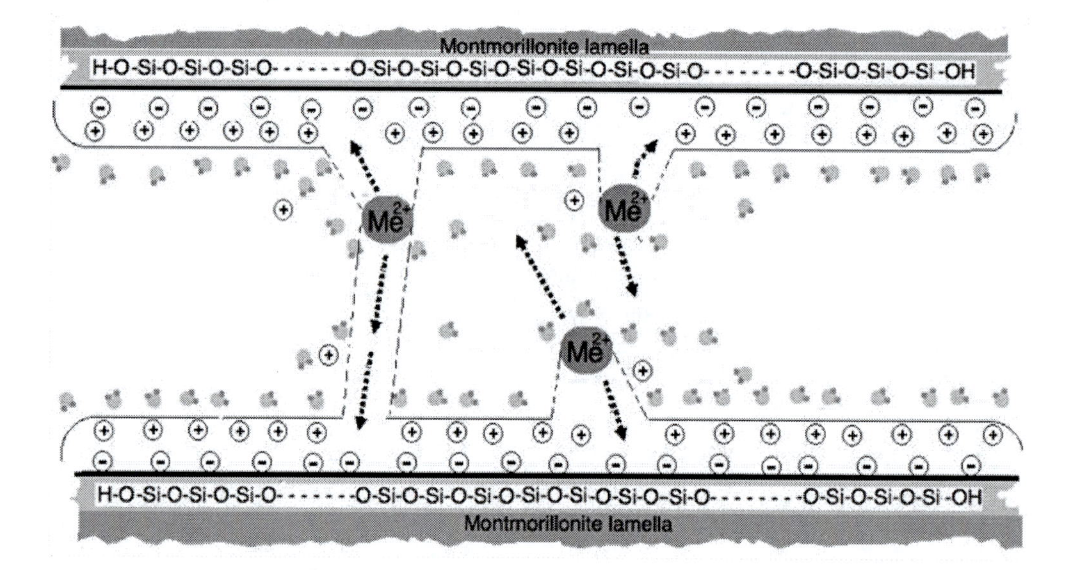

Figure 13. Destabilization via ion-exchange with bivalent cation.

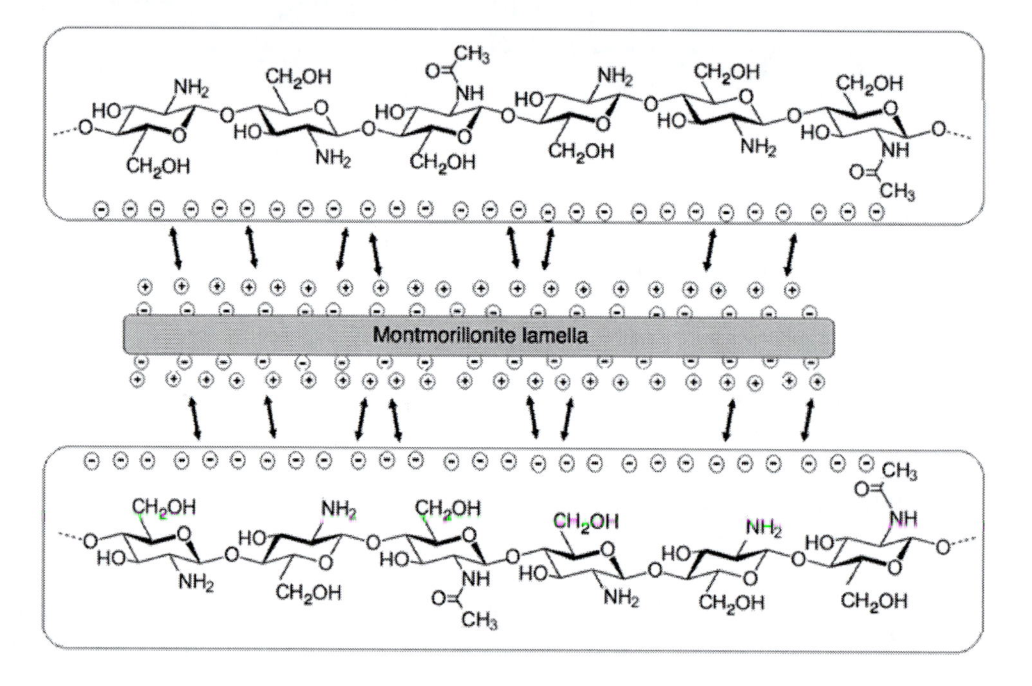

Figure 14. Interactions between chitosan macromolecules and dispersed clay sheets.

During coagulation in the presence of polymers [73], divalent cations are supposed to provide bridges between negatively charged surfaces, either between two montmorillonite sheets or between a clay lamella and an organic chain. Addition of a polymer such as chitosan to a clay suspension enhance the attraction force (Figure 14), leading to mutual coagulations of clay and chitosan, as in the case of an electrolyte [74].

As for simple organic molecules, the process is initialized by H^+ ion-exchange or cations interaction with the polymer. In the specific case of a chitosan-montmorillonite mixture, the acid-base interaction between an amino group of chitosan with montmorillonite is identical to an ion-exchange process.

According to some authors [75], the interaction between chitosan and montmorillonite in acid environment may first occur by ion-exchange. In this case, chitosan exerts strong electrostatic bonds on montmorillonite, compared to the weak hydrogen bonds. These bonds will not be affected by ion-exchange, more particularly when the capacity of ion exchange (CEC), i.e. the number of ion-exchange sites is higher than that of amino groups available.

The amino groups of the glucosamine monomer of chitosan may also be involved in the formation of complexes with exchangeable metal cations on the clay mineral surface [76]. At low pH, part of the amino groups can be transformed into ammonium, and undergo ion-exchange with metal cations therein. Because this process takes place by ion-exchange in the interlayer space, each divalent cation will be replaced by two amino groups (Figure 13).

The interaction by ion-exchange between chitosan and montmorillonite occurs mainly at acidic pH [77]. Nonetheless, under acidic conditions, amino groups are protonated, and the whole molecule of chitosan will behave as a poly-cation (Figure 15). However, it appears that positively charged polysaccharides are first retained (adsorbed) by cation exchange [78]. For this reason, they will have a greater affinity with montmorillonite sheets than with uncharged particles. In addition, the whole chitosan chain is stuck at the clay sheet surface, as shown in figure 15.

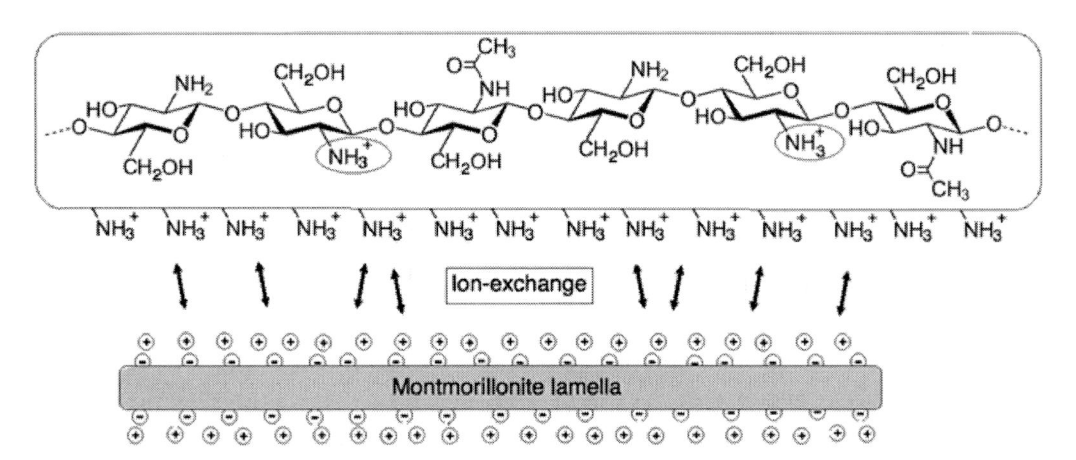

Figure 15. Ion-exchange scheme between a chitosan macromolecule acting as poly-cation and a montmorillonite lamella.

Such an ion-exchange phenomenon is strongly dependent on the pH level and behavior of the bulk water. Consequently, when amino groups are transformed into ammonium, chitosan binds to the clay sheet surface by exchanging the charge-compensating cations. This ion-exchange process is strongly correlated to both pH and behavior of water.

4.3. Role of Water

Water can act as donor and receiver of protons, and can promote or hinder the formation of hydrogen bridges depending on pH. The presence of water at the surface of a clay particle can either promote or inhibit the interaction with a macromolecule of chitosan. Since the oxygen atoms of the structure of the clay layer are weak electron donors [40,79-81], and display a weak base character, the interactions of water with the surface via hydrogen bridges will be weak. As a result, water molecules in the clay surface will be more associated with the exchangeable cations than with clay lamellae surface.

Currently, it is widely accepted that the mutual coagulation between montmorillonite and chitosan arises from inter-particle bridges formed during the adsorption of macromolecules of chitosan on the clay sheets [82]. Water may contribute by providing additional hydrogen bridges [83]. To argue this last statement, it should be reminded that at relatively high pH, chitosan exerts weak flocculating effect upon montmorillonite suspensions, because chitosan molecules curl in on themselves at alkaline pH, reducing thereby the number of sites available for bridging.

Low number of interactions between chitosan macromolecules and high numbers of amino or ammonium sites available on the chitosan chains are essential requirements for promoting inter-particle bridges between a clay sheet and an chitosan macromolecule, and effective removal of water turbidity. These two conditions can be fulfilled by using very low amounts of chitosan as compared to that of the clay mineral, in agreement with other works [41,56]. Anyway, the number of amino groups, protonated or not, necessary for the formation of inter-particle bridges, is almost always be higher than the ion-exchange sites on a clay lamella.

As an amino-biopolymer, chitosan displays a wide variety of physicochemical and biological properties [84]. The latter can be modified by applying suitable technique, and this is why so far diverse applications have been envisaged so far, in numerous fields like water treatment, cosmetics, biomedical engineering, pharmaceuticals, ophthalmology, biotechnology, agriculture, textiles, oenology, food processing and nutrition, and in nanotechnologies. Nowadays, nanotechnologies have focused growing interest, more particularly in the synthesis of chitosan-based materials. Most of these materials have exhibited improved interaction with both inorganic or organic solutes and suspensions, and turned out to be useful coagulants and flocculants in wastewater treatment. The presence of an initial turbidity is often regarded as being advantageous for the coagulating properties of chitosan in removing metal cations [41] or bacteria [85].

5. Applications Based on Interactions with Inorganic Species

5.1. Application of Chitosan – Clay Mineral Combinations

Chitosan is well known to favorably interact with clay minerals, inducing their coagulation and flocculation. Many applications, more particularly in pharmaceutical (drug controlled release) and dairy industries (removal of lactic acid and lactic seeds culture) are based on chitosan-clay interactions. These interactions determine the porosity and density of the stacking of the clay lamellae or entanglement of the chitosan polymeric chains.

In this regard, the carboxy-methylated chitosan has shown influence upon the microstructure, rheology, and the water retention capacity of clay minerals [53]. This opens new prospects to obtain more or less hydrophilic or hydrophobic organo-clays for specific purposes.

Similar research for obtaining montmorillonite-carboxymethylated chitosan composites [54] showed that more or less compact flakes may result depending on the strength of the interaction forces occurring between the chemical species involved. The more homogenous the mixture of chitosan fibers and clay lamellae, the stronger the cohesion of the flakes microstructure may be. To achieve such a pseudo-perfect homogeneity, chitosan fibers must be completely dispersed, and the clay mineral should be fully exfoliated and dispersed into single sheets.

It appears that prior dispersion of montmorillonite in a solution of potassium persulfate [55] can result in very high degree of exfoliation of the clay mineral. By contacting the suspension with an acid solution of chitosan, it was found that the persulfate salt reacts instantly with the chitosan, resulting in the dispersion of the latter and adsorption of clay layers in parallel positions along the polymer chains. Such an intimate contact between the clay and chitosan allows protecting chitosan chains against possible chemical or biochemical attacks, giving, thereby, rise to a quite rigid and chemically inert composite material.

In acidic media, the protonation of amino groups promote coagulation and only very low quantities of chitosan are sufficiently active to induce quick and effective settling of the colloidal particles [41]. In the neutral pH vicinity, flocculation prevails and higher amounts of chitosan are needed [86].

One of the most interesting applications of these interactions between chitosan and clay minerals is undoubtedly the synthesis of chitosan-clay nanocomposites. Such hybrid materials are bi-dimensional nanostructures based on the intercalation of Na-montmorillonite by chitosan. Clay minerals can interact with other substances in aqueous media by means of different types of sites, namely: 1. "broken edge" sites and accessible aluminol and silanol groups; 2 isomorphous incorporated atoms other than Al and Si; 3. charge-compensating cations; 4. hydrophobic silanol surfaces; 5. hydration water molecules around the exchangeable cations; 6. hydrophobic centers on adsorbed organic compounds [87]. The centers involve interactions between clay minerals and organic compounds dispersed or dissolved in aqueous media. Changes on the clay surface can induce specific interactions with different types of molecules. For instance, cationic clay minerals such as montmorillonite have negatively charged surface that can easily bind attract cations, but will repel negative or non polar hydrophobic particles. Ion-exchange with organic cations like ammonium based compounds (chitosan and derivatives, ammonium organic salts,....) might generate organoclays with hydrophobic moieties that attract non polar chemicals. Excessive organic cations loading should induce not only charge reversal and interactions with negatively charged molecules, but also rearrangements of the clay lamellae, inasmuch as expanded scattered structures are yielded.

When the amount of chitosan inserted is higher than the cationic exchange capacity (CEC) of the clay, chitosan reportedly adsorbs as a bilayer on the clay sheet. Consequently, the resulting organoclays will display anionic exchange properties, owing to the ($-NH3^{+}X^{-}$) sites belonging to chitosan [88]. The high affinity of chitosan towards montmorillonite appears to prevent from desorption or degradation. Incorporation of graphite particles, confers electrical conductibility to these materials, making them to be used as electrodes or electrochemical sensors. The latter can be employed in the selective determination of monovalent anions, more particularly for nitrate detection in waste-waters.

The mono- or bilayer chitosan adsorptions seem to be strongly dependent on the amount ratio [chitosan:CEC] of the clay. [89]. Cation-exchange appears to favor monolayer adsorption, while adsorption in the acetate salt form may lead to bilayer insertion.

All these changes can be more or less selectively induced through suitable modifications of the clay mineral and/or chitosan according to the purpose envisaged. The physicochemical properties of clay-chitosan composites are strongly dependent on the relative proportion of the components, and can open new prospects for obtaining a wide range of compositions for diverse purposes. For instance, the permeability to humidity can significantly be improved by insertion of clay in chitosan matrix, and an increase in the clay proportion seems to induce a decrease in the solubility in water [90].

The very structure and rheology of the bulk clay-chitosan matrix in both the hydrated and dry states are modified according the components amount ratio [91,92]. Indeed, thyxotropy properties for the hydrated state, exfoliation and increase in the surface roughness for the dry state seem to occur through incorporation of small quantities of Na-montmorillonite in chitosan. Other features of chitosan films are also modified, since an enhancement of the tensile strength and a decrease of the elongation-at-break take place. Insertion of 3 wt % clay mineral and 0.4 wt % carbon nanotubes also appears to significantly improve the tensile strength [93].

These findings open new prospects for producing special and high performance nanocomposites containing nanofillers of different sizes. In this regard, in other attempts [94], it was found that the composite structure is strongly dependent on the clay to chitosan ratio. It appears that the formation of an intercalated-and-exfoliated structure at low clay mineral content and an intercalated-and-flocculated structure at high clay mineral amount. Increasing clay loading improves the thermal stability of the composite. Here also, the tensile strength of a chitosan film by clay incorporation up to 2 wt% and elongation-at break decreased with addition of clay mineral into the chitosan matrix.

Researches for the development of biodegradable materials with controlled mechanical properties have progressively become a major priority for some scientists [95]. In this regard chitosan-based organoclays are still focusing glowing interest, owing to availability of the components of such composites. The latter are generally biodegradable, and display higher thermal stability, improved mechanical features and other interesting properties. The type of clay turned out to be a key factor regarding the composite feature. Rigid but breakable composites have been obtained with Na-bentonite, but better features have been provided by Cloisite 30B or Nanocor I-24 and addition of 5 % w/w polyethylene glycol.

Ternary combination of components can also generate interesting composite materials with improved mechanical and thermal feature [96]. In this regard, short chain composites have been synthesized via electrospinning technique using poly(vinyl alcohol), chitosan oligosaccharide [COS, (1→4)2-amino-2-deoxy-β-d-glucose], and montmorillonite in aqueous media. Here also, the results of this research reveals that the structure and properties of the composites are strongly dependent on the relative proportions of the three components.

Insertion of chitosan between clay lamellae produces unavoidably increase in the basal spacing of the clay mineral, and the magnitude of this change must strongly correlate to the chitosan to clay ratio. Here also, the thermal stability of chitosan is also improved thanks to the strong electrostatic interaction between the cationic chitosan chains and anionic clay sheets. Such materials seem to display synergy in the antimicrobial activity against Escherichia coli and Staphylococcus aureus [97].

For other medical purposes, some scientists [98] have synthesized novel composites based on Silver/Montmorillonite/Chitosan blends under UV exposure. This combination of three components must involve ternary interactions. These bio-nanocomposites seem to offer promising antibacterial and medical applications.

Furthermore, growing interest is now focused towards controlled drug release by means of chitosan-clay composite. Drug release from the nanocomposite particle carrier involves decomposition of the carrier matrix to its individual components or derived nanostructures with different compositions [99]. In chitosan–clay nanocomposite carriers, the slow drug release can be explained by the fact that the chemotherapeutic drug consists of positively charged molecules that strongly bind to the negative charges on the clay sheet surface. Such attraction forces must be strongly dependent on both the chitosan/clay ratio and pH.

Besides, chitosan-clay composites show interesting performances as membranes, more particularly for pervaporation. High separation level of the ethanol-water azeotrope mixture, has been registered at low water content, low temperature and low permeate pressure [100].

Other attempts and other researches are still in progress all over the word for obtained clay-chitosan composites, modified chitosan or chitosan-based bends for specific purposes. Thus, novel pervaporation membranes based on poly(vinylpyrrolidone) (PVP) and chitosan have been designed for tetrahydrofuran (THF) extraction from water and further dehydration [101]. Selectivity towards THF purification and mechanical properties have been improved by blending or cross-linking chitosan with glutaraldehyde or sulfuric acid in methanol/ sulfuric acid mixture.

5.2. Applications Based on Chitosan-Zeolites Interactions

Chitosan-silica hybrid films have been manufactured via sol-gel process starting tetraethoxysilane as precursor [102]. The microstructure of such composites consists of silica nanoparticles (2–7 nm), uniformly dispersed in the polymer matrix. Here, once again, it appears that the insertion of silica improves the thermal stability of chitosan.

Incorporation of mordenite into a blend containing chitosan and poly(acrylic acid) produces polyelectrolyte membranes for separation of ethylene-glycol from water via pervaporation [103]. The highest separation factor was obtained with membranes containing

4 wt% mordenite. The membranes show high selectivity towards water, because the diffusion of the water molecules through the membrane needs much less energy than that of ethylene-glycol.

Other membranes for the separation isopropanol from water via pervaporation have been synthesized by dispersion of NaY zeolite into a chitosan matrix [104].

Increasing amount of incorporated zeolite improves the flux and selectivity, presumably due to an enhancement of the hydrophilic character. The chitosan matrix acquires the zeolite microporosity, contributing, thereby, to the selectivity improvement.

Similar hybrid membranes but containing HY-faujasite have been designed the separation of the ethanol–water mixtures [105]. Strong interactions are supposed to occur between chitosan and the HY zeolite. This is presumably why increase in the zeolite content affects the crystalline structure of chitosan. The highest separation performances are obtained for a zeolite content of 20 wt%.

Dispersion of surface-modified Y-faujasite in a chitosan matrix yields hybrid membranes with higher methanol permeability than pure chitosan, more particularly at high methanol concentrations [106].The Y-faujasite surface is modified using silane coupling agents, namely 3-aminopropyl-triethoxysilane (APTES) and 3-mercaptopropyl-trimethoxysilane (MPTMS), to improve the compatibility towards the organic character of chitosan. Grafting [–SO3H] groups onto the zeolite surface causes an increase in the conductivity, making this type of hybrid membrane suitable for other purposes.

Direct hydrothermal synthesis of zeolite beta particles within a chitosan matrix gives rise composite membranes for direct methanol fuel cell [107]. The lowest methanol permeability has been obtained by incorporation of zeolite particles of about 800 nm in size. Grafting of sulfonic groups appears to reduce even more the methanol permeability.

Others hybrid membranes for the manufacture of direct methanol fuel cell have been prepared by insertion of zeolites like 3A, 4A, 5A, 13X, mordenite, and HZSM-5 into a chitosan matrix [108]. Chitosan interacts with zeolites via hydrogen bonds, and the hybrid membranes showed improved thermal and mechanical stabilities. Increase in both the free volume cavity size and methanol permeability, and a decrease in the methanol diffusion resistance are registered by incorporation of hydrophilic zeolites, and the reverse phenomena are observed after insertion of hydrophobic zeolites. The membrane performance seems to be strongly dependent on the sizes of the particles and pore, content, and hydrophilic/ hydrophobic character of the zeolites incorporated.

5.3. Applications Based on Interactions with Other Inorganic Compounds

In other researches, electrostatic interactions are involved in the preparation of biocompatible chitosan- carbon nanotube composites for biomedical purposes [109]. Their preparation uses nano-sized and micron-sized positively charged chitosan particles and carbon nanotubes bearing negatively charged chemical functions. The latter have been prepared by chemical oxidation method. Depending on the chitosan particle size, two kind of microspheres can be obtained: 1. grains with a carbon nanotube core coated by chitosan nanoparticles or cores of chitosan microsphere coated by carbon nanotubes.

Stronger interactions seem to be involved in the the synthesis of the so-called quantum dots, more particularly those composite nanoparticles obtained by incorporating cadmium sulfide (CdS) in a chitosan matrix [110]. Strong interactions seem to bind chitosan to CdS improving, thereby, the thermal stability of the chitosan matrix. The mere presence of CdS produces also significant improvement of the solubility in water.

6. Applications Involving Interactions with Organic Compounds

6.1. Chitosan Behavior in the Presence of Organic Chemical Species

Chitosan can be used for organic matter removal thanks to its interaction with organic compounds dispersed or dissolved in aqueous effluents. Removal of harmful micro-algae from waste-waters is a good example in this regard. During tertiary biological treatment of algae-containing waters, chitosan turned out to be more efficient than certain inorganic coagulating agents ($Al_2(SO_4)_3$, $KAl(SO_4)_2$, $Ca(OH)_2$). Indeed, almost 95% of algae can be removed using chitosan concentrations of 30 mg L-1 at pH 7.0 [111]. This performance seems to depend on the algae physiological state.

Chitosan also shows appreciable performances in the coagulation of solid suspensions in animal and vegetal food waste-waters [112]. Thanks to its high clotting capacity, chitosan can remove over 90% of organic wastes, reducing thereby significantly the chemical oxygen demand (COD).

Chitosan can also be used in the treatment of dairy industry waste-waters, and attempts to reduce the organic matter have been conducted in Germany [113]. pH seems to be a key factor, inasmuch as the best performances were achieved in the vicinity of the value of 6.5. Effective precipitation of organic matter, consisting mainly of milk protein, fat globules and lactose, occurs by coagulation-flocculation. Per cycle, nearly 50% of the organic matter and 25% of lactose could be eliminated by such treatments by chitosan. Owing to its affinity towards organic substances, chitosan even gave promising performances in biochemistry, as degreasing and clarification agent of protein suspensions [114].

6.2. Ternary Interactions with Cations and Organic Molecules

The presence of organic substances seems to improve the clotting properties of chitosan. Thus, chitosan and psyllium displayed high effectiveness in removing acidic and so called direct types of anionic organic dyes from waste-waters [115]. Together, these two compounds exert synergistic effect when combined with poly-aluminum chloride (PAC). Nonetheless, they turned out to be almost ineffective in eliminating the so-called reactive dyes with anthraquinone-like chromophorous groups. Here, pH, concentration and temperature appear to play key roles

Thanks to these synergy phenomena, a new application implying the combined effects of chitosan and the presence of organic matter in suspension for metal-loaded waters can be envisaged. Very complex interactions may take place in the presence of chitosan, cations and organic substrates, simultaneously dispersed in the same aqueous media. Indeed, one can expect that all three chemical species present exert not only individual effects, but also binary and ternary interactions. For instance, precipitation of a metal cation by chitosan should be greatly influenced by the presence of organic impurities. Conversely, coagulation of an organic substrate should depend on the ionic strength of the liquid media and thereby on the presence of metal cations.

An example in this regard is the elimination of uranium as uranyl ions in the presence of various organic compounds [116]. Removal of uranyl ions involves not only their adsorption on the chitosan macromolecules, but also precipitation, depending on the acidity or alkalinity of the organic compound present. For example, benzaldehyde promotes selective adsorption of uranyl ions on chitosan chains, while phenol, humic and benzoic acids appear to display negligible effect. An increased content of humic acid even seems to inhibit retention the uranyl cation. This must be due to a competitive adsorption, inasmuch as both uranyl cations and humic acid adsorb on chitosan, which displays base character in the presence of these cations.

To confirm this last statement, it would be judicious to compare the interactions of acidic cations with chitosan on the one hand, or with polymers bearing acidic groups on the other. In this regard, as previously discussed, a study for removing ReO_4^+ cations was comparatively carried out with chitosan, poly-galacturonic acid (PGA) and poly-styrene sulfonate (PSS) [37]. The results showed that PGA and PSS do not retain rhenium cations, because unlike chitosan, these two polymers contain rather carboxyl (-COOH) and sulfonate ($-SO_3H$) acidic sites. The latter do not interact with the acidic perrhenate ions (ReO_4^+), which adsorb more easily on the amino groups of chitosan.

Anyway, coagulation-flocculation attempts upon mud and sludge of rivers [117] provide evidence that even more complex quaternary interactions may be involved when using chitosan to reduce turbidity in waters containing cations, clay suspensions and organic matter. Explanation of such interaction can be of great interest for the combined use of chitosan and river silts in waste-waters treatments.

6.3. Applications via Interactions with Polymers, Dyes or Surfactants

Ternary interactions are involved between chitosan and organic dyes in the treatment of wool fibers [118]. At low chitosan contents, a 1:1 stoichiometry occurs between protonated amino groups and sulfonate acid groups on the dye ions, giving rise to an insoluble chitosan/dye chemical combination. Excessive chitosan content in the solution induces dye dispersion between the chitosan chains and solubilization chitosan/dye blend in the solution. This change is explained in terms of adsorption-desorption of the dye molecules and their redistribution between chitosan and the wool fibers, but the interactions between chitosan and acidic dye still remain to be elucidated.

The interactions occurring between chitosan and anionic species, more particularly anionic surfactants like sodium dodecyl sulfate seem to be highly exothermic, and insoluble chemical combinations are generated [119]. Nonetheless, the interaction strength dramatically decreases by change in the ionic force, e.g. in the presence of NaCl. This suggests the electrostatic nature of this interaction. This result could be of great interest for producing chitosan-based food ingredients with specific purposes like cholesterol lowering, fat replacement and others.

Interesting nano-aggregates comprised of a chitosan-poor core coated by almost pure chitosan shell have been obtained at high concentrations of surfactants like saturated sorbitan mono-laurate or unsaturated sorbitan mono-oleate [120]. Such structure assemblies in aqueous media must involve the hydrophillic character of chitosan and the hydrophobic character of sorbitan esters. Reportedly, the size and arrangement of the chitosan/surfactant assemblies is governed by the type and amount of the incorporated surfactant.

The possibility to improve the performances in drug release from chitosan pellets containing naproxen sodium, an anionic drug, by addition of use of anionic κ-carrageenan and nonionic hydroxy-propyl-methyl-cellulose (HPMC) has been envisaged [121]. The drug:chitosan ratio and pH appear to influence the water uptake, matrix erosion, and drug release. Pure chitosan shows rapid cohesiveness and, consequently quick release of naproxen sodium, but addition of optimum κ-carrageenan amount slows the drug release. The improvement of the pellets integrity has been explained in terms of the presence of HPMC. The latter seems to produce a viscous gel barrier that prevent from ionic interactions occur between chitosan chains. The retarded release of naproxen sodium from the chitosan matrices at different pH should also involve the low drug solubility in water, the drug interaction with [NH_3^+] groups of chitosan, along with the interaction of these amino groups with ionized sulfate groups of κ-carrageenan.

Organic films comprised of chitosan/polyvinyl alcohol and chitosan/polyethylene oxide mixture have synthesized through coagulation from aqueous solutions [122].

Chitosan has also been employed as a strengthening additive to improve in the pulp and paper technology [123,124]. His advantageous effect upon the wet network in paper must be due to its pH dependent adsorption behavior. Cellulose might be coated by chitosan in the absence of electrostatic attraction. The wet net strength improvement has been explained in terms of both increased wet adhesion between chitosan coated cellulose fibers at high pH and covalent bonding.

6.4. Applications via Interactions with Bio-Organic Substrates

Chitosan already turned out to be a performant gene carrier and pellets additive for drug delivery, but its contribution to lipid bilayer destabilization has scarcely been tackled. In this regard, attempts have been made to study the interactions occurring between chitosan and dipalmitoyl-Sn-glycero-3-phosphocholine (DPPC) bilayer [125]. It was found that the direct hydration of the DPPC/chitosan mixture produces bulkier DPPC multilamellar vesicles (MLV), and pure chitosan also induced fusions of individual MLV. Decrease in pH induces an increase of the number of protonated amines on the chitosan chain and affects the membrane structure. Prior to hydration, when DPPC and chitosan are mixed in an organic medium, the hydrophobic interactions between the two molecules are enhanced.

Novel materials comprised of coated silica beads with specific sites capable of hemoglobin recognition have be developed [126]. Here, chitosan acts is grafted on silica particles via phase inversion process. The exposed [NH_2] groups behave as reactive sites for incorporation aldehyde groups. Further, hemoglobin is covalently immobilized via imine bonds with the aldehyde groups, and acrylamide is then polymerized onto chitosan-coated silica grains to generates recognition sites. These sites favor selective adsorption of hemoglobin.

6.5. Chitosan and Derivatives for Enzyme Immobilization

Chitosan turned out to be promising adsorbent for enzymes and other bio-substrates. In this regard, chitosan nano-particles prepared by ions gelation of chitosan and tri-polyphosphate display amino and hydroxyl group capable to efficiently retain neutral proteinase from aqueous solution [127].

Gold nano-clusters insertion via electrostatic interaction on clay lamella surface can give rise to clay-chitosan-gold nanocomposites, which be useful for enzyme immobilization or to modify glassy carbon electrode (GCE) [128]. Immobilization of an enzyme, namely horseradish peroxidase, between the Clay/AuCS film and another clay layer shows that the original conformation of the enzyme is preserved. This result is of great importance for further enzyme immobilization without affecting the enzyme performances.

Enzymes such as α-amylase, β-amylase, and glucoamylase have been successfully immobilized on support based on blends containing equal weights of chitosan and activated clay and further cross-linked with glutaraldehyde [129]. Immobilization improves significantly the enzymes activity and lifetime after 50 times of repeated uses, within large pH and temperature ranges. Similar clay-chitosan composites grains turned out to be suitable supports for beta-glucosidase immobilization with higher performances than the free enzyme [130].

6.6. Applications of Chitosan Modified aith Organic Compounds

Knowledges about the chemical equilibrium and kinetics of the cation uptake mechanisms are very useful to improve the intrinsic physicochemical properties of chitosan through adequate modifications, even leading to the design of novel chitosan derivatives and combinations. For instance, changes in the physical form of chitosan into films, fibers, gel, beads, hollow particles and other forms can improve diffusion properties, while chemical modifications such as cross-linking, grafting of new chemical groups, or combinations with other substrates can significantly enhance selectivity, and even induce new features.

For instance, radionuclides and toxic metals are efficiently removed from aqueous effluents by chitosan or N-carboxymethyl-chitosan. For higher efficiency, chitosan may be combined with clay minerals like bentonite, attapulgite or volcanic ash [131].

Mercury in the Hg(II) form can be retained by chitosan membranes, but the amount of retained Hg(II) ions is significantly increased by cross-linking chitosan with epichlorohydrin or glutaraldehyde, presumably due to imino bonds, unreacted aldehyde terminals along with unreacted hydroxyl groups from starting chitosan [132]. The cation uptake is maximum at pH 6.0, for both epichlorohydrin-cross-linked and glutaraldehyde-cross-linked chitosan membranes. Cation desorption is optimal in the presence of NaCl in the solution.

A novel coagulation system for chitosan, in aqueous alcohol solution of calcium chloride or acetate seems to be found [133]. Via calcium chelation with amino groups during coagulation, such a system appears to favor an interesting arrangement of chitosan macromolecules into spun filaments with high solubility in water and improved coagulation capacity.

Chitosan derivatives obtained by modifying their functional groups (carboxyl, amido, sulfonate, N-vinylimidazole) or cross-linked have shown increased capacity for dye adsorption and improved mechanical features [134].

A new application in medicine consists in using modified chitosan to stop bleeding. In this regard, chitosan modified with aldehydes or its cross-linked with 4-azidobenzoic acid under the action of UV rays has shown improved clotting properties [135]. Later, it has been demonstrated [136] that any modification of chitosan to increase the number of amino groups would improve the coagulating capacity of chitosan. This finding is of great importance for opening new prospects in using chitosan in the treatment of food industries.

7. Conclusion

The chitosan particles have partial electrical charges in both isolated polymer chains or macromolecules clusters forms. These electrical charges make chitosan to be reactive towards any chemical species dispersed in aqueous media (cations, various molecules) that show a capacity to reduce these electrical charges. The nature and strength of the interactions involved strongly depend on the type of dispersed chemical particles, and confer to chitosan clotting properties and affinity to combine to a wide variety of chemical compounds. Besides, chitosan displays certain biodegradability and recyclability, and turns out to the best alternative to conventional aluminum-based coagulants. The latter are rather expensive for water treatment and possibly harmful towards human health. In addition, the sludges resulting from chitosan precipitation are also recyclable. Judicious chemical modifications of chitosan can improve the coagulation capacity, and may open new prospects in diversifying its purposes.

To date, several mechanisms of interactions occurring between metal cations and chitosan have been proposed in literature. These interactions appear to involve the acidic behavior of the cations and the base character of the amino groups. Regardless to their nature, most of the solid particles suspended in water can be trapped by chitosan settling. This can take place if chitosan adsorbs on these particles or vice versa and if adsorption attenuates the particle interactions with water. Clay minerals can behave as colloidal solids producing persistent turbidity in water, but the use of chitosan as coagulating agent can allow overcoming this drawback.

Interactions may also take place between chitosan and organic substrates dispersed in aqueous media. This makes chitosan to be an interesting bleeding inhibitor in medicine, and increasing the number of amino groups through suitable modifications of chitosan should not only improve the clotting ability, but also generate unexpected properties that can be very useful for new purposes, more particularly in obtaining novel chitosan-based composites.

This statement is of major importance for the chemistry of chitosan. Prospective applications of chitosan can also be envisaged, not only as coagulating agents to treat waters containing heavy metals, but also as co-coagulating agent by its combination with organic or inorganic polymers. This open new directions in obtaining hybrid matrixes chitosan-polymer for diverse purposes, more particularly in the manufacture of membranes for ultrafiltration, or excipients for the controlled release of drugs or ingredients in food technologies.

References

[1] Yang, T.C.; Zall, R.R. *Industrial and Engineering Chemistry Product Research and Development* 1984, 23, 168-172.

[2] Kawamura, Y.; Mitsuhashi, M.; Tanibe, H. *Industrial and Engineering Chemistry Research* 1993, 32, 386-391.

[3] Majeti, N.V.; Kumar, R. *Reactive and Functional polymers* 2000, 46, 1-27.

[4] Onsoyen, E, Skaugrud, O. *Journ Chem Technol Biotechnology* 1999, 49, 395-404.

[5] Ngah, W.S.W.; Isa, I.M. *Journal of Applied Polymer Science* 1998, 67, 1067–1070.

[6] Wu, F.C.; Tseng, R.L.; Juang, R.S. *Industrial and Engineering Chemistry Research*1999, 38, 270-275.

[7] Cheung, W.H.; Jcy, N.; Mckay, G. *Journal of Chemical Technology & Biotechnology* 2002, 78, 562–571.

[8] Nagy, M.N.; Konya, J. *Colloids and Surfaces A: Physicochemical and Engineering Aspects* 1998, 137, 231-242.

[9] Altin, O.; Ozbelge, O.H.; Dogu, T. *Journal of Chemical Technology & Biotechnology* 1999, 74, 1131-1138.

[10] Domard, A.; Rinaudo, M.; Terrassin, C. *J. Appl. Polymer Science* 1989, 38, 1799-1806.

[11] Dehonor-Gomez, M.; Hernandez-Esparza, M.; Ruiz-Trevino, F.A.; Contreras-Reyes, R. *Macromolecular Symposia* 2003, 197, 277-288.

[12] Huang, C.; Chen, S.C.; Pan, R.J.; *Water Research* 2000, 34(3), 1057-1062.

[13] Hollander, A.P.; Hatton, P.V. *Biopolymer methods in tissue engineering*; Humana Press: Bristol, UK, 2004; pp. 257.

[14] Howard, W.T.M. *Biomimetic materials and design: biointerfacial strategies, tissue engineering, and targeted drug delivery*; Marcel Dekker: New York, 2002; Chap.10, pp. 679.

[15] Shepherd, R.; Reader, S.; Falshaw, A. *Glycoconjugate Journal* 1997, 14(4), 535-542.

[16] Kurita, K.; Tomita, K.; Tada, T.; Ishii, S.; Nishimura, S.I.; Shimoda, K. *J. Polymer Sci. Part A: Polymer Chem.* 1993, 31, 485-91.

[17] Takagi, H.; Kadowaki, K. *Agric. Biol. Chem.* 1985, 49, 3051-57.

[18] Rinaudo M.; Domard. A. *In Chitin and Chitosan*; Skjak-Braek, G.; Anthonsen, T.; Sandford, P.; Eds.; Elsevier Science Publishing: New York; USA,1989; pp. 71-83.

[19] Beaudry, JP. *Traitement des eaux; Le Griffon d'argile*: Quebec; QC, 1984; pp. 231.

[20] Desjardins, R. *Le traitement des eaux; 2ème éd..;* École Polytechnique de Montréal: Montreal, QC, 1988; pp. 304.

[21] Benefield.; L.D.; Judkins.; J.F.; Weand.; B.L. *Process Chemistry for Water and Wastewater Treatment;* Prentice-Hall: Englewood Cliffs.; NJ., 1982; pp. 510.

[22] Bechac, JP.; Boutin, P.; Mercier B.; Nuer, P. *Traitement des Eaux Usées; Eyrolles* Ed.:Paris, 1984; pp. 281.

[23] Montgomery, J,M. *Water treatment principles and design*; J. Wiley: New York, 1977; pp. 696.

[24] Edeline, F. *L'épuration physico-chimique des eaux : théorie & technologie;* 2ème éd.; Lavoisier: Paris, 1992; pp. 282.

[25] Ruehrwein, RA.; Ward DW. *Soil Science* 1952, 73, 485-492.

[26] LaMer, V.K.; Healy, T.W. *Reviews of Pure and Applied Chemistry* 1963, 13, 112-133.

[27] Franchi, A.; O'Melia, C.R. *Environmental Science and Technology* 2003, 37(6), 1122-1129.
[28] Weber, W.J. *Physicochemical Processes for Water Quality Control;* J. Wiley: New York, 1972; pp. 640.
[29] Teramoto, T. *Deep Ocean Circulation - Physical and Chemical Aspects*; Elsevier: Amsterdam, 1993; pp. 382.
[30] Juang, RS.; Shao HJ. *Adsorption* 2002, 8, 71–78.
[31] Kaminski, W.; Modrzejewska, Z. *Separation Science and Technology* 1997, 32, 2659-2668.
[32] Oyrton, A.C.; Monteiro, J.; Airoldi, C. *Journal of Colloid and Interface Science* 1999, 212, 212-219.
[33] Guan, H.M.; Cheng X.S. *Polymers for Advanced Technologies* 2004, 15, 89-92.
[34] Inoue, K.; Yoshizuka, K.; Baba Y. *Polymeric Materials- Science and Engineering* 1992, 66, 346-347.
[35] Inoue, K.; Baba, Y.; Yoshizuka, K. *Bulletin of Chemical Society of Japan* 1993, 60, 444.
[36] Babel, S.; Kurniawan T.A.; *Journal of Hazardous Materials* 2003, 97 (1-3), 219-243.
[37] Kim, E.; Benedetti, M.F.; Boulègue, *J. Water Research* 2004, 38(2), 448-454.
[38] Piron, E.; Accominotti, M.; Domard, A. *Langmuir* 1997, 13(6), 1653-1658.
[39] Guibal, E. *Separation and Purification Technology* 2004, 38 (1), 43-74.
[40] Azzouz, A. *In Chitine et Chitosane – Du biopolymère à l'application;* Crini, G.; Badot, P.M.; Guibal, E.; Eds.; Presses Universitaire de Franche-Comté: Besançon, France, 2009; Chapter 12, pp. 231-255.
[41] Assaad, E.; Azzouz, A.; Nistor, D.; Ursu, A-V.; Sajin, T.; Miron, D.N.; Monette, F.; Niquette, P.; Hausler, R. *Applied Clay Science* 2007, 37, 258–274.
[42] Pan, JR.; Huang, C.; Chen, S.C.; Chung, Y.C. Colloids and Surfaces *A: Physicochemical and Engineering Aspects* 1999, 147(3), 359-364.
[43] Azzouz, A. *Physicochimie des Tamis Moléculaires*; Office des Publications Universitaires: Alger, DZ, 1994; pp. 25,47,103.
[44] Bergaya, F.; Theng, B.K.G.; Lagaly, G. *Handbook Of Clay Science*; Elsevier, 2006; pp. 1-150.
[45] Aubert, G.; Pierrot, R.; Guillemein, C. *Précis de minéralogie*; Masson: Paris, FR, 1978; pp. 335.
[46] Déribéré, M.; Esme, A. *La bentonite;* Dunod: Paris, FR, 1951; pp. 224.
[47] Hendrik, V.O. *An introduction to clay colloid chemistry: for clay technologists, geologists and soil scientists;* Interscience Publishers: New York, 1963; pp. 301.
[48] Janice, L.B.; Carle, M.P.; John O.P. *Clays and Clay Minerals* 1994, 42, 702-716.
[49] Huang, C.; Chen, Y. *Journ Chem Technology & Biotechnology* 1999, 66(3), 227–232.
[50] Roussy, J.; Van Vooren, M.; Guibal, E. *Journal of dispersion science and technology* 2004, 25(5), 663-677.
[51] Sridharan, A.; Jayadeva, M.S. *Australian Journal of Soil Research* 1980, 18(4), 461-466.
[52] Darder, M.; Colilla, M.; Ruiz-Hitzky, E. *Chemistry of materials* 2003, 15(20), 3774-3780.
[53] Joyce, M.K.; Gilbert, R.D.; Khan, S.A. *Tappi journal* (Technical Association of the Pulp and Paper Industry journal) 1997, 80(5), 185-190.

[54] Wang, S.F.; Chen, L; Tong, Y.J. *Journal of polymer science.* Part A. Polymer Chemistry 2006, 44(1), 686-696.

[55] Lin, K.F.; Hsu, C.Y.; Huang, T.S.; Chiu, W.Y.; Lee, Y.H.; Young, T.H. *Journal of applied polymer science* 2005, 98(5), 2042-2047.

[56] Assaad, E. *Étude du processus de coagulation-floculation du système montmorillonite-chitosane dans l'élimination de métaux de transition.* Master Thesis, Chemistry, Université du Québec à Montréal, january 2006.

[57] Celis, R.; Hermosin, M.C.; Cornejo, J. *Environmental Science and Technology2000*, 34, 4593-4599.

[58] Mercier, L.; Detellier, C. *Environmental Science and Technology* 1995, 29, 1318-1323.

[59] Lagaly, G. Coagulation *and Flocculation. Theory and Applications*; B Dobias.; Ed. Marcel Dekker Inc.: New York, USA Inc. 1993; pp. 427–494.

[60] Lagaly, G.; Ziesmer, S. *Advance in Colloid and Interface Science* 2003, 100, 105–112.

[61] Mermut, A.R.; Lagaly, G. *Clays and Clay Minerals* 2001,49, 393–397.

[62] Abollino, O.; Aceto, M.; Malandrino, M.; Sarzanini, C.; Mentasti, E. *Water Research* 2003, 37, 1619–1627.

[63] Egozy, Y. *Clays and Clay Minerals* 1980, 28(4),. 311-318.

[64] El-Batouti, M.; Sadek, O.M.; Assaad, F.F. Journal *of Colloid and Interface Science* 2003, 259, 223-227.

[65] Ghanayem, M.F. Removal *of heavy metals from aqueous solutions by soils*; PhD Thesis; The University of Texas at Dallas, USA, 1989; pp. 191.

[66] Puls,R.W. *Adsorption of heavy metals on soil clays*; PhD Thesis; The University of Arizona, USA, 1986; pp. 141.

[67] Vengris, T.; Binkiene, R.; Sveikauskaite, A. *Applied Clay Science* 2001, 8, 183–190.

[68] Lin, S.H.; Juang, R.S. *Journal of Hazardous Materials* 2002, B 92, 315–326.

[69] Kraepiel, A.M.L.; Keller, K.; Morel, F.M.M. *Journal of Colloid and Interface Science* 1999, 210, 43–54.

[70] Bradbury, M.H.; Baeyens, B. *Geochimica et Cosmochimica* Acta 1999, 63, 325–336.

[71] Viallis-Terrisse, H. *Interaction des Silicates de Calcium Hydratés.; principaux constituants du ciment.; avec les chlorures d'alcalins. Analogie avec les argiles*; PhD thesis; Université de Bourgogne, France, 2000; pp. 256.

[72] Van Olphen, H. *An Introduction to Clay Colloid Chemistry*; 2nd edition; Wiley-Interscience: New York, 1977; pp. 1.

[73] Sollins, P.; Homann, P.; Caldwell, B.A. *Geoderma* 1996, 74, 65-105.

[74] Klug, H.P.; Alexander, L.E. *X-ray diffraction procedures for polycrystalline and amorphous materials*; John Wiley: New Tork, 1967; pp.716.

[75] Clapp, C.E.; Emerson, W.W. *Soil Science* 1971, 114, 210-216.

[76] Pusino, A.; Micera, G.; Premoli, A.; Gessa, C. Clays *and Clay Minerals* 1989, 37, 377-380.

[77] Parfitt, R.L. *Soil Science* 1972, 113, 417-421.

[78] Theng, B.K.G. *Formation and properties of clay-polymer complexes*; Elsevier: Amsterdam, 1979; pp. 362.

[79] Garfinkel-Shweky, D.; Yariv, S. *Journal of Colloid and Interface Science* 1997, 188(1), 168–175.

[80] Yariv, S.; Michaelian, K.H. *Schriftenr. Angew. Geowiss* 1997, 1, 181–190.

[81] Shoval, S.; Yariv, S.; Michaelian, K.H.; Lapides, I.; Boudeuille, M.; Panczer, G. *Proceedings of The Israel Mineral Science and Engineering Association, Beer Sheva* 14th Conference, 14–16 December 1998; pp. 101–109.
[82] Benjelloun, M.; Cool, P.; Linssen, T.; Vansant, E.F. *Microporous and Mesoporous* Materials 2001, 49, 83-94.
[83] Chen, L.; Chen, D.; Wu, C. *Journal of Polymers and the Environment* 2003, 11, 87–92.
[84] Renault, F.; Sancey, B.; Badot, P.-M.; Crini, G.; *European Polymer Journal* 2009, 45 (5), 1337-1348).
[85] Bina, B.; Mehdinejad, M.H.; Nikaeen, M.; Movahedian Attar H. Iran. *J. Environ. Health. Sci. Eng.* 2009, 6(4), 247-252.
[86] Jean Roussy; Maurice Van Vooren; Eric Guibal.; *Journal of Dispersion Science and Technology* 2005, 25 (5), 663 – 677.
[87] Rytwo, G. *Revista de la sociedad española de mineralogía - macla* 2008, 9, 15-17.
[88] Darder, M.; Colilla, M.; Ruiz-Hitzky, E. *Applied Clay Science* 2005, 28(1-4), 199-208.
[89] Darder, M.; Colilla, M.; Ruiz-Hitzky, E.; *Chem. Mater.* 2003, 15 (20), 3774–3780.
[90] Casariego, A.; Souza, B.W.S.; Cerqueira, M.A.; Teixeira, J.A.; Cruz, L.; Dı´az, R.; Vicente, A.A. *Food Hydrocolloids* 2009, 23(7), 1895–1902.
[91] Xu, Y.; Ren, X.; Hanna, M.A. *Journal of Applied Polymer Science* 2006, 99 (4), 1684-1691.
[92] Liang, S.; Liu, L.; Huang, Q.; Yam, K.L. *J. Phys. Chem. B.* 2009,113(17), 5823-8.
[93] Tang, C.; Xiang, L.; Su, J.; Wang, K.; Yang, C.; Zhang, Q.; Fu, Q. *J. Phys. Chem. B,* 2008, 112 (13), pp 3876–3881.
[94] Altinisik, A.; Seki, Y.; Yurdakoç, K. *Polymer Composites* 2008, 30 (8), 1035-1042.
[95] Nunes, M.R.S.; Silva, R.C.; Silva Jr., J.G.; Tonholo, J.; Ribeiro, A.S. *Preparation and morphological characterization of chitosan/clay nanocomposites.* ICAM-2009, 11th International Conference on Advanced Materials, Rio de Janeiro, Brasil, sept. 20-25.
[96] Park, J.H.; Lee, H.W.; Chae, D.K.; Oh, W.; Yun, J.D.; Deng, Y.; Yeum, J.H. *Colloid & Polymer Science* 2009, 287 (8), 1435-1536.)
[97] Han, Y.-S.; Lee, S.-H.; Choi, K.H. In Park, Preparation and characterization of chitosan-clay nanocomposites with antimicrobial activity;15th International Symposium on Intercalation Compounds - ISIC 15 May 10-15, 2009, Beijing, *Journal of Physics and Chemistry of Solids*, 2010, 71 (4), 464-467.
[98] Bin Ahmad, M.; Shameli, K.; Darroudi, M.; Yunus, W.M.Zin.W.; Ibrahim, n.a. *American Journal of Applied Sciences* 2009, 6 (12), 2030-2035.
[99] Yuan, Q.; Shah, J.; Hein, S.; Misra, R.D.K. *Acta Biomaterialia* 2010, 6(3), 1140-1148.
[100] Nawawi, M.; Ghazali, M.; Niza, S.A.; Gi, T.G. *Jurnal Teknologi* 2008, 49, 179-188.
[101] Anjali Devi, D.; Smitha, B.; Sridhar, S.; Aminabhavi, T.M. *Journal of Membrane Science* 2006, 280 (1-2), 45-53.
[102] Al-Sagheer, F.; Muslim, S. *Journal of Nanomaterials* 2010, Article ID 490679, doi:10.1155/2010/490679.
[103] Hu, C.; Guo, R.; Li, B.; Ma, X.; Wu, H.; Jiang, Z.; *Journal of Membrane Science* 2007, 293 (1-2), 142-150.
[104] Kittur, A.A.; Kulkarni, S.S.; Aralaguppi, M.I.; Kariduraganavar, M.Y. *Journal of Membrane Science* 2005, 247 (1-2), 75-86.
[105] Chen, X.; Yang, H.; Gu, Z.; Shao, Z. *Journal of Applied Polymer Science* 2001, 79, 1144-1149.

[106] Wu, H.; Zheng, B.; Zheng, X.; Wang, j.; Yuan, w.; Jiang, Z. *Journal of Power Sources* 2007, 173 (2), 842-852.

[107] Wang, Y.; Yang, D.; Zheng, X.; Jiang, Z.; Li, J. *Journal of Power Sources* 2008, 183 (2), 454-463.

[108] Wang, J.; Zheng, X.; Wu, H.; Zheng, B.; Jiang, Z.; Hao, X.; Wang, B. *Journal of Power Sources* 2008, 178 (1), 9-19.

[109] Baek, S.-H.; KIM, B.; SUH, K.-D. Colloids and surfaces. A, Physicochemical and engineering aspects 2008, 316 (1-3), 292-296.

[110] Li, z.; Du, Y.; Zhang, Z.; Pang, D. *Reactive and Functional Polymers* 2003, 55 (1), 35-43).

[111] Lavoie, A.; De la Noüe, J.; Sérodes, J.B. *Can. J. Civ. Eng*. 1984, 11(2), 266–272.

[112] Bough, W.A.; Landes, D.R. *Proceedings of The First lnternational Conference on Chitin I Chitosan;* MIT Sea Grant Report MITSG 78-7; Muzzarelli, R.A.A.; Pariser, E.R., Ed.; 1978; pp. 218-230.

[113] Ordolff, D. Kieler Milchwirtschaftliche Forschungsberichte 1995, 47(4), 339-346.

[114] Novikov, V.Y.; ukhin., V.A. *Applied Biochemistry and Microbiology* 2001, 37, 629-634.

[115] Bhattacharya, B.; Sanghi, R. *Water Quality Research Journal of Canada: General Issue* 2005, 40 (1), 97-101.

[116] Gerente, C.; Andres, Y.; Le Cloirec, P. *Environmental Technology* 1999, 20(5), 515-521.

[117] Divakaram, R.; Pillai, V.N. *Water Research* 2001, 36, 2414-2418.

[118] Jocic, D.; Vílchez, S.; Topalovic, T.; Navarro, A.; Jovancic, P.; Julià, M.R.; Erra, P. *Carbohydrate Polymers* 2005, 60 (1), 51-59.

[119] Thongngam, M.; McClements, D.J. J. Agric. *Food Chem*. 2004, 52 (4), 987–991.

[120] Grant, J.; Lee, H.; Liu, R.C.W.; Allen, C. *Biomacromolecules*, 2008, 9 (8), 2146–2152.

[121] Bhise, K.S.; Dhumal, R.S.; Chauhan, B.; Paradkar, A.; Kadam, S.S. AAPS *PharmSciTech*. 2007, 8(2), E110-E118.

[122] Sashina, E.S.; Vnuchkin, A.V.; Novoselov, N.P. *Russian Journal of Applied Chemistry* 2006, 79 (10), 1608-3296; Original Russian Text published in Zhurnal Prikladnoi Khimii, 2006, 79(10), 1664–1667.

[123] Myllytie, P. *Interactions of Polymers with Fibrillar Structure of Cellulose Fibres: A New Approach to Bonding and Strength in Paper*; PhD Thesis, Helsinki University of Technology (Espoo, Finland), 18th of December, 2009, ISBN 978-952-248-229-7, Dissertation ISBN 978-952-248-228-0.

[124] Myllytie, P.; Salmi, J.; Laine, J. *BioResource*s 2009, 4(4), 1647-1662.

[125] Fang, N.; Chan, V.; Mao, H.-Q.; Leong, K.W. *Biomacromolecules*, 2001, 2 (4), 1161–1168.

[126] Xia, Y.-Q.; Guo, T.-Y.; Zhao, H.-L.; Song, M.-D.; Zhang, B.-H.; Zhang, B.-L. *Journal of Biomedical Materials Research Part A,* 2008, 90A (2), 326-332.

[127] Tang, Z.-X.; Shi, L.-E. *Biotechnol. & Biotechnol. Eq*. 2007, 21 (2), 223-228.

[128] Zhao, X.; Mai, Z.; Bang, X.; Zou, X. *Biosens. Bioelectron*. 2008, 23(7), 1032-1038.

[129] Chang, M.-Y.; Juang, R.-S. *Enzyme and microbial technology* 2005, 36 (1), 75-82.

[130] Chang, M.Y.; Kao, H.C.; Juang, R.S. *Int. J. Biol. Macromol*. 2008, 43(1), 48-53.

[131] Cuero, R.G.; McKay, D.S. *US Patent 7309437, Issued on december 18, 2007, Application No. 11031088 filed on 01/06/2005.*

[132] Vieira, R.S.; Beppu, M.M. *Colloids and Surfaces A: Physicochemical and Engineering Aspects* 2006, 279 (1-3), 196-207.

[133] Tamura, H.; Tsuruta, Y.; Itoyama, K.; Worakitkanchanakul, W.; Rujiravanit, R.; Tokura, S. *Carbohydrate Polymers* 2004, 56(2), 205-211.

[134] Kyzas, G.Z.; Kostoglou, M.; Lazaridis, NK. (2010). Relating Interactions of Dye Molecules with Chitosan to Adsorption Kinetic Data. *Langmuir* 2010, Mar 19. [Epub ahead of print].

[135] Yusuke, S.; Keisuke, O.; Satoru, I.; Yoshihiro, Y.; Masaki, T.; Mitsuji, Y. *Nippon Kagakkai Koen Yokoshu* 2005, 85 (1), 479.

[136] Yusuke, S.; Mitsuji, Y. *Nippon Kagakkai Koen Yokoshu* 2006, 86 (2), 874.

In: Handbook of Chitosan Research and Applications
Editors: Richard G. Mackay and Jennifer M. Tait
ISBN 978-1-61324-455-5

Chapter 4

CHITOSAN AS A BINDING AGENT FOR MODIFICATION OF ENDOTOXIN BIOLOGICAL ACTIVITY

V. N. Davydova, T. F. Solov'eva, G. A. Naberezhnykh, V. I. Gorbach, and I. M. Yermak
Pacific Institute of Bioorganic Chemistry,
Far East Branch of Russian Academy of Sciences,
Vladivostok, Russia

ABSTRACT

The ability of chitosan to form specific complexes with polyanions opens up broad opportunities for its application as a drug and gene delivery vector as well as a constituent component of biospecific adsorbents and composites. By virtue of its polycationic origin, chitosan can bind to endotoxin (LPS) and can be selected as a specific endotoxin ligand.

Date about the interaction of chitosan and its derivatives with enterobacterial LPS obtained by using of a broad range of advanced methods and techniques are considered. The process of forming the stable complexes chitosan with LPS, their stoichiometry and morphology are described. The influence of a number of factors (temperature, ionic strength, concentration) as well as the structure of LPS, chitosan and its derivatives on the binding process is discussed. Along with the binding of chitosan with the isolated LPS is considered the interaction of chitosan with the endotoxin in the bacterial cell. The three-dimensional structure models of complexes of LPS and chitosans and their derivatives are presented.

Modification of biological activities of endotoxin by chitosan in experiments *in vitro* and *in vivo* is considered. The protective effect of chitosan on the damaging effect of endotoxin in mice with bacterial intoxication is shown. The mechanisms of modulation of biological activity of LPS by chitosan are discussed.

INTRODUCTION

Appearance in circulation of lipopolysaccharide (LPS, endotoxin), the invariant structural component of Gram-negative bacterial outer membranes, as sequence of systemic bacterial infections frequently leads to a plethora of symptoms termed "endotoxic shock" or "sepsis". Endotoxic shock is characterized by hypotension, coagulation abnormalities, and multiple organ failure, treatment of which remains non-specific and supportive. Hence, mortality due to shock syndrome remains to be high (40-60%). The mechanisms leading to endotoxic shock are mediated by a cascade of host responses to LPS, important among which are the elaboration of proinflammatory cytokines including tumor necrosis factor and various interleukins by monocytes. These mediators at low concentrations are beneficial for regular functions of the host immune system against invading microorganisms while cytokines at high concentrations cause toxic effects such as irreversible shock [1, 2]. Much progress toward an understanding of mechanisms by which endotoxin activates the target cells opens the way to several experimental strategies for treating sepsis. These include neutralization of endotoxin by bio-molecules which are capable to its binding at high affinity that prevent endotoxin interaction with the susceptible cells.

Due to a high negative charge, the LPS macromolecules represented an important target for polycationic compounds which are discussed as new potential antiseptic drugs. In recent years, the search of drugs for the treatment of sepsis among the cationic proteins and peptides, and synthetic polycations, is actively explored [3]. However, to date none of the drugs was not introduced into clinical practice. The development of specific drugs against the syndrome of endotoxic shock remains an important problem.

As known, chitosan represented a widespread natural polycation which used in various biochemical specimens due to its availability and resistance [4, 5]. The ability of chitosan to form specific complexes with polyanions opens up broad opportunities for its application as a drug and gene delivery vector as well as a constituent component of biospecific adsorbents and composites [6]. In particular, chitosan was used for removing endotoxins from biological liquids [7]. Together with this, chitosan possesses antimicrobial activity [8] which is usually representative for the substances that is capable to bind endotoxins [9]. These data, as well as the positive charge of chitosan allows us to consider the polycation as a potential antiendotoxic agent.

Some aspects of interaction chitosan with LPS and its influence on biological activity of LPS are discussed at this article.

SOME ASPECTS OF INTERACTION OF CHITOSAN WITH ENDOTOXINS

LPS molecule consist of a carbohydrate and a lipid, called lipid A, portions. The carbohydrate part includes a relatively well-conserved negatively charged core hetero-oligosaccharide and an O-antigen-specific polymer of repeating oligosaccharide units, the composition of which is highly varied among Gram-negative bacteria. The O-polysaccharides have various lengths depending on the bacterial strains and their growth conditions. The lipid A is a negatively charged β-D-linked-glucosamine disaccharide phosphorylated and acylated by ester and amide-bound fatty acids [10]. The last as endotoxic constituent of LPS was shown to be responsible for toxicity in mammals [1]. LPS as an amphiphilic substance tends to form aggregates in aqueous solution [11] depending on the chemical structure of the endotoxin and ambient conditions [12, 13].

Chitosan is linear polysaccharide of β-1, 4 – linked residues of D-glucosamine and N-acetyl-D-glucosamine, which may vary in degree of *N*-acetylation (DA) and degrees of polymerization [14].

Influence of the ambient conditions on the interaction LPS with chitosan. The interaction of chitosan with endotoxin was studied by the analytical centrifugation, velocity sedimentation in sucrose or glycerol gradient as also isopicnic centrifugation. The process was shown to be dependent on time and reaction temperature. Upon mixing of the solutions of *Yersinia pseudotuberculosis* LPS and chitosan at 25°C as well as on incubation of polymers mixture at 37°C for 12 hours no complex was formed. Complex formation took place only after preincubation of LPS and chitosan solutions at 37°C and the subsequent incubation of their mixture at 37°C [15]. Such a strong effect of temperature on LPS interaction with chitosan points to the prominent role of the macromolecular structure of endotoxin in this process. The binding sites on LPS appear to be inaccessible at lower temperatures when the aggregate LPS molecules are more closely packed. The elevation of temperature up to 37°C that induce disaggregation of LPS leading to 20-fold decrease in aggregates molecular mass [16] and LPS phase transition accompanied by melting of the acyl chains of lipid A [13] appears to disorder the packing of LPS molecules in aggregate and as a result the ability of LPS to bind chitosan. It should however be emphasized that the process of LPS dissociation, which proceeds at an elevated temperature and seems to promote the release of binding sites, must precede the LPS interaction with chitosan. The effect of LPS phase state on its binding ability was observed in the experiments with the cationic peptide [17].

The process of LPS complexation with chitosan is accompanied by an additional disaggregation of LPS. The molecular weight of LPS/chitosan complex, prepared with preincubation of the initial components and subsequent incubation of their mixture (37°C), measured by analytical centrifugation (Archibald's method) was 160 kDa, whereas the molecular weight of the initial LPS held at the same temperature was 300 kDa [15]. Disaggregation of LPS was also noted at its binding with albumin and hemoglobin [18, 19].

It is interesting that elevation of temperature to 37°C to lead a change of the hydrodynamic parameters of chitosan (a decrease in viscosity and an increase in sedimentation coefficient) that appears to be result of conformation changes in the chitosan molecules [20].

The concentration of the components in reaction solution was shown to effect on stoichiometry of the LPS-chitosan complexes. The lower is the concentration of LPS, the higher is its amount bound to chitosan during complexing. At the saturation point there are on the average 9-11 and 200 glucosamine residues per mole of LPS on the average in the complex formed in the dilute solutions (LPS concentration <0.1 mg/ml) [21] and in concentrated solutions (LPS concentration 1 mg/ml) [22], respectively. Such a concentration dependence of stoichiometry of the complexes seems to be determined by the aggregate state of LPS, which depends on its concentration in the solution [23]. It is possible that the complexes formed in the solutions with different concentrations were structurally distinct.

The interaction of a cationic polymer, such as chitosan, with highly anionic LPS undoubtedly has an electrostatic component involving ionic binding between negatively charged groups on LPS and positively charged amino groups on chitosan. It is known that in the LPS molecule both lipid A and inner part of core oligosaccharide carry negatively charged groups. The phosphate substitutes of LPS, which are located on lipid A fragment, appear to be major contributor to the total negative charge of LPS and major LPS binding sites for chitosan. As was shown both of these fragments of *Escherichia coli* LPS molecule inhibit rather effectively (by 60-80%) chitosan binding to LPS [24]. However, the inhibitory activity of lipid A is higher than that of core. The O-specific polysaccharide, the uncharged LPS fragment, does not inhibit the interaction LPS with chitosan. The result obtained is an indirect evidence for the role of charged residues of LPS in its interaction with chitosan. The fact that the lipopolysaccharide may displace the anionic dye Tropaeolin 000-2 from its complex with chitosan [21] supports the conclusion that LPS binding occurs on amino groups of the polycation.

The interaction between chitosan and LPS strong depends on pH value which determines the ionization degree of charged groups involved in the complex formation. It was found that LPS–chitosan complexes are formed in the pH range from 4.0 to 7.0 [25]. The essential role of electrostatic forces in the interaction of endotoxin with chitosan is supported by the fact that the stability complexes decreases with increasing ionic strength. However in 1 M NaCl were found not only the free components – LPS and chitosan, but the stable complex as well. This clearly indicates the involvement of the different types of the interaction in the formation of complexes. Based on the chemical nature of these substances containing carbohydrate components, it is believed that, in addition to the electrostatic forces, hydrogen bonds are involved in formation of complex of LPS with chitosan [25].

Studies of various effects of urea on the formation of LPS–chitosan complexes provided conclusive evidence in favor of this hypothesis. Stipulating that ultracentrifugation in 6 M urea induces partial dissociation of the complexes, it was assumed that LPS molecules have several chitosan-binding sites, some of which are involved in ionic interactions, while in other sites this interaction involves hydrogen bonds and hydrophobic forces.

Influence of the structure chitosan and LPS on the interaction between them. Studies on interaction between chitosan and LPS are complicated the fact that both polymers have variable fine structures which may have a pronounced effect on their binding properties.

To gain a better insight into the mechanism of the interaction between LPS and chitosan knowledge of impact of the fine structures of the components on the complex formation between them is a necessity.

Comparison studies on binding the chitosan samples differing in molecular weight with LPS showed that the complexation depended on the degree of polymerization of the chitosan. Using sedimentation velocity experiments and analytical centrifugation the ratios LPS/chitosan for forming complex in solution with concentration of components (1-5 mg/ml) were established. The monomodal distribution was observed for the parent materials and their mixtures at ratios from 1:1 to 1:5 (w/w), indicating a complete binding of LPS *Y. pseudotuberculosis* with low molecular weight chitosan (LM-Ch, 20 kDa, 4% DA) within the above LPS/chitosan ratios of the constituents [22]. In the case of high molecular weight chitosan (HM-Ch, 140 kDa, 4% DA) the amount of polycation required for complete binding of LPS was less than for LM-Ch. The free HM-Ch was observed in the mixture of LPS/chitosan in ratios 1:5 (w/w) [25].

Table 1. Parameters of the *Y. pseudotuberculosis* LPS binding with chitosan [21]

Complex		Number of repeating units in LPS O-polysaccharide	K_bx10^5 M^{-1}	Number of LPS moles per chitosan monomer
	R-LPS	-	0,791	0,26
LM-Ch	LPS-I	2.4	1,009	0,19
	LPS-II	13	1,753	0,11
	LPS-III	24	6,015	0,04
	R-LPS	-	0,326	0,17
HM-Ch	LPS-I	2.4	0,594	0,13
	LPS-II	13	0,504	0,09
	LPS-III	24	2,820	0,05

The effect of the length of the O-specific polysaccharide endotoxin on its interaction with the polycation and binding constants was determined using the LPS replacement of the anionic dye - Tropaeolin 000-2 in its complex with chitosan and recording the released dye in diluted solutions (concentration of LPS <0.1 mg/ml) [21]. Using of LPSs isolated from the same strain of *Y. pseudotuberculosis* allowed obtaining a number of the samples with the same chemical structure, which were different only in length of the O-chains (table 1): number of repeating unit – 0 (R-LPS); 2.4 (LPS-I); 13 (LPS-II) and 24 (LPS-III) . The increasing of degree of LPS hydrophobicity (namely, reducing of the O-specific polysaccharide length) complicated the character of its interaction with chitosan [25, 21]. The more hydrophobic LPS, short-chained or free of O-chains, had to be incubated in the mixture with chitosan for a significantly longer time than LPS with the long O-chains. The increase in the LPS molecular mass caused by elongation of the O-specific polysaccharide was accompanied by an increase of the association constants from 0.3 to 6.0 x10^5 M^{-1} (table 1). The order of magnitudes (10^5 M^{-1}) is in consistence with those for LPS complexes with other polycations: human serum albumin [18] and NK-lysine [26].

An increase of the number of repeating units in O-polysaccharide is attend with a decrease in the number of binding sites on the endotoxin molecule from 0.26 to 0.04 as calculated per monosaccharide residue of chitosan (table 1). This result can derive from the fact that the long O-specific chains on the one hand, shield the binding sites on LPS, and on the other hand sterically prevent the approach of ligands during their binding to the adjacent sites. This model allows the explanation of a relatively low saturation level of the binding sites on LPS possessing the long O-polysaccharide chains [21].

The determinate dependences of the stoichiometry and affinity of LPS-chitosan interaction on O-polysaccharide moiety length were right for both LM-Ch and HM-Ch (table 1). However at the saturation point HM-Ch bound less LPS as calculated per chitosan monomer and with lower affinity than LM-Ch. These chitosan samples differed in the character of interaction with LPS. Analysis of the data shown that the binding of LM-Ch occurred at independent sites of one type with absence of cooperativity during the interaction. HM-Ch interacts with LPS with negative cooperativity [21]. It is suggested that it can be a result of the mutual repulsion of ligands. Large ligands sterically prevent approach of other molecules to the binding sites near-by. The approach of ligands during their binding to the adjacent sites can be thermodynamically disadvantageous. In the limiting case, the binding of two ligand molecules to adjacent sites can be completely excluded (model with binding site exclusion) [27].

Specific features of the LM-Ch and HM-Ch conformations can also influence the manifestation of these effects. The LM-Ch molecule is rather flexible [20] and its conformation in solution presents a randomly coiled sphere, whereas the HM-Ch molecule is rather rigid and elongated. Due to the chain flexibility, the binding sites on chitosan can spatially shift one relative to another and lower steric hindrances at the binding of LM-Ch ligands can occur that is impossible in the case of rigid elongated molecules of HM-Ch. This seems to explain the fact that the number of binding sites per LM-Ch monomer is higher than in HM-Ch.

LPSs isolated from various microorganisms have different amounts of negatively charged groups [10], which can be an important factor in their binding with polycations. Furthermore, LPSs are amphiphylic compounds and can form aggregates with various parking types depending on their primary chemical structure and ambient conditions. These properties in their turn can affect the varied availability of the negatively charged groups to bind with chitosan.

The influence of these aspects on the process of endotoxin interactions with chitosan was analyzed by using LPSs from three different microorganisms, *Y. pseudotuberculosis*, *E. coli*, and *Proteus vulgaris* [28]. These LPSs possess different chemical structures: the number of negatively charged groups in the lipid A-core moiety, the degree of acylation of lipid A, and the various lengths of the O-specific chains that may influence on the sizes [12] and packing densities of the acyl chains in aggregates of LPS [13]. In addition, lipid A of *P. vulgaris* LPS has an aminoarabinose group bound to phosphate on nonreducing end of molecule and only single free phosphate group [29].

The stable complex formation for all the investigated LPSs was also registered after preliminary incubation of the samples at 37°C. The complexation of *P. vulgaris* LPS and of *Y. pseudotuberculosis* LPS has only been noted on additional incubation of mixtures of LPS and chitosan at 37°C [15, 28]. At the same time, LPS from *E. coli* was shown to form a stable complex with chitosan after incubation of the mixture at 25°C that indicates the importance of lipid A structure of LPS to the binding process [28].

The complexes of chitosan with P. *vulgaris* and *Y. pseudotuberculosis* LPS, which have short O-specific chains showed a multimodal particle size distribution that determinating by the quasi-elastic light scattering. The main fraction of the complex contains particles with sizes in the range of 70-1100 nm. Whereas for complex of chitosan with *E. coli* LPS with long O-specific chains the narrow size distribution was noted [28].

The measurements of ζ-potential showed that interactions of LPS from different microorganisms with chitosan at the same w/w ratio of components (1:1) resulted in the formation of complexes in which the negative charges of LPS were neutralized (*E. coli* LPS) or overcompensated (*Y. pseudotuberculosis* and *P. vulgaris*). The difference observed in the electrokinetic properties of the complexes suggests that the availability of negatively charged groups on LPS molecules is completely different and depends on the aggregate structure of the endotoxins [28].

E. coli LPS with a long O-specific chain formed monodispersed aggregates with the lowest ζ-potential among the studied LPSs. After binding with chitosan the complex with the lowest surface charge was formed, which enabled to suggest that chitosan penetrated to anionic sites of LPS. It should be noted that the narrow size distribution as well as the low surface charge of this complex exclude bridging the LPS aggregate with chitosan as a possible mechanism of complexation. This agrees with the data of the dye displacement assay: 1 mole *E. coli* LPS is bound with a small portion of the chitosan amino groups. *E. coli* LPS possibly can present steric hindrances to interaction with the polycation [28]. A similar relationship was observed for *Y. pseudotuberculosis* LPS with the long O-specific chains [21].

In contrast to *E. coli* LPS with a long O-specific chain, the LPSs from *Y. pseudotuberculosis* and *P. vulgaris* have short O-chains. According to the chemical structures of the LPS it is possible to expect that it will form densely packed aggregates of larger size [12, 13]. According to earlier data [20], chitosan is a sufficiently rigid linear polymer. Its penetration into the more densely packed aggregates of LPS was difficult. In this regard, we can probably consider the complex formation as binding on the surface of LPS aggregate. In this case the interaction of the densely packed LPS aggregates with chitosan can be very schematically presented as the interaction between a highly charged polyelectrolyte and a charged sphere, which is known to occur with surface overcharging upon polyelectrolyte adsorption [30].

Influence of chitosan on LPS aggregate structure. The rearrangement of the structure of the endotoxin aggregates after binding with chitosan was established by the method of electron microscopy [31, 32].

Negative-stained preparation of *E. coli* LPS (concentration of 1 mg/ml) shown polymorphism and formed heterogeneous particles of the irregular membrane-like structure and the spherical structures with diameters of 18 nm combined in aggregates of various sizes and shapes. At the low concentration (0.25 mg/ml) there were weakly associated particles. Chitosan alone appeared as small, uniform discs with diameters of approximately 6–13 nm combined in a chain-like structure [31]. Incubation of diluted solution of LPS (0.25 mg/ml) with HM-Ch at 37°C (4:1 w/w) led to the formation of morphologically homogeneous spherical particles. With increasing the content of polycation in the complex to the ratio of LPS - chitosan (1:1 w/w) were observed heterogeneous in sizes spherical structures with diameters of 5 to 17 nm. The higher concentration of complex component led to the formation of polymorphic structures. Thus, at LPS concentration 1 mg/ml and ratio LPS/chitosan 1:1 w/w spherical particles with a diameter of 5 to 17 nm as well as rare membrane-like particles were detected [31].

Y. pseudotuberculosis LPS formed two types of structures: ribbon-like structures which consisted from of close-packed long filaments and the spherical structures with diameters of 18 nm combined in aggregates. As in the case of *E. coli* LPS, in a mixture of chitosan with *Y. pseudotuberculosis* LPS observed spherical particles various sizes, with a diameter of 6 to 14 nm, or ellipsoidal particles and the relatively short thin filaments [31].

P. vulgaris existed in a number of highly aggregated forms, including different freely branching ribbon-like structures (thickness: 5–13 nm). The LPS–chitosan mixture consisted of a number of particles with various forms. There were heterogeneous particles of irregular form and/or membrane-like structures [32].

So binding of LPS from different microorganism with chitosan led to changes in the morphology and sizes of endotoxin aggregates that depened on concentration and ratio of initial components. More homogeneous structure formed at a low content of chitosan at the complex, while increasing the content of chitosan in the complex leads to the formation of polymorphic structures.

Interaction of N-acylated derivatives of chitosan with LPS. It might be supposed from the literature data on the interaction of lipopolyamines with LPS [3, 33] that water-soluble hydrophobic derivatives of chitosan are promising for high-affinity LPS binding. In this connection chitosans with different degree of N-acetylation and chitooligosaccharides N-acylated with long-chain fatty acid were obtained and the LPS-binding activity of these compounds was explored.

The chitosan molecule has hydrophobic substituents, namely acetate groups, which partially substitute amino groups in the polymer. At present, there is no data on direct involvement of N-acetates in chitosan binding with LPS; although a possibility exists of interaction between these groups and acyl domain of lipid A. Moreover, N-acetate groups can indirectly influence the binding via their effect on both the conformation and physical properties of chitosan.

Chitosan specimens with various acetylation degrees (from 1.6 to 15%) but of equal molecular masses and random character of distribution of N-acetates along the polysaccharide chain were taken for the comparative study. This allowed to eliminate of the effect of chemical structure and chain length on the LPS-binding activity of chitosans. The increase in the acetylation degree of the chitosan molecule (from 1.5 to 9.0%) was found to result in decrease in efficacy of its binding to LPS in spite of the increase in general hydrophobicity of the polymer molecule. Further increase in the acetylation degree (up to 15%) does not lead to change in LPS-binding activity of chitosans. This may be due to rather high hydrophobicity of chitosan specimen with 15% of acetates, which can enhance its affinity to hydrophobic sites of LPS molecule and thereby compensate the decrease in contribution to interaction of electrostatic component resulting from the increase in amount of N-acetates [24].

The mechanism of a decrease in affinity of interaction between LPS and chitosan when N-acetylation degree of the latter increases appears to be complex. One cause of the observed dependence can be the decrease in linear density of charges and flexibility of chitosan polymer chain resulting from the increase in the content of N-acetate groups in the molecule [34]. In fact, the force of electrostatic interaction between polyelectrolytes (including chitosan) and oppositely charged amphiphilic compounds decreases with the increase in rigidity of polymer molecule and decrease in the charge density. The higher linear charge density in chitosan with acetylation degree of 1% than that in chitosan with 12% acetates is presumed to be the cause of stronger binding with negatively charged detergent micelles [35]. The linear charge density in the chitosan molecule and the extent of its flexibility can probably influence the size and localization of chitosan-binding sites on the LPS molecule, as it has been observed earlier when studying the interactions between polyelectrolytes and proteins [36]. As has been revealed earlier, an increase in the content of N-acetate groups in chitosan results in an increase in association between molecules in aqueous media [37]. The contribution of this process to alteration of binding affinity that is observed under alteration of chitosan acetylation degree can not be completely excluded.

Thus, the elevation of hydrophobicity of the chitosan molecule due to the increase in N-acetylation degree cannot serve as an effective way to increase its LPS-binding activity. It has been demonstrated that long-chain aliphatic groups, when presented in a polycation, stabilize its complexes with LPS through hydrophobic interactions of the lipohilic substituents with fatty acid lipid A domains [38]. A chemical modification of chitosan was performed to increase the affinity of LPS–chitosan binding. Hydrophobic chitosan derivatives substituted at the amino group by 3-hydroxytetradecanoic acid were synthesized. The solubility of these compounds in aqueous media was achieved via utilization of low molecular weight chitosan derivatives and low degree of their acylation. Mono-N-acyl-chitooligosaccharides (di–tetra) and mono-N-acyl-oligochitosan (5.5 kDa) produced had acyl substituent at the reducing terminus of the molecule [24]. As mentioned earlier [3], amphiphilic compounds of such type with segregated hydrophobic and hydrophilic cationic regions inside the same molecule interact preferentially with the lipid A fragment, which is important for neutralization of LPS. Comparison dissociation constant values showed that the affinity of the studied chitosan derivatives binding with LPS fell in the order: N-acyl-oligochitosan > N-acyl-chitotetraose > N-acyl-chitotriose > N-acyl-chitobiose > oligochitosan > chitotetraose. Elongation of the carbohydrate chain of the chitosans (from 2 to 30 of monomer residues) also leads to increase in LPS-binding activity—both of substituted and non-substituted chitosan derivatives. Analysis of the data shows that introduction of long acyl substituent into low molecular weight chitosan derivatives significantly (ten-fold) increases their LPS-binding activity [24].

Thus, water-soluble derivatives of chitosan with hydrophobic substituents and low substitution extent of amino groups are promising compounds for binding endotoxin.

Modeling of chitosans and their complexes with LPS. The study on the *in silico* interaction between LPS and chitosan is a difficult task. The method of molecular docking is used generally for prediction of the structure of a complex between protein (receptor, enzyme) and ligand. In this case, a binding site localized in a spatially limited inner space of the protein molecule is only regarded. In the case of LPS-chitosan complexes, the molecular surface of receptor (LPS) is exposed to the solvent to greater extent, and the ligand (chitosan) possesses substantially greater rotary and conformational freedom due to the absence of strict steric restrictions. At the same time the modeled complexes of chitosan and LPS are a useful tool to understand the chitosan interaction with LPS at atomic level and they might be used in structure based antiendotoxin drug design.

Theoretical models of spatial structures of low molecular weight chitosans (polymerization degree is 30) containing different amounts of acetate groups and di-, tri-, tetra-, and octa-chitooligosaccharides and their derivatives N-acylated with 3-hydroxytetradecanoic acid on the reducing terminus were generated using the carbohydrate and molecular construction modules of the MOE software package. Crystal structure of *E. coli* K12 LPS was obtained from the PDB database. The predicted structures after energy minimization in water were used in experiments on molecular docking of chitosan and LPS. The complexes were selected for each chitosan specimen with the strongest binding between the components (minimum free energy of the interaction). Generally, the models for LPS complexes with three types of chitosan derivatives were constructed: for chitosans with various contents of acetamide groups, chitooligosaccharides, and N-acyl-chitooligosaccharides [24].

The analysis of contacts between components in the predicted complexes showed that the structure of complexes was stabilized by several hydrogen and ionic bonds. The ionic, hydroxyl, and carbonyl groups of monosaccharide and fatty acid residues of core and lipid A moieties of LPS was demonstrated to take part in the interaction with chitosans and chitooligosaccharides. In case LPS complexes with N-acylated chitooligosaccharides the fatty acid residues were incorporated in polyacyl domain of lipid A and parallels to the fatty acid chains of lipid A that enforces their hydrophobic interactions and increases affinity of acylated chitosan binding to LPS. Increase in amount of acetyl groups in chitosans led to decrease in bonds stabilizing the complex. The putative binding sites on the LPS molecule were not the same for the chitosan specimens with different content of N-acetates. The elongation of the chitooligosaccharides carbohydrate chains (from 2 to 8 residues) led to multipoint binding on several anionic sites localized on both core and lipid A of the LPS molecule. Such character of the interaction probably explains experimentally found an increase of binding affinity with increase of chitosan polymerization degree. It is significant that with increase of length of the chitosan carbohydrate chain the conformational sites appear with the distance 1.23 nm between amino groups, which is close to the distance 1.4 nm between the phosphate groups on lipid A. These chitosan conformational sites should be involved in high-affinity binding to lipid A moiety of LPS [24]. As was shown the presence of two protonable amino groups positioned about 1.4 nm apart in a linear molecule of polyamine was sufficient to expect high-affinity binding to LPS [38].

LPS interaction with chitosan was estimated also quantitatively by computer modeling based on semiempirical quantum-chemical AM1 method [39]. It was found that $GlcNH_2$-Re-LPS complexes were characterized by an equilibrium distance 0.17 nm between interacting sites and binding energy 0.5 eV. Among three anionic binding sites of Ra-LPS (the carboxyl group on the KDO residue and the phosphate groups on lipid A), two sites are inactive. The involvement of only one phosphate group at reducing ends of lipid A in this binding is the most probable, but interaction with the other two negatively charged groups is energetically inefficient.

The results obtained indicate that computational predictions may be feasible also for the very difficult system of chitosan interacting with LPS.

The Interaction of Chitosan with the Endotoxin in the Bacterial Cell

Recent studies have suggested that LPS of outer membrane is the first protective layer that actually controls cationic antimicrobial peptides binding and insertion into Gram-negative bacteria [40, 41]

Although antimicrobial activity is one of the attractive features of chitosan, exact mode of antibacterial action polycation remained obscure. There are strong grounds to believe that the antibacterial action of chitosan has a double mechanism, affecting both the cell wall (outer membrane in gram-negative bacteria) and the membrane [42]. Outer membrane acts as an important protective barrier against chitosan which, once destroyed, would lead to a physiological situation in which the antibacterial activity of chitosan would be favored. The amino groups (NH_3^+) as active functional groups were found to be essential to the antibacterial activity of chitosan [42]. Interaction between positively charged chitosan molecules and negatively charged microbial cell membranes leads to the leakage of intracellular constituents [43, 44]. It was also shown by electron microscopy that chitosan caused extensive cell surface alterations, bound to the outer membrane and caused the loss of the barrier function of the outer membrane. It is important that that interaction of chitosan with *E. coli* and the Salmonellae involved no release of LPS or other membrane lipids into cell medium [45].

It is interesting to note that the same features affect both the interaction of chitosan with bacteria and the binding with the isolated LPS. Actually, chitosan shows its antibacterial activity only in an acidic medium. It at pH 5.3 induced significant uptake of the hydrophobic probe 1-N-phenylnaphthylamine (NPN) in *E. coli*, *Pseudomonas aeruginosa* and *Salmonella typhimurium*. The effect was reduced *(E. coli*, *S. typhimurium*) or abolished (*P. aeruginosa*) by $MgCl_2$. No NPN uptake was observed during exposure of the *S. typhimurium* to chitosan at pH 7.2. The interaction of chitosan with bacteria depended on concentration. At a lower concentration (<0.2 mg/mL), the polycationic chitosan does probably bind to the negatively charged bacterial surface to cause agglutination, while at higher concentrations, the larger number of positive charges may have imparted a net positive charge to the bacterial surfaces to keep them in suspension [43, 44].

As in the case with the LPS-binding properties the antimicrobial activity of chitosan varies depending on its molecular characteristics. The activity of chitosan is influenced by its DA: chitosan with 5% DA had higher antibacterial activity than chitosan with 25% DA [46]. Chitosan with a molecular weight ranging from 10 kDa to 100 kDa would be helpful in restraining the growth of bacteria. In addition, chitosan with an average molecular weight of 9.3 kDa was effective in restraining *E. coli*, while that with a molecular weight of 2.2 kDa accelerated growth [47]. The LM-Ch (5.5 kDa) poorly affected the permeability of cell walls of bacteria [48]. With binding a hydrophobic fatty acid residue to a molecule of the chitosan sample, the damaging effect of chitosan on membranes of bacteria was enhanced. Such a modification of chitosan was presumed to strengthen its interaction with components of the bacterial outer membrane, increasing its permeability. The assumption is supported by the fact the N-acylatyed chitosan is much more effectively bound to isolated LPS, major component of outer membrane, than its non-acylated parent. On the whole, the antibacterial activities of chitosans were shown to decrease in the order: N-acylatyed chitosan (about 5.5 kDa) > HM-Ch (80 kDa) > LM-Ch (5.5 kDa) [48].

N-, O-Carboxypropyl chitosan (65% substitution degree of amino and hydroxyl groups) differing from parent chitosan by lesser charge and greater hydrophobicity retained antibacterial activity but was less active than non-substituted chitosan [42]. The related result was obtained for carboxymethylated chitosans [49]. Their antibacterial activities were also found to be increasing in the order N-, O-carboxymethylated chitosan, chitosan, and O-carboxymethylated chitosan.

The results of chitosan antibacterial activity investigations support the hypothesis that the chitosan target in the cell wall of gram-negative bacteria is LPS.

Modification of Biological Activity of Endotoxin by Chitosan

As has been shown by us, LPS from different sources interact with chitosan to form complexes of different stoichiometry. Interaction of LPS with chitosan appeared to be connected with the lipid A region represented a toxophoric center of endotoxin. Therefore, an alteration of the LPS toxicity in the process of complex formation may be suggested.

Really, LPS-chitosan complexes possessed much lower toxicity in a comparison with the parent LPS at injection to mice in the similar concentration. This effect depends on the ratio of components in the complex and the molecular weight of chitosan. In the series of complexes of chitosan with *E. coli* LPS at different ratios of initial components (5:1, 1:1 and 1:5 w/w) the toxicity decreased with increasing chitosan content in the complex [50]. Decreasing toxicity also registered for complex of as HM-Ch as LM-Ch with *Y. pseudotuberculosis* [25] and *P. vulgaris* LPS [32]. The LD_{50} of LPS within complexes with chitosan was 20 times higher in comparison with *Y. pseudotuberculosis* endotoxin and 10 times higher in comparison with *E. coli* LPS and *P. vulgaris* LPS [22]. A substantial reduction of the LPS toxicity in the LPS-Ch complex may be explained due to chitosan as a constituent of the complex may prevent attachment of the LPS to specific cellular receptors or directly neutralize LPS toxicity alternating the molecular charge and/or structure of LPS aggregates.

As known, endotoxins induce a biosynthesis of different mediators such as interleukins, TNF-α and others in the immuncompetent cells. The LPS-chitosan complex was shown to maintain an ability to induce of IL-8 and TNF-α. The production of IL-8 and the TNF-α by the LPS-chitosan complex was found to be 70% (IL-8) and slightly lower (TNF-α) than that by the parent LPS [22].

In context of these results interest is in the data obtained by Chou et al. [51]. They showed that chitosan added simultaneously with LPS inhibited the pro-inflammatory cytokine formation in RAW 264.7 macrophages. The mechanisms by which chitosan might significantly inhibit the pro-inflammatory cytokine formation in LPS-treated macrophages and the cytokine-inducing activity of LPS in LPS-chitosan complexes are still unclear.

As known LPS induce the production of cytokines in monocytes/macrophages thought TLR-4 receptors [52]. In response to LPS MD2 protein is required for TLR4-dependent cellular activation [53]. Using human embryonal kidney cells (HEK 293 cells) transfected with TLR4 in combination with MD2, shown that the LPS-chitosan complex, similarly to the parent LPS, stimulated the formation of the IL-8 in these cells in the dose-dependent manner [22]. Chitosan was unable activate the IL-8 realization in the same cells. Furthermore after the pretreatment with chitosan the cells expressing TLR-4 is remained susceptible to activation by LPS. These data appear to indicate that the chitosan-induced secretion of the pro-inflammatory cytokines in macrophages is not dependent on TLR4 and chitosan do not block the TLR4-receptor for LPS binding and LPS signal transduction in the cell. However the possibility of some inhibition of LPS - TLR4 interaction in consequence of binding chitosan to the lipid A portion LPS is not to be excluded [22]. As known lipid A is responsible for the stimulation of the cytokine secretion via TLR4 receptor [10, 54] and its blocking by chitosan might contribute in the suppression of biological activity of LPS. The down-modulation of LPS cytokine-inducing activity might be also due to the competition between LPS and chitosan for CD14 receptor. CD14 has been recognized as is common binding receptor on monocytes both for LPS and for chitosan which is involved in the transduction of the cytokine-inducing signal [55].

A comparison of modulating effects of LPS, chitosan and their complexes on cell mediated delayed type hypersensitivity and humoral immunity established that chitosan in complex with LPS stimulated phagocytic activity of macrophages more effectively than free LPS by virtue of their strong ability to absorb *Y. pseudotuberculosis* [50]. Chitosan-bound LPS are deprived of the ability for partial suppression of delayed type hypersensitivity

It is known that endotoxins can directly activate blood platelets, which play a triggering role in pathogenesis of hemocoagulation disorders. Studies of aggregation of normal human platelets induced by single injections of *E. coli* LPS and LPS-chitosan complexes and possibility of elimination of damaging effects of LPS on blood cells revealed that chitosan protects platelets against ADP, widely known as a universal inducer of platelet aggregation. Five-minute incubation of platelets with LPS–chitosan complexes decreased the aggregation by 8% in comparison with control and by 14% in comparison with LPS- and ADP-induced aggregation [56, 57].

It was a great interest to exanimate the abilities of chitosan to neutralize LPS toxic effects *in vivo*. The activities of LM-Ch (5.5 kDa) and its derivative N-acylatyed by 3-hydroxytetradecanoic acid were evaluated at D-galactosamine-sensitized mouse model of endotoxic shock [58]. These compounds afford partial protection (up to 80% diminished lethality) when it was administrated simultaneously or 2 hours after of LPS injection. It is to be noted that partial protection is also observed when N-acylatyed chitosan but not parent LM-Ch is administrated 2 h hours prior to LPS challenge. The reason may be that non-acylated chitosan is more rapidly cleared from circulation than acylated one.

The capability of endotoxins to induce pathophysiological changes in various systems of host is now well established. The increase of nonspecific resistance by chitosan under induced by a single intraperitoneal dose of *E. coli* LPS (1 mg/kg) endotoxemia in mice was shown. Chitosan protects LPS-inducing hypertrophy of adrenal glands, thymus involution and changes of thyroid hormones and corticosterone in serum, activation of glycolysis and glycogenolysis, lipid peroxidation in liver cells [59].

Although the mechanisms responsible for chitosan effects need further elucidation, the data obtained provide strong evidence that chitosan increases the resistance of the host organism to bacterial LPS. Our studies from the past decade showed that LPS binding to chitosan gives stable complexes of different stoichiometry and reflects the general principle of endotoxin interactions with polycations and depends on numerous factors, e.g., LPS structure, chitosan polymerization, temperature, pH, ionic strength, etc. The mechanisms responsible for modification of biological activities of LPS–chitosan complexes are not completely understood and demand further verification and experiment.

CONCLUSION

The use of endotoxins of different structure, on the one hand, and polycations, on the other hand allowed elucidating certain details of the molecular mechanisms of their interaction.

It is safe to say that chitosan has the capability of forming the stable, water-soluble and variable in structure complexes with LPS.

The currently available evidence suggests demonstrates that the chitosan interaction with LPS is complicated process and which is significantly governed by the complex behavior of LPS in aqueous medium solutions. The external environmental factors such as the temperature, pH value, ionic strength influencing on the size and structure LPS aggregates and chitosan amino groups ionization have a profound effect on the chitosan binding with LPS. The interaction affinity and stoichiometry of the LPS−chitosan complexes are essentially affected by the structure of components and their concentration in the reaction mixture. The length of the O-specific polysaccharide as well as molecular weight and DA of chitosan are critical in determining of the complex structure. The presences of appropriable positioned hydrophobic groups in chitosan serve to future enhance binding affinity and stabilize the resultant of LPS-chitosan complex.

The formation of LPS-chitosan complex is accompanied by the essential decrease of LPS toxicity and modification of LPS immunological properties. The mechanisms responsible for modification of biological activities of LPS–chitosan complexes are not completely understood and demand further verification and experiment. The growing structural knowledge is opening pathways for using of chitosan as tool of binding and detoxication of LPS in accessory therapy of endotoxemia and endotoxic shock.

REFERENCES

[1] Rietschel, ET; Brade, H; Holst, O; Brade, L; Muller-Loennies, S; Mamat, U; Zahringer, U; Beckmann, F; Seydel, U; Brandenburg, K; Ulmer, AJ; Mattern, T; Heine, H; Schletter, J; Loppnow, H; Schonbeck, U; Flad, H.D; Hauschildt, S; Schade, UF; DiPadova, F; Kusumoto, S; Schumann, RR. Bacterial endotoxin: chemical constitution, biological recognition, host response, and immunological detoxification. *Curr.Top. Microbiol. Immunol.*, 1996, v. 216, pp. 39-81.

[2] Brandenburg, K; Andra, J; Muller, M; Koch, M; Caridel, P. Physicochemical properties of bacterial glycopolymers in relation to bioactivity. *Carbohydr. Res.*, 2003, v. 338, pp. 2477-2489.

[3] David, J. Towards a rational development of anti-endotoxin agents: novel approaches to sequestration of bacterial endotoxins with small molecules. *J. Mol. Recognit.*, 2001, v. 14, pp. 370–387.

[4] Kumar, MN; Muzzarelli, RA; Muzzarelli, C; Sashiva, H; Domb, AJ. Chitosan chemistry and pharmaceutical perspective. *Chem. Rev.*, 2004, v. 104, pp. 6017-6084.

[5] Singla, AK; Chawla, M. Chitosan: some pharmaceutical and biological aspects an update. *J. Pharm. Pharmacol.*, 2001, v. 53, pp. 1047-1067.

[6] Il'ina, AV. Chitosan-based polyelectrolyte complex: a review. *Appl. Biochem. Microbiol.*, 2005, v. 41, pp 9-16.

[7] Minobe, S; Watanable, T; Sato, T; Tosa, T. Characteristics and applications of adsorbents for pirogen removal. *Biotechnol. Appl. Biochem.*, 1988, v. 10, pp. 143-149.

[8] No, HR; Kim, SH; Lee, SH; Park, NY; Prinyawiwatkul, W. Stability and antibacterial activity of chitosan solutions affected by storage temperature and time. *Carbohyd. Polym.*, 2006, v. 65, pp. 174-178.

[9] Rabea, EI; Badawy, ME-T; Stevens, CV; Smagghe, G; Steurbaut, W. Chitosan as antimicrobial agent: applications and mode of action. *Biomacromolecules*, 2003, v. 4 pp. 1457–1465.

[10] Raetz, CRH. Biochemistry of Endotoxins. *Ann. Rev. Biochem.*, 1990, v. 59, pp. 129–170.

[11] Aurell, CA; Wistrom, AO. Critical aggregation concentrations of gram-negative bacterial lipopolysaccharides (LPS). *Biochem. Biophys. Res. Commun.*, 1998, v. 11, pp 230-236.

[12] Shands, JW. The physical structure of bacterial lipopolysaccharides of *Salmonella 99ewington.* In: Weinbaum, G; Kadis, S; Ajl, S., editors. *Microbial toxins,* v. 4. New York: Academic Press; 1971; pp. 127–144.

[13] Brandenburg, K; Funari, SS; Koch, MH; Seydel, U. Investigation into the acyl chain packing of endotoxins and phospholipids under near physiological conditions by WAXS and FTIR spectroscopy. *J. Struct. Biol.*, 1999, v. 128, pp. 175-186.

[14] Li, Q; Dunn, ET; Grandmaison, EW; Goosen, MFA. Applications and properties of chitosan. *J. Bioact. Compat. Polym.*, 1992, v. 7, pp. 370-397.

[15] Davidova, VN; Yermak, IM; Gorbach V.I; Solov'eva TF. The effect of temperature on the interaction of *Yersinia pseudotuberculosis* lipopolysaccharide with chitosan. *Membr. Cell. Biol.*, 1999, v. 13, pp. 49-58.

[16] Yermak, IM; Naberezhnykh, GA; Solov'eva, TF; Drozdov, AL; Ovodov, YuS. The effect of temperature on supramolecular structure and antigenic activity of *Yersinia pseudotuberculosis* endotoxin. *J. Biochem. Organization*, 1994, v. 1, pp. 295-304.

[17] Brandenburg, K; David, A; Howe, J; Koch, MH; Andrä, J; Garidel, P. Temperature dependence of the binding of endotoxins to the polycationic peptides polymyxin B and its nonapeptide. Biophys. J., 2005, v. 88, pp. 1845-1858.

[18] Mathison, JC; Tobias, PS; Wolfson, E; Ulevitch, RJ. Plasma lipopolysaccharide (LPS)-binding protein. A key component in macrophage recognition of gram-negative LPS. *J. Immunol.*, 1992, v. 149, pp. 200-206.

[19] Roth, RI. Hemoglobin enhances the binding of bacterial endotoxin to human endothelial cells. Thromb. Haemost., 1996, v. 76, pp. 258-262.

[20] Davydova, VN; Ermak, IM; Gorbach, VI; Drozdov, AL; Solov'eva, TF. Comparative study of the physico-chemical properties of chitosans with varying degree of polymerization in neutral aqueous solutions. *Biofizika*, 2000, v. 45, pp. 641–647.

[21] Davydova, VN; Naberezhnykh, GA; Yermak, IM; Gorbach; VI; Solov'eva, TF. Determination of binding constants of lipopolysaccharides of different structure with chitosan. *Biochemistry (Mosc.).*, 2006, v. 71, pp. 332–339.

[22] Yermak, IM; Davidova, VN; Gorbach, VI; Luk'yanov, PA; Solov'eva, TF; Ulmer, AJ; Buwitt-Beckmann, U; Rietschel, ET; Ovodov, YuS. Forming and immunological properties of some lipopolysaccharide-chitosan complexes. *Biochimie*, 2006, v. 88, pp. 23-30.

[23] Yermak, IM; Iadykina, GM; Solov'eva, TF; Ovodov, IuS. Study of aqueous solutions of the lipopolysaccharide-protein complex from *Yersinia pseudotuberculosis* using hydrodynamic methods. Biofizika, 1988, v. 33, pp. 288-292.

[24] Naberezhnykh, GA; Gorbach, VI; Likhatskaya, GN; Davidova, VN; Solov'eva, TF. Interaction of N-acylated chitosans with lipopolysaccharide. *Biochemistry (Mosc)*, 2008, v. 73, pp. 432-441.

[25] Davydova, VN; Yermak, IM; Gorbach, VI; Krasikova, IN; Solov'eva, TF. Interaction of bacterial endotoxins with chitosan. Effect of endotoxin structure, chitosan molecular mass, and ionic strength of the solution on the formation of the complex. *Biochemistry (Mosc)*, 2000, v. 65, pp. 1082–1090.

[26] Andersson, M; Giorard, R; Cazenave, P. Interaction of NK-lysine, a peptide produced by cytolytic lymphocytes, with endotoxin. *Infect. Immun.*, 1999, v. 67, pp. 201-205.

[27] Cantor, Ch; Shimmel, P. *Biophysical Chemistry* [Russian translation] Moscow: Mir; 1985.

[28] Davydova, VN; Bratskaya, SYu; Gorbach, VI; Solov`eva, TF; Kaca, W; Yermak, IM. Comparative study of electrokinetic potentials and binding affinity of lipopolysaccharides - chitosan complexes. *Biophys. Chem.*, 2008, v. 136, pp. 1-6.

[29] Sidorczyk, Z; Zahringer, U; Rietschel, ET. Chemical structure of the lipid A component of the lipopolysaccharide from a *P. mirabilis* Re-mutant. *Eur. J. Blochem.*, 1983, v. 137, pp. 15-22.

[30] Shin, YW; Roberts, JE; Santore, MM. The relationship between polymer/substrate charge density and charge overcompensation by adsorbed polyelectrolyte layers. *J Colloid Interface Sci.*, 2002, v. 247, pp. 220-230.

[31] Yermak, IM; Reunov, AV; Lapshina, LA; Davydova, VN; Solovyeva, TF. Electron-microscopic study of LPS from gram-negative bacteria and their complexes with chitosan. *Biologicheskie Membrany* (Rus.), 2005, v. 22, pp. 117-122.

[32] Arabski, M; Davydova, VN; Wasik, S; Reunov, AV; Lapshina, LA; Yermak, IM; Kaca, W. Binding and biological properties of lipopolysaccharide *Proteus vulgaris* O25 (48/57)–chitosan complexes. *Carbohydr. Polym.*, 2009, v. 78, pp. 481-487.

[33] Millera, EVK; Kumarb, S; Wooda, SJ; Cromera, JR; Dattab, A; David, SA. Lipopolysaccharide sequestrates: structural correlates of activity and toxicity in novel acylhomospermines. *J. Med. Chem.*, 2005, v. 48, pp. 2589-2599.

[34] Anthonsen, MW; Varum, KV; Smidsrod, O. Solution properties of chitosans: conformation and chain stiffness of chitosans with different degrees of *N*-acetylation. *Carbohydr. Polym.*, 1993, v. 22, pp. 193-201.

[35] Kayitmazer, AB; Shaw, D; Dubin, PL. Role of polyelectrolyte persistence length in the binding of oppositely charged micelles, dendrimers, and protein to chitosan and poly(dimethyldiallyammonium chloride). *Macromolecules*, 2005, v. 38, pp. 5198-5204.

[36] Kayitmazer, AB; Seyrec, E; Dubin, PL; Staggemeier, BA. Influence of chain stiffness on the interaction of polyelectrolytes with oppositely charged micelles and proteins. *J. Phys. Chem. B*, 2003, v. 107, pp. 8158-8165.

[37] Yanagisawa, M; Kato, Y; Yoshida, Y; Isogai, A. SEC-MALS study on aggregates of chitosan molecules in aqueous solvents: Influence of residual N-acetyl groups. *Carbohydr. Polym.*, 2006, v. 66, pp. 192-198.

[38] Andra, J; Gutsmann, Th; Garidel, P; Brandenburg, K. Mechanism of endotoxin neutralization by synthetic cationic compounds. *J. Endotoxin Res.*, 2006, v. 12, pp. 261-277.

[39] Zavodinsky, VG; Gnidenko, AA; Davydova, VN; Yermak, IM. Computer modeling of interaction of bacterial endotoxin with a chitosan polycation. *Khimiya I komp'yuternoe modelirovanie. Butlerovskie soobshcheniya* (Rus.), 2003, v. 2, pp. 12–14.

[40] Papo, N; Shai, Y. A molecular mechanism for lipopolysaccharide protection of Gram-negative bacteria from antimicrobial peptides. *J. Biol. Chem.*, 2005, v. 280, pp. 10378-10387.

[41] Balakrishna, R; Wood, SJ; Nguyen, TB; Miller, KA; Kumar, EVK; Datta, A; David, SA. Structural correlates of antibacterial and membrane-permeabilizing activities in acylpolyamines. *Antimicrob. Agents Chemother.*, 2006, v. 50, pp. 852–861.

[42] Chung, Y-C; Chen, CY. Antibacterial characteristics and activity of acid-soluble chitosan. *Bioresour. Technol.*, 2008, v. 99, pp. 2806-2814.

[43] Sudarshan, NR; Hoover, DG; Knorr, D. Antibacterial action of chitosan. *Food Biotechnol.*, 1992, v. 6, 257-272.

[44] Papineau, AM; Hoover, DG; Knorr, D; Farkas, DF. Antimicrobial effect of water-soluble chitosans with high hydrostatic pressure. *Food Biotechnol.*, 1991, v. 5, pp. 45-57.

[45] Helander, IM; Nurmiaho-Lassila, E-L; Ahvenainen, R; Rhoades, J; Roller, S. Chitosan disrupts the barrier properties of the outer membrane of Gram-negative bacteria. *Int. J. Food Microbiol.* 2001, v. 71, pp. 235 – 244.

[46] Chung, YC; Su, YP; Chen, CC; Jia, G; Wang, HL; Wu, CG; Lin, JG. Relationship between antibacterial activity of chitosan and surface characteristics of cell wall. *Acta Pharmacol. Sin.,* 2004, v. 27, pp. 932-936.

[47] Tokura, S; Miuray, Y; Johmen, M; Nishi, N; Nishimura, SI. Induction of drug specific antibody and the controlled release of drug by 6-*O*-carboxymethyl-chitin. *J. Controlled Release*, 1994, v. 28, pp. 235-241.

[48] Naberezhnykh, GA; Bakholdina, SI; Gorbach, VI; Solov'eva, TF. New chitosan derivatives with potential antimicrobial activity. *Russ. J. Mar. Biol.*, 2009, v. 35, pp. 498–503.

[49] Liu, XF; Guan, YL; Yang, DZ; Li, Z; Yao, KD. Antibacterial action of chitosan and carboxymethylated chitosan. *J. Appl. Polym. Sci.*, 2001, v. 79, pp. 1324-1335.

[50] Yermak, IM; Davydova, VN; Gorbach, VI; Berdyshev, EL; Kuznetsova, TA; Ivanushko, IA; Gazha, AK; Smolina, TP; Zaporozhets, TS; Solov'eva, TF. Modification of biological activity of lipopolysaccharide in the complex with chitosan. Bull. Exp. Biol. Med., 2004, v. 137, pp. 379-381.

[51] Chou, T.-C; Fu, E; Shen, E.-C. Chitosan inhibits prostaglandin E_2 formation and cyclooxygenase-2 induction in lipopolysaccharide-treated RAW 264.7 macrophages. *Biochem. Biophys. Res. Com.*, 2003, v. 308, pp. 403-407.

[52] Tapping, RI; Akashi, S; Miyake, K; Godowski, PJ; Tobias, PS. Toll-like receptor 4, but not Toll-like receptor 2, is a signaling receptor for *Escherichia* and *Salmonella* lipopolysaccharides. *J. Immunol.*, 2000, v. 165, pp. 5780-5787.

[53] Shimazu, R; Akashi, S; Ogata, H; Nagai, Y; Fukudome, K; Miyake, K. MD-2, a molecule that confers lipopolysaccharide responsiveness on Toll-like receptor 4. *J. Exp. Med.*, 1999, v. 189, pp. 1777-1782.

[54] Fenton, MJ; Golenbock, DT. LPS-binding proteins and receptors. *J. Leukocyte Biol.*, 1998, v. 64, pp. 25-32.

[55] Otterlei, M; Varum, KM; Ryan, L; Espevik, T. Characterization of binding and TNF-α-inducing ability of chitosans on monocytes: the involvement of CD 14. *Vaccine*, 1994, v. 12, pp. 825-832.

[56] Polyakova, AM; Astrina, OS; Bakhtina, YuA; Maleev, VV; Barabanova, AO; Yermak, IM. Possibility of correction of functional state of human trombocytes with natural polysaccharides in patients with alimentary toxic infections and in experimental endotoxinamia, *Infektsionnye Bolezni* (Rus.), 2005, v. 3, pp. 44–46.

[57] Polyakova, AM; Yermak, IM; Astrina, OS; Maleyev, VV; Gorbach, VI; Solovyeva, TF. Effects of biologically active natural polysaccharides on functional activity of platelets in experimental endotoxemia, *Infektsionniye Bolezni* (Rus.), 2003, v. 1, pp. 80–82.

[58] Ivanushko, LA; Soloviova, TF; Zaporozhets, TS; Somova, LV; Gorbatch, VI. Antibacterial and antitoxic properties of chitosan and its derivatives. *Pacific Medical Journal* (Rus.), 2009, v. 3, pp. 76-79.
[59] Khasina, EI; Sgrebneva, MN; Davydova, VN; Yermak, IM. Chitosan and nonspecific resistance. *Efferentnaya Terapiya* (Rus.), 2006, v. 12, pp. 32–35.

In: Handbook of Chitosan Research and Applications
Editors: Richard G. Mackay and Jennifer M. Tait
ISBN 978-1-61324-455-5

Chapter 5

RADIATION CROSSLINKED CHITOSAN GRAFTED WITH POLYANILINE

***A. T. Ramaprasad*[•] *and Vijayalakshmi Rao*[*]**
Materials Science Department, Mangalore University,
Mangalore, Karnataka, India

ABSTRACT

The chapter deals with the crosslinking of crosslinked chitosan using 8MeV electron beam and γ-rays and grafting of polyaniline on to the crosslinked chitosan. The crosslinked chitosan is characterized by dissolution, Fourier transform infrared spectroscopy (FTIR), X-ray diffraction (XRD), scanning electron microscopy (SEM) and differential scanning colorimetry (DSC). Mechanical properties such as elastic modulus and hardness increase with increase in the irradiation dose due to the increase in degree of crosslinking. E-beam crosslinked chitosan shows better mechanical properties compared to gamma irradiated samples.

Conducting polymer polyaniline has been grafted onto crosslinked chitosan to improve the processsibility and mechanical properties of polyaniline. Chitosan grafted with polyaniline has got potential applications such as biosensors, actuators, polymer battery electrode etc. But for these applications grafted polyaniline should be insoluble in the whole range of pH. Hence the chemical grafting of polyaniline onto the radiation crosslinked chitosan is carried out. Grafted polymer is characterized by dissolution, swelling, UV–vis–NIR spectroscopy, FTIR, XRD, SEM, TGA, DC conductivity and nanoindentation studies. From TGA it is observed that grafted polymer is thermally stable. DC conductivity of the grafted polymer is improved after doping with 1M HCl. The change in volume conductivity is from 10^{-11} to 10^{-5} S/cm and surface conductivity from 10^{-10} to 10^{-2} S/cm

The conductivity, mechanical properties and stability of grafted polymer under ambient condition are sufficient to make it an interesting material for diverse applications. Yang et al. (1989) has reported polyaniline grafted chitosan with highest surface conductivity of 10^{-2} S/cm. But they were able to get flexible film of graft polymer having surface conductivity of maximum10^{-4} S/cm. In the present study flexible film of

• e-mail: ramaprasadat@rediffmail.com.
* corresponding author: e-mail: vijrao@yahoo.com.

polyaniline grafted radiation crosslinked chitosan with surface conductivity of 10^{-2} S/cm is obtained. Present graft polymer can be used for applications, such as sensors and electrodes of polymeric batteries, since it is insoluble in whole range of pH.

1. INTRODUCTION

Polysaccharides such as cellulose, starch, chitin/chitosan and their water soluble derivatives have a variety of applications in many fields owing to their unique structure, distinctive properties, safety and biodegradability (Dumitriu et al., 1996). The solubility of chitosan in acid medium has several advantages for processing in various commercial applications such as in medicine, cosmetics, textile, paper, food and many other industrial branches (Wasikiewicz et al., 2005). However, for applications such as actuators, chromatographic separation and effluent treatment, it is necessary to crosslink chitosan to make it insoluble in the entire pH range. Chemical crosslinking of chitosan using glutaraldehyde, chlorohydrin, etc has been reported by Rorrer et al (1993). The resultant chitosan is insoluble in acid media, unlike normal chitosan.

Chemical modifications of polysaccharide has been widely studied, but only a very few reports on radiation crosslinking of polysaccharides are available. Radiation crosslinking of polysaccharide has been reported by Yoshii et al., for the first time (Yoshii et al., 2003). They showed that water soluble polysaccharide derivatives such as Carboxymethylcellulose (CMC), Carboxymethylstarch (CMS), Carboxymethylchitin (CMCT) and Carboxymethyl-chitosan (CMCTS) lead to radiation crosslinking at high concentrated aqueous solution. Chitosan like other natural polymers such as cellulose, carrageenan and algenates undergo predominantly degradation when exposed to ionizing radiation (Ulanski and Rosiak, 1992; Wenwei et al., 1993; Ershov et al., 1987). However chitosan crosslinks in the presence of sensitizers. Recently γ-ray induced crosslinking of chitosan in the powder form in the presence of CCl_4 has been reported by Ramnani et al., (2004).

The search for polymeric organic conductors is a very active and productive research area, which has been reviewed extensively (Pud et al., 2005). A central goal of the field is to prepare air-stable polymers with high conductivity which can be formed readily (processed) into freestanding materials. Considerable progress towards this goal has been achieved by electrochemically synthesizing thin films that can be removed from electrodes (Diaz et al., 1980; Street et al., 1986; Angelopoulos et al., 1987; Tourillon et al., 1983; Cushman et al., 1986), by preparing substituted polymers which have solubility imparted by these substituents and by identifying special solvent systems (Andreatta et al., 1988). These materials include polypyrrole, polythiophene, PANI etc. and are quite promising. PANI is an especially interesting air-stable conductive material. Unfortunately, it has poor mechanical and general physical properties.

In order to overcome these problems blending/grafting of PANI onto water soluble natural polysaccharide have been carried out in recent years (Yang et al., 1989; Kim et al., 2005(a); 2005(b); 2006; Tiwari, 2007a; 2007b; 2008; Tiwari and Singh, 2008). In general, incorporation of conducting polymers such as PANI into flexible biopolymer matrix could result in good processability, electrical conductivity and other requisite properties like chemical stability towards dopants and solubility under readily accessible conditions. Chitosan bears two types of reactive groups that can be grafted. First, the free amino groups on deacetylated units and secondly, the hydroxyl groups on the C_3 and C_6 carbon of acetylated or deacetylated units. The chemical grafting of PANI on to polysaccharide is first reported by Yang et al. (1989). PANI is grafted onto chitosan in solution phase and conductivity is found to be in the range of $\approx 10^{-2}$ to 10^{-5} S/cm depending on the ratio of initial concentration of chitosan to aniline. However they were not able to get flexible films having high conductivity i.e.; $\approx 10^{-2}$ S/cm. Later on Kim et al., (2005) have made an attempt on the blending of PANI with chitosan. They mixed two solutions of chitosan in acetic acid and PANI in NMP and then crosslinked the film using glutraldehyde to obtain semi interpenetrating network of chitosan and PANI. The conductivity of the blend was observed to be $\approx 10^{-4}$S/cm.

Many researchers have demonstrated the potential applications of chitosan/PANI composite such as, artificial muscles (Ismail et al., 2008), breast cancer sensor electrode (Tiwari et al., 2009) and immunosensor to detect ochratoxin-A (Khan et al., 2009). For some potential applications of chitosan-PANI composite such as, actuators, sensors and in polymeric batteries, these polymers should be insoluble in the whole range of pH. Hence PANI is chemically grafted onto the radiation crosslinked chitosan to get conducting flexible films insoluble in the whole range of pH. A Part of the results presented in this chapter is published in our earlier publications (Ramaprasad et al., 2008; 2009).

2. Radiation Crosslinked Chitosan

2.1. Preparation

Chitosan films were soaked in a mixture of CCl_4, water and methanol overnight. These soaked films were irradiated with electron beam and γ-rays. Specifications of electron beam used for irradiation is as follows: beam energy 8 MeV, beam current 25–30 mA, pulse repetition rate 50 Hz, pulse width 2.5 μs, dose rate of 1 kGy/min (Fricke dosimetry), and sample is kept at a distance of 30 cm. Gamma irradiation was carried out using Co^{60} gamma chamber having radiation dose rate of 5 kGy/h determined using Fricke dosimetry. Films were taken out and first washed with methanol to remove any traces of CCl_4 and other radiolytic product formed during the irradiation, followed by washing with water. The resultant films were yellow in colour and films were treated with NaOH to remove chlorine. So obtained brown coloured films were washed with water followed by acetone and then dried. Percentage of cross linking has been calculated from the gel fraction (Table 1). From the table it is clear that for e- beam crosslinked chitosan, the percentage of crosslinked portion is higher compared to γ-ray crosslinked chitosan. This may be due to the difference in the dose rate. If we compare the dose rate of e-beam and γ-rays, e-beam delivers dose in seconds where as γ-rays takes several hours to deliver the same dose (Chmielewski et al., 2005). This long residence time in the case of gamma rays leads to more oxidative degradation. During irradiation, chitosan undergoes crosslinking in presence of CCl_4, but at the same time it undergoes oxidative degradation due to the free radicals produced. Hence extent of crosslinking is less in the case of γ-ray crosslinked chitosan compared to electron beam crosslinked chitosan.

Table 1. Gel fraction of crosslinked Chitosan

Dose (kGy)	Gel fraction (%)	
	γ- ray	e-beam
25	16	29
50	55	67
70	62	76

2.2. Characterization

2.2.1. XRD and SEM

The XRD pattern of chitosan and e-beam crosslinked chitosan are shown in Figure 1.

The intensity of peak at ~2θ= 19.7° for cross linked chitosan decreases and gets broadened with irradiation dose. This clearly indicates that the crystalline structure of chitosan changes during irradiation in presence of methanol, CCl_4 and water mixture due to crosslinking. This is further supported by SEM (Figure 2) of irradiated chitosan samples. Surface and cross sectional view of uncrosslinked chitosan is shown in Fig 2(a and c) respectively, which shows very tightly bound structure with few voids. Surface and cross sectional view of crosslinked chitosan is shown in Figure 2(b and d) respectively. Crosslinked chitosan shows interlinked structure with more voids.

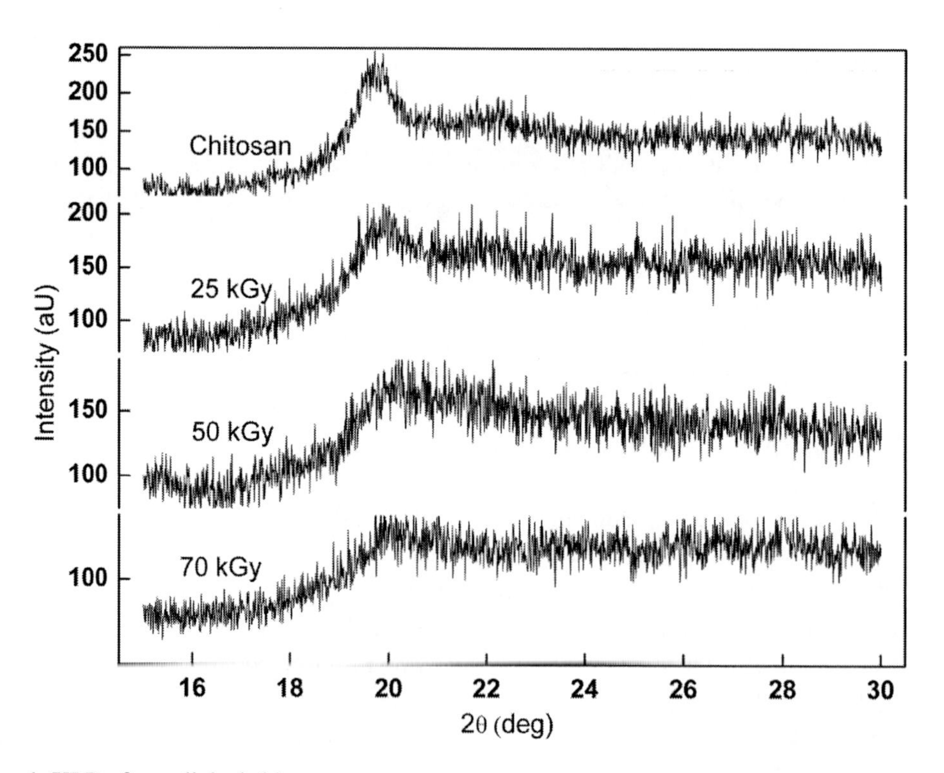

Figure 1. XRD of crosslinked chitosan.

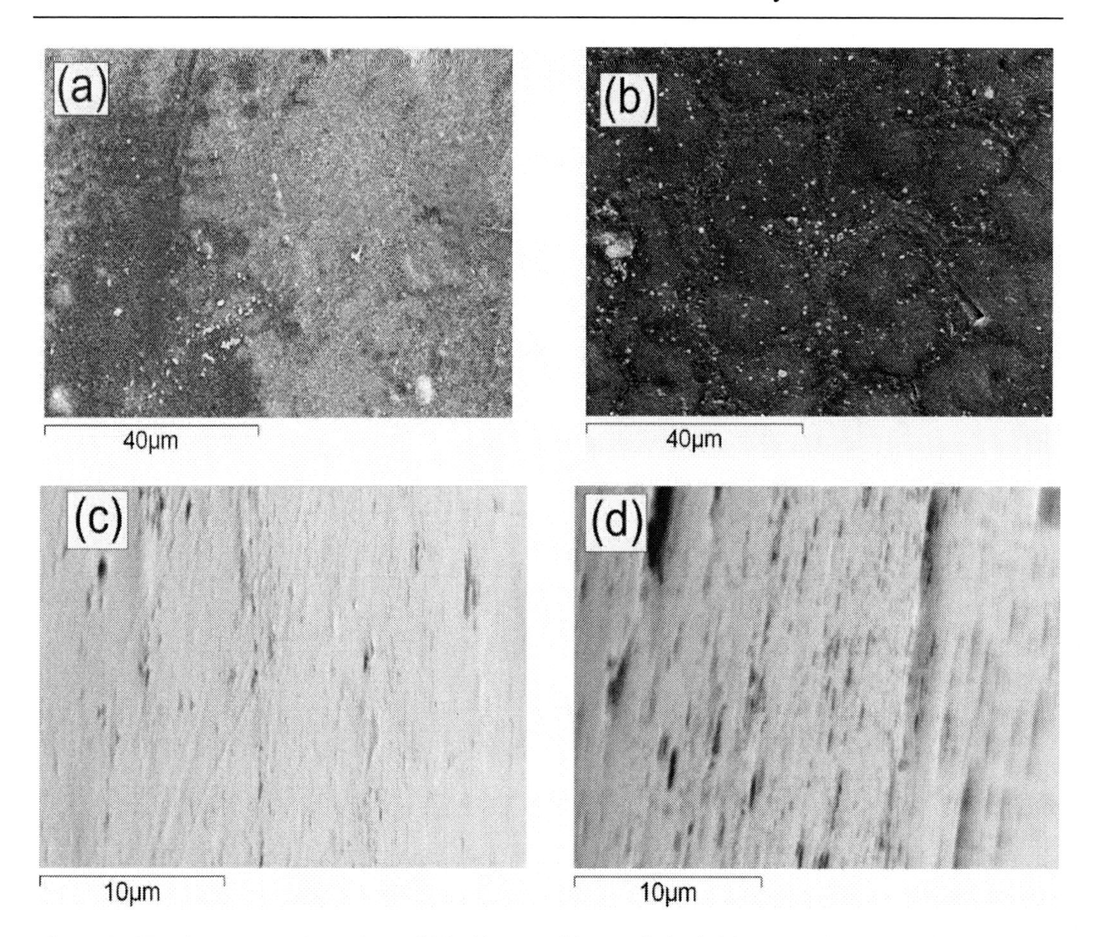

Figure 2. SEM images: surface view of (a) chitosan, (b) crosslinked chitosan and cross sectional view of (c) chitosan, (d) crosslinked chitosan.

Thus it seems that when chitosan is irradiated in the presence of a mixture of CCl_4, water and methanol, the polymer chains get crosslinked with each other, resulting in changes at the micro structural level, as evident from XRD and SEM results. Ramnani et al., (2004) have reported the XRD and SEM analysis of γ-ray irradiated chitosan powder in the presence of solvent. They observed lower intensity in XRD and more voids in SEM compared to the present e-beam irradiated films. That may be due to the difference in the form of chitosan taken, in the present case chitosan film is used for irradiation.

2.2.2. Thermal Analysis

The DSC thermograms of chitosan and crosslinked chitosan (Figure 3) show one endothermic peak at 80°C, due to the presence of moisture in the sample (Nogales et al., 1997) and a small endothermic peak at ~205°C, attributed to T_g of chitosan which vanishes in the case of crosslinked sample. In the case of chitosan, the endothermic peak corresponding to Tg observed at 205°C is small due to the presence of the rigid 2-amino-2-deoxy-D-glucopyranose or glucosamine residue (Clelend et al., 2003) and on crosslinking, motion of these chains at Tg is further hindered. The exothermic peak at 280°C is due to the degradation of chitosan. Figure 4 shows the thermogravimetric analysis of chitosan and crosslinked chitosan. Initial weight loss from room temperature to 150°C may be due to the moisture present in the sample. The weight loss from 250 to 350°C is due to the thermal degradation of chitosan. From figure one can see that there is no significant change in the weight loss pattern after crosslinking.

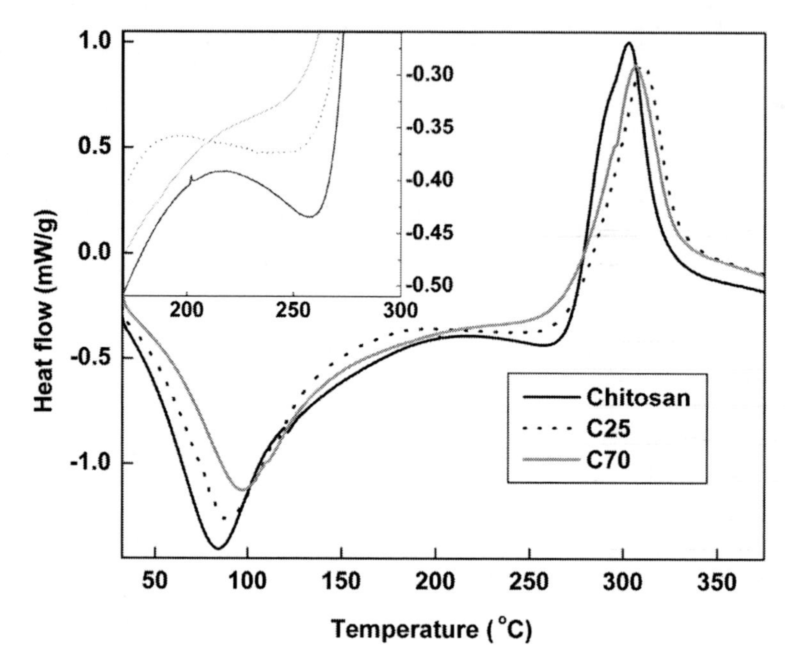

Figure 3. DSC thermograms of chitosan and crosslinked chitosan.

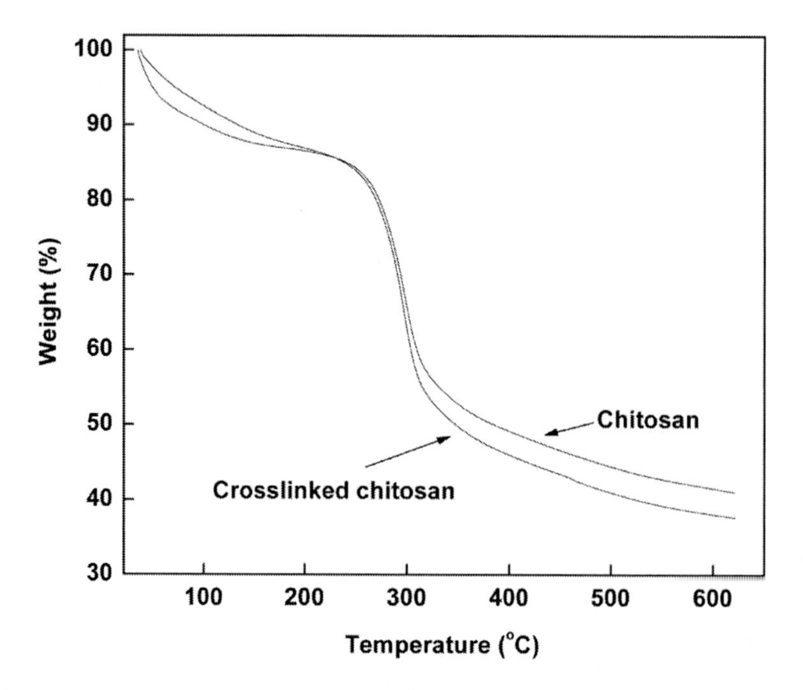

Figure 4. TGA thermograms of chitosan and crosslinked chitosan.

2.2.3. Mechanical Properties

The mechanical properties of the prepared polymers are studied using nanoindentation technique using a conical tip.

Figures 5 (a and b) show the AFM images of e-beam crosslinked chitosan before and after indentation respectively. Figure 5 (c) shows the load versus displacement curve for e-beam irradiated chitosan. The displacement decreases as the irradiation dose increases. It is due to the increase in crosslinking density. Elastic modulus and hardness of crosslinked chitosan are calculated from the unloading portion of the displacement curve.

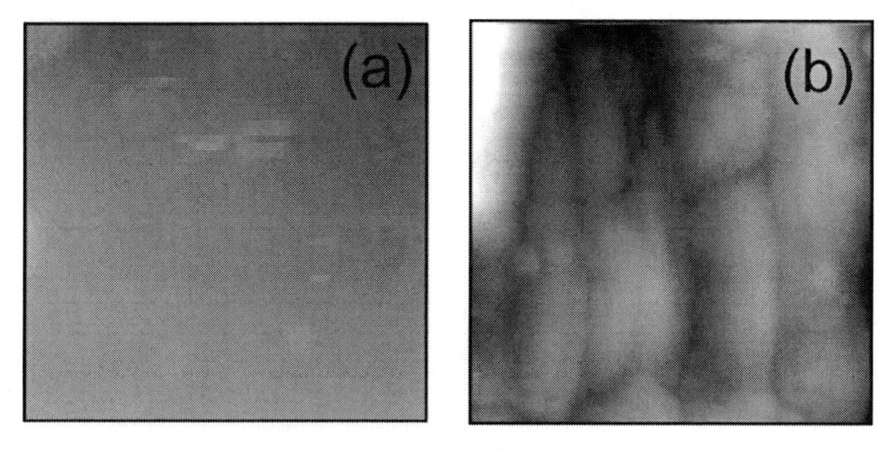

Figure 5. AFM image: (a) before indentation and (b) after indentation.

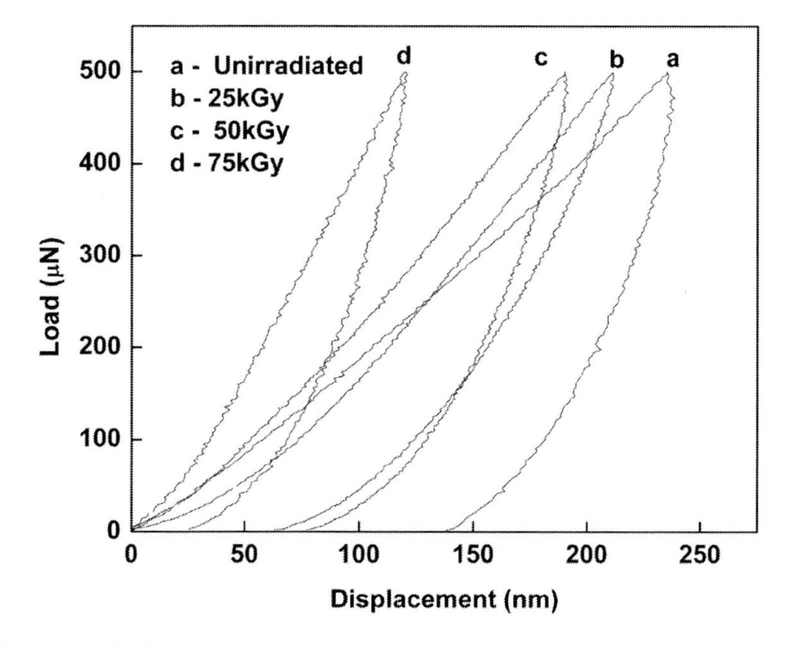

Figure 5. (c) Load vs. displacement curve.

Table 2. Nanoindentation Results

Dose (kGy)	Modulus (GPa)		Hardness (GPa)	
	e-beam	γ-rays	e-beam	γ-rays
0	4.82 ± 0.07	4.82 ± 0.07	0.19 ± 0.012	0.19 ± 0.012
25	4.93 ± 0.10	4.86 ± 0.09	0.28 ± 0.010	0.22 ± 0.009
50	5.36 ± 0.17	5.17 ± 0.14	0.34 ± 0.011	0.27 ± 0.013
70	7.85 ± 0.12	6.68 ± 0.11	0.37 ± 0.009	0.31 ± 0.010

As the irradiation dose increases hardness and modulus of the polymer increase, due to increase in the crosslinking fraction (Table 2). But modulus and hardness values of gamma irradiated sample are less compared to electron beam irradiated sample, due to the difference in the degree of crosslinking .

2.3. Mechanism of Crosslinking

When chitosan in the solid state is irradiated with ionizing radiation, many free radicals are produced on the back bone of the chain. These radicals generated preferably on the position 1 and 4 of the 2-deoxy-D-glucose residue results in chain scission at the 1, 4 glycosidic bond, thus leading to degradation (Lim et al., 1998). When chitosan is exposed to ionizing radiation in the presence of CCl_4, following changes are noticed.

- The irradiated chitosan films swell in 1% aqueous acetic acid solution in which chitosan otherwise dissolves.
- The product formed is dark yellow compared to unirradiated chitosan.
- After hydrolysis with 1% NaOH, it becomes brown in colour and still it swells in 1% acetic acid.

It is well known that any polymer undergoing crosslinking reaction swells in the solvent in which, it otherwise dissolves. The insolubility of irradiated films in 1% acetic acid solution indicates that either chitosan has undergone a crosslinking reaction or NH_2 groups of chitosan are involved in some interaction during irradiation. However FTIR spectrum of crosslinked chitosan is similar to that of chitosan showing the presence of NH_2 group and that NH_2 group is not involved in any interaction (Figure 6). The broad peak from 2600-3600 cm^{-1} is assigned to overlapping of – OH and – NH_2 groups. A small peak at 1669 cm^{-1} is attributed to the -C=O stretching in -$NHCOCH_3$ arising from a small portion of pristine chitin. Peaks at 1656 cm^{-1} and 1030cm^{-1} are due to N-H and O-H bending respectively. The peak at 1076 cm^{-1} is due to C-O stretching.

Further the presence of NH_2 group is confirmed by the estimation of ionic capacity of chitosan samples. The ionic capacity of uncrosslinked and crosslinked chitosan is almost same for all samples. (Table 3).

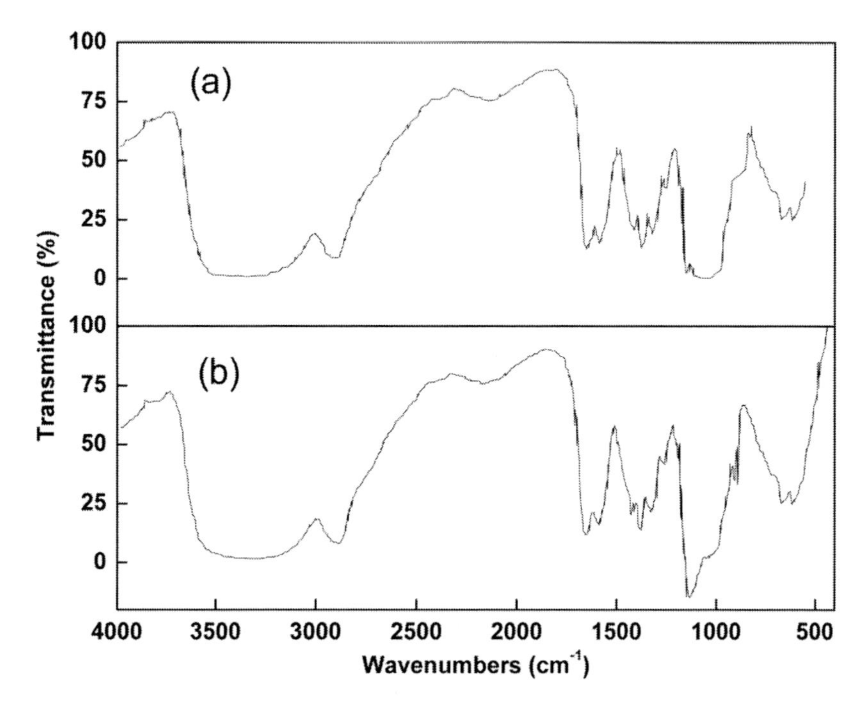

Figure 6. FTIR spectrum of (a) chitosan and (b) crosslinked chitosan.

Table 3. Ionic capacity of irradiated chitosan

Dose (kGy)	Ionic Capacity (meq g^{-1})	
	γ-ray	e-beam
0	4.75	4.75
25	4.78	4.76
50	4.74	4.71
70	4.73	4.80

The dark yellow colour of crosslinked chitosan suggests the incorporation of chlorine in crosslinked chitosan, as organochloride compounds are normally yellow in colour. Qualitative analysis is performed to confirm the presence of chlorine in crosslinked chitosan. When unhydrolysed crosslinked chitosan is treated with 1% NaOH and filtrate is acidified with aqueous nitric acid, the filtrate gave white precipitate with silver nitrate, where as when the same procedure is repeated with uncrosslinked chitosan, did not give white precipitate with silver nitrate. This clearly indicates the incorporation of Cl in the crosslinked chitosan. The presence of chlorine also indicates the role of CCl_4 in the crosslinking of chitosan. Third observation, i.e.; swelling of crosslinked chitosan in acetic acid even after hydrolyzing with 1% NaOH suggests that during hydrolysis chlorine atoms present in the crosslinked chitosan are removed and even after the removal of Cl atoms the crosslinking structure is retained.

When chitosan is irradiated with e-beam in the presence of mixed solvents 30:35:5 parts by weight of water, methanol, and CCl_4, respectively, energy is absorbed by both chitosan and solvent. The free radicals generated on the back bone of chitosan, particularly at the radicals sites on the 1 and 4 carbon atom, are responsible for scission of the 1, 4 glycosidic bond. It is well known that CCl_4 reacts efficiently with organic radicals. Also irradiation effect of CCl_4 is well established in literature (Yoshii et al., 2003). Radiolysis of CCl_4 results in the formation of $\bullet CCl_3$ radicals by the following reaction:

$$CCl_4 \xrightarrow{\text{e- beam}} {}^{\bullet}CCl_3 + Cl^{\bullet} \quad (1)$$

Since C-Cl bond has lower dissociation energy than the C-H bond, more radicals are produced upon irradiation of CCl_4 than chitosan (Wasikiewicz et al., 2005). In the presence of CCl_4, the radical $\bullet CCl_3$ formed may either react with the radicals formed on the chitosan backbone, resulting in the attachment of the $\bullet CCl_3$ group to the chitosan chain or it may react with chitosan and result in the formation of radical sites on the chitosan backbone. Also $Cl^{\bullet}$ radical reacts with chitosan to produce radicals in the chitosan backbone (eqn.(2)).

$$\left.\begin{aligned} R^{\bullet} + {}^{\bullet}CCl_3 &\longrightarrow R - CCl_3 \\ RH + {}^{\bullet}CCl_3 &\longrightarrow R^{\bullet} + CHCl_3 \\ Cl^{\bullet} + RH &\longrightarrow R^{\bullet} + HCl \end{aligned}\right\}$$

RH - Chitosan R$^{\bullet}$ -Chitosan radical (2)

This $\bullet CCl_3$ attached chitosan chain will react with adjacent chains having radicals, resulting in the formation of crosslinked chitosan structure (eqn.(3)).

$$R - CCl_3 + R^{\bullet} \longrightarrow R - CCl_2 - R + Cl^{\bullet} \quad (3)$$

Hydrogen bonds are disrupted in the formation of R-CCl_2-R. When chitosan chains are crosslinked, a bulky Cl-C-Cl group links the two adjacent chains with each other; for such crosslinks to occur, the chitosan chains have to orient themselves in such a way as to accommodate the Cl-C-Cl group. Thus an ordered arrangement of chains that exist in the crystalline region is affected, resulting in the broadening of the XRD peak and decrease in the intensity. When crosslinked chitosan is treated with NaOH two Cl atoms of –CCl_2- bridge are replaced by -OH groups and two -OH groups on same carbon atom is unstable and undergoes dehydration resulting in loss of water molecule, with formation of carbonyl group(>C=O). The colour of crosslinked chitosan changes from dark yellow to brown after hydrolysis, confirming the formation of C=O group. This results in restoration of hydrogen bonding network. Therefore crosslinked chitosan shows no appreciable change in decomposition temperature in TG analysis. All the above results are depicted in Figure 7.

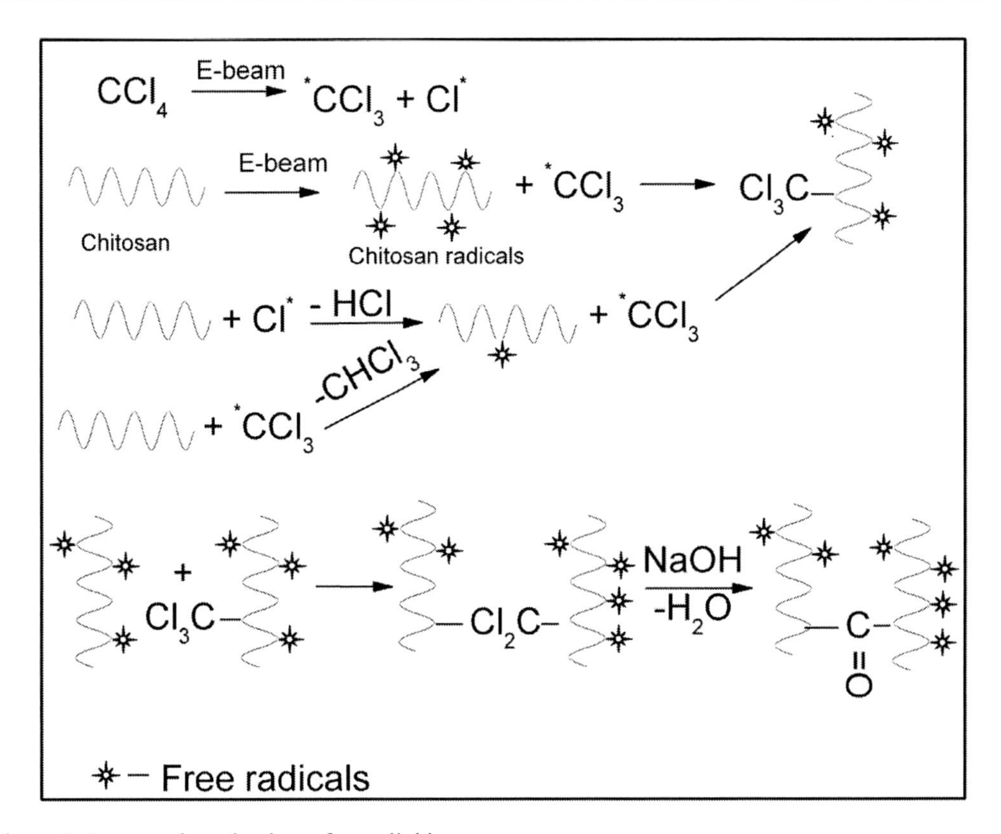

Figure 7. Suggested mechanism of crosslinking.

3. Grafting of Polyaniline onto Radiation Crosslinked Chitosan

3.1. Preparation

Grafting of PANI onto the e-beam and γ-ray crosslinked chitosan films is done chemically by polymerizing aniline in the presence of APS. Grafting of aniline is carried out as follows. A weighed chitosan film is soaked in precooled (0-5°C) 0.25M aniline prepared in 1M HCl. After 24hrs of soaking, pre-cooled 0.25M APS solution prepared in 1M HCl is added drop wise with in 20 minutes interval and reaction mixture is maintained at 0-5°C for 4hrs. 0.5N NaOH solution is added to the reaction mixture until colour of the film turned from green to blue and then washed with deionized water until it was free from NaOH and dried at 60°C for 24 hrs. So obtained films are treated with DMA for 2 hrs in order to remove homopolymerized PANI and again washed with deionized water and dried. So obtained polymer films are doped with 1M HCl for different time intervals and used for further characterization.

3.2. Characterisation

3.2.1. Grafting Percentage

From the weight of the crosslinked chitosan films before and after grafting, percentage of grafting is calculated and is presented in Table 4.

From Table 4 it is clear that as the radiation dose or percentage of crosslinking fraction increases, percentage of grafting decreases. This may be due to the decrease in penetration of monomer in to the crosslinked chitosan matrix. Gamma crosslinked chitosan shows slightly higher grafting percentage compared to e-beam crosslinked one. This is because e-beam crosslinked chitosan has got higher crosslinked fraction compared to gamma crosslinked chitosan due to the longer residence time of chitosan inside the gamma chamber for a particular dose, which can be delivered in a fraction of seconds in the case of e-beam irradiation. Hence penetration of monomer is more in the case of gamma crosslinked chitosan compared to e-beam crosslinked chitosan. But during the doping of grafted films with 1M HCl, handling of γ-ray crosslinked sample was difficult compared to e-beam crosslinked one due to lower crosslinking fraction and therefore further studies have been carried out only with e-beam crosslinked samples.

Table 4. Percentage of PANI grafted onto radiation crosslinked chitosan

Sample Code*	Dose (kGy)	Monomer concentration (M)	% of grafting	
			e-beam	γ-rays
$C_{25}P_{0.1}$	25	0.1	48.58	54.5
$C_{25}P_{0.25}$	25	0.25	71.24	88.2
$C_{25}P_{0.5}$	25	0.5	148.2	160.5
$C_{50}P_{0.1}$	50	0.1	28.7	36.3
$C_{50}P_{0.25}$	50	0.25	49.56	57
$C_{50}P_{0.5}$	50	0.5	117.6	132
$C_{70}P_{0.1}$	70	0.1	22.6	27.8
$C_{70}P_{0.25}$	70	0.25	41.8	44.2
$C_{70}P_{0.5}$	70	0.5	88.3	94.7

*Subscript of C and P stands for dose and aniline concentration respectively.

3.2.2. UV-vis-NIR and FTIR Spectroscopy

Figure 8 shows the UV-vis-NIR spectra for insitu doped PANI grafted crosslinked chitosan. Absorption peaks are observed at ≈330, ≈600 and a small peak at ≈1951nm. Since there is no significant absorption peak for chitosan in this region and these peaks are similar to the absorption peaks observed for PANI, the absorption peak at ≈330nm corresponds to π-π^* transition of the benzenoid ring. On doping, peaks are observed at ≈330, ≈750, ≈1465, ≈1951 and absorption edge at ≈440nm. These additional peaks are due to the increase in polaronic concentration (Yang et al., 1989; Jeevananda et al., 2001; Xia and Wang, 2003). Further intensity of peak at ≈750nm increases and shifts to higher wavelength and intensity of absorption edge at ≈440nm increases up to 3 hrs of doping and there after not much change in the intensity of the peaks even after 24 hrs of doping. This shows the saturation of doping level after 3 hrs of doping.

Figure 9 shows the FTIR spectrum of chitosan grafted with PANI. FTIR spectrum of grafted samples show all significant peaks corresponding to chitosan and PANI. The peak at ≈3286-3468cm^{-1} is due to free O-H stretching and NH_2 stretching with hydrogen bonded secondary amino groups; ≈2925 and ≈2858 cm^{-1} due to aliphatic C—H stretching; ≈1655cm^{-1} due to –NH bending; ≈1636cm^{-1} due to C=O stretching of carbonyl group of chitin moiety present in chitosan; ≈1571cm^{-1} is attributed to C=C and C=N stretching of quinoid rings of PANI; ≈1459cm^{-1} is assigned to C=C and C=N stretching of benzenoid rings of PANI; ≈1301cm^{-1} is assigned for aromatic stretching (Yang et al., 1989; Kim et al., 2005; Jeevananda et al., 2001; Tiwari and Singh, 2007).

The transmission band at ≈1214cm^{-1} is attributed to N=Q=N bending vibration of pure doped PANI, but in the case of grafted polymer it is shifted to ≈1167cm^{-1} due to the steric effect of chitosan (Tiwari and Singh, 2007). In the doped grafted polymer peak at ≈825cm^{-1} is attributed to para substituted aromatic rings (Yang et al., 1989) and peak at ≈1598cm^{-1} is due to proton doping. The N-H band observed around ≈1655cm^{-1}in chitosan is shifted to lower wave number in the case of grafted chitosan indicating strong interaction between chitosan and PANI. Aniline polymerizes at the active sites (2 and 5) of chitosan (Muzzralli, 1977; Tiwari, 2007), that is –OH and –NH_2, which is supported from FTIR results.

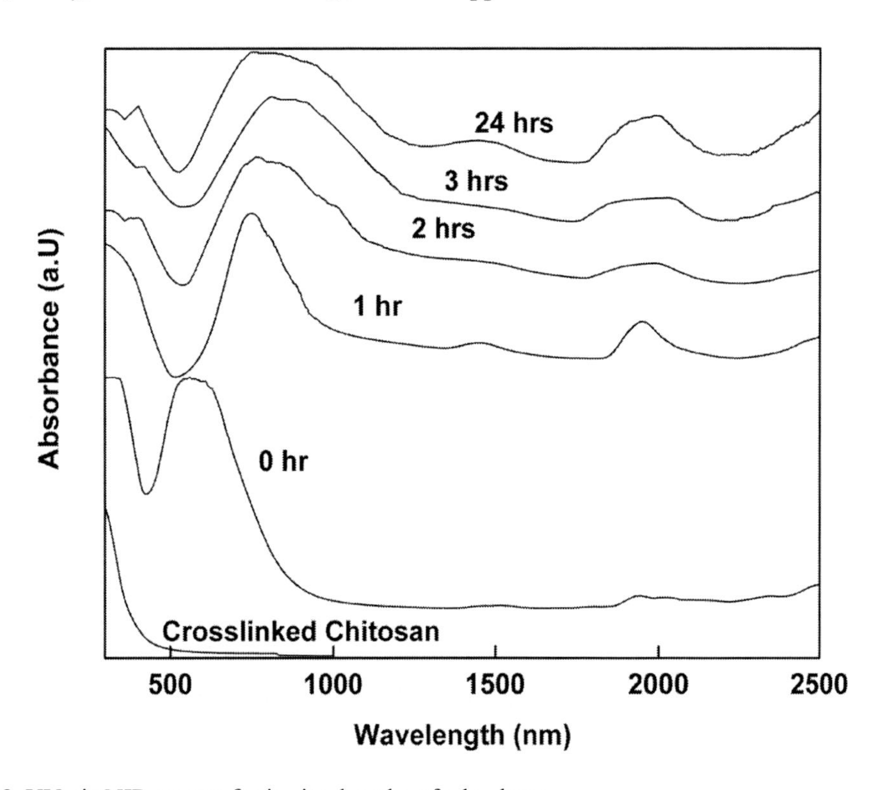

Figure 8. UV-vis-NIR spectra for in situ doped grafted polymer.

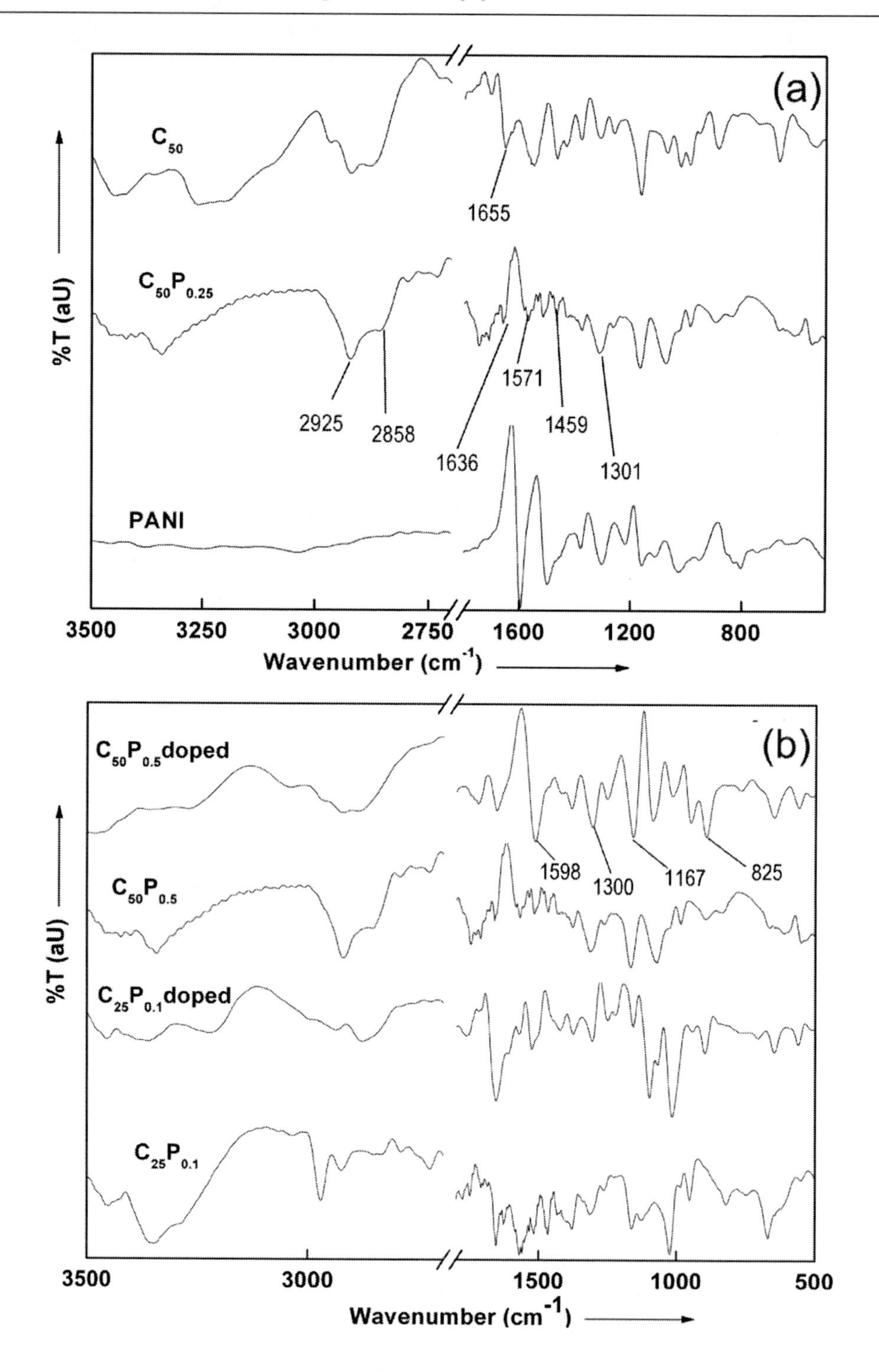

Figure 9. FTIR spectrum of (a) PANI, chitosan crosslinked at 50kGy (C_{50}) and crosslinked chitosan film grafted with PANI ($C_{50}P_{0.25}$) and (b) undoped and doped grafted polymers.

3.2.3. XRD and SEM

The X-ray diffractograms of the grafted polymer show distinct crystalline peaks of chitosan and PANI, indicated by the presence of a broad peak at 11.2°, (Figure 10) in agreement with the characteristic diffractogram of the original chitosan and PANI. The peak, appearing at about 20°, is assigned to the overlapping of chitosan and PANI. The peak at 20° became sharp with increasing irradiation dose. Crystallinity of crosslinked chitosan decreases with increase in the percentage of crosslinking. Further on grafting with PANI, which is an amorphous polymer, crystallinity of grafted polymer decreases. As the percentage of crosslinking increases, the amount of PANI (amorphous phase) grafted on to chitosan decreases. In other words percentage of PANI grafting is more in the case of chitosan crosslinked at 25 kGy dose compared to the 50 kGy. Hence crystallinity of grafted polymer increases with percentage of crosslinking.

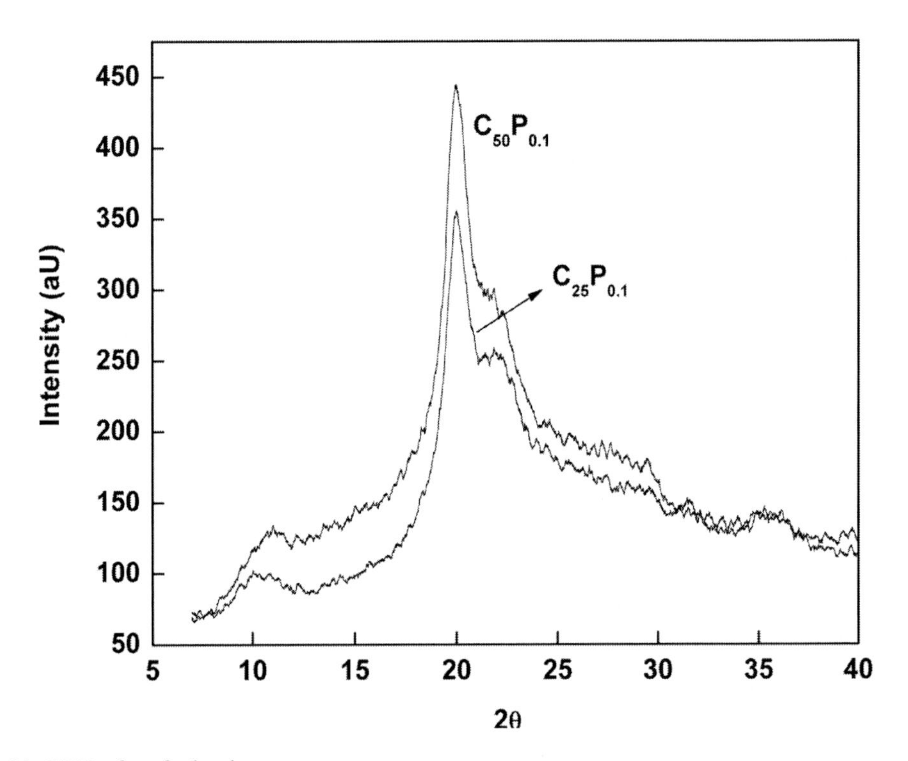

Figure 10. XRD of grafted polymers.

Surface and cross sectional view of SEM pictures are shown in Figure 11 and 12 respectively, cross section of the sample in the image is marked by black line. Surface view of chitosan (Figure 11(a)) shows tightly bound structure with a very few voids and crosslinked chitosan shows interlinked structure with more voids (Figure 11(b)).

After grafting of PANI, smooth surface with fewer voids is observed (11(d)). But cross sectional view (Figure 12 (a) shows two separate phases, near surface it is continuous with less voids and in bulk larger voids with pillar like structure. This continuous phase near the surface increases as the monomer concentration increases (Figure 12 (a, c and d)). Further if irradiation dose is increased this continuous phase near the surface decreases (Figure 12 (a, e and f)). This can be explained as follows: as the crosslinking fraction of chitosan increases, penetration of aniline monomer to the bulk reduces. Hence there are two separate phases in the film. Surface of the grafted film is rich in PANI and bulk poor in PANI. This is reflected in the conductivity values. The surface conductivity is found to be more than the volume conductivity, which is discussed in section 3.4. The concentration of PANI in the bulk can be increased by increasing the concentration of monomer (Fig 12 (a, c and d)). On doping these films with 1M HCl, pillar like structure is improved, but micro cracks are observed in the bulk of the grafted film (Figure 12 (b). But surface view shows rough surface without crack (Figure 11 (e)). As the degree of crosslinking increases these micro cracks propagate to the surface of the film (Figure 11(f) and films become brittle and break easily.

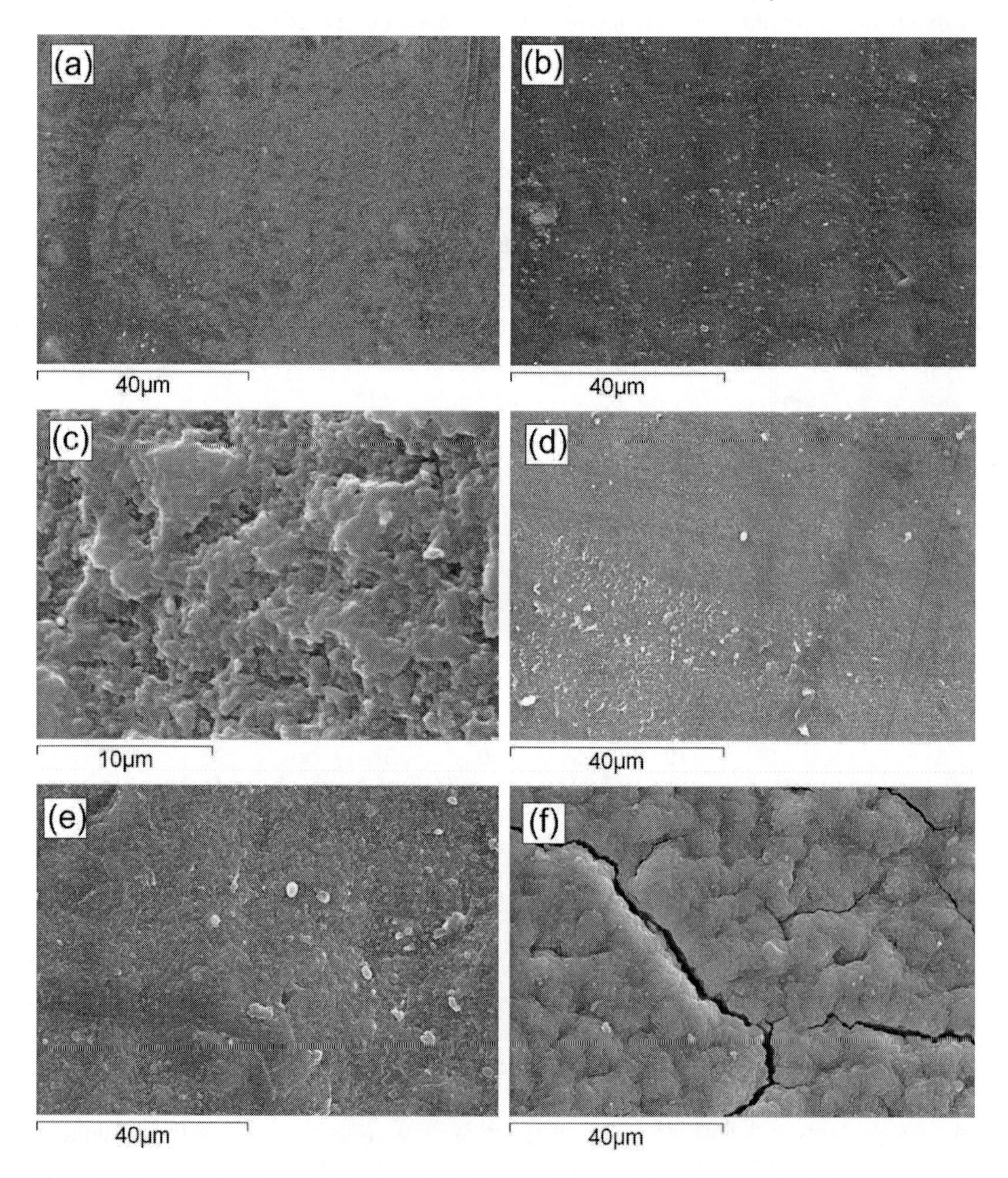

Figure 11. Surface view of SEM pictures of (a) chitosan, (b) crosslinked chitosan (at 25 kGy), (c) PANI (EB), (d) $C_{25}P_{0.1}$, (e) $C_{25}P_{0.1}$ doped with 1M HCl and (f) $C_{70}P_{0.1}$ doped with 1M HCl.

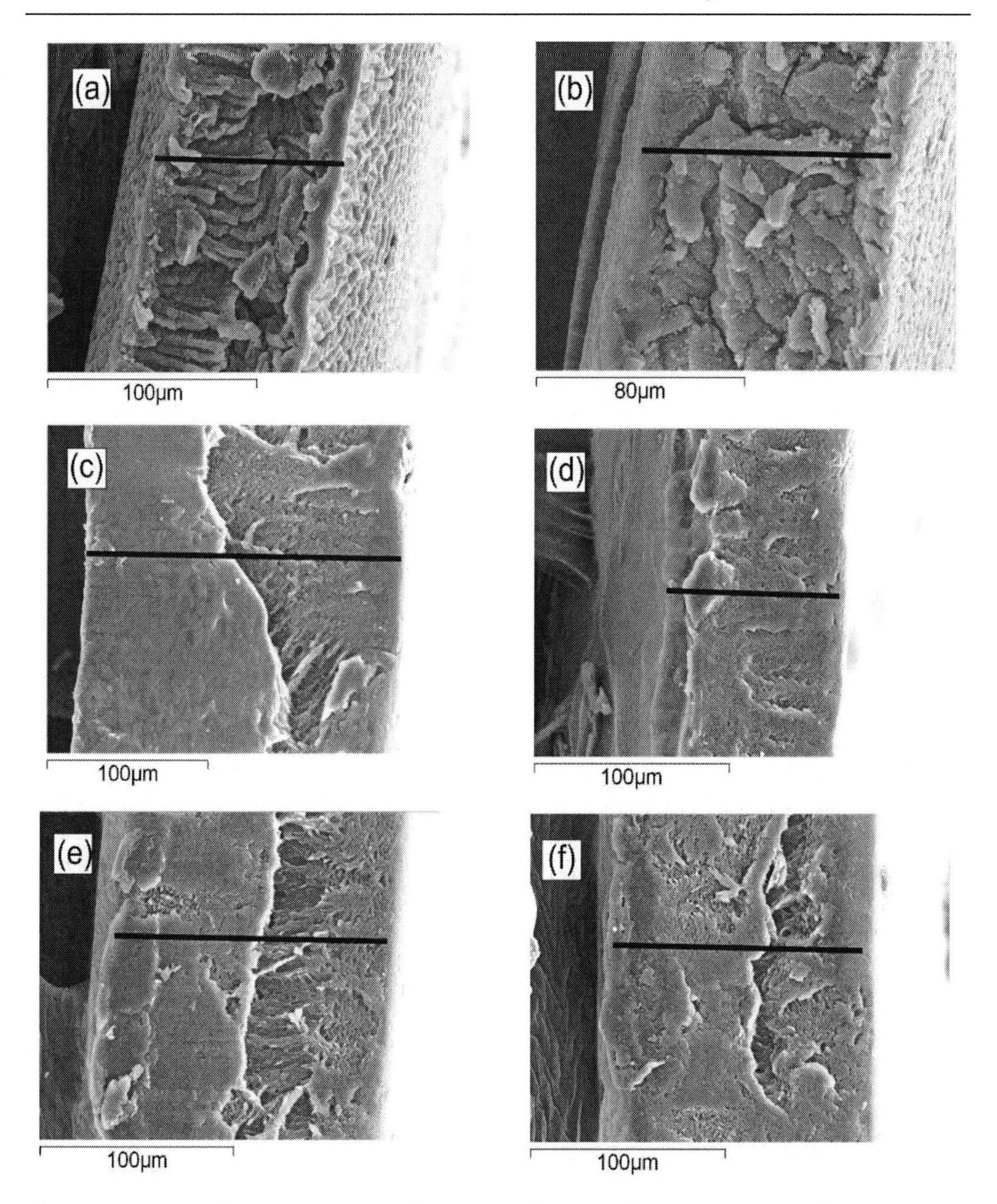

Figure 12. Cross-sectional SEM pictures of (a) $C_{25}P_{0.1}$, (b) $C_{25}P_{0.1}$ doped with 1M HCl, (c) $C_{25}P_{0.25}$, (d) $C_{25}P_{0.5}$, (e) $C_{50}P_{0.1}$ and (f) $C_{70}P_{0.1}$.

3.2.4. Thermal Analysis

All grafted polymers, doped / undoped, are stable at room temperature for several months. Thermogravimetric analysis (TGA) studies also show that, these materials are stable up to 250°C in air with out decomposition (Figure 13). Initial weight loss up to 100°C is due to the loss of residual solvent and moisture. There is not much difference in the thermal behavior of crosslinked chitosan and grafted chitosan.

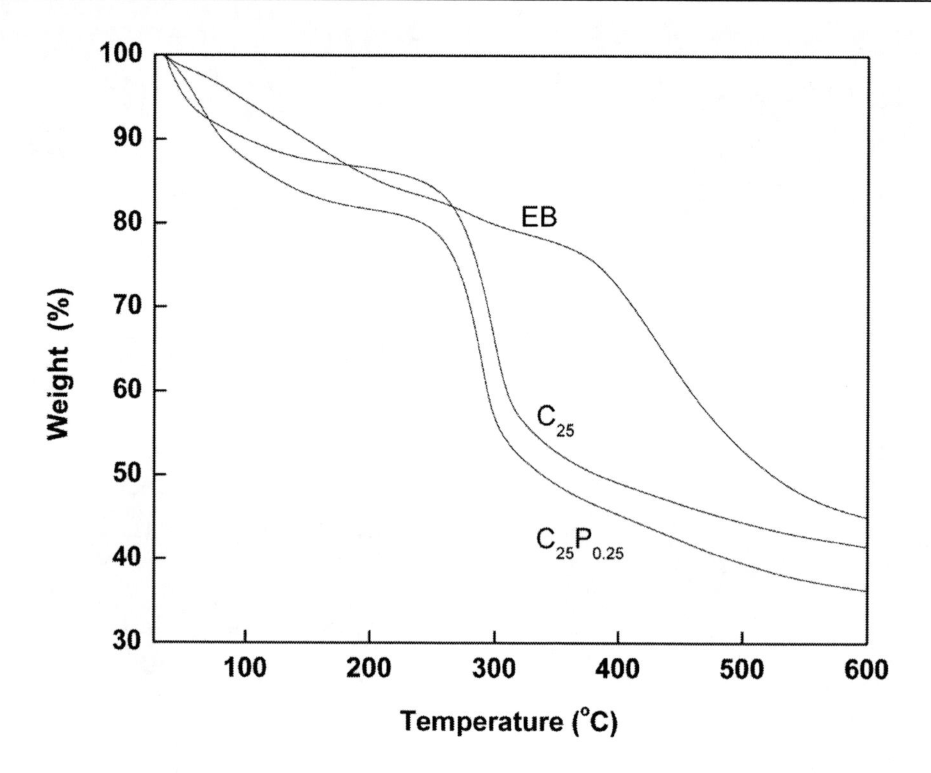

Figure 13. TGA of grafted polymer.

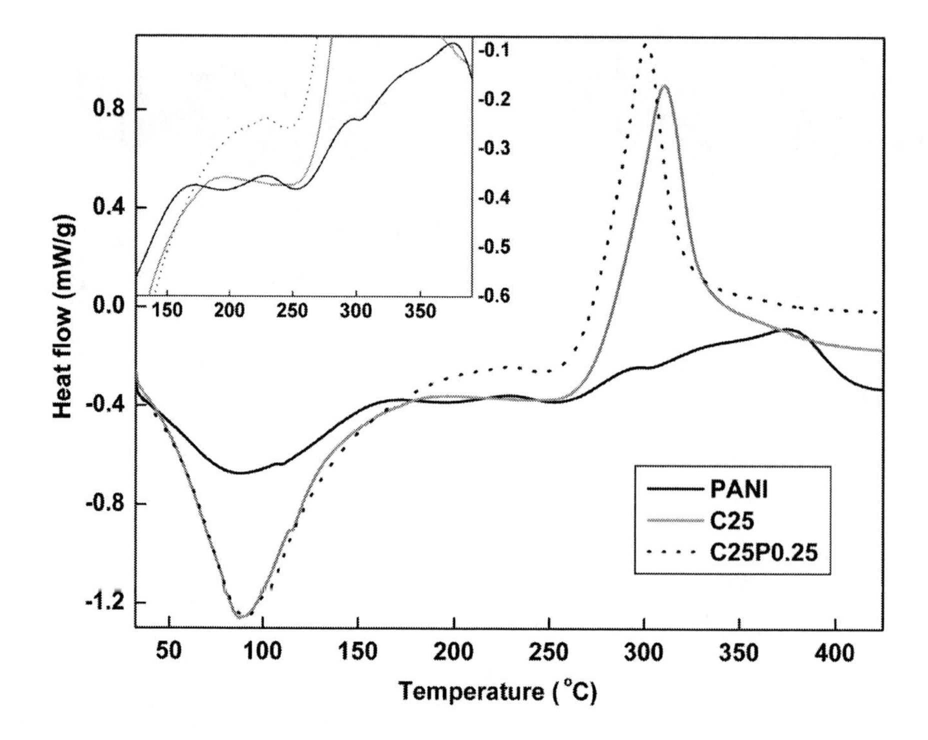

Figure 14. DSC theromograms.

Table 5. Nanoindentation results

Sample ID	Hardness (Gpa)	Modulus (Gpa)
Chitosan	0.19 ± 0.012	4.82 ± 0.07
C_{25}	0.28 ± 0.010	4.93 ± 0.10
C_{50}	0.34 ± 0.011	5.36 ± 0.17
C_{70}	0.37 ± 0.009	7.85 ± 0.12
$C_{25}P_{0.1}$	0.12 ± 0.006	4.16 ± 0.02
$C_{25}P_{0.1}$ (doped)	0.11 ± 0.008	4.12 ± 0.016
$C_{50}P_{0.1}$	0.17 ± 0.014	4.97 ± 0.04
$C_{50}P_{0.1}$ (doped)	0.14 ± 0.010	4.66 ± 0.06
$C_{70}P_{0.1}$	0.25 ± 0.013	5.90 ± 0.037

Figure 14 shows the DSC thermograms of PANI, crosslinked chitosan and grafted polymer. All polymers show a broad endothermic peak at 80°C due to the traces of solvent or water present in the polymer. Crosslinked chitosan shows a small endothermic peak around 205°C, which can be attributed to the Tg of chitosan. In the case of chitosan, the endothermic peak corresponding to T_g is small due to the presence of the rigid 2-amino-2-deoxy-D-glucopyranose or glucosamine residue (Clelend, et al., 2003) and on crosslinking, motion of these chains at Tg is further hindered. The exothermic peak at 280°C is due to the degradation of chitosan. From modulated DSC and from DMA, T_g of the PANI is reported at around 120-135°C, which is a very small deflection. So transition corresponding to T_g can not be seen in DSC. Exothermic peak at around 160°C corresponds to recrystallization of PANI (Bhadra et al., 2008). Endothermic peak at 200°C may be due to the loss of bound water or low MW oligomeric units (Bhadra et al., 2008). Broad exothermic peak around 375°C for pure PANI is due to the thermal crosslinking of PANI. DSC thermogram of grafted polymer is similar to the DSC of crosslinked chitosan. Endothermic peak corresponding to T_g of chitosan is further reduced due to the grafting of PANI.

3.2.5. Mechanical Properties

The calculated values of elastic modulus and hardness are presented in the Table 5. As the irradiation dose increases hardness and modulus of chitosan increases, due to increase in the crosslinking fraction. There is a direct correlation between chain flexibility of polymer and measured values of modulus and hardness. Modulus and hardness value of chitosan increase as the crosslinking fraction increases due to the reduction in the chain flexibility of crosslinked chitosan. Hence as the irradiation dose increases chain flexibility reduces. On grafting crosslinked chitosan with PANI, modulus and hardness reduces. This is due to the incorporation of PANI having poor mechanical properties. The reduction in the modulus value is less compared to the reduction in the hardness value. As crosslinking fraction increases, elastic modulus and hardness of grafted polymer increase. Hence the flexibility of grafted polymer reduces as the crosslinking dose increases, in other words grafted polymer becomes hard and brittle. Further on doping the elastic modulus and hardness decrease, but decrease in the elastic modulus is very small.

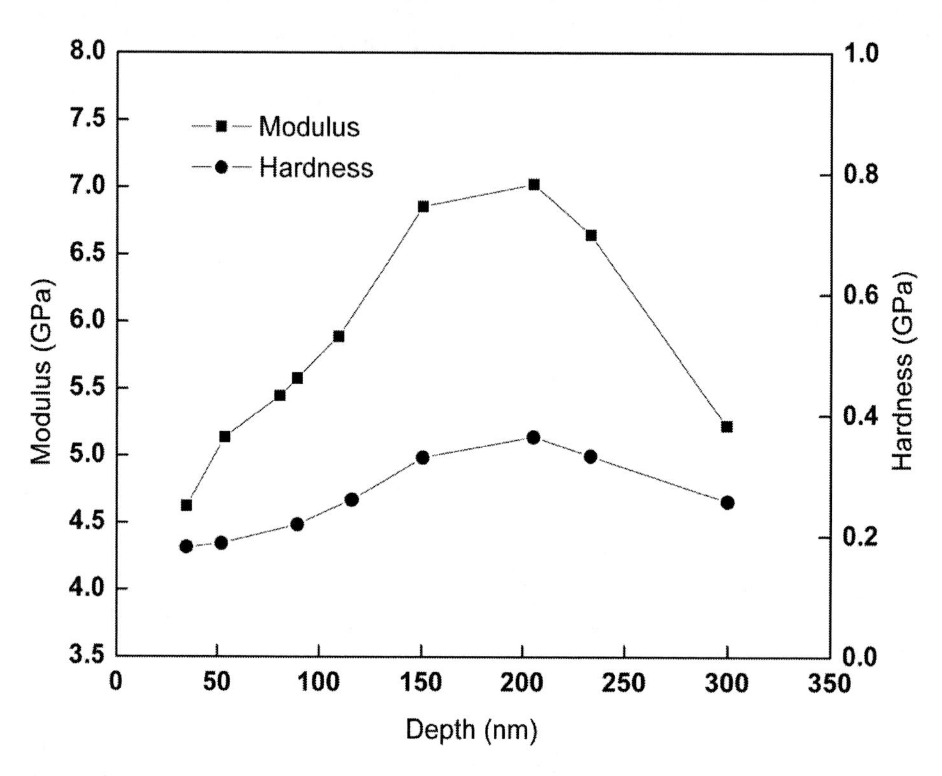

Figure 15. Variation of modulus and hardness with penetration depth for $C_{70}P_{0.1}$.

SEM micrograph shows heterogeneous distribution of the PANI from surface to bulk. Hence in order to understand the variation of elastic modulus and hardness across the cross section, a plot of hardness and elastic modulus with penetration depth is drawn for grafted polymer (Figure 15). Hardness and elastic modulus of the grafted polymer slightly increases as the penetration depth increases and at 200 nm depth modulus decreases. As we go from surface to bulk, concentration of PANI (having poor mechanical property) decreases (Figure 12), in other words chitosan starts dominating in bulk, hence improvement in mechanical property up to 200 nm. After that again PANI concentration in the film increases.

3.3. Grafting Mechanism

Many researchers have exploited the free radical copolymerization of biopolymers with vinyl monomers using ammonium peroxy disulphate (APS) initiator (Singh et al., 2004; 2005a; 2006; Tiwari, 2007; Tiwari and Singh, 2008). A chain mechanism (Singh et al., 2005b) is involved due to the formation of sulfate ion radicals ($SO_4^{-\cdot}$)(Scheme 1), which are well-known ion chain carriers for the graft copolymerization. At the same time, APS stimulates the oxidative polymerization reaction of aniline (Gospodinova and Terlemezyan, 1998) to form PANI radicals. Further APS reacts with chitosan backbone resulting in the formation of chitosan radicals (Prashanth and Tharanathan, 2007; Tian et al., 2004; Gospodinova and Terlemezyan, 1998). The detailed mechanism of grafting is shown in the scheme 2. The possible molecular structure of the grafted polymer is shown in the scheme 3.

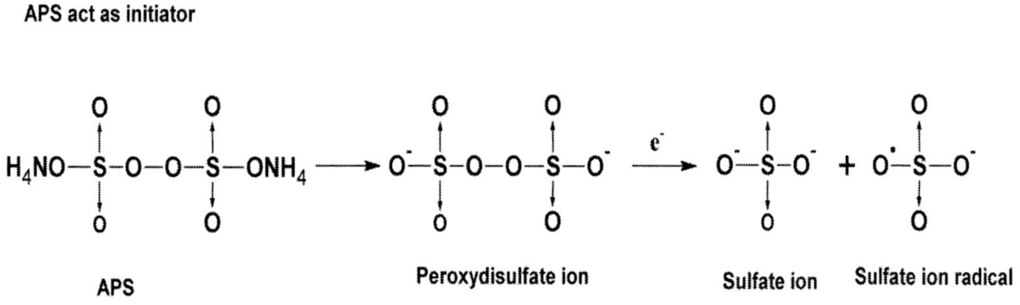

APS act as initiator

$$H_4NO-\overset{O}{\underset{O}{S}}-O-O-\overset{O}{\underset{O}{S}}-ONH_4 \longrightarrow {}^{-}O-\overset{O}{\underset{O}{S}}-O-O-\overset{O}{\underset{O}{S}}-O^{-} \xrightarrow{e^{-}} {}^{-}O-\overset{O}{\underset{O}{S}}-O^{-} + {}^{\bullet}O-\overset{O}{\underset{O}{S}}-O^{-}$$

APS | Peroxydisulfate ion | Sulfate ion | Sulfate ion radical

Scheme 1. Generation of sulfate ion radicals.

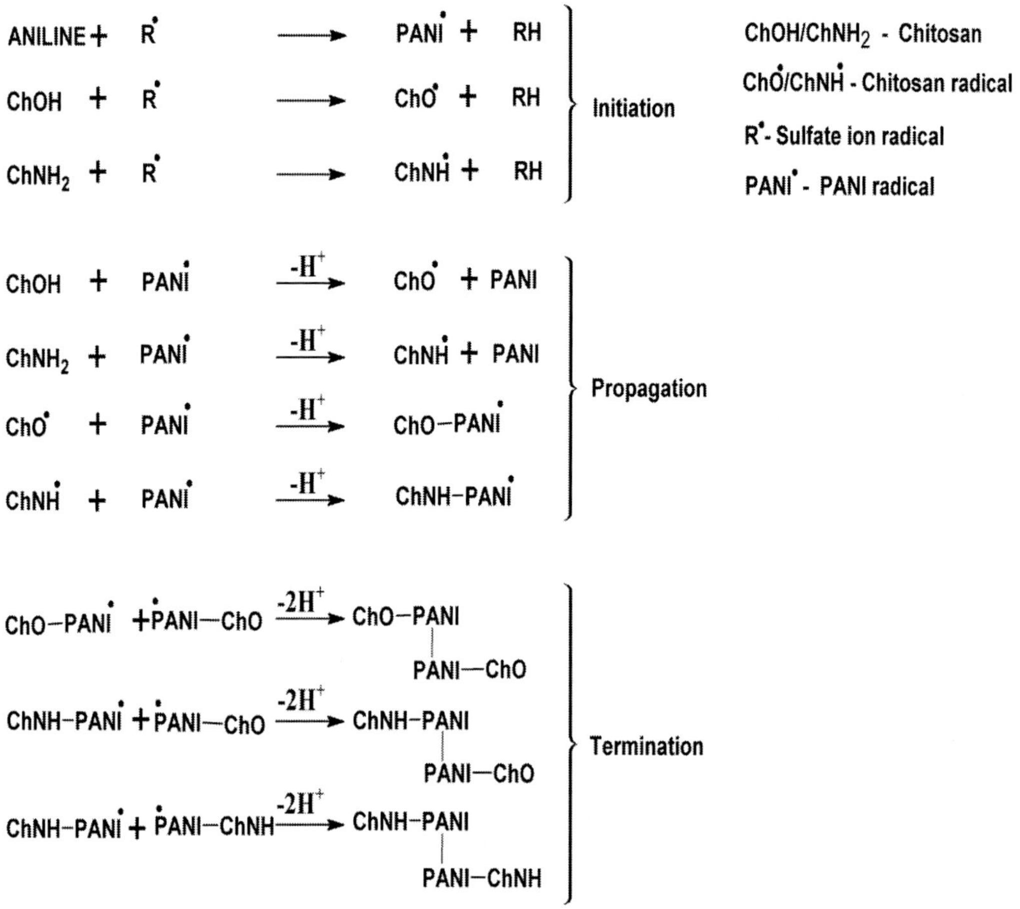

ChOH/ChNH$_2$ - Chitosan
ChO$^\bullet$/ChNH$^\bullet$ - Chitosan radical
R$^\bullet$- Sulfate ion radical
PANI$^\bullet$ - PANI radical

Initiation

$$\text{ANILINE} + R^\bullet \longrightarrow \text{PANI}^\bullet + RH$$
$$\text{ChOH} + R^\bullet \longrightarrow \text{ChO}^\bullet + RH$$
$$\text{ChNH}_2 + R^\bullet \longrightarrow \text{ChNH}^\bullet + RH$$

Propagation

$$\text{ChOH} + \text{PANI}^\bullet \xrightarrow{-H^+} \text{ChO}^\bullet + \text{PANI}$$
$$\text{ChNH}_2 + \text{PANI}^\bullet \xrightarrow{-H^+} \text{ChNH}^\bullet + \text{PANI}$$
$$\text{ChO}^\bullet + \text{PANI}^\bullet \xrightarrow{-H^+} \text{ChO-PANI}^\bullet$$
$$\text{ChNH}^\bullet + \text{PANI}^\bullet \xrightarrow{-H^+} \text{ChNH-PANI}^\bullet$$

Termination

$$\text{ChO-PANI}^\bullet + {}^\bullet\text{PANI-ChO} \xrightarrow{-2H^+} \text{ChO-PANI(-PANI-ChO)}$$
$$\text{ChNH-PANI}^\bullet + {}^\bullet\text{PANI-ChO} \xrightarrow{-2H^+} \text{ChNH-PANI(-PANI-ChO)}$$
$$\text{ChNH-PANI}^\bullet + {}^\bullet\text{PANI-ChNH} \xrightarrow{-2H^+} \text{ChNH-PANI(-PANI-ChNH)}$$

Scheme 2. Graft copolymerization of PANI onto chitosan.

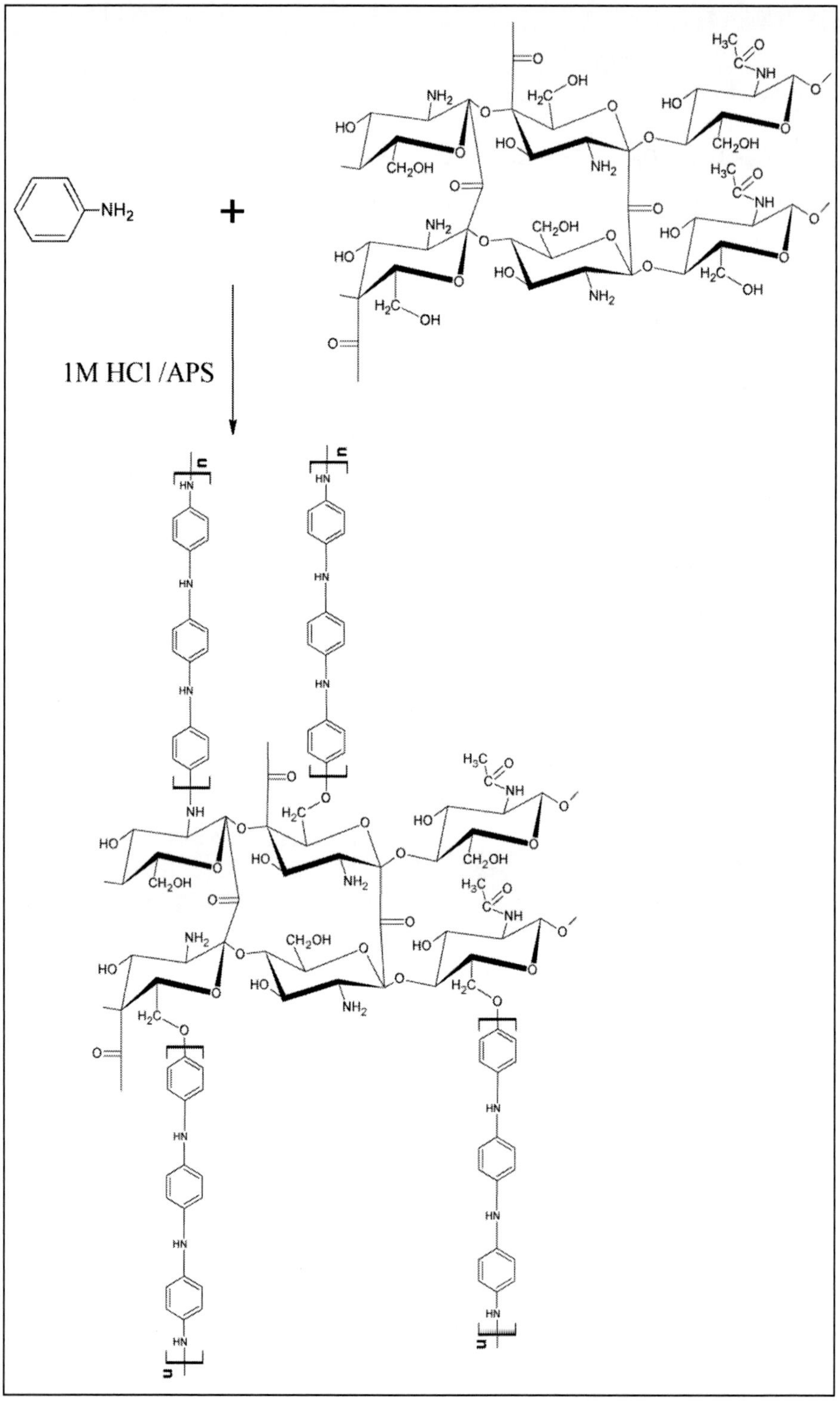

Scheme 3. Grafting reaction.

Table 6. Surface and Volume conductivity of grafted polymer doped with 1M HCl for different time intervals

Sample code	Surface Conductivity (S/cm2)					Volume Conductivity (S/cm)				
	0 hr	2 hrs	4 hrs	6 hrs	24 hrs	0 hr	2 hrs	4 hrs	6 hrs	24 hrs
C25P0.1	4.86×10-11	5.05×10-3	6.57×10-3	7.92×10-3	8.69×10-3	4.81×10-10	6.86×10-7	9.08×10-7	2.69×10-6	4.00×10-6
C25P0.25	2.5×10-11	8.79×10-3	9.57×10-3	1.28×10-2	1.49×10-2	3.89×10-10	1.49×10-6	2.82×10-6	2.83×10-6	4.72×106
C25P0.5	1.72×10-11	1.06×10-2	2.38×10-2	2.85×10-2	3.53×10-2	2.92×10-10	2.86×10-6	4.57×10-6	8.83×10-6	1.17×10-5
C50P0.1	2.74×10-11	1.13×10-3	1.53×10-3	4.54×10-3	--	3.84×10-10	1.81×10-6	2.29×10-6	2.42×10-6	--
C50P0.25	3.03×10-11	1.05×10-3	9.87×10-3	1.14×10-2	--	6.06×10-10	1.05×10-6	6.85×10-6	9.87×10-6	--
C50P0.5	2.04×10-11	4.55×10-3	6.71×10-3	7.90×10-3	--	2.24×10-10	2.69×10-6	6.71×10-6	9.09×10-6	--

3.4. Electrical Conductivity Studies

3.4.1 DC Conductivity

D.C electrical conductivity values of undoped and doped grafted polymers are recorded by two probe method using Keithley electrometer (model 6517A). The surface and volume conductivity of grafted chitosan increases with increase in the percentage of grafting. Also conductivity decreases as the crosslinking density increases (Table 6). Highest surface conductivity observed is 3.53×10^{-2} S/cm for $C_{25}P_{0.5}$. But volume conductivity is found to be 3 orders less than that of surface conductivity. That may be due to decrease in the penetration of monomer on to chitosan matrix due to increase in the crosslinking fraction , supported by the SEM results.

In order to understand the conduction mechanism, I-V characteristic data at different temperatures for grafted polymers are analyzed on the basis of variable range hopping model. Figure 16 (a-d) shows the ln I vs. ln V plots at various temperatures for crosslinked chitosan and grafted polymer undoped / doped with HCl for different time intervals.

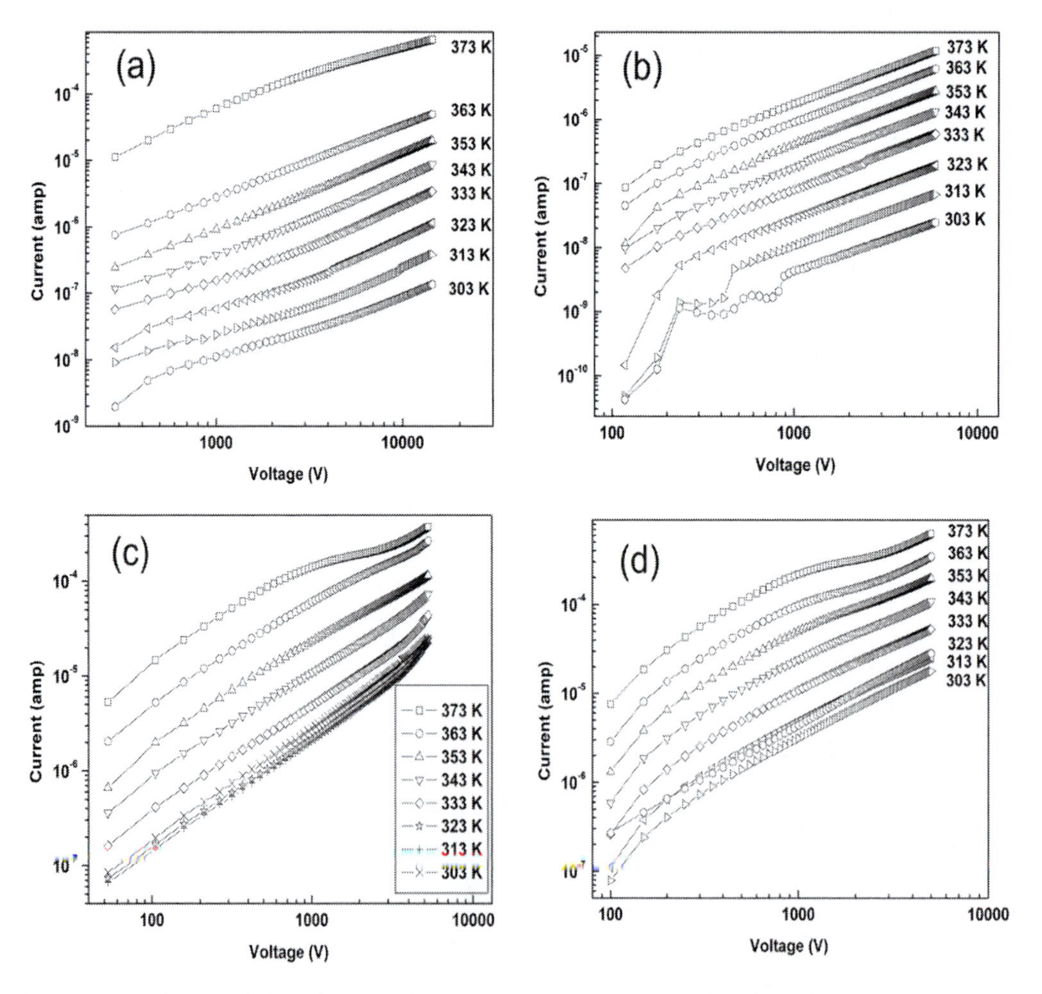

Figure 16. I-V characteristics of (a) C_{25}, (b)undoped $C_{25}P_{0.5}$,(c) $C_{25}P_{0.5}$ doped for 2 hrs and (d) $C_{25}P_{0.5}$ doped for 24 hrs.

The current increases nonlinearly with the applied voltage and slope of the plots are more than one clearly indicating that conduction is non-ohmic. Further from Tables 6, it is seen that conductivity of grafted polymer increases from insulating region to semiconducting region on doping with 1M HCl. Hence possibility of ohmic conduction is ruled out. In addition to this UV-vis – NIR spectra show the formation of charge carriers polarons and bipolarons. These charge carriers can not move so easily as electron moves, in other words these charges can hop along the back bone of polymer with the assistance of phonons (Mott, 1969; Chandrasekhar, 1999). Charge carrier hopping mechanism is well established in the case of PANI in literature (Chandrasekhar, 1999). Since these graft polymers can be considered as disordered organic semiconductors, variable range hopping conduction mechanism is analyzed. Figures 17(a-c) show 1D-VRH, 2D-VRD and 3D-VRH plots respectively.

Figure 17 shows linear behavior for all samples, from least square fitting analysis it is found that crosslinked chitosan and undoped grafted polymer follow 1D-VRH and doped grafted polymer 3D-VRH mechanism. Grafted polymers are semi-crystalline in nature, which is confirmed from the XRD analysis. Hence long range order is missing in the grafted polymer and due to this charge carriers are localized. In random electric fields, charge transport takes place via phonon assisted hopping between localized sites.

From the slope of the VRH plot Mott's characteristic temperature T_o is determined (Table 7). T_o is a parameter representing the potential barrier for electron to hop between the chains and a measure of disorder in the system.

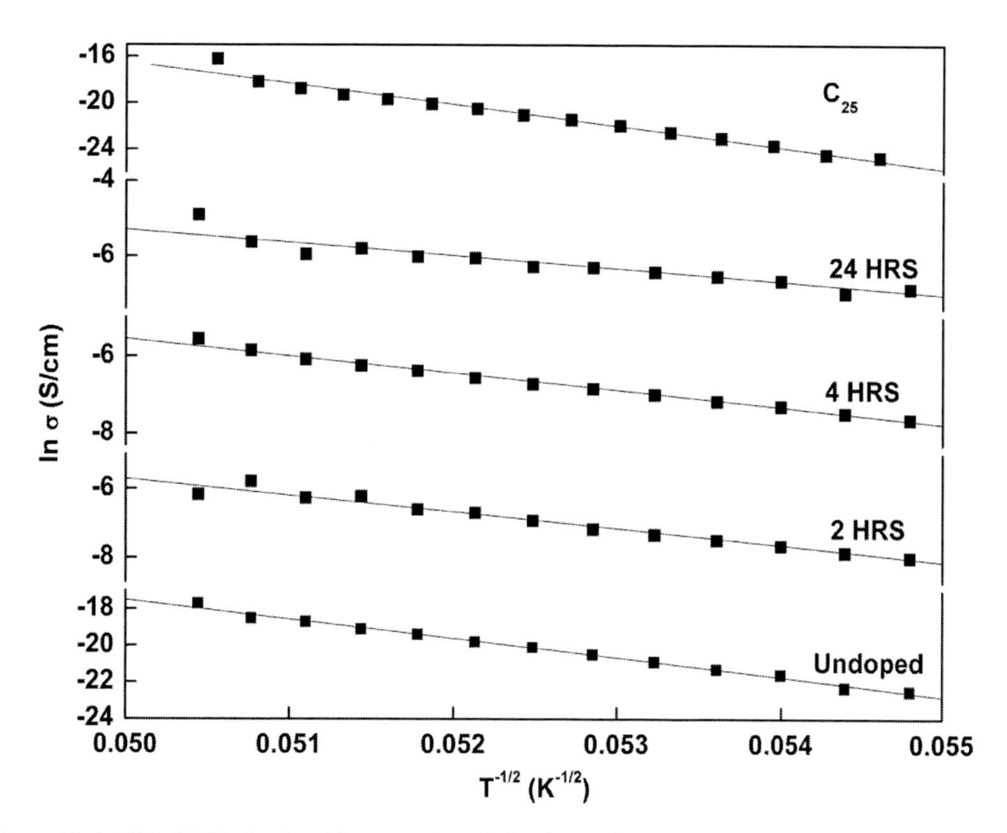

Figure 17. (a) 1D - VRH plot for chitosan and grafted polymer ($C_{25}P_{0.5}$).

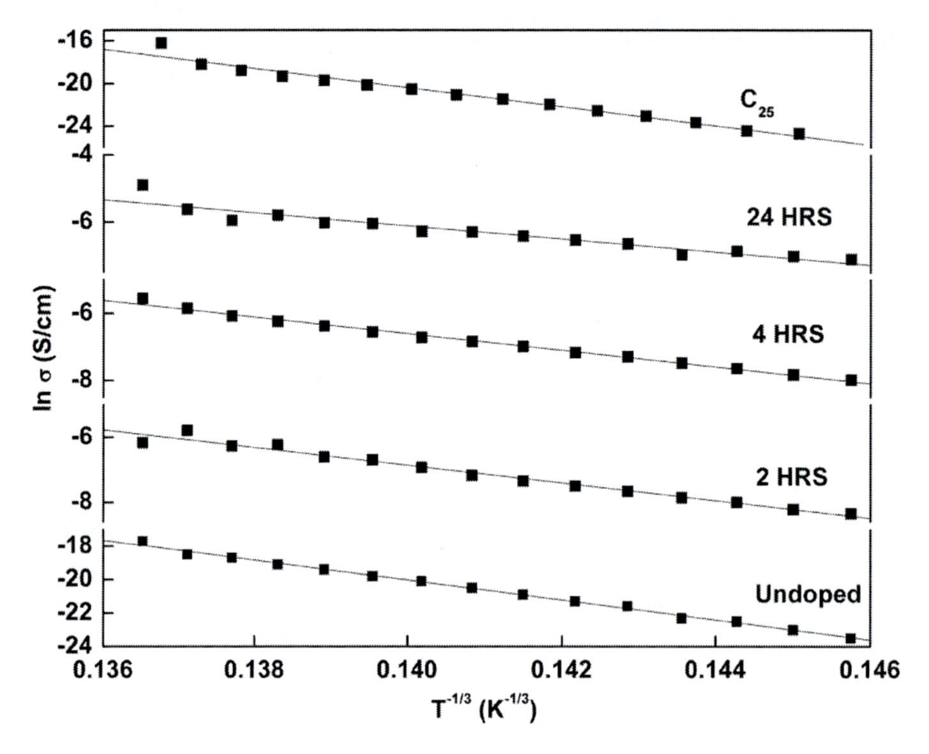

Figure 17. (b) 2D-VRH plot for grafted polymer ($C_{25}P_{0.5}$).

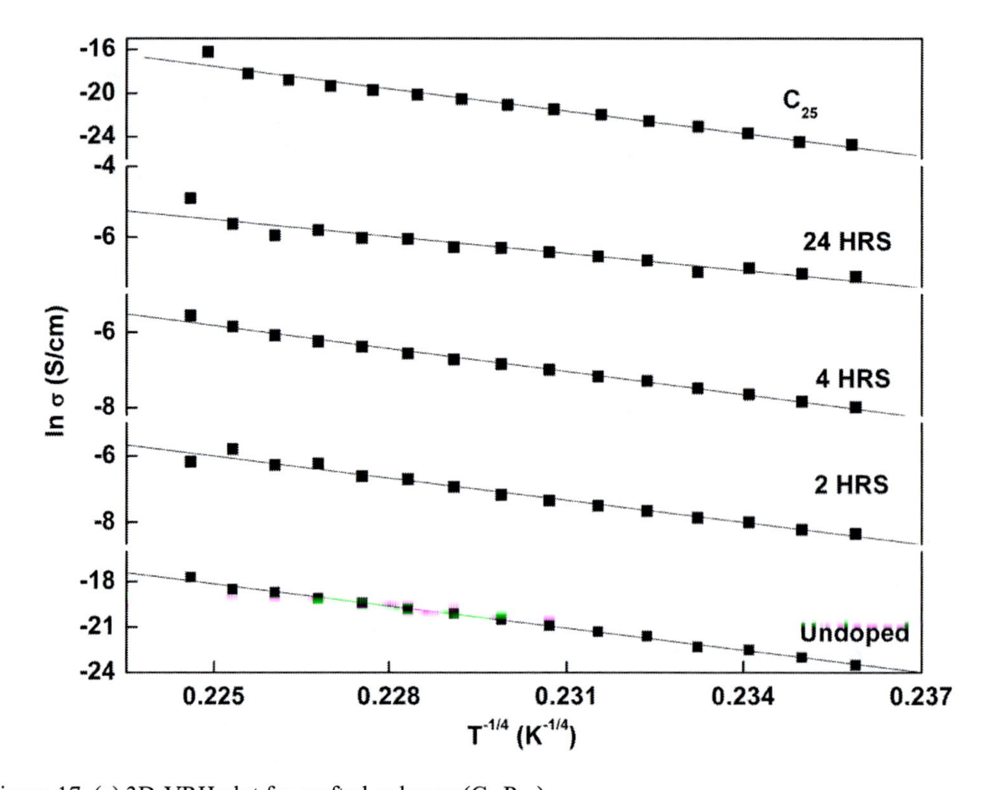

Figure 17. (c) 3D-VRH plot for grafted polymer ($C_{25}P_{0.5}$).

Table 7. Different physical parameters of grafted polymers for various doping time and different monomer concentrations

Sample	T_0 (K)	γ	W_M (eV)
$C_{25}P_{0.5}$ (undoped)	1.12×10^6	1/2	0.315
$C_{25}P_{0.5}$ (2hrs doped)	4.93×10^4	1/4	0.239
$C_{25}P_{0.5}$ (4 hrs doped)	4.08×10^4	1/4	0.255
$C_{25}P_{0.5}$ (24 hrs doped)	2.53×10^4	1/4	0.243
C25P0.1 (undoped)	1.36×10^6	1/2	0.29
C25P0.1 (24hrs doped)	1.59×10^4	1/4	0.286
C25P0.25 (24hrs doped)	4.45×10^4	1/4	0.245
C50P0.1 (24 hrs doped)	2.10×10^4	1/4	0.244
C50P0.25 (24hrs doped)	1.72×10^4	1/4	0.210
C50P0.5 (24hrs doped)	1.21×10^4	1/4	0.25
C_{25}	1.72×10^6	1/2	0.494

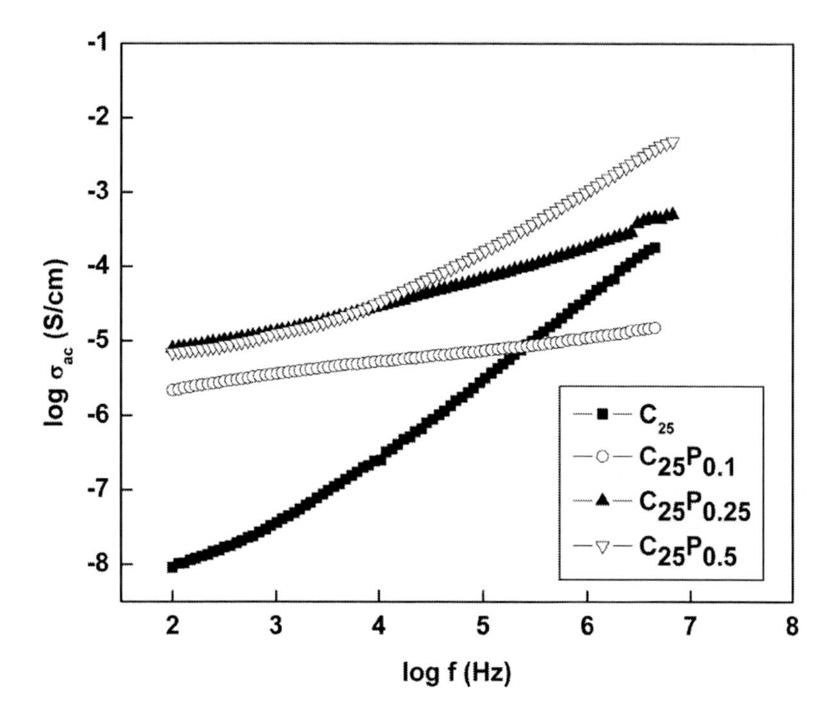

Figure 18. Variation of ac conductivity with frequency for different monomer concentrations.

T_o is inversely proportional to localization length. T_o value is very high for undoped or insulating grafted polymers. Value of the T_0 decreases as the doping time increases or as the conductivity of grafted polymer increases.

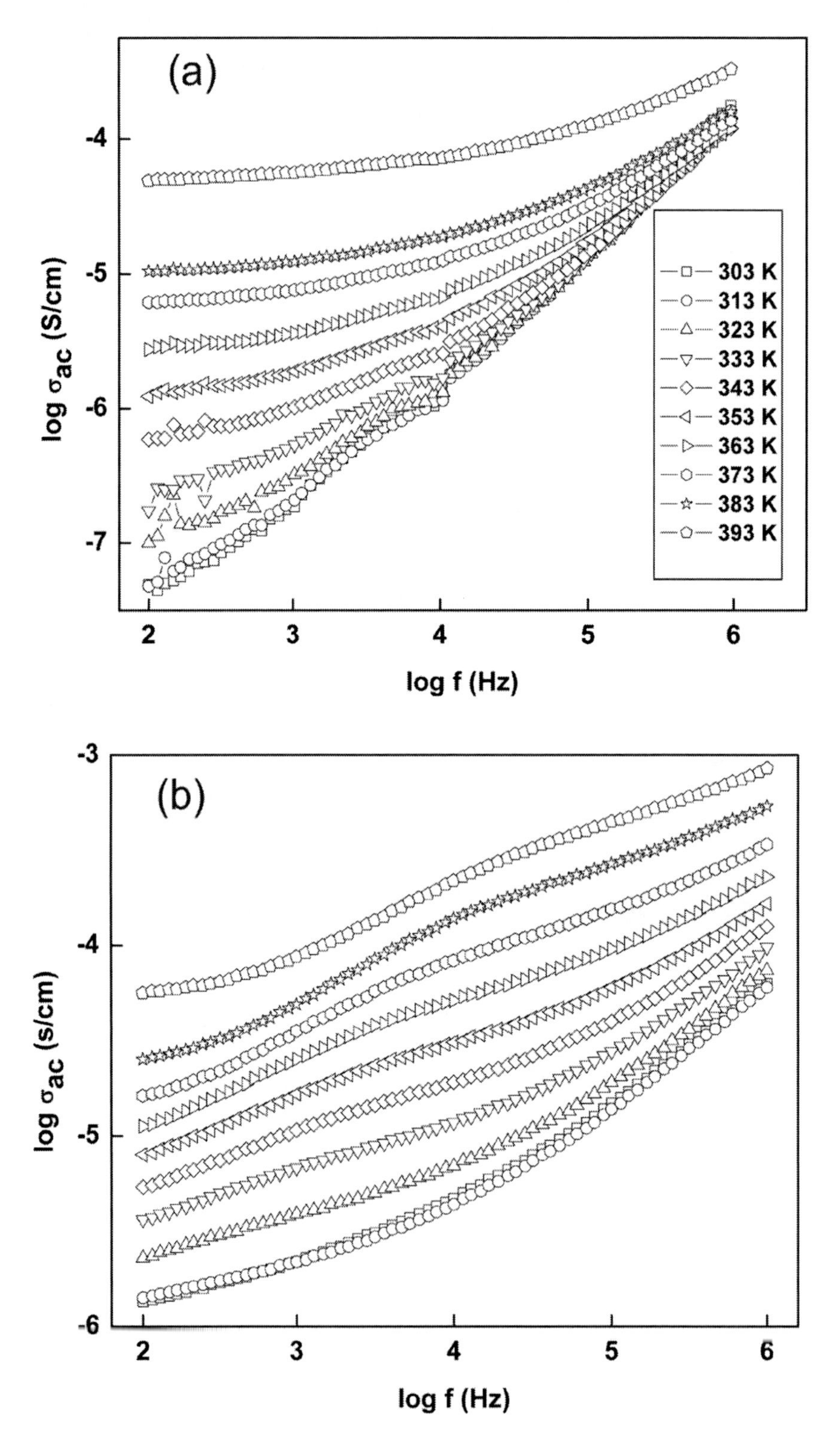

Figure 19. Variation of ac conductivity with frequency for various temperatures (a) undoped and (b) doped grafted polymer ($C_{25}P_{0.5}$).

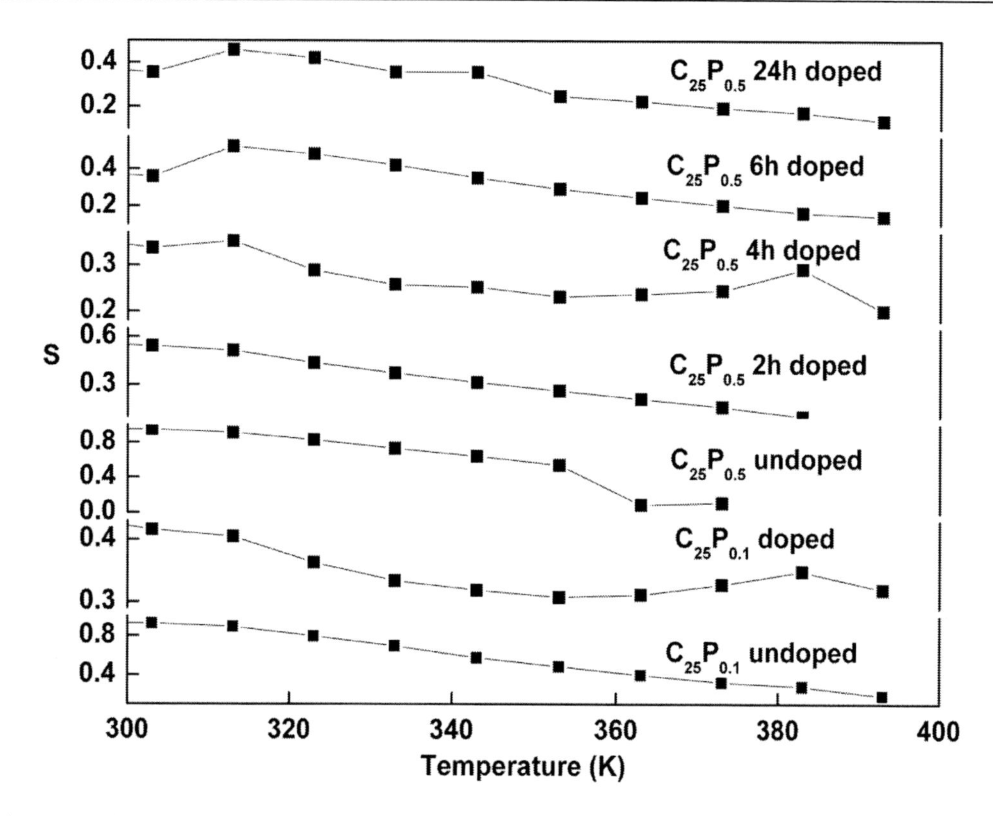

Figure 20. The variation of '*S*' with temperature for different grafted polymers.

3.4.2. AC Conductivity

The frequency dependent conductivity of grafted polymer is presented in the Figure 18. Conductivity increases monotonically with frequency. As the concentration of aniline increases, conductivity of the grafted polymer increases due to the increase in the polaron and bipolaron concentration in the grafted polymer. Figure 19 represents the frequency dependent conductivity of undoped and doped grafted polymer ($C_{25}P_{0.5}$) for various temperatures. Log-log plot shows increase in ac conductivity as the frequency increases. At low frequency, increase in the conductivity is slow and as frequency increases, conductivity increases at a faster rate. At higher temperatures the variation of ac conductivity with frequency becomes more linear. The conductivity values converge at higher frequency in the case of undoped grafted polymer. But for doped grafted polymer ac conductivity linearly increases with frequency as the temperature increases and does not converge.

The total frequency dependent conductivity at a given temperature can be expressed as

$$\sigma(f) = \sigma_{dc} + \sigma_{ac}(f) \tag{4}$$

where σ_{dc} is the dc electrical conductivity and $\sigma_{ac}(f)$ the ac conductivity.

The frequency variation of $\sigma_{ac}(f)$ at a particular temperature for a disordered semiconductor obeys the following power law

$$\sigma_{ac}(f) = Af^{s} \tag{5}$$

where A is a constant dependent on temperature and s is the index which is characteristic of the type of conduction mechanism/relaxation mechanism in the material and $s \leq 1$ (Elliott, 1987). The value of s at various temperatures can be determined from the slope of log $\sigma(f)$ versus log f plot. The calculated '*s*' values for various grafted polymer at different temperatures is presented in the Figure 20. '*s*' value decreases as temperature increases. The temperature dependence of '*s*' is very complex. Various theoretical models for ac conductivity have been applied in order to explain the temperature dependence of '*s*' (Elliott, 1987). The value of '*s*' is less than 1 for all samples. The electron tunneling model suggests that *s* is independent of *T*, but dependent on f (Efros and Philos, 1981). In the case of small polaron tunneling process, *s* increases (Elliott, 1987), whereas for large polaron tunneling, '*s*' decreases up to a certain temperature and then increases for further increase of temperature (Long, 1982). However the value of '*s*' decreases with increase of temperature. Thus these models are not suitable to explain the observed results.

On the other hand, in the correlated barrier-hopping model (CBH) (Elliott, 1987), the charge carrier hops between the sites over the potential barrier separating them. The frequency exponent *s* for such a model is

$$s = 1 - \frac{6kT}{W_M - kT \ln\left(\frac{1}{\omega \tau_0}\right)} \tag{6}$$

where W_M is the effective barrier height or polaron binding energy, τ_o is the characteristic relaxation time which is in the order of an atom vibrational periodic $\tau_o \sim 10^{-13}$ s. For large values of W_M/kT, '*s*' is nearly unity. Also above equation predicts that s decreases with increasing temperatures at large W_M/kT. The above demonstrations lead to the approach

$$1 - s = {6kT}/{W_M} \tag{7}$$

From the slope of variation of '*s*' with temperature (Figure 20) the value of W_M, has been calculated (Table 7). In addition to this an increase in ac conductivity with frequency and temperature indicates that charge carriers may be transported by hopping through the defect sites along the polymer chain (Jonscher, 1967). This temperature and frequency dependence of ac conductivity suggests that the CBH conduction mechanism is dominating in the present system.

Figure 21 shows the frequency dependence of dielectric constant for grafted polymer for various aniline concentrations at room temperature. At low frequencies dielectric constant is high and as the frequency increases, dielectric constant decreases. Frequency dependence of dielectric constant is more pronounced at low frequencies because interfacial polarization plays an important role at lower frequencies. This polarization will arise only when the phases with different conductivities are present (Prakash et al., 1996).

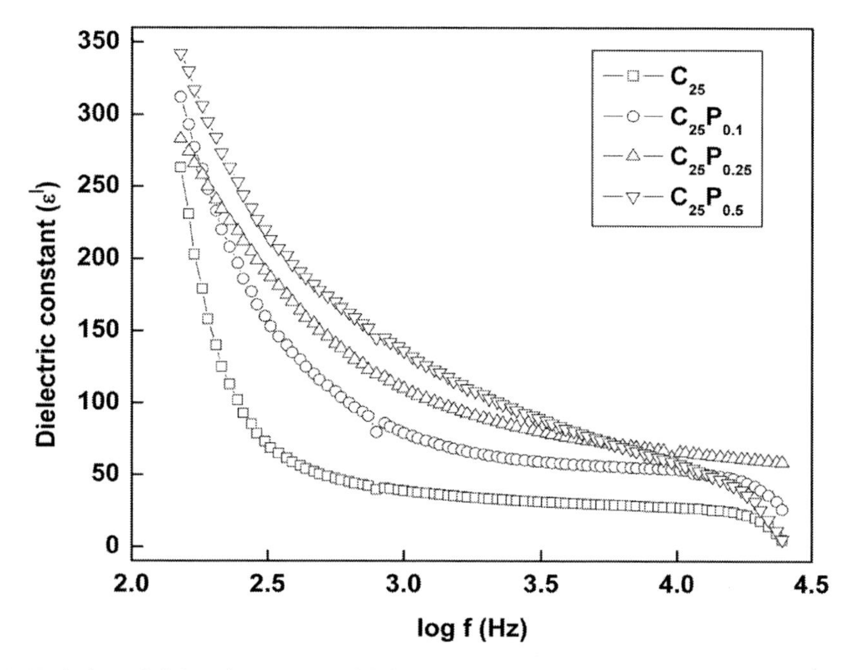

Figure 21. Variation of dielectric constant with frequency for different monomer concentrations.

A relatively high dielectric constant at low frequency and fast decrease with frequency are characteristics of a conducting polymer and consistent with earlier reports (Matveena, 1996; Lian et al., 1995). This may be due to the large effective size of metallic islands and easy charge transfer through well ordered polymer chains in disordered regions as suggested by Joo et al. (1998). As the aniline concentration increases the dielectric constant of the grafted polymer increases due to the increase in the grafting percentage of PANI onto chitosan.

Figure 22 and 23 show dependence of dielectric constant with frequency and temperature respectively for undoped and doped grafted polymer. At low temperature dielectric constant is almost independent of frequency, where as at high temperature it increases as frequency decreases, which reveals that the system exhibits strong interfacial polarization at low frequency. The dielectric permittivity increases with increase of temperature and at a particular temperature it decreases monotonically with increasing frequency. This may be due to the tendency of induced dipoles in macromolecules to orient themselves in the direction of the applied field when the frequency of alternation is low. However, at high frequencies the induced dipoles will hardly be able to orient themselves in the direction of the applied field and hence the dielectric permittivity decreases (Reicha et al., 1991). In order to understand the relaxation in the system, tanδ with frequency and temperature (inset of Figure 22 and 23) plot is presented and no significant peak is observed. Further it should be noted that on doping, dielectric constant increases due to increase in polaron and bipolaron concentration in the grafted polymer. This behavior points to a Debye-type dielectric dispersion (Cde and Cole, 1941; Frohlieh, 1958) characterized by a relaxation temperature. The magnitude of the dielectric dispersion depends on the temperature because the dielectric dispersion is related to σ_{ac}, which in turn depends on the temperature. Further the dielectric constant and ac conductivity as a function of temperature do not show a maximum, which confirms the 3D-VRH mechanism for doped grafted polymers.

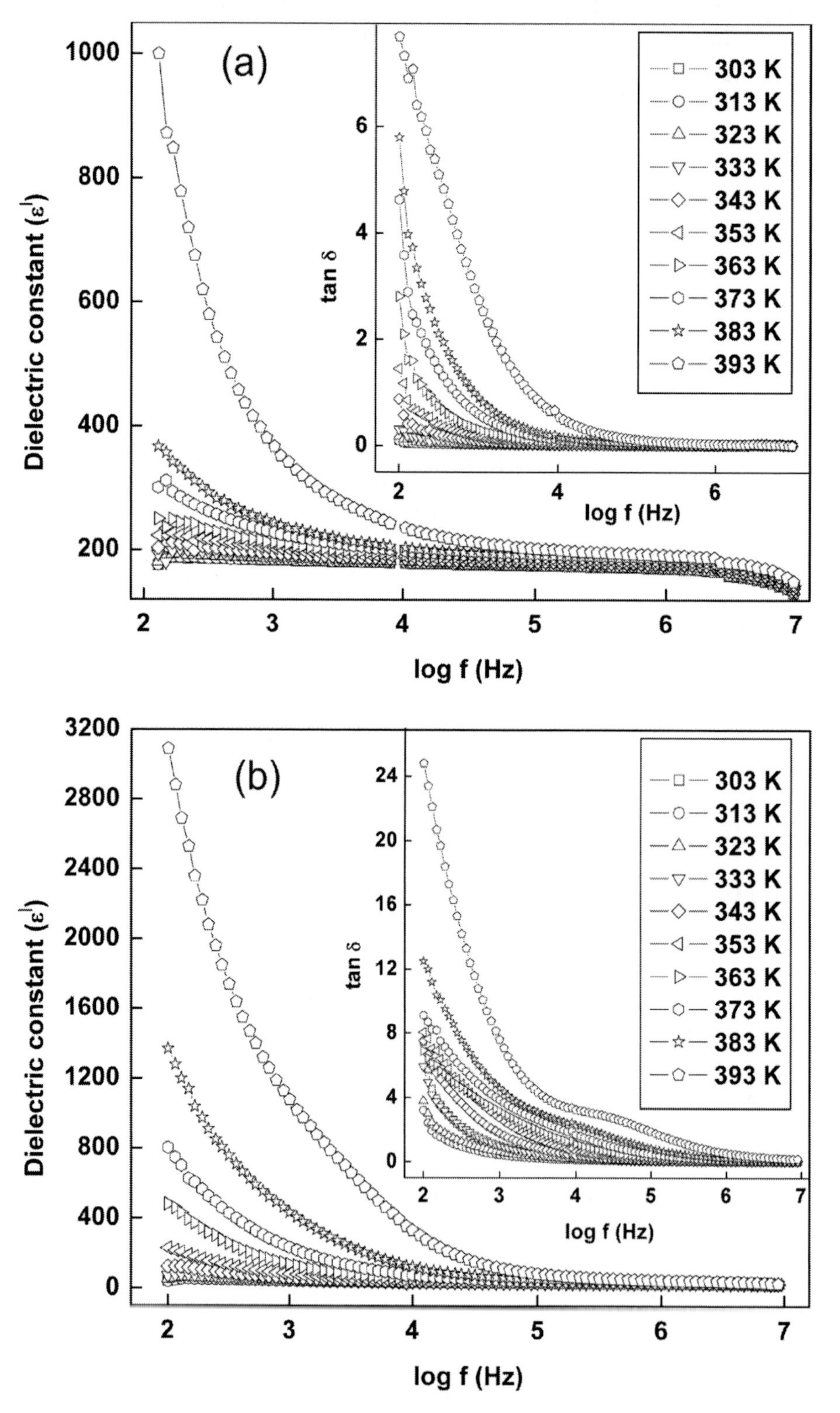

Figure 22. Variation of dielectric constant and tan δ with frequency for (a) undoped grafted polymer ($C_{25}P_{0.5}$) (b) 1 M HCl doped grafted polymer ($C_{25}P_{0.5}$).

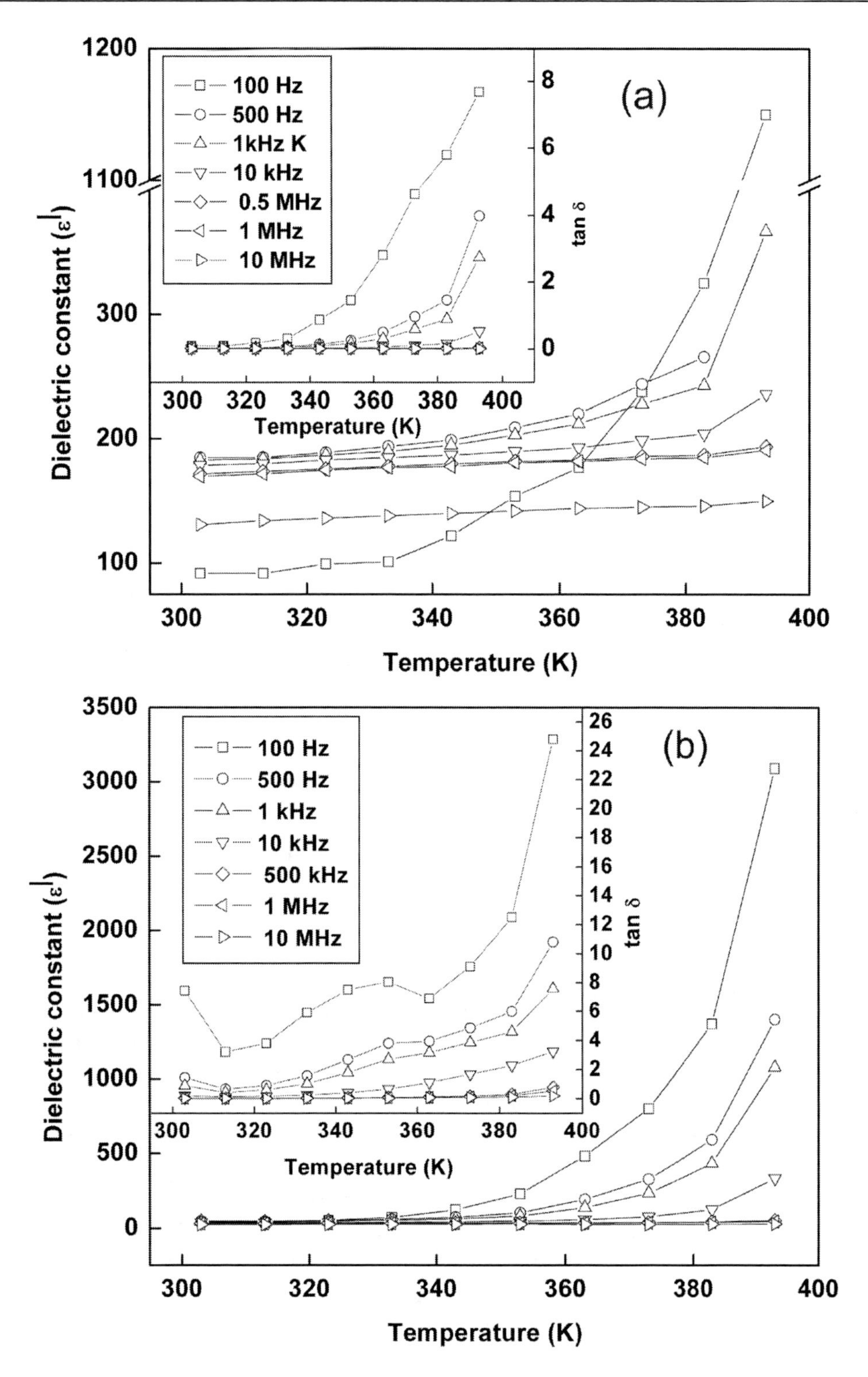

Figure 23. Variation of dielectric constant and tan δ with temperature for (a) undoped grafted polymer ($C_{25}P_{0.5}$) (b) 1 M HCl doped grafted polymer ($C_{25}P_{0.5}$).

4. CONCLUSION

Chitosan which normally undergoes degradation, when exposed to e-beam irradiation and gamma rays can be crosslinked in the presence of CCl_4. From the dissolution study it is seen that electron beam crosslinked chitosan has better crosslinked fraction compared to gamma ray crosslinked chitosan.Ionic capacity measurement confirms the presence of NH_2 group, which is further supported by FTIR results. Mechanical properties of crosslinked chitosan are increased with irradiation dose due to the increase in crosslinking fraction. Electron beam irradiated sample shows better mechanical properties compared to gamma irradiated samples. Hence electron beam is more efficient in crosslinking chitosan and has got better mechanical properties compared to γ-ray corsslinked chitosan.

Chemical grafting of PANI on to the radiation crosslinked chitosan is carried out for varying irradiation doses and concentration of aniline. Grafting percentage increases as the monomer concentration increases and decreases with increase in crosslinking fraction. This may be due to the decrease in the penetration of monomer on to the chitosan matrix with increase in crosslinking density. Further it is found that percentage of grafting is more in the case of gamma crosslinked chitosan compared to electron beam crosslinked chitosan. But it is difficult to handle gamma crosslinked chitosan due to its poor mechanical property. UV-vis-NIR studies show, absorption peaks at ≈330nm, ≈600nm and small peak at ≈1951nm in the undoped grafted polymer. On doping additional peaks are observed at ≈750nm, ≈1465nm and an absorption edge at ≈440nm due to increase in polaronic concentration. This is verified from SEM (cross sectional view) of grafted polymer. FTIR studies on grafted polymer show all characteristic peaks of chitosan as well as PANI. The band at ≈825 cm^{-1}, ≈1300 cm^{-1}, ≈1500 cm^{-1} and 1568 cm^{-1} in doped and undoped grafted polymer is due to PANI. A new band appears at ≈1598 cm^{-1} in the case of doped polymer due to protonic doping. The N-H band belonging to chitosan is observed around ≈1655cm^{-1}in chitosan and is shifted to lower wave number in the case of grafted polymer indicating strong interaction between chitosan and PANI.

Electrical properties of grafted polymer improved after doping with 1M HCl. The change in volume conductivity is from 10^{-11} to 10^{-5} S/cm and surface conductivity from 10^{-10} to 10^{-2} S/cm. This may be due to the lower penetration of monomer onto crosslinked chitosan matrix. This is supported by SEM studies; grafting of PANI takes place more on the surface, than in the bulk. Nanoindentation studies show good mechanical properties for grafted polymer. X-ray diffractogram shows all characteristics peaks of chitosan and PANI. From the I-V characteristics, conduction mechanisms have been predicted and results obtained suggest, 1D-VRH conduction mechanism for undoped graft polymer and 3D-VRH conduction mechanism for doped grafted polymer. The ac conductivity of grafted polymer obeys power law. The frequency exponent's' decreases with increase in temperature. Such behavior of 's' has been interpreted by the CBH model. At low temperature dielectric constant shows a small variation while a large change is observed at higher temperature. At constant temperature, the dielectric constant increases with decreasing frequency, which is consistent with the Debye type dispersion. The dielectric constant and ac conductivity as a function temperature do not show a maximum, which further confirms the 3D-VRH mechanism of doped grafted polymers.

To conclude, the conductivity, mechanical properties and stability of grafted polymer under ambient condition is sufficient to make it an interesting material for diverse applications. All other previous reported PANI grafted chitosan has highest surface conductivity of 10^{-2} S/cm. But they were able to get flexible film of grafted polymer having surface conductivity of maximum10^{-4} S/cm only. In the present case, flexible film of PANI grafted radiation crosslinked chitosan with surface conductivity of 10^{-2} S/cm is obtained. Present grafted polymer can be used for applications, such as sensors and electrodes of polymeric batteries, since it is insoluble in the whole range of pH.

REFERENCES

Angelopoulos, M; Ray, A ; MacDiarmid, A.G. *Synth. Met.* 1987, 21, 21-30.

Andreatta, A; Cao, Y; Chiang, J.C; Heeger, A.J; Smith, P. *Synth. Met.* 1988, 26, 383-389.

Bhadra, S; Singha, N.K; Khastgir, D. *J. Appl. Polym. Sci.* 2008a, 107, 2486-2493.

Cole, K.S; Cole, R.H. *J. Chem. Phys.* 1941, 9, 341-352.

Chandrasekhar, P., "*Conducting polymers, fundamentals and applications: Apractical approach*", Kluwer Academic Publishers, USA, 1999.

Chmielewski, A.G; Haji-Saeid, M; Shamshad, A. *NIMB.* 2005, 236, 38-43.

Clelend, M.R; Parks, L.A ; Cheng, S. *NIMB.* 2003, 208, 66-73.

Cushman, R. J; McManus, P.M; Yang, S. C. *J. Electroanal. Chem.* 1986, 291,335-346.

Dumitriu, S et al., "*Polysaccharides in Medical Applications*", Marcel Dekker Inc, New York, 1996.

Diaz, A. F; Logan, J. A. *J. Electroanal. Chem.* 1980, 111,111-114.

Elliott, S. R. *Adv. Phys.* 1987, 36,135-218.

Ershov, B. G; Isakova, O. V; Rogoshin, S. V; Gamazazade, A. I; Leonova, E. *U. Dokl. Akad. Nauk. SSSR.* 1987, 295, 152 (in Russian).

Frohlieh, A. "*Theory of Dielectrics*",Oxford University Press, London, 1958.

Fros, A.L. *Philos. Mag. B.* 1981, 43,829.

Gospodinova, N; Terlemezyan, L. *Prog. Polym. Sci.* 1998, 23,1443-1484.

Ismail, Y.A; Shin, S. R; Shin, K. M; Yoon, S. G; Shon, K; Kim, S. I; Kim, S. J. *Sensors and Actuators B-Chemical.* 2008, 129, 834-840.

Jeevananda; Siddaramaiah; Annadurai, V; Somashekar, R. *J. Appl. Poly. Sci.* 2001, 82,383-388.

Joo, J; Long, S.M; Pouget, J.P; Oh, E. J; MacDiramid, A. G; Epstein, A. J. *Phys. Rev. B.* 1998, 57, 9567-9580.

Jonscher, A.K. *Thin Solid Films.* 1967, 1, 213-234.

Khan, R; Dhayal, M. *Biosensor and Bioelectronics.* 2009, 24,1700-1705.

Kim, S.J; Shin, S.R; Spinks, G.M; Kim, I.Y; Kim, S.I. *J. App. Polym. Sci.* 2005a, 96,867-873.

Kim, S.J; Kim, M.S; Shin, S.R; Kim, I.Y; Kim, S.I; Lee, S.H; Lee, T.S; Spinks, G.M. *Smart. Mater. Struct.* 2005b, 14, 889-894.

Kim, S.J; Kim, M.S; Kim, S.I; Spinks, G.M; Kim, B.C; Wallace, G.G. *Chem. Mater.* 2006, 18, 5805-5809.

Lim, L.Y; Khor, E; Koo, O. *J. Biomed. Mater. Res. (Appl Biomater).* 1998, 43, 282-290.

Lian A; Besner S; Dao L H. *Synth. Met.* 1995, 74, 21-27.

Long, A.R. *Adv. Phys.* 1982, 31, 553-637

Matveeva E S. *Synth. Met.* 1996, 79, 127-139.

Mott, N.F. *Advan. Solid. State. phys*. 1969, 9, 22.

Muzzarelli, R.A.A. *"Chitin"*, Pergamon Press: Oxford, UK, 1977.

Oliver, W.C; Pharr G M. *J. Mater. Res.* 1992, 7, 1564-1580.

Prakash, O; Mandal, K.D; Christopher, C.C; Sastry, M.S; Kumar, D. *J. Mater. Sci.* 1996, 31, 4705.

Prashanth, K.V.H; Tharanathan, R.N. *Tren. Food Sci. Tech.* 2007, 18, 117-131.

Prigodin, V.N; Samukhin, A.N; Epstein, A. J. *Synth. Met.* 2004, 141, 155-164.

Pud, A; Ogurtsov, N; Korzhenko, A; Shapoval, G. *Progr. Polym. Sci.* 2003, 28, 1701-1753.

Ramaprasad, A.T; Rao, V; Ganesh Sanjeev; Ramnani, S.P; Sabharwal S. *Synth. Met.* 2008, 159, 1983-1990.

Ramaprasad, A.T; Rao, V; Praveena, M; Ganesh Sanjeev; Ramnani, S.P; Sabharwal, S. *J. Appl. Poly. Sci.* 2009, 111, 1063-1068.

Ramnani, S.P; Chaudhari, C.V; Patil, N.D; Sabharwal, S. *J. Polym. Sci. Part A: Polym. Chem.* 2004, 42, 3897-3909.

Reicha, F.M; Hiti, M.E; Sonbati, A.Z.E; Diab, M.A. *J. Phys. D: Appl.Phys.* 1991, 24, 369-374.

Rorrer, G.A.F; Hsien, T. Y ; Way, J.D. *Industrial and Engineering Chemistry Research.* 1993, *32, 2170.*

Singh, V; Tiwari, A; Tripathi, D. N; Sanghi, R. *J. Appl. Polym. Sci.* 2004, 92, 1569-1575.

Singh, V; Tiwari, A; Sanghi, R. *J. Appl. Polym. Sci.* 2005, 98, 1652-1662.

Singh, V; Tiwari, A; Singh, S.P; Shukla, P.K; Sanghi, R. *React. Funct. Polym.* 2006, 66,1306-1318.

Sixou, B; Travers, J.P; Barthet, C; Guglielmi, M. *Phys. Rev. B.* 1997, 56, 4604-4613.

Sneddon, I.N, *Int. J. Engng. Sci.* 1965, 3, 47

Street, G.B., in *"Handbook of Conducting Polymers";* Skotheim T A (ed.), Marcel Dekker, New York, 1986, p. 265

Tian, F; Liu, Y; Hu, K; Zhao, B. *Carb. Polym.* 2004, 57, 31-37.

Tiwari, A. *J. Macr. Sci.- A., Pure and Applied Chemistry* 2007, 44, 735-745.

Tiwari, A; Singh V. *Express Polym. Lett.* 2007, 1, 308-311.

Tiwari, A. *J. Polym. Res.* 2008, 15, 337-342.

Tiwari, A; Singh S.P. *J. Appl. Polym. Sci.* 2008, 108,1169-1177.

Tiwari, A; Gong S. *Talanta* 2009, 77,1217-1222.

Tourillon, G; Garnier F. *J. Phys. Chem.* 1983, 87, 2289.

Ulanski, P ; Rosiak, J.M. *Int. J. Radiat. Appl. Instrum. Part C. Radiat. Phy.s Chem.* 1992, 39, 53.

Wang, Z.H; Scherr, E. M ; MacDiarmid, A.G; Epstein, A.J. *Phys. Rev. B.* 1992, 45, 4190

Wasikiewicz, J.M; Nagasawa N; Tamada M; Mitomo H; Yoshii F. *NIMB.* 2005, 236, 617-623.

Wenwei, Z; Xiaoguang, Z; Li, Y; Yuefang, Z; Jiazhen, S. *Polym. Degrad. Stab.* 1993, 41, 83-84.

Xia, H; Wang, Q. *J. Appl. Poly. Sci.* 2003, 87, 1811-1817.

Yang, S; Tirmizi, S.A; Burns, A; Barney, A.A; Risen, W.M. *Synth. Met.* 1989, 32,191-200.

Yoshii, F; Zhao, L; Wach, R.A; Nagasawa, N; Mitomo, H; Kume, T. *NIMB.* 2003, 208, 320-324

In: Handbook of Chitosan Research and Applications
Editors: Richard G. Mackay and Jennifer M. Tait
ISBN 978-1-61324-455-5

Chapter 6

BIO-APPLICATIONS OF CHITOSAN

Magdy Elnashar*
Laboratory of Advanced Materials and Nanotechnology,
Polymers Dept., Centre of Excellence for Advanced Sciences,
National Research Centre, Cairo, Egypt

ABSTRACT

Efficient commercial carriers suitable for the immobilization of enzymes are economically expensive. Contrarily, the immobilization techniques would enable the reusability of enzymes for tens of times, thus significantly reducing both enzyme and product costs.

To prepare a novel carrier for enzyme immobilization it is, accordingly advantageous, that the utilizable starting materials are among those already, permitted for pharmaceutical or food industrial use. Carrageenans and chitosans are two commercially available polysaccharide families, belonging to this category having in addition, diverse features and reasonable costs. Unfortunately, chitosan is not individually manipulated, as it has low physical/mechanical stability, while carrageenan, in addition to its low thermal stability, is lacking the active functionalities required to covalently bind enzymes.

In our laboratory, we have succeeded in assembling a combination of the two biopolymers, in such a way to gain the benefits of both, such as the abundance of active functional amino (NH2) groups of chitosans, and being shapeable, while having good thermal stability for carrageenan gel.

Carrageenan gels were treated with protonated polyamines "chitosan" to form a polyelectrolyte complex, which was then followed by glutaraldehyde treatment. The newly developed carrier revealed an outstanding gel's thermal stability as it was augmented from 35 to 95 °C. The novel gel incorporating the aldehydic chemical functionality has been efficaciously manipulated to covalently immobilize enzymes.

FTIR techniques, as well as Schiff's base color development were used to elucidate the structure of the newly grafted carrier biopolymer. FTIR, equally confirmed the incorporation of the aldehydic carbonyl group to the carrageenan coated chitosan at via tracing the band 1720 cm^{-1}. Interestingly, the operational stability retained 97% of the enzyme activity, even after 15 time uses. In brief, the newly developed immobilization

* E-mail: magmel@gmail.com.

methodology is simple and the carriers are economically favored vis à vis the commercially available Eupergit C® or Agaroses®, yet effective and utilizable for the immobilization of other enzymes.

2. INTRODUCTION

Chitosan is a cationic polyamine naturally occurring polymer, obtainable via deacetylation of chitin. The later is the most abundant biopolymer in nature, following cellulose. [1, 2] It is the major constituent of the exoskeleton of the aquatic crustaceous animals (crabs, shrimps, *etc.*) and cell walls of fungi. This natural non-toxic biodegradable polysaccharide is hydrophilic in nature, while morphologically available as solution, flakes, fine powder, beads and fibers. The difference between chitin and chitosan however, reside in the degree of deacetylation (%DA), since the %DA of chitin ranges from 70 to 95%. Most publications use the term chitosan when the degree of deacetylation is more than 70%.

The amino group in chitosan has a pKa value of ~ 6.5; thus, such acid soluble biopolymer is positively charged having a pH value < 6, with a charge density dependent on pH and the %DA-value. This makes chitosan a bio-adhesive in nature that readily binds to negatively charged surfaces, as mucosal membranes. Chitosan, being hypo-allergenic, having a natural anti-bacterial properties, is further manipulated in medical uses, as bandages. [3] The compound is also used for diet, blood clotting and industrial applications as surface coatings and immobilization of enzymes.

The industrial worldwide figure of enzymes in billions (except in medicines) is about € 1.5. However, immobilized enzymes, on solid supports, are economically more favorable than free enzymes (liquid state), since they offer the possibility of continuous flow processing, in such a way that easy recovery and reutilization of the immobilized enzyme, by simple filtration is practically feasible. The reusability of enzymes for tens of times tends to significantly reduce the costs of the enzyme, as well as, the enzymatic products.

Several techniques have been used for enzyme immobilization, including entrapment [4], cross-linking [5, 6], adsorption [7], or an alternative combination of these methods. [8] Industrial surfaces such as glass beads [9], nylon-6 [10], chitosan [11], Eupergit C® (epoxy-activated acrylic beads) and Agaroses® [12] have been variably used for this purpose. However, efficient commercial carriers suitable for the immobilization of enzymes are fairly expensive [13]. For examples, available carriers such as Eupergit C® or Agarose® are sold for € 6,000 and € 3,250 /kg, respectively. Accordingly, the need for a cheaper carrier is still an industrial dream to compensate the high costs of their entire inputs.

To prepare a new carrier for enzyme immobilization it is, consequently, comprehensively advantageous if the startings are initially permitted for use in the pharmaceutical or food industries.

Carrageenans and chitosans are two polysaccharide families belonging to this category that are commercially available and at a reasonable cost (Figure 1a, b). [14]

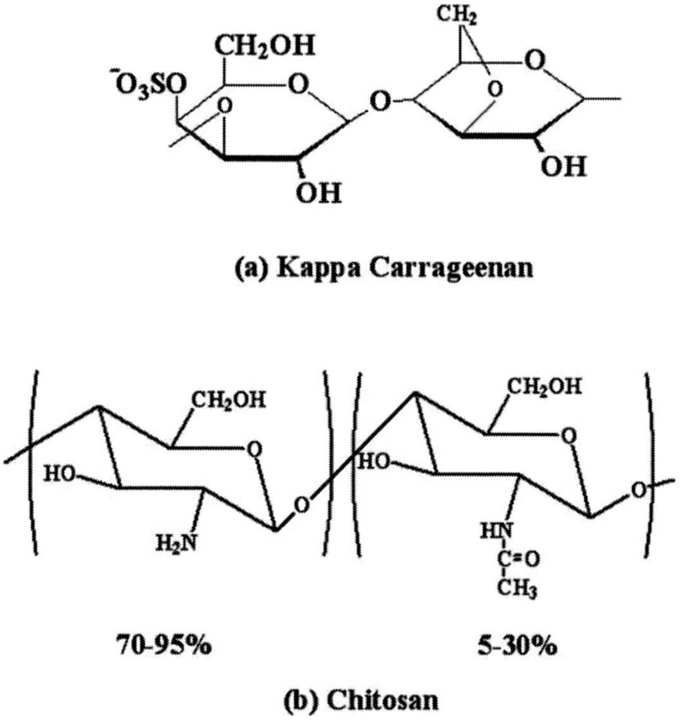

(a) Kappa Carrageenan

70-95% 5-30%

(b) Chitosan

Figure 1. Chemical structures of (a) κ-carrageenan and (b) chitosan.

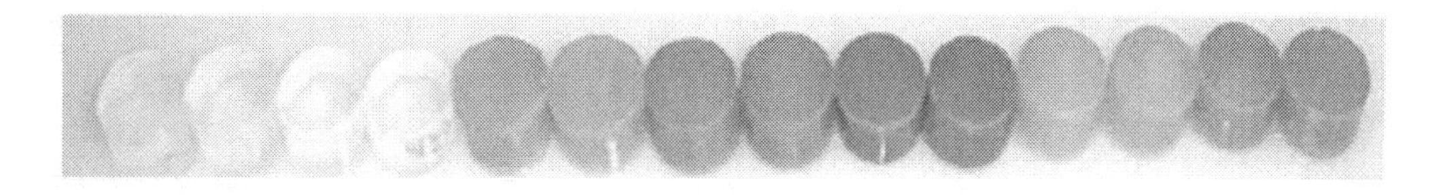

Figure 2. Modified carrageenan gel disks using the parallel plates equipment.

According to many authors, κ-carrageenan is one of the main supports used for cell and enzyme immobilization, namely via entrapment. [15-17]. κ-Carrageenan gel is a naturally abundant polyanion biopolymer, isolated from red seaweed polysaccharides (carrageenose 4′-sulphate). [18] Such polyanion contains one sulfate group (– OSO3-) / molecule, (Figure 1a). A gel can be easily formed by dissolving the carrageenan powder in hot water then setting the solution at room temperature. According to Wang and Kiang (1999), [19] Gel formation involves a coil, to helix, to gel conformational transition changes of the molecule. This advantage enables the carrageenan gel to be shaped in any form, for example, as uniform gel disks, as shown in Figure 2. [20] However, a known drawback of carrageenan is its low thermal stability [21] and lacking of active functional groups. [15-17]

The cationic properties of chitosan are one of its main advantages, since most other polysaccharides are anionic. Typical commercial chitosan is usually revealing an abundance of amino groups (70-95%) (Figure 1b). The chitosan possesses primary amine groups (NH_2) that could be protonated (NH_3^+) in acidic medium thus, to harden the carrageenan gel, by forming a polyelectrolyte complex.

To our knowledge, carrageenan-chitosan systems were never used for covalent immobilization of enzymes, yet were usually used to entrap enzymes, [15-17] which have the major problem of enzyme leakage. For example, Boadi *et al.*, 2001 [22] used alginate and carrageenan to entrap tannase, then crosslinked the gel beads with chitosan, followed by glutaraldehyde. The entrapment technique practically limits their industrial use as supports for enzyme immobilization due to enzyme leakage. Consequently, efforts to immobilize enzymes on novel type of carriers, especially with covalent bonds, are still undertaken in several laboratories. [23-26]

In this chapter, which is based on our experience in the field of immobilization of enzymes, [27-30] we are presenting a breakthrough in the field of carrageenan-chitosan complex, as we succeeded to covalently immobilize an important enzyme, namely, β-galactosidase, on a novel, thermally stable carrageenan coated with chitosan [31, 32] in manipulating glutaraldehyde as a crosslinker and as a mediator.

Chitosan has been chosen for this Treatment Due to Two Important Advantages

The first, is based on the study made by Chao *et al.*, 1986 [33], where he used different amines to harden carrageenan and conclude that only polyamines substantially improved the thermal stability of carrageenan gels.

The second, is based on its chemical structure, as it abundantly encompasses free amino groups (NH_2). Chitosan's amino groups have been partially acidified, so that its protonated NH_3^+ groups form a strong polyelectrolyte complex with the anionic-counterpart of carrageenan ($-OSO_3^-$), whereas, the free $-NH_2$ groups of chitosan could be used for *covalent immobilization* of enzymes via glutaraldehyde residues .

β-galactosidase has been immobilized as a model of industrial enzymes. The enzyme is used to hydrolyze lactose in whey and milk to afford both glucose and galactose. Knowing that consumption of foods with a high content of lactose presents the etiology of medical problems for almost a *70% of the world population*, especially in the developing countries, as the naturally occurring enzyme, in the human intestine loses its activity during lifetime. [34]

Unfortunately, *there is no cure to lactose intolerance*. This fact, together with the relatively low solubility and sweetness of lactose, has led to an increasing interest in the development of industrial processes to hydrolyze the dairy lactose (milk and whey) with both free and immobilized enzymes. [35] The hydrolysis of lactose affords two monosaccharides, namely, glucose and galactose, which are four times sweeter than lactose, while more soluble, digestible [36] and safely consumed by the 'lactose-intolerance' people. [37]

In view of the fact that the β-galactosidase is covalently attached to the gel, it will enable the immobilized enzyme to convert lactose to both glucose and galactose without being detached from the gel or migrate to food, and thus it is unlikely to be consumed. Another advantage could be that the hydrolysis of lactose present in whey permeate, will produce lactose free syrup thus solving an aquatic pollution problem, since whey is commonly thrown in water.

In Brief, the Purpose of this Chapter is to focus on

1. Providing valuable information on the chemical and physical modifications of the newly developed carrageenan coated with chitosan as a carrier for covalent immobilization of enzymes. The gel's structure was elucidated using techniques such as FTIR, DSC, Schiff's base formation and the gel's thermal stability.
2. Exploring the factors affecting the enzyme loading capacity in terms of chitosan pH and concentration variations.
3. Investigating the factors affecting the free and immobilized enzyme activity and stability searching for the optimum process conditions.

<u>This includes the effect of</u>

a) pH,
b) temperature,
c) Michaelis–Menten kinetics parameters,
d) enzyme operational stability.

METHODOLOGIES

3.1. Preparation of κ-Carrageenan Gel Disks

κ-Carrageenan gels were prepared as previously reported by Elnashar *et al.*, 2005[20] .Carrageenan at a concentration of 2 % (w/v) was dissolved in distilled water at 70 °C using an overhead mechanical stirrer, until complete dissolution. Glass parallel plates equipment designed by Elnashar *et al.*, 2005[20] , as shown in Figure 3, with 10 mm gaps were then immersed into the hot molten gel to produce uniform gel sheets.

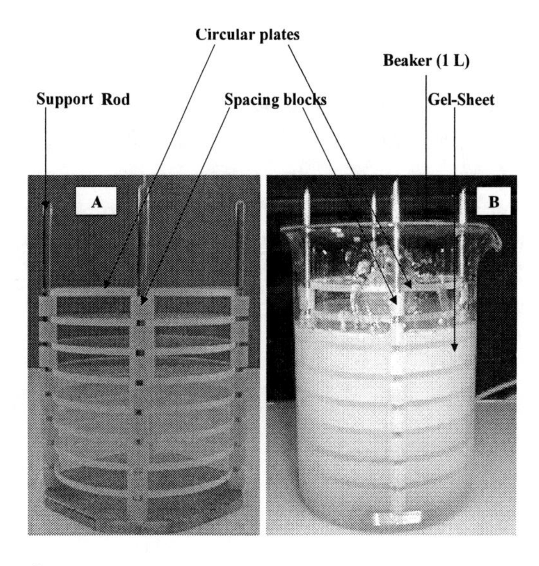

Figure 3. Parallel plates equipment.

The 10 mm thick gel-sheets were cut into disks using cork borer for enzyme immobilization. Typically, ~4 mm diameter gel disks, of 145 mg average weight, were thus produced to be ready for immobilization Panels A and B demonstrate the parallel plates equipment while empty and filled with gel respectively. The circular plates are of 2 mm thick and 8.6 cm in diameter, whilst the spacing blocks are 10 mm deep with 8 mm and 4 mm outer diameter and inner diameter. The stainless steel support rods are 4 mm diameter.

3.2. Carrageenan Coated with Chitosan and Schiff's Base Formation

In this experiment, the carrageenan gel was hardened using 0.3 M KCl solution as a control, [20] and with a series of chitosan at different pHs. The gels were filtered off and the supernatant was left to determine the optimum chitosan pH using Cibacron Brilliant Red® reagent.

The modified gel disks were thoroughly washed with distilled water then soaked in glutaraldehyde.

In this step, glutaraldehyde acted as a chitosan crosslinker, thus improving the gels thermal stability [33] while functioning as a spacer arms, offering free reactable aldehyde groups with the enzyme's amino groups, via Schiff's base formation.

The modified gel was then thoroughly washed with distilled water to get rid of the unbound glutaraldehyde.

3.3. Colorimetric Determination of Chitosan

A colorimetric method for the determination of chitosan in an aqueous solutions was described by Muzzarelli (1998). [38] A solution of Cibacron brilliant Red® 3B-A dye was prepared by dissolving 150 mg of the powder in 100 ml bi-distilled water. Five milliliters of the stock solution was then diluted to 100 ml with 0.1M glycine hydrochloride buffer. The final concentration of the dye solution was 75μg/ml. To prepare the standard curve of chitosan, 15, 30, 45, 60, 80, 100, 150, 200 and 250μl of 0.5 g/L chitosan were filled into test tubes, followed by the addition of different volumes of buffer to reach 300 μl. 3ml aliquots of dye solution were then added to each tube. The absorbance values were spectrophotometrically measured at 575 nm. The aqueous chitosan solutions were tested before and after the incubation of the gel disks. The difference in the amount of chitosan before and after incubation was the amount of chitosan bound to the carrageenan gel disks. The chitosan surface density coat per one wet gel disk was also calculated knowing that the surface area of one gel disk is equal to 1.256 cm^2 using the following equation:

$$Surface\ area\ of\ a\ disk = \pi dl = 3.14 * 0.4 * 1 \tag{1}$$

3.4. Thermal Stability of Carrageenan Gel Formulations

Two formulations of gel disks were used for this test: namely, the control gel and the modified gel of 2 % (w/v) κ-carrageenan/KCl soaked in 0.75 % (w/v) chitosan at pH 4.0, followed by glutaraldehyde.

The gel disks were incubated for 1 h in distilled water at 25-95 °C. The appearance of the gels was then visually inspected to check whether the gel disks remained solid or dissolved.

3.5. Fourier Transform Infrared Spectroscopy

The infrared spectra of all formulations were recorded with Fourier transform infrared spectroscopy (FTIR-8300, Shimadzu, Japan). The spectra were taken in the wavelength region 4000 to 400 cm^{-1} and at ambient temperature.

3.6. Differential Scanning Calorimetry (DSC) and Gel's Thermal Stability in Solutions

The DSC was performed to prove the formation of a strong polyelectrolyte complex between carrageenan and chitosan followed by glutaraldehyde. The thermal behavior of the different gel components (carrageenan powder, chitosan powder, carrageenan/chitosan powders were physically mixed and the "carrageenan /chitosan/glutaraldehyde gel" was characterized by differential scanning calorimetry (SDT 600, TA Instruments, USA). Approximately 3 to 6 mg of the dried gels were weighed into an alumina pan. The samples were heated from 25 to 350 °C, at a heating rate of 10 °C/min.

3.7. Immobilization Efficiency and Soluble Protein Determination

β-Galactosidase was immobilized onto the control and modified gel. Six disks of control and modified gel (10 mm height * 4 mm diameter) were thoroughly washed with distilled water and were incubated into 10 ml of enzyme solution (3.4 U/mL) prepared in 100 mM citrate-phosphate buffer at pH 4.5. The immobilized enzyme was thoroughly washed with a buffer solution containing Tris-HCl to block any free aldehyde group and to remove any unbound enzyme. The immobilized enzyme was stored at 4 °C for further assays. The supernatant and the wash were kept for soluble protein assay via Lowery assay using bovine serum albumin (BSA) as a standard. The amount of protein immobilized onto and into the gel carrier Pg (mg/g) was calculated using the following equation:

$$Pg = \frac{CoVo\_C_fV_f}{w} \quad (2)$$

where Co is the initial protein concentration (mg/ml), C_f, the protein concentration of the filtrate (mg/ml), Vo the initial volume of the enzyme solution (ml), V_f the volume of filtrate (ml) and w the weight of used gel carrier (g).

To increase the percent immobilization efficiency (% I.E.) as in equation 3, six disks of control and modified gels were incubated into 10 ml of enzyme solution of 2.0 U/ml instead of 3.4 U/ml, prepared in 100 mM citrate-phosphate buffer, pH 4.5.

$$\% I.E.\% = \frac{MoVo - M_f Vo}{w} * 100 \quad (3)$$

where Mo is the initial enzyme concentration (U/ml), M_f the enzyme concentration of the filtrate (U/ml), Vo the initial volume of the enzyme solution (ml), V_f the volume of filtrate (ml) and w the weight of the used carrier material (g)

3.8. Determination of β-Galactosidase Activity

β-Galactosidase activity was determined by the rate of glucose formation in the reaction medium. A known amount of immobilized or free enzyme were incubated into 10 ml of 200 mM lactose solution in citrate-phosphate buffer (100 mM, pH 4.5) for 3 h at 37 °C and at 100 rpm [33]. At the end of the time, 50 µl of reaction mixture was added to 950 µl buffer and boiled for 10 min, to inactivate the enzyme and analysis for glucose content using the glucose kits. One enzyme unit (IU) was defined as the amount of enzyme that catalyzes the formation of 1 µmol of glucose/ minute, under the specified conditions.

Glucose concentration was spectrophotometrically determined with a glucose test based on the Trinder reagent [39] .Glucose is transformed to gluconic acid and hydrogen peroxide by glucose oxidase (GOD). The formed hydrogen peroxide was reacted in the presence of peroxidase (POD) with 4-aminoantipyrine and p-hydroxybenzene sulfonate to form a quinoneimine dye as shown in Scheme 1.

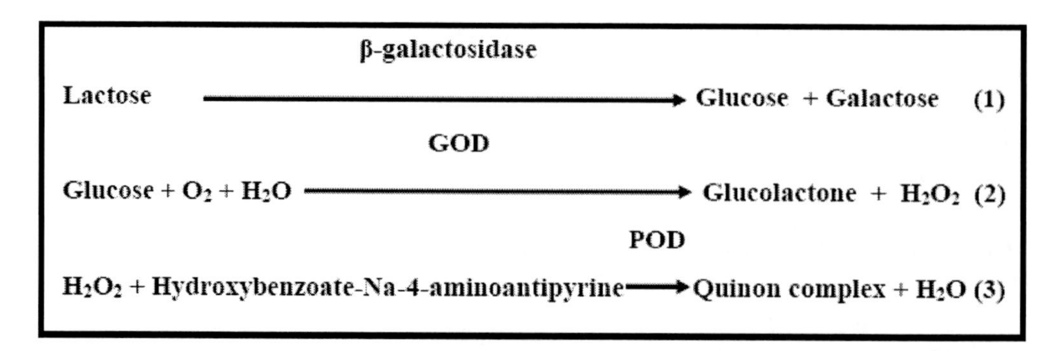

Scheme 1. Hydrolysis of lactose by β-galactosidase and glucose determination using a mixture of enzymes, glucose oxidase (GOD) and peroxidase (POD).

The intensity of the color produced is directly proportional to the glucose concentration in the sample. The assay was performed by mixing 30 μl of a sample of unknown concentration and 3 ml of Trinder® reagent. The reaction was let to proceed for 20 min at room temperature, and the absorbance of the unknown concentration was read at 510 nm.

3.9. Enzyme Incubation Period

In this experiment, six disks of gels were incubated in an enzyme solution of 34 U (10 ml of 3.4 U/ml) for an incubation time of 0.5 to 32 h, to determine the optimum time for immobilization to reach the maximum enzyme loading capacity (E.L.C.).

$$\%E.L.C. = \frac{M_o - M_f}{M_o} * 100 \quad (4)$$

where M_o is the initial enzymatic activity (U), M_f the enzyme activity of the filtrate (U) after immobilization.

3.10. Temperature Effect on β-Galactosidase Activity

The optimum temperature for the free and immobilized enzyme was examined. The free and immobilized β-galactosidase were incubated into 10 ml of 250 mM lactose at pH 4.5 and temperatures from 30 to 70 °C, for 3h. The data were normalized to 100% activity at 37 °C, at which the activities of the respective enzyme preparations were determined. The relative activity at each temperature is expressed as a percentage of the 100 % activity.

3.11. pH Effect on β-Galactosidase Activity

The effect of pH on the activity of free and immobilized enzyme was examined. The free and immobilized β-galactosidase were incubated into 10 ml of 250 mM lactose at pH 3-6.5 at 50 °C, for 3h. The data were normalized to 100% relative activity at pH 4.5 at which activities of the respective enzyme preparations were determined. The relative activity at each pH is expressed as a percentage of the 100 % activity.

3.12. *K*m and *V*max of Free and Immobilized Β-Galactosidase

To obtain the adequate Michaelis-Menten kinetic models for the description of the hydrolysis of lactose by the free and the immobilized enzyme, the apparent *Km* and *V*max of free and immobilized β-galactosidase were determined for lactose using the Hanes–Woolf plot method, equation 5.

$$\frac{[S]}{Vo} = \frac{1}{V\max} * [S] + \frac{Km}{V\max} \qquad (5)$$

where, [S] is the substrate concentration (lactose), Vo is the initial enzyme velocity, V max is the maximum enzyme velocity and Km is the Michaelis constant and is defined only in experimental terms and equals the value of [S] at which Vo equals ½ Vmax). Experimentally, the Km from the plot is equal to –[S], whereas the Vmax is equal to 1/slope. The assay mixture was comprised of 5 U of free and immobilized enzyme and substrate concentration of 20–300 mM at 37 °C and pH 4.5 for 3h.

3.13. Conversion of Lactose to Glucose

Forty units of free and immobilized enzyme were incubated into 10 ml of 250 mM lactose at pH 5.5 and 50 °C. Samples were withdrawn at time intervals ranging from 15 min to 7h and analyzed for glucose content.

3.14. Operational Stability

The reusability of immobilized enzyme was studied using the modified gel disks. [40] Forty units of immobilized enzyme were incubated into 10 ml of 250 mM lactose at pH 5.5 and 50 °C for 3 h and the substrate solution was assayed for glucose content. The same gel-disks were then washed with distilled water and re-incubated with another substrate solution. The procedure was repeated 15 times and the starting operational activity was considered as 100% relative activity.

4. Results and Discussion

4.1. Structure Elucidation of Carrageenan Coated with Chitosan

Carrageenan gel sheets were prepared using the parallel plates apparatus [20]. The sheets were cut into disks and modified using partially protonated chitosan. The protonated chitosan amino groups (NH_3^+) formed a polyelectrolyte complex (PEC) with the $–OSO_3^-$ of the carrageenan gel to increase the gel's thermal stability [28] and to incorporate free amino groups new functionality. [27] The free chitosan amino groups ($-NH_2$) were used to form Schiff's base and covalently immobilize β-galactosidase via glutaraldehyde as a mediator as shown in Scheme 2.

The FTIR bands of the carrageenan, chitosan and the modified gel (Carrageenan/ Chitosan/Glutaraldehyde) were shown in Figure 4. The data were tabulated in Table 1. The modified gel revealed the presence of carbonyl group, at 1720 cm^{-1}, attributed to glutaraldehyde free aldehydic groups. The presence of glutaraldehyde -C=O was also confirmed visually as it reacted with the chitosan $–NH_2$ group forming *Schiff's base* accompanied with a color change from pale-yellow to brown-red according to the chitosan's pH. [31]

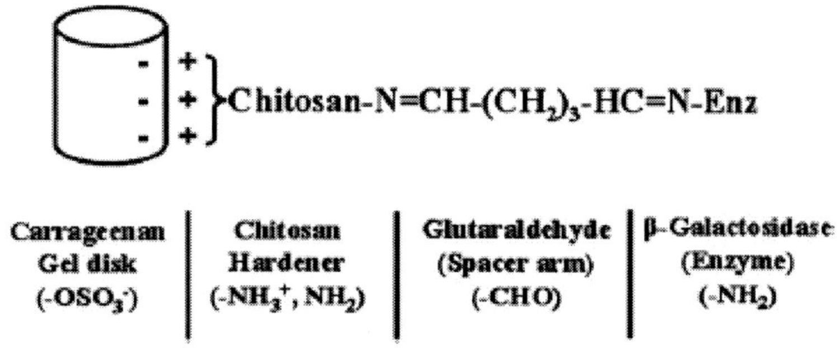

- + Chitosan-N=CH-$(CH_2)_3$-HC=N-Enz

Carrageenan Gel disk ($-OSO_3^-$)	Chitosan Hardener ($-NH_3^+$, NH_2)	Glutaraldehyde (Spacer arm) (-CHO)	β-Galactosidase (Enzyme) ($-NH_2$)

Scheme 2. Carrageenan coated with chitosan followed by glutaraldehyde (GA) as a spacer arm to immobilize covalently β-galactosidase (Enz) to the chitosans amino groups via Schiff's base formation.

Table 1. Analysis of the FTIR data of carrageenan, chitosan, and carrageenan-coated chitosan followed by GA

Polymer	Wavenumber (cm^{-1})	
	$-OSO_3^-$	-C=O
Carrageenan	1446	---
Chitosan	---	1670
Carrageenan/Chitosan/GA	1390	1720, 1670

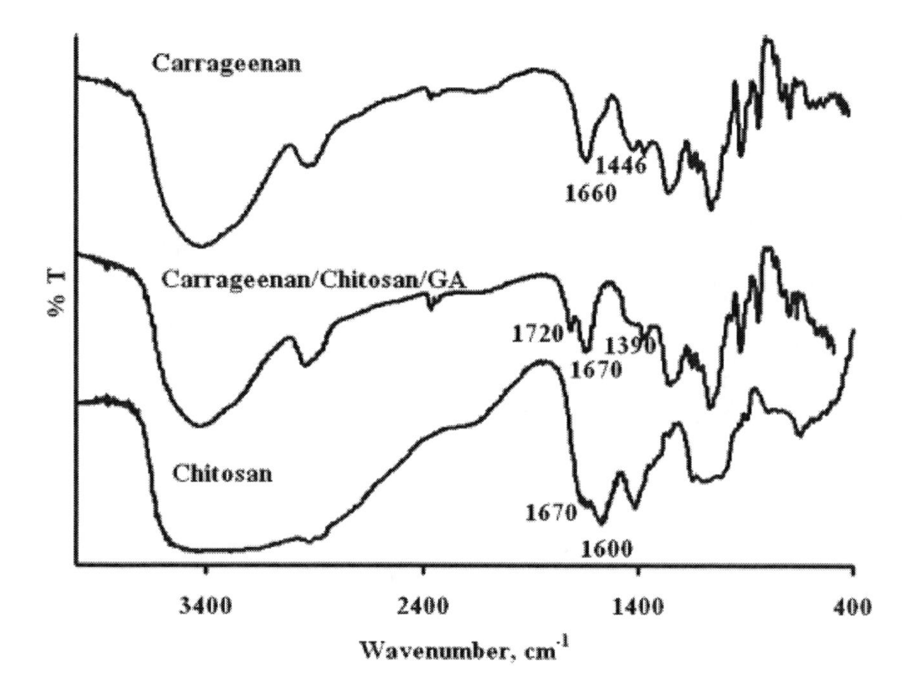

Figure 4. FTIR of carrageenan, chitosan, and the carrageenan coated with chitosan followed by glutaraldehyde (GA).

The FTIR bands also revealed a decrease in intensity and a shift of the carrageenan $-OSO_3^-$ absorption band from 1446 cm^{-1} to 1390 cm^{-1} after reaction with chitosan, which was derived from the ionic interaction between the carrageenan and the chitosan. This interaction evidenced the formation of strong polyelectrolyte complexes. [41]

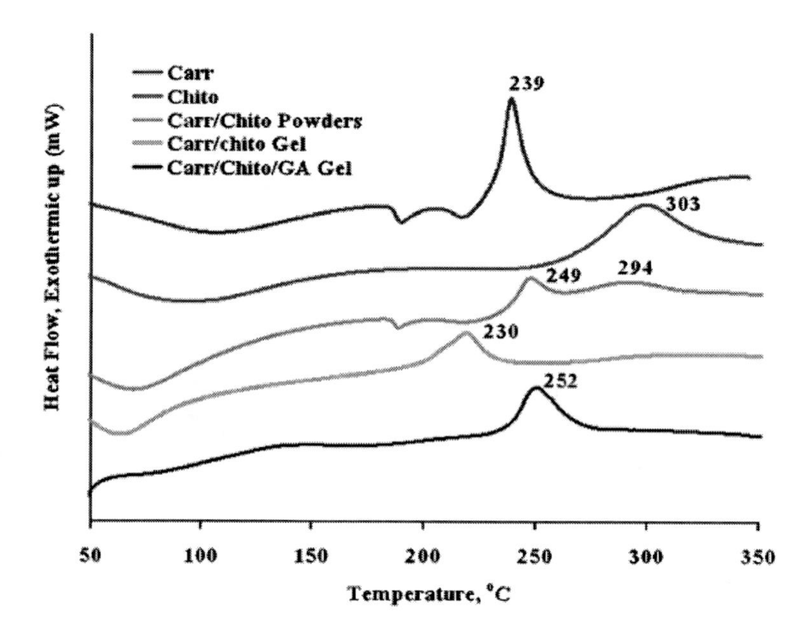

Figure 5. DSC thermogram of the pure carrageenan and chitosan powders, the carrageenan/chitosan powders physically mixed, the carrageenan/chitosan and the carrageenan/chitosan/glutaraldehyde.

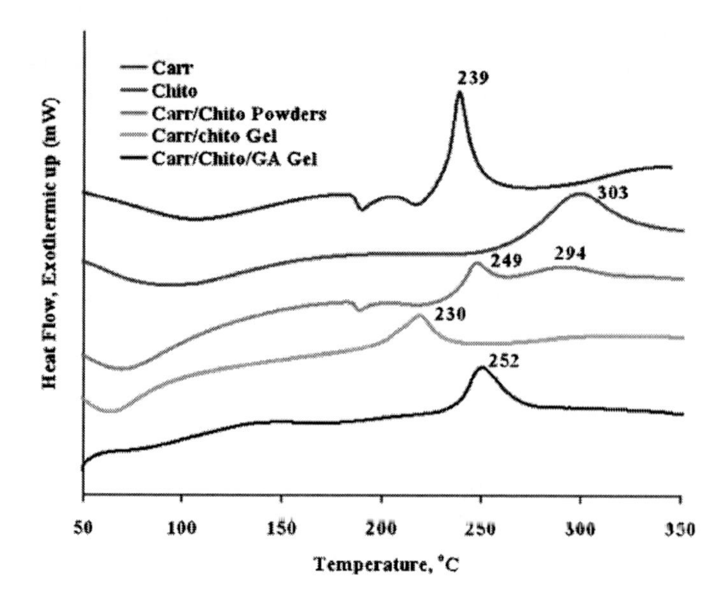

Figure 5. DSC thermogram of the pure carrageenan and chitosan powders, the physically mixed, carrageenan/chitosan powders, the carrageenan/chitosan and the carrageenan/chitosan/glutaraldehyde.

Table 2. Thermal stability of control carrageenan gel and carrageenan coated with chitosan. The appearance of the gel was inspected visually, where (-) means turbid solution (gel dissolved), and (+) refers to a clear solution (gel remained solid)

Chitosan Concentration % (w/v)	0	0.25	0.5	0.75	1.0
Thermal Stability at 30 °C	+	+	+	+	+
Thermal Stability at 35 °C	-	+	+	+	+
Thermal Stability at 95 °C	-	+	+	+	+

The DSC thermograms of pure carrageenan, chitosan powders, the physically mixed carrageenan/chitosan powders and the carrageenan/chitosan/glutaraldehyde gel were shown in Figure 5. The pure carrageenan and chitosan powders revealed exothermic peaks at 239 and 303 °C, respectively, which represent their degradation. The physically mixed chitosan/carrageenan powders showed two exothermic peaks at 249 and 294 °C, which are in between those of the pure carrageenan and chitosan respectively. However the carrageenan/chitosan/glutaraldehyde gel showed a single peak at 252 °C. [42] This finding indicated formation of a strong polyelectrolyte complex due to ionic interaction between the polyanion (carrageenan $-OSO_3^-$) and the polycation (chitosan – NH_3^+) [41, 43]. The modification of carrageenan using chitosan and glutaraldehyde greatly improved the carrageenan thermal stability from 239 °C to 252 °C.

4.2. Thermal Stability of Carrageenan and Carrageenan Coated with Chitosan

It is well known that a disadvantage of carrageenan gel is its low thermal instability. [21] The thermal stability of the carrageenan coated with chitosan revealed greater stability over the control gel, as revealed in Table 2, where the control gel dissolved at 35 °C and the modified gel remained intact at 95 °C. This fact was supported by Chao *et al.*, 1986 [33] who reported that only polyamines substantially improved the carrageenan gels thermal stability. This outstanding thermal stability improvement in the modified gels could be the result of the strong polyelectrolyte complexation between the polyanion, carrageenan gel, and the chitosan polycation, [43] and to the reinforcement of the carrageenan gel's network using chitosan through multiple points, [19] as shown in Scheme 2. This result could also be supported by the fact that hardening of hydrogels by polyelectrolyte complexation is an interesting alternative to covalently crosslinked hydrogels. [44]

4.3. Optimization of Chitosan pH and Concentration

In this Section, the optimum pH and concentration of chitosan were studied to find out the maximum enzyme loading capacity. Lowry assay showed that the enzyme activity did not decrease by the immobilization process, i.e almost 100% retention of enzyme efficiency. The units of enzyme / milligram protein were found to be 4 U/mg and not 11.8 U/mg, as stated by the manufacturer.

Table 3. Optimization of chitosan pH. Carrageenan gels hardened with 0.3 M KCl for 3 h were soaked in a series of 0.5% (w/v) chitosan at pH 2-6 for 3 h. The chitosan concentration was determined using Cibacron® reagent

Chitosan pHs	mg Chitosan/ g wet gel disk	mg Chitosan/cm^2 g wet gel disk	Chitosan/Carrageenan % (w/w) dry gel
pH2	4.0	3.2	20.2
pH3	4.0	3.2	20.2
pH4	*5.7*	*4.5*	*28.5*
pH5	4.2	3.3	20.9
pH6	3.6	2.9	18.2

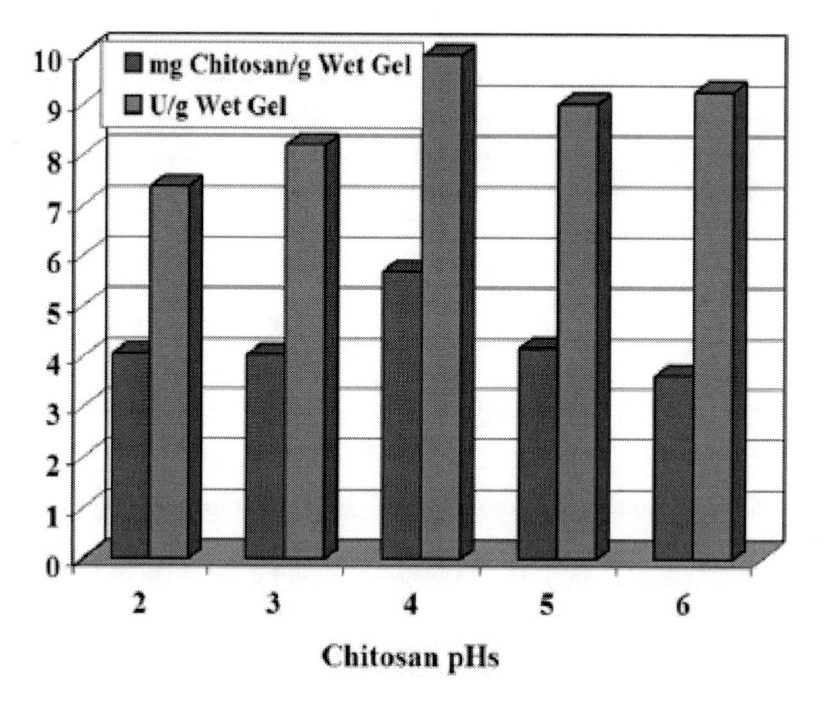

Figure 6. Effect of chitosan pH(s) on the amount of bound chitosan and the units of enzymes immobilized per gram wet gel. Carrageenan gel disks hardened with KCl were soaked in 0.5% (w/v) chitosan at pH 2-6 for 3 h afterward in 2% (v/v) glutaraldehyde for 3 h then in 34 U of β-galactosidase for 16 h.

4.3.1. Optimization of Chitosan pH

Chitosan was prepared at pH 2 to 6 to find out the best pH for maximum chitosan coating to the carrageenan gel by forming a polyelectrolyte complex between the chitosan ($-NH_3^+$) and the carrageenan ($-OSO_3^-$). Since the pK of chitosan is ~ 6.3 and the sulphates groups of carrageenan have negative charges in pure water (~ pH 5.5), [44, 45] we were expecting the chitosan and carrageenan to be ionized, reaching their maximum binding at low pHs(2-3). By increasing the chitosan pHs, to 4-6, carrageenan/chitosan polyelectrolyte complex formation was expected to gradually decline due to deprotonation of chitosan.

Table 4. Optimization of chitosan concentration. Carrageenan gels hardened with 0.3 M KCl were soaked in a series of chitosan concentrations of 0.25-1.0% (w/v) at pH 4 for 3 h. The chitosan concentration was determined using Cibacron reagent

Chitosan Conc.	mg Chitosan/ g wet gel disk	mg Chitosan/cm^2 g wet gel disk	Chitosan/Carrageenan % (w/w) dry gel
0.25	4.9	3.9	24.3
0.50	5.7	4.5	28.5
0.75	*6.0*	*4.8*	*30.0*
1.0	5.4	4.3	27.1

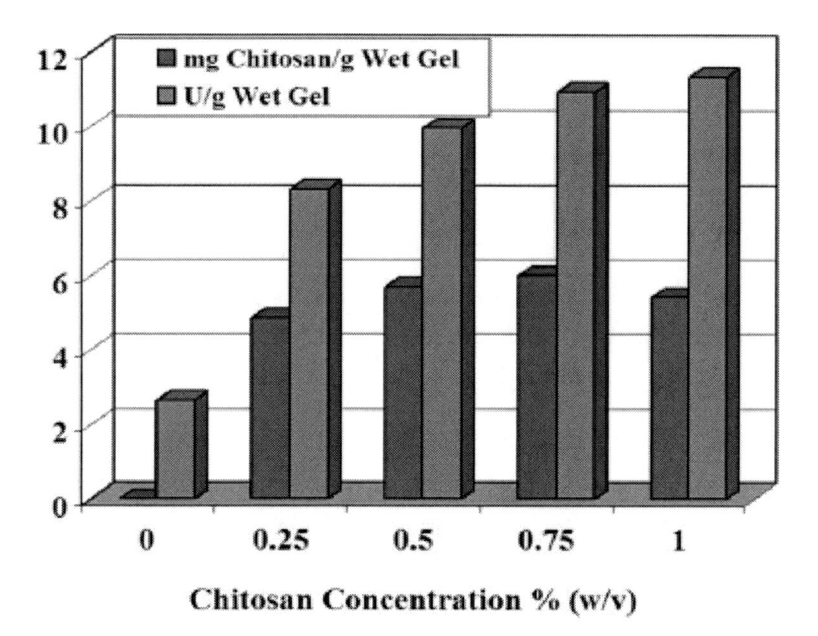

Figure 7. Effect of chitosan concentration on the amount of bound chitosan and the units of enzymes immobilized per gram of wet gel. Carrageenan gel disks hardened with KCl were soaked in 0.25-1.0% (w/v) chitosan at pH 4 for 3 h afterward in 2% (v/v) glutaraldehyde for 3 h then in 34 U of β-galactosidase for 16 h.

Amazingly, as presented in Table 3, the maximum chitosan /carrageenan binding of 5.7 mg chitosan / g of wet gel disks, was reached at pH 4, instead of pH 2, then it decreased afterwards to 3.6 mg chitosan/ g wet gel disk at pH 6. The low carrageenan/chitosan polyelectrolyte binding at pH 2-3 could be regarded, as the fact that κ-carrageenan under pH ~2.6 is susceptible to hydrolytic degradation. Degradation of carrageenan resulting in the formation of low molecular weight material provides an obstacle in the regulation of the polyelectrolyte complex process. [14]

In terms of chitosan surface density, it reached its maximum of 4.5 mg chitosan/ cm^2 g wet gel disks, knowing that chitosan coat was almost on the carrageenan surface due to its high molecular weight and viscosity. The results in Figure 6 were in agreement with that of Table 3, where the maximum enzyme loading of 9.6 U/g wet gel was achieved, as well at pH 4. This could be attributed to that the more bound chitosan to the carrageenan, the more free amino groups to bind enzyme to the gel via glutaraldehyde as a mediator. [27]

4.3.2. Optimization of Chitosan Concentration

The optimum formulation of Section 4.2.1 was chosen for this experiment, where chitosan formulations of 0.25-1.0% (w/v) at pH 4 were used to coat the carrageenan gel. The data in Table 4 revealed an increase in the maximum binding to 6.0 mg chitosan/ g wet gel disks and the maximum surface density to 4.8 *mg* chitosan/cm^2 g wet gel disks using a chitosan solution of 0.75%. In terms of dry gel, the chitosan to carrageenan percentage (w/w) increased to 30%. By increasing the chitosan concentration to 1.0%, the binding efficiency decreased. This may be due to the high viscosity of chitosan solution, which caused difficulty for the chitosan molecules to penetrate the carrageenan gel pores.

The results in Figure 7 were in accordance with that of Table 4, where the amount of immobilized enzyme was gradually increased by increasing chitosan concentration from 0.25 to 0.75%, reaching its maximum of 10.92 U/g wet gel (2.73 mg protein/g wet gel) at 0.75% chitosan, which is more than four folds that of the control gel, 2.65 U/g wet gel.

However, at 1.0% (w/w) chitosan, the amount of enzyme units immobilized slightly increased, but owing to the high viscosity of chitosan and the difficulty of preparation of 1.0% (w/w) chitosan, we have chosen the concentration of 0.75% (w/w) chitosan for future experiments.

4.4. Enzyme Incubation Period

The optimum immobilization time for β-galactosidase was found to be 16 h, at when 28% of the initial enzyme solution were immobilized, after which a plateau was attained as shown in Figure 8.

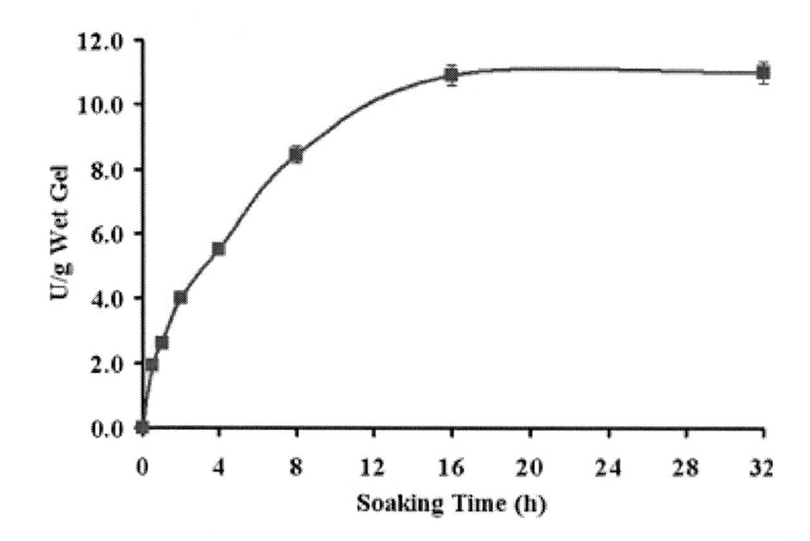

Figure 8. β-Galactosidase incubation period. Carrageenan gels disks (Carrageenan/Chitosan/ Glutaraldehyde) were soaked in 10 ml of 3.4 U/ml of enzyme solution at an incubation time of 0.5 h to 32 h.

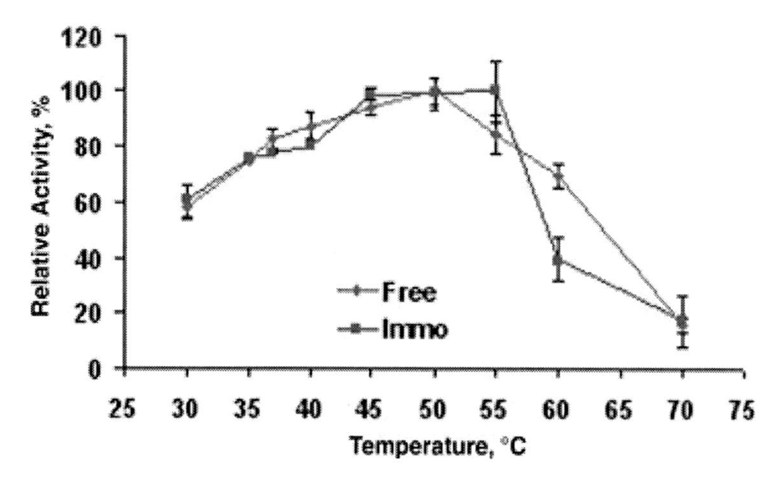

Figure 9. Temperature-activity profile of free and immobilized β-galactosidase.

By increasing the time of immobilization from 0.5 to 16 h, the immobilized enzyme loading capacity was gradually increased from 1.9 to 10.92 U/g of wet gel respectively. A longer enzyme incubation time has no effect on the immobilization efficiency, which could be regarded to all aldehyde groups have been engaged with the enzymes after 16 h.

4.5. Immobilization Efficiency

In order to increase the immobilization efficiency percentage (I.E. %), 1g gel was incubated in 20 U of enzyme (10 ml of 2 U/ml) instead of 34 U , as previously used by the authors. [27] The results showed the immobilization of ~11 U/g wet gel, which revealed the increase of I.E. % from 28% to 50% using equation 3.

4.6. Temperature-Activity Profile of Free and Immobilized β-Galactosidase

Figure 9 illustrates the temperature-activity profile of the free and immobilized β-galactosidase. The enzymatic activity is dependent on the temperature, as there is an optimum pH and temperature above which the enzymatic activity decreases due to the denaturation of the enzyme protein.

As shown from Figure 9, the immobilized enzyme was found to be stable at a wider range of temperature (45 – 55 °C), compared to the free enzyme (50 °C). The shift of the optimum temperature towards higher temperatures when the biocatalyst is immobilized indicates that the enzyme structure is strengthened by the immobilization and the formation of a molecular cage around the enzymatic protein molecule was found to enhance the enzyme thermal stability. [48] The increase of the immobilized enzyme temperature tolerance may also be due to diffusional effects, where the reaction velocity is more likely to be diffusion limited, so that improvements in thermal diffusion would correspondingly result in proportionally higher reaction rates.

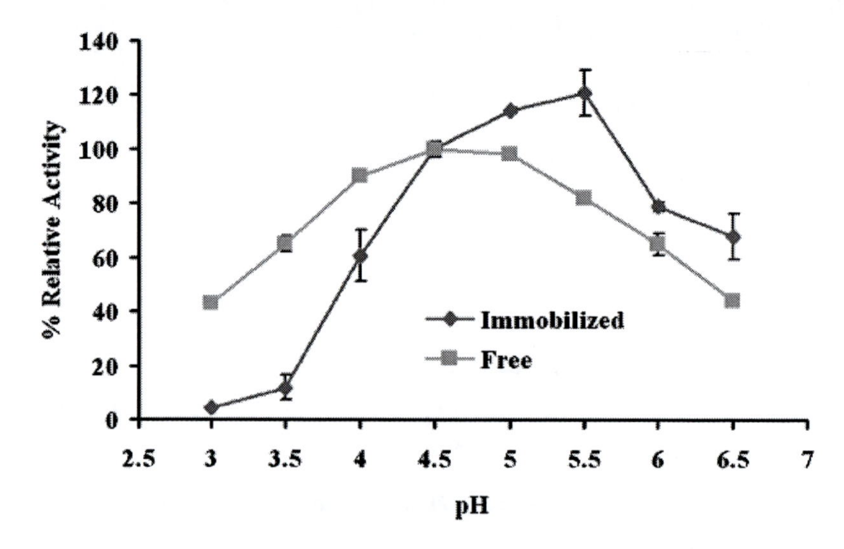

Figure 10. pH–activity profile of free and immobilized β-galactosidase.

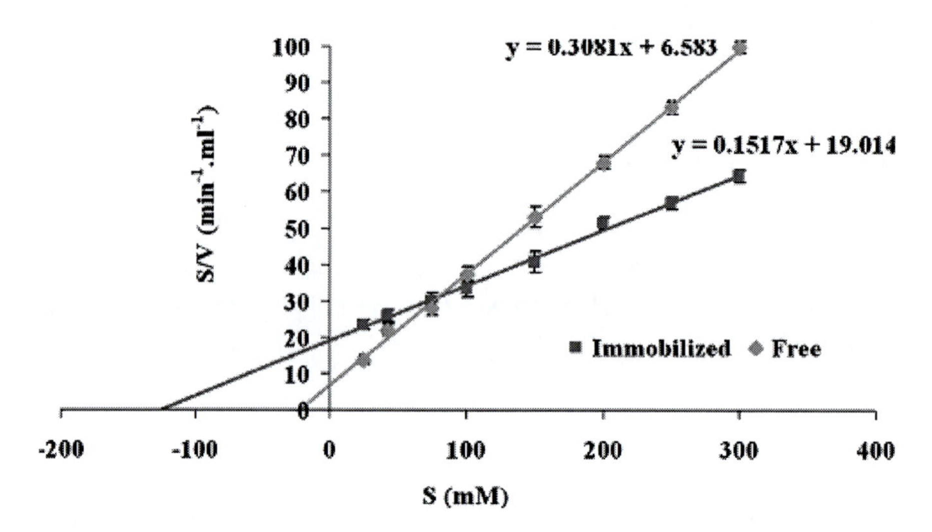

Figure 11. Kinetic parameters of free and immobilized β-galactosidase using Hanes–Woolf plot.

4.7. pH-Activity Profile of Free and Immobilized β-Galactosidase

Figure 10 illustrates the pH-activity profile of the free and immobilized β-galactosidase. The optimum pH values for free and immobilized enzyme were 4.5–5 and 5-5.5, respectively, which indicated that the immobilized enzyme was more stable at higher pH. [46]

These properties could be very useful for lactolysis in sweet whey permeate, which has a pH range of 5.5–6. [47] The shift in the pH-activity profile of the immobilized β-galactosidase may be attributed to the partition effects that were arising from different concentrations of charged species in the microenvironment of the immobilized enzyme and in the domain of the bulk solution. [48]

Table 5. Kinetic constants of free and immobilized β-galactosidase

β-Galactosidase form	Kinetic Constants	
	K_m (mM)	V_{max} (μmol/min)
Free	13.2 ± 0.03	3.2 ± 0.005
Immobilized	125 ± 0.025	6.6 ± 0.007

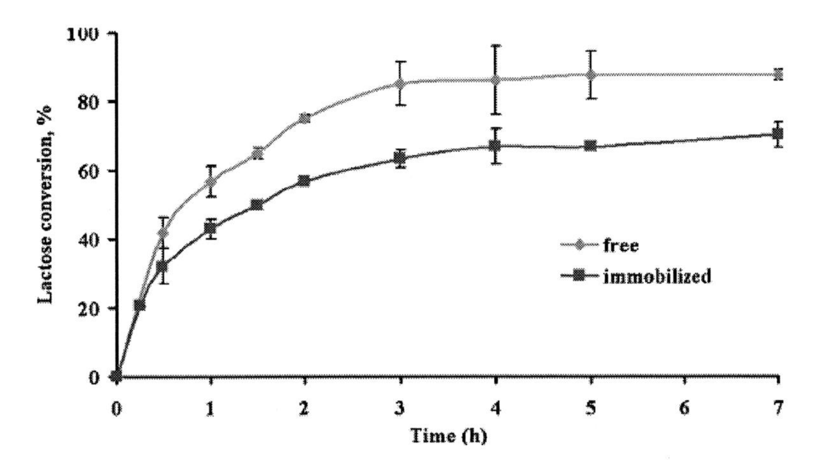

Figure 12. Effect of time on lactose conversion to glucose using free and immobilized β-galactosidase.

4.8. Kinetic Constants of Free and Immobilized β-Galactosidase

The kinetic constants of free and immobilized β-galactosidase were calculated using Hanes–Woolf plot method as shown in Figure 11. The calculated values are shown in Table 5.

The apparent *Km* after immobilization, estimated as 125 mM, is higher than that of the free enzyme, 13.2 mM, indicating that a higher concentration of substrate is needed for the immobilized enzyme compared to the free enzyme. Higher *Km* result (~ 187 mM) for the same immobilized enzyme was obtained by other authors. [40] It should be taken into consideration that the modification of enzyme's structure may result in the loss of its specificity towards lactose. However, no substrate or product inhibition by the increase of substrate concentration up to 300 mM could be observed during our experiment, as shown by the straight line of the Hanes-Wolf representation (Figure 11).

The maximum reaction velocity "*V*max" values for the immobilized enzyme was astounding, since it was doubled relative to the free enzyme, i.e., it increased from 3.2 μmol/min to 6.6 μmol/min.

The apparent *Km* after immobilization, 125 mM, is higher than that of the free enzyme, 13.2 mM, indicating that a higher concentration of substrate is needed for the immobilized enzyme compared to the free enzyme. Higher *Km* value (~ 187 mM) for the same immobilized enzyme was obtained by other authors. [40] It should be taken into consideration that the modification of enzyme's structure may result in the loss of its specificity towards lactose. However, no substrate or product inhibition by the increase of substrate concentration up to 300 mM could be detected during our experiment as shown by the straight line of the Hanes-Wolf representation (Figure 11). The maximum reaction velocity "*V*max" values for the immobilized enzyme was astounding, it was found to double that of the free enzyme, i.e., it increased from 3.2 μmol/min to 6.6 μmol/min.

4.9. Lactose Hydrolysis Using Free and Immobilized β-Galactosidase

This experiment has been carried out so that the immobilized enzyme could attain its maximum efficiency and act with its highest velocity using double the *K*m = 125 mM concentration, namely, 250 mM. The high substrate, concentration(250 mM) ,was used in this study as this enzyme was supposed to be suitable for hydrolysis of higher lactose concentrations found in mammal milk (88 – 234 mM lactose) [46] and whey permeate (85% lactose). [47] The free enzyme reached a plateau and conversion of 87% of lactose after 3 h, whereas the immobilized enzyme reached 63% of conversion at that time and it increased to 70% after 7 h, as shown in Figure 12.

The decrease in substrate conversion after three hours could be regarded to diffusion limitation. [49] Perceptively, in order to increase the enzyme substrate affinity and decreasing the *K*m value, our future trial will be carried out using smaller gel particles with larger surface areas and more gel pores to decrease the substrate/product diffusion limitation.

4.10. Operational Stability of Immobilized β-Galactosidase

The major advantage of enzyme immobilization is the realized easy separation and reusability. The data shown in Figure 13 indicate that almost no decrease in the enzyme activity for nine consecutive uses. [50] However, after the ninth use, the relative enzyme activity started to gradually decrease to attain 97%. The loss in activity was attributed to inactivation of enzyme due to continuous use. [51]

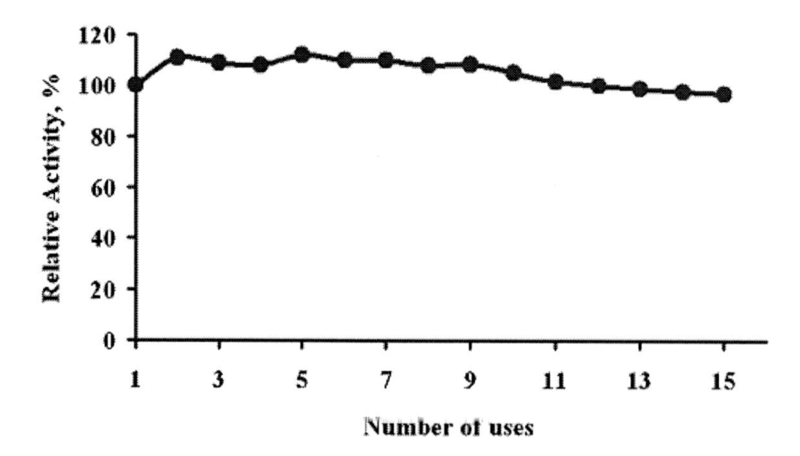

Figure 13. Operational stability of immobilized β-galactosidase onto the modified gel (Carrageenan/Chitosan/Glutaraldehyde).

5. CONCLUSION

The modification of carrageenan with chitosan imparts two extra benefits to carrageenan. The first is the creation of a new amino group's functionality, (free NH_2); while the second, is the amelioration of the carrageenan gel's thermal stability by forming a polyelectrolyte complex (PEC) between the carrageenan $-OSO_3^-$ and the chitosan $-NH_3^+$.

The free chitosan amino groups ($-NH_2$) were then activated with glutaraldehye to covalently immobilize β-galactosidase. The two individual biopolymers have the advantage of being already, permitted for use in the pharmaceutical or food industries. The new hybrid carrier increased the thermal stability of the carrageenan gel from 35 to 95 °C, as well as the enzyme temperature and pH stabilities from 50 to 45-55 °C, while from pH 4.5-5.0 to 5.0-5.5, respectively. The immobilized enzyme showed a lactose conversion of 70% at 7h compared to 87% for the free enzyme. Additionally, the reusability of the immobilized enzyme for tens of times tremendously reduces the enzyme cost. For example, the operational stability retained 97% of the enzyme activity after 15 uses.

In brief, the simplicity and effectiveness of the newly developed method for covalent immobilization of β-galactosidase, in addition to the promising results for lactose hydrolysis are encouraging to be applied for lactose sources such as whey and milk.

ACKNOWLEDGMENT

The author would like to thank the Centre of Excellence for Advanced Sciences, and the Science and Technology Development Fund (STDF/IMC) for supporting this work, and highly appreciates the efforts of Mrs Joanne Maryam Yachou and Professor Mohamed Abo-Ghalia for their contribution towards editing.

REFERENCES

[1] Roberts, G. (1992). Structure of chitin and chitosan, In: Roberts G. A. F. (ED.), *Chitin Chemistry*, MacMillan Press, London, 1-35.

[2] Hejazi, R. and Amiji, M. (2003). Chitosan. *J. Cont. Rel*, 89, 151.

[3] Kevin, McCue. (2003). New Bandage Uses Biopolymer. Chemistry.org (American Chemical Society). Retrieved 2006-07-10.

[4] Betancor, L., Luckarift, R., Seo, H. and Brand, O. (2008). Three-dimensional immobilization of **β**-galactosidase on a silicon surface. *Biotechnol. Bioeng*, 99, 261.

[5] Wang, Y., Xu, J., Luo, G. and Dai, Y. (2008). Immobilization of lipase by ultrafiltration and cross-linking onto the polysulfone membrane surface. *Biores. Technol*, 99, 2299.

[6] Jozef, S. and Sylwia W. (2006). Immobilization of thermostable β-glucosidase from *Sulfolobus shibatae* by cross-linking with transglutaminase. *Enz. Microb. Technol*, 39, 1417.

[7] Salah, S., Srimathi, S., Gulnara, S., Ikuo, S. and Bengt, D. (2008). Hydroxyapatite as a novel reversible *in situ* adsorption. Talanta. 77, 490.

[8] D'Souza, S. F. (1999). Immobilized enzymes in bioprocess. *Current Sci*, 77, 69.

[9] Kahraman, V., Bayramoglu, G., Kayaman-Apohan, N. and Atilla, G. (2007). α-Amylase immobilization. *Food Chem,* 104, 1385.

[10] Shweta P., Shamsher, K., Ghanshyam, C. and Reena, G. (2008). Glutaraldehyde activation of polymer. *Biores. Technol*, 99, 2566.

[11] Emese, B., Ágnes, N., Csaba, S., Tivadar, F. and János, G. (2008). Preparation. *J. Biochem. Biophys. Meth*, 70, 1240.

[12] Peppler, H. and Reed, G. (1987). Enzymes in food and feed processing, in: H. J. Rehm, G. Reed (Eds.), *Biotechnology*, Vol. 7a, VCH, Weinheim, 578.

[13] Bickerstaff GF. (1995). Impact of genetic technology on enzyme technology. *Genet Eng. Biotechnol.* 15, 13.

[14] Hugerth, A., Caram-Lelham, N. and Sundeliir, L. (1997) The effect of charge density. *Carbohyd. Polym*, 34, 149.

[15] Chang, J., Cho, C. and Chen, S. (2001). Decolorization of azo dyes. *Proc. Biochem*, 36, 757.

[16] Moon, S. and Parulekar, S. J. (1991). Characterization of .kappa.-carrageenan gels used for immobilization. *Biotechnol. Prog*, 7, 516.

[17] Tramper, J. and Grootjen, D. R. (1986). Operating performance of *Nitrobacter agilis* immobilized in carrageenan. *Enz. Microb. Technol*, 8, 477.

[18] Gerhard, A. and Brian, R. (1997) Carrageenan biotechnology. *Trends in Sci. Technol*, 8, 389.

[19] Wang, J. and Qiang, Y. (1999). Microbial degradation. *Chemosphere,* 38, 3109.

[20] Elnashar, M., Millner, P., Johnson, A. and Gibson, T. (2005). Parallel plate equipment for preparation of uniform gel sheets. *Biotechnol. Lett*, 27, 737.

[21] Luong, J. (1985). Cell immobilization in κ-carrageenan for ethanol production *Biotechnol. Bioeng*, 27, 1652.

[22] Boadi, D. and Neufeld, R. (2001) Encapsulation of tannase for the hydrolysis. *Enz. Microb. Technol.*, 28, 590.

[23] Eberhardt, A. M. Pedroni, V. Volpe, M. and Ferreira, M. L. (2004). Immobilization of catalase from *Aspergillus niger* on inorganic and biopolymeric supports for H_2O_2 decomposition. *Appl. Cat* B: Environm, 47, 153

[24] Hidalgo, A. Betancor, L. Lopez-Gallego, F. Moreno, R. Berenguer, J. Fernandez-Lafuente, R. and Guisan, J. M. (2003) Design of an immobilized preparation . *Enz. Microb. Technol*, 33, 278.

[25] Magnan, E. Catarino, I. Paolucci-Jeanjean, D. Preziosi-Belloy, L. and Belleville, M. P. (2004). Immobilization of lipase on a ceramic. *Membr. Sci*, 241, 161.

[26] Rocchietti, S. Urrutia, A. S. Pregnolato, M. Tagliani, A. Guisán, J. M. Fernández-Lafuente, R. and Terreni, M. (2002). Influence of the enzyme. *Enz. Microb. Technol*, 31, 88.

[27] Elnashar M., Yassin, A. and Kahil, T. (2008). Novel thermally and mechanically stable hydrogel for enzyme immobilization of penicillin G acylase via covalent technique. *Appl. Polym. Sci*, 109, 4105.

[28] Mansour, M., Elnashar, M. and Hazem, M. (2007). Amphoteric hydrogels using template polymerization technique. *Appl. Polym Sci*, 106, 3571.

[29] Elnashar. M., Danial, E. and Awad, G. (2009). Novel Carrier of Grafted Alginate for Covalent Immobilization of Inulinase. *Indus. Eng. Chem. Res*, 48, 9781.

[30] Danial, E., Elnashar, M. and Awad, G. (2010). Immobilized Inulinase on Grafted Alginate Beads Prepared by the One-Step and the Two-Steps Methods. *Indus. Eng. Chem. Res. Published online in March 2010.*

[31] Elnashar, M. and Yassin, A. (2009). Covalent immobilization of β-galactosidase on carrageenan coated chitosan. *Appl. Polym. Sci*, 114, 17.
[32] Elnashar, M. and Yassin, A. (2009). Lactose Hydrolysis by β-Galactosidase Covalently Immobilized to Thermally Stable Biopolymers. *J. Appl. Biochem. Biotechnol*, 159, 426.
[33] Chao, K., Haugen, M. and Royer, G. P. (1986). Stabilization of κ-carrageenan gel with polymeric amines: Use of immobilized cells as biocatalysts at elevated temperatures. *Biotechnol. Bioeng* 28, 1289.
[34] Richmond, M., Gray, J. and Stine, C. (1981). *J. Dairy Sci*, 1759, 64.
[35] German, J. H. (1997). Applied Enzymology of Lactose Hydrolysis, in: Milk Powders for the Future, 81.
[36] Sungur, S. and Akbulut, U. (1994**)**. Immobilisation of **β**-galactosidase onto gelatin by glutaraldehyde and chromium (III) acetate**.** *J. Chem. Technol. Biotechnol*, 59, 303.
[37] Nijipels, H. (1981). Lactases and their applications, in: G.G. Birch, H. Blakebrough, K.J. Parker (Eds.), Enzyme and Food Processing, Applied Science Publishers, London, 89.
[38] Muzzarelli, R. (1998). Colorimetric Determination of Chitosan. *Anal. Biochem*, 260, 255.
[39] Trinder, P. (1969). Determination of Glucose in Blood using Glucose Oxidase with an Alternative Oxygen Acceptor. *Anal. Clin. Biochem,* 6, 24.
[40] Elnashar, M.M. (2010). Review article: "Immobilized Molecules Using Biomaterials and Nanobiotechnology". *J. Biomat. Nanobiotechnol.* 1, 61-76.
[41] Tapia, C., Escobara, Z., Costab, E., Sapag-Hagara, J., Valenzuelaa, F., Basualtoa, C., Gaib, M. and Yazdani-Pedramc, M. (2004). Comparative studies on polyelectrolyte complex. *Eur. J. Pharma Biopharma*, 57, 65.
[42] Maciel, J., Silva, D., Haroldo, C. and Paula, R. C. (2005). Chitosan. *Eur. Polym. J*, 41, 2726.
[43] Desai, P., Dave, A. and Devi, S. (2004). Entrapment of lipase into K-carrageenan beads and its use in hydrolysis. *J. Mol. Catal.* B: Enz, 31, 143.
[44] Berger, J., Reist, M., Mayera, J., Felt, O., Peppas, N. and Gurny, R. (2004). Structure and interactions in covalently and ionically crosslinked chitosan. *Eur. J. Pharma Biopharma*, 57, 19.
[45] Mitsumataa, T., Suemitsua, Y. Fujiib, K. Fujiic, T. and Taniguchia, T. (2003) PH-response. *Polym*, 44, 7103.
[46] Tanriseven, A. and Dog˘an, S. (2002). A novel method for the immobilization. Process Biochem, 38, 27.
[47] Szczodrakr, J. (2000). Hydrolysis of lactose. *J. Mol. Catal.* B: Enz, 10, 631.
[48] Kennedy, J. and Cabral, M. (1987). In Enzyme Technology; Kennedy, J, Ed.; VCH: *Germany; Biotechnology; Rehm, J.,* Reed, G, Eds.; Vol. 7a, Chapter 7.
[49] Mohy Eldin, M. S., Portaccio, M. Diano,N. Rossi, S. Bencivenga, U. D'Uva, A. Canciglia, P. Gaeta, F. S. and Mita, D. G. (1999). influence of microenvironment on the activity of enzymes immobilized on teflon membranes grafted by γ radiation. *J. Mol. Catal.* B: Enzy. 7, 251.
[50] Haider, T. and Husain, Q. (2007). Preparation of lactose-free milk by using salt-fractionated almond (*Amygdalus communis*) **β**-galactosidase. *J. Sci. Food Agri,* 87, 1278.

[51] Nakane, K., Ogihara, T., Ogata, N. and Kurokawa, Y. (2001). Entrap-immobilization of invertase on composite gel fiber of cellulose acetate and zirconium alkoxide by sol-gel process. *J. Appl. Polym. Sci*, 81, 2084.

In: Handbook of Chitosan Research and Applications
Editors: Richard G. Mackay and Jennifer M. Tait
ISBN 978-1-61324-455-5

Chapter 7

CHITOSAN-INDUCED MODULATION OF SIGNAL PATHWAYS FOR VALUE-ADDITION TO HORTICULTURAL COMMODITIES

G. Manjunatha, V. Lokesh, Nandini P. Shetty and N. Bhagyalakshmi*

Plant Cell Biotechnology Department,
Central Food Technological Research Institute, Mysore -570 020, India

ABSTRACT

Once known as a mere surfactant, chitosan is now recognized as a marvelous signal-inducer capable of eliciting a wide array of secondary and primary metabolic networks in plants, imparting tremendous improvements to field crops and harvested flowers, fruit and vegetables. The presence of amino group in C_2 position of chitosan provides major functionality allowing various modifications to meet the needs of horticultural, biotechnological and food applications. Jasmonic acid (JA), calcium (Ca^{++}), ethylene (Et), abscisic acid (ABA), nitric oxide (NO) etc., in plants act as chief signal molecules modulating the networks involved in regulation and turn-over of shikimates and terpenoids implicated in defense, and polyamine pools needed for supporting most of such biochemical pathways. Such chitosan-induced regulations point towards developing better synergies for desirably eliciting specific type of biochemical changes. Chitosan-induced JA alteration also modulates genes of ethylene-responsive networks that affect senescence and other biochemical, sensorial and keeping quality of fruits and vegetables. Chitosan induced enzymatic alterations accomplished through different signal transducers alter shikimic acid pathways modulating the transcription activation and translational regulations of phenylalanine ammonia-lyase (PAL), tyrosine ammonia-lyase (TAL), peroxidase (POD), polyphenol oxidase (PPO), catalase (CAT) etc., resulting in conferring beneficial changes to fruit and vegetables such as conversions of phenolic compounds to more desirable ones, increase in soluble solids, ascorbic acid accumulation, pigment biosynthesis, while simultaneously eliciting phytoalexins, defense compounds and anti-

*(A constituent laboratory of the Council of Scientific and Industrial Research, New Delhi). E-mail: blakshmi_1999@yahoo.com; pcbt@cftri.res.in.

oxidants imparting resistance against decay of fresh harvests. The present review highlights progress made recently in chitosan-induced elicitation of signal networks for accomplishing desirable changes in horticultural commodities.

INTRODUCTION

Plants are a chief source of a variety of secondary metabolites and many of them are of potential economic importance. Fruit and vegetables are most popular nutritional components of daily life which are time and again proven to offer properties of anti-ageing and prevention/delaying of chronic ailments. Losses of harvested fruit, vegetables and cut flowers account for 25% in tropical countries, although this number is much smaller in temperate zones of developed countries. Thus, enormous research has taken place in developing gentle preservation methods to retain their freshness for extended periods. One of the promising approaches to prevent such losses is by the use of chitosan either as an elicitor of *in situ* defense functions or as a semi-permeable edible coating. Chitosan is derived from the natural product, chitin, which is a natural polysaccharide found as exoskeleton of crustaceans and cell walls of fungi. Chitosan is a low acetyl derivative of chitin and is composed primarily of glucosamine, 2-amino-2-deoxy-β-D-glucose, known as (1-4)-2-amino-2-deoxy-(D-glucose). Three types of reactive functional groups, an amino group as well as both primary and secondary hydroxyl groups at the C-2, C-3 and C-6 positions forms biologically active chitosan (Furusaki et al., 1996) (Figure1). The opportunities for altering these reactive groups make this molecule versatile for numerous biotechnological applications.The property of cross linking chitosan with other food materials has been widely applied for enzyme immobilization and adsorption process (Shantha and Harding, 2002).

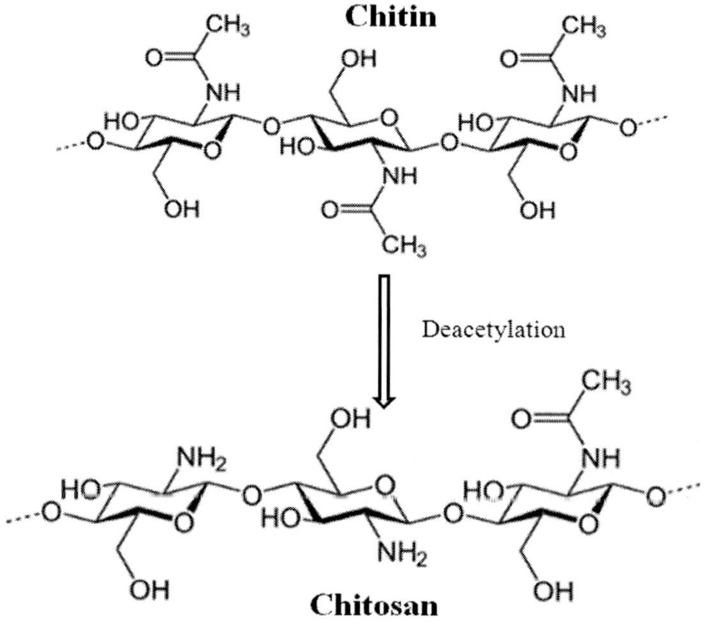

Figure 1. Structure of chitin and the chitosan derived from deacetylation of chitin.

Chitosan is widely used to alter pre-harvest physiologies of many crops, particularly to impart growth promoting effects (El-Sawy et al., 2010). Chitosan application enhanced the growth of roots and shoots of radish (*Raphanussativus*var.*longipinnatus*) (Tsugitaet al., 1993) and induced early and higher flowering in orchids (Ohta et al., 2004). Such flowering enhancement was also found to increase the number of vascular bundles containing silica, probably effecting silica metabolism (Limpanavech et al., 2008). When used for minimizing the water loss in field crops, chitosan was found to regulate stomata openings by modulating the ABA through hydrogen peroxide (H_2O_2), and protect the photosynthetic apparatus against photo damage. Chitosan treatment enhanced the lipase activity, gibberellic acid (GA) and indole acetic acid (IAA) in peanut (Zhou et al., 2002). Furthermore, chitosan has been widely used as a coating to enhance the shelf life of fruits. Chitosan coating delayed ripening of sugar apple (*Annonasquamosa*), softening of tomatoes by interfering with ethylene biogenesis (Chunprasert et al., 2006) and reduced decay of peach, pear and kiwifruit probably by eliciting cellular defense compounds (Jianming et al., 1997). In many vegetables and fruit, chitosan altered the activities of enzymes such as chitinase, peroxidase (POD), β-1,3-glucanase and chitosanase protecting firmness. Chitosan has been used for preventing the discoloration and delaying colour loss and browning of fruit litchi, longan and Chinese water chestnut (Rinaudo et al., 1989). Maintenance of water in the fruits and vegetables at appropriate level is very important for keeping metabolic homeostasis and turgidity for appropriate preservation of nutrients, which is often achieved by chitosan coatings (Iriti et al., 2009). Refrigeration is widely practiced for preservation of fresh and processed products, although quite often it imparts stress on live materials. Circumventing such stress by various physicochemical factors is an interesting approach, finding a wide application. In maize under low temperature stress, chitosan enhanced germination index, reduced the mean germination time (MGT), and increased shoot height, root length, dry weights of shoot and root, declined malondialdehyde (MDA), relative permeability of the plasma membrane and increased the concentrations of soluble sugars and proline, a stress combating amino acid (Guan et al., 2009). Benefits offered to horticultural commodities by chitosan through alterations of various biochemical pathways are discussed under separate headings hereafter.

*Chitosan as elicitor:*The foremost mechanism by which chitosan imparts beneficial effects is by its elicitation actions. Use of biotic and abiotic elicitor to enhance secondary metabolites *in planta* is an interesting strategy (Vasconsueloand Boland, 2007). Elicitation is the process of triggering cellular biochemical events leading to the activation of signaling cascades from outside the cell to specific gene activation, thereby inducing expression of enzymes responsible for enhanced synthesis of secondary metabolites. Elicitors are compounds of mainly microbial or insect origin or non-biological origin, which have been successfully applied for higher plant cells to trigger the increased production of pigments, flavones, phytoalexins and other defense related compounds (Gomez-Vasquez et al., 2004). It has been demonstrated that chitin and its derivatives elicit a wide range of defense compounds (Manjunatha et al., 2008). This knowledge is exploited to induce resistance in plants, inhibition of physiological disorders and enhancement of the shelf life of fruit by delaying ripening, ripening inhibition in vegetables, lengthening floral displays and overproduction of secondary metabolites of food and pharmaceuticals applications. Treatment of hairy root cultures of *Brugmansia candida* and*Hyoscyamusmuticus*with chitosan enhanced scopolamine and hyoscyamine, which are valuable anticholinergic drugs (Sevonet al., 1992). Application of chitosan to the culture medium of *Petroselinumcrispum* enhanced the production of hernandulcin, a minor constituent of the essential oil obtained from the aerial parts of *Lippiadulcis* (Sauerweinet al., 1991). In *Capsicum* cell cultures, chitosan significantly enhanced accumulation of capsaicin and other phenyl proponiod pathway (PPP) compounds (Johnson et al., 1990), the formation of scopolin and chlorogenic acid in suspension cultures of *Nicotiana*sp. and *Eschscholtziacalifornica (*Brodeliuset al., 1989). Indirubin production was enhanced in the presence of chitosan along with calcium chloride, and indole production in *Polygonumtinctorium* (Kim et al., 1997). The chitosan application at a low level of 50 mg l^{-1} produced a 5-fold enhancement of oleic acid accumulation (Wiktorowska et al., 2010). Chitosan caused conversion of pulegone to 40 fold higher menthol than control in *Menthapiperita*cell culture (Chang et al.,1998), enhanced vanillin in *Vanilla planifolia*(Funk and Brodelius, 1990) and several secondary metabolites in *Glycine max* and *Phaseolus vulgaris* (Young et al., 1982).

BIOCHEMICAL PATHWAYS ELICITED BY CHITOSAN

Shikimic acid pathway (SAP): A large number of PPP compounds synthesized through SAP participate in cellular defense, and hence are also responsive to elicitors. Upon chitosan treatment the synthesis of phenolic acids was stimulated in parsley where ferulic acid synthesis was linear with chitosan concentration (Cornath et al., 1989; Pearce and Ride, 1982). The lignin content of primary leaves also showed a similar pattern. The synthesis of precursors of lignin such as *p*-coumaric, ferulic, sinapic and other phenolic acids having antimicrobial activity were also elicited. Direct correlations between chitosan levels and β-1, 3-glucan synthase activity leading to callose deposition in the cell wall was found. Such callose deposition was implicated in changing protein phosphorylation, cellular calcium increase, decrease of potassium, and internal pH, and alterations in JA production (Kauss, 1994). Chitosan elicited synthesis of phytoalexin and callose formation in soybean cells occurring as a Ca^{2+} dependent process during membrane perturbation, resulting in electrolyte leakage from cells (Kohle et al., 1985), which probably played a defensive role.

In many plants, it has been demonstrated that chitosan triggers PAL and TAL, the key enzymes of PPP and further altered precursors of secondary metabolites progressing towards lignin formation and synthesis of flavonoid pigments and phytoalexins known for their roles in plant-pathogen interactions (Morrison and Buxton, 1993; Khan, 2003; Kim et al., 2005) of SAP. In grapefruit, chitosan elicited the above enzymes, resulting in increased total soluble solids and decreased titratable acidity (Meng et al., 2010). Chitosan enhanced biosynthesis of phenolic volatiles such as vanillin in suspension cultures of *Vanilla planifolia* (Funk and Brodelius, 1990), and accumulation of *p*-hydroxybenzoic acid in endosperm cell suspension cultures of *Cocosnucifera* (Chakraborty et al., 2009).

Coating *Litchi chinensis* withvarious concentrations of chitosan dissolved in glutamic acid delayed degradation of anthocyanins, flavonoids, and total phenolics (Zhang and Quantick, 1997), delayed the increase in PPO activity and partially countered POD activity (Jiang et al., 2005). Such effects of chitosan coating were also observed with peeled litchi fruit (Dong et al., 2004), longan fruit (Jiang and Li, 2001), and fresh-cut Chinese water chestnut vegetable (Rinaudo et al., 1989; Pen and Jiang 2003). Chitosan also influenced the litchi browning and water content (Caro and Joas, 2005), its pH, and dehydration rate during storage (Joas et al., 2005). In maize, activities of POD and CAT were found modulated by chitosan treatment during chilling (Guan et al., 2009). Although chitosan also induced browning through PAL induction, such sensitivities of plant cells was affected by the amount of POD present in the cells. High POD levels in the potato cells resulted in a rapid degradation of H_2O_2 and suppressed PAL after chitosan treatment. Different degrees (reversible, irreversible) and locations (tonoplast, plasma membrane) of permeabilization of cell membranes led to polyphenol (PP) production, which correlated with reaction rates of PPO and increased pressure treatments. Chitosan activated the octadecanoid pathway via different recognition events, which further induced the expression of defense genes (Doares et al., 1995) as well as various bio-molecules (polyamines, flavonoids, isoprenes etc.,) functioning as precursors for the synthesis of secondary metabolites.

*Ethylene:*Majority of the fruit, vegetables and flowers, immediately after harvest, are characterized by a surge of ethylene production, which is associated with progression of senescence. Various approaches are now adopted to reduce ethylene formation and prevent postharvest losses, one of which is the use of different modified atmosphere packaging systems (MAP) in conjuction with refrigiration (Annese et al., 1997). MAP is known to reduce ethylene and respiration due to low O_2 availability and higher CO_2 accumulation in the fruit surroundings. Slowing down respiration and ethylene delays all physiological processes, including those associated with senescence. Edible coatings are also used to create MAP for enhancing functional biomolecules in fruit and vegetables such as antioxidants, flavors, colors, antimicrobial agents, and nutraceuticals (Kester and Fennema,1986; Diab et al., 2001; Han et al., 2004). Thus chitosan coating has been successfully used for reducing the respiration rate and ethylene production, control of decay and retention of firmness in apples (Hwang et al., 1998; Davies et al., 1989), banana (Kittur et al., 2001), citrus (Chien and Chou, 2006; Chien et al., 2007a), mango (Kittur et al., 2001; Chien et al., 2007b), peach (Li and Yu, 2000), carrots (Durango et al., 2006), and lettuce (Devlieghere et al., 2004). Apples coated with N,O-carboxymethyl chitosan films and placed in cold storage could be kept in fresh condition for extended periods (Davies et al., 1989). Coating fruit with semipermeable film has also been shown to retard ripening by modifying the endogenous CO_2, O_2, and ethylene levels of fruits (El Ghaouth et al., 1991). Tissue exposure to low levels of O_2 and/or high levels of CO_2 suppresses C_2H_4 biosynthesis in climacteric fruit (Kader, 1986). Under anaerobic conditions, conversion of 1-aminocyclopropane-1-carboxylic acid (ACC) to C_2H_4 can be completely inhibited because O_2 is a co-substrate in the oxidation of ACC to C_2H_4 by ACC oxidase (ACCO; Yang, 1985). The combination of low O_2 and elevated CO_2 can have a synergestic effect in suppressing C_2H_4 formation(Gorny and Kader, 1996) functioning through alerations in expression levels of key C_2H_4 biosynthesis enzymes - ACC synthase (ACC-S) and ACC-Oxidase (Yip et al., 1988; Poneleit and Dilley, 1993; Gorny and Kader, 1996). In general, non stressful CO_2 enriched atmospheres were helpful in reducing C_2H_4 biosynthesis in numerous climacteric fruit (Kerbel et al., 1988). Similarly, elevated CO_2 levels could enhance/reduce, or had no effect on C_2H_4 biosynthesis in fruit tissue, depending up on the tissue and amount of CO_2 present (Kader, 1986). Chitosan coating is likely to modify the internal atmosphere without switiching on anaerobic respiration, since chitosan films are more selectively permeable to O_2 than to CO_2 (Bai et al., 1988). Therefore, chitosan coating has a potential to prolong storage life by ethylene supression through elicitation of other signal molecules (Manjunatha et al., 2010).

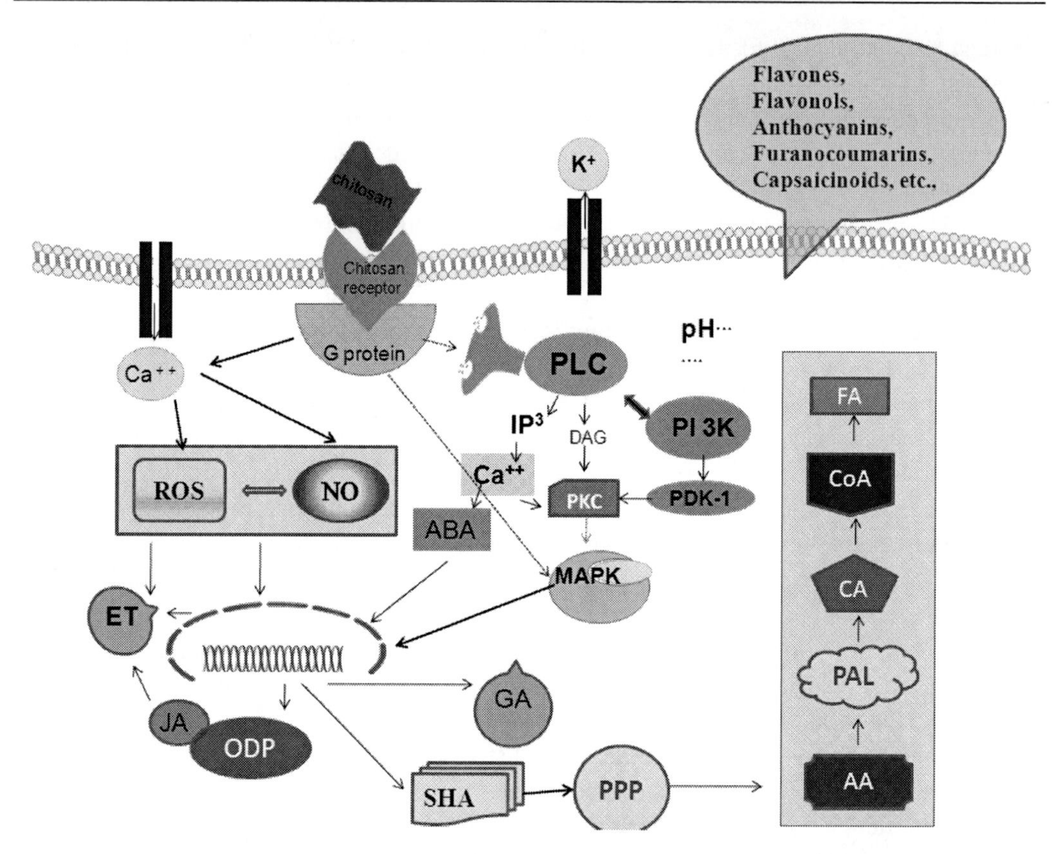

Figure 2. Signaling cascades of various biosynthetic pathways induced by chitosan. Binding of chitosan to receptors localized in the plasma membrane coupled to G-protein activation stimulates phospholipase-C (PLC) increasing secondary messengers levels, diacylglycerol (DAG), inositol 1,4,5-trisphosphate (IP3), releasing calcium for activation of their target kinases (PKC). Chitosan also leads to changes in cytoplasmic Ca^{2+} fluxes through plasma membrane or intracellular reservoirs leading to activation of phosphoinositide 3-kinase (PI3K) regulating PKC through phosphoinositide-dependent kinase (PDK-1). Whilst, jasmonic acid (JA), ethylene (ET), nitric oxide (NO), reactive oxygen species (ROS) are chief signal molecules elicited by chitosan. This cascade stimulates the mitogen activated protein kinase (MAPK), changes in the extent of phosphorylation leading to transcriptional activation of genes coding for secondary metabolites such as shikimic acid (SHA), arachidionic acid (AA), phenyl alanine ammonialyase (PAL), coumaric acid (CA), acetyl CoA, ferulic acid (FA) and other phenyl proponoid intermediates (flavones, flavonols, anthocyanins, furanocoumarins, isoflavonoids, capsaicinoids etc.), resulting in synthesis of phenolic compounds, callose and lignin having defense functions. Abscisic acid (ABA) and Gibberilic acid (GA) are plant hormones regulated by chitosan treatment.

Activation of signal transduction: Perception of soluble and particulate chitosan by plant cells results in a variety of cellular responses such as cellular calcium regulation, hypersensitive response, programmed cell death (Zuppini et al., 2003), membrane H^+-ATPase inhibition (Amborabe, 2008), MAPK activation (Lizama Uc et al., 2007), development of callose for evading pathogens (Conrath et al., 1989), elicitation of reactive oxygen species (Lin et al., 2005), regulation of abscisic acid (Iriti and Faoro, 2008), jasmonate (Doares et al., 1995), synthesis of phytoalexins (Chakraborthy et al., 2008), induction of pathogenesis related proteins and systemic acquired resistance (Manjunatha et al., 2009). Chitosan also controlled intracellular proteins through regulation of phosphorylation at various levels (Lee et al., 1992) (Figure 2).

In the elicitor –mediated signal transduction, the foremost event is the perception of the elicitor through identification of membrane bound receptors and their numbers being different in different species. In *Brassica campestris* (sp. *Chinensis)* chitosan interaction with a membrane-bound receptor and a specific chitosan-binding glycoprotein having a mass of 53 kDa was isolated (Chen and Xu, 2005). In rice a similar kind of protein was shown to be involved in chitin signaling which is suspected functioning as chitosan receptor, although precise confirmation is needed through a knock-out mutant of the chitosan binding protein. In rice, the chitin-binding protein CEBiP was found to contain two extracellular LysM and a transmembrane domain (Kaku et al., 2006) and since CEBiP does not have an obvious intracellular domain, it needs a partner such as receptor-like kinase (RLK) for signal transduction from the plasma membrane into the cytoplasm. Chitosan differentially expressed some genes which had a similarity to a RLK gene pathway in calli of *Cocos nucifera.* Chitosan probably binds to this RLKmembrane integral protein that switches on signal transduction (Lizama et al., 2007) (Figure 2).

Cytosolic alkalinization is a common and early messenger which preceds the production of ROS and NO signals during activation of various pathways by chitosan (Gonugunta et al., 2009), which might be similar with other signals like abscisic acid (ABA) and methyl jasmonate (MJ). Elicitation of pathways for reducing the water stress in bean plants through induction of antitranspirant activity, mediated by ABA and driven by H_2O_2 imparted changes for decrease of stomatal conductance and transpiration rate (Iriti et al., 2009). Complementing the above process, chitosan is suspected to up-regulate the 9-*cis*-epoxycarotenoid dioxygenase, a key enzyme of ABA biosynthesis through the carotenoid pathway (Qin and Zeevaart, 1999; Thompson et al., 2000) and this ABA-induced stimulation was also known to protect the photosynthetic apparatus against photo-damage (Ivanov et al., 1995; Sharma et al., 2002). The addition of calcium - a well reputed signal molecule to chitosan coating formulation, not only enhanced the nutritional value of strawberry by increasing the uptake of calcium (Hernandez-Munoz et al., 2006), but also helped delaying softening, suppress mold growth and decrease the incidence of physiological disorders (Poovaiah, 1986).

Guanine nucleotide-binding proteins (G-proteins) are important signaling molecules in eukaryotes. Heterotrimeric G proteins, which consist of three subunits, α, β, and γ make-up, one of the major classes of G proteins involved in many signaling pathways. Thus, plant G-proteins have been implicated in various cellular processes linked to growth and development, hormone signaling and secondary metabolism (Assmann et al., 2002; Pandey and Assmann, 2004). Several studies have demonstrated the role of G-protein in response to elicitors and in elicitation process (Park et al., 2000; Kelly et al., 2003). For example the involvement of heterotrimeric G proteins in *Nod* factor signaling has been shown (Pingret et al., 1998). The putative receptor might function by coupling to a G-protein. Furthermore, Vasconsuelo et al., (2006) showed that chitosan stimulated anthraquinone synthesis in *Rubiatinctorum* cells through elicitation of heterotrimeric G-proteins. Although, many details related to plant G-proteins are still unclear, undoubtedly both heterotrimeric and small G-proteins are involved in the biosynthesis of various plant secondary metabolites elicited by chitosan.

MAPK cascades are major components downstream of receptors and sensors that transduce external signals into intracellular responses (Shetty et al., 2008). Plant MAPKs are activated by a variety of biotic and abiotic stimuli (Stratmann et al., 1997). This activation may affect other pathways or specific genes where MAPK is also known to participate in chitosan signaling (Figure 2). Thus, MAPK cascade was activated in *Rubiatinctorum* when anthroquinone production was initiated after chitosan application where the signaling was mediated by the MAPK to the nucleus. It was suggested that the MAPK phosphorylated transcription factors leading to increased expression of genes coding for PPP and ODP enzymes, play an important role in the biosynthetic pathway of anthroquinones(Vasconsuelo et al., 2006). A similar signal cascade was reported in coconut calli (Lizama et al., 2007) and during saponin biosynthesis in cell cultures of ginseng (Hu et al., 2004). The first step being the rapid activation of plasma membrane H^+-ATPase in all types of cells and the upstream signaling through activation of PLC mediated MAPK have been prevalent in animal cells where PLC acts as an upstream activator of MAPK. Whereas, upstream activation of PKC was involved in chitosan stimulation in plant cells with no significant contribution towards elicitation by cAMP (Amborabe et al., 2008). In a microarray study in *Brassica napus*, a JA synthase gene, a JA-mediated ethylene receptor gene, and two ethylene responsive element binding protein (EREBP) genes were found induced by chitosan-elicited JA/ET signaling pathways (Yin et al., 2006).

Clearly, chitosan may affect several signaling processes in plants and that the events occurring are highly complex and inter-connected (Figure 2). Chitosan may induce different signaling pathways in soybean cells depending on its concentration that at lower doses it activated cell death pathway and the higher doses just led to the membrane disruption (Zuppini et al., 2003). The ability of eliciting certain pathways may depend on the plant system rather than the elicitor. Hence further studies in different plants are necessary to draw conclusions on the number of chitosan activated pathways. Furthermore, it is also necessary to understand the relationship between different molecular weights of chitosan elicitors and the signaling pathways as well as whether different receptor molecules are involved. To ensure this, different molecular modification of chitosan were made and their effects on value addition to horticultural commodities are discussed below.

MODIFICATION OF CHITOSAN FOR ELICITATION

Chitosan is a linear molecule composed of randomly distributed β-(1-4)-linked D-glucoseamine (deacetylated units) and N-acetyl-D-glucosamine (acetylated units) and is processesed by removing acetyl groups (CH_3-CO) for the molecule to be soluble in diluted acids. This process of deacetylation, releases amine groups (NH_2) making chitosan cationic (Furusaki et al., 1996).

The degree of acetylation (DA) of chitosan can affect its biological activities, particularly elicitation and its ability to bind polygalacturonate, a component of plant cell walls responsible for the activation or inhibition of enzymes, the inactivation of viruses, bacterial growth and agglutination. Low level of acetylation (DA-1%) caused complete growth inhibition of three strains of *Phytophthroaparasitica,* whereas DA with 36.5% was inefficient in completely inhibiting such strains (Falcon et al., 2008).

DA of chitosan also affected its spatio-temporal functions in triggering enzyme production in plants. Application of chitosan to roots of tobacco systemically enhanced POD and PAL activities in leaves where different DA's showed specific effects (Falcon-Rodríguez et al., 2009). Similarly, the antibacterial activity of chitosan was influenced by its DA; the polycationic chitosan was found to bind to the negatively charged bacterial surface causing agglutination of bacterial cells, while higher concentrations resulted in higher positive charge causing a net positive charge at the bacterial surfaces keeping them in suspension (Sudarshan et al., 1992). Highly deacetylated chitosan oligomers induced PAL activation, H_2O_2 synthesis and cell death in *Arabidopsis thaliana* cell suspensions, and the progressive reacetylation progressively impaired the ability to enhance H_2O_2 accumulation and cell death, but without affecting PAL (Cabrera et al., 2006). DA also influenced growth promoting properties (El-Sawy et al., 2010).

Since DA influenced elicitation to such an extent that altering chitosan DA, polymerization/molecular weight were the major aspects defining its application. Irradiation of chitosan with different doses (2–20 kGy) of Co-60 γ rays decreased its molecular weight, enhancing its antioxidant activity. For example chitosan of molecular weight 2.1×10^3 which had been irradiated at 20 kGy exhibited high reductive capacity and expressed good inhibition of linoleic acid peroxidation. At a concentration of 0.1 mg/mL ray γ-irradiated chitosan could scavenge 74.2% of superoxide radical (Feng et al., 2008). Hence, the DA and polymerization, require much experimentation to make this unique molecule applicable to enhance the nutritional components in food products. LMWC (low molecular weight chitosan) exhibited stronger scavenging activity toward DPPH radicals, superoxide anion radicals and H_2O_2, compared to either MMWC (medium molecular weight chitosan) or HMWC (higher molecular weight chitosan). At a concentration of 0.8 mg/ml in apple juice, the LMWC chitosan exhibited 88.2%, 99.8% and 93.0% scavenging activities toward DPPH radicals, H_2O_2 and superoxide anion radicals, respectively (Po-jung et al., 2007). Increasing the DA of chitosan from 5 to 30% appreciably increased the elicitation of chitinase, peroxidase, and β-1,3-glucanase activities in cucumber and bean plants (Ben-Shaldm et al., 2002). LMWC improved fruit firmness, titratable acidity, ascorbic acidity and water content of fruit, with no such effects by HMWC (Chienet al., 2007). In addition, the fruit coated with LMWC exhibited greater and prolonged protection from decay than the treatment with standard fungicide thiabendazole. Chitosan coating that was formed due to the coacervation either by the addition of ethanol or sodium hydroxide could increase the retention of amino acid in the fat prills. CAT immobilized on chitosan beads exhibited better thermal and pH stabilities than the free ones, and tyrosinase, glucosidase, and acid phosphatase activities were higher showing improved stability (Cetinus and Oztop, 2003; Wu et al., 2001). Chitosan coating in combination with ascorbic acid in pears increased beneficial nutrients and also helped to retain a much higher ascorbic acid content and the activities of antioxidative enzymes (superoxide dismutase (SOD), CAT and ascorbate POD), reducing the respiration rate and inhibiting the senescence process (Luo and Barbosa-Canovas, 1996; Lin et al., 2008).

Due to its excellent compatibility with other functional substances, chitosan has been widely used in different forms with amalgamation of other bioactive compound and nutraceuticals. Several workers have attempted to incorporate calcium (Han et al., 2004; Hern´andez-Mu˜noz et al., 2006), vitamin E (Han et al., 2004, 2005), potassium (Park et al., 2005), or oleic acid (Vargas et al., 2006) into chitosan film formulation to prolong the shelf life and to enhance the nutritional value of fruits. Chitosan based coatings containing a mixture of calcium lactate and calcium gluconate and DL-α-tocopheryl acetate applied to strawberries (*Fragaria ananassa*) and red raspberries (*Rubus ideaus*) proved to extend the shelf life. Chitosan-based coatings with high concentration of calcium salts added countered the ion interactions among adjacent molecular structures with small mineral ions acting as fillers, thus resulting in decreased diffusivity of water vapor through the film matrix and an increase in the hydrophobicity of chitosan films. Similarly, oleic acid and acetic acid impregnation enhanced antimicrobial activity and water vapor resistance of chitosan (1% in 1% acetic acid) in strawberries (Han et al., 2004).

Conclusions

A vast number of studies have shown that chitosan affects plant secondary metabolism by modulating their rates of biosynthesis, accumulation, turnover and degradation. However, the exact mechanism by which the secondary metabolism is elicited is only partly understood. In general, activation of any plant secondary metabolite depends on the complex interactions between elicitor and biosynthetic machinery of plant cell. Signal perception by receptors is the first step in the elicitor signal transduction cascade, where different molecular modifications imparted to chitosan appear to dramatically change cellular perceptions. Therefore, it is important to test the molecular nature and concentration of chitosan for inducing specific functions in plants. Chitosan based polymeric materials that can be formed into fibers, films, gels, sponges, beads or even nano-particles are employed for pre- and post-harvest treatment as well as used to develop edible/eco-friendly packaging material.

References

Amborabe BE, Bonmort J, Fleurat-Lessard P, Roblin G. Early events induced by chitosan on plant cells. *J. Exp. Bot.* 2008; 59: 2317-24.

Annese M, Manzano M, Nicoli MC. Quality of minimally processed apple slices using different modified atmosphere conditions. *J. Food Quality.* 1997; 20: 359-370.

Assmann SM. Heterotrimeric and unconventional GTP binding proteins in plant cell signaling. *Plant Cell Supl.* 2002; S355-S373.

Bai RK, Huang MY, Jiang YY. Selective permeabilities of chitosan-acetic acid complex membrane and chitosan-polymer complex membranes for oxygen and carbon dioxide. *Polym. Bull.* 1988; 20: 83-88.

Ben-shaldm N, Cuneyt A, Ruti A, Riki P. Elicitation effects of chitin oligomers and chitosan sprayed on the leaves of cucumber (*Cucumissativus*) and bean (*Phaseolus vulgaris*) plants. *Israel J. Plant Sci.* 2002; 50: 199-206.

Brodelius PC, Funk A, Hainer M, Villegas. A procedure for the determination of optimal chitosan concentration of cultured plant cells. *Phytochem.* 1989; 28: 2651-2654.

Cabrera JC, Johan M, Pierre C, Pierre VC. Size, acetylation and concentration of chito-oligosaccharide elicitors determine the switch from defense involving PAL activation to cell death and water peroxide production in Arabidopsis cell suspensions. *Physiol. Plant.* 2006; 56: 127-44.

Caro Y, Joas J. Postharvest control of litchi pericarp browning (cv. Kwai Mi) by combined treatments of chitosan and organic acids. II. Effect of the initial water content of pericarp. *Postharvest Biol. Technol.* 2005; **38**: 137- 44.

Cetinus SA, Oztop HN. Immobilization of catalase into chemically cross-linked chitosan beads. *Enzyme Microbial Technol.* 2003; 32: 889–894.

Chakraborty M, Karun A, Mitra A. Accumulation of phenyl propanoid derivatives in chitosan-induced cell suspension culture of *Cocosnucifera. J. Plant Physiol.* 2009: 166; 63-71.

Chang JH, Shin JH, Chung IS, Lee HJ. Improved menthol production from chitosan-elicited suspension culture of *Mentha piperita. Biotechnol. Lett.* 1998; 20: 1097-1099.

Chen HP, Xu LL. Isolation and characterization of a novel chitosan-binding protein from non-heading Chinese cabbage leaves. *J. Integrative Plant Biol.* 2005; 47: 452-456.

Chiang YW, Wang TH, Lee WC. Chitosan coating for the protection of amino acids that were entrapped within hydrogenated fat. *Food Hydrocolloids* 2009; 23: 1057-1061.

Chien PJ, Chou CC. Antifungal activity of chitosan and its application to control post-harvest quality and fungal rotting of Tankan citrus fruit (*Citrus tankan* Hayata). *J. Sci. Food Agric.* 2006; 86: 1964-1969.

Chien PJ, Sheu F, Lin HR. Coating citrus (*Murcotttangor*) fruit with low molecular weight chitosan increases postharvest quality and shelf life. *Food Chem.* 2007; 100: 1160-1164.

Chien PJ, Sheu F, Yang FH. Effects of edible chitosan coating on quality and shelf life of sliced mango fruit. *J. Food Eng.* 2007; 78: 225-9.

Chunprasert A, Uthairatanakij A, Wongs-Aree C. Storage quality of ′neang′ sugar apple treated with chitosan coating and MAP. Proceedings of the IVth international conference on managing quality in chains MQUIC 2006.

Conrath U, Domard A, Kauss H. Chitosan-elicited synthesis of callose and of coumarin derivatives in parsley cell suspension cultures. *Plant Cell Rep.* 1989; 8: 152-155.

Davies DH, Elson CM, Hayes ER. N,O- Carboxymethyl chitosan, a new water soluble chitin derivative. In: Skjak-Bræk G, Anthonsen T, Sandford P, editors. Chitin and chitosan: sources, chemistry, biochemistry, physical properties and applicatins. London, U.K.: Elsevier. 1989: p 467-72.

Devlieghere F, Vermeulen A, Debevere J. Chitosan: antimicrobial activity, interactions with food components and applicability as a coating on fruit and vegetables. *Food Microbiol.* 2004; 21: 703-14.

Diab T, Biliaderis CG, Gerasopoulos D, Sfakiotakis E. Physicochemical properties and application of pullulan edible films and coatings in fruits preservation. *J. Sci. Food Agric.* 2001; 81: 988-1000.

Doares SH, Syrovets T, Weiler EW, Ryan CA. Oligogalacturonides and chitosan activate plant defence genes through the octadecanoid pathway. *PNAS* 1995; 92: 4095-4098.

Dong H, Cheng L, Tan J, Zheng K, Jiang Y. Effects of chitosan coating on quality and shelf life of peeled litchi fruit. *J. Food Eng.* 2004; 64: 355-358.

Durango AM, Soares NFF, Andrade NJ. Microbiological evaluation of an edible antimicrobial coating on minimally processed carrots. *Food Control* 2006; 17: 336-41.

El Ghaouth A, Arul J, Benhamou N, Asselin A, Belanger RR. Effect of chitosan on cucumber plants: suppression of *Pythium aphanidermatum* and induction of defense reactions. *Phytopathol.*1994; 84:313-320.

El-Sawy NM, Abd El-Rehim HA, Elbarbary AM, El-Sayed A. Hegazy. Radiation-induced degradation of chitosan for possible use as a growth promoter in agricultural purposes. *Carbohydrate Polymers* 2010; 79: 555-562.

Falcon AB, Cabrera JC, Costales D, Ramırez MA, Cabrera G, Toledo V, Angel M, Martınez-Tellez. The effect of size and acetylation degree of chitosan derivatives on tobacco plant protection against *Phytophthora parasitica* var. *nicotianae. World J. Microbiol. Biotechnol.* 2008; 24: 103-112.

Falcon-Rodríguez AB, Cabrera JC, Martinez-Tellez MA. Concentration and physicochemical properties of chitosan derivatives determine the induction of defense responses in roots and leaves of tobacco (*Nicotiana tabacum*) plants. *American J. Agri. Biol. Sci.* 2009; 4: 192-200.

Feng T, Du Y, Li J, Hu Y, Kennedy JF. Enhancement of antioxidant activity of chitosan by irradiation. *Carbohydrate Polymer* 2008; 73:126-132.

Funk C, Brodelius P. Influence of growth regulators and an elicitor on phenyl propanoid metabolism in suspension culture of *Vanilla planifolia. Phytochem.* 1990; 29: 845-848.

Furusaki E, Ueno Y, Sakairi N, Nishi N, Tokura S. Facile preparation and inclusion ability of a chitosan derivative bearing corboxymethyl-beta-cyclodextrin. *Carbohydrate polymer* 1996; 9: 29-34.

Gómez-vásquez R, Day R, Buschmann H, Randles S, Beeching JR, Cooper RM. Phenylpropanoids, phenylalanine ammonia lyase and peroxidases in elicitor-challenged Cassava (*Manihot esculenta*) suspension cells and leaves. *Annals Botany* 2004; 94: 87-97.

Gonugunta V, Nupur S, Agepati S, Raghavendra. Cytosolic alkalinization is a common and early messenger preceding the production of ROS and NO during stomatal closure by variable signals, including abscisic acid, methyl jasmonate and chitosan. *Plant Signal Behaviour* 2009; 4: 561-564.

Gorny JR, Kader AA. Controlled atmosphere suppression of ACC synthase and ACC oxidase in 'Golden Delicious' apples during long term cold storage. *J. Amer. Soc.Hort Sci.* 1996; 121: 751-755.

Gorny JR. Summary of CA and MA requirements and recommendations for fresh-cut fruits and vegetables. In J. R. Gorny (Ed.), CA 97 Proceedings: Fresh cut fruits and vegetables and MAP, Davis, CA: University of California 1997; 5: 30–33.

Guan Ya-jing, Hu J, Wang Xian-ju, Shao Chen-xia. Seed priming with chitosan improves maize germination and seedling growth in relation to physiological changes under low temperature stress. *J. Zhejiang Univer. Science* 2009; 10: 427-433.

Han C, Lederer C, McDaniel M, Zhao Y. Sensory evaluation of fresh strawberries (*Fragariaananassa*) coated with chitosan-based edible coatings. *J. Food Sci.* 2005; 70: S172-178.

Han C, Zhao Y, Leonard SW, Traber MG. Edible coatings to improve storability and enhance nutritional value of fresh and frozen strawberries (*Fragaria× ananassa*) and raspberries (*Rubusideaus*). *Postharvest Biol. Technol.* 2004; 33: 67-78.

Hernandez-Munoz P, Almenar E, Jose OM, Gavara R. Effect of calcium dips and chitosan coatings on postharvest life of strawberries (*Fragaria*x *ananassa*). *Postharvest Biol. Technol.* 2006; 39: 247–253.

Hwang YS, KimYA, Lee JC. Effect of postharvest application of chitosan and wax, and ethylene scrubbing on the quality changes in stored 'Tsugaru' apples. *J. Kor. Soc.Hort Sci.* 1998; 39: 579-82.

Iriti M, Faoro F. Abscisic acid mediates the chitosan-induced resistance in plant against viral disease. *Plant Physiol.Biochem.* 2008; 46: 1106-11.

Iriti M, Picchib V, Rossonia M, Gomarascac S. Nicola L, Marco G, Franco F, Chitosan antitranspirant activity is due to abscisic acid-dependent stomatal closure. *Environ. Experim. Bot.* 2009; 66: 493-500.

Ivanov AG, Krol M, Maxwell D, Huner NPA. Abscisic acid induced protection against photoinhibition of PSII correlates with enhanced activity of the xanthophyll cycle. *FEBS Lett.*1995; 371: 61-64.

Jiang Y, Li J, Jiang Weibo. Effects of chitosan coating on shelf life of cold-stored litchi fruit at ambient temperature. *LWT* 2005; 38: 757-761.

Jiang Y, Li Y. Effects of chitosan coating on postharvest life and quality of longan fruit. *Food Chem.* 2001; 73: 139–43.

Jianming D, Hiroshi, Shuichi. Effect of chitosan coating on the storage of peach, Japanese pear and Kiwifruit. *J. Japan Soc. Hort Sci.* 1997; 66: 15-22.

Joas J, CaroY, Ducamp MN, Reynes M. Postharvest control of pericarp browning of litchi fruit (*Litchi chinensis* Sonn cv Kwai Mi) by treatment with chitosan and organic acids. Effect of pH and pericarp dehydration. *Postharvest Biol. Technol.* 2005; **38**: 128–36.

Johnson TS, Ravishankar GA, Venkataraman LV. *In vitro* capsaicin production by immobilized cells and placental tissues of *Capsicum annum* L. grown in liquid medium. *Plant Sci.* 1990; 70: 223-229.

Kader AA. Biochemical and physiological basis for effects of controlled and modified atmosphere on fruits and vegetables. *Food technol.* 1986; 40: 99-104.

Kaku H, Nishizawa Y, Ishii-Minami N, Akimoto-Tomiyama C, Dohmae N, Takio K, Minami E, Shibuya N. Plant cells recognize chitin fragments for defense signaling through a plasma membrane receptor. *PNAS* 2006; 103: 11086-11091.

Kauss H. Systemic signals condition plant cells for increased elicitation of phenylpropanoid defence responses. *Biochem. Soc. Symp*, Portland Press: 1994, 60, 95.

Kelly MN, Irving HR. *Nod* factors activate both heterotrimeric and monomeric G-proteins in *Vignaunguiculata* (L.) Walp. Planta 2003; 216: 674-685.

Kerbel EL, Kader AA, Romani RJ. Effects of elevated CO_2 concentrations on glycolysis in intact 'Bartlett' pear fruit. *Plant Physiol.* 1988; 86: 1205-1209.

Kester JJ, Fennema OR. Edible films and coatings: a review. Food Technol. 1986; 60: 47–59.

Khan W, Prithiviraj B, Smith DL. Chitosan and chitin oligomers increase phenylalanine ammonia-lyase and tyrosine ammonia-lyase activities in soybean leaves. *J. Plant Physiol.* 2003; 160: 859-63.

Kim HJ, Chen F, Wang X, Rajapakse NC. Effect of chitosan on the biological properties of sweet basil (*OcimumbasilicumL.*). *J. Agri. Food Chem.* 2005; 53: 3696-3701.

Kim JH, Shin JH, Lee HJ, Chung IS, Lee HJ. Effect of chitosan on indirubin production from suspension culture of *Polygonum tinctorium. J. Ferment. Bioeng.* 1997; 83: 206-208.

Kittur FS, Saroja N, Habibunnisa, Tharanathan RN. Polysaccharide-based composite coating formulations for shelf-life extension of fresh banana and mango. *Eur. Food Res. Technol.* 2001; 213: 306-11.

Kohle H, Jeblick W, Poten F, Blaschek W, Kauss H. Chitosan-elicited callose synthesis in soybean Cells as a Ca^{2+} dependent process. *Plant Physiol.* 1985; 77: 544-551.

Lee CH, Park JC, Choi K, Chae Q. The chitosan effect on the phosphorylation of intracellular proteins of ginseng callus cell. *Korean Biochem. J.* 1992; 25: 587-593.

Li H, Yu T. Effect of chitosan on incidence of brown rot, quality and physiological attributes of postharvest peach fruit. *J. Sci. Food Agric.* 2000; 81: 269-74.

Limpanavech P, Chaiyasuta S, Vongpromek R, Pichyangkura R, Khunwasi C, Chadchawan S, Lotrakul P, Bunjongrat R, Chaidee A, Bangyeekhun T.Chitosan. *Sci. Hort.* 2008; 116: 65-72.

Lin L, Wang B, Wang M, Cao J, Zhang J, Wu Y, Jiang W. Effects of a chitosan-based coating with ascorbic acid on post-harvest quality and core browning of 'Yali' pears (*Pyrus bertschneideri* Rehd.) *J. Sci. Food Agric.* 2008; 88: 877-884.

Lin W, Hu X, Zhang W, Rogers WJ, Cai W. Hydrogen peroxide mediates defense responses induced by chitosans of different molecular weights in rice. *J. Plant Physiol.* 2005; 162: 937-44.

Lizama G, Estrada-Mota IA, Caamal-Chan MG, Souza-Perera R, Oropeza-Salìn C, Islas-Flores I, Zuñiga-Aguillar JJ. Chitosan activates a MAP-kinase pathway and modifies abundance of defence-related transcripts in calli of *Cocusnucifera*L. *Physiol. Mol. Plant Pathol.* 2007; 70: 130-41.

Luo Y, Barbosa-Canovas GV. Preservation of apple slices using ascorbic acid and 4-hexylr esorcinol. *Food Sci. Technol. Intern.* 1996; 2: 315–321.

Manjunatha G, Roopa KS, Geetha NP, Shetty HS. Chitosan enhances systemic resistance in pearl millet against downy mildew disease caused by Scelrosporagraminicola and defense related enzyme activation. *Pest Management Sci.* 2008; 64: 1250-1257.

Manjunatha G, Lokesh V, Bhagyalakshmi N. Nitric oxide in fruit ripening-Emerging challenges and opportunities. *Biotech. Adv.* 2010; 28: 489-499.

Manjunatha G, Niranjan Raj S, Deepak S, Geetha NP, Amruthesh KN, Shetty HS. Nitric oxide is involved in chitosan induced systemic resistance in pearl millet against downy mildew disease. *Pest Manag. Sci.* 2009; 65: 737-743.

Meng X, Qin G, Tian S. Influences of preharvest spraying *Cryptococcus laurentii* combined with postharvest chitosan coating on postharvest diseases and quality of table grapes in storage. LWT - *Food Sci. Technol.* 2010; 43: 596-601.

Morrison TA, Buxton DR. Activity of Phenylalanine Ammonia-Lyase, Tyrosine Ammonia-Lyase, and Cinnamyl Alcohol Dehydrogenase in the Maize Stalk. *Crop. Sci.* 1993; 33: 1264-1268.

Ohta K, Morishita S, Suda K, Kobayashi N, Hosoki T. Effects of chitosan soil mixture treatment in the seedling stage on the growth and flowering of several ornamental plants. 2004; 73: 66-68.

Pandey S, Assmann SM. The Arabidopsis putative G protein–coupled receptor GCR1 interacts with the G protein a subunit GPA1 and regulates abscisic acid signaling. *Plant Cell* 2004; 16: 1616-1632.

Park J, Choi HJ, Lee S, Lee T, Yang Z, Lee Y. Rac-related GTP-binding protein in elicitor-induced reactive oxygen generation by suspension-cultured soybean cells, *Plant Physiol.* 2000; 124: 725-732.

Park SI, Stan SD, Daeschel MA, Zhao Y. Antifungal coatings on fresh strawberries (*Fragaria× ananassa*) to control mold growth during cold storage. *J. Food Sci.* 2005; 70: M202-207.

Pearce RB, Ride JP. Chitin and related compounds as elicitors of the lignification response in wounded wheat leaves. *Physiol. Plant Pathol.* 1982; 20: 119-123.

Pen LT, Jiang YM. Effects of chitosan coating on shelf life and quality of fresh-cut Chinese water chestnut. *LWT* 2003; 36: 359-364.

Pingret JL, Journet EP, Barker DG. Rhizobium nod factor signaling. Evidence for a g protein-mediated transduction mechanism. Plant Cell. 1998; 10: 659-672.

Po-Jung C, Fuu S, Wan-Ting H, Min-Sheng S. Effect of molecular weight of chitosans on their antioxidative activities in apple juice. *Food Chem.* 2007; 102: 1192-1198.

Poneleit LS, Dilley D. Carbon dioxide activation of 1-amonocyclopropane-l-carboxylate (ACC) oxidase in ethylene synthesis, *Postharvest Biol. Technol.* 1993; 3: 191-199.

Poovaiah BW. Role of calcium in prolonging storage life of fruits and vegetables. *Food Technol.* 1986; 40: 86-88.

Qin XQ, Zeevaart JAD. The 9-*cis*-epoxycarotenoid cleavage is the key regulatory step of abscisic acid biosynthesis in water stresse bean. *PNAS* 1999; 96: 15354-15361.

Rinaudo M, Domard A. In *Chitin and Chitosan*; Skjak-Braek G, Anthonsen T, Sandford P. Eds.; Elsevier Applied Science: London, 1989; pp 71-86.

Sauerwein M, Flores HM, Yamazaki T, Shimomura K. *Lippia dulcis* shoot cultures as a source of the sweet sesquiterpene hernandulcin. *Plant Cell Rep.*1991; 9, 663-666.

Sevon N, Hiltunen R, Oksman-Caldentey KM. Chitosan increases hyoscyamine content in hairy root cultures of *Hyoscyamus muticus*. *Pharma Pharmacol. Lett.* 1992; 2: 96-99.

Shantha KJ, Harding DRK. Synthesis and characterization of chemically modified chitosan microspheres.*Carbohydrate Polymers* 2002; 48: 247-253.

Sharma PK, Sankhalkar S, Fernandes Y. Possible function of ABA in protection against photodamage by stimulating xanthophyll cycle in sorghum seedlings. *Curr. Sci.* 2002; 82: 167-171.

Shetty NP, Jorgensen HJL, Jensen JD, Collinge DB, Shetty HS. Roles of reactive oxygen species in interactions between plants and pathogens. *Europ. J. Plant Pathol.* 2008; 121: 267-280.

Stratmann JW, Ryan C. Myelin basic protein kinase activity in tomato leaves is induced systemically by wounding and increases in response to systemin and oligosaccharide elicitors. *PNAS* 1997: 94: 11085-11089.

Sudarshan NR, Hoover DG, Knorr D. Antibacterial action of chitosan. *Food Biotechnol.*1992; 6: 1532-4249.

Thompson AJ, Jackson AC, Parker RA, Morpeth DR, Burbidge A, Taylor IB. Abscisic acid biosynthesis in tomato: regulation of zeaxanthin epoxidase and 9-*cis*-epoxycarotenoid dioxygenase mRNAs by light/dark cycles,water stress and abscisic acid. *Plant Mol. Biol.* 2000; 42: 833-845.

Tsugita T, Takahashi K, Muraoka T, Fukui H. The application of chitin/chitosan for agriculture. *7th Symposium on Chitin and Chitosan*, Fukui, Japan, 1993, pp 21-22.

Vargas M, Albors A, Chiralt A, Gonzalez-Martınez C. Quality of cold-stored strawberries as affected by chitosan-oleic acid edible coatings. *Postharvest Biol. Technol.* 2006; 41: 164-71.

Vasconsuelo A, Boland R. Molecular aspects of the early stages of elicitation of secondary metabolites in plants. *Plant Sci.* 2007; 172; 861-875.

Vasconsuelo A, Morelli S, Picotto G, Giulietti AM, Ricardo B. Intracellular calcium mobilization: A key step for chitosan-induced anthraquinone production in Rubiatinctorum L. *Plant Sci.* 2005; 169:712-720.

Wiktorowska E, Dlugosz M, Janiszowska W. Significant enhancement of oleanolic acid accumulation by biotic elicitors in cell suspension cultures of *Calendula officinalis. Enzyme Microbial Technol.* 2010; 46: 1: 14-20.

Wu FC, Tseng RL, Juang RS. Enhanced abilities of highly swollen chitosan beads for color removal and tyrosinase immobilization. *J. Hazard Mater.* 2001; 81: 167-177.

Yang SF. Biosynthesis and action of ethylene. Hortscience 1985; 20: 41-45.

Yin H, Li S, Zhao X, Du Y, Ma X. cDNA microarray analysis of gene expression in *Brassica napus* treated with oligochitosan elicitor. *Plant Physiol.Biochem.* 2006;1644: 910-916.

Yip W, Liao X, Yang SF. Dependence of *in vivo* ethylene production rate on 1-amino cyclopropane-1-carboxylic acid content and oxygen concentrations. *Plant Physiol.* 1988; 88: 553-558.

Young HD, Harald K, Kauss H. Effect of chitosan on membrane permeability of suspension cultured *Glycine max* and *Phaseolus vulgaris* cells. *Plant physiol.* 1982; 70: 1449-1454.

Zhang DL, Quantick PC. Effects of chitosan coating on enzymatic browning and decay during postharvest storage of litchi (*Litchi chinensis*Sonn.) fruit. *Postharvest Biol. Technol.* 1997; 12: 195-202.

Zhou YG, Yang YD, Qi YG, Zhang ZM, Wang XJ, Hu XJ. Effects of chitosan on some physiological activity in germinating seed of peanut. *J. Peanut Sci.* 2002; 31: 22-25.

Zuppini A, Baldan B, Millioni R, Favaron F, Navazio L, Mariani P. Chitosan induces Ca^{2+} mediated programmed cell death in soybean cells. *New Phytol.* 2003; 161: 557-68.

In: Handbook of Chitosan Research and Applications
Editors: Richard G. Mackay and Jennifer M. Tait

ISBN 978-1-61324-455-5

Chapter 8

CHITOSAN EDIBLE FILMS AND COATINGS – A REVIEW

Henriette M.C. Azeredo[1], Douglas de Britto[2] and Odílio B. G. Assis[2]

[1]EMBRAPA Tropical Agroindustry, Fortaleza, CE, Brazil;
[2]EMBRAPA Agricultutal instrumentation, São Carlos, SP, Brazil

ABSTRACT

The use of conventional food packaging materials is usually effective in terms of barrier. On the other hand, their non-biodegradability creates serious environmental problems, motivating researches on edible biopolymer films and coatings to at least partially replace synthetic polymers as food packaging materials. Chitosan is a biopolymer obtained by N-deacetylation of chitin, which is the second most abundant polysaccharide on nature after cellulose. Chitosan forms clean, tough and flexible films with good oxygen barrier, which may be employed as packaging, particularly as edible films or coatings, enhancing shelf life of a diversity of food products. One of the main applications of the film forming properties of chitosan is to develop coatings to decrease respiration rates and moisture loss in fresh fruits and vegetables. Chitosan has some advantages over other biomaterials because of its antimicrobial activity against a wide variety of microorganisms. The main purpose of this review is to summarize the literature on development and applications of chitosan films and coatings, and their properties/performance as affected by intrinsic factors (molecular weight and degree of deacetylation), chemical/physical treatments, and incorporation of nanoreinforcements.

INTRODUCTION

Packaging is important in post-harvest preservation of fruits and vegetables, and to extend shelf life of processed foods. The main purpose of food packaging is to act as a barrier against environmental factors, thus reducing the rates of quality changes, including microbiological, chemical, and physical deterioration. Microbiological changes are typically related to food safety, while chemical and physical changes are usually (but not only) associated with loss of sensory quality (e.g., rancidity, color changes, loss of crispiness). Regarding the barrier properties, the critical compounds that can deteriorate food quality are water vapor and oxygen.

The use of conventional food packaging materials such as synthetic polymers is usually effective in terms of barrier. On the other hand, they are non biodegradable, which create serious environmental problems. Approximately 40,000,000 tons of plastic packaging is used every year worldwide, and most of this is put to one time use and then discarded (Srinivasa and Tharanathan, 2007). This has motivated researches on edible films and coatings to at least partially replace synthetic polymers as food packaging materials. Edible films and coatings, elaborated from biopolymers (e.g., polysaccharides, proteins) can act as an adjunct to synthetic packaging for extending food stability, ans possibly improving economic efficiency of packaging materials (Kester and Fennema, 1986). Moreover, edible films represent non-polluting alternative packaging options.

Chitosan is one of the most promising biomaterials to replace the synthetic ones, particularly for food and packaging applications. Chitosan is a linear polysaccharide consisting of (1,4)-linked 2-amino-deoxy-β-D-glucan, and commercially prepared by chemical alkali deacetylation of chitin, which is considered as the second most abundant polysaccharide in nature after cellulose (Dutta et al., 2009; Aranaz et al., 2010). The deacetylation is usually not complete, so chitosan is generally a copolymer comprised of D-glucosamine along with N-acetyl-D-glucosamine with various fractions of acetylated units (Aranaz et al., 2010). Chitosan is a polycation whose charge density depends on the degree of deacetylation and pH. It is soluble in diluted aqueous acidic solutions due to the protonation of $-NH_2$ groups at the C2 position (Aranaz et al., 2010). Commercially, chitosan is available from a number of suppliers in different grades of purity, molecular weight and degree of deacetylation.

Antimicrobial Properties of Chitosan

Chitosan has some advantages over other biomaterials because of its antimicrobial activity against a wide variety of microorganisms including fungi, algae, and some bacteria (Tsai and Su, 1999; Entsar et al., 2003; Wu et al., 2005).

Recent data in literature has the tendency to characterize chitosan as bacteriostatic rather than bactericidal (Coma et al., 2002), although the exact mechanism is not fully understood and several other factors may contribute to the antibacterial action (Raafat et al., 2008).

Several mechanisms have been proposed, the most acceptable being the electrostatic interactions between the positively charges of the amino groups (NH_3^+) at pH values lower than 6.3 (the pKa of chitosan) and the negatively charged surface of bacteria. This electrostatic interaction results in twofold interference: i) by promoting changes in the properties of membrane wall permeability, thus provoking internal osmotic imbalances and consequently inhibiting the microbial growth (Shahidi et al., 1999), and ii) by the hydrolysis of the peptidoglycans in the microorganism wall, leading to the leakage of intracellular electrolytes such as potassium ions and some low molecular weight constituents (e.g. nucleic acids, glucose, and lactate dehydrogenase) (Helander et al., 2001; Liu et al., 2004; Dutta et al., 2009). Corroborating this proposition, Fernandez-Saiz et al. (2009) recently demonstrated that only the soluble protonated glucosamine fractions that are released from the solid chitosan film have antimicrobial activity.

The antimicrobial activity varies with several factors concerning the type of chitosan, such as degree of deacetylation (DD) and molecular weight (MW). Chitosans with lower MW have been reported to have greater antimicrobial activity than native chitosans with high MW (Liu et al., 2006; Dutta et al., 2009), because of their higher water solubility, favoring the reaction with the active sites of the microorganisms (Aider and de Halleux, 2010). On the other hand, Zheng and Zhu (2003) reported different effects of chitosan MW on *Staphyloccoccus aureus* (Gram-positive) and *Escherichia coli* (Gram-negative). For the former, the antimicrobial activity increased on increasing chitosan MW, while for the latter it was the other way around. Antimicrobial effectiveness is also favored by a highly DD, since it increases chitosan solubility and charge density, improving chitosan adhesion to the microbial cells (Aider and de Halleux, 2010).

The antimicrobial activity is also affected by the conditions of the medium. At low pH values, the higher solubility and protonation of chitosan enhances its effectiveness. Some solutes of the medium can react with chitosan cations, thus blocking the reactivity of the amine groups and decreasing antimicrobial activity (Aider and de Halleux, 2010).

The target microrganisms have also an important role on chitosan antimicrobial effectiveness. The charge density on the microorganism cell surface is a determinant factor to establish the amount of adsorbed chitosan. More adsorbed chitosan would evidently result in greater changes in the structure and in the permeability of the cell membrane. Yeasts and moulds are generally most sensitive, followed by Gram-positive and Gram-negative bacteria (Aider and de Halleux, 2010).

Moreover, chitosan is effective in chelating transition metal ions (Chen et al., 2002), since the $-NH_2$ groups are involved in interactions with metals (Rinaudo, 2006). It is well known that some ions such as Ag^+, Cu^{2+}, and Zn^{2+} are antimicrobial agents. Wang et al. (2004, 2005) have reported a wide spectrum antimicrobial activity of chitosan-metal complexes. When compared with free chitosan and metal salts, the complexes showed much better antimicrobial activities. According to Wang et al. (2005), it was concluded that the higher antimicrobial activities of chitosan-metal complexes were due to the stronger positive charge after complexation. The antimicrobial activities were affected by the chelate ratio, being favored by increasing content of metal ions. The complexes were more effective against bacteria than fungi. Additionally, it is supposed that chitosan molecules in bacteria surroundings might complex metals and block some essential nutrients to flow, also contributing to cell death (Kumar et al., 2005). Nevertheless, this is not a determinant antimicrobial action, since the sites available for interaction are limited, and the complexation reaches saturation as a function of metal concentration.

Chitosan can be combined with other compounds in order to enhance its antimicrobial activity. Indeed, the combination of chitosan or its derivatives with essential oils (Wilson et al., 2003) or diluted solution of organic acids such as acetic, sorbic, propionic, lactic, and glutamic acid (Wilson and El-Ghaouth, 2002) have been demonstrated to improve the antimicrobial effect, and may represent an actual alternative to the use of synthetic fungicides such as thiabendazole.

CHITOSAN FILMS AND COATINGS

Like most polysaccharide based films, the strongly hydrophilic character of chitosan films provides them with good barrier properties against gases and lipids but a poor barrier against water vapor (Bordenave et al., 2007; Sebti et al., 2005), which limits chitosan films uses (Caner et al., 1998), since an effective control of moisture transfer is desirable in most food packaging applications. The water vapor barrier can be improved by incorporating lipid compounds such as waxes and fatty acids into chitosan-based films. Increased resistance to water vapor transmission through chitosan films has been obtained by addition of saturated (Wong et al., 1992; Srinivasa et al., 2007) or unsaturated fatty acids (Vargas et al., 2009), and tocopherols (Park and Zhao, 2004). on the other hand, the incorporation of lipid compounds usually impairs the mechanical properties of chitosan films (Park and Zhao, 2004; Vargas et al., 2009).

Brittleness is an inherent quality attributed to the complex/branched primary structure and weak intermolecular forces of natural polymers (Srinivasa et al., 2007). Chitosan films are brittle (Suyatma et al., 2004), which is attributed to the high glass transition temperature (T_g) of the polymer (203°C, according to Sakurai et al., 2000). Due to the stiffness of the backbone and the molecule configuration, the glass transition (T_g) of chitosan was attributed by Quijada-Garrido et al. (2008) to torsional oscillations between two glucosamine rings across glucosidic oxygen and a cooperative reordering of hydrogen bonds. By reducing intermolecular forces and thus increasing the mobility of the polymer chains, plasticizers such as glycerol, sorbitol and polyethylene glycol lower the glass transition temperature of films and improve film flexibility, elongation and toughness (Banker, 1966; Domjan et al., 2009). Suyatma et al. (2005) reported that a glycerol concentration of 20% (w/w) was sufficient to improve flexibility of chitosan films. Some theories have been proposed to explain mechanisms of plasticization action (Di Gioia and Guilbert, 1999). According to the lubrication theory, interspersion of plasticizers make them act as internal lubricants by reducing frictional forces between polymer chains. The gel theory postulates that the rigidity of polymers comes from 3-dimensional structures, and plasticizers act by breaking polymer–polymer interactions, causing adjacent chains to move apart and so increasing flexibility. On the other hand, the main drawback of plasticizers is the increasing permeability rates resulting from decreased film cohesion (Gontard et al., 1993). The most effective plasticizers are those whose structure is similar to that of the polymer, so hydrophilic plasticizers such as polyols are best suited to polysaccharide films (Sothornvit and Krochta, 2005). As expected, the presence of glycerol have been reported to result in chitosan films with higher elongation, but decreased strength and modulus, and increases permeability rates (for both oxygen and water vapor) of chitosan films (Caner et al., 1998; Lazaridou and Biliaderis, 2002; Ziani et al., 2008; Azeredo et al., 2010).

Chitosan coatings have been demonstrated to be effective in enhancing shelf life of a variety of foods, such as carrots (Cheah et al., 1997), mangoes (Srinivasa, et al., 2002), strawberries (Park et al., 2005) and apples (Assis, 2004). The effects of chitosan have been reported to be related to reduced rates of several changes such as weight/water loss (Jeon et al., 2002; Bhale et al., 2003; Dong et al., 2004; Hernández-Muñoz et al., 2006, 2008; Chien et al., 2007a, b, c), respiration/ripening (Jiang and Li, 2001; Hernández-Muñoz et al., 2008), lipid oxidation (Jeo et al., 2002), pigment degradation (Jiang et al., 2005), decline in sensory quality (Dong et al., 2004), ascorbic acid losses (Chien et al., 2007a, b, c), and enzyme activity (Jiang and Li, 2001; Pen and Jiang, 2003; Dong et al., 2004; Jiang et al., 2005).

Film formation as a coating is dependent not only on cohesion (attractive forces between film polymer molecules), but also on adhesion (attractive forces between film and substrate) (Sothornvit and Krochta, 2005). A coating process involves wetting of the food by the coating solution, followed by a possible adhesion between them (Casariego et al., 2008). The amino groups of chitosan backbone, when protonated in acid solution, plays important role when adsorption and adhesion is dominantly governed by ionic attraction between oppositely charged surfaces. In any case, the wettability of the coating solution is a key factor to the effectiveness of edible coatings, since it affects the thickness and uniformity (Hong et al., 2004). Sometimes a surfactant is necessary to reduce the surface tension of the coating solution and enhance its wettability (Choi et al., 2002; Casariego et al., 2008). Choi et al. (2002) used Tween 80 as a surfactant to reduce the surface tension of a chitosan coating solution and enhanced its wettability on apple skin. Casariego et al. (2008) reported that the wettability and adhesion coefficients of chitosan coatings in tomato and carrot surfaces was decreased by increasing concentrations of chitosan and hydrophilic plasticizers.

Characteristics of the chitosan, mainly MW and DD, can affect the film properties. Films from high MW chitosans tend to present higher tensile strength (Chen and Hwa, 1996; Hwang et al., 2003) and better barrier properties (Chen and Hwa, 1996) than those from low MW. Moreover, the higher the DD, the higher the tensile strength of chitosan films (Chen et al., 1994; Hwang et al., 2003), since chitosans of higher DD are more flexible and have a higher tendency to form inter- or intra-chain hydrogen bonds (Chen et al., 1994).

The nature of acidic solvent can also influence the properties of chitosan films. The solutions of choice have usually been based on aqueous acetic acid at 1%, but it imparts a strong acidic flavor to the foods (Casariego et al., 2008), then some studies have been conducted to compare its effectiveness with those of other acids. Caner et al. (1998) studied the effects of four acids (propionic, lactic, formic, and acetic) on mechanical and barrier properties of chitosan films, and reported films elaborated with lactic acid presented the best oxygen barrier, while acetic acid produced the lowest water vapor permeability. The highest elongation was obtained with lactic acid, while formic and acetic acids resulted in higher tensile strength.

Besides the mechanical and barrier properties of chitosan, which are important in any material intended to be used to elaborate edible films, its antimicrobial properties are those which differentiates it from most other biopolymers. The application of films and coatings based on an antimicrobial polymer is one promising approach to prevent both contamination of pathogens and growth of spoilage microorganisms on the surface of food, where the antimicrobial effects are most needed (Ouattara et al., 2000). Coating a food surface with an antimicrobial film has important advantages over the alternative ways of adding antimicrobial agents to food, i.e., their incorporation into food formulations or their direct application on food surfaces without a polymer matrix. The incorporation of antimicrobial agents into food formulations may result in partial inactivation of the agent by food components, being expected to have only limited effect on surface. Also, a direct application of the antimicrobial by coating the food surface without a polymer matrix has little effect because of its rapid diffusion within the bulk of food (Torres et al., 1985; Siragusa and Dickinson, 1992).

The antimicrobial activity of chitosan has motivated its use as edible films or coatings. Direct application of chitosan formulations onto food surfaces can be attained by spraying or dipping, rendering tough, flexible and transparent films (Tharanathan, 2003). Several studies have proved the antimicrobial effectiveness of chitosan films in different culture media (Coma et al., 2002, 2003; Möller et al, 2004; Sebti et al., 2005). Indeed, chitosan coatings have proved to control microbial growth in fresh fruits and vegetables (Chien et al., 2007a; Liu et al., 2007; Badawy and Rabea, 2009), minimally processed fruits and vegetables (Devlieghere et al., 2004; Durango et al., 2006; Hernández-Muñoz et al., 2006; Chien et al., 2007b, c; Campaniello et al., 2008; Geraldine et al., 2008), seafood (Jeon et al., 2002), cheeses (Coma et al., 2003), and meats (Ouattara et al., 2000; Zivanovic et al., 2005; Beverly et al., 2008).

Chemical and Physical Treatments to Improve Properties of Chitosan Films

Some chemical modifications have been made in order to improve mechanical and gas barrier properties of chitosan films, as well as their water solubility. Although chitosans have been confirmed as attractive biomacromolecules with relevant antimicrobial properties, applications are somewhat limited due to its pH-dependent solubility. Moreover, the low pH required of the solvent to form chitosan gel plays an important role on food surface reactions. During coating, the acidic degree of chitosan solutions directly affects fruit color as a result of fast oxidation, mainly on fruit cut surfaces (Assis and Pessoa, 2004). Quaternization of the nitrogen atoms of amino groups is a modification aimed to introduce permanent positive charges along the polymer chains, providing the molecule with a cationic character independent on the aqueous medium pH (Curti et al., 2003; Aranaz et al., 2010). Quaternization is usually carried out by *N*-methylation, producing a trimethylated quaternary salt, *N,N,N*-trimethylchitosan (TMCh) (Britto and Campana-Filho, 2004; Britto and Assis, 2007). Films obtained from *N*-methylated chitosan have been reported to present higher plasticity, but lower strength and elastic modulus (Britto et al., 2005; Britto and Assis, 2007). Moreover, quaternization has been associated with improvements in antimicrobial effects (Jia et al., 2001; Guo et al., 2007), because the groups usually grafted to chitosan are electronegative, strengthening the charge density of the cation (Guo et al., 2007).

Alkylation can be a good strategy to improve mechanical properties (especially elongation) and to change the hydrophilic character of chitosan (Desbrières et al., 1996). The alkyl moiety may reduce inter- and intra-chain hydrogen bonds, plasticizing the molecule, and improve its water vapor barrier (Muzzarelli and Muzzarelli, 2005). Moreover, chitosan antimicrobial activity increases with increasing the chain length of the alkyl substitute (Jia et al., 2001; Rabea et al, 2003), because of the increased lipophilic properties of the derivatives (Kenawy et al., 2007).

The quaternary salts of chitosan were tested against several bacteria such as *Escherichia coli* and *Staphylococcus aureus* (Kim et al., 1997; Jia et al., 2001; Sun et al., 2006). The most import finding is that the antibacterial activity of quaternary ammonium chitosan was stronger in acetic acid medium than in pure water. Indeed, its antibacterial activity increased as the concentration of acetic acid increases (Jia et al., 2001). It was also found that the antibacterial activities of the chitosan derivatives, mainly those containing quaternary ammonium sites, increased with increasing chain length of the alkyl substituent in acetate buffer (pH = 6.0).

From these findings it is evident that the number of positive sites along the polymer chain, which are higher in acid than in neutral medium, strongly affects the antibacterial activity. There is no published data regarding the action of chitosan quaternary salts against fungal action, but other derivatives such as N,O-carboximethyl chitosan (El-Ghaouth et al., 1992) and glycolchitosan (El-Ghaouth et al., 2000a,b) have been tested. It was discovered that the N,O-carboximethylchitosan, a water soluble derivative, was less effective than chitosan in reducing the radial growth of the fungi *B. cinerea* and *Rhizopus stolanifer*, while glycolchitosan, an acid soluble derivative, was more effective than chitosan against the fungi *B. cinerea* and *P. expansum*. Thus, as already mentioned, it would appear that the fungicide effect depends on the charge density of the derivative chain.

The application of electric fields has been suggested to improve properties of chitosan films. The application of electric fields resulted in decreased permeability of chitosan films to water vapor (García et al., 2009; Souza et al., 2009), oxygen and carbon dioxide (Souza et al., 2009), as well as increased tensile strength and elongation at break (García et al., 2009; Souza et al., 2010). Images from atomic force microscopy (AFM) suggested that the application of electric fields improved the uniformity of the film surfaces, which was attributed to a more uniform gel structure leading to the improvements in barrier properties (Souza et al., 2009).

ADDITION OF NANOREINFORCEMENTS TO CHITOSAN FILMS

The addition of reinforcing fillers to biopolymers such as chitosan has been reported to enhance their thermal, mechanical, and barrier properties. The smaller the filler particles, the better is the filler-matrix interaction (Ludueña et al., 2007). Fillers with at least one dimension in the nanometric range (< 100 nm) are called nanoreinforcements, and their composites with polymers are nanocomposites (Alexandre and Dubois, 2000). A uniform dispersion of nanofillers leads to a very large matrix/filler interfacial area, which changes the molecular mobility, and favors the thermal and mechanical properties of the material (Vaia and Wagner, 2004).

Azeredo et al. (2010) reinforced chitosan films with cellulose nanofibers, which increased the T_g of the films and improved their tensile strength and water vapor barrier, although the elongation was decreased. A nanocomposite film with 15% cellulose nanofibers was comparable to some synthetic polymers in terms of strength and stiffness.

Han et al. (2010) prepared chitosan-clay nanocomposites by an ion exchange reaction between oligomeric chitosan and a Na^+-montmorillonite (NaMMT). The thermal stability of chitosan was improved by combination with NaMMT, which was attributed to a thermal insulation effect of the clay and strong electrostatic interactions between cationic chitosan molecules and anionic silicate layers. The antimicrobial activity was also enhanced by the clay, which seems contradictory, since the anionic silicate layers tend to neutralize the positive charges of chitosan. The authors attributed this effect to a synergy between the components of the nanocomposite, because of the high dispersion of the chitosan molecules through the inorganic matrix.

Chitosan Nanoparticles in Films

De Moura et al. (2009) incorporated chitosan/tripolyphosphate nanoparticles (CsN) to hydroxypropyl methylcellulose (HPMC) films, and observed improvements in mechanical properties and water vapor barrier of the films. The water vapor permeability was affected by the size of the CsN – the smaller the particles, the lower the permeability. According to the auhors, CsN tended to occupy the empty spaces in the pores of the HPMC matrix, thereby improving tensile and barrier properties. The thermal stability of the films was also favored by addition of CsN.

Chang et al. (2010) added CsN to a glycerol plasticized starch matrix. The tensile strength, storage modulus, glass transition temperature, water vapor barrier and thermal stability were improved by low contents of CsN, when the particles were well dispersed in the matrix. Such effects were attributed to close interactions between CsN and the matrix, due to their chemical similarities. However, higher CsN loads (8%, w/w) resulted in aggregation of CsN in the nanocomposites.

Final Considerations

Considering the problems related to disposal of non-biodegradable materials, the use of biopolymers as food packaging materials is an important tendency. Among the known biopolymers, chitosan is especially promising because of its antimicrobial properties. The antimicrobial activity is related to the cationic character of chitosan, and has been successfully explored to extend shelf life of a variety of foods. Introducing alkyl chain and promoting the quaternization of chitosan it is possible to obtain water soluble films with different hydrophilic character, having potential application for minimally processed food industry. The beneficial use of water soluble quaternary salt of chitosan is devoted to its improvement on decreasing browning effect, increasing the fungicidal action and represents an advance in overcoming consumer acceptance.

References

Aider, M., and de Halleux, D. (2010). Chitosan application for active bio-based films production and potential in the food industry: review. LWT – Food Science and Technology, *in press.*

Alexandre, M., and Dubois, P. (2000). Polymer-layered silicate nanocomposites: preparation, properties and uses of a new class of materials. *Materials Science and Engineering*, 28, 1-63.

Aranaz, I., Harris, R., and Heras, A. (2010). Chitosan amphiphilic derivatives. Chemistry and applications. *Current Organic Chemistry*, 14, 308-330.

Assis, O. B. G. (2008). The effect of chitosan as a fungistatic agent on cut apples. *Revista Iberoamericana de Tecnologia Postcosecha,* 9 (2), 148-152.

Assis, O. B. G., and Pessoa, J.D.C. (2004). Preparation of thin-film of chitosan for use as edible coating to inhibit fungal growth on sliced fruits. *Brazilian Journal of Food Science and Technology*, 7, 17-22.

Azeredo, H. M. C., Mattoso, L. H. C., Avena-Bustillos, R. J., Ceotto Filho, G., Munford, M. L., Wood, D., and McHugh, T. H. (2010). Nanocellulose reinforced chitosan composite films as affected by nanofiller loading and plasticizer content. *Journal of Food Science*, 75(1), N1-N7.

Badawy, M. E. I., and Rabea, E. I. (2009). Potential of the biopolymer chitosan with different molecular weights to control postharvest gray mold of tomato fruit. *Postharvest Biology and Technology,* 51, 110-117.

Banker, G. S. (1966). Film coating theory and practice. *Journal of Pharmaceutical Sciences*, 55(1), 81-89.

Beverly, R. L., Janes, M. E., Prinyawiwatkul, W., and No, H. K. (2008). Edible chitosan films on ready-to-eat roast beef for the control of *Listeria monocytogenes*. *Food Microbiology*, 25, 534-537.

Bhale, S., No, H. K., Prinyawiwatkul, W., Farr, A. J., Nadarajah, K., and Meyers, S. P. (2003). Chitosan coating improves shelf life of eggs. *Journal of Food Science*, 68(7), 2378-2383.

Bordenave, N., Grelier, S., and Coma, V. (2007). Water and moisture susceptibility of chitosan and paper-based materials: Structure-property relationships. *Journal of Agricultural and Food Chemistry*, 55(23), 9479–9488.

Britto, D., and Assis, O. B. G. (2007). Synthesis and mechanical properties of quaternary salts of chitosan-based films for food application. *International Journal of Biological Macromolecules,* 41, 198-203.

Britto, D., Campana-Filho, S. P., and Assis, O. B. G. (2005). Mechanical properties of N,N,N-trimethylchitosan chloride films. *Polímeros: Ciência e Tecnologia*, 15(2), 142-145.

Britto, D., and Campana-Filho, S. P. (2004). A kinetic study on the thermal degradation of *N,N,N*-trimethylchitosan. *Polymer Degradation and Stability*, 84, 353-361.

Campaniello, D., Bevilacque, A., Sinigaglia, M., and Corbo, M. R. (2008). Chitosan: Antimicrobial activity and potential applications for preserving minimally processed strawberries. *Food Microbiology*, 25, 992-1000.

Caner, C., Vergano, P. J., and Wiles, J. L. (1998). Chitosan film mechanical and permeation properties as affected by acid, plasticizer, and storage. *Journal of Food Science*, 63(6), 1049-1053.

Casariego, A., Souza, B. W. S., Vicente, A. A., Teixeira, J. A., Cruz, L., and Díaz, R. (2008). Chitosan coating surface properties as affected by plasticizer, surfactant and polymer concentrations in relation to the surface properties of tomato and carrot. *Food Hydrocolloids*, 22, 1452-1459.

Chang, P. R., Jian, R., Yu, J., and Ma, X. (2010). Fabrication and characterisation of chitosan nanoparticles/plasticized-starch composites. *Food Chemistry*, 120, 736-740.

Cheah, L. H., Page, B. B. C., and Shepherd, R. (1997). Chitosan coating for inhibition of sclerotinia rot of carrots. *New Zealand Journal of Crop and Horticultural Science*, 25, 89-92.

Chen, R. H., and Hwa, H. -D. (1996). Effect of molecular weight of chitosan with the same degree of deacetylation on the thermal, mechanical, and permeability properties of the prepared membrane. *Carbohydrate Polymers*, 29, 353-358.

Chen, R. H., Lin, J. H., and Yang, M. H. (1994). Relationships between the chain flexibilities of chitosan molecules and the physical properties of their casted films. *Carbohydrate Polymers*, 24, 41-46.

Chen, X.-G., Zheng, L., Wang, Z., Lee, C.-Y., and Park, H.-J. (2002). Molecular Affinity and Permeability of Different Molecular Weight Chitosan Membranes. *Journal of Agricultural and Food Chemistry*, 50(21), 5915–5918.

Chien, P. -J., Sheu, F., and Lin, H. -R. (2007a). Coating citrus (Murcott tangor) fruit with low molecular weight chitosan increases postharvest quality and shelf life. *Food Chemistry*, 100, 1160-1164.

Chien, P. -J., Sheu, F., and Lin, H. -R. (2007b). Effects of edible chitosan coating on quality and shelf life of sliced mango fruit. *Journal of Food Engineering*, 78, 225-229.

Chien, P. -J., Sheu, F., and Lin, H. -R. (2007c). Quality assessment of low molecular weight chitosan coating on sliced red pitayas. *Journal of Food Engineering*, 79, 736-740.

Choi, W. Y., Park, H. J., Ahn, D. J., Lee, J., and Lee, C. Y. (2002). Wettability of chitosan coating solution on ′Fuji′ apple skin. *Journal of Food Science,* 67(7), 2668-2672.

Coma, V., Deschamps, A., and Martial-Gros, A.(2003). Bioactive packaging materials from edible chitosan polymer – Antimicrobial activity assessment on dairy-related contaminants. *Journal of Food Science*, 68(9), 2788-2792.

Coma, V., Martial-Gros, A., Garreau, S., Copinet, A., Salin, F., and Deschamps, A. (2002). Edible antimicrobial films based on chitosan matrix. *Journal of Food Science*, 67(3), 1162-1169.

Curti, E., Britto, D., Campana-Filho, S. P. (2003). Methylation of chitosan with iodomethane: Effect of reaction conditions on chemoselectivity and degree of substitution. *Macromolecular Bioscience*, 3, 571-576.

De Moura, M. R., Aouada, F. A., Avena-Bustillos, R. J., McHugh, T. H., Krochta, J. M., and Mattoso, L. H. C. (2009). Improved barrier and mechanical properties of novel hydroxypropyl methylcellulose edible films with chitosan/tripolyphosphate nanoparticles. *Journal of Food Engineering*, 92(4), 448–453.

Desbrières, J., Martinez, C., Rinaudo, M. (1996). Hydrophobic derivatives of chitosan: Characterization and rheological behaviour. *International Journal of Biological Macromolecules,* 19, 21–28.

Devlieghere, F., Vermeulen, A., and Debevere, J. (2004). Chitosan: antimicrobial activity, interactions with food components and applicability as a coating on fruits and vegetables. *Food Microbiology*, 21, 703-714.

Domjan, A., Bajdik, J., and Pintye-Hodi, K. (2009). Understanding of the plasticizing effects of glycerol and PEG 400 on chitosan films using Solid-State NMR Spectroscopy. *Macromolecules*, 42, 4667-4673.

Di Gioia, L., and Guilbert, S. Amphipilic plasticizers. *Journal of Agricultural and Food Chemistry,* 47, 1254–1261, 1999.

Dong, H., Cheng, L., Tan, J., Zheng, K., and Jiang, Y. (2004). Effects of chitosan coating on quality and shelf life of peeled litchi fruit. *Journal of Food Engineering*, 64, 355-358.

Durango, A. M., Soares, N. F. F., and Andrade, N. J. (2006). Microbiological evaluation of an edible antimicrobial coating on minimally processed carrots. *Food Control*, 17, 336-341.

Dutta, P. K., Tripathi, S., Mehrotra, G. K., and Dutta, J. (2009). Perspectives for chitosan based antimicrobial films in food applications. *Food Chemistry*, 114(4), 1173-1182.

El-Ghaouth, A., Arul, J., Asselin, A., and Benhamou, N. (1992). Antifungal activity of chitosan on post-harvest patogens: induction of morphological and cytological alterations in *Rhizopus stolonifer*. *Mycological Research*, **96(**9), 769–779.

El-Ghaouth, A., Smilanick, J. L., and Wilson, C. L. (2000). Enhancement of the performance of *Candida saitoana* by the addition of glycolchitosan for the control of postharvest decay of apple and citrus fruit. *Postharvest Biology and Technology*, 19, 103-110.

El Ghaouth, A., Smilanick, J.L., Brown, G. E., Ippolito, A., Wisiewski, M., and Wilson, C. L. (2000). Applications of *Candida saitoana* and glycolchitosan for the control of postharvest diseases of apple and citrus fruit under semi-commerical conditions. *Plant Disease*, 84, 243-248.

Entsar, I. R., Badawy, M. E. T., Stevens, C. V., Smagghe, G., and Walter, S. (2003). Chitosan as antimicrobial agent: Application and mode of action. *Biomacromolecules*, 4, 1457-1465.

Fernandez-Saiz, P., Lagaron, J. M., Ocio, M. J. (2009). Optimization of the biocide properties of chitosan for its application in the design of active films of interest in the food area. *Food Hydrocolloids*, 23, 913-921.

García, M. A., Pinotti, A., Martino, M., and Zaritzky, N. (2009). Electrically treated composite films based on chitosan and methylcellulose blends. *Food Hydrocolloids*, 23, 722-728.

Geraldine, R. M., Soares, N. F. F., Botrel, D. A., and Gonçalves, L. A. (2008). Characterization and effect of edible coatings on minimally processed garlic quality. *Carbohydrate Polymers*, 72, 403-409.

Gontard, N., Guilbert, S., and Cuq, J. -L. (1993). Water and glycerol as plasticizers affect mechanical and water vapor barrier properties of and edible wheat gluten film. *Journal of Food Science,* 58(1), 206-211.

Guo, Z., Xing, R., Liu, S., Zhong, Z., Ji, X., Wang, L., and Li, P. (2007). The influence of the cationic of quaternized chitosan on antifungal activity. *International Journal of Food Microbiology,* 118, 214-217.

Han, Y.-S., Lee, S.-H., Choi, K. H., and Park, I. (2010). Preparation and characterization of chitosan-clay nanocomposites with antimicrobial activity. *Journal of Physics and Chemistry of Solids, in press.*

Helander, I. M., Nurmiaho-Lassila, E. L., Ahvenainen, R., Rhoades, J., and Roller, S. (2001). Chitosan disrupts the barrier properties of the outer membrane of Gram-negative bacteria. *International Journal of Food Microbiology*, 71(2-3), 235-244.

Hernández-Muñoz, P., Almenar, E., Del Valle, V., Velez, D., and Gavara, R. (2008). Effect of chitosan coating combined with postharvest calcium treatment on strawberry (*Fragaria* x *ananassa*) quality during refrigerated storage. *Food Chemistry*, 110, 428-435.

Hernández-Muñoz, P., Almenar, E., Ocio, M. J., and Gavara, R. (2006). Effect of calcium dips and chitosan coatings on postharvest life of strawberries (*Fragaria* x *ananassa). Postharvest Biology and Technology*, 39, 247-253.

Hong, S., Han, H., and Krochta, J. M. (2004). Optical and surface properties of whey protein isolate coating on plastic films as influenced by substrate, protein concentration and plasticizer type. *Journal of Applied Polymer Science*, 92, 335–343.

Hwang, K. T., Kim, J. T., Jung, S. T., Cho, G. S., and Park, H. J. (2003). Properties of chitosan-based biopolymer films with various degrees of deacetylation and molecular weigths. *Journal of Applied Polymer Science*, 89, 3476-3484.

Jeon, Y. -J., Kamil, J. Y. V. A., and Shahidi, F. (2002). Chitosan as an edible invisible film for quality preservation of herring and Atlantic cod. *Journal of Agricultural and Food Chemistry,* 50, 5167-5178.

Jia, Z., Shen, D., and Xu, W. (2001). Synthesis and antibacterial activities of quaternary ammonium salt of chitosan. *Carbohydrate Research*, 333, 1-6.

Jiang, Y., and Li, Y. (2001). Effects of chitosan coating on postharvest life and quality of longan fruit. *Food Chemistry*, 73, 139-143.

Jiang, Y., Li., and Jiang, W. (2005). Effects of chitosan coating on shelf life of cold-stored litchi fruit at ambient temperature. LWT – *Food Science and Technology*, 38, 757-761.

Kenawy, E. -R., Worley, S. D., and Broughton, R. (2007). The chemistry and applications of antimicrobial polymers: A state-of-the-art review. *Biomacromolecules,* 8(5), 1359-1384.

Kester, J. J., and Fennama, O. R. (1986). Edible films and coatings: a review. *Food Technology,* 40(12), 47-59.

Kim, C. H., Choi, J. W., Chun, H. J., and Choi, K. S. (1997). *Synthesis of chitosan derivatives with* quaternary ammonium salt and their antibacterial activity. *Polymer Bulletin, 38, 387*–393.

Kumar, A. B. V., Varadaraj, M. C., Gowda, L. R., and Tharanathan, R. N. (2005). Characterization of chito-oligosaccharides prepared by chitosanolysis with the aid of papain and Pronase, and their bactericidal action against *Bacillus cereus* and *Escherichia coli. Biochemical Journal*, 391, 167–175.

Lazaridou, A., and Biliaderis, C. G. (2002). Thermophysical properties of chitosan, chitosan-starch and chitosan-pullulan films ncar thc glass transition. *Carbohydrate Polymers*, 48, 179-190.

Liu, H., Du, Y., Wang, X., and Sun, L. (2004). Chitosan kills bacteria through cell membrane damage. *International Journal of Food Microbiology*, 95(2), 147-155.

Liu, J., Tian, S., Meng, X., and Xu, Y. (2007). Effects of chitosan on control of postharvest diseases and physiological responses of tomato fruit. *Postharvest Biology and Technology,* 44, 300-306.

Liu, N., Chen, X. -G., Park, H. -J, Liu, C. -G., Liu, C. -S., Meng, X. -H., and Yu, L. -J. (2006). Effect of MW and concentration of chitosan on antibacterial activity of *Escherichia coli. Carbohydrate Polymers*, 64, 60-65.

Ludueña, L. N., Alvarez, V. A., and Vasquez, A. (2007). Processing and microstructure of PCL/clay nanocomposites. *Materials Science and Engineering*: A, 460-461, 121-129.

Möller, H., Greiler, S., Pardon, P., and Coma, V. (2004). Antimicrobial and physicochemical properties of chitosan-HPMC-based films. *Journal of Agricultural and Food Chemistry*, 52, 6585-6591.

Muzzarelli, R. A. A., and Muzzarelli, A. (2005). Chitosan chemistry: relevance to the biomedical sciences. *Advances in Polymer Science*, 186, 151-209.

Ouattara, B., Simard, R. E., Piette, G., Begin, A., and Holley, R. A. (2000). Inhibition of surface spoilage bacteria in processed meats by application of antimicrobial films prepared with chitosan. *International Journal of Food Microbiology*, 62(1–2), 139–148.

Park, S. -I., Stan, S. D., Daeschel, M. A., and Zhao, Y. Y. (2005). Antifungal coatings on fresh strawberries (*Fragaria x ananassa*) to control mold growth during cold storage. *Journal of Food Science*, 70(4), M202-M207.

Park, S. -I., and Zhao, Y. (2004). Incorporation of a high concentration of mineral or vitamin into chitosan-based films. *Journal of Agricultural and Food Chemistry*, 52, 1933–1939.

Pen, L. T., and Jiang, Y. M. (2003). Effects of chitosan coating on shelf life and quality of fresh-cut Chinese water chestnut. Lebensmittel Wissenschaft und Technologie, 36, 359–364.

Quijada-Garrido, I., Iglesias-González, V., Mazón-Arechederra, J. M., and Barrales-Rienda, J. M. (2008). The role played by the interactions of small molecules with chitosan and their transition temperatures. Glass-forming liquids: 1,2,3-propantriol (glycerol). *Carbohydrate Polymers*, 68(1), 173–86.

Raafat, D.; von Bargen, K.;.Haas, A.; Sahl, H. -G. (2008). Insights into the mode of action of chitosan as an antibacterial compound. *Appl. Environ. Microbiol.*, 74, 3764–3773.

Rabea, E. I., Badawy, M. E. T., Stevens, C. V., Smagghe, G., and Steurbaut, W. (2003). Chitosan as amtimicrobial agent: Applications and mode of action. *Biomacromolecules,* 4(6), 1457-1465.

Rinaudo, M. (2006). Chitin and chitosan: Properties and applications. *Progress in Polymer Science,* 31, 603-632.

Sakurai, K., Maegawa, T., Takahashi, T. (2000). Glass transition temperature of chitosan and miscibility of chitosan/poly(*N*-vinyl pyrrolidone) blends. *Polymer,* 41(19), 7051-7056.

Sebti, I., Martial-Gros, A., Carnet-Pantiez, A., Grelier, S., and Coma, V. (2005). Chitosan polymer as bioactive coating and film against *Aspergillus nige*r contamination. *Journal of Food Sciences*, 70(2), M100–M104.

Shahidi, F., Arachchi, J., and Jeon, Y. -J. (1999). Food applications of chitin and chitosans. *Trends in Food Science and Technology*, 10, 37–51.

Siragusa, G. A., and Dickson, J. S. (1992). Inhibition of *Listeria monocytogenes* on beef tissue by application of organic acids immobilized in a calcium alginate gel. *Journal of Food Science,* 57, 293–296.

Sothornvit, R., and Krochta, J. M. (2005). Plasticizers in edible films and coatings. In J.H. Han (Ed.), Innovations in food packaging (pp. 403-33). London, UK: Academic Press.

Souza, B. W. S., Cerqueira, M. A., Casariego, A., Lima, A. M. P., Teixeira, J. A., and Vicente, A. A. (2009). Effect of moderate electric fields in the permeation properties of chitosan coatings. *Food Hydrocolloids*, 23, 2110-2115.

Souza, B. W. S., Cerqueira, M. A., Martins, J. T., Casariego, A., Teixeira, J. A., and Vicente, A. A. (2010). Influence of electric fields on the structure of chitosan edible coatings. *Food Hydrocolloids*, 24, 330-335.

Srinivasa, P. C., Baskaran, R., Ramesh, M. N., Prashanth, K. V. H., and Tharanathan, R. N. (2002). Storage studies of mango packed using biodegradable chitosan film. *European Food Research and Technology*, 215, 504-508.

Srinivasa, P. C., Ramesh, M. N., and Tharanathan, R. N. (2007). Effect of plasticizers and fatty acids on mechanical and permeability characteristics of chitosan films. *Food Hydrocolloids,* 21, 1113-1122.

Srinivasa, P. C., and Tharanathan, R. N. (2007). Chitin/chitosan – Safe, ecofriendly packaging materials with multiple potential uses. *Food Reviews International*, 23, 53-72.

Sun, L., Du, Y., Fan, L., Chen, X., and Yang, J. (2006). Preparation, characterization and antimicrobial activity of quaternized carboxymethyl chitosan and application as pulp-cap. *Polymer*, 47(6), 1796-1804.

Suyatma, N. E., Copinet, A., Coma, V., and Tighzert, L. (2004). Mechanical and barrier properties of biodegradable films made from chitosan and poly(lactic acid) blends. *Journal of Polymers and the Environment*, 12, 1–6.

Suyatma, N. E., Tighzert, L., and Copinet A. (2005). Effects of hydrophilic plasticizers on mechanical, thermal, and surface properties of chitosan films. *Journal of Agricultural and Food Chemistry*, 53(10), 3950–3957.

Tharanathan, R. N. (2003). Biodegradable films and composite coatings: Past, present and future. *Trends in Food Science and Technology*, 14(3), 71–78.

Torres, J. A., Motoki, M., and Karel, M. (1985). Microbial stabilization of intermediate moisture food surfaces. I. Control of surface preservative concentration. *Journal of Food Processing and Preservation,* 9, 75–92.

Tsai, G. J., and Su, W. H. (1999). Antibacterial activity of shrimp chitosan against *Escherichia coli. Journal of Food Protection*, 62(3), 239–243.

Vaia, R. A., and Wagner, H. D. (2004). Framework for nanocomposites. *Materials Today*, 7, 32–37.

Vargas, M., Albors, A., Chiralt, A., and González-Martínez, C. (2009). Characterization of chitosan-oleic acid composite films. *Food Hydrocolloids*, 23, 536-547.

Wang, G. H. (1992). Inhibition and inactivation of five species of foodborne pathogens by chitosan. *Journal of Food Protection*, 55(11), 916–919.

Wang, X., Du, Y., Fan, L., Liu, H., and Hu, Y. (2005). Chitosan-metal complexes as antimicrobial agent: Synthesis, characterization and structure-activity study. *Polymer Bulletin,* 55, 105-113.

Wang, X., Du, Y., and Liu, H. (2004). Preparation, characterization and antimicrobial activity of chitosan-Zn complex. *Carbohydrate Polymer*, 56, 21-26.

Wilson, C. L.; El-Ghaouth, A.; Wisniewski, M. E. Synergistic combinations of natural of compounds that control decay of fruits and vegetables and reduce contamination by foodborne human pathogens. US Patent Application 20030113421 A1, 2003.

Wilson, C. L.; El-Ghaouth, A. Biological coating with a protective and curative effect for the control of postharvest decay. US Patent 6423310 B1, 2002.

Wong, D., Gastineau, F., Gregorski, K. S., Tillin, S. J., and Pavlath, A. E. (1992). Chitosan–lipid films: Microstructure and surface energy. *Journal of Agriculture and Food Chemistry,* 40, 540–544.

Wu, T., Zivanovic, S., Draughon, F. A., Conway, W. S., and Sams, C. E. (2005). Physicochemical properties and bioactivity of fungal chitin and chitosan. *Journal of Agricultural and Food Chemistry*, 53(10), 3888–3894.

Zheng, L. Y., and Zhu, J. F. (2003). Study of antimicrobial activity of chitosan with different molecular weight. *Carbohydrate Polymers*, 54(4), 527–530.

Ziani, K., Oses, J., Coma, V., and Maté, J. I. (2008). Effect of the presence of glycerol and Tween 20 on the chemical and physical properties of films based on chitosan with different degree of deacetylation. LWT – *Food Science and Technology*, 41, 2159-2165.

Zivanovic, S., Chi, S., and Draughon, A. F. (2005). Antimicrobial activity of chitosan films enriched with essential oils. *Journal of Food Science*, 70(1), M45-M51.

In: Handbook of Chitosan Research and Applications ISBN 978-1-61324-455-5
Editors: Richard G. Mackay and Jennifer M. Tait

Chapter 9

ENCAPSULATION OF FUNCTIONAL FOOD INGREDIENTS USING CHITOSAN-BASED MATERIALS

Lu-E Shi[1*] ***and Zhen-Xing Tang***[2, 3]

[1]. College of Life and Environmental Sciences,
Hangzhou Normal University, Hangzhou,
Zhejiang, China
[2]. Key Laboratories for Food Bioengineering Research,
Wahaha Research Institute,
Hangzhou Wahaha Group Co. Ltd,
Hangzhou, Zhejiang, China
[3]. Guelph Food Research Center,
Agriculture and Agri-Food Canada,
Guelph, Ontario, Canada

ABSTRACT

Consumers prefer food products that are tasty, healthy and convenient. Encapsulation, a process to entrap active agents into particles, is an important way to protect them during processing does not affect their availability in the body. For example, this technology may allow taste and aroma differentiation, mask bad tasting or bad smelling components, stabilize food ingredients and/or increase their bioavailability. Chitosan is a natural polysaccharide prepared by the N-deacetylation of chitin. It has been widely used in food industrial and biomedical aspects, including in encapsulating active food ingredients, in enzyme immobilization, and as a carrier for controlled drug delivery, due to its significant biological and chemical properties such as biodegradable, biocompatible, bioactive, and polycationic. In this chapter chitosan-based carriers used to encapsulate for food active ingredients are reviewed.

* Corresponding author: Lu-E Shi, E-mail: shilue@126.com.

I. INTRODUCTION

The emergence of dietary compounds with benefits offers an excellent opportunity to improve public health. Known as functional food ingredients, this category of compounds has received much attention in recent years from the scientific community, consumers and food manufacturers. The list of potential functional food ingredients (vitamins, probiotics, bioactive peptides and enzymes, antioxidants, et al.) is endless, and scientific evidence to support the concept of health-promoting food ingredients is growing steadily. Even though the nature of the involvement of functional food ingredients in physiological functions is not yet fully understood, it is now well recognized that their addition to foods, as simple and direct means of decreasing the incidence of illnesses, holds much promise for benefit to society. The scientific community can therefore contribute to the development of innovative functional foods that bring physiological benefits or reduce long-term risks of developing diseases [1].

The effectiveness of functional food in preventing diseases depends on preserving the bioavailability of the active ingredients. This represents a formidable challenge, given that only a small proportion of these remains available following oral administration, because of insufficient gastric residence time, low permeability and/or low solubility within the gastrointestinal tract, as well as instability under conditions encountered during food processing (temperature, oxygen, light) or in the gut (pH, enzymes, presence of other nutrients), all of which limit the activity and potential health benefits of functional molecules [2]. The delivery of these molecules will therefore require food formulators and manufacturers to provide protective mechanisms that both maintain the active molecular form until the time of consumption, and deliver this form to the physiological target within the organism.

Polymer-based delivery systems that trap molecules of interest within networks have been developed extensively for the biomedical and pharmaceutical sectors to protect and transport bioactive compounds to target functions [3, 4]. In spite of the success of carrier systems such as hydrogels, micro- and nanoparticles designed for enhanced delivery of medicinal drugs, drug targeting or time-controlled release, most of these polymer-based systems, cannot be used for food applications that require wall materials generally recognized as safe (GRAS).

Chitosan, the principal derivative of chitin, is obtained by *N*-deacetylation to a varying extent that is characterized by the degree of deacetylation, and is consequently a copolymer of *N*-acetyl-d-glucosamine and d-glucosamine. Chitosan possesses distinct chemical and biological properties [5-20]. In its linear polyglucosamine chains of high molecular weight, chitosan has reactive amino and hydroxyl groups, amenable to chemical modifications [5, 9, 10,14]. Additionally, amino groups make chitosan a cationic polyelectrolyte (p*K*a ≈ 6.5). Both the solubility in acidic solutions and aggregation with polyanions impart chitosan with excellent gel-forming properties. Chitosan offer an extraordinary potential in a broad spectrum of areas, such as medicine, biotechnology and food engineering [7, 11-13, 20], owing to the unparlleled biological properties that include biocompatibility, biodegradability to harmless products, nontoxicity, physiological inertness, remarkable affinity to proteins. This chapter focused on discussion the potential of chitosan-based matrices to serve as carriers for encapsulation of functional food components.

II. Encapsulation of Food Active Ingredients

Chitosan-based delivery systems have found wide and rapidly increasing applications in the food industry because they can be precisely designed for use in many food formulations and virtually any ingredient can be encapsulated, whether hydrophobic, hydrophilic, or even bacterial. The following section discusses a few chitosan-based delivery systems for functional food ingredients.

Vitamins

Vitamins are naturally occurring compounds present in most foods. They are essential in small quantities for all body functions including growth, repair of tissues, and the maintenance of health. However, they are generally sensitive to light, pH and oxidation.

Chen and Subirade [21] prepared core-shell nanoparticles (100-200 nm hydrodynamic diameters) for delivery of riboflavin by crosslinking chitosan with sodium tripolyphosphate (TPP), followed by coating the nano-gel-beads with either native or denatured 3-lactoglobulin.

Microcapsules made of chitosan and gelatin, were also developed for delivery of bioactive molecules (riboflavin) by ingestion [22]. Chitosan, gelatin and the bioactive molecule solutions were mixed at a low pH (using acetic acid) and 37 °C, and then dropped into cold sesame-seed oil to induce gelation of the gelatin. Then, ionotropic crosslinking was done at pH 7.0 to improve the mechanical strength of the beads, using either sulphate or citrate. Citrate and sulfate crosslinked beads, released about 60 % and 70 % of the riboflavin in 24 h in SIF, respectively, and completely released it in less than 5 h in SGF. The studies discussed also exemplify the important role that multivalent counterions may play in electrostatic interactions between biopolymers, in particular when encapsulation and controlled release is sought.

In another study, sequentially coated multilayered beads were prepared by complex coacervation of calcium alginate and chitosan layers [23]. The beads containing vitamin B_2 were studied for the release at varying pH levels at 37 °C. Nearly 54 ± 4 % of the vitamin were released in the first 3 h in a SGF (pH 1.0) while the remaining amount was released in the SIF (pH 7.4) within 6 h.

Vitamin C, a representative water-soluble vitamin, has a variety of biological, pharmaceutical, and dermatological functions. Vitamin C is widely used in various types of foods as a vitamin supplement and as an antioxidant. Hence, sustained release carriers of vitamin C have been prepared by using cross-linked chitosan as a wall material by spray-drying technique [24-26]. Chitosan was cross-linked with nontoxic cross-linking agent TPP. Vitamin C–encapsulated chitosan microspheres of different size, surface morphology, loading efficiency, and zeta potential with controlled-release property could be obtained by varying the manufacturing parameters (inlet temperature, flow rate) and using the different molecular weight and concentration of chitosan. Vitamin C–encapsulated chitosan microcapsules were spherical in shape with a smooth surface as observed by scanning electron microscopy.

Jiang and Lee [27] investigated the stability and characteristics of Vitamin C-loaded chitosan nanoparticles during heat processing in aqueous solutions. Vitamin C-loaded chitosan nanoparticles were prepared by ionic gelation of chitosan with TPP anions. The smallest chitosan nanoparticles (170 nm) were obtained with a chitosan concentration of 1.5 mg/ml and a TPP concentration of 0.6 mg/ml. During heat processing at various temperatures, the size and zeta potential of the particles decreased rapidly in the first 5 min and then slowly fell to the regular range. At the beginning of the release profiles, the burst release-related stability of the surface increased with the temperature. Then, the release of the internal Vitamin C was constantly higher with a longer release time. Consequently, it was confirmed that the stability of Vitamin C-loaded chitosan nanoparticles was affected by temperature but that the internal stability was greater than the surface stability. These results demonstrate the stability of chitosan nanoparticles for during heat processing and suggest the possible use of Vitamin C-loaded chitosan nanoparticles to enhance antioxidant effects because of the continuous release of Vitamin C from chitosan nanoparticles in food processing.

Probiotics

Most probiotics in the food supply are used in fermented milks and dairy products. Probiotics can be defined as living microbial supplements which can improve the balance of intestinal microorganisms [28]. The probiotic effect has been attributed to the production of acid and/or bacteriocins, competition with pathogens and enhancement of the immune system. Claimed benefits include controlling serum cholesterol levels, preventing intestinal infection, improving lactose utilization in persons who are lactose intolerant, and possessing anticarcinogenic activity.

Good probiotic viability and activity are considered essential for optimal functionality [29, 30]. Furthermore, the ability of microorganisms to survive and multiply in the host strongly influences their probiotic benefits. The bacteria in a product should remain metabolically stable and active, surviving passage through the upper digestive tract in large numbers sufficient enough to produce beneficial effects when in the host intestines [31]. Suggested beneficial minimum level for probiotics in yogurt is 10^6 cfu/ml [32, 33] or the daily intake should be about 10^8 cfu/ml. Earlier studies indicated that some strains of probiotics, especially *Bifidobacterium* spp., lack the ability to survive gastrointestinal conditions [34, 35].

Other studies have also reported low viability of probiotics in dairy products such as yogurt and frozen dairy desserts [36-38] due to the concentration of lactic acid and acetic acid [39], low pH [40, 41], the presence of hydrogen peroxide [42], and the oxygen content [43]. To overcome this problem, encapsulation has been investigated for improving the viability of microorganisms in both dairy products and the intestinal tract [44-46].

Droplets of a chitosan solution suspended in an oil phase can be hardened by cross-linking with glutaraldehyde via solvent evaporation or by the addition of polyvalent anions such as TPP. The stirring rate, temperature, level of the gelling agent, concentration of the surfactant polymer, and the viscosities of the phases were reported to affect the size and morphology of the particles [47, 48].

Coating alginate beads with chitosan has been studied extensively for encapsulating probiotics [49-51]. Calcium alginate–chitosan microcapsules can be made by one- or two-step processes, based on the presence or absence of Ca^{2+} in the chitosan solution [52]. Capsules' mechanical strength and permeability strongly depend on the process of capsule preparation

[53, 54]. In the one-step process (in the absence of Ca^{2+} in chitosan solution), chitosan is located only at the interface, as a thin-alginate–chitosan membrane with a weak mechanical resistance. The capsules were much stronger when the two-step protocol was used. This difference between two protocols of capsule formation is due to the ability of chitosan to penetrate through the membrane [52]. The kinetics of membrane formation and the capsule parameters (like thickness, permeability and mechanical strength) depend on the concentration of components, molar masses of both, alginate and chitosan, reaction time, pH and ionic strength. Sprayed particles coated with chitosan are also recommended as impressively effective vehicles in delivering viable bacterial cells to the colon and stable shells during refrigerated storage.

Krasaekoopt et al [54] compared the survival of microencapsulated probiotics using different coating materials and found that chitosan-coated alginate beads provide better protection for *Lactobacillus acidophilus* and *Lactobacillus casei* than did poly-L-lysine-coated alginate beads in 0.6 % bile salts. And chitosan-coated alginate beads provided the best protection for the strains *L. acidophilus* 547 and *L. casei* 01, while sensitive *B. bifidum* ATCC 1994 did not survive the acidic conditions of gastric juice. Zhou et al [55] reported that suspending alginate capsules in a low molecular weight chitosan solution reduced cell release by 40 %. Low molecular weight chitosan diffuses more readily into the calcium alginate gel matrix resulting in a denser membrane than with high molecular weight chitosan [56]. On the contrary, Lee et al [57] indicated that high molecular weight chitosan coating resulted in the highest survival for *Lactobacillus bulgaricus* in simulated gastric juice and better stability at 22 °C. Lee et al indicated that microencapsulation of freeze dried *L. bulgaricus* by chitosan-coated calcium alginate greatly improved the viability of probiotics in simulated gastric and intestinal juices. Krasaekoopt et al [58] studied the survival of probiotics encapsulated in chitosan-coated alginate beads in yogurt and found that the survival of the encapsulated probiotic bacteria was higher than free cells by approximately 1 log cycle.

Bifidobacterium animalis subsp. *lactis* was entrapped in alginate (as control), alginate-chitosan, alginate-chitosan-Sureteric and alginate-chitosan-Acryl-Eze [59]. Survival and in vitro release of *bifidobacteria* from the microparticles were investigated under conditions simulating gastrointestinal fluids covering the pH range from 1.5 to 7.5, with and without pepsin, pancreatin, and bile. All types of microcapsules protected *B. animalis*, but the use of chitosan and enteric polymers in the formulation of the beads, especially Acryl-Eze, enhanced the beneficial effects of the microencapsulation technique. Besides promoting the controlled release of bifidobacteria in simulated gastrointestinal juices, the microencapsulation with enteric polymers improved the survival rate of these microorganisms.

Unsaturated Fatty Acids

Unsaturated fatty acids are important nutrients involved in many body functions, including neuro-protective, antioxidant, anti-inflammatory effects and cardiovascular health. However, their development as nutritional supplements is limited by their high susceptibility to oxidative rancidity, which leads to the formation of off-flavors and potentially toxic compounds. Chitosan is very effective at stabilizing oil-in-water emulsions by adsorbing to the oil droplet surface, lowering interfacial tension, forming protective membranes around the lipid. Unsaturated fatty acid encapsulated in chitosan-based carrier can be converted to a stable powder form that is more resistant to oxidation.

The use of a single layer composed of sodium caseinate cross-linked with transglutaminase could not inhibit lipid oxidation [60]. The use of two layers, which were composed of lecithin and chitosan, resulted in a slight inhibition of oxidation [61, 62]. These emulsions were prepared by first making a so-called primary emulsion with lecithin, and then mixing this emulsion with a solution of a positively charged chitosan [62, 63]. These two-layered emulsions could be freeze or spray-dried after addition of corn syrup solids to improve the oxidative stability [63-65].

Tuna oils are a good source of ω -3 polyunsaturated fatty acids, especially the long-chain ω-3 fatty acids EPA and DHA. Klinkesorn et al [66] investigated physicochemical and oxidative stability of the spray-dried tuna oil-in-water emulsions droplets with a coating of chitosan and lecithin multilayer systems. Their results shown spray-dried tuna oil-in-water emulsions stabilized by chitosan-lecithin membranes were more oxidatively stable than buck oils and thus have excellent potential as ω-3 fatty acid ingredients for functional foods.

Shaw et al [67] also investigated the stability of spray-dried multilayer emulsion. Lecithin-chitosan multilayered emulsions were spray-dried with corn syrup solids. The lecithin-chitosan multilayer interfacial membrane remained intact on the emulsion droplets on reconstitution into an aqueous system. Lecithin-chitosan emulsions were more oxidatively stable than lecithin emulsions. These studies suggest that a microencapsulated multilayered emulsion system could be used as a delivery system for improved oxidative stability.

Park et al [68] investigated the influence of interfacial composition on the in vivo digestibility of lipids coated with chitosan. Their results suggest that encapsulation of lipids by chitosan does not inhibit their in vivo digestibility. Consequently, it should be possible to use chitosan to microencapsulate lipids and lipid-solube components without compromising their bioavailability, although further human studies are needed to confirm it.

Klaypradit and Huang [69] studied the feasibility of chitosan-based encapsulation of fish oil by the new ultrasonic atomizer encapsulation technique. Emulsion preparation variables such as concentration of wall materials (chitosan, maltodextrin and whey protein isolate) and tuna oil were optimized. At 20 g/100 g tuna oil, the optimum ratios of chitosan to maltodextrin and was 1:10. The combination of chitosan and maltodextrin giving the smallest particle size had the highest emulsion stability. The EPA and DHA content (240 mg/g) of the encapsulated powder were slightly higher than commercial specification (100 mg/g) and they had low moisture content and water activity, acceptable appearance and encapsulation efficiency. The ultrasonic technology used in this study could lead to application in the food industry improving the stability of tuna and other oils.

Bustos et al [70] has demonstrated the feasibility of using an emulsion phase separation technique involving an ionotropic gelation method to be used to prepare chitosan microcapsules stabilizing fish oil such as krill oil rich in both EPA and DHA. Emulsions were prepared from a mixture of chitosan, Tween 20 as an emulsifier and krill oil and then slowly added into a stirred solution of sodium hydroxide in ethanol for hardening of the wall matrix of dispersed emulsion particles. Chitosan microcapsules prepared were reported to be not significantly effective in stabilizing the encapsulated krill oil against oxidation compared to the corresponding emulsion form, suggesting some modifications of the properties of wall materials to improve oxidative stability.

Phytochemicals

Phytochemicals have received much attention in recent years from the scientific community, consumers, and food manufacturers due to their potential in lowering blood pressure, reducing cancer risk factors, regulating digestive tract activity, strengthening immune systems, regulating growth, controlling blood sugar concentration, lowering cholesterol levels and serving as antioxidants [71]. Although the use of phytochemicals in capsules and tablets is abundant, their effect is frequently diminished or even lost due to their lack of solubility in water, vegetable oils or other food-grade solvents. In addition, insufficient gastric residence time, low permeability and solubility within the gut, as well as instability under conditions encountered in product processing (temperature, oxygen, light) or in the gastro-intestinal tract (pH, enzymes, presence of other nutrients) limit the activity and potential health benefits of phytochemical molecules [2]. To overcome instability, poor water solubility and bioavailability of phytochemicals, encapsulation techniques have been employed to bring about effective amounts of the intact active component to desired target sites in the body. Ideally, actives such as phytochemicals should be stable and intact under stomach acidic conditions, but readily available under prevailing alkaline conditions of the small intestines [72-74].

Hu et al [75] investigated the process in fabricating chitosan-TPP nanoparticles intended to be used as carriers for delivering tea catechins. The effects of modulating conditions including contact time between chitosan and tea catechins, molecular mass, chitosan concentration, chitosan-TPP mass ratio, initial pH value of chitosan solution, and concentration of tea catechins on encapsulation efficiency and the release profile of tea catechins in vitro were examined systematically. The study found that the encapsulation efficiency of tea catechins in chitosan-TPP nanoparticles ranged from 24 to 53 %. In addition, FT-IR analysis showed that the covalent bonding and hydrogen bonding between tea catechins and chitosan occurred during the formation of chiotsan-TPP nanoparticles loaded with tea catechins. Furthermore, studies on the release profile of tea catechins in vitro demonstrated that the controlled release of tea catechins using chitosan-TPP nanoparticles was achievable.

Bao et al [76] incorporated tea polyphenol-loaded chitosan nanoparticles into gelatin film. The antioxidant effect of tea polyphenol-loaded chitosan nanoparticles against the oxidation of fish oil was investigated, as well as the release of tea polyphenol from nanoparticles in the film. The incorporation of tea polyphenol-loaded chitosan nanoparticles greatly decreased the tensile strength and oxygen permeability but increased the water vapour permeability and affected the transparency of gelatin film. Scanning electron microscopy images indicated that extensive interference was embedded with tea polyphenol-loaded chitosan nanoparticles in the film microstructure. The radical-scavenging activity of films incorporated with tea polyphenol-loaded chitosan nanoparticles was higher than that of control films, and increasing radical-scavenging activity of the former was observed during the storage period. The antioxidant activity and properties of gelatin film were affected by tea polyphenol-loaded chitosan nanoparticles incorporation and storage time. Also, the oxidation of fish oil was retarded. The addition of tea polyphenol-loaded chitosan nanoparticles could improve the antioxidant activity of gelatin film, and the release of tea polyphenol from nanoparticles in the film was achieved during the storage period.

Carotenoids may protect cells against photosensitization and work as light-absorbing pigments during photosynthesis. Some carotenoids may inhibit the destructive effect of reactive oxygen species [77]. Due to the antioxidative properties of carotenoids, many investigations regarding their protective effects against cardiovascular diseases and certain types of cancers, as well as other degenerative illnesses, have been carried out in the last years. Astaxanthin one kind of carotenoids, exist in salmon, lobster, fish eggs, crabs, trout, fish eggs, and flamingo. Its molecule has two cyclic end groups with polar groups, similar to lutein and zeaxanthin. Kittikaiwan et al [78] have developed a novel encapsulation method using chitosan (80 % deacethylation) to encapsulate *Haematococcus pluvialis*, a rich source of astaxanthin. *H. pluvialis* beads were coated five times with chitosan solution forming multilayers of chitosan film. Chitosan capsules containing the algal have shown uniform shape with diameters from 0.35 to 0.5 cm and 100 mm film thickness. Storage studies on chemical stability have proven that chitosan as wall material can prevent degradation of astaxanthin. Higuera-Ciapara et al (79) used solvent evaporation method to microencapsulate astaxathin in a chitosan matrix cross-linked with glutaraldehyde. Microcapsules were chemically stable over a storage time of 8 weeks.

Flavors

Flavor is one of the most important characteristics of a food product, since people prefer to eat only food products with an attractive flavor [80]. Flavor can be defined as a combination of taste, smell and/or trigeminal stimuli. Aroma consists of many volatile and odorous organic molecules. Most of the aroma molecules are lipophilic, but some are hydrophilic. Aroma molecules can be either added to food products, produced during processing of the food product or are formed during cooking of the food product. Aroma can reach the nose cavity directly when the food is not yet in the mouth or via the oral cavity. Aroma release from food products before and after eating is controlled by both thermodynamic and kinetic parameters, which depends on the aroma characteristics and on the composition and the physical state of the matrix. These parameters will determine the volatility of the flavor compounds and their resistance to mass transfer between different phases, especially from the product to air [81-83]. Proper choice of food composition and food microstructures may thus control aroma release during food product preparation and consumption. Encapsulation might be one of the tools in such a design [84].

Exact information about the annual sales of aroma delivery systems is not available. The flavor and fragrance industry had a turnover of about US$20.5 billion in 2008. Ubbink and Schoonman [85] assumed that about 20–25 % of all flavors are estimated to be sold in an encapsulated form. However, Campanile [86] estimated that the total market for aroma delivery systems was only half of this value, i.e. at about $750 m in 2002. About 80–90 % of the encapsulates are spray-dried ones, 5–10 % are prepared by spray-chilling, 2–3 % are prepared by melt extrusion and 2% are prepared by melt injection [85, 87-89].

Hsieh et al [90] emulsified citronella oil in the presence of 0.2 % chitosan and increased the pH with 0.1–1.5 % sodium hydroxide to form a shell of chitosan around the emulsion droplets. After washing twice with distilled water, the dispersion of the encapsulates was improved by incubation with coconut oil amphoteric surfactant for 10 days. The resulting microcapsules were dried in a vacuum oven. The aroma released out of dried microcapsules within a few hours (depending on the temperature). Interestingly, the authors suggest that the structure of the chitosan wall shrinked once the microcapsules were thermally pre-treated at 80°C.

Citral and limonene are the major flavor components of citrus oils. Both of these compounds can undergo chemical degradation leading to loss of flavor and the formation of undesirable off-flavors. Djordjevic et al [91] examined if citral and limonene were more stable in emulsions stabilized with a sodium dodecyl sulfate (SDS)-chitosan complex than gum arabic. Citral degraded less in GA-stabilized than in SDS-chitosan-stabilized emulsions at pH 3.0. However, SDS-chitosan-stabilized emulsions were more effective at retarding the formation of the citral oxidation product, p-cymene, than GA stabilized emulsions. Limonene degradation and the formation of limonene oxidation products, limonene oxide and carvone, were lower in the SDS-chitosan- than GA-stabilized emulsions at pH 3.0. The ability of an SDS-chitosan multilayer emulsifier system to inhibit the oxidative deterioration of citral and limonene could be due to the formation of a cationic and thick emulsion droplet interface that could repel prooxidative metals, thus decreasing prooxidant-lipid interactions.

Ponce et al [92] studied the use of chitoan coating enriched with rosemary and olive oleoresins applied to butternut squash did not produce a significant antibacterial effect, however antioxidant effects were observed during the first day, exerting peroxidise inhibition up to 5 days of storage. The use of chitosan enriched rosemary and olive did not introduce deleterious effects on the sensorial acceptability of squash. Chitosan enriched with rosemary and olive improved the antioxidant protection of the minimally processed squash offering a great advantage in the prevention or browning reactions which typically result in quality loss in fruits and vegetables.

III. Conclusions

Encapsulation of functional food ingredients, such as vitamin, probiotics, flavors, phytochemicals and unsaturated fatty acids, using chitosan-based material was reviewed. Due to chitosan biological and chemical properties, it is possible more and more chitosan-based delivery systems for functional food ingredients will be studied and developed. Until now, most of these researches are limited to lab level. More studies are needed before the industrial application of chitosan-based delivery systems. Hopefully, more chitosan-based delivery systems can be applied in food relevant fields in the future.

References

[1] Elliott, R., Ong, T. J. Science, medicine, and the future of nutritional genomics. *British Medicine Journal*, 2002(324), 1438-1442.

[2] Bell, L. N. Stability testing of nutraceuticals and functional foods. In: Wildman, R.E.C., *Handbook of Nutraceuticals and Functional Foods*, New York: CRC press; 2001, 501-516.

[3] Peppas, N. A., Buers, P., Leobandung, W., Ichikawa, H. Hydrogels in pharmaceutical formulations. *European Journal of Pharmaceutics and Biopharmaceutics*, 2000(50), 27-46.
[4] Lang, R., Peppas, N. A. Advances in biomaterials, drug delivery, and bionanotechnology. *AIChE Journal*, 2003(49), 2990-3006.
[5] Peter, M. Applications and environmental aspects of chitin and chitosan. *Journal of macromolecular science. Pure and Applied Chemistry*, 1995(32), 629–640.
[6] Hudson, S. M., Smith, C. Polysaccharides: chitin and chitosan: chemistry and technology of their use as structural materials. In: Kaplan, D.L., *Biopolymers from Renewable Resources*, Berlin: Springer; 1998, 96–118.
[7] Khor, E. Chitin: a biomaterial in waiting. *Current Opinion in Solid State and Materials Science*, 2002(6), 313–317.
[8] Felse, P.A., Panda, T. Studies on applications of chitin and its dervatives. *Bioprocess Engineering*, 1999(20), 505–512.
[9] Tharanathan, R.N., Kittur, F.S. Chitin—the undisputed biomolecule of great potential. *Critical Reviews in Food Science and Nutrition*, 2003(43), 61–87.
[10] Kurita, K. Controlled functionalization of the polysaccharide chitin. *Progress in Polymer Science*, 2001(26), 1921–1971.
[11] Paul, W., Sharma, C.P. Chitosan, a drug carrier for the 21st century: a review. *STP Pharmaceutical Sciences*, 2000(10), 5–22.
[12] Felt, O., Buri, P., Gurny R. Chitosan: a unique polysaccharide for drug delivery. *Drug Development and Industrial Pharmacy*, 1998(24), 979–993.
[13] Singla, A.K., Chawla, M. Chitosan: some pharmaceutical and biological aspects—an update. *Journal of Pharmacy and Pharmacology*, 2001(53), 1047–1067.
[14] Dutta, P.K., Ravikumar, M.N.V., Dutta, J. Chitin and chitosan for versatile applications. *Journal of Macromolecular Science*, 2002(42), 307–354.
[15] Shahidi, F., Arachchi, J.K.V., Jeon, Y.J. Food applications of chitin and chitosan. *Trends in Food Science and Technology*, 1999(10), 37–51.
[16] Agboh, O.C., Qin, Y. Chitin and chitosan fibers. *Polymer for Advanced Technologies*, 1996(8), 355–365.
[17] Kawamura, Y., Yoshida, H., Asai, S., Kurahashi, I., Tanibe, H. Effects of chitosan concentration and precipitation bath concentration on the material properties of porous crosslinked chitosan beads. *Separation Science and Technology*, 1997(32), 1959–1974.
[18] Kubota, N., Kikuchi, Y. Macromolecular complexes of chitosan. In: Dumitriu S., *Polysaccharides*, New York: Dekker; 1998, 595–628.
[19] Krajewska, B. Chitin and its derivatives as supports for immobilization of enzymes. *Acta Biotechnologica*, 1991(11), 269–277
[20] No, H. K., Meyers, S.P. Application of chitosan for treatment of wastewaters. *Reviews of Environmental Contamination and Toxicology*, 2000(163), 1–28.
[21] Chen, L. Y., Subirade, M. Chitosan/ß-lactoglobulin core-shell nanoparticles as nutraceutical carriers. *Biomaterials*, 2005(26), 6041-6053.
[22] Shu, X. Z., Zhu, K. J. Controlled drug release properties of ionically cross-linked chitosan beads: the influence of anion structure. *International Journal of Pharmaceutics*, 2002(233), 217-225.

[23] Bajpai, S.K., Tankhiwale, R. Investigation of dynamic release of vitamin B_2 from calcium alginate/chitosan multilayered beads: Part II. *Reactive and Functional Polymers*, 2006(66), 1565-1574.

[24] Desai, K.G.H. Park, H.J. Encapsulation of vitamin C in tripolyphosphate crosslinked chitosan microspheres by spray drying. *Journal of Microencapsulation*, 2005(22), 179–192.

[25] Desai, K.G.H., Park, H.J. Effect of manufacturing parameters on the characteristics of vitamin C encapsulated tripolyphosphate-chitosan microspheres prepared by spray drying. *Journal of Microencapsulation*, 2006(23), 91-103.

[26] Desai, K.G.H., Liu, C., Park, H.J. Characteristics of vitamin C encapsulated tripolyphosphate-chitosan microspheres as affected by chitosan molecular weight. *Journal of Microencapsulation*, 2006(23), 79-90.

[27] Jang, K. I., Lee, H. G. Stability of Chitosan Nanoparticles for L-Ascorbic Acid during Heat Treatment in Aqueous Solution, *Journal of Agricultural and Food Chemistry*, 2008(56), 1936-1941.

[28] Fuller, R. *Probiotics: The Scientific Basis*. London: Chapman and Hall; 1992.

[29] Mattila-Sandholm, T., Myllärinen, P., Crittenden, R., Mogensen, G., Fonden, R., Saarela., M. Technological challenges for future probiotic foods. *International Dairy Journal,* 2002(12), 173–182.

[30] Champagne, C. P., Gardner, N. J. Challenges in the addition of probiotic cultures to foods. *Critical Reviews in Food Science and Nutrition,* 2005(45), 61–84.

[31] Gilliland, S. E. Acidophilus milk products: A review of potential benefits to consumers. *Journal of Dairy Science,* 1989(72), 2483–2494.

[32] Robinson, R. K. Survival of *Lactobacillus acidophilus* in fermented products. *Suid Afrikaans Tydskrif Vir Suiwelunde,* 1987(19), 25–27.

[33] Kurman, J. A., Rasic, J. L. The health potential of products containing *Bifidobacteria*. In: Robinson, R. K., editor. *Therapeutic Properties of Fermented Milk.* London: Elsevier Application Food Science Series; 1991; 115–117.

[34] Berrada, N., Lemeland, J. F., Laroche, G., Thouvenot, P., Piaia, M. *Bifidobacterium* from fermented milks: Survival during gastric transit. *Journal of Dairy Science,* 1991(74), 409–413.

[35] Lankaputhra, W. E. V., Shah, N. P. Survival of *Lactobacillus acidopilus* and *Bifidobacterium* spp. in the presence of acid and bile salts. *Cultured Dairy Products Journal,* 1995(30), 2–7.

[36] Iwana, H., Masuda, H., Fujisawa, T., Suzuki, H., Mitsuoka, T. Isolation and identification of *Bifidobacterium* ssp. in commercial yogurt sold. *Europe Mitsuokifidobacteria Microflora,* 1993(12), 39–45.

[37] Shah, N. P., W. Lankaputhra, E. V. Improving viability of *Lactobacillus acidophilus* and *Bifidobacterium* ssp. in yogurt. *International Dairy Journal,* 1997(7), 349–356.

[38] Schillinger, U. Isolation and identification of lactobacilli from novel-type probiotic and mild yogurts and their stability during refrigerated storage. *International Journal of Food Microbiology,* 1999(47), 79–87.

[39] Samona, A., Robinson, R. K. Effect of yogurt cultures on the survival of *Bifidobacteria* in fermented milks. *Journal of the Society of Dairy Technology,* 1994(47), 58–60.

[40] Maitrot, H., Paquin, C. Lacroix, C., Champagne, C. P. Production of concentrated freeze-dried cultures of *Bifidobacterium longum* in *k*-carrageenan-locust bean gum gel. *Biotechnology Techniques,* 1997(11), 527–531.

[41] Klaver, F. A. M., Kingma, F., Weerkamp, A. H. Growth and survival of *bifidobacteria* in milk. *Netherlands Milk and Dairy Journal,* 1993(47), 151–164.

[42] Lankaputhra, W. E. V., Shah, N. P. Survival of *Bifidobacteria* during refrigerated storage in the presence of acid and hydrogen peroxide. *Milchwissenschaft,* 1996(51), 65–70.

[43] Dave, R. I., Shah, N. P. Effectiveness of ascorbic acid as an oxygen scavenger in improving viability of probiotic bacteria in yoghurts made with commercial starter cultures. *International Dairy Journal*, 1997(7), 435–443.

[44] Prevost, H., Divies, C. Continuous pre-fermentation of milk by entrapped yogurt bacteria. I. Development of the process. *Milchwissenschaft,* 1988(43), 621–625.

[45] Lacroix, C., Paquin, C., Arnaud, J. P. Batch fermentation with entrapped growing cells of *Lactobacillus casei.* I. Optimization of the rheological properties of the entrapment. *Applied Microbiology and Biotechnology,* 1990(32), 403–408.

[46] Champagne, C. P., Gaudy, D. P., Neufeld, R. J. *Lactococcus lactis* release from calcium alginate beads. *Applied and Environmental Microbiology,* 1992(58), 1429–1434.

[47] Peniche, C.W., Argüelles-Monal, H. P., Acosta, N. Chitosan: An attractive biocompatible polymer for microencapsulation. *Macromolecular Bioscience,* 2003(3), 511–520.

[48] Groboillot, A. F., Champagne, C. P., Darling, G. F., Poncelet, D. Membrane formation by interfacial crosslinking of chitosan for microencapsulation of *Lactococcus lactis. Biotechnology and Bioengineering,* 1993(42), 1157–1163.

[49] Cui, J., Goh, J., Kim, P., Choi, S., Lee, B. Survival and stability of *Bifidobacteria* loaded in alginate poly-Llysine microparticles. *International Journal of Pharmaceutics,* 2000(210), 51–59.

[50] Canh, L., Millette, M., Mateescu, M., Lacroix, M. Modified alginate and chitosan for lactic acid bacteria immobilization. *Biotechnology and Applied Biochemistry,* 2004(39), 347–354.

[51] Smidsrod, O., Skjak-Braek, G.. Alginate as immobilization matrix for cells. *Trends in Biotechnology,* 1990(8), 71–78.

[52] Lacík, I. Polyelectrolyte complexes for microcapsule formation. In: Nedović, V., Willaert, R.G., editors. *Fundamentals of cell immobilisation biotechnology*. Kluwer: Dordrecht; 2004; 103–120.

[53] Gaserod, O., Smidsrod, O., Skjak-Braek, G. Microcapsules of alginate-chitosan I. A quantitative study of the interaction between alginate and chitosan. *Biomaterials*, 1998(19), 1815–1825.

Gaserod, O., Sannes, A., Skjak-Braek, G. Microcapsules of alginate-chitosan II. A study of capsule stability and permeability. *Biomaterials*, 1999(20), 773–783.

[54] Krasaekoopt, W., Bhandari, B., Deeth, H. The influence of coating materials on some properties of alginate beads and survivability of microencapsulated probiotic bacteria. *International Dairy Journal*, 2004(14), 737–743.

[55] Zhou, Y., Martins, E., Groboillot, A., Neufeld, R. J. Spectrophotometric quantification of lactic bacteria in alginate and control of cell release with chitosan coating. *Journal of Applied Microbiology,* 1998(84), 342–348.

[56] McKnight, C.A., Ku, A., Goosen, M.F. Synthesis of chitosan-alginate microencapsule membrane. *Journal of Bioactive and Compatible Polymers*, 1988(3), 334–354.

[57] Lee, J. S., Cha, D. S., Park, H. J. Survival of freeze-dried *Lactobacillus bulgaricus* KFRI 673 in chitosan-coated calcium alginate microparticles. *Journal of Agricultural Food Chemistry,* 2004(52), 7300–7305.

[58] Krasaekoopt, W., Bhandari, B., Deeth, H. C. Survival of probiotics encapsulated in chitosan-coated alginate beads in yoghurt from UHT- and conventionally treated milk during storage. *LWT,* 2006(39), 177–183.

[59] Liserre, A. M., Re, M. I., Franco, B. D.G.M. Microencapsulation of *bifidobacterium animalis subsp. Lactis* in modified alginate-chitosan beads and evaluation of survival in simulated gastrointestinal conditions, *Food Biotechnology*, 2007(21), 1-16.

[60] Ogawa, S., Decker, E.A., McClements, D.J. Influence of environmental conditions on the stability of oil in water emulsions containing droplets stabilized by lecithin-chitosan membranes. *Journal of Agricultural and Food Chemistry*, 2003(51), 5522–5527.

[61] Klinkesorn, U., Sophanodora, P., Chinachoti, P., McClements, D.J., Decker, E.A. Increasing the oxidative stability of liquid and dried tuna oil-in-water emulsions with electrostatic layer-bylayer deposition technology. *Journal of Agricultural and Food Chemistry*, 2005(53), 4561–4566.

[62] Klinkerson, U., Sophanodora, P., Chinachoti, P., Decker, E.A., McClements, D.J. Encapsulation of emulsified tuna oil in two-layered interfacial membranes prepared using electrostatic layer by layer deposition. *Food Hydrocolloids*, 2005(19), 1044–1053.

[63] Klinkerson, U., Sophanodora, P., Chinachoti, P., Decker, E.A., McClements, D.J. Characterization of spray-dried tuna oil emulsified in two-layered interfacial membranes prepared using electrostatic layer-by-layer deposition. *Food Research International*, 2006(39), 449–457.

[64] McClements, D.J., Decker, E.A. Biopolymer encapsulation and stabilization of lipid systems and methods for utilisation thereof, *WO2005/086976*, 2005.

[65] McClements, D.J, Decker, E.A. Encapsulated emulsions and methods of preparations, *WO2007/038616*, 2007.

[66] Klinkerson, U., Sophanodora, P., Chinachoti, P., McClements, D.J., Decker, E.A. Stability of spray-dried tuna oil emulsions encapsulated with two-layered interfacial membranes. *Journal of Agricultural and Food Chemistry*, 2005(53), 8365–8371.

[67] Shaw, L.A., McClements, D.J., Decker, E.A. Spray-dried multilayered emulsions as a delivery method for ω-3 fatty acids into food systems. *Journal of Agricultural and Food Chemistry*, 2007(55), 3112–3119.

[68] Park, G. Y., Mun, S., Park, Y., Rhee, S., Decker, E. A., Weiss, J., McClements, D. J., Park, Y. Influence of encapsulation of emulsified lipids with chitosan on their in vivo digestibility. *Food Chemistry*, 2007(104), 761-767.

[69] Klaypradit, W., Huang, Y. W. Fish oil encapsulation with chitosan using ultrasonic atomizer. *LWT*, 2008(41), 1133-1139.

[70] Bustos, R., Romo, L., Yáñez, K., Díaz, G., Romo, C. Oxidative stability of carotenoid pigments and polyunsaturated fatty acids in microparticulate diets containing krill oil for nutrition of marine fish larvae. *Journal of Food Engineering*, 2003(56), 289-293.

[71] Wildman, R. E. C., Kelly, M. Nutraceuticals and Functional Foods. In: Wildman, R. E. C. *Handbook of Nutraceuticals and Functional Foods*. New York: CRC Press; 2001.

[72] Ho, C. T., Lee, C. Y., Huang, M. T. *Phenolic Compounds in Food and their Effects on Health. I: Analysis, Occurrence, and Chemistry. ACS Symp. Ser. 506*, Washington, D.C.: American Chemical Society, 1992.

[73] Salah, N., Miller, N. J., Paganga, G. Polyphenolic flavanols as scavengers of aqueous phase radicals and as chain breaking antioxidants. *Archives of Biochemistry and Biophysics*, 1995(322), 339–346.

[74] Havsteen, B. Flavonoids, a class of natural products of high pharmacological potency. *Biochemistry Pharmacology*, 1983(32), 1141–1148.

[75] Hu, B., Pan, C., Sun, Y., Hou, Z., Ye, H., Hu, B., Zeng, X. X. Optimization of fabrication parameters to produce chitosan-tripolyphosphate nanoparticles for delivery of tea catechins. *Journal of Agricultural and Food Chemistry*, 2008(56), 7451-7458.

[76] Bao, S. B., Xu, S. Y., Wang, Z. Antioxidant activity and properties of gelatin films incorporated with tea polyphenol-loaded chtosan nanoparticles. *Journal of the Science of Food and Agriculture*, 2009(89), 2692-2700.

[77] Shu, B., Yu, W., Zhao, Y., Liu, X. Study on microencapsulation of lycopene by spray-drying. *Journal of Food Engineering*, 2006(76), 664–669.

[78] Kittikaiwan, P., Powthongsook, S., Pavasant, P., Shotipruk, A. Encapsulation of *Haematococcus pluvialis* using chitosan for astaxanthin stability enhancement. *Carbohydrate Polymers*, 2007(70), 378–385.

[79] Higuera-Ciapara, I., Felix-Valenzuela, L., Goycoolea, F.M., Argüelles-Monal, W. Microencapsulation of astaxanthin in a chitosan matrix. *Carbohydrate Polymers*, 2004(56), 41–45.

[80] Voilley, A., Etiévant, P. *Flavor in Food*. Cambridge: Woodhead Publishing Limited; 2006.

[81] Druaux, C., Voilley, A. Effect of food composition and microstructure on volatile flavour release. *Trends of Food Science and Technology*, 1997(8), 364–368.

[82] Van Ruth, S., Roozen, J.P. Delivery of flavor from food matrices. In: Taylor, A.J. *Food flavour technology*. UK: Sheffield Academic Press; 2002; 190-206.

[83] De Roos, K.B. Effect of texture and microstructure on flavor retention and release. *International Dairy Journal*, 2003(13), 593–605.

[84] Porzio, M.A. Flavor delivery and product development. *Food Technology*, 2007(1), 22–29.

[85] Ubbink, J., Schoonman, A. Flavor delivery systems. In: Othmer, K. *Kirk-Othmer encyclopedia of chemical technology*. New York: Wiley-Interscience; 2003; 527-562.

[86] Campanile, F. The flavor saviour? *Food Processing*, 2004(73), 14.

[87] Porzio, M. Flavor encapsulation: a convergence of science and art. *Food Technology*, 2004(58), 40–47.

[88] Porzio, M. Spray drying. An in-dept look at the steps in spray drying and the different options available to flavorists. *Perfumer Flavorist*, 2007(32), 34–39.

[89] Porzio, M. Melt extrusion and melt injection. *Perfumer Flavorist*, 2008(33), 48–53.

[90] Hsieh, W.C., Chang, C.P., Gao, Y.L. Controlled release properties of chitosan encapsulated volatile citronella oil microcapsules by thermal treatments. *Colloids and Surface B: Biointerfaces*, 2006(53), 209–214.

[91] Djordjevic, D., Cercaci, L., Alamed, J., McClements, D. J., Derker, E. A. Chemical and physical stability of citral and limonene in sodium dodecyl sulphate-chitosan and gum

Arabic-stabilized oil-in-water emulsions. *Journal of Agricultural and Food Chemistry*, 2007(55), 3585-3591.

[92] Ponce, A. G., Roura, S. I., del Valle, C. E., Moreira, M. R. Antimicrobial and antioxidant activities of edible coatings enriched with natural plant extracts: *In vitro* and *in vivo* studies. *Postharvest Biology and Technology*, 2008(49), 294-300.

In: Handbook of Chitosan Research and Applications ISBN 978-1-61324-455-5
Editors: Richard G. Mackay and Jennifer M. Tait

Chapter 10

CHITOSAN NANOPARTICLES AND NANOFIBERS: PREPARATION AND APPLICATION FOR ENZYME IMMOBILIZATION

Zhen-Xing Tang,[1,3]* Lu-E Shi[2] and Qing-Bin Guo[3]

[1]. Food Bioengineering Research Laboratory,
Research and Development Center,
Hangzhou Wahaha Group Co. Ltd,
Hangzhou, Zhejiang, China
[2]. College of Life and Environmental Sciences,
Hangzhou Normal University, Hangzhou,
Zhejiang, China
[3]. Guelph Food Research Center,
Agriculture and Agri-Food Canada,
Guelph, Ontario, Canada

ABSTRACT

Enzyme immobilization has attracted continuous attention in the fields of fine chemical engineering and bio-chemical engineering. The performance of immobilized enzyme largely depends on the supports. Compared with traditional enzyme immobilized carriers, nano-structured carriers show many advantages including surface area, mass transfer resistance, and effective enzyme loading. Nano-structured carriers are believed to be able to retain the catalytic activity as well as ensure the immobilization efficiency of enzyme to a high extent. Various nanomaterials, such as nanoparticles, nanofibers and nanoporous matrices, have shown potential for revolutionizing the preparation and application of enzyme immobilized carriers. Chitosan is a natural polysaccharide prepared by the N-deacetylation of chitin. It has been widely used in many industrial and biomedical aspects, including in wastewater treatment, in enzyme immobilization, and as a carrier for controlled drug delivery, due to its significant biological and chemical properties such as biodegradable, biocompatible, bioactive, and polycationic. This chapter mainly discussed the recent advances in using chitosan nanoparticles and

* Corresponding author: Zhen-Xing Tang, E-mail: tangzhenxing@126.com.

nanofibers supports as hosts for enzyme immobilization. Preparation of chitosan nanoparticle and nanofibers, our work we achieved in our group on this topic were introduced. Some problems encountered with chitosan nanoparticles and nanofibers carriers for enzyme immobilization were discussed together with the future prospects of such systems.

I. Introduction

Enzymes are powerful catalysts that can be applied be various fields including biodegradation, biosensors, food processing, and the development of biofuel cells because they are highly specific and active [1]. Although a lot of merits exist in using enzymes for different application, their use is limited by their short active lifetimes. In general, enzymes become unstable over time due to changes in the vital three-dimensional structure. In addition, the reuse of enzymes is difficult to achieve because it is not easy to separate the enzyme from reaction by-products.

To overcome these disadvantages, many researchers use immobilization enzyme techniques [2-4]. The immobilization of enzymes allows their reuse as well as an extended life of activity. Conceptually, three ways are often used to immobilize enzymes, i.e., adsorption, covalent binding, and encapsulation. The performance of immobilized enzyme largely depends on the carriers or matrices.

Nanobiocatalysis, in which enzymes are incorporated into nanostructured materials, has emerged as a rapidly growing area. Various nanomaterials, particularly nanoparticles and nanofibers, are available for enzyme immobilization applications [5-9]. Typically, smaller particles provide a larger surface area for the attachment of enzymes [10] and a shorter diffusional path for the substrates. Thus, nanoparticles are utilized as carriers for enzyme immobilization in many studies [11-13]. Nanofibers have large surface areas because they are produced in a small string-like structure. A supporting carrier with a large surface-to-volume ratio is desirable because a high catalytic efficiency can be achieved [14]. Thus, it is assumed that nanofibers, with their high surface-to-volume ratios, will be good supports for enzyme immobilization, whereas substrate diffusion limitations, an inherent problem when enzymes are trapped within nanoporestructured materials, are not as likely because the enzymes are located on the surface of the support [15]. Furthermore, enzyme–nanofiber composites can be recovered easily and used repeatedly. It is also reported that an enzyme aggregate coating on nanofibers improves the enzyme activity and stability [14]. The α-chymotrypsin-coated nanofibers, formed via cross-linking and aggregation, show negligible loss of the enzyme activity for more than 1 month.

Chitosan is a partially deacetylated polymer of acetyl glucosamine obtained through alkaline deacetylation of chitin. It consists of β-(1-4)-linked-D-glucosamine residue with the 2- hydroxyl group being substituted by an amino or acetylated amino group [16]. The amine groups and –OH endow chitosan with many special properties, making it applicable in many areas and readily available for chemical reactions. Chitosan is safe, non-toxic and can interact with polyanions to form complexes and gels [17, 18].

Although it is recognized that there is no universal support for all enzymes, a number of desirable characteristics should be common to any material considered for immobilizing enzymes. These include: high affinity to proteins, availability of reactive functional groups for direct reactions with enzymes and for chemical modifications, hydrophilicity, mechanical stability and rigidity, regenerability, and ease of preparation in different geometrical configurations that provide the system with permeability and surface area suitable for a chosen biotransformation [19]. Understandably, for food, pharmaceutical, medical and agricultural applications, nontoxicity and biocompatibility of the materials are also required. Furthermore, to respond to the growing public health and environmental awareness, the materials should be biodegradable, and to prove economical, inexpensive.

Of the many carriers that have been considered and studied for immobilizing enzymes, organic or inorganic, natural or synthetic, chitosan have the most of the above characteristics for enzyme immobilized carriers. In recent years, for chitosan nanoparticles and nanofibers, various preparation methods and applications for enzyme immobilization have been reported. In this chaper, preparation of chitosan nanoparticle and nanofibers and their application in enzyme immobilization were reviewed.

II. Preparing Chitosan Nanoparticles and Nanofibers

Preparing of Chitosan Nanoparticles

Nanoparticles are defined as particulate dispersions or solid particles with a size in the range of 1-1000 nm. The enzyme is dissolved, entrapped, encapsulated or attached to a nanoparticle matrix. In recent years, biodegradable polymeric nanoparticles including chitosan have been used as potential drug delivery carriers [20]. Until now, most of researcher work on application of chitosan nanoparticles is focused on drug encapsulation and delivery. Here we summaried the preparation of chitosan nanoparticles firstly.

(a) Emulsion Method

A water-oil (w/o) emulsion is prepared by emulsifying chitosan aqueous solution in oil [21-23]. Aqueous droplets are stabilized using a suitable surfactant. The stable emulsion is cross-linked by using an appropriate cross-linking agent such as glutaraldehyde to harden the droplets. Ohya et al [21] reported for the first time the preparation of chitosan-gel nanospheres (average diameter 250 nm) using w/o emulsion method followed by glutaraldehyde crosslinking of the chitosan amino groups. These pioneering studies demonstrated the feasibility of synthesizing stable and reproducible chitosan nanoparticles that could bind and deliver drugs and enzymes. Unfortunately, this method is seldom used for enzyme carrier preparation. Liu et al [24] prepared linoleic acid chitosan nanoparticles using o/w emulsification method with methylene chloride. However, methylene chloride was used as an oil phase, which is not also suitable for enzyme immobilization carrier.

Instead of cross-linking the stable droplets, precipitation is induced by allowing chitosan droplets to interact with NaOH droplets [25]. First, a stable emulsion containing aqueous solution of chitosan along with drug is produced in liquid paraffin oil and then, another stable emulsion containing chitosan aqueous solution of NaOH is produced in the same manner. When both emulsions are mixed under high-speed stirring, droplets of each emulsion would collide at random and coalesce, thereby precipitating chitosan droplets to give small size particles.

Gadopentetic acid-loaded chitosan nanoparticles have been prepared by this method [26]. Particle size depends upon the type of chitosan. As deacetylation degree of chitosan decreased, particle size increased. Particles produced using 100 % deacetylated chitosan had the mean particle size of 452 nm. Since gadopentetic acid is a bivalent anionic compound, it interacts electrostatically with the amino groups of chitosan, which would not have occurred if a cross-linking agent is used that blocks the free amino groups of chitosan. Thus, it was possible to achieve higher gadopentetic acid loading by using the emulsion-droplet precipitation method compared to the simple emulsion crosslinking method.

(b) Ionic Gelation Method

The conjugation of oppositely charged macromolecules is another method for preparing chitosan nanoparticles. Tripolyphosphate (TPP) was often used to prepare chitosan nanoparticles, because TPP is nontoxic, multivalent and able to form gelate through ionic interaction between positively charged amino groups of chitosan and negatively charged TPP. The interaction could be controlled by the charge density of TPP and chitosan, which is dependent on the pH of solution.

Calvo et al [27] first prepared the nanoparticles using TPP cross-linking method, which were used as protein carrier. This formulation was also evaluated as a carrier for therapeutic peptides and food ingredients [28-30]. Nasti et al [31] studied the influence of a number of orthogonal factors (pH, concentrations, ratios of components, different methods of mixing) on the preparation of chitosan/TPP nanoparticles. Lin et al [32] investigated the relationship of free amino groups on the surface and the characteristics of chitosan nanoparticles prepared by ionic gelation method. The result showed that the surface free amino groups reduced, the average size, zeta potential, stability of nanoparticles, was not affected with the increase of TPP concentration. With the increase of pH, the free amino groups could be deprotonated and the ionizable level was stepped down, correspondingly the particle size and zeta potential of chitosan nanoparticles decreased. The amount and ionizable level of free amino groups on the surface are affected by the gelation degree and pH, which further affected the volume phase transitions of chitosan nanoparticles. The properties of chitosan nanoparticles have correlation with the surface free amino groups.

Keresztessy et al [33] prepared chitosan/poly-γ-glutaic acid nanoparticles via an ionotropic gelation method. Solubility, surface charge and size distribution of these nanoparticles were depended on the concentration and ratio of biopolymers, pH environment as well as addition. Their results shown chitosan/poly-γ-glutaic acid nanoparticles have the potential for building mutifuctional nanocarriers for biochemical engineering application.

A new and simple method to prepare magnetic Fe_3O_4-chitosan nanoparticles by cross-linking with TPP, precipitation with NaOH and oxidation with O_2 in hydrochloric acid aqueous phase containing chitosan and $Fe(OH)_2$ was developed by Wu et al [34]. TEM showed that the diameter of the magnetic Fe_3O_4 chitosan nanoparticles with a diameter of 20 nm. The adsorption capacity of lipase onto nanoparticles could reach 129 mg/g; and the maximal enzyme activity was 20.02 μmol min^{-1} mg^{-1} (protein), and activity retention was as high as 55.6% at a certain loading amount.

(c) Reverse Micellar Method

Brunel et al [35] used a reverse micellar method to prepare chitosan nanoparticles. The lower the molar mass of chitosan, the better the control over particle size and size distribution, probably as a result of either a reduction in the viscosity of the internal aqueous phase or an increase in the disentanglement of the polymer chain during the process. These new and safe nanoparticles offer wide perspectives of development in further applications.

Mitra et al [36] have encapsulated doxorubicin–dextran conjugation in chitosan nanoparticles prepared with reverse micellar method. The surfactant sodium bis (ethyl hexyl) sulfosuccinate (AOT) was dissolved in n-hexane. To AOT solution, 0.1 % chitosan in acetic acid, doxorubicin–dextran conjugate, liquid ammonia and 0.01 % glutaraldehyde were added with continuous stirring at room temperature. This method produced chitosan nanoparticles with doxorubicin–dextran conjugate inside.

Mansouri et al [37] prepared BSA-loaded chitosan nanoparticles using reverse micellar method. The size of particles was obtained in range 143 to 428 nm. FT-IR spectrum analysis indicated that the BSA was successfully encapsulated into BSA loaded chitosan nanoparticles. The chitosan concentration and theoretical BSA loading had no obvious effect on the diameters of nanoparticles. The chitosan concentration and BSA loading, therefore, played an important role in the release of BSA. Increasing the chitosan solution concentration retained the release of BSA, both in 10 % and 20 % BSA loading, which decreasing the BSA loading, accelerated the release of BSA both in 0.1 and 0.2 % chitosan concentration.

The magnetic chitosan nanoparticles were prepared by reversed phase suspension method using Span-80 as an emulsifier, glutaraldehyde as cross-linking reagent [38]. The results shown the nanoparticles were spherical and almost superparamagnetic. The laccase was immobilized on nanoparticles by adsorption and subsequently by cross-linking with glutaraldehyde. The immobilized laccase exhibited an appreciable catalytic capability and had good storage stability and operation stability. The K_m of immobilized and free laccase for ABTS were 140.6 and 31.1 μM in phosphate buffer (0.1 M, pH 3.0), respectively. The immobilized laccase is a good candidate for the research and development of biosensors based on laccase catalysis.

(d) Self-Assembling Method

There has been an increasing interest in the self-assembling of polymeric amphiphiles to prepare nanoparticles. There were several reports on chitosan hydrophobical modification and nanoparticle formation through self-assembly in aqueous solution.

Liu et al [39] studied trypsin was immobilized on linolenic acid (LA) modified chitosan nanoparticles using glutaraldehyde as crossing-linker. Their results indicated the activity of trypsin immobilized onto linolenic acid modified chitosan nanoparticles increased with increasing concentration of glutaraldehyde up to 0.07 % and then decreased with increasing amount of glutaraldehyde. On the other hand, particle size increased (from 523 to 1372 nm) with the increasing concentration of glutaraldehyde (from 0.03 to 0.1 %). The enzyme catalytic characteristics of nanoparticle solution were also studied. The kinetic constant value (K_m) of immobilized enzyme (71.9 mg/ml) was higher than that free enzyme (50.2 mg/ml). However, the thermal stability and optimum temperature of immobilized enzyme improved, which make it more attractive in the application aspect.

Chen et al [40] modified chitosan by coupling it with LA through the EDC-mediated reaction. The fluorescence spectra indicated that the self-aggregation of LA-chitosan occurred at the concentration of 1.0 g L^{-1} or above. The micelles of LA-chitosan formed nanosized particles ranging from 200 to 600 nm. The LA chitosan nanoparticles encapsulated the lipid soluble model compound, retinal acetate, with 50 % efficiency.

Hu et al [41] obtained hydrophobically modified chitosan nanoparticles containing 5.4 stearic acid groups per 100 anhydroglucose units synthesized by an EDC-mediated coupling reaction. Mean diameter of self-aggregates in pH 7.0 PBS was 25.0 ± 14.7 nm with a unimodal size distribution. The diameter, as well as the zeta potential of self-aggregates increased when the pH value of dispersion medium decreased. Bovine serum albumin (BSA) was further enveloped in the interface of different single self-aggregate and formed nanoparticles. The size of BSA-loaded stearic acid modified nanoparticles depended on the pH values of the dispersed aqueous vehicle, and the size diminished when the pH values of the dispersed aqueous vehicle decreased, whilst, the BSA encapsulation efficiency enhanced. The nanoparticles were characterized by TEM. BSA release from stearic acid modified nanoparticles decreased when the pH values of the delivery media decreased, in the range from 7.2 to 5.8.

Lee et al [42] modified chitosan hydrophobically by making it contain 5.1 deoxycholic acid groups per 100 anhydroglucoses with EDC-mediated coupling reaction. Deoxycholic acid is a main component of bile acid, which is biologically the most detergent-like molecule in the body. Since bile acid can assemble in water, the deoxycholic acid-modified chitosan also self-assembles to form micelles of a mean diameter of 160 nm.

Preparation of Chitosan Nanofibers

Electrospinning was first described as a fabrication technology by Rayleigh in 1897 [43]. However, a large industrial use was not seen until the advent of nanotechnology around 1990. Since then, electrospinning has found increasing use as studies have shown that a large variety of polymers can be electrospun [44]. However, most processes to date focus on the electrospinning of synthetic polymers since the electrospining of biopolymers is associated with considerable difficulties due to the fact that biopolymers, (polysaccharides and proteins) often have distributed molecular weights or have complex chemical structures.

One of the principal parameters that impacts the electrospinning process is the viscosity of the polymer solution, which depends on the polymer concentration and the solvent type. Both these properties thus have a profound influence on the resulting morphology of the fibers [45]. Viscosity is a key parameter in electrospinning because it is related to the extent of the polymer molecule chain entanglement within the solution [45]. Electrospinning is also influenced by some other parameters such as the operating voltage, the tip-to-target distance, the temperature, the pressure, and the flow rate. Of these parameters, the applied voltage is the most important one since it determines the degree of electrostatic interaction forces that induce the expulsion of a polymer jet. In some cases an increase in fiber diameter combined with a decrease in the number of bead defects can be observed with increased voltage [46, 47]. Lower applied voltages may sometimes cause a decrease in the fiber diameter since a decreased flight speed may allow the jet to split [48]. Further, an increase in voltage can cause a broadening of the fiber diameter distribution [49]. Finally, if the applied voltages are too high, charges may be directly discharged onto the plates effectively shortening the circuit and resulting in a breakdown of the electric field. An increase in the flow rate or a decrease in the syringe tip to the collector plate distance usually leads to the formation of fibers with larger diameters [50]. In the case of bead formation, increasing flow rates or decreasing the syringe tip to the collector plate distances also increase the size of the beads. Increasing the temperature reduces the viscosity allowing for more polymer to be added to the solution. In addition, the solvent may evaporate more rapidly at high temperatures. Conversely at higher humidity, the driving force for solvent evaporation is reduced, resulting again in the formation of junction zones between the deposited fibers.

An electrospun non-woven fabric of chitosan was successfully prepared by Ohkawa et al. [51]. Their study focuses on the effect of the electrospinning solvent and the chitosan concentration on the morphology of the resulting non-woven fabrics. The solvents tested were diluted hydrochloric acid, acetic acid, formic acid and trifluoroacetic acid (TFA). TFA is the main constituent in the successful solvent system for chitosan because the amino groups of the chitosan can form salts [52] with the TFA, which can effectively destroy the rigid interactions between the chitosan molecules thus facilitating electrospinning. As the chitosan concentration was increased, the morphology of the deposition on the collector changed from spherical beads to interconnected fibrous networks. The addition of dichloromethane to the chitosan–TFA solution has improved the homogeneity of the electrospun chitosan fiber. Under optimized conditions, homogenous chitosan fibers with a mean diameter of 330 nm were obtained. Ohkawa et al [53] also idealized the viscosity of chitosan solutions [54, 55] in order to decrease the average fiber diameter. It was determined that fiber diameter and polymer concentration have an inverse relationship.

Schiffman and Schauer [56] reported, chitosans with a low (70 kDa), medium (190–310 kDa), and high (500 kDa) molecular weight, as well as a commercial chitosan (190–375 kDA) sample were dispersed in 2.7 % TFA. With TFA, all solutions formed uniform nanofibers with a diameter of 74 ± 28 nm, 77± 29 nm, 108 ± 42 nm, and 58 ± 20 nm for low, medium, high, and commercial grade chitosan, respectively. The higher molecular weight chitosan produced larger fibers. Subsequent crosslinking of chitosan nanofibers with glutaraldehyde led to a substantial increase in the fiber size for all chitosans, e.g., the mean diameter of fibers for low, medium, high molecular weight, and commercial chitosan increased to 387 ± 183 nm, 172 ± 75 nm, 137 ± 59 nm, and 261 ± 160 nm, respectively. Thus, fibers with initially smaller sizes had the greatest relative increase in the mean diameter after crosslinking. The crosslinked fibers were insoluble in various media including NaOH, acetic acid, and ultrapure water for at least 72 h. This illustrated how the processing parameters may be adjusted to find the most suitable working conditions to efficiently and effectively produce nanofibers.

Torres-Giner et al [57] developed electrospun chitosan nanofibers using TFA and dicholomethane solvent. In addition, Torres-Giner et al [58] also developed porous electrospun chitosan nanofibers using pure trichloromethane solvent. Besides electrospinning chitosan in TFA and trichloromethane, the second solvent system that has been demonstrated to effectively produce nanofibers is concentrated acetic acid. Geng et al [59] reported the manufacture of chitosan nanofibers in concentrated aqueous acetic acid solutions varying in acetic acid concentration. Chitosan with low (30 kDa), medium (106 kDa), and high molecular weight (398 kDa) formed beads and thin fibers (40 nm) at an acetic acid concentration of 30 %. However, using neat acetic acid did not dissolve enough chitosan for a successful experiment. Increasing the acetic acid concentration to 90 % led to the surface tension being decreased and the net charge of polymers being increased, both conditions favoring continuous fiber formation. However, of all the polymers and polymer concentrations tested, only 106 kDa chitosan produced bead free fibers with an average diameter of 130 nm from 7 % solutions. The low molecular weight chitosan solutions of 9.5–

10 % yielded large beads and very fragile fibers, while the high molecular weight chitosan solutions of 2.5–3 % produced very fine and fragile fibers with rough surfaces and an average diameter of 60 nm. This again highlights the need to balance the molecular weight and the polymer concentration requirements such that sufficient polymer chain entanglement occurs while maintaining a low enough viscosity to allow the processing of the solution. The solvent type plays a crucial role in modulating both the entanglement and the solution viscosity.

De Vrieze et al [60] reported success in producing pure chitosan fibers for chitosans similar to the one studied above [56] using acetic acid as a solvent. Chitosan (3 %) solutions in 90 % aqueous acetic acid were found to produce fibers with a mean average fiber diameter of 70 ± 45 nm, while using 1 % hydrochloric acid, 90–100 % formic acid or less concentrated acetic acid and other concentrations of polymer only produced beads instead of fiber which was attributed to excessive solution viscosities.

Chitosan nanofibers were also produced by electrospinning of 5 % chitin solutions with a molecular weight of 910 kDa using 1,1,1,3,3,3-hexafluoro-2-propanol (HFIP) as the spinning solvent [61]. Fiber diameters ranged between 50 and 460 nm, but most fibers were smaller than 200 nm. Subsequent deacetylation in sodium hydroxide yielded chitosan fibers with a degree of deacetylation of approximately 85 %. The deacetylation apparently had no significant influence on the morphology and the fiber diameter.

In another study, Homoyoni et al [62] developed electrospinning of chitosan. The problem of chitosan high viscosity, which limits its spinability, is resolved through the application of an alkali treatment which hydrolyzes chitosan chains and so decreases their molecular weight. The alkali treated chitosan in aqueous 70–90 % acetic acid produces nanofibers with appropriate quality and processing stability. Decreasing the acetic acid concentration in the solvent increases the mean diameter of the nanofibers. Optimum sizes of nanofibers were achieved with chitosan, which was hydrolyzed with alkali for 48 h. The diameter of these nanofibers was strongly affected by the electrospinning conditions as well as the concentration of the solvent.

Shauer and Schiffman [63] patented preparation of chitosan nanorfibers. In their patent, the size of nanofibers is from 50 to 150 nm. Protein or enzymes can be immobilized on chitosan nanofibers cross-linked with glutarldehyde. Ye et al [64] studied poly (acrylonitrile-co-maleic acid)-chitosan naofibers was used for lipase immobilization. Their results show that the pH and thermal stabilities of lipases increased upon immobilization. The residual activities of the immobilized lipases are 55 % on chitosan composite nanofibers after 10 uses.

III. OUR WORK ABOUT CHITOSAN NANOPARTICLES

Reverse miscellar and ionic gelatation method were tried to prepare chitosan nanoparticles in our group. Reverse micelles are thermodynamically stable liquid mixtures of water, oil, and surfactant. Macroscopically, they are homogeneous and isotropic, structured on a microscopic scale into aqueous and oil microdomains separated by surfactant-rich films. One of the most important aspects of reverse micelle systems is their dynamic behavior. Preparation of ultrafine polymeric nanoparticles with narrow size distribution can be achieved using reverse micellar medium [65]. Their particle size was so small that they could not be collected through the supercentrifugal effect. Moreover, a plentiful amount of solvent was needed. The use of complexation between oppositely charged macromolecules to prepare chitosan microspheres has attracted much attention because the process is very simple and mild [66, 67]. In addition, reversible physical crosslinking by electrostatic interaction, instead of chemical crosslinking, has been applied to avoid the possible toxicity of reagents and other undesirable effects. TPP is a polyanion, which can interact with cationic chitosan by electrostatic forces [68, 69]. Figure 1 presents scanning electron micrographs of nanoparticles prepared by different methods. The nanoparticles were about 50–80 nm in size. However, in Figure 1B there is some oil phase, which indicates that the product was washed insufficiently.

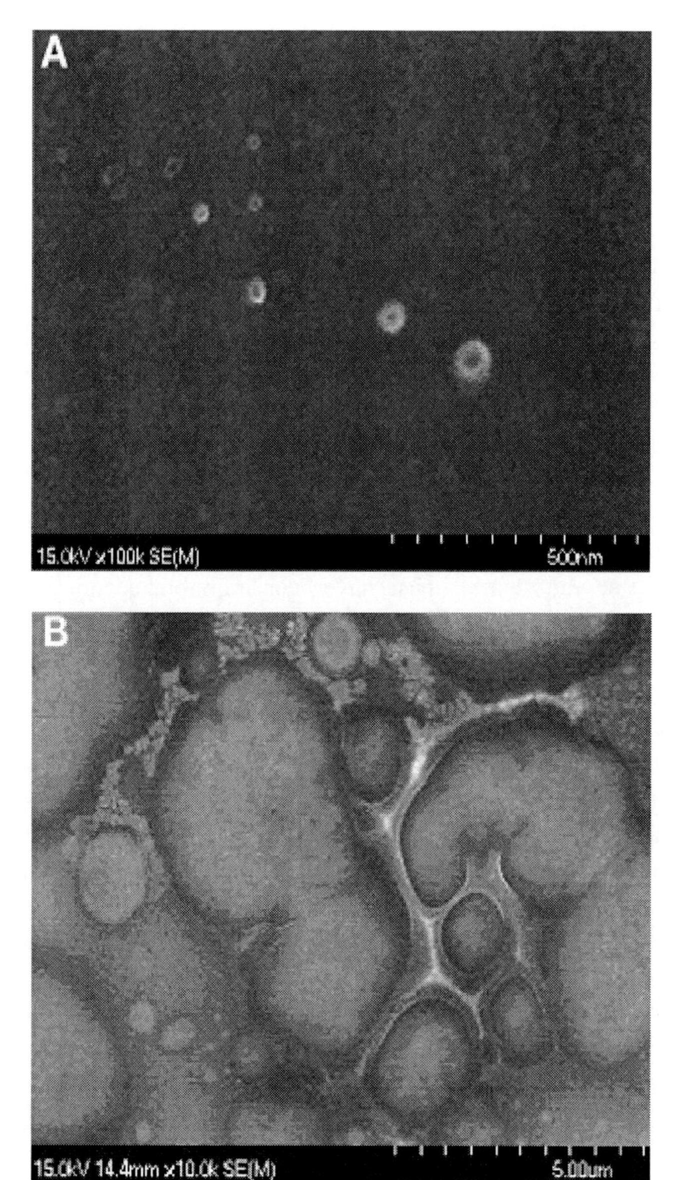

Figure 1. Scanning electron micrographs of (A) nanoparticles made by ionic gelatation method and (B) nanoparticles made by reverse micelle method.

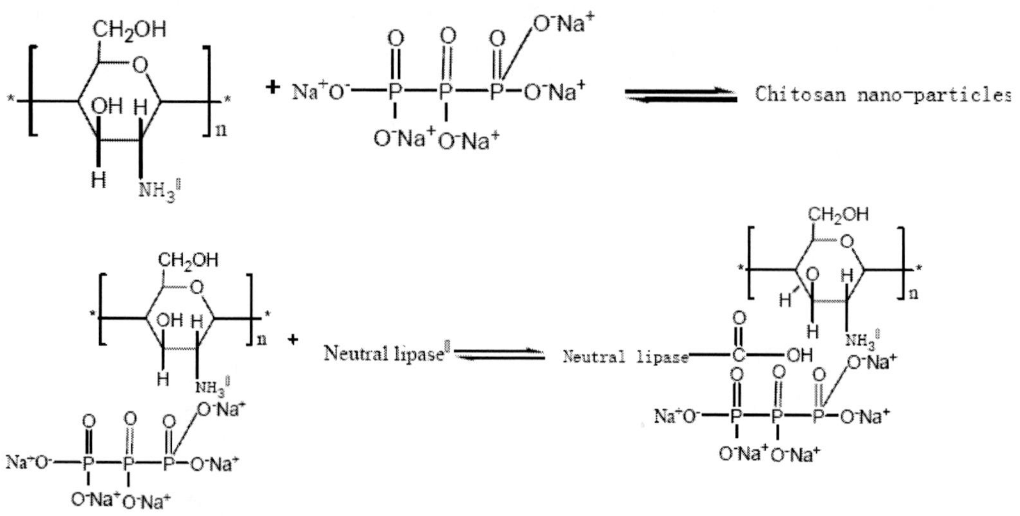

Figure 2. Hypothesis of adsorption mechanism.

The effect of factors such as molecular weight of chitosan, chitosan concentration, TPP concentration, and solution pH on the size of chitosan nanoparticles was studied [70]. Based on these results, response surface methodology was employed to optimize the size of chitosan nanoparticles. The results showed that solution pH, TPP concentration, and chitosan concentration significantly affected the size of chitosan nanoparticles. The adequacy of the predictive model equation for predicting the magnitude orders of the size of chitosan nanoparticles was verified effectively by the validation data. The minimum particles size was about 42 ± 5 nm under the optimized conditions.

Neutral proteinase and neutral lipase have been immobilized on chitosan nanoparticles prepared through chitosan/TPP method, respectively [71, 72]. For neutral proteinase, the optimal conditions of immobilization were as follows: one milligram of neutral proteinase was immobilized on chitosan nanoparticles for about 15 min at 40 °C. Under the optimized conditions, the enzyme activity yield was 84.3 %. Enzyme exhibited remarkably improved properties, such as thermal stability, reuse, and storage stability after it has been immobilized on chitosan nanoparticles. In addition, the stability of immobilized enzyme in the acidic solution was improved after immobilization.

Sorption kinetics and sorption mechanism for neutral lipase and nuclease p1 were studied [73-74]. The factors, such as enzyme concentration, chitosan nano-particles solution concentration, adsorption temperature, size of chitosan nano-particles, stirring rate, solution pH, could affect their adsorption kinetics. For example, adsorption of neutral lipase on chitosan nano-particles was fitted into Lagergren first-order equation at initial neutral lipase of 3.0 mg/ml (pH 7.0). The first-order constant for neutral lipase was 23.34 h^{-1}. When neutral lipase concentration was controlled under certain region, it was fitted into Freundlich isothermal linear equation. Mechanism of adsorption for neutral lipase was presumed by analyzing IR spectra. The hydrogen in the carboxyl group was connected with electronegative oxygen. Electron pair was attractive to oxygen, so hydrogen atom became cation and proton. When electronegative neutral lipase was closed to chitosan nano particle, the hydrogen bond might be formed. It was the main force between hydroxyl group, NH_2 group and neutral lipase (Figure 2).

IV. CONCLUSIONS

This chapter summarized the preparation of chitosan nanoparticles and nanofibers and their application in enzyme immobilization. From this chapter, it is concluded that chitosan nanoparticles and nanofibers have been used as enzyme immobilization carriers. They have the potential for the development of novel enzyme immobilization carriers. Due to their favourable biological properties such as non-toxicity, biocompatibility, biodegradability and antibacterial ability, they are also promising candidates for drug delivery carrier and cell proliferation enhancer. However, most of these studies are still at the laboratory level. Additional studies are necessary before the industrial application of chitosan nanoparticles and nanofibers. We hope more chitosan- based nanoparticle and nanofibers can be developed and applied in biochemical engineering field in the near future.

REFERENCES

[1] Lee, J. H., Hwang, E. T., Kim, B. C., Lee, S. M., Sang, B. I., Choi, Y. S., Kim, J., Gu, M. B. Stable and continuous long-term enzymatic reaction using an enzyme-nanofiber composite. *Applied Microbiology and Biotechnology*, 2007(75), 1301-1307.

[2] Fernandez-Lorente, G., Palomo, J.M., Mateo, C., Munilla, R., Ortiz, C., Cabrera, Z., Guisan, J. M., Fernandez-Lafuente, R. Glutaraldehyde cross-linking of lipases adsorbed on aminated supports in the presence of detergents leads to improved performance. *Biomacromolecules*, 2006(7), 2610–2615.

[3] Morana, A., Mangione, A., Maurelli, L., Fiume, I., Paris, O., Cannio, R., Rossi, M. Immobilization and characterization of a thermostable beta-xylosidase to generate a reusable biocatalyst. *Enzyme* and *Microbial Technology*, 2006 (39), 1205–1213.

[4] Tardioli, P.W., Zanin, G.M., De Moraes, F.F. Characterization of Thermoanaerobacter cyclomaltodextrin glucanotransferase immobilized on glyoxyl-agarose. *Enzyme and Microbial Technology*, 2006(39), 1270–1278.

[5] Jia, H.F., Zhu, G.Y., Vugrinovich, B., Kataphinan, W., Reneker, D.H., Wang, P. Enzyme-carrying polymeric nanofibers prepared via electrospinning for use as unique biocatalysts. *Biotechnology Progress*, 2002(18), 1027–1032.

[6] Kim, J., Grate, J.W. Single-enzyme nanoparticles armoured by a nanometer-scale organic/inorganic network. *Nano Letters*, 2003(3), 1219–1222.

[7] Lee, J., Han, S., Hyeon, T. Synthesis of new nanoporous carbon materials using nanostructured silica materials as templates. *Journal of Material Science*, 2004(14), 478–486.

[8] Sawicka, K., Gouma, P., Simon, S. Electrospun biocomposite nanofibers for urea biosensing. *Sensors and Actuators B: Chemical*, 2005(108), 585–588.

[9] Wei, Y., Xu, J., Feng, Q., Dong, H., Lin, M. Encapsulation of enzymes in mesoporous host materials via the nonsurfactant-templated sol–gel process. *Materials Letters*, 2000(44), 6–11.

[10] Jia, H., Zhu, G., Wang, P. Catalytic behaviors of enzymes attached to nanoparticles; the effect of particle mobility. *Biotechnology and Bioengineering*, 2003(84), 406–414.

[11] Govardhan, C.P. Crosslinking of enzymes for improved stability and performance. *Current Opinion in Biotechnology*, 1999(10), 331–335.
[12] Haring, D., Schreier, P. Cross-linked enzyme crystals. *Current Opinion in Chemical Biology*, 1999(3), 35–38.
[13] Xu, K., Klibanov, A.M. pH control of the catalytic activity of crosslinked enzyme crystals in organic solvents. *Journal of the American Chemical Society,* 1996 (118), 9815–9819.
[14] Kim, B.C., Nair, S., Kim, J., Kwak, J.H., Grate, J.W., Kim, S.H., Gu, M.B. Preparation of biocatalytic nanofibers with high activity and stability via enzyme aggregate coating on polymer nanofibers. *Nanotechnolgy*, 2005 (16), S382–S388.
[15] Kim, J., Lee, J., Na, H.B., Kim, B.C., Youn, J.K., Kwak, J.H., Moon, K., Lee, E., Kim, J., Park, J., Dohnalkova, A., Park, H.G., Gu, M.B., Chang, H,N., Grate, J.W., Hyeon, T. Magnetically-separable and highlystable enzyme system based on composites of enzymes and magnetic nanoparticles in mesoporous silica. *Small*, 2005(1), 1203–1207.
[16] Sevda, S., McClureb, S. J. Potential applications of chitosan in veterinary medicine. *Advanced Drug Delivery Review,* 2004(56), 1467-1480.
[17] Sunil, A. A., Nadagouda, N. M., Tejraj, M. Recent advances on chitosan-based micro- and nanoparticles in drug delivery. *Journal of Controlled Release,* 2004(100), 5-28.
[18] Se, K. K., Niranjan, R. Enzymatic production and biological activities of chitosan oligosaccharides (COS): a review. *Carbohydrate Polymers,* 2005(62), 357-368.
[19] Barbara, K. Application of chin- and chitosan-based materials for enzyme immobilization: a review. *Enzyme and Microbial Technology*, 2004(35), 126-139.
[20] Mohanraj, V. J., Chen, Y. Nanoparticles – a review. *Tropical Journal of Pharmaceutical Research*, 2006(5), 561-573.
[21] Ohya, Y., Shiratani, M., Kobayashi, H., Ouchi, T. Release behavior of 5-fluorouracil from chitosan-gel nanospheres immobilizing 5-fluorouracil coated with polysaccharides and their cell specific cytotoxicity. *Pure and Applied Chemistry,* 1994 (31), 629-642.
[22] Yokohama, M., Fukushima, S., Uehara, R., Okamoto, K., Kataoka, K. Characterization of physical entrapment and chemical conjugation of adriamycin in polymeric micelles and their design for *in vivo* delivery to a solid tumor. *Journal of Controlled Release*, 1998(50), 79-92.
[23] Kataoka, K., Matsumoto, T., Tokohama, M., Okano, T., Sakurai, Y. Doxorubicin-loaded poly(ethylene glycol)-poly(b-benzyl-L-aspartate) copolymer micelles: their pharmaceutical characteristics and biological significance. *Journal of Controlled Release*, 2000(64), 143-153.
[24] Liu, C. G., Desai, K. G. H., Chen, X. G., Park, H. J. Linolenic acid-modified chitosan for formation of self-assembled nanoparticles. *Journal of Agricultural and Food Chemistry*, 2005(53), 437-441.
[25] Tokumitsu, H., Ichikawa, H., Fukumori, Y. Chitosangadopentetic acid complex nanoparticles for gadolinium neutron capture therapy of cancer: preparation by novel emulsiondroplet coalescence technique and characterization. *Pharmaceutical Research,* 1999(16), 1830-1835.
[26] Van der Lubben, I. M., Verhoef, J. C., Borchard, G., Junginger, H. E. Chitosan for mucosal vaccination. *Advanced Drug Delivery Review,* 2001(52), 139-144.

[27] Calvo, P., Remunan-Lopez, C., Vila-Jato, J. L., Alonso, M. J. Novel hydrophilic chitosan-poly ethylene oxide nanoparticles as protein carriers. *Journal of Applied Polymer Science,* 1997(63), 125-132.
[28] Xu, Y. M., Du, Y. M. Effect of molecular structure of chitosan on protein delivery properties of chitosan nanoparticles. *International Journal of Pharmaceutics,* 2003 (250), 215-226.
[29] Janes, K. A.; Alonso, M. J. Depolymerized chitosan nanoparticles for protein delivery: preparation and characterization. *Journal of Applied Polymer Science,* 2003 (88), 2769-2776.
[30] Ohya, Y., Cai, R. Nishizawa, H. Hara, K. Ouchi, T. Preparation of PEG-grafted chitosan nano-particle for peptide drug carrier. *Proceedings* International Symposium *on* Controlled Release *of Bioactive Materials.* 1999(26), 655-656.
[31] Nasti, A., Zaki, N. M., Leonardis, P. D., Ungphaiboon, S., Sansongsak, P., Rimoli, M.G., Tirelli, N. Chitosan/TPP and chtiosan/TPP-hyaluronic acid nanoparticles: systematic optimization of the preparative process and preliminary biological evaluation, *Pharmaceutical Research*, 2009(26), 1918-1930.
[32] Lin, A. H., Liu, Y. M., Ping, Q. N. Free amino groups on the surface of chitosan nanoparticles and its characteristics. *Yao Xue Xue Bao*, 2007(42), 323-328.
[33] Kereszessy, Z., Bodnar, M., Ber, E., Hajdu, I., Zhang, M., Hartmann, J. F., Minko, T., Borbely, J. Self-assembling chitosan/poly-γ-glutamic acid nanoparticles for targeted drug delivery, *Colloid and Polymer Science*, 2009(287), 759-765.
[34] Wu, Y., Wang, Y. J., Luo, G. S., Dai, Y. Y. In situ preparation of magnetic chitosan nanoparticles for lipase immobilization by cross-linking and oxidation in aqueous solution. *Bioresource Technology*, 2009(100), 3459-3464.
[35] Brunel, F., Veron, L., David, L., Domard, A., Delair, T. A Novel synthesis of chitosan nanoparticles in reverse emulsion. *Langmuir*, 2008 (24), 11370-11377.
[36] Mitra, S., Gaur, U., Ghosh, P. C., Maitra, A. N. Tumor targeted delivery of encapsulated dextran-doxorubicin conjugate using chitosan nanoparticles as carrier. *Journal of Controlled Release*, 2001(74), 317-323.
[37] Mansouri, M., Khorram, M., Samimi, A., Osfouri, Sh. preparation of bovine serum albumin loaded chitosan nanoparticles using reverse micelle method. *Journal of Controlled Release*, 2003(91), 135-145.
[38] Fang, H., Huang, J., Ding, L. Y., Li, M. T., Chen, Z. Preparation of magnetic chitosan nanoparticles and immobilization of laccase. *Journal of Wuhan University of Technology*, 2009(24), 42-47.
[39] Liu, C. G., Chen, X. G.,Park, H. J. Self-assembled nanoparticles based on linoleic-acid modified chitosan: Stability and adsorption of trypsin. *Carbohydrate Polymers,* 2005(62), 293-298.
[40] Chen, X. G., Lee, C. M., Park, H. J. O/W emulsification for the self-aggregation and nanoparticle formation of linoleic acids modified chitosan in the aqueous system. *Journal of Agricultural and Food Chemistry,* 2003(51), 3135-3139.
[41] Hu, F. Q., Ren, G. F., Yuan, H., Du, H. Z., Zeng, S. Shell cross-linked stearic acid grafted chitosan oligosaccharide self-aggregated micelles for controlled release of paclitaxel. *Colloids and Surfaces. B, Biointerfaces*, 2006(50), 97-103.

[42] Lee, K. Y., Kim, J. H., Kwon, I. C., Jeong, S. Y. Self-aggregates of deoxycholic acid-modified chitosan as a novel carrier of adriamycin. *Colloid and Polymer Science*, 2001(278), 1216-1219.

[43] Burger, C., Hsiao, B. S., and Chu, B. Nanofibrous materials and their applications. *Annual Review of Materials Research,* 2006(36), 333-368.

[44] Reneker, D. H., Chun, I. Nanometre diameter fibres of polymer, produced by electrospinning. *Nanotechnology,* 1996 (7), 216-223.

[45] Ramakrishna, S, Fujihara, K., Teo, W. E., Lim, T. C., Ma, Z. H. *An Introduction to Electrospinning and Nanofibers.* Singapore: World Scientific; 2005.

[46] Jiang, H. L., Zhao, P. C., Zhu, K. J. Fabrication and characterization of zein-based nanofibrous scaffolds by an electrospinning method. *Macromolecular Bioscience.* 2007(7), 517-525.

[47] Yao, C., Li, X. S., Song, T. Y. Electrospinning and crossfinking of Zein nanofiber mats, *Journal of Applied Polymer Science,* 2007(103), 380-385.

[48] Zhao, S. L., Wu, X. H., Wang, L. G., Huang, Y. Electrospinning of ethyl-cyanoethyl cellulose/tetrahydrofuran solutions. *Journal of Applied Polymer Science,* 2004(91), 242-246.

[49] Ki, C. S., Baek, D. H., Gang, K. D., Lee, K. H., Um, I. C., Park, Y. H. Characterization of gelatin nanofiber prepared from gelatin-formic acid solution. *Polymer,* 2005(46), 5094-5102.

[50] Han, S. O., Son, W. K., Youk, J. H., Park, W. H. Electrospinning of ultrafine cellulose fibers and fabrication of poly(butylene succinate) biocomposites reinforced by them. *Journal of Applied Polymer Science,* 2008(107), 1954-1959.

[51] Ohkawa, K., Cha, D. I., Kim, H., Nishida, A., Yamamoto, H. Electrospinning of chitosan. *Macromolecular Rapid Communications,* 2004(25), 1600-1605.

[52] Hasegawa, M., Isogai, A., Onabe, F., Usuda, M., Atalla, R.H. Characterization of cellulose-chitosan blend films. *Journal of Applied Polymer Science*, 1992(45), 1873–1879.

[53] Ohkawa, K., Minato, K. I.,Kumagai, G., Hayashi, S., Yamamoto, H. Chitosan nanofiber. *Biomacromolecules,* 2006(7), 3291-3294.

[54] Baumgarten P.K. Electrostatic spinning of acrylic microfibers. *Journal of Colloid and Interface Science,* 1971(36), 71–79.

[55] Fridrikh, S.V., Yu, J.H., Brenner, M.P., Rutledge, G.C. Controlling the fiber diameter during electrospinning. *Physical Review Letter*, 2003(90), 144502.

[56] Schiffman, J. D., Schauer, C. L. One-step electrospinning of crosslinked chitosan fibers. *Biomacromolecules,* 2007(8), 2665-2667.

[57] Torres-Giner, S., Gimenez, E., Lagaron, J.M. Characterization of the morphology and thermal properties of Zein Prolamine nanostructures obtained by electrospinning. *Food Hydrocolloids*, 2008(22), 601–14.

[58] Torres-Giner, S., Ocio, M.J., Lagaron, J.M. Development of active antimicrobial fiber based chitosan polysaccharide nanostructures by electrospinning. *Engineering in Life Science*, 2008(8), 303–314.

[59] Geng, X.Y., Kwon, O. H., Jiang, J. H. Electrospinning of chitosan dissolved in concentrated acetic acid solution. *Biomaterials*, 2005(26), 5427-5432.

[60] De Vrieze, S., Westbroek, P., Van Camp, T., and Van Langenhove, L. Electrospinning of chitosan nanofibrous structures: feasibility study. *Journal of Materials Science,* 2007(42), 8029-8034.

[61] Min, B.M., Lee, S.W., Lim, J.N., You, Y., Lee, T.S., Kang, P.H. Chitin and chitosan nanofibers: electrospinning of chitin and deacetylation of chitin nanofibers. *Polymer*, 2004(45), 7137–7142.

[62] Homoyoni, H., Ravandi, S.A.H., Valizadeh, M. Electrospinning of chitosan nanofibers: processing optimization. *Carbohydrate Polymers*, 2009(77), 656–661.

[63] Shauer, C. L., Schiffman, J. D. Fibrous mats containing chitosan nanofibers, WO 2009/011944

[64] Ye, P., Xu, Z. K., Wu, J., Innocent, C., Seta, P. Nanofibrous poly(acrylonitrile-co-maleic acid) membranes functionalized with gelatine and chitosan for lipase immobilization. *Biomaterials*, 2006(27), 4169-4176.

[65] Leong, Y. S., Candau, F. Inverse microemulsion polymerization, *Journal of Physical Chemistry,* 1982(86), 2269–2271.

[66] Polk, A., Amsden, B., Yao, K. D., Peng, T., Goosen, M. F. A. Controlled Release of Albumin from Chitsan-Alginate Microcapsules. *Journal of Pharmaceutical Sciences,* 1994(83), 178–185.

[67] Liu, L. S., Liu, S. Q., Ng, S. Y., Froix, M., Ohno, T., Heller, J. Controlled release of interleukin-2 for tumour immunotherapy using alginate/chitosan porous microspheres. *Journal of Controlled Release*, 1997(43), 65–74.

[68] Kawashima, Y., Handa, T., Takenaka, H., Lin, S. Y., Ando, Y. Novel method for the preparation of controlled-release theophylline granules coated with a polyelectrolyte complex of sodium polyphosphate-chitosan. *Journal of Pharmaceutical Sciences*, 1985(74), 264–268.

[69] Kawashima, Y., Handa, T., Kasai, A., Takenaka, H., Lin, S. Y. The effects of thickness and hardness of the coating film on the drug release rate of theophylline granules coated with chitosan-sodium tripolyphosphate complex. *Chemical and Pharmaceutical Bulletin*, 1985(33), 2469–2474.

[70] Tang, Z.X., Qian, J.Q., Shi, L.E. Preparation. *Applied Biochemistry and Biotechnology*, 2007(136), 77-96.

[71] Tang, Z.X., Qian, J.Q., Shi, L.E. Characterizations of immobilized neutral. *Materials Letters*, 2007(61), 37-40.

[72] Tang, Z.X., Qian, J.Q., Shi, L.E. Characterizations of immobilized neutral. *Process Biochemistry*, 2006(41), 1193-1197.

[73] Shi, L.E., Tang, Z.X. Adsorption of chitosan nanoparticles. *Brazilian Journal of Chemical Engineering*, 2009(26), 435-443.

[74] Tang, Z. X., Shi, L.E., Qian, J.Q. Neutral lipase from aqueous solutions. *Biochemical Engineering Journal*, 2007(34), 217-223.

In: Handbook of Chitosan Research and Applications
Editors: Richard G. Mackay and Jennifer M. Tait ISBN 978-1-61324-455-5

Chapter 11

RESEARCH PROGRESS ON THE PREPARATION AND APPLICATION OF AMPHOTERIC CHITOSAN*

Hu Yang,• Yaobo Lu and Rongshi Cheng
Key Laboratory for Mesoscopic Chemistry of MOE,
Department of Polymer Science and Technology,
School of Chemistry and Chemical Engineering,
Nanjing University, Nanjing, P.R.China

ABSTRACT

Chitosan is one of the high-performance polysaccharide materials, which has been already applied widely in various fields such as biotechnology, biomedicine, wastewater treatment and food process. However, some disadvantages, such as inactive chemical properties and bad solubility, always limit chitosan in practical application. In order to improve the performances of chitosan, amphoteric chitosan, containing both cation and anion on the chain backbone, have been prepared by chemical modifications, such as etherification, esterification and graft copolymerization. The amphoteric chitosan derivatives present some interesting features and distinguishing performances. Therefore, based on the ionic strength of groups on the amphoteric chitosan, the preparation and applications of amphoteric chitosan have been reviewed in this paper.

Keywords: *Amphoteric Chitosan; Preparation; Application*

---

* Supported by National Natural Science Foundation of China (Grant No.: 50633030 and 51073077).
• E-mail address: yanghu@nju.edu.cn.

1. INTRODUCTION

Natural polymer materials, coming from animals, plants, microorganisms and so on, are a kind of resources abundant in nature, which are believed to be nontoxic and environment-friendly materials, since they are facile to degrade into water, carbon dioxide and so on after being disused. Another important characteristic of natural polymers is that they are also a kind of reproducible and inexhaustible materials fully independent of petroleum resources, which peculiarities make the natural polymer materials applicable widely[1, 2].

Chitosan, poly-β-(1→4)-2-amino-2-deoxy-D-glucose, is one of the high-performance polysaccharide materials, which is prepared from deacetylation of natural chitin, and chitin is the second most abundant natural polymers in the world. In addition, chitosan has many prominent characteristics such as low toxicity, and high biocompatibility. Furthermore, chitosan presents abundant free amino groups along the chain backbone that are cationically charged in a wide range of physiological pHs, and show prominent performances in practical applications. So chitosan has been already widely applied in many fields such as biotechnology, biomedicine, wastewater treatment and food process [3-5].

However, chitosan also have some defects in real applications, such as low molecular weight, inactive chemical properties and poor solubility. To improve these performances further, modified chitosan materials have been manufactured. Among the chitosan derivatives, amphoteric chitosan [6-10], containing both cation and anion on the chain backbone, present some interesting features and distinguishing performances, which bear better solubility in a wide range of pH and excellent salt-resistant properties. Furthermore, amphoteric chitosan still bear excellent chelating effect for the abundant free cationic and anionic groups, and could be used as adsorbent to absorb impurities, such as metal ions, humic acids and synthetical surfactants, from the water. On the other hand, strong cationic chitosan having quaternary ammonium salt also show good anti-bacterial performance. Therefore, considerable attentions have been paid to the amphoteric chitosan materials recently. In this paper, based on the ionic strength of groups on the amphoteric chitosan, the preparation methods of amphoteric chitosan have been reviewed, and their applications in various fields have been discussed also.

2. PREPARATION OF AMPHOTERIC CHITOSAN

In acid medium, the amino (-NH_2) group on glucose ring of chitosan was facile to combine a proton to -NH_3^+, being a kind of weak cationic polyelectrolytes, so it was the easiest method to prepare amphoteric chitosan materials by introducing anionic groups into chitosan chains. Furthermore, strong cationic chitosan derivatives could be also obtained by further quaternization.

2.1. Weak-Cation/Weak-Anion Type Amphoteric Chitosan(WWAC)

As mentioned above, chitosan itself was a kind of weak cationic polyelectrolytes, so weak-cation/weak-anion type amphoteric chitosan could be prepared only by introducing weak anionic groups into molecular chains. Among WWAC materials, carboxymethyl chitosan (CMC) was the simplest one, many researches and applications of which have been made in recent years.

CMC was usually prepared by etherification of chitosan with chloroacetic acid in alkali solutions [6, 11-12]. The reaction process was shown in scheme 1.

In comparison to 3-OH and 2-NH_2 groups on glucose ring of chitosan, 6-OH was much more reactive, being smaller steric hindrance and higher electronegativity of oxygen, which usually made etherification reaction carry out mostly on 6-OH groups. However, caboxymethyl groups could also substitute the protons at 3-OH and 2-NH_2, respectively, under various modifying conditions, such as different alkalization time, alkalization temperature and various proportions of alkali and isopropanol as reaction medium[13-15]. Compared to chitosan, the solubility of CMC in aqueous solution was improved remarkably, since the carboxymethyl groups have effectively destroyed the original regular structure of chitosan, and reduced the degree of crystallinity of chitosan. On the other hand, H-form CMC could transit into salt-form CMC by neutralization, which conferred CMC much better solubility. In addition, with the change of the substitution degree, CMC also showed various isoelectric points (IEP) [6, 11-15].

Zhou et al.[17,18] selectively oxidized the hydroxymethyl group of 6-CH_2OH to carboxyl group with NO_2 to form amphoteric chitosan of 6-carboxy chitosan by one step so as to give chitosan a new bioactivity. In the optimal conditions, the degree of carboxylation was around 52.5 %, which IEP was 4.9. Besides NO_2 as oxidant, partially dicarboxylated chitosan was prepared by the ring-opening oxidation of periodate and chlorite in two steps, which showed good biodegradability and pH-sensitivity [9, 19].

Furthermore, graft copolymerization could also achieve to introduce the weak anionic groups on chitosan chain. For example, acrylic acid (AA), containing both unsaturated carbon-carbon double bonds and carboxyl groups, could graft onto chitosan by radical polymerization to form amphoteric chitosan [20-23]. In addition, Li et al. [24] reported a kind of novel full-biodegradable amphoteric chitosan materials of chitosan-g-poly(L-lactic acid). Poly(L-lactic acid) as macromonomer was grafted onto chitosan selectively at the hydroxyl group of 6-OH through the protection-graft-deprotection route for 2-NH_2 by phthalic anhydride. Yao et al. [25] prepared a novel amphoteric chitosan hydrogel of chitosan-*g*-poly(L-lactic-co-citric acid) by the reaction of chitosan with oligomer of citric acid and L-lactic acid. Here the oligomer took the effects not only of its carboxyl group supplier but also of the crosslinker. Batista et al. [26] grafted peptide acid onto chitosan chain to obtain a kind of chitosan derivatives with high solubility in a wide range of pH around 2-10.

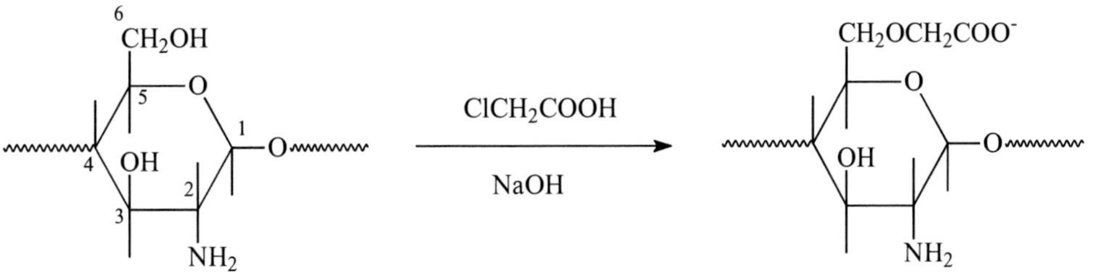

$$\text{chitosan unit (6: } CH_2OH\text{; 3: OH; 2: } NH_2\text{)} \xrightarrow[NaOH]{ClCH_2COOH} \text{unit with } CH_2OCH_2COO^- \text{, OH, } NH_2$$

Scheme 1. Etherization Reaction Formula for preparation of carboxymethyl chitosan.

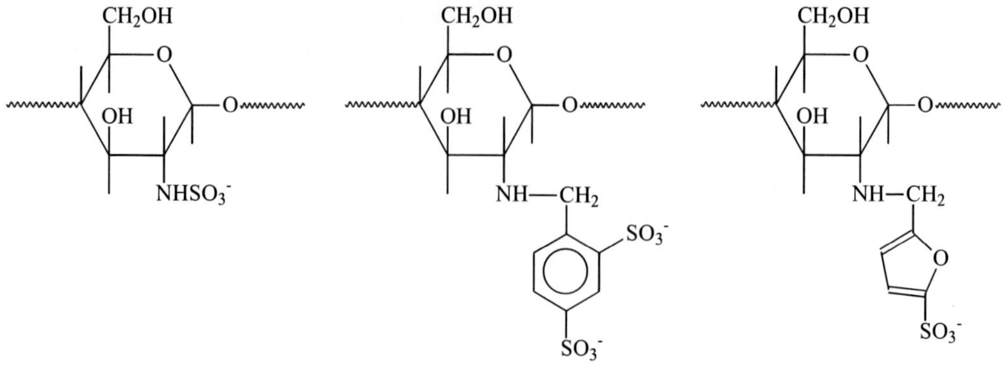

(a)[7,27] (b)[28] (c)[29]

(a) *N*-sulfonated chitosan; (b) *N*-Benzyl sulfonate chitosan; (c) *N*-sulfofurfuryl chitosan.

Scheme 2. The chemical structures of several common *N*-sulfonated chitosan.

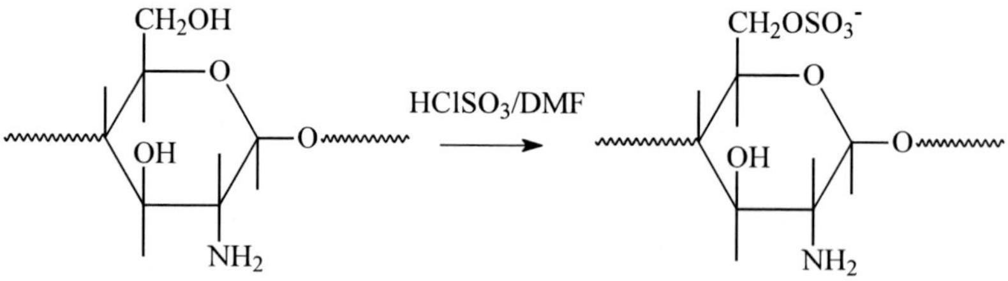

Scheme 3. Reaction Formula for preparation of *O*-Sulfated Chitosan.

2.2. Weak-Cation/Strong-Anion Type Amphoteric Chitosan(WSAC)

Schatz et al. [7] and Holme et al.[27] synthesized a kind of *N*-sulfonated chitosan, as shown in scheme 2(a), using trimethylamine-sulfur trioxide complex as sulfonation reagent and $NaCO_3$ aqueous solution as medium. A series of *N*-sulfonated chitosan with various degree of sulfonation from 0.4 to 0.86 were obtained through adjusting the reaction time and the concentration of $NaCO_3$[27]. The experimental results indicated that all of *N*-sulfonated chitosan were soluble in water, and had better solubility at pH=8.0 than pH=6.0, furthermore, the increase of degree of sulfonation would be favorable for improvement of solubility.[27]

In addition, many researchers [28, 29] also tried to prepare various forms of the sulfonated chitosan derivatives, as shown in scheme 2 (b) and (c), by aldehyde ammonia condensation using the aromatic or heterocyclic aldehyde containing sulfonic groups as sulfonation reagents. Firstly, the aromatic or heterocyclic aldehyde formed Schiff's bases with amino groups of chitosan, and the sulfonic groups could be introduced on the chitosan chain indirectly; then, the Schiff's bases transited into the final sulfonated chitosan derivatives by hydrogenation. Similarly, the novel sulfonated chitosan derivatives could improve the solubility of chitosan effectively, especially in alkaline aqueous solution.

In comparison to *N*-substituted sulfonated chitosan, *O*-substituted sulfonated chitosan could be also prepared by various modifying conditions. Miao et al. [30, 31] reported a homogeneous method to prepare *O*-substituted sulfonated chitosan in the mixture of chlorosulfonic acid and dimethylformamide (DMF), as shown in scheme 3. The degree of sulfonation was around 38.1 wt%. Furthermore, the experimental results indicated that the products had good performances in solubility, anticoagulant, inhibit inflammation and film forming.

Recently, Amaral et al. [32] supplied a method to prepare amphoteric chitosan by phosphorylation. Phosphorylation was carried out at room temperature in the $H_3PO_4/Et_3PO_4/P_2O_5$/butanol system, and the phosphate groups were connected to 6-OH of chitosan through esterification. The materials were promising in orthopaedic applications, since phosphate groups were easy to chelate with calcium ions.

2.3. Strong-Cation Type Amphoteric Chitosan

As for the strong-cation chitosan materials, it was the popular method to introduce quaternary ammonium salt to the chain backbone of chitosan [33-37]. Based on the origin of the quaternary ammonium salt, two routes for preparation of strong-cation chitosan have been summarized: one was to quaternize 2-NH_2 on glucose ring of chitosan directly; the other was that chitosan was modified using some materials containing the quaternary ammonium salt by etherification or esterification indirectly. The large steric hindrance and stronge hydration capacity of quaternary ammonium salt weakened the inter-molecular hydrogen bonding of chitosan effectively, which made the solubility of quaternized chitosan derivatives in aqueous solutions improve remarkably.

However, there were only a few reports related to the strong-cation type amphoteric chitosan until now [8, 38-44]. In fact, the method for preparation of strong-cation type amphoteric chitosan was quite similar to that of strong-cation chitosan as mentioned above, and the strong-cation type amphoteric chitosan could be prepared by quaternization of weak-cation type amphoteric chitosan.

Recently, Yao et al. [38] reported a kind of novel strong-cation type amphoteric chitosan. Firstly, chitosan was dispersed in *N*-methyl-2-pyrrole alkyl ketone; then, sodium hydroxide (NaOH), methyl iodide (CH_3I) and potussium iodide (KI) were added successively at room temperature, which quaternized the 2-NH_2 on glucose ring of chitosan into *N*-trimethyl chitosan quaternary ammonium salt directly; at last, the quaternized chitosan derivative was reacted with chloroacetic acid through etherification further, which resulted in a kind of strong-cation / weak-anion amphoteric chitosan derivatives. The whole preparation process was showed in Scheme 4. Furthermore, the experimental results of FTIR spectra and ^1H NMR spectra both indicated that the carboxymethyl groups mainly substituted the protons of 6-OH and 3-OH on glucose ring of chitosan.

CH_2OH, OH, NH_2 → (CH_3I / KI, NaOH) → CH_2OH, OH, $N(CH_3)_3^+$ → ($ClCH_2COOH$, NaOH) → $CH_2OCH_2COO^-$, OH, $N(CH_3)_3^+$

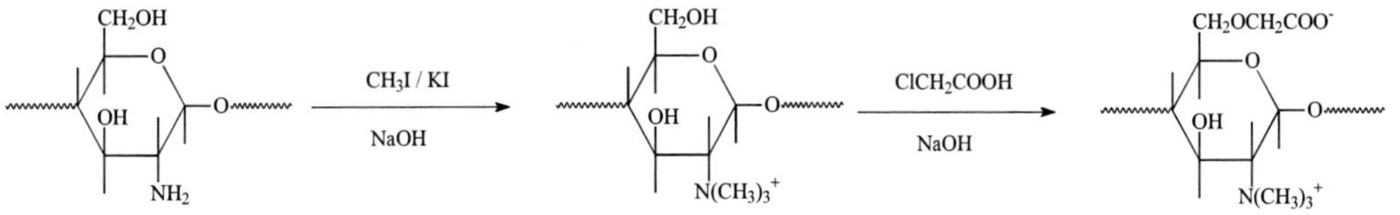

Scheme 4. Reaction Formulas for preparation of O-carboxymethyl-N-trimethyl chitosan quaternary ammonium salt.

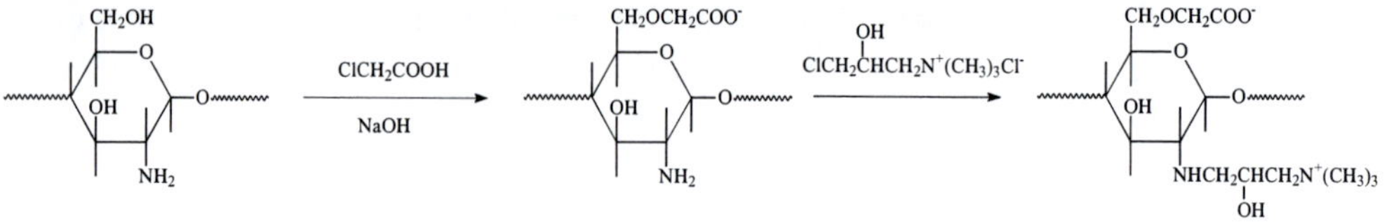

Scheme 5. Reaction Formulas for preparation of quaternized carboxymethyl chitosan.

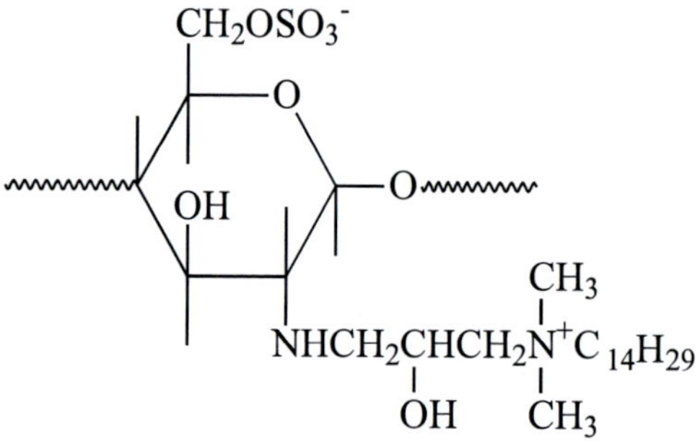

Scheme 6. The chemical structure of *O*-sulfonic-*N*-2-hydroxypropyl dimethyl tetradecane ammonium chloride chitosan.

Except for the self-quaternization, the strong-cation type amphoteric chitosan could be prepared by introducing additional quaternary ammonium salt. Cai et al. [8, 39] synthesized a kind of quaternized carboxymethyl chitosan through the grafting reaction of CMC with 3-chloro-2-hydroxypropyl trimethylammonium chloride (CTA) as a quaternizing agent in 2-propanol medium under alkaline condition, preparation process of which was showed in Scheme 5. The final product could be applied in wastewater treatment. Recently, Cai et al. [43, 44] also prepared a series of quaternized amphoteric chitosan using 3-chloropropionic acid as anionic modifying agent, and 2,3-epoxypropyl trialkyl ammonium chlorides with various alkyl groups as cationic modifying agents. The experimental results indicated that these materials had strong antimicrobial activity. And their antimicrobial activity increased with the increase of the substitution degree of the quaternary ammonium salt, as well as with increasing chain length of the alkyl in the quaternary ammonium salt. Furthermore, among the quaternized amphoteric chitosan materials[43,44], 2-Hydroxypropyl dimethylbenzylammonium *N,O*-(2-carboxyethyl) chitosan chloride showed the best performances in antimicrobial activity.

In addition, Tang et al. [40] obtained a kind of strong-cation/strong-anion type amphoteric chitosans (MTGA) by grafting glycidyl retradecyl bimethlammonium chloride on chitosan, and then reacted with chlorosulfonic acid, and the structure of which was shown in Scheme 6. The experimental results indicated that the grafting efficiency of MTGA was affected remarkably by the type of solvents and the reaction time. Compared to ethanol, isopropanol as reaction solvent was much easier to break the inter-molecular hydrogen bonds of chitosan. Furthermore, it was great helpful to improve the solubility of MTGA by adding some water into isopropanol. At the mixing ratio of isopropanol to water near 4:1, the grafting efficiency reached the highest one up to around 50 %, while the grafting efficiency was only kept below 30 % without adding water. In addition, MTGA showed good hygroscopicity and surface activity in applications.

3. APPLICATIONS OF AMPHOTERIC CHITOSAN

Amphoteric chitosan were of great current interest due to not only their high biocompatibility, nontoxicity and biodegradable, but also the existence of cationic and anionic groups on the backbone of chitosan chain conferring the amphoteric chitosan with improved solubility in aqueous solutions. Compared to chitosan, the hygroscopicity and anti-bacterium of strong-cation amphoteric chitosan were also improved remarkably because of the quaternary ammonium salt. Therefore, amphoteric chitosan had a bright future in applications at various fields, such as biomedicine, membrane separation and wastewater treatment.

3.1. Applications in Biomedical Field

It was well known that CMC was one of very important amphoteric chitosan derivatives. It could be used as immunity auxiliary agent in medicine for nontoxicity of CMC. Furthermore, CMC possessed not only antitumor activity without any hurt to other healthy cells simultaneously, but also biological activity in cells growth-promoting and antiarrhythmia etc. In addition, the excellent coordination of CMC with metal ions, such as Ca^{2+}, Fe^{2+} and Zn^{2+}, made it an ideal ligand to prepare the trace elements' supplier. Furthermore, *N,O*-carboxymethyl chitosan had good effect on prevention of pericardial adhesion after cardiac operation; *N,N*-bis (carboxymethyl) chitosan calcium phosphate showed good performances in the promotion of injured bone restore and regeneration.

Moreover, the carboxymethyl groups had effectively improved pH-sensitivity of CMC as controlled release drug carrier. Chen et al. [12] made a systematical study of swelling behaviors of CMC hydrogels with the change of pH, which indicated that the hydrogels showed typical amphoteric characteristics for responding to pH of the external medium. At the IEP, the hydrogel shrunk mostly, on the contrary, when the pH deviated from IEP, the swelling degree increased. Chen et al. [12] also tried to study the drug release behaviors of CMC hydrogels as carriers for a kind of protein drug under the simulated gastric (pH=2.1) and intestinal (pH=7.4) circumstances, respectively.

Wang et al. [23] prepared a series of biodegradable heparin/amphoteric chitosan (HP/ACS) complexes by electrostatic interaction for pH-sensitive release of protein drugs, and ACS was synthesized by chitosan grafted by acrylic acid (AA). Bovine serum albumin (BSA) was selected as a model protein. The experimental results indicated that the entrapment efficiency of BSA was very high almost near 100%, and the BSA release was extremely pH-dependent. Furthermore, the transition of BSA release could occur within a rather narrow pH range (<0.4 unit). However, the IEP of HP/ACS was dependent on the substitution degree of AA, which could be modulated by the AA/chitosan feed ratio. Therefore, the critical pH of transition for drug release could be controlled well by the substitution degree of AA. In addition, the complex had a good performance in the sustained release of BSA, and the release duration could be up to 15 days.

Recently, Liang et al. [41] reported a modified approach to prepare novel amphoteric octadecyl-quaternized carboxymethyl chitosan(QACMC), in which CMC was made to react with glycidyl octadecyl dimethylammonium chloride; thus the octadecyl quaternary ammonium salt was introduced on CMC. The crystalline properties of QACMC were perfect, which had a high degree of crystallinity around 72.3 %. QACMC also had good thermal stability when the temperature was lower than 200° C. Affected by both the molar mass and the hydrophobic side chains of long alkyl moieties, the moisture-absorption and retention abilities of QACMC were lower than that of hyaluronic acid and CMC, which resulted from no synergistic effect between the carboxymethyl and quaternary ammonium salt. Furthermore, QACMC showed a good solubility in both water and organic solvents, which made the product promise to be a high-potential delivery vector for lipophilic drugs and soluble drugs with a remarkably higher efficiency than chitosan and CMC. For Minocycline hydrochloride, the loading efficiency could reach 10.9 wt%.

However, considerable attentions have been paid to amphoteric chitosan derivatives containing sulfonate groups currently, since the molecular structure of this kind of chitosan derivatives was quite similar to that of heparin, and sulfonated chitosan derivatives possessed high anticoagulant activity and no side-effect. Moreover, sulfonated chitosan derivatives were much cheaper than heparin, so it may be one of the best substitutes for heparin in medical and pharmaceutical applications[45,46].

Vongchan et al.[47] sulfonated chitosan, a degree of acetylation of 0.21, with chlorosulfonic acid in DMF under semi-heterogeneous conditions to give 87% of water-soluble *N,O*-sulfonated chitosan with substitution degree of 2.13. Gel filtration on the sulfonated chitosan got three fractions with different average molecular weights. The three sulfonated chitosan all showed strong anticoagulant activities, and followed the same mechanism of action: inhibiting the activity of thrombin while enhancing the antithrombin III-mediated inhibition of FXa.

N-sulfofurfuryl chitosan, an amphoteric watersoluble derivative, was synthesized by Amiji et al. [29], as mentioned above. The product appeared to possess non-thrombogenic properties, and may be suitable for some blood-contacting applications. As compared with an average of more than 73.0 platelets per 25000 mm^2 on unmodified chitosan, only 4.50 platelets were present on the sulfonated chitosan derivative. The extent of platelet activation was also significantly reduced on the *N*-sulfofurfuryl chitosan surface. From the SEM micrographs, it could be found that all of the adherent platelets which presented on the sulfonated chitosan derivative had retained the discoid shape, which was due to that the presence of sulfonate prevented platelet adhesion and activation completely. In addition, Miao et al. [30, 31] reported that *O*-sulfonated chitosan also had excellent performances in anticoagulant and inflammation-inhibition.

3.2. Applications in Membrane Separation

Recently, CMC/poly(ethersulfone) composite microfiltration membranes were prepared by Zhao et al. [13, 14], and the IEP of CMC was around 5.5. Studies indicated that the composite membranes possessed strong pH-sensitivity: a weak positively charged characteristic at low pHs but a strong negative charged characteristic at high pHs. It was further observed that the composite membranes had high adsorption capacities of BSA at lower pH around 3.0-4.7, but low adsorption capacities at higher pH around 6.0-8.0. Therefore, the composite membranes may be suitable for resistant to protein fouling at high pHs or protein adsorption separations at low pHs applications.

Feng et al. [48] reported a kind of macroporous amphoteric natural polymeric membrane—chitosan/CMC blend membrane, and the adsorption and desorption properties of two model proteins: lysozyme and ovalbumin, which had large different IEP, on the amphoteric membrane, have been studied. Experimental results indicated that the adsorption capacity of proteins varied with the change of environmental pH, and the maxima of ovalbumin and lysozyme were found at pH 4.8 and 9.2, respectively. Both lysozyme and ovalbumin could be effectively separated from their mixtures only by adjusting the pH of feed and desorption solution. Furthermore, the adsorption capacity on the membrane was also affected by initial protein concentration and the content of CMC etc.

The sulfonated chitosan/polysulfone (SCS/PSF) nanofiltration (NF) membranes were prepared by Miao et al. [30, 31], using sulfonated chitosan aqueous solution as casting solution of the active layer, the glutaraldehyde as cross-linking agent and the polysulfone ultrafine membrane as support membrane, respectively. NF, which was a new type of separation membrane technology actively developed in recent years, was an appealing field of membrane science and technology. Compared to ultrafiltration (UF) and reverse osmosis (RO) technologies, the operated pressure was considerably lower, which reduced the operating cost much, and was applied widely in many industrial fields. Based on the structure of the membrane, the high efficiency of separation of ions at low pressure was ascribed to the electric charge on the membrane. In addition, the separation mechanisms of NF membrane included convection, diffusion and electromigration effects. The rejections for the various electrolytes by NF membranes were also investigated by Miao et al. [30, 31] in detail. According to the experimental results, the rejections to the electrolytes, which all contained identical anion of Cl^-, decreased in the order of K^+, Na^+, and Mg^{2+}, but comparing the electrolytes of $NaSO_4$ and NaCl, both containing identical cation of Na^+, the rejection of $NaSO_4$ was much higher than that of NaCl. The active layer of the membranes acquired a negative surface charge distribution by adsorbing anions from the electrolyte solution. Therefore, the membranes had stronger repulsion to SO_4^{2-} than Cl^-. As for the separation behaviors of KCl, NaCl and $MgCl_2$, it might be due to the different size of the cations, which were in the decreasing order of K^+, Na^+, and Mg^{2+}. The smaller ionic size was unfavorable to increase the effective surface charge of membranes, which would decrease the separation performances of NF membranes.

3.3. Applications in Wastewater Treatment

In addition to the applications in the fields of biomedicine and membrane separation, amphoteric chitosan also showed excellent flocculating and chelating properties in wastewater treatment. Compared to chitosan, amphoteric chitosan containing abundant both cationic and anionic groups on the chain backbone, could dissolve in aqueous solutions with various charge, which showed excellent salt-resistant properties. Furthermore, amphoteric chitosan were charged in a wide range of physiological pHs, and showed prominent flocculating effect. On the other hand, the excellent chelating effect of amphoteric chitosan could be removal of metal ions, humic acids and synthetical surfactants efficiently for formation of complex precipitation.

Recently, An et al. [49] prepared a series of water soluble *N*-substituted CMC with various substitution degrees from full deacetylated chitosan, and the structures of the products were characterized by IR, ^{1}H, ^{13}C and ^{1}H–^{13}C NMR-HSQC spectra. Based on the NMR analysis, the substitution degrees of CMC including *N*-mono- and *N,N*-di- substitution have been detected, respectively. The *N*-CMC was also applied for adsorption of Cu^{2+} in nitrate aqueous solution, and the experimental results indicated that the optimal adsorption condition was pH 6.5, temperature of 30° C, for 60–90 min and the substituted chitosan derivative having substitution degree of *N*-monosubstitution of 0.16 and that of *N,N*-disubstitution of 0.81. The maximum adsorption capacity for Cu^{2+} was 192 mg/g.

Bratskaya et al. [50] obtained a series of *N*-carboxyethyl chitosan derivatives with different substitution degrees, which IEP were various between 3.55 and 6.30. The derivatives have been tried to applied for removal of Cu^{2+}, $Zn^{2+}$and $Ni^{2+}$from aqueous solutions, respectively. Studies found that the derivatives had the best adsorption performance for Cu^{2+}, and the worst one was for Ni^{2+}, which might be due to the strongest chelating effects for Cu^{2+} with carboxyl groups.

Generally, cationic flocculants had poor flocculating effects on metal cationic ions and colloidal particles with positive charges, while that of anionic flocculants on Cr(VI) and colloidal particles with negative charge was also bad, which limited their applications in wastewater treatment. In order to overcome these disadvantages, Cai et al. [8, 39] prepared a kind of quaternized amphoteric chitosan by the grafting reaction of CMC with CTA, as mentioned above. And the flocculating properties of the products for Cd^{2+} and $Cr_2O_7^{2-}$ were studied in detail, respectively. The experimental results indicated that the product had good performances for removal of the ions of Cd^{2+} and $Cr_2O_7^{2-}$. For Cd^{2+}, at the optimal conditions: the pH of 8.3, and the mass concentration of the product around 120-140 mg/L, 99.7 % of Cd^{2+} could be removed successfully, which may be ascribed to the stronge electrostatic attraction between Cd^{2+} and salt-form carboxyl groups in weak alkaline condition. As for Cr(VI), it was found that the flocculating properties for Cr(VI) showed strongly pH-dependence. When the pH was around 5.0, the removal ratio for Cr(VI) could reach the highest value near 94.4 %. However, under higher or lower pH, the removal ratio for Cr(VI) was both low, which might result from the various forms of Cr(VI) in water mainly $HCrO_4^-$ and CrO_4^{2-}, respectively. The form of Cr(VI) at pH 5.0 was $Cr_2O_7^{2-}$ dominantly. It was partly ascribed to different charge density and the content of Cr(VI) among the various forms of Cr(VI), and CrO_4^{2-} or $HCrO_4^-$ was unfavorable to be removal of, but $Cr_2O_7^{2-}$ could be chelated with ionic groups of amphoteric chitosan effectively. In addition, the study of the isotherm adsorption behaviors for Cd^{2+} and $Cr_2O_7^{2-}$ showed that they both obeyed Langmuir isotherm adsorption.

Weltrowski et al. [28] tried to apply *N*-benzyl sulfonate chitosan derivatives as adsorbents for removal of several heavy metals, such as Cd^{2+}, Zn^{2+}, Ni^{2+}, Pb^{2+}, Cu^{2+}, Fe^{3+} and Cr^{3+}. The experimental results indicated that the synthesized sulfonate chitosan derivatives were especially adapted to the adsorption of heavy metals from the acidic industrial effluents, and the adsorption capacities of disulfonate compounds were better than that of monosulfonate compounds.

Recently, the application performances of CMC used as soaping agents for removal of reactive dyes were studied by Liu et al. [51]. The experimental results showed that CMC could remove the loose color more effectively and have higher soaping efficiency than anionic chitosan derivatives. The optimal conditions for removal of dyes was the concentration of CMC around 0.25-0.5 g/L under weak alkaline condition (pH=9.0) at 90° C for 15 min, in which the soaping agents would have great effects to form micelles with degraded dyes.

As mentioned above, amphoteric chitosan had good performances in the adsorption of heavy metals and removal of organic pollutants. Therefore, amphoteric chitosan had a bright prospect in wastewater treatment.

4. Perspective

Above all, amphoteric chitosan materials, one kind of chemical modified chitosan derivatives, possessed much better performances than chitosan in practical applications. Therefore, the preparation and applications of amphoteric chitosan had more extensive foregrounds in the future. However, the current technologies for preparation of amphoteric chitosan were too complicated to apply for real industrial productions. One of the urgent matter for development of amphoteric chitosan materials was to explore a simple, efficient and inexpensive preparation technology [10].

Moreover, the research works about the mechanisms in practical applications of amphoteric chitosan were not adequate until now, which would limit their applications severely. For example, chitosan could be used as flocculants in wastewater treatment, however, the flocculating mechanism of chitosan simply ascribed to that of polyacrylamide: charge neutralization and bridging flocculating mechanism. In fact, based on the actural flocculating properties, chitosan, a weak cationic polyelectrolyte, possessed similar flocculating ability to that of the strong cationic polyelectrolyte; Furthermore, although the molecular weight of chitosan was usually kept much lower than that of synthetic polymer flocculants, such as polyacrylamide, chitosan still had excellent flocculating performances. It was obviously that the flocculating mechanisms of chitosan could not be explained simply by charge neutralization and bridging flocculating mechanism, which might be related to the special structures of the chitosan in solutions[52]. Furthermore, amphoteric chitosan, containing the characteristics of both polysaccharide and polyeletrolyte, would bear more distinguished structures in solutions. So it was of great significance to investigate the structures and conformations of amphoteric chitosan materials for understanding well their mechanisms in actual applications.

References

[1] Zhang L N. Natural Polymer Modified Materials and Application, Beijing: Chemical Industry Press, 2006

[2] Ge J J. Biodegradable Polymer Materials and Their Applications, Beijing: Chemical Industry Press, 2002

[3] Jiang T D. Chitosan, 1st Edn., Beijing: Chemical Industry Press, 2001

[4] Rinaudo M. Chitin and chitosan: Properties and applications, *Progress in Polymer Science*, 2006, 31(7), 603-632

[5] Muzzarelli R A A, Muzzarelli C. Chitosan chemistry: Relevance to the biomedical sciences, *Advances in Polymer Science*, 2005, 186, 151-209

[6] Chen X G, Park H J. Chemical characteristics of O-carboxymethyl chitosans related to the preparation conditions, *Carbohydrate Polymers*, 2003, 53(4), 355-359

[7] Schatz C, Bionaz A, Lucas J M, Pichot C, Viton C, Domard A, Delair T. Formation of Polyelectrolyte Complex Particles from Self-Complexation of N-Sulfated Chitosan, *Biomacromolecules*, 2005, 6(3), 1642-1647

[8] Cai Z S, Song Z Q, Shang S B, Yang C S. Study on the flocculating properties of quaternized carboxymethyl chitosan, *Polymer Bulletin*, 2007, 59, 655-665

[9] Matsumura S, Yokochi E, Winursito I, Toshima K. Preparation of novel biodegradable polyampholyte: Partially dicarboxylated chitosan, *Chemistry Letters*, 1997, 3, 215-216

[10] Yang H, Shang Y B, Yuan B, Jiang Y X, Lu Y B, Qin Z Q, Cheng R S, Chen A M. ZL200810124282.0 [P]

[11] Carolan C A, Blair H S, Allen S J, Mckay G. N,O-Carboxymethyl Chitosan, A Water-Soluble Derivative and Potential Green Food Preservative, *Chemical Engineering Research and Design*, 1991, 69(3), 195-196

[12] Chen L Y, Tian Z G, Du Y M. Synthesis and pH sensitivity of carboxymethyl chitosan-based polyampholyte hydrogels for protein carrier matrices, *Biomaterials*, 2004, 25(17), 3725-3732

[13] Zhao Z P, Wang Z, Ye N, Wang S C. A novel N,O-carboxymethyl amphoteric chitosan/poly(ethersulfone) composite MF membrane and its charged characteristics, *Desalination*, 2002, 144(1), 35-39

[14] Zhao Z P, Wang Z, Wang S C. Formation, charged characteristic and BSA adsorption behavior of carboxymethyl chitosan/PES composite MF membrane, *Journal of Membrane Science*, 2003, 217(1), 151-158

[15] Wu G, Shen Y H, Xie A J, Lin H Y, Guo W H. Synthesis and Studies on Properties on N,O-carboxymethylchitosan, *Chinese Journal of Chemical Physics*, 2003, 16(6), 499-503

[16] de Abreu F R, Campana-Filho S P. Characteristics and properties of carboxymethylchitosan, *Carbohydrate Polymers*, 2009, 75, 214-221

[17] Zhou Y G, Yang Y D, Wang D J, Liu X M. Preparation and Characterization of 6-Carboxychitosan, *Chemistry Letters*, 2003, 32(8), 682-683

[18] Zhou Y G, Yang Y D, Chuo H M, Liu X M. Preparation and Rheological Behaviors of 6-Carboxy-Chitosan, *Journal of Applied Polymer Science*, 2004, 94(3), 1126-1130

[19] Ohya Y, Okawa K, Murata J, Ouchi T. Preparation of Oxidized 6-O-Glycolchitosan, pH Sensitivity of its Aqueous Solution and of its Cross-Linked Hydrogel, *Angewandte Makromlekulare Chemie*, 1996, 240, 263-273

[20] Mahdavinia G R, Pourjavadi A, Hosseinzadeh H, Zohuriaan M J. Modified chitosan 4.Superabsorbent hydrogels from poly(acrylic acid-co-acrylamide) grafted chitosan with salt- and pH-responsiveness properties, *European Polymer Journal*, 2004, 40, 1399-1407

[21] Yazdani-Pedram M, Retuert J, Quijada R. Hydrogels based on modified chitosan, 1-Synthesis and swelling behavior of poly(acrylic acid) grafted chitosan, *Macromolecular*

Chemistry and Physics, 2000, 201(9), 923-930

[22] Zhang L Y, Li Z W, Ni M L, Zeng Q X. Preparation of N-carboxyethyl chitosan hydrogel and its pH-sensitivity, *Chemical Industry and Engineering Progress*, 2008, 27(4), 585-589

[23] Wang Y J, Jiang H L, Hu Y Q, Zhu K J. Biodegradable Heparin/Ampholytic Chitosan Complexes for pH-Sensitive Release of Proteins, *Acta Polymerica Sinica*, 2005, 4, 524-528

[24] Li G, Zhuang Y L, Mu Q, Wang M, Fang Y. Preparation, characterization and aggregation behavior of amphiphilic chitosan derivative having poly(L-lactic acid) side chains, *Carbohydrate Polymers*, 2008, 72(1), 60-66

[25] Yao F L, Chen W, Liu C, Yao K D. A Novel Amphoteric, pH-Sensitive, Biodegradable Poly[chitosan-g-(L-lactic-co-citric) acid] Hydrogel, *Journal of Applied Polymer Science*, 2003, 89(14), 3850-3854

[26] Batista M K S, Pinto L F, Gomes C A R, Gomes P. Novel highly-soluble peptide–chitosan polymers: Chemical synthesis and spectral characterization, *Carbohydrate Polymers,* 2006, 64, 299-305

[27] Holme K R, Perlin A S. Chitosan N-sulfate. A water-soluble polyelectrolyte, *Carbohydrate Research*, 1997, 302, 7-12

[28] Weltrowski M, Martel B, Morcellet M. Chitosan N-Benzyl Sulfonate Derivatives as Sorbents for Removal of Metal Ions in an Acidic Medium, *Journal of Applied Polymer Science,* 1996, 59(4), 647-654

[29] Amiji M M. Platelet adhesion and activation on an amphoteric chitosan derivative bearing sulfonate groups, *Colloids and Surfaces B: Biointerfaces,* 1998, 10(5), 263-271

[30] Miao J, Chen G H, Gao C J. A novel kind of amphoteric composite nanofiltration membrane prepared from sulfated chitosan (SCS), *Desalination*, 2005, 181, 173-183

[31] Miao J, Chen G H, Gao C J, Lin C G, Wang D, Sun M K. Preparation and Characterization of Sulfated chitosan(SCS)/Polysulfone(PSF) Composite Nanofiltration Membrane Cross-linked by Glutaraldehyde, *Journal of Chemical Engineering of Chinese Universities,* 2007, 21(2), 227-232

[32] Amaral I F, Granja P L, Barbosa M A. Chemical modification of chitosan by phosphorylation: an XPS, FT-IR and SEM study, *Journal of Biomaterials Science-Polymer Edition,* 2005, 16(12), 1575-1593

[33] Domard A, Rinaudo M, Terrassin C. New Method for the Quaternization of Chitosan, *International Journal of Biological Macromolecules*, 1986, 8, 105-107

[34] Jia Z S, Shen D F, Xu W L. Synthesis and antibacterial activities of quaternary ammonium salt of chitosan, *Carbohydrate Research*, 2001, 333, 1-6

[35] Thanou M, Florea B I, Geldof M, Junginger H E, Borchard G. Quaternized chitosan oligomers as novel gene delivery vectors in epithelial cell lines, *Biomaterials*, 2002, 23, 153-159

[36] Bratskaya S, Schwarz S, Laube J, Liebert T, Heinze T, Krentz O, Lohmann C, Kulicke W M. Effect of Polyelectrolyte Structural Features on Flocculation Behavior: Cationic Polysaccharides vs Synthetic Polycations, *Macromolecular Materials and Engineering*, 2005, 290, 778-785

[37] Chi W L, Qin C Q, Zeng L T, Li W, Wang W. Microbiocidal Activity of Chitosan-N-2-hydroxypropyl Trimethyl Ammonium Chloride, *Journal of Applied Polymer Science*, 2007, 103, 3851-3856

[38] Yao P J, Li S Q, Wei Y A. Synthesis and structural characterization of the quaternary ammonium salt of O-carboxymethyl-N-triMethylchitosan, *Journal of Guangxi University*, 2006, 31(3), 208-211

[39] Cai Z S, Song Z Q, Yang C S, Wang J C, Xu Q. Preparation of Amphoteric Chitosan and Characterization of Its Structure, *Chemistry and Industry of Forest Products*, 2007, 27(2), 123-126

[40] Tang Y G, Jiang G B, Xie G D. Synthesis of A Novel Type of Chitosan Amphoteric Polymer Surfactant, *Hunan Chemical Industry*, 2002, 30(2), 30-33

[41] Liang X F, Wang H J, Tian H, Luo H, Chang J. Synthesis, Structure and Properties of Novel Quaternized Carboxymethyl Chitosan with Drug Loading Capacity, *Acta Physico-Chimica Sinica*, 2008, 24(2), 223-229

[42] Tang H L, Chen Y, Tan H M. Primary Study on Preparation and Water Absorption Property of the CMCTS-g-(PAA-co-PDMDAAC) Polyam Pholytic Superabsorbent Polymer. *Polymer Materials Science and Engineering*, 2006, 22(3), 115-118

[43] Cai Z S, Song Z Q, Yang C S, Shang S B, Yin Y B. Synthesis, characterization and antibacterial activity of quaternized N,O-(2-carboxyethyl) chitosan, *Polymer Bulletin*, 2009, 62, 445-456

[44] Cai Z S, Song Z Q, Yang C S, Shang S B, Yin Y B. Synthesis of 2-Hydroxypropyl Dimethylbenzylammonium N,O-(2-carboxyethyl) Chitosan Chloride and Its Antibacterial Activity, *Journal of Applied Polymer Science*, 2009, 111, 3010-3015

[45] Doczi J, Fischman A, King J A. Direct Evidence of the Influence of Sulfamic Acid Linkages on the Activity of Heparin-like Anticoagulants, *Journal of the American Chemical Society,* 1953, 75, 1512-1513

[46] Wolfrom M L, Shen T M, Summers C G. Sulfated Nitrogenous Polysaccharides and Their Anticoagulant Activity, *Journal of the American Chemical Society*, 1953, 75, 1519-1519

[47] Vongchan P, Sajomsang W, Subyen D, Kongtawelert P. Anticoagulant activity of a sulfated chitosan, *Carbohydrate Research*, 2002, 337(13), 1239-1242

[48] Feng Z C, Shao Z Z, Yao J R, Chen X. Protein adsorption and separation with chitosan-based amphoteric membranes, *Polymer,* 2009, 50, 1257-1263

[49] An N T, Thien D T, Dong N T, Dung P L. Water-soluble N-carboxymethylchitosan derivatives: Preparation, characteristics and its application, *Carbohydrate Polymers*, 2009, 75, 489-497

[50] Bratskaya S Y, Pestovb A V, Yatluk Y G, Avramenko V A. Heavy metals removal by flocculation/precipitation using N-(2-carboxyethyl)chitosans, *Colloids and Surfaces A-Physicochemical and Engineering Aspects*, 2009, 339, 140-144

[51] Liu P, Xue X T, He J X. Research on application of amphoteric chitosan derivative NAOC in wash--off for reactive dyeing, *Textile Auxiliaries*, 2008, 25(8), 31-35

[52] *Cai Z C,* Dai J, *Yang H, Cheng R S, Study on the interfacial properties of viscous capillary flow of dilute acetic acid solutions of chitosan, Carbohydrate Polymers, 2009, 78(3), 488-491.*

In: Handbook of Chitosan Research and Applications
Editors: Richard G. Mackay and Jennifer M. Tait
ISBN 978-1-61324-455-5

Chapter 12

CHITOSAN REDUCES DRY ROT OF POTATO TUBER: FUNGISTATIC EFFECTS AND INDUCE RESISTANCE

Yongcai Li [1,2]*, Yang Bi[1], Yonghong Ge[1], Yi Wang[1] and Di Wang[2]
[1]. College of Food Science and Engineering,
[2]. Gansu Key Laboratory of Crop Genetic and Germplasm Enhancement,
Gansu Agricultural University,
Lanzhou, P.R. China

ABSTRACT

Potato dry rot caused by *Fusarium sulphureum* is one of the most important postharvest diseases restricting potato production and incidence of disease is up to 30% during storage in Gansu, China. Fungistatic characteristic and induced disease resistance of chitosan on dry rot of potato tuber (cv. Atlantic) were studied. *In vitro* tests, spore germination and mycelial growth of *F. sulphureum* were inhibited by chitosan treatment and the inhibitory effect was highly correlated with chitosan concentration. Morphological changes such as intertwisting hyphal, distortion, and swelling with excessive branching were observed by scanning electron microscopy (SEM) observation. Transmission electron microscopy (TEM) observation further indicated the ultrastructural alterations of hyphae. These changes included abnormal distribution of cytoplasma, non-membraneous inclusion bodies assembling in cytoplasm, considerable thickening of the hyphal cellular walls, and very frequent septation with malformed septa. Application of chitosan at higher concentration caused serious damage to fungal hyphae, including cellular membrane disorganisation, cell wall disruption, and breaking of inner cytoplast. New hyphae (daughter hyphae) inside the collapsed hyphal cells was often detected in the cytoplasm of chitosan-treated hyphae. Chitosan at 0.25% increased the activities of peroxidase (POD) and polyphenoloxidase (PPO), and the contents of flavonoid compounds and lignin in tissues. Increased activities of β-1, 3-glucanase (GLU) and

* E-mail: liyc_2005@yahoo.com, Tel.: +86 931 7632256; fax: +86 931 7631201.

phenylalanine ammonialyase (PAL) were observed, but there were no significant differences between the treated and the control. This suggests chitosan is promising as fungicide to reduce potato tuber disease and to partially substitute for the utilization of synthetic fungicides in fruit and vegetables.

Keywords: *chitosan, potato tuber, dry rot, Fusarium sulphureum, fungistatic characteristic, induced disease resistance*

INTRODUCTION

Dry rot caused by *Fusarium* spp. is one of the most important diseases of stored potato, affecting tuber storage and seed pieces after planting. *Fusarium sulphureum* Schlechlendah is one of the major causal agents in Gansu province, the largest potato production area in China (He *et al.* 2004). As *Fusarium* spp. cannot penetrate the periderm of the tubers, infection can only occur through wounds or breaks in the periderm. Therefore, *Fusarium* infection is generally controlled by limiting wounding during harvest, transport and storage or storing potato tubers under the optimal conditions for wound healing (Secor and Gudmestad 1999). Primary control of pathogens is reported by postharvest application of fungicides, such as thiabendazole. However, the resistance of *Fusarium* spp. to thiabendazole (Desjardins *et al.* 1993), and public concern over food safety require an investigation on the control methods with potentially less harmful to human health and the environment (Tripathi and Dubey 2004).

Chitosan (poly-β-(1→4) *N*-acetyl-d-glucosamine) is currently obtained from the outer shell of crustaceans such as crabs, krills and shrimps (Sandford and Hutchings 1987; Sandford 1989). The chemical agent is a low acetyl form of chitin mainly composed of glucosamine, and 2-amino-2-deoxy-β-D-glucose. The positive charge of chitosan confers to this polymer numerous and unique physiological and biological property with great potential in a wide application range. Chitosan has been proven to control effectively postharvest diseases on various horticultural commodities such as apples, kiwifruit, pears, tomato, table grape strawberries, raspberries and others (El Ghaouth *et al.* 1991, 1992a, b; Zhang and Quantick 1998; Bautista-Banos *et al.* 2004a; Du *et al.* 1997; de Capdeville *et al.* 2002; Liu *et al.* 2007; Meng *et al.* 2008; Sun *et al.* 2008). The fungicidal activity of chitosan has been well documented by *in vitro* and *in vivo* studies (Bautista-Banos *et al.* 2006). The level of inhibition of fungal growth is highly correlated with chitosan concentration. Recent studies have shown that chitosan treatment is not only effective in halting pathogen growth (Ben-Shalom *et al.* 2003; Xu *et al.* 2006; Liu *et al.* 2007; Meng *et al.* 2008), but also results in marked morphological changes, structural alterations and molecular disorganisation of the fungal cells (El Ghaouth *et al.* 1999; Ait Barka *et al.* 2004).

In addition to its direct microbial activity, other studies strongly suggest that chitosan induces a series of defence reactions correlated with enzymatic activities. Chitosan has been shown to increase the production of glucanohydrolases, phenolic compounds and synthesis of specific phytoalexins with antifungal activity (Bautista-Banos *et al.* 2006). Postharvest chitosan treatment could increase the activities of the peroxidase (POD), chitinase (CHT), and β-1, 3-glucanase (GLU) in tomatoes, strawberries, and raspberries (Liu *et al.* 2007; Zhang et al. 1998). Besides, it has been reported that the activities of lipoxygenase (LOX) and phenylalanine ammonialyase (PAL), and the content of lignin have been elicited in wheat leaves treated with chitosan (Vander *et al.* 1998).

The objective of this study was to determine fungistatic characteristic and induced disease resistance of chitosan on dry rot of potato tuber, and to investigate its possible mechanism.

MATERIALS AND METHODS

Potato Tubers

Potato (cv. Atlantic) was harvested from Zhongchuan Farm, Yongdeng County in Gansu, China, and then the potato tubers were packed in net bags (15 kg/bag) and transported to the laboratory within 24 h. Tubers were sorted by size while tubers without physical injuries or visible infection were stored at 5-8°C. Before treatment, tubers were surface-disinfected with 2% sodium hypochlorite for 3 min, then rinsed with tap water, and finally air-dried.

Pathogen

Fusarium sulphureum isolate, the pathogen of dry rot of potato (He *et al.* 2004), was provided by the Institute of Plant Protection, Gansu Academy of Agricultural Sciences.

Chitosan

Chitosan (edible level, deacetyl degree 90.2%, granularity 80) was obtained from Ocean Bioengineering Limited Company, China.

Optimum Solvent for Chitosan Selecting

Mycelial disks (5mm in diameter) from the 2-week-old culture of the fungus were placed in the centre of 90 mm Petri dishes containing 20 mL PDA containing 0.5% chitosan dissolved in 0.5 M formic acid, acetic acid or lactic acid at pH 5.6 adjusted with 0.1 M NaOH and then incubated at 25°C in the dark. Mycelial growth was determined by measuring the colony diameter after 7 days of inoculation. PDA containing each solvent was used as a control. Each treatment was replicated three times and the experiment was repeated twice.

Measurements of Spore Germination and Mycelial Growth of *Fusarium Sulphureum In Vitro*

To assess the effects of chitosan on spore germination of *F. sulphureum*, 20-μL aliquots of a conidial suspension of 1×10^6 spores mL^{-1} were plated on 1% agar plugs (10 mm in diameter) containing chitosan at 0, 0.25, 0.5 or 1%. Agar plugs were incubated in Petri dishes at 25°C. After 8 hours of incubation, germination rates were recorded under the light microscope. The germination rate was measured for approximately 200 spores per treatment. Each treatment was replicated three times and the experiment was repeated twice.

The effects of chitosan on mycelial growth of *F. sulphureum* were assessed according to Yao and Tian (2005). Mycelial disks (5 mm in diameter) from the 2-week-old culture of the fungus were placed in the centre of 90 mm Petri dishes containing 20 mL PDA containing chitosan at 0, 0.25, 0.5 or 1% and then incubated at 25°C in the dark. Mycelial growth was determined by measuring the colony diameter after 7 days of inoculation. Each treatment was replicated three times and the experiment was repeated twice.

Hyphae Morphological Change Observation with Scanning Electron Microscopy (SEM)

Hyphal samples excised from the 7-day-old culture of the fungus treated with 0 and 0.5% chitosan were fixed, sputter-coated and observed according to the method of Benhamou *et al.* (1999). Micrographs were taken by CCD (charged coupled device)-Camera (America Gatan Company, USA). The experiment was repeated three times on two replicate plates for each treatment. For each replicate, 10 agar blocks were examined using SEM.

Hyphae Ultrastructure Alteration Observation with Transmission Electron Microscopy (TEM)

Mycelial samples (2 mm) were excised from the 7-day-old PDA culture of the fungus grown with 0.25 and 0.5% chitosan were fixed, dehydrated, cut and observed according to the method of Benhamou *et al.* (1999). Micrographs were taken by CCD-Camera (America Gatan Company, USA). Average of five samples from each sampling time were examined using five sections per sample.

Effect of Chitosan on Lesion Diameters of Tuber Inoculated with *F. Sulphureum*

Following the method of Bi *et al.* (2006), with some modifications, the tuber was dipped in 0.25, 0.5, and 1% chitosan (pH 5.6 ± 0.2), which was predissolved in 0.5% lactic acid for about 10 min, and then dried in air for 2 h. A uniform wound (3 mm deep and 3 mm wide) was made at the equator of each tuber using a sterile dissecting needle after 72 h of treatment, and then a 20-μL conidial suspension of *F. sulphureum* was inoculated into each wound after 1 h of wounding. Inoculated tubers were incubated in the boxes, covered with plastic film at room temperature (25°C ± 2). Lesion diameters were evaluated after 15 days of inoculation. Each treatment was applied to three replicates of 10 tubers. The experiment was repeated thrice.

Effect of Chitosan on Lesion Diameters of Slices Inoculated with *F. Sulphureum*

According to the method of Ray and Hammers (1998), Slices (20 mm in diameter and about 5 mm in thickness) were made with a sterile knife, rinsed in sterile water, sterilized with 75% ethanol and flame, placed on wet sterile filter paper, and incubated for 4 h in darkness. A 200-μl chitosan solution at 0.25, 0.5 or 1% was scrawled on discs. The control was treated with water. The pathogen was inoculated after 24, 48, 72, and 96 h of treatment. The inoculums, 4 mm in diameter, were obtained from seven-day-old *F. sulphureum* grown on PDA, and then placed hyphae side down on slices. The slices were incubated for 3 days at 23-25°C in the dark. The lesion diameter was recorded. Each treatment was applied to three replicates of five slices. The experiment was repeated twice.

Mechanism of Induced Disease Resistance in Potato Tuber by Chitosan and Challenged Inoculation

Assay of Defense Enzymes

Tubers were removed after 1, 2, 3, 4, and 5 days of treatment with 0.25 % chitosan. Approximately 3 g tissue samples were taken from 1-2 mm below the treated side with a stainless steel cork borer. Tissue samples inoculated with *F. sulphureum* were taken from 3-4 mm below the treated side with a stainless steel cork borer, and were removed at 3, 4, and 5 days, respectively. Each sample was packed and frozen in liquid nitrogen, and finally kept at -80°C prior to crude enzyme extraction.

POD activity was determined according to the methods of Venisse *et al.* (2001), with some modifications. Guaiacol was used as the substrate. Oxidation of guaiacol to tetraguaiacol was monitored spectrophotometrically at 470 nm for 2 min at 30°C. The enzyme activity was expressed in unit (U) per microgram protein. The enzyme unit was defined as $\Delta A470$ of 0.01 per minute.

PPO activity was determined according to the method of Jiang *et al.* (2002), by adding 0.5 ml of enzyme preparation to 3 mL of 500 mM catechol (100 mM sodium phosphate, pH 6.4), as a substrate. An absorbance was measured at 420 nm. The activity was expressed in the unit (U) per microgram protein. One unit was defined as $\Delta A420$ of 0.01 per minute.

β-1,3-Glucanase was assayed by measuring the amount of reducing sugar released from the substrate using the method of Ippolito *et al.* (2000), with some modifications. Enzyme preparation (0.25 mL) was incubated with 250 μL of 0.5% laminarin (w/v) for 60 min at 37°C and 50 μL reactive mixture was then taken out. The reaction mixture was diluted 1:4 with sterile distilled water. For the control, the same mixture was similarly diluted at zero incubation time. The reaction was stopped by adding 250 μL of 3, 5-dinitrosalicylate and boiling for 5 min on a water bath. The solution was diluted with 4 ml of distilled water, and the amount of reducing sugars was measured spectrophotometrically at 630 nm using a UV-160 spectrophotometer (Shimadzu, Japan). Specific activity was expressed as the formation of 1 μmol glucose equivalent per hour per milligram protein.

PAL activity was assayed referring to the method of Assis *et al.* (2001), with some modifications. One milliliter of enzyme extract was incubated with 2 ml of borate buffer (50 mM, pH 8.8) and 1 ml of L-phenylalanine (20 mM) for 60 min at 37°C. The reaction was stopped with 1 ml of 1 M HCl. PAL activity was determined by the production of cinnamate, which was measured at 290 nm absorbance. The control mixture was stopped by adding 1 ml of 1 M HCl immediately after mixing the crude enzyme preparation with L-phenylalanine. Specific enzyme activity was defined as nanomoles cinnamic acid per hour per milligram protein.

Flavonoid compounds were measured according to the methods of Pirie and Mullins (1976). One gram of frozen tissue was homogenized with 5 mL of 1% HCl-methanol ice-cold solution and extracted for 2 h and then centrifuged at 4°C for 30 min at 14 000 × g. The supernatant was collected and absorbance measured at 325 nm, with flavonoid compounds expressed as OD325 g^{-1} FW.

Lignin was assayed according to the method of Morrison (1972), with some modifications. Frozen tissue of 0.5 g was homogenized with 5 ml of 95% ice-cold ethanol solution and then centrifuged at 4°C for 30 min at 14 000 × g. The collected pellet was washed with 95% ethanol thrice, and then washed thrice with ethanol : hexane = 1:2 (V/V). Pellets was collected for drying (24 h at 60°C) and transferred to tubes, and then 1 ml of 25% bromized acetyl-acetic acid was added. The tubes were incubated in a water bath at 70°C for 30 min, and then 1 ml of 2 mol L^{-1} NaOH was added to stop the reaction. Ice acetic acid of 2 mL and 0.1 mL 7.5 M hydroxylamine hydrochloric acid were added to each tube and centrifuged. Supernatant solution of 0.5 mL was removed. The absorbance of the supernatant was measured at 280 nm with a spectrophotometer. Lignin compounds were expressed as OD280 g^{-1} FW.

Protein content was determined according to the method of Bradford (1976), using bovine serum albumin as a standard.

Statistical Analysis

All statistical analyses were performed using SPSS. To test the effect of the treatment, the data were analyzed by one-way analysis of variance (ANOVA). Mean separations were performed using the least significant difference method. The data were also expressed as the means (±SE).

RESULTS

Effect of Various Solvents on Antifungal Activity of Chitosan Against *F. Sulphureum*

Chitosan dissolved in formic acid, acetic acid and lactic acid had different antifungal activity (Figure 1-A). Chitosan in lactic acid exhibited the best inhibitory effect, with the inhibitory rate being up to 80%. Moreover, direct inhibition of lactic acid on mycelial growth of *F. sulphureum* was significantly lower than other two solvents (Figure 1-B). Thus, lactic acid was the best solvent for enhancing antifungal activity of chitosan.

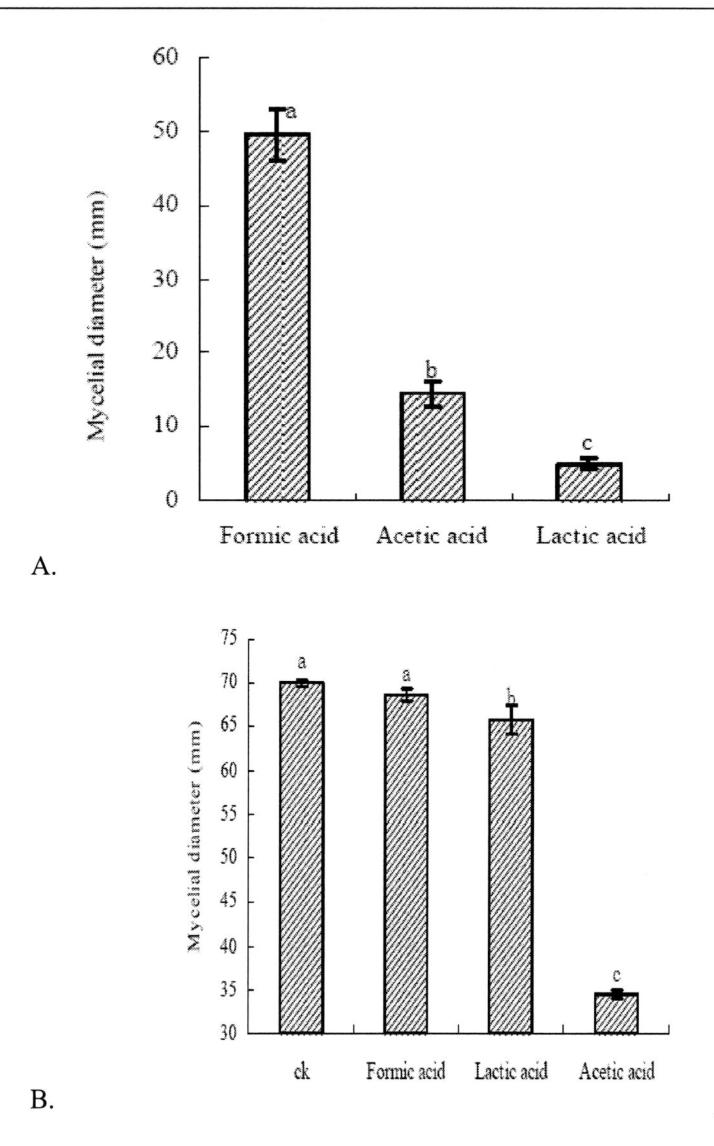

Figure 1. Effects of chitosan (0.5%) dissolved in different solvents (A) and different solvents alone (B) on mycelial growth of *F. sulphureum*. Different letters show significant levels at 5%. The same as below. Vertical bars represent the standard error.

Effect of Chitosan Treatment on Spore Germination and Mycelial Growth of *F. Sulphureum*

Spore germination of *F. sulphureum* was significantly inhibited by chitosan at different concentrations and inhibitory effects enhanced with increasing concentration (Figure 2-A). Chitosan at 1% completely inhibited spore germination of *F. sulphureum*. Chitosan at different concentrations also inhibited markedly mycelial growth of *F. sulphureum* (Figure 2-B). The inhibitory rate of *F. sulphureum* treated with 0.5 and 1% chitosan was up to 86 and 89%, respectively.

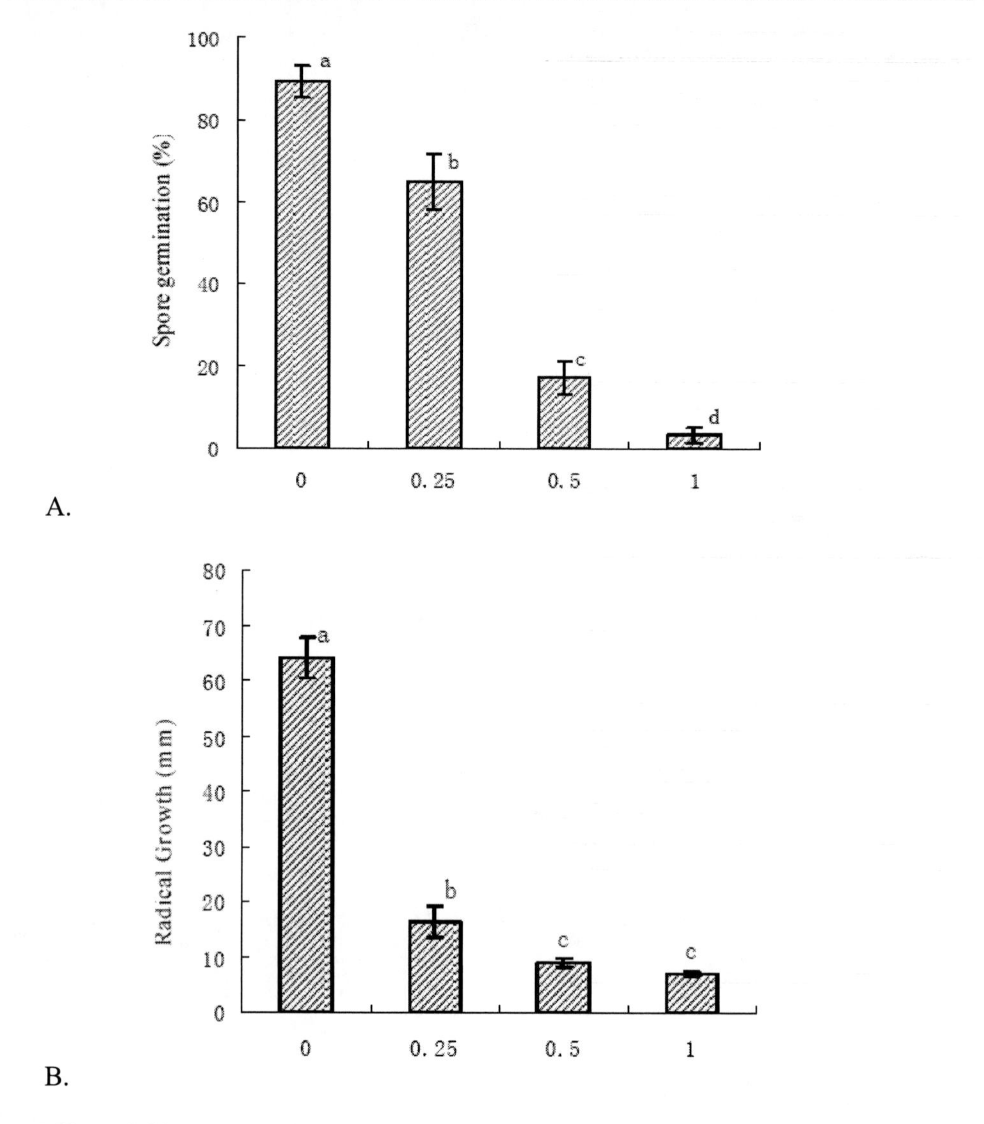

Figure 2. Effects of chitosan concentration on spore germination (A) and mycelial growth (B) of *F. sulphureum*. Different letters show significant levels at 5%. Vertical bars represent the standard error.

Effect of Chitosan Treatment on Hyphal Morphology of *F. Sulphureum*

SEM observations of *F. sulphureum* treated with chitosan revealed the effects of the morphology of the hyphae (Figure 3). Hyphae of the control sample were even (Figure 3-A) and the hyphal branching occurred at some distance from the hyphal tip (Figure 3-B). The growth of hyphae treated with chitosan was strongly inhibited and the chitosan-treated hyphae was tightly twisted and formed rope-like structures (Figure 3-C and D). Spherical or club-shaped abnormally inflated ends were observed on the twisted hyphae (Figure 3-E). At the edge of the twisted hyphae, hyphae were swollen with excessive branching and the inflated mycelium became blasted (Figure 3-F).

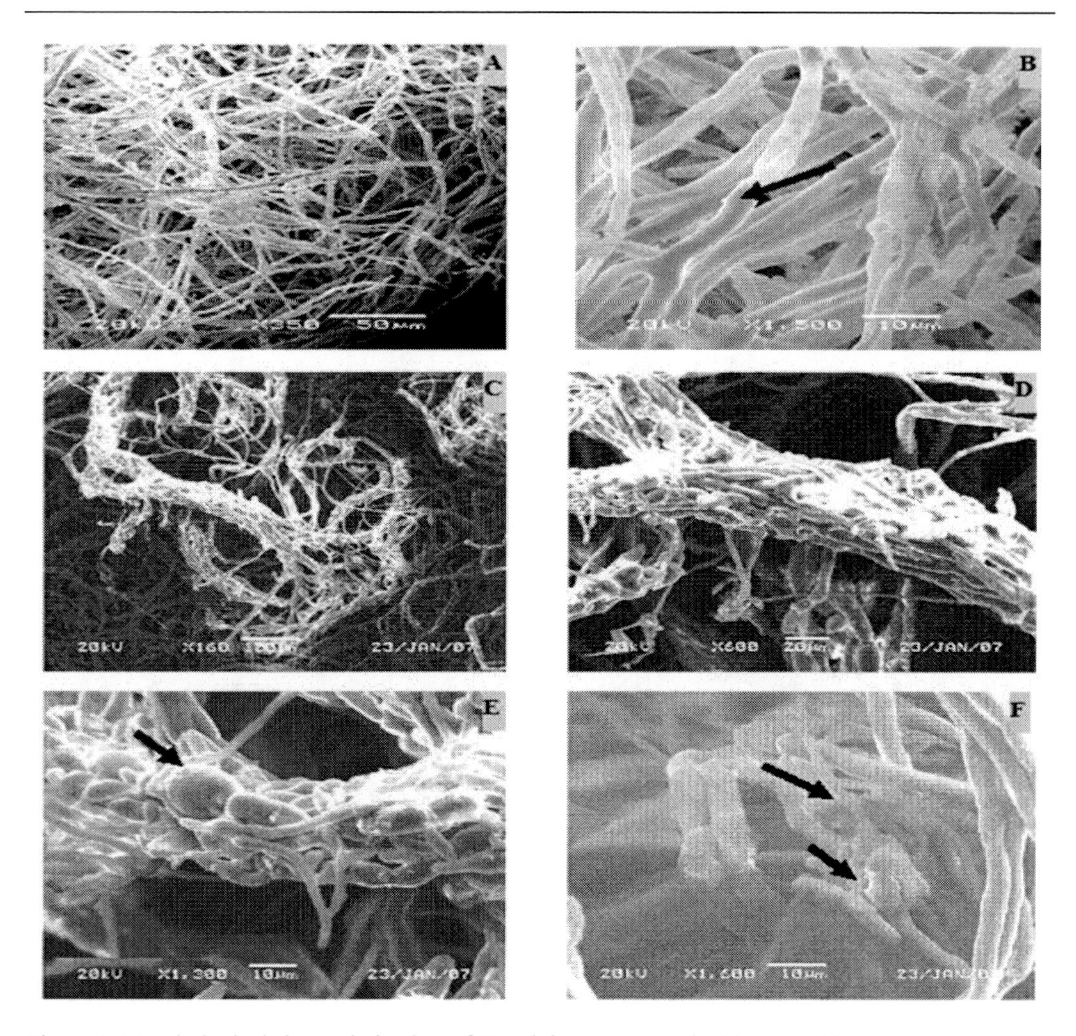

Figure 3. Morphological change in hyphae of *F. sulphureum* treated with 0.5% chitosan (C, D, E and F) and untreated control (A and B) removed from 7-day-old PDA cultures as revealed by SEM. Hyphae were even (A) and the hyphal branching occurred at some distance from the tip (B) in untreated control. Tightly twisted hyphae (C and D), spherical or club-shaped abnormally inflated ends (arrowhead) (E), excessive branching and blasted inflated mycelium (arrowheads) (F) appeared in chitosan treated pathogen cultures.

Effect of Chitosan Treatment on Hyphae Ultralstructure of *F. Sulphureum*

The cytoplasm of hyphae in the control sample was dense, and hyphal and septal walls were thin and uniform (Figure 4-A, B). Marked damages were induced in hyphae by chitosan treatment. In a longitudinal section of a hypha treated with 0.25% chitosan, the cytoplasm was uneven while the hyphal cell walls were irregularly thickened (Figure 4-C), and some parts of the cell wall were considerably thickened so that the cytoplasm was distorted (Figure 4-D). Thickened, malformed and incomplete septae and some inclusion bodies were detected in the cytoplasm of a cross section (Figure 4-E).

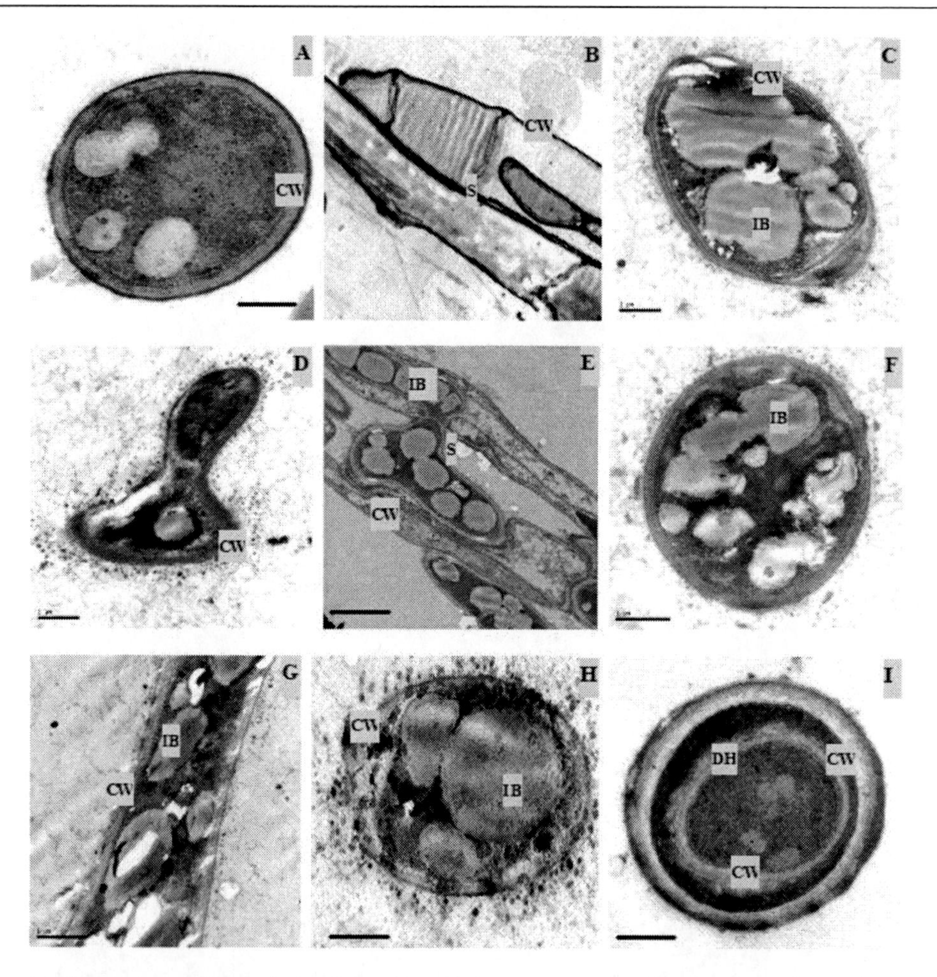

Figure 4. Ultrastructural change in hyphae of *F. sulphureum* treated with 0.25% (C, D and E), 0.5% chitosan (F, G, H and I) and untreated control (A and B) removed from 7-day-old PDA cultures as revealed by TEM. Hyphal walls and septae walls were thin and uniform in untreated controls (A and B). Irregularly thickened cell walls (C), distorted cytoplasm (D), thickened and malformed, and incomplete septae (E) and non-membraneous inclusion bodies (C, D and E) was observed in 0.25% chitosan treated hyphae. Inclusion bodies increased (F and G), and cell wall disruption, leakage of inner cytoplast, and some dark spots (H) appeared in hyphae treated with 0.5% chitosan. A daughter hypha (DH) was formed within the original degenerated hypha (I). Scale bars represent 0.5μm in (A), 5 μm in (B), 2 μm (E), 6 μm (I), and 1 μm in (C), (D), (F), (G) and (H). CW, hyphal cell wall; S, Septum; IB, Inclusion bodies.

Inclusion bodies increased in the cytoplasm of hyphae treated with chitosan at 0.5% (Figure 4-F, G) and parts of the cellular plasmalemma were considerably damaged resulting in cytoplasm overflowing, and some dark spots were observed throughout the cells (Figure 4-H). A daughter hypha (DH) was formed within the original degenerated hypha, and the cytoplasm of the daughter hypha looked very dense and the cellular walls irregularly thickened (Figure 4-I).

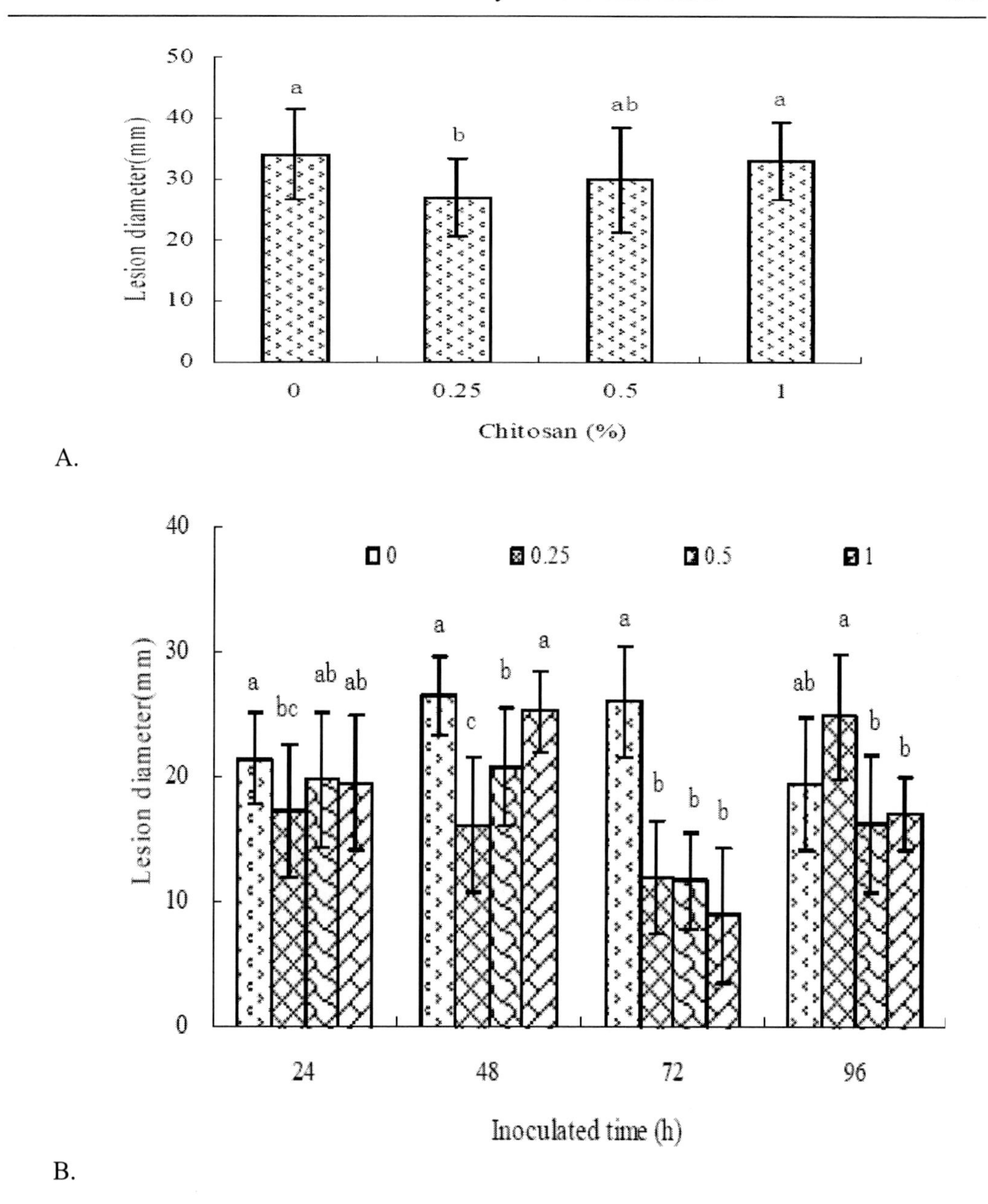

Figure 5. Effect of chitosan on lesion diameters of tuber (A) and slices (B) inoculated with *F. sulphureum*. Different letters show significant levels at 5%. Vertical bars represent the standard error.

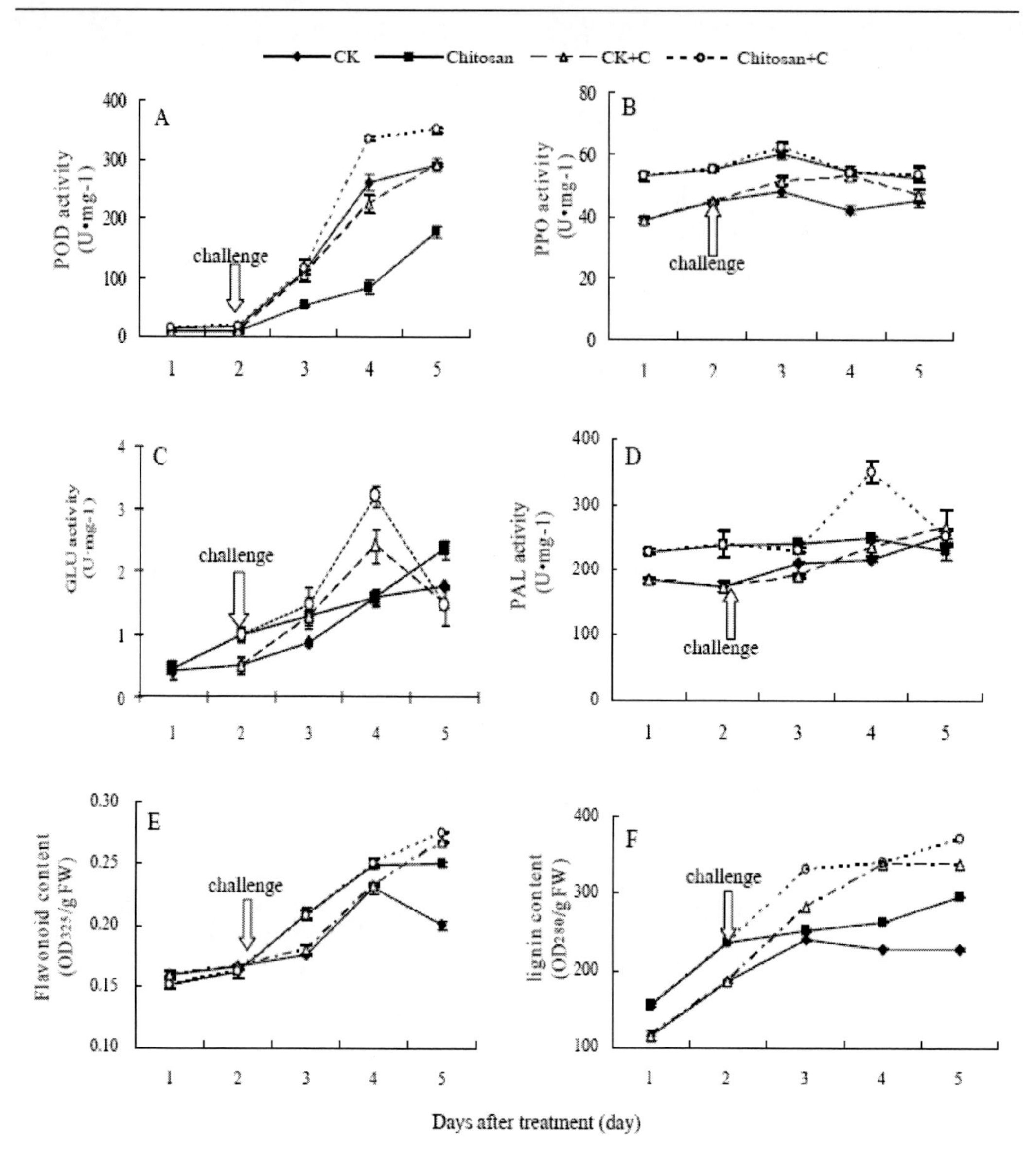

Figure 6. Effect of chitosan treatment and challenge inoculation on POD (A), PPO (B), GLU (C) and PAL (D) activities and content of flavonoids (E) and lignin (F). CK+C: the control + challenge inoculation, Chitosan+C: chitosan treatment + challenge inoculation. Bar shows ±SE.

Effect of Chitosan on Lesion Diameters of Tuber and Slices Inoculated with *F. Sulphureum*

Chitosan treatment at 0.25% significantly reduced the lesion diameter of a tuber inoculated with *F. sulphureum* (Figure 5A). However, no difference was found between high concentration treatment (at 1%) and the control. Although the treatment at 0.5% had less lesion diameter, there was no noticeable difference when compared with the control.

Lesion diameters of slices inoculated with *F. sulphureum* varied with chitosan concentration and incubation time (Figure 5-B). Chitosan at 0.25% showed an effective reduction of lesion diameters at 24, 48, and 72 hours of inoculation time. The treatment at 0.5 and 1% reduced lesion diameters at 48 and 72 hours, and 24 and 72 hours of inoculation time, respectively. The smallest lesion diameters after treatments were observed at 72 hours of inoculation time. No significant difference was found between the treatments and the control when the pathogen was inoculated 96 hours after treatment.

Effect of Chitosan Treatment and Inoculation on POD and PPO Activities

Chitosan at 0.25% enhanced POD and PPO activities in slices (Figure 6). No significant difference of POD was observed during the 2 days of treatment. The activity of POD increased gradually after 2 days of treatment. The maximum POD activity was found 5 days after treatment, with the activity 3.8 times in the slices treated with chitosan than in the control (Figure 6-A). However, the activity of PPO kept mostly stable after treatment (Figure 6-B). Chitosan treatment at 0.25% caused a more progressive and significant increase in POD activity on tissues challenged by *F. sulphureum*. The maximum POD activity was found 5 days after inoculation, with the activity 1.5 times in the slices treated with chitosan than in the control (Figure 6-A). A little increase of PPO activity was observed in the treated slices challenged by *F. sulphureum* (Figure 6-B).

Effect of Chitosan Treatment and Inoculation on GLU and PAL Activities

Chitosan at 0.25% enhanced GLU and PAL activities in the slices (Figure 6). The activity of GLU increased gradually after a one-day treatment (Figure 6-C). However, the activity of PAL kept stable after treatment (Figure 6-D). Chitosan treatment caused a more progressive and significant increase in GLU and PAL activities on tissues challenged by *F. sulphureum*. The maximum GLU and PAL activities were found 4 days after inoculation, with the activities 1.3 and 1.5 times in treated slices than in the control, respectively (Figure 6).

Effect of Chitosan Treatment and Inoculation on Flavonoid Compounds and Lignin Contents in Tissue

Chitosan at 0.25% increased the content of flavonoids and lignin in slices (Figure 6). The content of flavonoids enhanced gradually after 2 days of treatment (Figure 6-E). However, the content of lignin increased continually during the period of experiment (Figure 6-F). The maximum content of flavonoids and lignin was found 5 days after treatment, with the contents 1.2 times and 1.3 times in treated slices than in the control, respectively. Chitosan caused a progressive and significant increase in lignin content in tissues challenged by *F. sulphureum*. The maximum content of lignin was found 5 days after inoculation, with the activity 1.1 times in treated slices than in the control (Figure 6-E). A little increase of flavonoids was observed in treated slices challenged by *F. sulphureum* (Figure 6-F).

DISCUSSION

Chitosan directly inhibited spore germination and mycelial growth of *F. sulphureum*. The study confirmed other findings on the antifungal activity of chitosan on spore germination, germ tube elongation and mycelial growth of many plant pathogens (Ben-Shalom *et al.* 2003; El Ghaouth *et al.* 1992b; Xu *et al.* 2006). The mode of action of chitosan may be related to its ability to interfere with the negatively charged residues of macromolecules exposed on the fungal cell surface resulting in the leakage of intracellular electrolytes and proteinaceous constituents (Leuba and Stossel 1986; Xu *et al.* 2006). Another explanation may be the interaction of the diffused hydrolysis products with microbial DNA, which affects mRNA and protein synthesis (Hadwiger *et al.* 1986; Zakrzewska *et al.* 2005).

SEM observations showed that chitosan induced morphological change. Similar observations in this study were obtained in other studies on fungi such as *F. oxysporum* f. sp. *radicis-lycopersici, Rhizopus* stolonifer and *Sclerotinia* sclerotiorum that showed excessive mycelial branching, abnormal shapes, swelling, and reduction in hyphal size in the presence of chitosan (Benhamou, 1992; El Ghaouth *et al.* 1992a, c; Cheah *et al.* 1997). Chitosan also caused morphological changes such as large vesicles or empty cells devoid of cytoplasm in the mycelium of *Botrytis cinerea* and *F. oxysporum* f. sp. *albedinis* (Ait Barka *et al.* 2004; El Hassni *et al.* 2004). This ultrastructural study confirmed alterations to hyphae of *F. sulphureum* by chitosan. In this study, abnormal distribution of cytoplasm, non-membraneous inclusion bodies assembling in cytoplasma, considerable thickening of the hyphal cell walls, and very frequent septation with malformed septa were observed in chitosan-treated hyphae, and high concentration of chitosan caused fungal hyphae cell to be seriously damaged, including cell membrane disorganization, cell wall disruption and leakage of the inner cytoplast. This result is consistent with the report of Bautista-Banos *et al.* (2004b) who found swellings and hyphal convolutions, surrounded by loosened cell walls. The formation of the daughter hyphae within the original degenerated hypha with very dense cytoplasm and irregularly thickened cellular walls may be due to the defense response of the pathogen hyphae to chitosan.

Different induced effects were found at different times, after treatment with the same concentration of chitosan. The result suggested that defense response induced by chitosan was associated with incubation time, which may be needed for host reception, signal transduction, and related gene repression. It was also found that the high concentration treatments have worse induced effect in this experiment. However, an opposite result was reported in tomato fruits (Liu *et al.* 2007). Liu *et al.* (2007) found that chitosan at 0.5 and 1% could significantly decrease gray mould and blue mould caused by *B. cinerea* and *P. expansum* in tomato fruit ($P < 0.05$) and controlled effects of chitosan on both diseases were enhanced when the concentration of chitosan was increased from 0.5 to 1%. The phenomenon was that high chitosan concentration could stimulate the sensitivity of potato tissue to pathogen.

Induced defense reactions in plants were highly correlated with enzymatic responses. The enzymes participated in the first defensive line and inhibited the incidence of pathogens. It has been reported that chitosan induces the occurrence of defense marks. In this experiment, chitosan treatment and inoculation has increased the activities of POD and PPO in potato tissue. This supports reports of chitosan-enhanced activities of POD and PPO in tomato fruit (Liu *et al.* 2007). POD participates into the cell wall reinforcement and is involved in the final steps of lignin biosynthesis and in the cross-linking of cell-wall protein (Graham 1991). When inoculated with the pathogen, the structural barriers strengthen or limit the activities of the pathogen. PPO is involved in the oxidation of polyphenols into quinines, which can restrict the growth of the plant-pathogen. It has also been found that activities of GLU and PAL in inoculated tuber tissue treated with chitosan have been increased. Several studies have demonstrated that chitosan treatment induces activities of GLU and PAL, and the formulation of lignin (Smith 1996; Greyerbiehl 2002). The increased activity of GLU has been thought to be a common phenomenon in the control of plant disease in resistance systems (Venisse 2002). No significant induction of GLU activity by chitosan can be because of other factors such as plant variety or the method of the experiment. PAL is the first enzyme of the phenylpropanoid pathway and is involved in the biosyntheses of phenolics, phytoalexins, and lignin (Pellegrini *et al.* 1994). Therefore, increased PAL activity can increase disease resistance of plant. In the present study, the levels of flavonoid compounds and lignin in the tuber tissue with chitosan treatment increase, suggesting that flavonoid compounds, as a type of phytoalexins, can directly kill the fungal pathogen (Shadle 2003). In other cases lignin deposition may actually contain the pathogen, by the lignification of the pathogen cell wall surrounding the pathogen or by the deposition of lignin around the penetrating pathogen (Greyerbiehl 2002).

In conclusion, the present study shows that chitosan could directly inhibit the growth of *Fusarium sulphureum* in vitro and potently induce defense reactions in potato tuber. This suggests that chitosan is promising as a natural fungicide, to partially substitute for the utilization of synthetic fungicides in fruits and vegetables. However, further study is needed on the mechanism of chitosan against fungal pathogens, as well as, appropriate application of postharvest disease control.

ACKNOWLEGEDGMENTS

This research was supported by the Gansu Agricultural Bio-Technology Foundation, China (GNSW-2005-08), and the R and D Special Funds for Public Welfare Industry (Agriculture) of Ministry of Agriculture of China (NYHYZX07-6).

REFERENCE

Ait Barka, E., Eullaffroy, P., Clement, C. and Vernet, G. (2004). Chitosan improves development, and protects *Vitis vinifera* L. against *Botrytis cinerea*. *Plant Cell Report,* 22, 608-614.

Assis, J. S., Maldonado, R., Muñoz, T., Escribano, M. I. and Merodio, C. (2001). Effect of high carbon dioxide concentration on PAL activity and phenolic contents in ripening cherimoya fruit. *Postharvest Biology and Technology*, 23, 33-39.

Bautista-Banos, S., DeLucca, A. J. and Wilson, C. L. (2004a). Evaluation of the antifungal activity of natural compounds to reduce postharvest blue mould of apples during storage. *Mexican Journal of Phytopathology*, 22, 362-369.

Bautista-Banos, S., Hernandez-Lauzardo, A. N., Velazquez-del Valle, M. G., Hernandez-Lopez, M., Ait Barka, E. and Bosquez-Molina, E. (2006). Chitosan as a potential natural compound to control pre and postharvest diseases of horticultural commodities. *Crop Protection,* 25, 108-118.

Bautista-Banos, S., Hernandez-Lopez, M. and Bosquez-Molina, E. (2004b). Growth inhibition of selected fungi by chitosan and plant extracts. *Mexican Journal of Phytopathology,* 22, 178-186.

Benhamou, N., Rey, P., Picard, K. and Tirilly, Y. (1999). Ultrastructural and cytochemical aspects of the interaction between the mycoparasite *Pythium oligandrum* and soilborne plant pathogens. *Phytopathology,* 89, 506-517.

Benhamou N. (1992). Ultrastructural and cytochemical aspects of chitosan on *Fusarium oxysporum* f. sp. *radicis-lycopersici,* agent of tomato crown and root rot. *Phytopathology,* 82, 1185-1193.

Ben-Shalom, N., Ardi, R., Pinto, R., Aki, C. and Fallik, E. (2003). Controlling gray mould caused by *Botrytis cinerea* in cucumber plants by means of chitosan. *Crop Protection*, 22, 285-290.

Bi, Y., Tian, S. P., Guo, Y. R., Ge, Y. H. and Qin, G. Z. (2006). Sodium silicate harmful effects on reduces postharvest decay on Hami melons: induced resistance and fungistatic effects. *Plant Disease*, 90, 279-283.

Bradford, M. M. (1976). A rapid and sensitive method for the quantitation of microgram quantities of protein utilizing the principle of protein-dye binding. *Analytical Biochemistry*, 72, 248-254.

Cheah, L. H., Page, B. B. C. and Sheperd, R. (1997). Chitosan coating for inhibition of Sclerotinia rot of carrots. *Zealand Journal of Crop and Horticultural Science*, 25, 89-92.

de Capdeville, G., Wilson, C. L., Beer, S. V. and Aist, J. R. (2002). Alternative disease control agents induce resistance to blue mold in harvested 'Red Delicious' apple fruit. *Phytopathology*, 92, 900-908.

Desjardins, A. E., Christ-Harned, E. A., McCormick, S. P. and Secor, G. A. (1993). Population structure and genetic analysis of field resistance to thiabendazole in *Gibberella pulicaris* from potato tuber. *Phytopathology*, 83, 164-170.

Du, J., Gemma, H. and Iwahori, S. 1997. Effects of chitosan coating on the storage of peach, Japanese pear and kiwifruit. *Journal of the Japanese Society for Horticultural Science*, 66, 15-22.

El Ghaouth, A., Arul, J., Asselin, A. and Benhamou, N. (1992c). Antifungal activity of chitosan on post-harvest pathogens: induction of morphological and cytological alterations in *Rhizopus stolonifer*. *Mycological Research*, 96, 769-779.

El Ghaouth, A., Arul, J. and Asselin, A. (1992b). Potential use of chitosan in postharvest preservation of fruits and vegetables. In *Advances in Chitin and Chitosan* ed. Brines CJ, Sandfors P A and Zikakis J P. pp. 440-452. London and New York: Elsevier Applied Science

El Ghaouth, A., Arul, J., Grenier, J. and Asselin, A. (1992a). Antifungal activity of chitosan on two postharvest pathogens of strawberry fruits. *Phytopathology*, 82, 398-402.

El Ghaouth, A., Ponnampalam, R. and Boulet, M. (1991). Chitosan coating effect on storability and quality of fresh strawberries. *Journal of Food Science*, 56, 1618-1621.

El Ghaouth, A., Smilanick, J. L., Brown, G. E., Wisniewski, M. and Wilson, C. L. (1999). Application of *Candida saitoana* and glycolchitosan for the control of postharvest diseases of apple and citrus fruit under semi-commercial conditions. *Plant Disease*, 84, 243-248.

El Hassni, M., El Hadrami, A., Daayf, F., Barka, E. A. and El Hadrami, I. (2004). Chitosan, antifungal product against *Fusarium oxysporum* f. sp. *albedinis* and elicitor of defence reactions in date palm roots. *Phytopathology Mediterr*, 43, 195-204

Graham, M. Y. and Graham, T. L. (1991). Rapid accumulation of anionic peroxidases and phenolic polymers in soybean cotyledon tissues following treatment with *Phytophthora megasperma* f. sp. *glycinea* wall glucan. *Plant Physiology*, 97, 1445-1455.

Greyerbiehl, J. A. (2002). Induction of resistance in potato tuber tissue to control *Fusarium sambucinum*. Ph D dissertation, Michigan State University, East Lansing, Michigan USA.

Hadwiger, L. A., Kendra, D. F., Fristensky, B. W. and Wagoner, W. (1986). Chitosan both activates genes in plants and inhibits RNA synthesis in fungi. In *Chitin in nature and technology* ed. Muzarelli G, Jeuniaus C, Graham G W. pp. 209-214. New York: Plenum Press.

He, S. Q., Jin, X. L., Wei, Z. Q., Zhang, T. Y., Du, X. and Luo, D. G. (2004). Isolation and identification of pathogens causing dry rot of potato tuber in Dingxi prefecture of Gansu Province. *Journal of Yunnan Agricultural University*, 19, 48-51.

Ippolito, A., El-Ghaouth, A., Wilson, C. L. and Wisniewski, M. (2000). Control of postharvest decay of apple fruit by *Aureobasidium pullulans* and induction of defense responses. *Postharvest Biology and Technology*, 19, 265-272.

Jiang, A. L., Tian, S. P. and Xu, Y. (2002). Effects of controlled atmospheres with high-O_2 or high-CO_2 concentrations on postharvest physiology and storability of "Napoleon" sweet cherry. *Acta Botanica Sinica*, 44, 925-930.

Leuba, J. L. and Stossel, P. (1986). Chitosan and other polyamines: Antifungal activity and interaction with biological membranes. In *Chitin in nature and technology* ed. Muzarelli G, Jeuniaus C, Graham GW. pp. 215-222 . New York: Plenum Press.

Liu, J., Tian, S. P., Meng, X. H. and Xu, Y. (2007). Effects of chitosan on control of postharvest diseases and physiological responses of tomato fruit. *Postharvest Biology and Technology*, 44, 300-306.

Meng, X. H., Li, B. Q., Liu, J. and Tian, S. P. (2008). Physiologic responses and quality attributes of table grape fruit to chitosan preharvest spray and postharvest coating during storage. *Food Chemistry*, 106, 501-508.

Morrison, I. M. (1972). A semi-micro method for the determination of lignin and its use in predicting the digestibility of forage crops. *Science Food Agricultural*, 23, 455-463.

Pellegrini, L., Rohfritsch, O., Fritig, B. and Legrand, M. (1994). Phenylalanine ammonia-lyase in tobacco. *Plant Physiology*, 106, 877-886.

Pirie, A., and Mullins M. G. (1976). Changes in anthocyanin and phenolics content of grapevine leaf and fruit tissues treated with sucrose, nitrate and abscisic acid. *Plant Physiology*, 58, 468-472.

Ray, H., and Hammerschmidt, R. (1998). Responses of potato tuber to infection by *Fusarium sambucinum. Physiological and Molecular Plant Pathology*, 53, 81-92.

Sandford, P. A. and Hutchings, G. P. (1987). Chitosan, a natural, cationic biopolymer: commercial applications. In *Industrial Polysaccharides Genetic Engineering, Structure/Property Relations and Applications* ed. Yapalma M. pp. 363-376. Amsterdam, the Netherlands: Elsevier Science Publishers BV.

Sandford, P. (1989). Chitosan: commercial uses and potential applications. In *Chitin and Chitosan. Sources, Chemistry, Biochemistry, Physical Properties and Applications*. ed. Skjak-Braek G, Anthosen T, Standford P. pp. 51-69. London and New York: Elsevier Applied Science.

Secor, G. A. and Gudmestad, N. C. (1999). Managing fungal diseases of potato. *Canadian Journal of Plant Pathology*, 21, 213-221.

Shadle, G. L., Wesley, S. V., Korth, K. L., Chen, F., Lamb, C. and Dixon, R. A. (2003). Phenylpropanoid compounds and disease resistance in transgenic tobacco with altered expression of L-phenylalanine ammonia-lyase. *Phytochemistry*, 64, 153-161.

Smith, C. J. (1996). Accumulation of phytoalexins: defence mechanism and stimulus response system. *New Phytology*, 132, 1-45.

Sun, X. J., Bi, Y., Li, Y. C., Han, R. F. and Ge, Y. H. (2008). Postharvest chitosan treatment induces resistance in potato against *Fusarium sulphureum*. Agricultural Sciences in China, 7, 615-621.

Tripathi, P. and Dubey, N. K. (2004). Exploitation of natural products as an alternative strategy to control postharvest fungal rotting of fruit and vegetables. *Postharvest Biology and Technology,* 32, 235-245.

Vander, P., Kjell, M. V., Domard, A., El-Gueddari, N. E. and Moerschbacher, B. M. (1998). Comparison of the ability of partially N-acetylated chitosans and oligosaccharides to elicit resistance in wheat leaves. *Plant Physiology*, 118, 1353-1359.

Venisse, J. S., Gullner, G. and Brisset, M. N. (2001). Evidence for the involvement of an oxidative stress in the initiation of infection of pear by *Erwinia amylovora*. *Plant Physiology*, 125, 2164-2172.

Venisse, J. S., Malnoy, M., Faize, M., Paulin, J. P. and Brisset, M. N. (2002). Modulation of defense responses of *Malus* spp. during compatible and incompatible interactions with *Erwinia amylovora*. *Molecular Plant-Microbe Interactions*, 15, 1204-1212.

Xu, J. G., Zhao, X. M., Han, X. W. and Du, Y. G. (2006). Antifungal activity of oligochitosan against *Phytophthora capsici* and other plant pathogenic fungi *in vitro*. *Pesticide Biochemistry and Physiology*, 87, 220-228.

Yao, H. J. and Tian, S. P. (2005). Effects of a biocontrol agent and methyl jasmonate on postharvest diseases of peach fruit and the possible mechanisms involved. *Journal of Applied Microbiology,* 98, 941-950.

Zakrzewska, A., Boorsma, A., Brul, S., Hellingwerf, K. J. and Klis, F. M. (2005). Transcriptional response of *Saccharomyces* to the plasma membrance-perturbing compound chitosan. *Eukaryotic Cell,* 4, 703-715.

Zhang, D. and Quantick, P. C. (1998). Antifungal effects of chitosan coating on fresh strawberries and raspberries during storage. *Journal of Horticultural Science and Biotechnology,* 73, 763-767.

In: Handbook of Chitosan Research and Applications
Editors: Richard G. Mackay and Jennifer M. Tait
ISBN 978-1-61324-455-5

Chapter 13

THE USE OF CHITOSAN IN SURGERY: PREVENTION OF ADHESIONS

Neusa Margarida Paulo[1] and Ângela Maria Moraes[2]

[1]. Department of Veterinary Medicine,
Federal University of Goiás, School of Veterinary,
Goiania, Goiás – GO - Brazil

[2]. Department of Biotechnological Processes,
School of Chemical Engineering,
State University of Campinas – Campinas,
SP - Brazil

ABSTRACT

Adhesions consist on fibrotic union among tissues and organs, as a consequence of trauma and/or tissue ischemia, infection, and foreign body reactions, among other causes. Even though the term adhesion has been classically employed to designate peritoneal fibrosis formation, it also refers to fibrosis reaction due to ophthalmic, orthopedic, cardiovascular (pericardium), neurological (dura mater or dura) and gynecological surgical procedures. The experimental research on adhesion formation has often been associated to the implantation of synthetic biomaterials as polypropylene meshes for the treatment of patients with hernia. It is known that when a macroporous biomaterial as a polypropylene mesh is put into direct contact with visceral peritoneum there is a risk of triggering complications as adhesions over the mesh, what may lead to intestine obstruction processes or fistulae development. Chitosan is a highly biocompatible polysaccharide that has drawn attention lately regarding adhesion prevention. This biopolymer and some of its derivatives can be employed as gels or films for this purpose, and some of the interesting results attained with their use will be discussed in this chapter.

POSTOPERATIVE PERITONEAL ADHESIONS

The study on adhesion formation and prevention must consider the function of parietal and visceral peritoneum, pleura and pericardium. These membranes are of the same embryological origin (from mesoderm) [1] and they are composed of a specialized mesothelial monolayer, constituted by predominantly flattened cells that rest over a basal membrane constituted by connective tissue, lymphatic and blood vessels, resident inflammatory cells and fibroblast-like cells [2-6]. The importance of peritoneal mesothelium for adherence prevention lies on its protective function against physical lesions caused by friction of serous surfaces [4]. Moreover, apart from being a mechanical barrier, the peritoneum is a biologically active structure that regulates responses to surgical lesions and other alterations of the peritoneal cavity. Failure of the peritoneum regulatory function due to injury may result in the formation of adhesions, despite these processes may also occur even if no peritoneum trauma is observed [7]. From the structural point of view, the organization of the peritoneum is similar in humans, rodents and other mammal species, justifying the choice of these animal species as experimental models for the study of adherence formation and prevention [8]. Even though adherence may occur in any region that is covered with a serous membrane, such as the thorax, most studies related to the formation of fiber bundles are concentrated in the peritoneal cavity, which includes the area covered by the mesothelial layer that extends from the diaphragm to most of the pelvis, including not only the anterior and posterior walls of the abdomen, but the abdominal and pelvic viscera [9]. The mesothelial layer is known by its high regenerative capacity and independently from the peritoneal lesion extension, its repair is complete within a period of 8 to 10 days [10], what has been attributed to different hypotheses. Although the exact mechanism of mesothelial repair remains uncertain, there is evidence pointing to a multifactorial mechanism [5]. Due to the phospholipids present in mesothelial cells, which exert a surfactant activity, there is a constant lubrication of the visceral surfaces. Furthermore, the integrity of the mesothelium protects the viscera against adhesions because the mesothelial cells themselves present fibrinolytic activity and secrete cytokines that act in tissue repair [11].

As pointed out by Werner et al. (2009) [12], abdominal adhesions are known since the year 440 BC, although studies about them have first appeared after 1880. Adhesions consist of abnormal fibrous bundles formed among tissues and organs (Figures 1 and 2) due to a general inflammatory response caused, for instance, by surgical approaches, causing serious problems for patients. This problem affects humans and other animal species, such as horses. Considering this species, adhesions are the most frequent cause of colic episodes [12-14]. Regardless of the type of animal studied, the complex mechanism of adhesion formation is triggered by cellular and biochemical responses that occur in injured organs usually due to laparotomy incisions or other traumas [15], but factors such as ischemia, hemorrhage, presence of foreign bodies and peritonitis can also start the adhesion process [11, 16]. Nevertheless, peritoneal trauma results in loss of plasminogen activator factor activity, culminating in impaired fibrinolysis [17].

For the occurrence of adhesions, it is necessary that the mesothelial layer of visceral peritoneum and parietal peritoneum are affected [13, 18] and that there is apposition between them [19]. Surgical procedures are related to the reduction of fibrinolytic activity of the peritoneum [20, 21], which allows the presence of a permanent fibrinous scaffold that will be invaded by fibroblasts and endothelial cells. Following, the deposition of collagen takes place, resulting in the formation of fibrosis [21]. In the beginning, adhesions are fibrinous, but within 10 days to two weeks after the injury, organization of these adhesions occurr, with the formation of a fibrovascular tissue that may contain nerve fibers [21]. Additionally, peritoneal adhesions may contain calcified nodules [19].

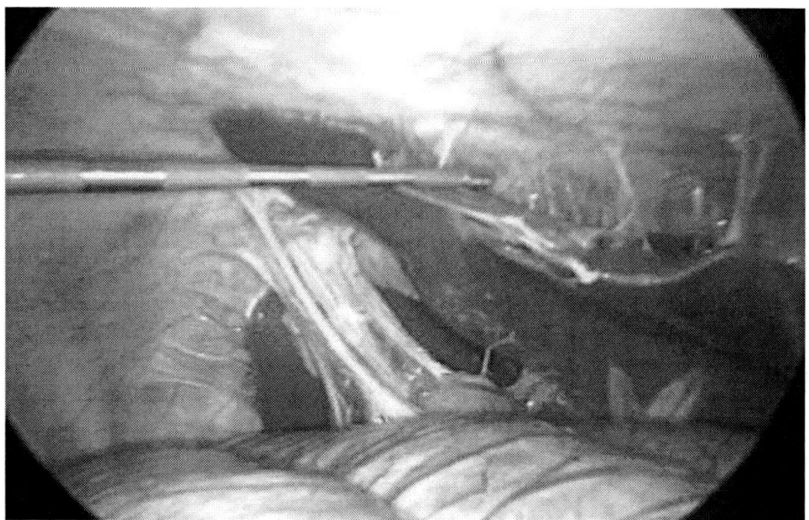

Figure 1. Videolaparoscopic image of adhesions between the polypropylene surface and the pig's liver. Photograph courtesy of A. F. Martins.

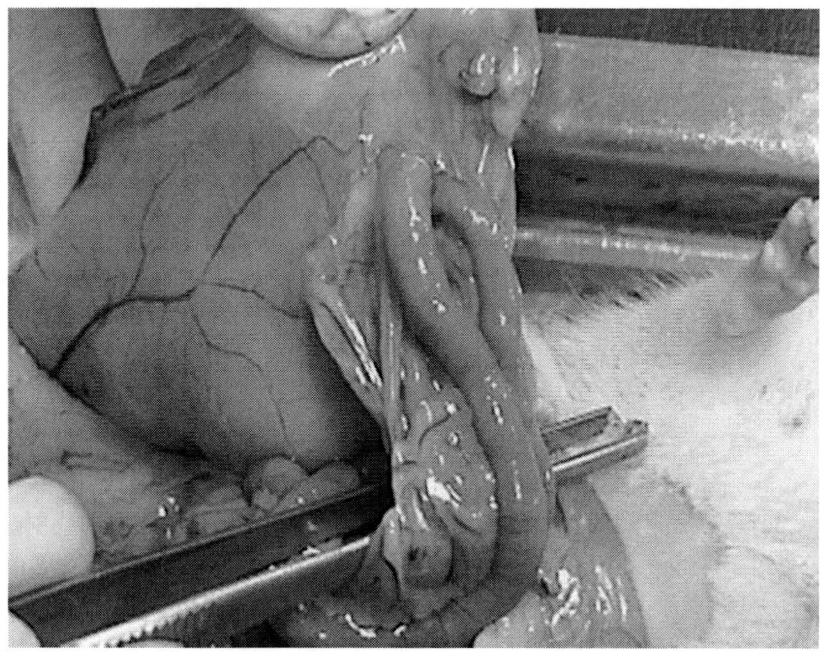

Figure 2. Fibrous adhesion between the polypropylene surface and a rat intestinal segment. Photograph of N. M. Paulo.

Since the serosa lining the chest cavity has the same origin of the one coating the peritoneal cavity, their behavior on inflammatory processes is identical. Therefore, the formation of postoperative adhesions is particularly important in this region. Thoracic adhesions involving the epicardium and underlying structures such as the pericardium, mediastinal fat, pleura and sternum, as well as peritoneal adhesions, are formed early after surgery, being both related to reduced fibrinolytic activity [22].

Obstruction of the bowel is one of the most frequent consequences of adhesions. In the case of equines, it can lead to fatal colic [14], and in humans, it may lead to abdominal pain and reoperation procedures for adhesiolysis. Another type of adhesion of special interest is that formed between the bowel and prosthetic material implanted in the abdominal wall (as shown in Figure 2), which can lead to formation of enterocutaneous fistulas [23]. Besides, thoracic adhesions turn reinterventions dangerous due to risk of injury to the structures attached to the adhesions, which can result in fatal bleeding, apart from the possibility of altering the thoracic architecture.

Peritoneal Adhesions Caused by the Presence of Biomaterials

One of the more interesting chapters of surgical practice focuses the study of abdominal wall defects. Since such defects are relevant to many animal species, including humans, similarity between species is often considered, both regarding the formation of the adhesions themselves and their complications, and in relation to adhesion prevention approaches.

Ideally, a biomaterial used to correct abdominal wall defects should present mechanical resistance similar to that of the tissue to be fixed, not propitiating infection or causing foreign body and inflammatory reactions, what could result in its removal from the body after implantation. In addition, the biomaterial should be rapidly incorporated by the body, without stimulating excessive fibrosis or causing adhesion to organs as liver, spleen or intestines. Non carcinogenicity is another important desired characteristic, as economical attractiveness and easy of sterilization and handling.

Polypropylene meshes attend many of these requirements and numerous studies have been performed focusing the evaluation of the effect of its thickness, pore diameter, mechanical resistance, rigidity, surface texture and electrical charge on patient recovery time and life quality. Despite these efforts, many of the biomaterials available in the market for correction of abdominal defects are not totally efficient regarding adhesion avoidance. Therefore, strategies that contribute to minimize or eliminate this problem are quite valuable.

To elucidate the many questions regarding the important role that adhesions play after abdominal wall reconstruction, animal experiments have been performed to assess different types of biomaterials that meet the demands not only of abdominal surgery, but also that of other areas covered by serosa. As it has been mentioned earlier, tissue ischemia, as well as the presence of foreign bodies may lead to the formation of peritoneal adhesions. Thus, it should be considered that meshes implanted in the abdominal wall are essentially foreign bodies, whose association with surgical trauma can lead to tissue ischemia and poor mesothelization. Such factors, associated with direct contact of the biomaterial with the visceral surface, have been considered important in the genesis of peritoneal adhesions [24]. This behavior seems to be connected to the structure and porosity of the material, which will influence the organization of neoperitoneum, which will be linear if the material is laminar, what is not observed with the use of polypropylene meshes, which are reticular or even macroporous [25]. There is a strong relationship between the roughness of the biomaterial surface and the intensification of mechanical irritation, what often leads to the formation of fibrous tissue.

Therefore, when using biomaterials for the reconstruction of the abdominal wall, the interface between this material and the peritoneum should be strongly considered. It is a key variable for the prevention of adhesions and fistulization, since the implant will be in direct contact with the viscera [26]. Suture material itself used for the fixation of the biomaterial to adjacent tissues can also contribute to the formation of adhesions [15, 27]. In an attempt to minimize the occurrence of adhesions associated to biomaterial implantation, strategies as combining the strength of polypropylene meshes with materials which do not promote cell adherence are quite useful, in addition to approaches based on the preservation of the parietal peritoneum, which may reduce adhesion incidence, severity and extent. Another alternative is the use of barrier systems, which are presented in the form of polymer solutions, gels or solid membranes. These systems, when deposited between peritoneal surfaces, separate them, preventing the deposition of fibrin-rich clots and fibroblast infiltration in the lesion site [28].

The use of barriers has been considered as the most effective method for the prevention of thoracic and abdominal adhesions. Its action is normally restricted to the injured site, not being observed in other areas. A barrier system must meet some important favorable characteristics, but unfortunately there has not yet reported an ideal barrier constitution [29].

CHITOSAN AND ADHERENCE PREVENTION

One of the highest human expectations is the possibility to restore, improve or substitute lesioned or lost tissues and organs. For that matter, biomaterials development require a multidisciplinary approach that involves knowledge on the biological phenomena associated to the implantation of clinical devices in live tissue, tissue engineering and materials science, among other areas. Moreover, with the increasing knowledge on the physiological mechanisms of tissue cicatrization and regeneration, there has also been a rise in the interest on bioactive materials capable of modulating tissue responses in addition to substituting a lost part of the organism.

Implanted biomaterials can be degradable or not, and in both cases it is necessary that their interaction with the tissues is not associated to an intense inflammatory reaction. It is well known that frequently the implantation of a biomaterial in the organism triggers a foreign body response initiated by an acute inflammatory reaction, culminating, in some cases, in chronic effects and in the development of a fibrous capsule around the material [28]. For this reason, biocompatibility criteria considering the simultaneous events which occur in the receiving tissue and in the implanted material should be taken into account. As a result, a given material may be biocompatible in one tissue but not in another one, and its biocompatibility may not be directly related to cytotoxicity [30]. Tissue response may be also dependent on the type of compounds released from the biomaterial during its permanence in the body or resulting from its degradation [30, 31], reinforcing the need of improvement in the materials science area.

Polysaccharides are among the biologically-originated molecules most used in studies focusing adhesion prevention and other clinical applications. Some of the general structural characteristics of these molecules that make them attractive are their high molecular weight, their tendency to form intra and interchain hydrogen bonds due to the high number of –OH groups, their structural stability dependency on temperature and presence of ions, the fact that they are chemically modifiable, and their stereoregularity, with semi-rigid character and helical conformation in solution.

Chitosan and its derivatives are polysaccharides that structurally mimic, to a certain extent, extracellular matrix glycoproteins. These polymers have been increasingly used in regenerative medicine due to their quite remarkable properties, as low or inexisting toxicity and possibility of degradation by lysozyme, an enzyme normally present in body tissues. In fact, the chitosan degradation products are quite safe, entering in the normal metabolic routes of the organism and being excreted as carbon dioxide [32]. Chitosan is well known for its antimicrobial activity, presenting low immunogenicity and good compatibility towards mixing with many different compounds, such as growth factors and cytokines. Its solubility is close to that of cellulose and it is possible to have chitosan presented in distinct forms, as powder, gel, cream, emulsion, film or fibers. These characteristics and others which are discussed ahead are the basis for the adequacy of the use of chitosan as a biomaterial component.

It is interesting to point that the term Chitosan is frequently used to identify a mixture of copolymers with distinct ratios of β(1→4)-glucosamine and N-acetyl-Dglucosamine residues soluble in dilute organic acids, instead of a single defined compound. It is of paramount importance, therefore that the relevant chemical characteristics of the chitosan employed are described along with the experimental results verified *in vitro* or *in vivo*. Moreover, many studies report data obtained with chemically modified forms of chitosan or with chitosan presented in different forms (as gels, creams, films) or yet with chitosan associated to a multitude of components, turning difficult the interpretation and correlation of experimental results. Finally, the animal model chosen to evaluate distinct experimental adhesion formation aspects may also contribute to the variability of the results.

Chitosan has been attracting attention lately because of its capacity to promote dermal regeneration and accelerate wound healing due to its stimulatory effect over macrophages and chemoattraction of neutrophils [33], inhibiting fibroplasia and facilitating the resolution of inflammatory processes [34]. Also, chitosan stimulates cell proliferation and hystoarchitectural organization [32]. showing hemostatic [35], and immunstimuling activity [36]. Another important chitosan characteristic is its hemocompatibility [37]. In the presence of chitosan, the formation of collagen with well defined organization can be noticed [32]. In view of these properties, chitosan can be considered as a functional material [38], what is quite important concerning tissue reconstruction.

Based on these characteristics, chitosan has shown potential use on numerous biomedical applications, especially when implantable materials are aimed. Particularly, the effects of chitosan and its derivatives have been considered in many experimental and clinical studies focusing adhesions resulting from peritoneal lesions in various animal species, including humans.

USE OF CHITOSAN IN THE FORM OF GELS AND FILMS AS BARRIERS

Chitosan can be employed in the form of films or gels as a strategy to prevent adhesion formation based on the principle that it acts as a mechanical barrier, separating the lesioned peritoneal surfaces. Furthermore, since this polysaccharide facilitates re-epithelization and improves skin wound healing, it could also affect post-surgical adhesion formation caused by the incisions and the anastomoses themselves without affecting the strength of the repaired tissue [34]. Also, as an additional advantage, due to the fact that chitosan is safely degraded by lyzozyme, there is no need of additional surgical procedures to remove it after implantation [39].

Numerous studies concerning the influence of chitosan and its derivatives on the formation of peritoneal adhesions have been performed, mainly in Japan and China. Some variations can be observed in the results, possibly attributable to differences in the physico-chemical characteristics of the chitosan types employed, since its composition, chain size, net charge and viscosity, among other properties, can differ depending on the chitin source and on the deacetylation process employed. In fact, the deacetylation degree of a chitosan molecule is a quite important characterization parameter. The higher the deacetylation degree, the higher the number of positive charges per molecule, mostly at pH conditions below 6.5. Since many of the chitosan biological effects are credited to its charge density, it is natural to expect impacts of this particular property on the performance of implantable devices produced with this compound. For instance, chitosan can be degraded *in vivo* by lysozyme, however, its degradation rate is low for highly deacetylated forms of chitosan [40, 41]. *In vitro* cytotoxicity also may vary with deacetylation degree, as pointed out for macrophages, but apparently this is not the case for nonphagocytic mesothelial cells [27].

Several adhesion formation models have been evaluated in different animal species. Some of the methods commonly used are: abrasion, puncture or incision of the ceccum or of the uterine horn, together with traumatic lesion of the parietal peritoneum, aortic anastomosis, talc deposition in the peritonial cavity and ligation of blood vessels. In addition, there can be an association of some of these methods with the suture of a fragment of polypropylene mesh or other related material, on the abdominal wall, with or without right lateral wall muscle ressection.

Despite the variation in results observed with the different animal species employed, and also due to chitosan distinct forms and chemical modifications, the results of its use in the peritoneal cavity are, overall, favorable, including even tests in women with post-operative adhesions. Chitosan is more frequently used in the form of solutions or gels. The N,O-carboxymethyl chitosan gel, for instance, presents antiadhesive activity attributed to the dilution of fibrin present in the inflammatory exudate, added to its barrier function, isolating the visceral surfaces and also the inflammatory cells and the factors released by them. In fact, the presence of a macromolecular compound in peritoneal or thoracic cavities might also inhibit the removal of the plasminogen activator factor derived from lesioned peritoneal areas [17, 42].

Other possible mechanisms of action of the N,O-carboxymethyl chitosan gel refer to blocking the complement activation process by the hydrophilic carboxymethyl groups [43] or preventing hydrophilic interactions between the extracellular matrix fibronectin molecules.

The advantage of the use of N,O-carboxymethyl chitosan over regular (not chemically modified) chitosan is that the first form is soluble in water, while the second requires, as already mentioned, the addition of organic acids, as acetic acid, to be efficiently solubilized. These acids may, by themselves, result in cytotoxic effects when instillated in lesioned abdominal cavities, turning more difficult the analysis of *in vivo* results.

Gels produced with regular chitosan seem to show a satisfactory effect on traumatic or ischemic peritoneal adhesion, but not on foreign body-induced peritoneal adhesion [44]. This behavior is attributed to the degradation time of the chitosan gel *in vivo*, estimated to be around two weeks. The foreign body (talc was used in this particular study) persisted after this period, inducing inflammatory reaction with formation of a granuloma and persistent massive fibroplasia, results which the chitosan gel was not able to adequately circumvent.

An UV-cross linked chitosan gel, on the other hand, when in contact with mesothelial cells and peritoneal macrophages to *in vitro* has not shown attractive interactions with the cells nor cell growth stimulation effects. However, when applied in the peritoneal cavity of rabbits, the same UV-cross-linked chitosan caused a granulomatous reaction with adhesion formation, which persisted up to 4 weeks after exposure [27]. The authors reported that unmodified chitosan also caused adhesions, stressing that *in vitro* cytotoxicity assays may not be sufficient to safely show if a given material would be or not biocompatible *in vivo*.

The few works that deal with the implantation of films produced with chitosan or its derivatives in the abdominal cavity also show somehow controversial results regarding adhesion prevention. In general, it seems that chitosan in the form of film may be more satisfactory than as a gel. As a solid film, migration of chitosan to regions away from the lesioned area would less likely occur, what could not be stated for the gel, which flows easily. Also, as a film, chitosan could more effectively act as a mechanical barrier before being completely degraded. There is still the possibility that the monomers resulting from chitosan degradation might also have antiadhesion effects, however, in this presentation form, chitosan does not prevent already formed fibrinous adherences. The slow degradation rate of chitosan *in vivo*, mostly when it is the form of films, can consist on a limitation of the intraabdominal use of this biomaterial, since the residual undegraded material can trigger the foreign body reaction, resulting in the formation of a fibrous capsule.

It is interesting to mention that similarly to what is observed for the poor description of chitosan composition or source, no details are frequently given on the tested biomaterials preparation or characterization methods, what turns difficult the comparison of results attained in *in vivo* experiments by different groups. The *in vivo* results attained by our research group with chitosan films, for instance, can be considered as quite promising. A straight forward method was employed by our group to prepare the films, which basically consisted of obtaining a 2% (w/w) 85% deacetylated chitosan solution by dissolution in a 1% (v/v) acetic acid aqueous solution at room temperature, followed by filtration, deaeration, drying at 50°C, neutralization with sodium hydroxide solution, washing with water and sterilization with ethylene oxide. The prepared films, when dry had tensile strength around 44 MPa, strain at break of 14% and thickness around 152 μm. When immersed in water, the films presented some change in behavior, with, in average, tensile strength of 6 MPa, strain at break of 86% and thickness of 202 μm. These films were capable to absorb up to 1.6 times their weight in blood serum, losing 5.6% of their weight after 8 days when incubated in the same liquid. The experimental model employed in the mentioned work showed that these films, when placed on top of conventional polypropylene meshes, allowed the peritonization of the meshes while successfully minimizing the formation of adhesions in rats [45]. Visceral adhesions were detected in the group implanted with the polypropylene meshes only, but not in the group of animals that received also the chitosan films. In addition, for the last group, no aggravation of inflammatory reaction associated to the peritoneal lesions was statistically demonstrated. A limitation of the use of this particular system was that the fixation of the chitosan film had to be performed independently of that of the polypropylene mesh, increasing the complexicity and the duration of the surgery. This issue can be easily bypassed by developing coats that bind tightly to the meshes, so only the implantation of the mesh is effectively required. This approach is currently under evaluation by our group.

Not only chitosan but also chitin can be used as peritoneal adhesion barrier [46]. A commercially available formulation, denominated Suprofim®, has presented adequate adhesion prevention activity without showing hepathotoxicity or nefrotoxicity.

Final Remarks

Despite the fact that chitosan and many of its derivatives seem to be effective in peritoneal adhesion prevention, there is still some controversy about the mechanisms involved in this result. And since discrepancy in results might be attributed to variations in the composition and physical form of the material used as well as on the procedures employed to prepare it, establishing reproducible and easily up-scalable experimental methodologies to obtain uniform devices produced from chitosan and related compounds is advisable, regardless if it is the form of a solution, a gel, a film, a sponge or even as a coating.

In this sense, the processing conditions which we recommend to be carefully controlled to improve *in vivo* data reproducibility are: polymer composition, type and concentration; solvent composition; type and concentration of cross-linking agent; mixing and drying conditions (particularly if solid devices are aimed), incorporation of other potentially active agents and sterilization method, among others. Variations in these processing conditions can affect properties as morphology, swelling degree, stability in body fluids, mechanical characteristics, release rate of incorporated agents, and finally biocompatibility, biodegradability and *in vivo* behavior. Developments in the biomaterials area are still an urgent need, and the clinical use of biomaterials which do not attend to the particular requirements related to adhesion prevention may deprive patients from better life quality and higher longevity.

References

[1] Van Der Wal, JBC; Jeekel, J. Biology of the peritoneum in normal homeostasis and after surgical trauma. *The Association of Coloproctology of Great Britain and Ireland,* 2007 9, 9-13.

[2] Foley-Comer, AJ; Herrick, SE; Al-Mishlab, T; Prêle, CM; Laurent, GJ; Mutsaers, SE. Evidence for incorporation of free-floating mesothelial cells as a mechanism of serosal healing. *Journal of Cell Science,* 2002 115, 1383-1389.

[3] Michailova, K; Wassilev, W; Wedel, T. Scanning and transmission electron microscopic study of visceral and parietal peritoneal regions in the rat. *Annals of Anatomy,* 1999 181, 253-260.

[4] Mutsaers, SE. The mesothelial cell. *The International Journal of Biochemistry and Cell Biology,* 2004 36, 9-16.

[5] Mutsaers, SE; Whitaker, D; Papadimitriou, JM. Mesothelial regeneration is not dependent on subserosal cells. *Journal of Pathology,* 2000 190, 86-92.

[6] Herrick, SE; Mutsaers, SE. Mesothelial progenitor cells and their potential in tissue engineering. *The international Journal of Biochemistry and Cell Biology,* 2004 36, 621-642.

[7] Stadlmann, S; Pollheimer, J; Renner, K; Zeimet, AG; Offner, FA; Amberger, A. Response of human peritoneal mesothelial cells to inflammatory injury is regulated by interleukin-1β and tumor necrosis factor-α. *Wound repair and regeneration,* 2005 14, 187-194.

[8] Trbojević, J; Nešić, D; Laušević, Ž; Obradović, M; Brajušković, G; Stojimirović, B. Histological characteristics of healthy animal peritoneum. *Acta Veterinaria,* 2006 56, 405-412.

[9] Ergul, E; Korukluoglu, B. Peritoneal adhesions: Facing the enemy. *International Journal of Surgery,* 2008 6, 253-260.

[10] Menzies, D. Peritoneal adhesions: incidence, cause, and prevention. *Surgery Annual,* 1992 24, 27-45.

[11] Ward, BC; Panitch, A. Abdominal adhesions: current and novel therapies. *Journal of Surgical Research,* 2009, 1-21. (Proof).

[12] Werner, M; Galecio, JS; Bustamante, H. Adherencias abdominales postquirúrgicas em equinos: patofisiología, prevención y tratamiento. *Archivos de Medicina Veterinaria,* 2009 41, 1-15.

[13] Kiliç, N. The effect of *Aloe vera* gel on experimentally induced peritoneal adhesion in rats. *Revue de Médicine Véténinaire,* 2005 156, 409-413.

[14] Proundman, C. Peritoneal adhesions in the horse: prevalence and prevention. *Adhesions News and Views,* 2004 5, 10-11.

[15] Whitfield, RR; Stills JR, HF; Huls, HR; Crouch, JM; Hurd, WW. Effects of peritoneal closure and suture material on adhesion formation in a rabbit model. *American Journal of Obstetrics and Gynecology,* 2007 197, 644. E1-644.e5.

[16] Torre, M; Favre, A; Prato, AP; Brizzolara, A; Martucciello; G. Histologic study of peritoneal adhesions in children and in a rat model. *Pediatric Surgery International,* 2002 18, 673-676.

[17] Kennedy, R; Costain, DJ; McAlister, VC; Lee, TDG. Prevention of experimental postoperative peritoneal adhesions by N,O-carboxymethyl chitosan. *Surgery,* 1996 120, 866-870.

[18] Lopes, JB; Dallan, LAO; Campana-Filho, SP; Lisboa, LAF; Guitierrez, PS; Moreira, LFP; Oliveira, SA; Stolf, NAG. Keratinocyte growth factor: a new mesothelial targeted therapy to reduce postoperative pericardial adhesions. *European Journal of Cardio-Thoracic Surgery,* 2009 35, 313-318.

[19] Di Zerega, GS; Campeau, JD. Peritoneal repair and post-surgical adhesion formation. *Human reproduction update,* 2001 7, 547-555.

[20] Holmdahl, L; Eriksson, E; Eriksson, BI; Risberg, B. Depression of peritoneal fibrinolysis during operation is a local response to trauma. *Surgery,* 1998 123, 539-544.

[21] Sullaiman, H; Gabella, G; Davis, C; Mutsaers, SE; Boulos, P; Laurent, GJ; Herrick, SE. Presence and distribution of sensory nerve fibers in human peritoneal adhesions. *Annals of Surgery,* 2001 234, 256-261.

[22] Iliopoulos, J; Cornwall, GB; Evans, Ron; Manganas, C; Thomas, KA; Newman, DC; Walsh, WR. Evaluation of a bioabsorable polylactide film in a large animal model for the reduction of retrosternal adhesions. *Journal of Surgical Research,* 2004 118, 144-153.

[23] Felemovicius, I; Bonsack, ME; Hagerman, G; Delaney, JP. Prevention of adhesions to polypropylene mesh. *Journal of American College of Surgeons,* 2004 198, 543-548.

[24] Jin, J; Voskerician, G; Hunter, SA; McGee, MF; Cavazzola, LT; Schomisch, S; Harth, K; Rosen, MJ. Human peritoneal membrane controls adhesion formation and host tissue response following intra-abdominal placement in a porcine model. *Journal of Surgical Research,* 2009 156, 297-304.

[25] Bellón, JM; Jurado, F; García-Honduvilla, N; López, R; Martín, ACS; Buján, J. The structure of a biomaterial rather than its chemical composition modulates the repair process at the peritoneal level. *The American Journal of Surgery,* 2002 184, 154-159.

[26] Bellón, JM; Rodríguez, M; García-Honduvilla, N; Pascual, G; GIL, VG; Bújan, J. Peritoneal effects of prosthetic meshes used to repair abdominal wall defects: monitoring adhesions by sequential laparoscopy. *Journal of Laparoendoscopic and Advanced Surgical Techniques,* 2007 17, 160-167.

[27] Yeo, Y; Burdick, JA; Highley, CB; Marine, R; Langer, R; Kohane, DS; Peritoneal application of chitosan and UV-cross-linkable chitosan. *Journal of Biomedical Materials Research Part A*, 2006 78A, 668-675.

[28] Babensee, JE; Anderson, JM; McIntire, LV; Mikos, AG. Host response to tissue engineered devices. *Advanced Drug Delivery Reviews,* 1998 33, 111-139.

[29] Attard, J-AP; MacLean, AR. Adhesive small bowel obstruction: epidemiology, biology and prevention. *Canadian Journal of Surgery,* 2007 50, 291-300.

[30] Kohane, DS; Langer, R. Polymeric biomaterial in tissue engineering. *Pediatric Research,* 2008 63, 487-491.

[31] Williams, DF. Biomaterials and tissue engineering in reconstructive surgery. *Sadhana - Academy Proceedings in Engineering Sciences,* 2003 28, 563-574.

[32] Muzzarellli, R; Baudassarre, V; Conti, F; Ferrara, P; Biagini, G. Biological activity of chitosan: ultrastructural study. *Biomaterials,* 1998 9, 247-252.

[33] Shi, C; Zhu, Y; Ran, X; Wang, M; Su, Y; Cheng, T. Therapeutic potential of chitosan and its derivatives in regenerative medicine. *Journal of Surgical Research,* 2006 133, 185-192.

[34] Koide, SS. Chitin-chitosan: properties, benefits and risks. *Nutrition Research,* 1998 18, 1091-1101.

[35] Athanasiadis, T; Beule, AG; Robinson, BH; Robinson, SR; SHI, Z; Wormald P-J. Effects of a novel chitosan gel on mucosal wound healing following endoscopic sinus surgery in a sheep model of chronic rhinosinusits. *Laringoscope*, 2008 118, 1088-1094.

[36] Song, S; Zhou, F; Nordquist, RE; Carubelli, R; Liu, H; Chen, WR. Glycated chitosan as a new non-toxic immunological stimulant. *Immunopharmacology and Immunotoxicology,* 2009 31, 202-208.

[37] Khor, E; Lim, LY. Implantable applications of chitin and chitosan. *Biomaterials,* 2003 24, 2339-2349.

[38] Majeti, NV, Kumar, R. A review of chitin and chitosan applications. *Reactive and Functional Polymers*, 2000 46, 1–27.

[39] Cárdenas, G; Anaya, P; Von Plessing, C. Rojas, C; Sepúlveda, J. Chitosan composite films. Biomedical applications. *Journal of Materials Science,* 2008 19, 2397-2405.

[40] Senel, S; Mcclure, SJ. Potential applications of chitosan in veterinary medicine. *Advanced Drug Delivery Reviews* 2004 56,1467–1480.

[41] Suh, JKF; Matthew, HWT. Application of chitosan-based polysaccharide biomaterials in cartilage tissue engineering: A review. *Biomaterials* 2000 21, 2589–2598.

[42] Cetin, M; Ak, D; Duran, B; Cetin, A; Guvenal, T; Yanar, O. Use of methylene blue and N,O-carboxymethylchitosan to prevent postoperative adhesions in a rat uterine horn model. *Fertility and Sterility,* 2003 80, 698-701.

[43] Krause, RJ; Zazanis, G; Malaesta, P; Solina, A. Prevention of pericardial adhesions with N-O carboxymethylchitosan in the rabbit model. *Journal of Investigative Surgery,* 2001 14, 93-97.

[44] Zhou, J; Liwski, RS; Elso, C; Lee, TDG. Reduction in postsurgical adhesion formation after cardiac surgery in a rabbit model using N,O-carboxymethy chitosan to block cell adherence. *The Journal of Thoracic and Cardiovscular Surgery,* 2008 135, 777-783.

[45] Paulo, NM; Silva, MSB; Moraes, AM; Rodrigues, AP; Menezes, LB; Miguel, MP; Lima, FG; Faria, AM; Lima, LML. Use of chitosan membrane associated with polypropylene mesh to prevent peritoneal adhesion in rats. *Journal of Biomedical Materials Research Part B: Applied Biomaterials,* 2009 91B, 221-227.

[46] Sahin, M; Cakir, M; Avsar, FM; Tekin, A; Kucukkartallar, T; Akoz, M. The effects of anti-adhesion materials in preventing postoperative adhesion in abdominal cavity (anti-adhesion materials for postoperative adhesions). *Inflammation*, 2007 30, 244-249.

In: Handbook of Chitosan Research and Applications
Editors: Richard G. Mackay and Jennifer M. Tait
ISBN 978-1-61324-455-5

Chapter 14

SOLID-PHASE PRODUCTION AND MODIFICATION OF CHITOSAN UNDER CONDITIONS OF SHEAR DEFORMATION

S. Z. Rogovina
Semenov Institute of Chemical Physics
of Russian Academy of Sciences,
Moscow, Russia

ABSTRACT

The development of new methods for production of chitosan and its derivatives is among the most important problems of the modern polysaccharide chemistry. The solid-phase processes carried out under conditions of joint action of high pressure and shear deformation is the promising methods of chitin and chitosan modification. The advantage of this method is a significant reduction of the process time, reagent consumption, and wastewater amounts that leads to the improvement of the economical and ecological parameters of the process. The reactions were carried out in a two-screw extruder, where the joint action of high pressure and shear deformation on material is realized. The properties and fractional composition of chitosan prepared by solid-phase deacetylation of chitin, as well as chitosan interaction with stearic, oxalic, malonic, and succinic acids and with phthalic, succinic and maleic anhydrides under conditions of shear deformation were investigated. Also, as a result of chitin interaction with monochloroacetic acid in the presence of sodium hydroxide under these conditions, carboxymethyl ether of chitosan was obtained. Based on the data of elemental analysis, IR spectroscopy, viscometry, potentiometry, and other experimental techniques, the formation of the corresponding chitosan derivatives is corroborated and the possible mechanisms of the reactions of modification are discussed.

INTRODUCTION

The interest to chitosan – polysaccharide formed from of glucosamine entities is explained by the unique properties of this polymer (nature origin, biocompatibility, absence of toxicity, etc.) and by its high capacity of chemical modification [1]. Traditionally, the chitosan modification is realized under heterogeneous conditions with preactivation of the polymer by grinding or swelling in organic liquids or in water-acid solutions [2, 3]. Solid-state modification of polysaccharides including chitin and chitosan under conditions of joint action of high pressure and shear deformation is a promising approach to the creation of waste-free procedures for preparation of their versatile derivatives. Under these conditions the reacting material is subjected to plastic flow with an unlimited strain. The physical principle underlying this method consists in the fact that the energy accumulated in the material after the application of pressure is consumed in forming new surface under the influence of shear deformation. In this case the processes are characteristic of the following features: 1) upon solid-phase reactions under these conditions, a very effective mixing of the reagents on the molecular level necessary for the reaction completion takes place, in spite of the fact that initially both reactants are the solid powders; 2) under such conditions, various chemical reactions can occur due to interaction of polymer functional groups in the zone of joint plastic flow of deformed system [4].

DEACETYLATION OF CHITIN AND PROPRTIES OF CHITOSAN OBTAINED BY SOLID-PHASE METHOD

The deacetylation of chitin results in the most famous chitin derivative – chitosan, which is usually prepared by treating chitin with concentrated (40-50%) aqueous NaOH solutions at elevated temperatures for 2-3 h. As chitin and chitosan are not soluble in alkali medium, the reaction proceeds under heterogeneous conditions. A high excess of alkali (not less then 10 mol NaOH per 1 mol chitin) is removed by water. The conversion of chitin to chitosan is characterized by the degree of deacetylation (DD), i.e. by degree of substitution of amino groups [5-7].

$$\xrightarrow[-\ CH_3COONa]{+\ NaOH}$$

The process of chitin deacetylation was carried out in a two-screw extruder, in which the joint action of high pressure and shear deformation on material is realized [8, 9]. A mixture of chitin and solid sodium hydroxide taken in different ratios was fed in the extruder. On the output of extruder the mixture of chitosan with sodium acetate and excess of NaOH was obtained. The mixture was washed by water to neutral reaction and dried at room temperature. The use of an extruder compared with other kinds of equipment, e.g., blenders has some advantages, such as the continuous action and a higher capacity.

The comparative characteristics of chitosan samples obtained by conventional suspension technique (sample 1) and by solid-phase chitin deacetylation (samples 2 and 3) are given in Table 1.

As folllows from these data the sample 2 (obtained by solid-phase synthesis) is rather identical to sample 1 (synthesized by the conventional suspension deacetylation) with respect to important characteristics such as the solubility and degree of deacetylation that allowed one to conclude that, upon production of chitosan by solid-phase method, the side reactions are absent.

Table 1. Conditions of synthesis and characteristics of chitosan samples

Sample	Method of synthesis	Molar ratio Chitin:NaOH:H_2O	T, °C	Reaction time	Solubility, % *	DD		[η], dl/g
						soluble fraction	insoluble fraction	
1	Suspension	1:20:43	150	2.0	98	0.82	0.18	7.0
2	Solid-phase**	1:5.0:11	180	0.1	96	0.81	0.19	1.2
3	Solid-phase	1:4.3:13	150	0.1	86	0.78	0.22	1.2

* Solubility in 2% acetic acid.

** The solid state of the system is virtually independent of the presence of water in the reaction mixture (10-13 mol H_2O per 1 mol chitin, or ~1g/g).

Table 2. Parameters of chitosan obtained by extrusion method

Process conditions		Parameters of chitosan samples		
Molar ratio of chitin:NaOH	Temperature, °C	DD	Solubility, %	MM, 10^{-4}
1:3	120	0,09	5,0	6,05
1:3	180	0,63	62,0	4,70
1:3*	180	0,73	70,0	4,50
1:3	200	0,70	65,0	4,52
1:5	25	0,33	36,0	6,72
1:5	100	0,40	39,0	4,65
1:5	160	0,59	60,0	7,22
1:5	180	0,81	78	7,75
1:5*	180	0,90	87	6,00
1:5	200	0,98	90,0	4,30
1:10	25	0,12	5,0	4,06
1:10	50	0,27	8,5	4,23
1:10	100	0,43	46,0	6,00
1:10	160	0,76	77,0	7,00

* - Repeated treatment of reaction mixture in extruder.

The results of chitin deacetylation under shear deformation at different initial ratios chitin: NaOH and different temperatures are presented in Table 2. As shown from table, the greater is the NaOH excess, the higher are the degree of deacetylation and the solubility of the obtained products.

The 5-fold molar excess of NaOH suffices to obtain a highly deacetylated product with the solubility about 90%, although in the common suspension procedure, at least 10 fold molar excess is used. So the alkali consumption is reduced in the proposed method up to 75-80%.

Temperature is the most important factor, which influences on the chitosan yield. The degree of deacetylation increased from 0.39 to 0.98 as temperature changed from 25 to 200°C at the molar ratio of reagents of 1:5.

The temperature range of 180-200° C and 5-fold excess of NaOH are probably the optimal conditions to obtain highly deacetylated and highly soluble chitosan on the equipment used. The great excess of NaOH leads obviously to blocking of reaction centres of chitin and to lower product yields; at high temperature the combustion of reaction mixture is possible.

MM of chitosan samples was ranged from 40000 –to 60000. Generally, the decrease of MW highly depends on temperature, being more intensive at temperature increase. The effective decrease of degree of polymerization of polysaccharide derivatives and other polymers obtained under conditions of joint action of high pressure and shear deformations is a general feature caused by the specificity of this kind of action. A decrease in molecular weight (and accordingly in viscosity) is associated with the intensive degradation of the initial chitin. A decrease in solubility is related to the difficulties in providing the uniform distribution of the reagents within a short reaction time (minutes) and results in molecular heterogeneity of the products.

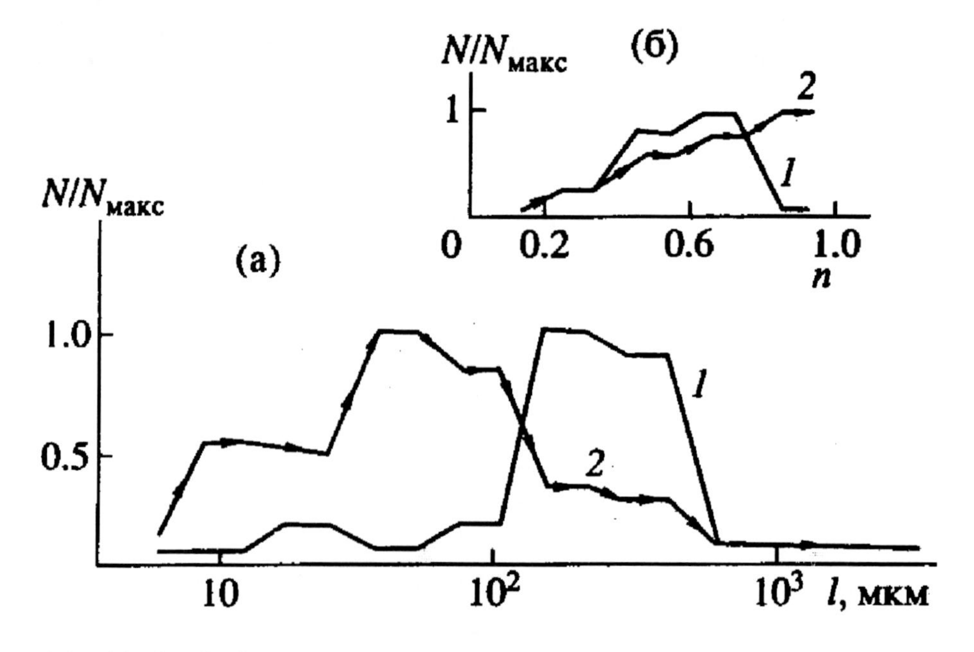

Figure 1. Particle size distribution (a) and form factor distribution (b) for suspension (1) and extrusion (2) chitosan samples.

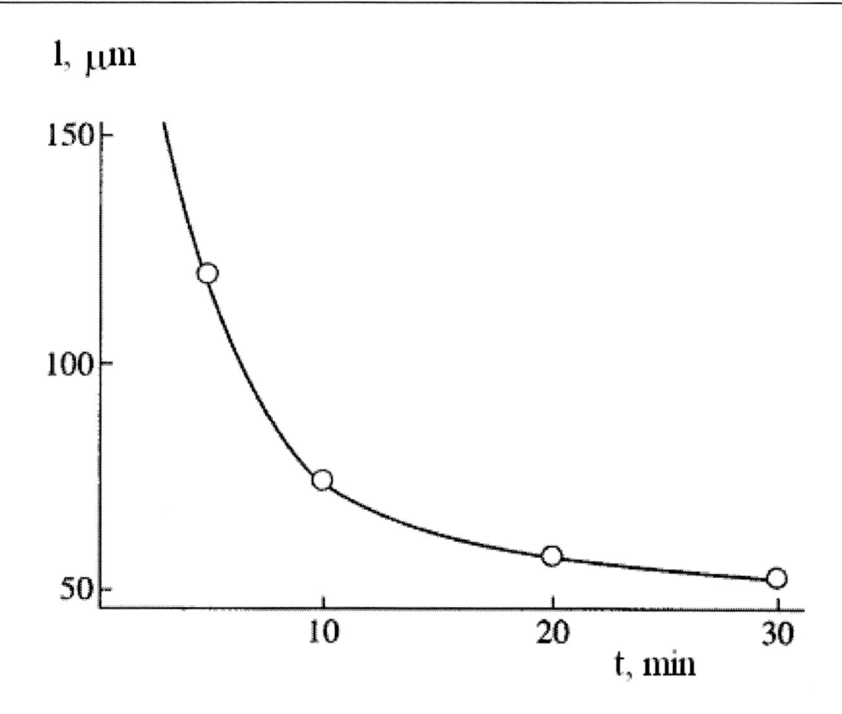

Figure 2. Average reduced particle size for sample 1 as a function of grinding duration at 180^0 C.

The different character of change in MW of chitosan in the presence of NaOH at varied initial chitin: NaOH ratios with temperature (decrease at ratio 1:3, nonmonotonic character at ratio 1:5 and increase at ratio 1:10) may be explained by different number of reaction centres of chitosan under these conditions. At the minimum excess of NaOH (1:3), the degradation becomes more intensive with temperature and MM decreases. At the maximum excess of NaOH (1:10) the reaction centres are blocked by NaOH and the degradation decreases with temperature (MM grows). The intermediate 1:5 ratio is characterized by nonmonotonic change in MM.

It is quite natural that the degrees of dispersibility of chitosan samples obtained by suspension and solid-phase methods are differ to a great extent. As can be seen from Figure1 sample 1 consists of large (130-500 μm) and asymmetric particles, whereas, in the extruded sample 3, the particles are smaller (10-130 μm) more uniform in size and round-shaped.

The effect of extrusion grinding conditions (time and temperature) on the properties of chitosan and the size of chitosan particles was studied especially. The most dramatic decrease in size of the chitosan particles takes place within the first 5-10 min of the process (Figure 2). Chitosan ground in the extruder for this period of time is similar be the particle size distribution to chitosan obtained by solid-phase method in extruder. Further prolongation of the grinding time (increasing number of extrusion runs) has almost no effect on the particle size, but predominantly increases their uniformity and decreases their asymmetry (oval particles become round). It is noteworthy that the highest grinding efficiency is attained at room temperature (Figure 3), since at elevated temperatures, the polymer becomes less brittle.

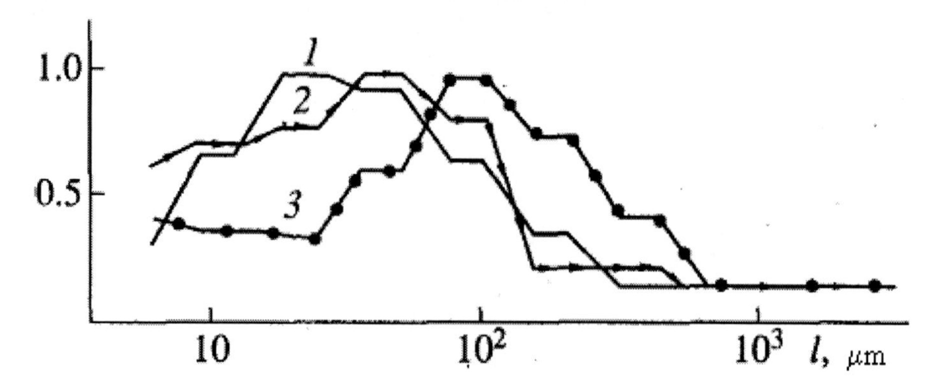

Figure 3. Particle size distribution for sample 1 ground at (1) 25°, (2) 100° and (3) 180° C.

Table 3. Characteristics of various fractions of samples 1 and 2 obtained upon fractionation in water - HCl-isopropyl alcohol system

V_{prec}/Vs*	Fraction yield, %	DD	[η], dl/g	$M\times10^{-3}$
Sample 1				
0	-	0.82	7.0	343
0.5	20	0.82	8.0	402
0.8	55	0.81	7.5	372
1.0	3.0	0.82	6.6	320
1.5	2.0	0.81	6.2	297.5
2.0	5.0	0.83	5.5	258
2.5	8.0	0.83	5.2	242
3.0	4.0	0.82	4.6	209
Sample 2				
0	-	0.81	1.2	43.1
0.8	4.0	0.80	1.8	69.4
1.0	8.0	0.79	1.6	60.5
1.4	12.0	0.82	1.4	51.7
1.6	20.0	0.80	1.2	43.1
2.0	35.0	0.83	1.1	39.0
3.0	10.0	0.82	0.9	30.7

*Volume ratio of precipitant and solution.

At high temperatures, a decrease in the molecular mass resulting from the grinding is slower. It is noteworthy that the variation of molecular masses with time is of nonmonotonic character. For example, after a single run at 25° C, molecular mass decreases by another 10%, whereas with the increasing number of runs when the zone of back mixing is formed, the molecular mass increases slightly. With an increase in the extrusion grinding time, the polymer solubility decreases; the decrease is most pronounced (50% loss of solubility) upon grinding at 180°C.

The obtained data suggest that the destruction of the chitosan structure during extrusion grinding is obviously accompanied by intermolecular crosslinking of the polymer; at 180° C, the crosslinking starts earlier.

Table 4. Characteristics of fractions of sample 3 obtained by variation of pH of chitosan acetate solution

PH	Fraction yield, %	DD	[η], dl/g	MM×10^{-4}
6.3	16	0.69	1.3	4.7
6.7	28	0.75	1.25	4.5
7.0	38	0.82	1.2	4.3
7.5	15	0.80	1.21	4.3
8.5	3	0.88	1.3	4.7
	100	0.78	1.20	4.3

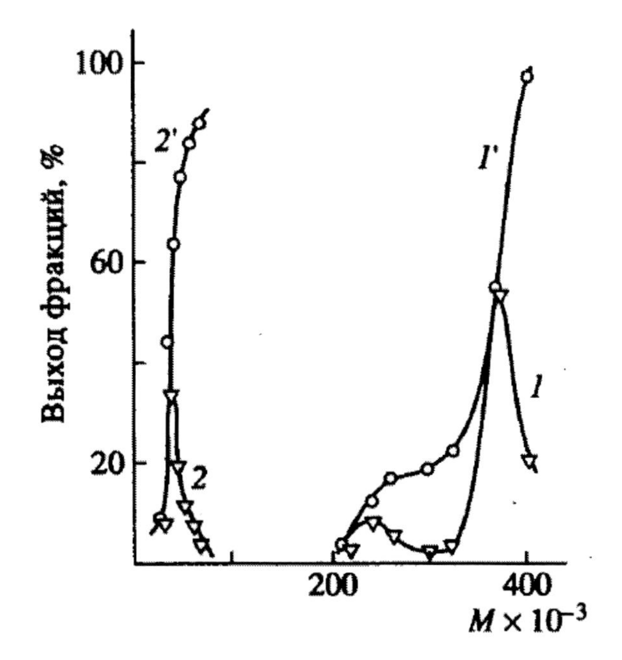

Figure 4. Differential (1, 2) and integral (1', 2') curves of molecular mass distribution of samples 1 (1, 1') and 2 (2. 2').

An important characteristic of chitosan obtained by suspension and solid-phase methods is the fractional composition, which is related to its physical and chemical inhomogeneity and determines the toxicity and probably the biological activity of chitosan-based polymeric drugs. This characteristic is especially important for chitosan obtained by solid-phase synthesis.

The fractionation of chitosan was performed by the method of fractional precipitation. It is believed that the addition of organic solvent leads to the precipitation of fractions differing mostly by the molecular mass, whereas the precipitation of the polymer by pH variation results infractions with different degrees of deacetylation. The latter assumption was based on the notion that a fraction with a lower DD value and hence a smaller degree density on macroions must exhibit a lower solubility at the given pH.

The characteristics of fractions obtained by fractionation of samples 1-3 are given in Tables 3 and 4 and Figure 4 shows the molecular mass distributions for these samples.

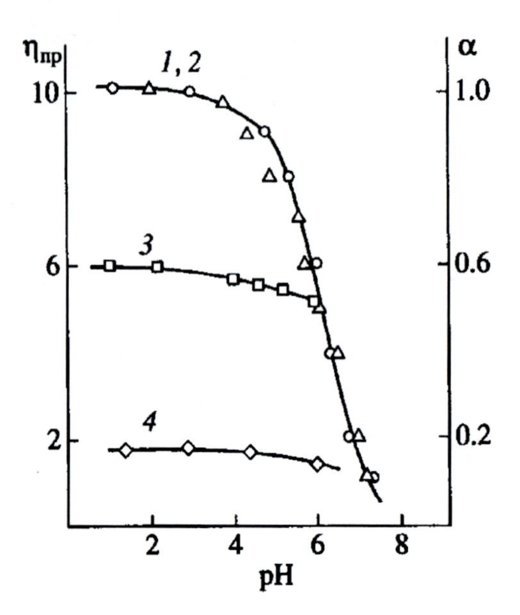

Figure 5. Ionization degree (1, 2) and viscosity (3, 4) vs. pH for 0, 05% solutions of samples 1 (1, 3) and 2 (2, 4) in 0.1M HCl.

As is seen from Table 3, the difference between isolated fractions of samples 1 and 2 is rather small and comparable with the measurement error. Therefore we can conclude that the partitioning procedure gives fractions differing in molecular mass and these data can be used for evaluating the molecular mass distribution of samples 1 and 2. The main difference between two samples was that sample 1 (obtained by the suspension method) has a bimodal molecular mass distribution while sample 2 (obtained by solid-phase deacetylation) has a unimodal distribution (Figure 4). Apparently, the mechanical degradation of polymers under conditions of the joint action of pressure and shear deformations has a less selective character leading to a low-molecular-mass product with a comparatively narrow mass distribution.

Evidently the significant decrease of MM chitosan obtained by solid-phase method does not result from the mechanical degradation. It is possible that macromolecular degradation due to deacetylation is related to thermooxidative processes accelerated by the interaction of polysaccharides with alkali. It should be note that decrease in their molecular mass may be useful and even necessary for specific practical applications of polymer leads to decreasing of drug toxity, increasing of solubility and decreasing of polymer solution viscosity. For this reason, a number of works [10-12] is dedicated to the development of process conditions for obtaining chitosan with reduced molecular masses.

The parameters of fractions obtained with sample 3 partitioned by method of pH variation in chitosan solution are listed in Table 4. As is seen, the chitosan fractions separated as a result of pH variation exhibit no significant difference in molecular mass, but show a clearly pronounced tendency of the degree of deacetylation to increase in the fractions precipitated at high pH values. At the same time, the DD values of the precipitated fractions differ by no more than 20%. Thus, despite the solid state of the initial components, the reaction of chitin deacetylation under the conditions of shear deformation results in the products with relatively homogeneous chemical properties.

As is known from literature [13], the viscosity of chitosan solutions depends slightly on the pH medium. To verify this effect, the values of η_{red} of diluted HCl solutions of chitosan at various pH for samples obtained by the suspension (1) and solid-phase methods (2) were determined. Figure 5 is shows the results of a comparative study of the degree of ionization of amino groups and the reduced viscosity η_{red} of hydrochloric solutions as a function of pH for samples 1 and 2. The α versus pH plots for the samples are absolutely identical (curves 1 and 2). The behaviour of viscosity (curves 3 and 4) shows that the polymer precipitates at the same pH value, for example, pH 6.1+0.2 for chitosan concentration of 0.05% and anionic strength about 0.1 mol/l, which corresponds to α=0.5. Although the viscosity of solutions of sample 2 is much lower as compared to that of sample 1, the η_{sp} versus pH curves are of rather similar character. In the range of pH 1-6, where α varies from 1 to 0.5) the drop of viscosity of both solutions did not exceed 10% which is probably caused by a relatively high rigidity of macromolecules of this polyelectrolyte.

The parameters of characteristics of supramolecular structure of chitosan obtained by solid-phase method is of special interest, because a decrease in MM promotes a higher mobility and results in formation of a more oriented structure, whereas the grinding leads to disordering phenomena. X-ray data showed that, in spite of polymer amorphization in the course of extrusion solid phase deacetylation the obtained chitosan has rather high degree of crystalinity (55%), which is evidently related to recrystalization processes occurring during the washing chitosan from alkali excess. Moreover, chitosan obtained by solid-phase method is characteristic an elevated degree of crystallite dispersity.

It is obviously that low-molecular highly dispersed and amorphized chitosan has a higher sorption capacity to water vapors and another sorbates, for example metal ions.

The films formed from solutions of chitosan obtained by solid-phase method do not differ visually from films formed from high-molecular chitosan solutions. They are characterized by the absence of nonhomogenity and are transparent in the dry form. The mechanical characteristics of these films are suitable for their use as separate membranes or films for bandaging.

Thus, the solid-phase chitosan production is characterized by the simultaneous grinding, amorphization, and increasing of molecular mass of polymer. The method advantages consist in decreasing of reagents consumption (NaOH and washing water) as well as in a significant reduction of a process time.

Chemical Transformations of Chitosan

The chemical transformations of chitin and chitosan continue to attract the attention of many researchers in different countries, because of their high capability for chemical modification due to the presence of hydroxyl and amino groups that makes it possible to form salts, complexes, etc. Among such derivatives, particularly ethers (e.g. carboxymethyl) and esters should be noted.

The solid phase modification of chitosan under conditions of the joint action of high pressure and shear deformation has the same advantages that the process of solid-phase synthesis of chitosan. Moreover upon chitosan modification in heterogeneous (suspension) media as a rule the stage of polymer preactivation by grinding or swelling in organic liquids or in water-acid solutions exists, while the proposed method allows one to combine the stage of activation with the chemical transformation.

CARBOXYLATION OF CHITIN AND CHITOSAN

The products of alkylation, in particular, carboxylation of chitin and chitosan are the promising materials for practical application as they are water-soluble, show high fiber- and film-forming ability ion exchange and complexing properties, along with a high physiological activity and nontoxicity. This combination of properties allows their successful use in medicine, pharmaceutical and cosmetic industries.

Carboxymethylation of chitosan is usually carried out under heterogenous conditions via its treatmentt with monochloroacetic acid (MCA) or its sodium salt in the presence of alkali in aqueous or organic-aqueous medium. As a result of this treatment both O-substituted (I) and N-substituted (II) carboxymethyl chitosan can be obtained.

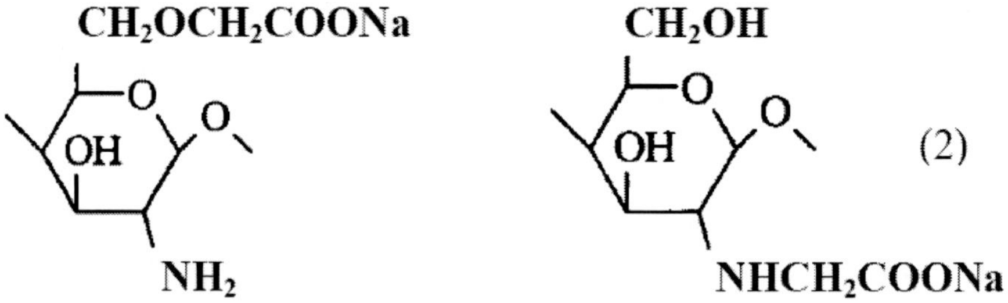

CH_2OCH_2COONa — O — O — OH — NH_2 — I

CH_2OH — O — O — OH — $NHCH_2COONa$ — II (2)

A simpler original method for the synthesis of selectively substituted N-carboxymethyl chitosan involves the preparation of N-carboxymethylidene chitosan and its subsequent reduction to N-carboxymethyl chitosan with sodium cyanoborohydride [14]. To obtain selectively substituted N-carboxymethyl chitosan, a procedure was developed based on the alkylation of chitosan after its reprecipitation and protection of amino groups with salicyl aldehyde [15]. This method is disadvantageous in that it requires large amounts of the alkylating agent; moreover the protection of amino groups and subsequent elimination of the protecting groups are laborious and time-consuming procedures. The advantage of using an alkylating agent (MCA in isopropyl alcohol) for activation of chitosan, is the change of the procedure used in synthesis of carboxymethyl cellulose namely: first, the treatment with MCA and then with NaOH [16]. This procedure leads to high-substituted and completely soluble products and reduces the consumption of reagents.

The method of preparation of carboxymethyl chitosan in solid phase [17] allows one to combine the stages of activation and chemical reaction with lower consumption of reagents (NaOH, organic solvents and water). The production of chitosan in an extruder was performed under different ratios of reagents at temperatures 60-120^0 C. When preparing carboxymethyl chitosan, it was also appropriate to combine in a single process the preparation of chitosan from chitin and subsequent carboxymethylation of chitosan without the stage of chitosan isolation. The comparative characteristics of the CMCh samples obtained by solid-phase and suspension methods are summarized in Table 5. The close values of the degrees of substitution with respect to all functional groups (x_1 and x_2 are O- and N-carboxymethyl, respectively, y – free amino groups, z – N-acetyl groups) calculated from the experimental data on elemental composition for the samples of CMCh obtained by solid-phase and suspension procedures suggested that their structures are almost identical (Table 5).

It should be noted that both samples are substituted nonselectively but still predominantly via the hydroxyl groups of elementary units (x_1 is significantly greater than x_2). It is also revealed that MM of the obtained products decreased dramatically with the process temperature increasing.

The X-ray difraction studies of CMCh obtained by suspension and extrusion methods and carboxymethylcellulose (CMC) obtained by extrusion methods revealed a number of differences in their structure (Figure 6).

Table 5. Data of analysis of CMCh samples [C6H7O2(OH)2x1 (OCH2COONa)x1 (NHCH2COONa)x2 (NH2)y (NHCOCH3)m]

Method	Content, %					Degree of substitution by functional groups				
	C	N_{total}	N_{acyl}	N_{NH2}	COONa	X_1	x_2	Y	Z	M
Solid-phase*	41.8	5.7	1.9	2.8	19.4	0.53	0.15	0.46	0.29	226.1
Suspensional*	41.2	5.4	1.2	3.2	21.4	0.55	0.18	0.52	0.20	226.4

*Molar ratio of chitin:NaOH:MCA= 1:5,25:2,5.

**Molar ratio of chitosan:NaOH:MCA = 1:4,2:2.

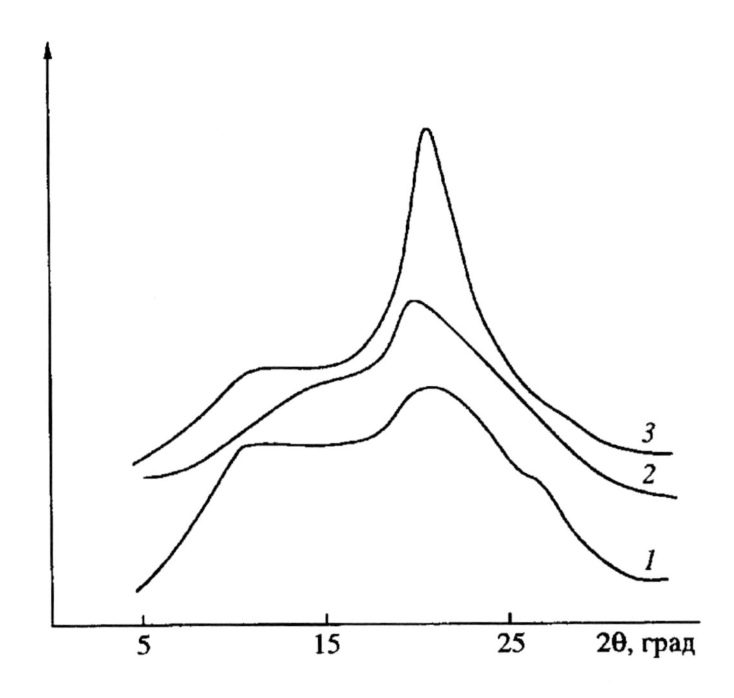

Figure 6. Diffraction patterns for carboxymethyl chitosan (1, 2) and carboxymethyl cellulose (3). Sample 1 was obtained by suspension method (DS = 1.1); samples 2, 3 were obtained by solid-phase extrusion (DS = 0.7 and 0.5, respectively).

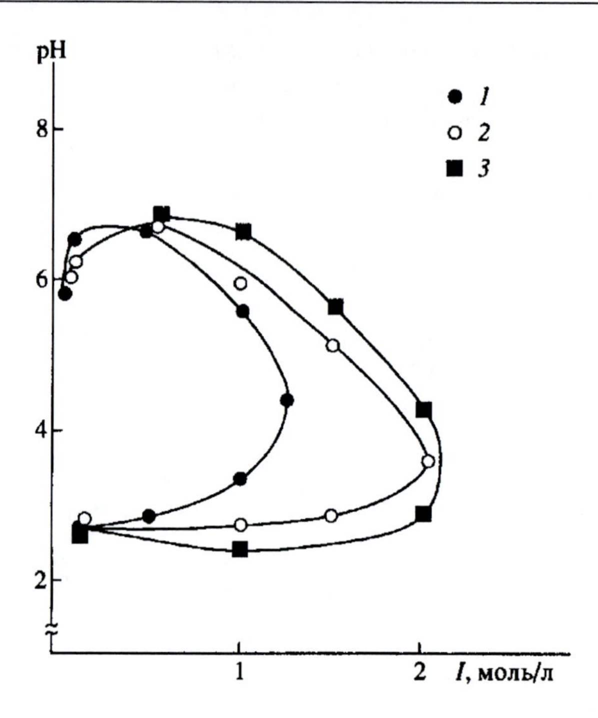

Figure 7. Phase separation boundaries in aqueous saline solution-CMCh system for samples with DS = 0.7 and MM × 10^{-3} = 6 (1), 40 (2) и 82 (3).

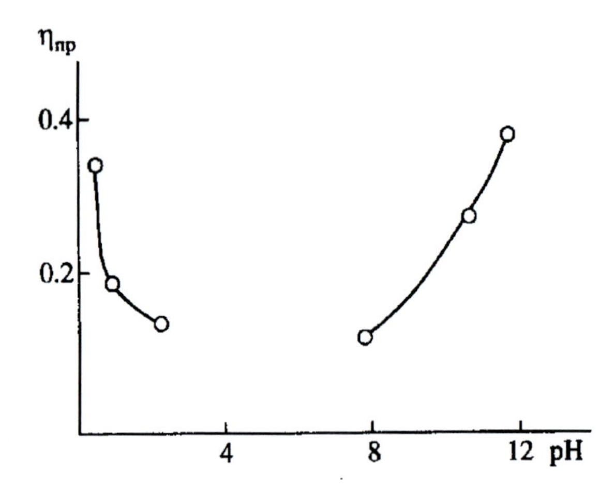

Figure 8. Reduced viscosity of 0.15 %- CMCh solution as a function of pH (DS = 0.7; M = 6×10^3).

The X-ray diffraction patterns of all three samples show a broad reflex at $2\theta = 20^0$ and a broad plateau at $2\theta = 10\text{-}15^0$. Such patterns with a small number of broad reflexes are characteristic of low-ordered structures with low crystallinity. The fact that the maximum is more pronounced in the diffraction pattern of CMC is obviously related to the lower degree of carboxylation of this sample as compared to chitosan derivatives; so the degree of crystallinity increases with an increase in the degree of substitution. It is also possible that cellulose derivatives recover their crystalline structure more readily during washing with polar solvents.

The solubility of CMC in water and the viscous properties depend on pH and ionic strength of solutions, that indicates on its polyampholytic nature. It was interesting to clarify whatever this ability was retained in low-molecular samples of CMCh. Figure 7 shows the phase separation boundaries for the system CMCh-aqueous saline solution, where the CMCh samples were obtained by different procedures and have different molecular masses.

As is seen from the Figure 8, even a sample with MM = $6x10^3$ (i.e., in the isoelectric region, where both the carboxyl and amino groups are ionized and the total charge on the molecule and its hydration decrease) precipitates at pH 3-6.5. In dilute solutions, as pH approaches the isoelectric region, the total charge of polyampholyte decreases, and, as is suggested by a reduced viscosity of the solution, this decrease is accompanied by a decrease in the asymmetry of macromolecules (Figure 7).

MODIFICATION OF CHITOSAN WITH CARBOXYLIC ACIDS AND ANHYDRIDES

As was mentioned above, polybase chitosan is not dissolved in water at pH 6, but is easily soluble in acid media, for example, in solvents of monobasic acids (formic, acetic, hydrochloric), in which the polymer amino groups transfer in ionic and hydrated state. In solutions of dicarboxylic acids (sulpfuric, oxalic) the protonation of amino groups and salt formation leads to crosslinking of macromolecules. The X-ray study of the structure of chitosan and inorganic acids (H_2SO_4, HNO_3 and halogenated acids) salts was carried out in [18].The salt form of chitosan obtained from solution (by precipitation or by lyophylization) and dried retains the solubility in water (so-called water-soluble chitosan), which can be used in cosmetics and agriculture.

The chitosan acid derivatives are usually obtained in liquid media [19-22]] The chitosan derivatives have a complex structure, where the fragments of acids and anhydrides may be bound to hydroxyl groups, or amino groups of polymer by ether- (I), amide- (II) and salt (III) bonds.

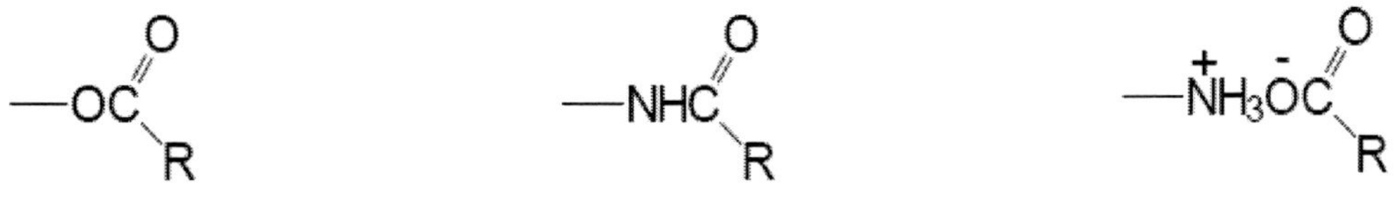

$$—OC(=O)R \qquad —NHC(=O)R \qquad —\overset{+}{N}H_3\overset{-}{O}C(=O)R$$

I II III

The chitosan interaction with some solid acids and anhydrides was studied thoroughly in [23]. In an extruder under conditions of shear deformation, the interactions of chitosan with solid dicarboxylic acids (oxalic, malonic and succinic) and with anhydrides of dicarboxylic acids (phthalic, succinic, maleic) and also fatty stearic acid were realized.

The modification of chitosan with stearic acid and another surfactants allows one to verify hydrophilic-lipophilic balance and to increase its surface activity and emulsifying abiliity. These important properties of chitosan and the derivatives allow their use as structure forming stable additives in food and cosmetic products. Chitosan modification by functional carbonyl containing compounds must lead to formation of cyclic products, as well as partial crosslinking due to intermolecular interaction.

To estimate the possibility of anhydrides addition to chitosan hydroxyl groups with formation of ether bonds under condition of shear deformation the interaction of phthalic anhydride interaction with cellulose having only hydroxyl functional groups was studied. The IR spectra of the products display weak characteristic bands, due to absorption of benzene ring at 1600 cm^{-1}, carbonyl groups at 1780, 1420 cm^{-1} and aromatic ethers at 1250 cm^{-1}. The results of the elemental analysis showed a high C content compared with the initial cellulose in samples synthesized at 100°C. Thus the obtained results allow one to confirm the possibility of O-acetylation reaction, but the degree of substitution was not higher than 10%.

Unlike the cellulose derivative, the IR spectra of the products of chitosan modification by phthalic and another anhydrides show intensive absorption bands due to carboxylic groups (1740-1710 cm^{-1} and 1550 cm^{-1}) and benzene ring (1600 cm^{-1}). In the case of chitosan modification by malonic and succinic acids and also by maleic anhydride the absorption bands of CH group at 2950 and 720-700 cm^{-1}. are observed (Figure 9, curves 2-4).

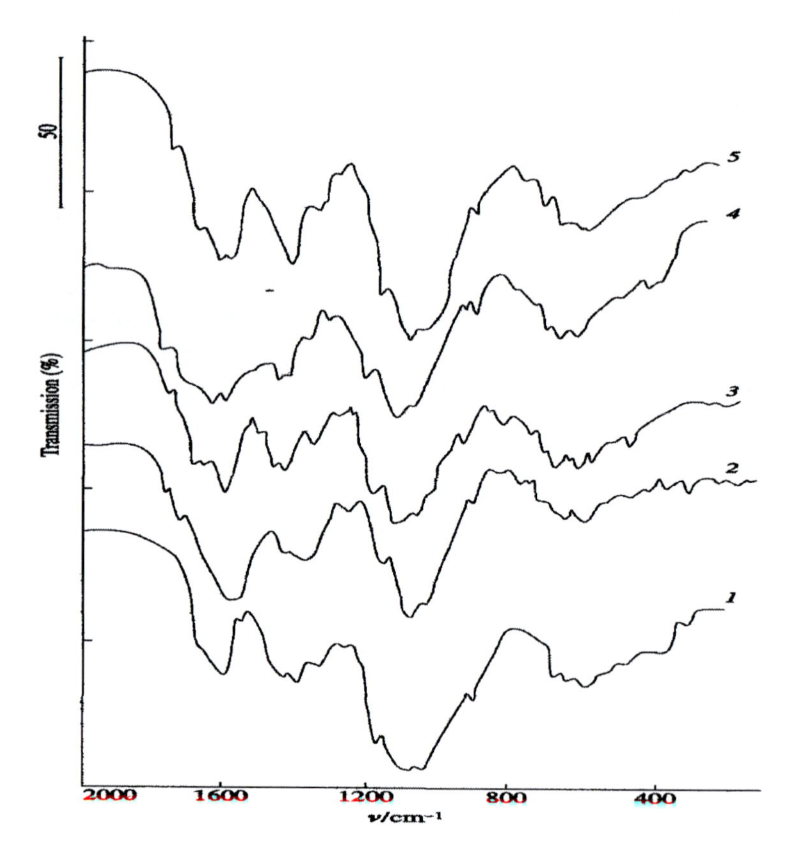

Figure 9. Regions of the IR spectra of chitosan (1) and products of its reactions with oxalic (2), malonic (3) and succinic (4) acids at 100^0 C and with maleic (5), succinic (6) and phthalic (7) anhydrides at 50^0 C and 100^0 C, respectively.

Table 6. Characteristics of the products obtained by solid-phase reaction between polysaccharides and carbonyl containing compounds

Reaction conditions			C/N	N_{aM}/N_{tot}	Degree of substitution			Solubility in 0.1 M HCl, %
Reagent	Molar excess of reagent	T, °C			Total	Ionic	Covalent	
Chitosan	-	-	5.56	0,90	-	-	-	100
Stearic acid	1.50 1.50	25 50	10.7 11.3	- -	0.16 0.22	0.16 0.22	- -	83 68
Oxalic acid	1.50 1.50	25 50	7.82 7.85	- -	0.42 0.43	0.34 0.23	0.08 0.20	98 94
Malonic acid	1.00 1.50 1.50	50 50 100	7.57 8.22 8.42	- - -	0.20 0.40 0.46	0.20 0.36 0.33	- 0.04 0.13	96 94 74
Succinic acid	1.50 1.50	50 100	7.33 7.47	- -	0.10 0.13	0.05 0.12	0.05 0.01	60 40
Maleic anhydride	0.25 0.50	25 50	7.33 8.30	0.90 0.81	0.25 0.50	0.07 0.36	0.18 0.14	58 31
Succinic anhydride	0.25 0.50	50 100	6.93 8.19	0.89 0.59	0.16 0.47	0.08 0.27	0.08 0.20	81 40
Phtalic anhydride	0.50 1.00 1.00	100 50 100	7.96 8.86 9.95	0.89 0.90 0.53	0.21 0.32 0.45	0.07 0.32 0.23	0.14 - 0.22	59 93 33

The characteristics of some products of chitosan interaction with carboxylic acids and anhydrides under condition of shear deformation in the temperature range of 25-100^0 C are given in Table 6. In the above temperature range the starting compounds do not enter into the reaction and remain unchanged. The data in Table 6 show that, under a significant excess of the reagents, the addition proceeds via a mixed (mainly ionic - salt) mechanism. The total amount of added substituents increased with the reagent excess and temperature. The temperature rise has strongest effect on formation of covalent bonds; by this the amide bonds are formed only under a significant excess of reagents, which testifies to the decrease of the ratio N_{am}/N_{tot} (where N_{am} is the nitrogen content in amino groups and N_{tot} the total nitrogen content) in products of the interaction of chitosan with phtalic anhydride and succinic anhydride obtained at 100^0 C. The characteristic feature of this process is it's dependence from aid strength (less strong highest polyhomologers have less reaction capacity) and do not depend on reaction temperature (50-100^0 C), that testifies the analogous with proceeding of these reactions in liquid phase and sequence the absent of diffusion limits for them in solid phase.

In the case of N- and O-acetylation of chitosan by anhydrides of dicarboxylic acids the reaction is accelerated with temperature and weakly depends on the nature of anhydride [23]. An unexpected activity of maleic anhydride compared with other anhydrides was so high that its interaction with chitosan proceeds quantitatively both at room and at elevated temperatures. Under conditions of solid-phase reaction carrying out at the same temperature and initial molar ratio of reactants, the reactivity of dicarboxylic acids and anhydrides changes as follows: oxalic acid > malonic acid > succinic acid; maleic anhydride > succinic anhydride > phthalic anhydride. The conversion of the most weak succinic acid is no more then 10%, and succinic anhydride as might be expected demonstrates a significantly higher reactivity.

The chitosan reaction with fatty stearic acid proceeds via the ionic mechanism and degree of substitution of chitosan stearates obtained at elevated temperatures is as high as 0.22. The hydrophobization of chitosan by stearic acid is confirmed also by a partial loss of solubility in diluted acids (by 60% in 2% acetic acid and by 30% in HCl). An insoluble and more hydrophobic fraction of the obtained products remains insoluble in organic solvents, such as alcohols, hexane, chloroform and acetone.

Thus, the chitosan modification by carboxylic acids and anhydrides does not make them soluble in water and organic solvents that might be expected taking into account a significant break of the structure regularity and the introduction of hydrophilic carboxylic and hydrophobic methyl groups. Obviously the obtained chitosan derivatives are characterized by a hydrophilic-lipophilic balance, which does not allow their complete dissolution in water-acid media, but simultaneously does not endow them with solubility in organic solvents. The considerable decrease in solubility of some products in acid-water media is evidently related to formation both intermolecular and intramolecular crosslinks.

The obtained derivatives of chitosan and carboxylic acids and their anhydrides have polyampholytic nature and can be used as selective sorbents and carriers of amino acids enzymes and other biologically active substances, and soluble fractions of chitosan hydrophobized by stearic acid can be employed as emulsifiers and stabilizers of foams and emulsions.

Thus, the possibility of chitosan modification by carboxylic acids and their anhydrides under conditions of joint action of pressure and shear deformation in solid phase for producing chitosan derivatives is demonstrated. Nevertheless, it should be note once more that the main practical interest is solid-phase method of production of chitosan as the widely used polysaccharide with unique properties.

REFERENCES

[1] M. Rinaudo, *Polym. Int.* 2008. V. 57. P. 397-430.
[2] R.A.A. Muzarelli Chitin. Oxford., New York, Toronto, Sydney, Paris, Frankfurt, Pergamon Press, 1977.
[3] Y.C. Wei, S.M.Hudson. *Macromolecules.* 1993. V. 26. P. 4151.
[4] E.V. Prut, A.N. Zeleneskii. *Uspechi chimii.* 2001. V. 70. № 2. P. 72-87.
[5] Mima S., Miya M., Iwamoto R., Yoshikawa S. *J. Appl. Polym. Sci.* 1983. V. 28. P. 1909-1917.
[6] Kurita K., Kamiya M., Niskimura S. *Carbohydr. Polym.* 1991. V. 16. P. 83-92.
[7] Pelletier A., Lemire A., Sygusch J., Chornet E., Overend R.P. *Alkaline Biotech. and Bioeng.* 1990. V. 36. P. 310-318.
[8] S.Z. Rogovina, T.A. Akopova, *Polymer Science*. Ser. A. 1994. V. 36. № 4. P. 4151.
[9] S.Z. Rogovina, T.A. Akopova, G.A. Vikhoreva. *J. Appl. Polym .Sci.* 1998. V. 70. P. 927.

[10] Hasegava M., Isogai A., Onabe F. *Carbohydr. Polym.* 1993. V. 20. № 4. P. 279-283.
[11] Tanioka S., Matsui Y., Irie T., Tanigava T., Tanaka Y., Shibata H., Sawa Y., Kono Y. *Biosci.Biotech. Biochem.* 1996. V. 60. № 12. P. 2001-2004.
[12] Chen R.H., Chang J.R., Shyar J.S. *Carbohydr. Res.* 1997. V. 29. № 4. P. 287-294.
[13] M.Rinaudo, A.Domard, «Chitin and Chitosan. Soures, Chemistry, Biochemistry, Physical Properties and Applications», Ed. by J. Skjak-Brak, T.Anthonsen, P.Sandford, London, New York. Elsevier. 1989.
[14] Muzzarelli R.A.A., Tanfani F., Emanuelli M., Mariotti S. *Carbohydr. Res.* 1982. V. 107. P. 199-214.
[15] Nud'ga L.A., Plisko E.A., Danilov S.N., *Zh. Obshch. Khim.* 1973. V. 13. № 12. P. 2752]
[16] Vikhoreva G.A., Gladyshev D.Yu, Bust M.F., Barkov V.V., Galbraich L.S. *Cellulose Chem. Techn.* 1992. V. 26. № 6. P. 663-674.
[17] Patent 2044741 RF. C08B37/08.
[18] Osgava K., Saeko N. *Carbohydr. Res.* 1987. V. 16. P. 425-433.
[19] Patent 802290 USSR. C08B37/08.
[20] Patent 325324 USSR C08B37/08.
[21] Argulles-Monal W., Penche-Corvas C. *Makromol. Chem. Rapid Commun.* 1993. V. 14. N 12. P. 735-740.
[22] Demargerandre S., Domard A. *Carbohydr. Polym.* 1994. V. 29. N 3. P. 211-219.
[23] Rogovina S.Z., Vikhoreva G.A., Akopova T.A., Gorbacheva I.N. *J. of Appl. Pol. Sci.* 2000. V. 76. P. 616-622.

In: Handbook of Chitosan Research and Applications
Editors: Richard G. Mackay and Jennifer M. Tait
ISBN 978-1-61324-455-5

Chapter 15

The Use of Chitosan in Basic Studies and Clinical Application in Urinary Bladder Epithelium

Andreja Erman and Peter Veranic
Institute of Cell Biology, Faculty of Medicine,
University of Ljubljana, Slovenia, EU

Abstract

Chitosan is a cationic polysaccharide mainly used as an absorption enhancer. Its known low toxicity for most tissues gave it a wide applicability in the local delivery of pharmaceuticals to selected tissues. The positive charge of the chitosan enables it to attach to the mainly negatively charged proteoglycans that cover the apical surface of most epithelial cells and allow a gradual release of the drugs trapped in this polysaccharide cage. The intravesical application of chitosan to urinary bladder recently revealed an unpredicted deteriorating effect on superficial epithelial cells. This discovery broadens the application of chitosan to basic research of urinary bladder regeneration and gave it a possible application in treating chronic bacterial cystitis and detachment of superficial urothelial tumors. The presented commentary examines our current understanding of the effects of chitosan on urinary bladder epithelial cells in normal rodent bladder, with its concentration and time dependent effects including reversible dysfunction of tight junctions at mildest conditions, over selective removal of superficial epithelial cells and at the severest conditions the complete eradication of the urinary bladder epithelium down to the basal lamina. Its local and very controllable effects make it a perfect model system for the studies of urothelial tissue regeneration without involving inflammatory side effects. The stimulated exfoliation of the bladder epithelium provides a new and very intriguing model for the removal of intracellular pools of uropathogenic E.coli responsible for the majority of recurrent uroinfections and possibly enables a noninvasive removal of superficial bladder tumors predictably with fewer side-effects in comparison to classical treatment.

INTRODUCTION

Chitosan is a polysaccharide obtained by the partial N-deacetylation of chitin, the second most abundant natural polysaccharide. Chitosan is a linear copolymer, composed of glucosamine and N-acetyl glucosamine. It is the main structural component of the cell walls of certain fungi and exoskeletons of insects and crustaceans, which are the main source of chitin. Chitosan polymers vary in molecular weights (50–2000 kDa), viscosity and degree of deacetylation (40–98%). These and other important physicochemical characteristics determine many properties of this polymer, including its influence on tissue permeability. Chitosan has a high density of positive charge and thus adheres to negatively charged surfaces and chelates metal ions. Due to its cationic nature, it forms complexes with negatively charged drugs and forms gels and precipitates. Chitosan acts also as an absorption enhancer, when its amino groups are protonated. This enables the interaction of chitosan with a negatively charged mucosal surface [1]. *In vivo* and *in vitro* studies proved that chitosan increases the transport of some substances across intestinal, nasal, ocular, buccal and vaginal mucosae as well as across the urinary bladder mucosae [2, 3, 4, 5, 6, 7]. Chitosan is evaluated as a biocompatible, biodegradable and non toxic polymer with bioadhesive properties. All these characteristics make chitosan a promising tool in a number of pharmaceutical applications such as controlled drug delivery, fast drug release, peptide delivery and gene delivery. Besides that, chitosan is widely used also in various biological applications as a fat reducer, bactericide and wound healing material [8, 9, 10].

Thus, chitosan is a main component of several products employed as dietary supplements for the control of overweight. Namely, it was reported that chitosan reduces fat absorption in the intestine by binding fatty acids, triglycerides and bile acids and increases their secretion from the body. However, the molecular mechanisms of these effects are still poorly understood, and some adverse effects such as deprivation of trace elements and liposoluble vitamins have appeared [11].

EFFECT OF CHITOSAN ON TIGHT JUNCTION PERMEABILITY

The mucoadhesive property of chitosan has been proposed as a novel potential enhancer for transmucosal drug delivery. Several studies have been undertaken to examine the effect of chitosan on the permeability of epithelial cells. It has been found that chitosan causes a dose and time dependent reduction in transepithelial electrical resistance (TEER) and translocation of tight junction proteins (ZO-1, occludin) from the plasma membrane to the cytoskeleton in the human intestinal Caco-2 cell line [12, 13]. Besides tight junction disruption, chitosan also causes changes in F-actin cytoskeleton. The mechanism by which chitosan increases the permeability of epithelia is probably a combination of its mucoadhesion and the effect on tight junctions. The binding of chitosan to cell membranes of epithelial cells is enabled by electrostatic interactions between positively charged chitosan and negatively charged proteoglycans on the surface of epithelial cells. Studies on interactions between chitosan and biomembrane models revealed that electrostatic interactions with negatively charged phospholipids are crucial for the actioning of chitosan [14]. By binding to negatively charged phosphate groups and alkyl chains of phospholipids, chitosan penetrates into the membrane and causes increased order in the phospholipid layer. Therefore it seems that the action of chitosan mainly depends on interaction with membrane lipids and not on removal of cholesterol from the membrane as previously thought. Chitosan has been already shown to increase permeability across several epithelia. Important studies were performed on intestinal epithelium because of its well known poor permeability for peptides and proteins in the case of peroral peptide drug delivery [2, 15]. In one of them it was proved that chitosan is able to substantially improve the intestinal absorption of the peptide drug buserelin *in vivo* in rats. The proven safety and effect of chitosan may therefore imply that peroral application of buserelin has promising prospects for therapeutic use in the future [15].

Another highly impermeable epithelium is the urinary bladder epithelium, the so-called urothelium, which represents the blood-urine permeability barrier in the body. It has long been aimed to treat the superficial cancers and chronic infections of urinary bladder by local application of drugs to the apical surface of this epithelium, by so sparing the rest of the body with the side effects of cytostatics or high doses of antibiotics. The extremely high impermeability of both apical membrane and the tight junctions of differentiated urothelial cells prevented the successful application of mucoadherent drug holders. The highly differentiated superficial urothelial cells, which cover two layers of less differentiated cells, the basal cells attached to the basal lamina and partly differentiated intermediate cells, are reported to have one of the highest transepithelial resistance (TEER) of 10000 - 75000 Ωcm^2, which highly exceeds the value of the small intestine with TEER lower than 100 Ωcm^2 [16, 17]. Tight junctions are located at the boundary between apical and basolateral membrane domains in superficial cells and prevent paracellular diffusion of substances from urine to blood. A major impact to the impermeability of superficial cells to the substances in the urine gives a specialized apical membrane. It is composed of proteins uroplakins, arranged in semi-crystaline structures called plaques. Since the bulk of mass of uroplakins lies in the luminal leaflet of the membrane, plaques look asymmetric in cross-section, hence the name "asymmetric unit membrane" [18]. Unique plaques of asymmetric unit membrane give a scalloped appearance of the apical surface and enclose specific fusiform vesicles in the cytoplasm of superficial cells. These plaques are thought to be stabilized by attachment to the highly condensed subapical network of cytokeratins. Plaques of asymmetric unit membrane and cytokeratins are unequivocal markers of urothelial cell terminal differentiation [19]. Besides tight junctions and apical membrane, the thin layer of glycosaminoglycans on the surface of superficial cells also offers an important permeability barrier [20].

Such highly efficient barrier function of bladder urothelium recently provoked several research studies with an intention to increase the drug delivery across the urothelium. Studies on an isolated pig urinary bladder wall revealed that chitosan at concentrations between 0.001 and 0.05% w/v increased the permeability by opening tight junctions. However, at high concentrations 0.5% w/v or even higher, the increase of the permeability of the urinary bladder wall was mainly due to induced massive desquamation of urothelial cells, which results in the removal of all components of the urothelial permeability barrier [7]. Therefore the mechanism of chitosan to the permeation of the drug across the pig urinary bladder wall was established to be time and concentration dependent [21].

Chitosan Induces Desquamation of Urothelial Cells

The ultrastructural and biomechanical studies proved that the urothelium is a very stable tissue, resistant to mechanical forces due to numerous intercellular junctions and a compact cytokeratin network characteristic for urothelial cells [22, 23]. Cytokeratins, especially cytokeratin 20 as a differentiation marker of superficial cells, are arranged into parallel cytokeratin tubes pointing towards the apical membrane, forming a trajectorial organization of cytokeratin network [24]. This type of cytoskeleton is connected to a dense array of desmosomes that become concentrated in the subapical area of lateral membranes of superficial cells and together with well-developed tight junctions tightly connect the neighbouring cells. Namely, besides maintaining of the barrier function, superficial cells have to bear extreme changes during cycles of contractions and distensions of the urinary bladder wall appearing as a part of the normal functioning of the bladder.

Inspite of the well known stability, urothelium could quickly respond to various factors with a massive desquamation. Desquamation of urothelial cells is a process that can be induced by a certain signal to which cells respond by disconnection of intercellular junctions or it can result from a mechanical or chemical damage inflicted to cells that directs them into necrosis. The first mechanism is characteristic for stress induced desquamation found after treatment with stress hormones, induction of the rise of stress hormones by disturbing the circadian rhythm with prolonged constant illumination of experimental animals or also as a part of tissue remodeling during normal postnatal development [25, 26]. As a result, detachment of viable cells was determined in urine of stressed animals. The second mechanism is characteristic for desquamation induced with for example instillation of bacterial lipopolysaccharide (LPS) that causes a massive necrosis and removal of morphologically obviously damaged cells [23]. Our recent studies intended to answer the question by which mechanism chitosan at concentrations above 0.5% w/v induces massive desquamation of urothelial cells.

Chitosan, which had first been shown to induce desquamation in *ex vivo* experiments on pig urinary bladder urothelium [21], was tested also as a possible detachment inducer in live animals. Namely, induced removal of cells is the most appropriate way of studying differentiation of the epithelia *in vivo*. Indeed, chitosan as a low-toxic biopolymer was attested as an optimal inducer of desquamation of the urothelium *in vivo* [27]. Intravesical application of chitosan at the concentration of 0.005% w/v induces desquamation of a complete superficial cell layer after 20 min of treatment. Most of the cells of this layer show signs of necrosis before detachment, but no ultrastructural changes are present in underlying urothelial cells. It seems that chitosan causes cell destruction and desquamation only of cells that are in close contact with this polymer. This is in agreement with the results of studies on isolated pig urothelium [21], where the response of urothelial cells to chitosan was strictly concentration dependent.

The main advantage of using chitosan as desquamation inducer comparing to other models for studying urothelial cell differentiation, as for example protamin sulphate [28] or cyclophosphamide [29], is a better control of desquamation process and quicker regeneration. Urothelial desquamation was usually induced by agents such as cyclophosphamide, which could cause extensive loss of urothelial cells, reversible hyperplastic state and delayed beginning of differentiation because of prolonged inflammatory response. Complete recovery of urothelium after cyclophosphamide treatment takes 10 to 14 days [30, 31]. Treatment with chitosan induces controllable removal of urothelial cells, which can be limited to only a single cell layer by a short time treatment (20 min) [27] or can induce complete eradication of the whole urothelium down to the basal lamina during longer period (60 min) of treatment. Urothelial regeneration after chitosan induced cell desquamation approaches the normal physiological regeneration process very accurately and with no inflammatory response. The intravesical instillation of 0.005% (w/v) chitosan dispersion for 20 min causes complete desquamation of superficial cell layer, where desquamated cells show typical morphological signs of necrosis. The first signs of differentiation during regeneration appear already 10 min after chitosan treatment as an instant formation of tight junctions and maturation of apical plasma membrane of new superficial cells, just exposed to the luminal surface. Both events contribute to quick restoration of the barrier function of the urinary bladder and they are therefore quite expected. What is really amazing and unexpected, is the remarkable speed of differentiation, which is completed within an hour after chitosan induced detachment of superficial cell layer. Namely, after restoration of tight junctions and apical membrane architecture, completely new synthesis and arrangement of the cytokeratin 20, as the last event in differentiation, is completed within 60 min after chitosan treatment. At that time, the urothelium once again achieved normal three-layered stage with highly differentiated superficial cells at the surface and fully developed tight junctions as two prerequisites for the functional recovery [27].

The mechanism for the fast differentiation of newly exposed superficial urothelial cells, in order to restore the barrier function, likely depends on highly prepared cells of intermediate cell layer to replace the desquamated superficial cells. The cells of intermediate layer were found to already contain the proteins of tight junctions i.e. ZO-1 proteins [32] at the lateral membranes and proteins of apical membrane, the uroplakins, already packed within vesicles [33]. After the right signal such as the exposure of intermediate cells to the luminal surface the already synthesized proteins of tight junctions could be quickly composed into tight junctions. Similarly, the removal of superficial cells is probably the signal for triggering intense exocytosis of uroplakin carrying vesicles, allowing fast reorganization and differentiation into typically scalloped apical plasma membrane [19].

Chitosan, which is generally considered as a biopolymer with a low-toxicity, shows some toxic effects on the urothelium in live animals, which is the rare exception of its functioning. However, the cytotoxic effect was revealed only in cells that were in close contact with a suitably high concentration of chitosan. No deteriorating effects of chitosan were found in the intermediate cells lying under the superficial cell layer. This result indicates for a specific interaction of that polymer with the apical membrane of highly differentiated superficial cells. The reason is most likely the very special structure of that apical membrane composed of two very different membrane domains, the rigid plaques and the fluid interplaque areas. Such limited toxicity gives chitosan a new and a very promising role not only for theoretical studies of cell detachment and subsequent regeneration but also for clinical application in the future.

CHITOSAN AND TREATMENT OF BACTERIAL CYSTITIS

A growing number of recent studies describe an efficient antibacterial and anti fungal properties of chitosan and its derivatives [34, 35]. The antibacterial activity of chitosan has been demonstrated almost exclusively *in vitro*, where the mechanism of this activity was studied. The antibacterial action was proposed to be due to crosslinking between polycations of chitosan and the anions on the bacterial surface [36]. The interaction between chitosan and bacteria alters cell permeability, causes the leakage of cell components and leads into death of the bacteria. It was found that more negatively charged cell surface have a greater interaction with positively charged chitosan, which means that antibacterial activity of chitosan is related with the surface characteristics of the bacterial cell wall. The density of negative charge on the cell surface of Gram-negative bacteria is higher than that on Gram-positive ones. This clearly explains why more chitosan is adsorbed to Gram-negative bacteria than Gram-positive bacteria and why most Gram-negative bacteria are sensitive to chitosan [37].

From such *in vitro* results it can be concluded that the bactericidal action of chitosan is very promising also in fighting against infection diseases caused by gram-negative bacteria, for example pathogenic *Escherichia coli* (*E. coli*). This bacterium is the most common cause of infections of the urinary tract including urinary bladder infections called cystitis [38]. Urinary tract infections (UTI) are a major human health concern in developed countries and worldwide and can result in significant costs and morbidity. Some of the main problems in urinary tract infections are the high frequency of infection and the increasing resistance on antibiotics. It has been reported that between 40 and 50% of adult women experience at least one UTI during their lifetime [39]. Even with antibiotic treatment, UTIs can recur in 25% of women within six months of an initial acute infection [40]. This clearly implies that another main problem in uroinfections is specific behaviour of uropathogenic *Escherichia coli* (UPEC) adapted to dwell inside urothelial cells where they are fairly protected from antibiotics and immune system. UPEC initiates infection by attacking to extracellular domain of uroplakin Ia in apical plasma membrane of superficial cells via the FimH adhesins capping the distal end of its fimbriae [41]. Attachment induces conformational changes in the entire uroplakin receptor complex, which transmit a signal to the cytoplasmic tails of uroplakins, trigger downstream signaling events and cause cytoskeletal rearrangements. FimH was found to play a role not only as the »adherin« but also as the »invasin« that enables UPEC to invade into the host cells, which is necessary for triggering host cell signaling cascades leading to bacterial invasion [41, 42]. Studies on infected mouse bladder models reveal that UPEC invasion subsequently leads to the formation of intracellular bacterial communities (IBC) in the superficial urothelial cells [43]. In the first phase, during the first three hours after infection, bacteria in the IBC are non motile, rod shaped (average length 3μm) and grow rapidly in loosely organized colonies in the cytoplasm of superficial cells. In the second phase, 6-8 hours after infection, the IBC mature into tighter community of bacteria than in early phase, forming a biofilm inside cytoplasm of superficial cells [44]. Bacteria are cocoid shaped, smaller (0.7μm) than in the first phase and they occupate of almost the entire superficial cell. This stage of infection coincides with the appearance of pods. They represent a unique, very dense and organized intracellular microbial community with a globular architecture, in which bacterial differentiation results in subpopulations with some survival advantages. In the third phase, 12 hours after infection, bacteria become highly motile and rod shaped again, dissociate from the IBC and even exit the superficial cells. Bacteria that flux out of the superficial cells are able to reenter the cells, hence colonize and reestablish infection of other superficial cells [43]. A subpopulation of the bacteria continue to grow but fail to septate, which results in the formation of filamentous bacteria with average length of 20 μm, representing the fourth phase of maturation process (20-48 hours after infection). Septation of filamentous bacteria results in release of rod-shaped daughter cells that could serve as a new population for invasion into other superficial cells. Filamentation provided an advantage to the bacteria in evading host immune cells, especially polymorphonuclear leukocytes and appears to be a critical step for bacterial survival. Despite the strong host antimicrobial defense mechanism, which causes massive exfoliation of infected superficial cells [45], still many intracellular bacteria survive. They form a persistent and quiescent reservoir undetectable within the urine, save from effectors of host immunity and serving as a seed for recurrent infections [46]. UTIs have a strong propensity to recur [47]. The bacteria, associated with a recurrent UTI often appear to be phenotypically or genotypically identical to bacterial strain that caused the initial infection. Hence, it is likely that many recurrent UTIs occur due to resurgence of UPEC from quiescent reservoirs within the bladder urothelium [48].

Recurrent infections are one of the most urgent problems associated with UTIs and so many basic studies are focused into searching of more efficacious antibacterial therapy than widespread antibiotic use is. The rise of antibiotic-resistant strains of UPEC threatens to make the situation even worse. Recent efforts are concentrated on improving current treatment of UTI and searching for new therapeutical methods, which include simultaneous use of antibiotics and direct removal of intracellular bacteria. One of the potential therapeutics could also be chitosan due to its antimicrobial properties, already mentioned at the beginning of this chapter. The ability of chitosan to bind the negatively charged cell walls of UPEC can directly affect bacteria attached to urothelial cells or those floating in urine. In addition to that, chitosan could also efficiently stimulate desquamation of infected urothelial cells and so remove intracellular pools of UPEC, responsible for recurrent infections [1, 32]. Therefore, chitosan treatment would be very useful for the elimination of intracellular pathogenic bacteria, which are not amenable to treatment with antibiotics. Intensive work has been done on chitosan used as a vehicle for incorporating antimicrobials such as quaternary ammonium cations (quats) or silver ions, which all have well known bacteriostatic and bactericidal properties. These derivatives of chitosan have an even amplified antimicrobial action [49, 50, 51, 52] and could represent a promising tool for application in treatment of chronic bacterial cystitis. Strict time and concentration dependent chitosan effects enable a very controllable application in removal of infected urothelial cells without induction of inflammation and with fewer side effects compared to prolonged antibiotic treatment that often result in secondary fungal infections. We predict that a dispersion of chitosan with silver-nanoparticles should be efficient enough to eradicate bacterial infection as a single dose treatment and so omit the necessity of repeated intravesical administration. As already proven in our previous work [27], the urothelial regeneration after chitosan treatment is extremely fast, so the recovery after treatment should not cause any problems to the patients.

Conclusion

Chitosan has a broad spectrum of properties, which makes it useful as a tool for safe and controllable removal of urothelial cells and hence as a tool for a simple and effective model of studying urothelial differentiation. Besides applicability in basic research of urothelium, chitosan could also have a great potential for clinical application as an auxiliary antimicrobial drug in the treatment of uroinfections. Another promising area of chitosan use is in verifying the efficiency of chitosan in the treatment of superficial bladder tumors by application of chitosan as a nanoparticulate carrier system for Mitomycin C to obtain a high concentration of the drug attached close to the tumor and enables controlled release and prevention of rapid drug loss by miction [53]. Further investigations of the application of chitosan and its derivatives for clinical purposes are to be encouraged as a counterbalance to chemical drugs and their widespread use in the world.

References

[1] Singla A.K. and Chawla M. (2001). Chitosan: some pharmaceutical and biological aspects - an update. *J. Pharm. Pharmacol,* 53, 1047-67.

[2] Lueßen H.L., Rentel C.O., Kotze A.F., Lehr C.M., de Boer A.B., Verhoef J.C., Junginger H.E. (1997). Mucoadhesive polymers in peroral peptide drug delivery. IV. Polycarbophil and chitosan are potent enhancers of peptide transport across intestinal mucosae *in vitro.* Carbomer and chitosan improve the intestinal absorption of the peptide drug buserelin in vivo. *J. Control Rel*, 45, 15-23.

[3] Alonso M.J. and Sanchez A. (2003). The potential of chitosan in ocular drug delivery. *J. Pharm. Pharmacol,* 55, 1451-63.

[4] Sinswat P. and Tengamnuay P. (2003). Enhancing effect of chitosan on nasal absorption of salmon calcitonin in rats: comparison with hydroxypropyl- and dimethyl-beta-cyclodextrins. *Int. J. Pharm*, 257, 15-22.

[5] Senel S. and Hincal A.A. (2001). Drug permeation enhacement via buccal route: possibilities and limitations. *J. Control Release*, 72, 133-44.

[6] Sandri G., Rossi S., Ferrari F., Bonferoni M.C., Muzzarelli C., Caramella C. (2004). Assessment of chitosan derivatives as buccal and vaginal penetration enhancers. *J. Pharm Sci*, 21, 351-9.

[7] Kerec M., Bogataj M., Veranič P., Mrhar A. (2005). Permeability of pig urinary bladder wall: the effect of chitosan and the role of calcium. *Eur. J. Pharm. Sci*, 25, 113-21.

[8] Kanauchi O., Deuchi K., ImasatoY., Shizukuishi M., Kobayashi E. (1995). Mechanism for the inhibition of fat digestion by chitosan and for the synergistic effect of ascorbate. *Biosci. Biotechnol. Biochem*, 59, 786-90.

[9] Rabea E.I., Badawy M.E.-T., Stevens C.V., Smagghe G., Steurbaut W. (2003). Chitosan as antimicrobial agent: applications and mode of action. *Biomacromolecules,* 4, 1457-65.

[10] Okamoto Y., Shibazaki K., Minami S., Matsuhashi A., Tanioka S., Shigemasa Y. (1995). Evaluation of chitin and chitosan in open wound healing in dogs. *J. Vet. Med. Sci*, 57, 851-4.

[11] Koide S. (1998). Chitin-chitosan: properties, benefits and risks. Nutr Res, 18, 1091-101.

[12] Smith J., Wood E., Dornish M. (2004). Effect of chitosan on epithelial cell tight junctions. *Pharm Res,* 21, 43-9.

[13] Ranaldi G., Marigliano I., Vespignani I., Perozzi G., SambuyY. (2002). The effect of chitosan and other polycations on tight junction permeability in the human intestinal Caco-2 cell line. *J. Nutr. Biochem*, 13, 157-67.

[14] Pavinatto F.J., Pacholatti C.P., Montanha E.A., Caseli L., Silva H.S., Miranda P.B., Viitala T., Oliveira Jr. O.N. (2009). Cholesterol mediates chitosan activity on phospholipid monolayers and Langmuir-Blodgett films. *Langmuir,* 25, 10051-61.

[15] Lueßen H.L., de Leeuw B.J., Langemeÿer M.W., de Boer A.B., Verhoef J.C., Junginger H.E. (1996). Mucoadhesive polymers in peroral peptide drug delivery. VI. Carbomer and chitosan improve the intestinal absorption of the peptide drug buserelin in vivo. *Pharm Res,* 13, 1668-72.

[16] Lewis S.A. (2000). Everything you wanted to know about the bladder epithelium but you were afraid to ask. *Am. J. Physiol. Renal Physiol*, 278, F867-F874.

[17] Ward P.D., Tippin T.K., Thakker D.R. (2000). Enhancing paracellular permeability by modulating epithelial tight junctions. *PSTT,* 3, 346-58.

[18] Hicks R.M. (1975). The mammalian urinary bladder: an accomodating organ. *Biol. Rev*, 50, 215-46.

[19] Veranič P., Romih R., Jezernik K. (2004). What determines differentiation of urothelial umbrella cells? *Eur. J. Cell Biol,* 83, 27-34.

[20] Lilly J.D. and Parsons C.L. (1990). Bladder surface glycosaminoglycans is a human epithelial permeability barrier. *Surg. Gynecol. Obstet*, 171, 493-6.

[21] Kerec Kos M., Bogataj M., Veranič P., Mrhar A. (2006). Permeability of pig urinary bladder wall: time and concentration dependent effect of chitosan. *Biol. Pharm Bull*, 29, 1685-91.

[22] Minsky B.D. and Chlapowski F.J. (1978). Morphometric analysis of the translocation of luminal membrane between cytoplasm and cell surface of transitional epithelial cells during the expansion-contraction cycles of mammalian urinary bladder. *J. Cell Biol*, 77, 685-97.

[23] Veranič P. and Jezernik K. (2000). The response of junctional complexes to induced desquamation in mouse bladder epithelium. *Biol. Cell*, 92, 105-13.

[24] Veranič P. and Jezernik K. (2002). Trajectorial organisation of cytokeratins within the subapical region of umbrella cells. *Cell Motil. Cytoskeleton*, 53, 317-25.

[25] Veranič P. and Jezernik K. (2001). Succesion of events in desquamation of superficial cells as a response to stress induced by prolonged constant illumination. *Tissue cell*, 33, 280-85.

[26] Erman A, Zupančič D., Jezernik K. (2009). Apoptosis and desquamation of urothelial cells in tissue remodeling during rat postnatal development. *J. Histochem. Cytochem*, 57, 721-30.

[27] Veranič P., Erman A., Kerec Kos M., Bogataj M., Mrhar A., Jezernik K. (2009). Rapid differentiation of superficial urothelial cells after chitosan-induced desquamation. *Histochem. Cell Biol*, 131, 129-39.

[28] Lavelle J., Meyers S., Ramage R., Doty D., Bastacky S., Apodaca G. Zeidel M. (2001). Protamine-sulfate-induced cystitis: a model of selective cytodestruction of the urothelium. *Urology,* 57, 113

[29] Philips F.S., Sternberg S.S., Cronin A.P., Vidal P.M. (1961). Cyclophosphamide and urinary bladder toxicity. *Cancer Res*, 21, 1577-89.

[30] Locher G.W. and Cooper E.H. (1970). Repair of rat urinary bladder epithelium following injury by cyclophosphamide. Invest Urol, 8, 116-23.

[31] Romih R., Koprivec D., Štiblar Martinčič D., Jezernik K. (2001). Restoration of the rat urothelium after cyclophosphamide treatment. *Cell Biol. Int,* 25, 531-37.

[32] Acharya P., Beckel J., Ruiz W.G., Rojas R., Birder L., Apodaca G. (2004). Distribution of the tight junction proteins ZO-1, occludin, and claudin-4, -8, and -12 in bladder epithelium. *Am. J. Physiol. Renal. Physiol*, 287, F305-318.

[33] Yu J., Manabe M., Wu X.R., Xu C., Surya B., Sun T.T. (1990). Uroplakin I: a 27-kD protein associated with the asymmetric unit membrane of mammalian urothelium. *J. Cell Biol,* 111, 1207-16.

[34] Fernandez J.C., Tavaria F.K., Soares J.C., Ramos O.S., Monteiro M.J., Pintado M.E., Malcata F.X. (2008). Antimicrobial effects of chitosan and chitooligosaccharides, upon Staphylococcus aureus and Escherichia coli, in food model systems. *Food Microbiol,* 25, 922-28.

[35] Lee D.S., Jeong S.Y., Kim Y.M., Lee M.S., Ahn C.B., Je J.Y. (2009). Antibacterial activity of aminoderivatized chitosans against methicilin-resistant Staphylococus aureus (MRSA). *Bioorg. Med. Chem*, 17, 7108-12.

[36] Tsai G.J. and Su W.H. (1999). Antibacterial activity of shrimp chitosan against Escherichia coli. *J. Food Prot,* 62, 239-43.

[37] Chung Y.-C., Su Y.-P., Chen C.-C., Jia G., Wang H.I., Wu J.C.G., Lin J.G. (2004). Relationship between antibacterial activity of chitosan and surface characteristics of cell wall. *Acta Pharmacol. Sin*, 25, 932-6.

[38] Svanborg C. and Godaly G. (1997). Bacterial virulence in urinary tract infection. *Infect. Dis. Clin. N. Am*, 11, 513-29.

[39] Kunin C.M. (1994). Urinary tract infections in females. *Clin. Infect. Dis*, 18, 1-10.

[40] Foxman B. (2002). Epidemiology of urinary tract infections: incidence, morbidity, and economic costs. *Am. J. Med*, 113, Suppl 1A, 5S-13S.

[41] Wang H., Min G., Glockshuber R., Sun T.-T., Kong X.-P. (2009). Uropathogenic E.coli adhesin-induced host cell receptor conformational changes: implications in transmembrane signaling transduction. *J. Mol. Biol*, 392, 352-61.

[42] Martinez J.J., Mulvey M.A., Schilling J.D., Pinkner J.S., Hultgren S.J. (2000). Type 1 pilus-mediated bacterial invasion of bladder epithelial cells. *EMBO J,* 19, 2803-12.

[43] Justice S.S., Hung C., Theriot J.A., Fletcher D.A., Anderson G.G., Footer M.J., Hultgren S.J. (2004). Differentiation and developmental pathways of uropathogenic Escherichia coli in urinary tract pathogenesis. *Proc. Natl. Acad. Sci. USA*, 101, 1333-8.

[44] Anderson G.G., Palermo J.J., Schilling Y.D., Roth R., Heuser J., Hultgren S.J. (2003). Intracellular bacterial biofilm-like pods in urinary tract infections. *Science*, 301, 105-7.

[45] Aronson M., Medalia O., Amichay D., Nativ O. (1988). Endotoxin-induced shedding of viable uroepithelial cells is an antimicrobial defense mechanism. *Infect. Immun*, 56, 1615-17.

[46] Mulvey M.A., Schilling J.D., Hultgren S.J. (2001). Establishment of a persistent Escherichia coli reservoir during the acute phase of a bladder infection. *Infect. Immun*, 69, 4572-9.

[47] Hooton T.M. and Stamm W.E. (1997). Diagnosis and treatment of uncomplicated urinary tract infection. *Infect. Dis. Clin. N. Am*, 11, 551-81.

[48] Brauner A., Jacobson S.H., Kuhn I. (1992). Urinary Escherichia coli causing reccurent infections - a prospective follow-up of biochemical phenotypes. *Clin. Nephrol*, 38, 318-23.

[49] Rhim J.-W., Hong S.-I., Park H.-M., Ng P.K.W. (2006). Preparation and characterization of chitosan-based nanocomposite films with antimicrobial activity. *J. Agric. Food Chem*, 54, 5814-22.

[50] Sanpui P., Murugadoss A., Prasad P.V.D., Ghosh S.S., Chattopadhyay A. (2008). The antibacterial properties of a novel chitosan-Ag-nanoparticle composite. *Int. J. Food Microbiol,* 124, 142-6.

[51] Sajomsang W., Tantayanon S., Tangpasuthadol V., Daly W.H. (2009). Quaternization of N-aryl chitosan derivatives: synthesis, characterization, and antibacterial activity. *Carbohydrate Res,* 344, 2502-11.

[52] Wei D., Sun W., Qian W., Ye Y., Ma X. (2009). The synthesis of chitosan-based silver nanoparticles and their antibacterial activity. *Carbohydrate Res*, 344, 2375-82.

[53] Bilensoy E., Sarisozen C., Esendagli G., Dogan A.L., Aktas Y., Sen M., Mungan N.A. (2009). Intravesicalcationic nanoparticles of chitosan and polycaprolactone for the delivery of Mitomycin C to bladder tumors. *Int. J. Pharmaceutics*, 371, 170-176

Reviewed by: prof.dr. Kristina Sepčič.,
Chair of Biochemistry, Department of Biology
Biotechnical Faculty, University of Ljubljana
Večna pot 111, 1000 Ljubljana, Slovenia
Phone: 00386 1 423 33 88; fax: 00386 1 257 33 90
kristina.sepcic@bf.uni-lj.si

In: Handbook of Chitosan Research and Applications
Editors: Richard G. Mackay and Jennifer M. Tait
ISBN 978-1-61324-455-5

Chapter 16

FUNCTIONALIZATION AND APPLICATIONS OF CHITOSAN

Congming Xiao*
College of Material Science and Engineering of
Huaqiao University, Quanzhou, China

ABSTRACT

Chitosan, the second abundant naturally occurring polysaccharide, is an N-deacetylated derivate of chitin. Owing to its unique properties such as biodegradability, biocompatibility, biological activity, and capacity of forming ployelectrolyte complex with anionic polyelectrolytes, chitosan has been widely applied in food, cosmetics, biomedical, pharmaceutical and the other fields. On the other hand, chitosan has its inherent limitations. Thus, many attempts to render more functions to chitosan via chemical or physical modifications have been made and reported. Herein, we describe several means to functionalize chitosan in order to improve its water-solubility or provide it some useful properties. The applications of these chitosan derivates are also involved.

Keywords: *chitosan; chemical; physical; functionalization; application*

1. INTRODUCTION

1.1. Origin and Structure of Chitosan

Chitosan is the second abundant natural biopolymer and has been receiving growing attention for decades. It is obtained by alkaline deacetylation from chitin. The characteristic structural unit of chitosan is 2-amino-2-deoxy-D-glucopyranose. Actually, it should be regarded as a copolymer of beta-(1,4')-linked 2-acetamido-2-deoxy-D-glucopyranose and 2-

* Corresponding author, xcm6305@yahoo.com.cn.

amino-2-deoxy-D-glucopyranose (Figure 1) as the deacetalation is hardly completed. The content of amino groups of chitosan varies with the degree of deacetalation. Chitosan is linear, rigid and semi-crystalline. In addition, there is strong intra- and intermolecular hydrogen bonding [1-3].

1.2. Advantages and Limitations of Chitosan

The backbone bears a lot of amino groups means that chitosan is a basic polysaccharide. Thus, it can dissolve in dilute acid at pH below its PKa (ca. 6.3) and form polycation. However, it is not soluble in neutral and alkali aqueous medium, neither in any organic solvents. This makes it difficult to process and will hinder some of its potential applications.

The positive attribute of chitosan is its excellent biocompatibility, admirable biodegradability, ecological safety, low toxicity, and versatile biological activity such as antimicrobial activity and low immuogenicity [2]. It is also cheap due to its replenishable resource. These make chitosan one of the most promising candidates for biomedical, food packaging, and eco-friendly applications [2-4].

The properties of chitosan are affected significantly with both the molecular weight and degree of decetylation.

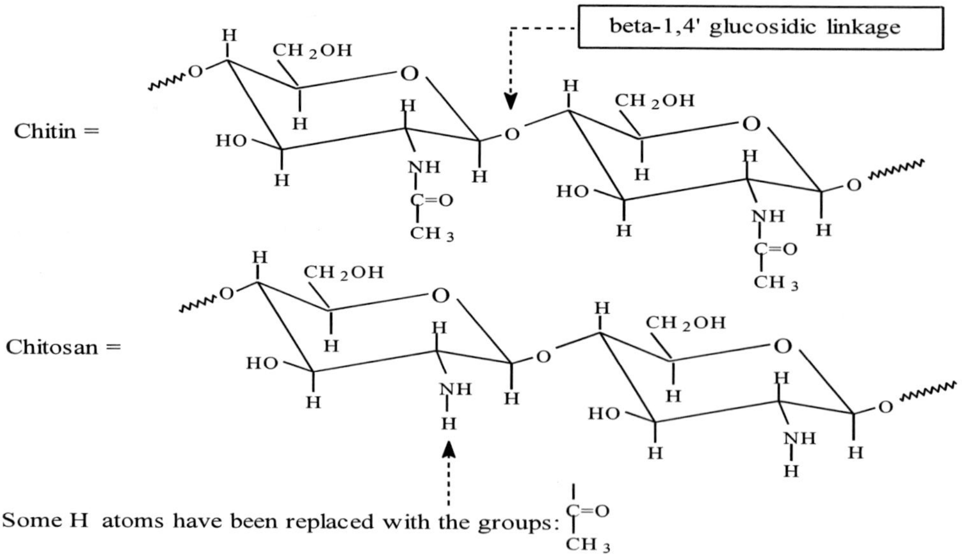

Figure 1. Structure of chitosan and its precursor.

2. Functionalization of Chitosan

Chitosan itself shows some functional properties. To overcome its inherent limitation and generate desired properties or functions, chitosan is modified or functionalized by different strategies. The presence of hydroxyl especially amino groups provides chemical reaction or physical interaction sites for chitosan. Modifications will not change the fundamental skeleton of chitosan. As a consequence, the original physicochemical and biochemical properties are kept while new or improved properties are imparted [5].

2.1. Functionalizing Ways

Functionalizing chitosan can be carried out by chemical reactions or physical combination (Figure 2). It can be molecular-level modification or just happen on the surface/interface, which will construct sophisticated architecture having various interesting properties.

Chemical reactions include oligomerization, alkylation, acylation, quternization, hydroxyalkylation, carboxyalkylation, thiolation, sulfation, phosphorylation, enzymatic modification, graft copolymerization, and covalently cross-linking. Physical modifications usually assume blend, ionic complex and composite.

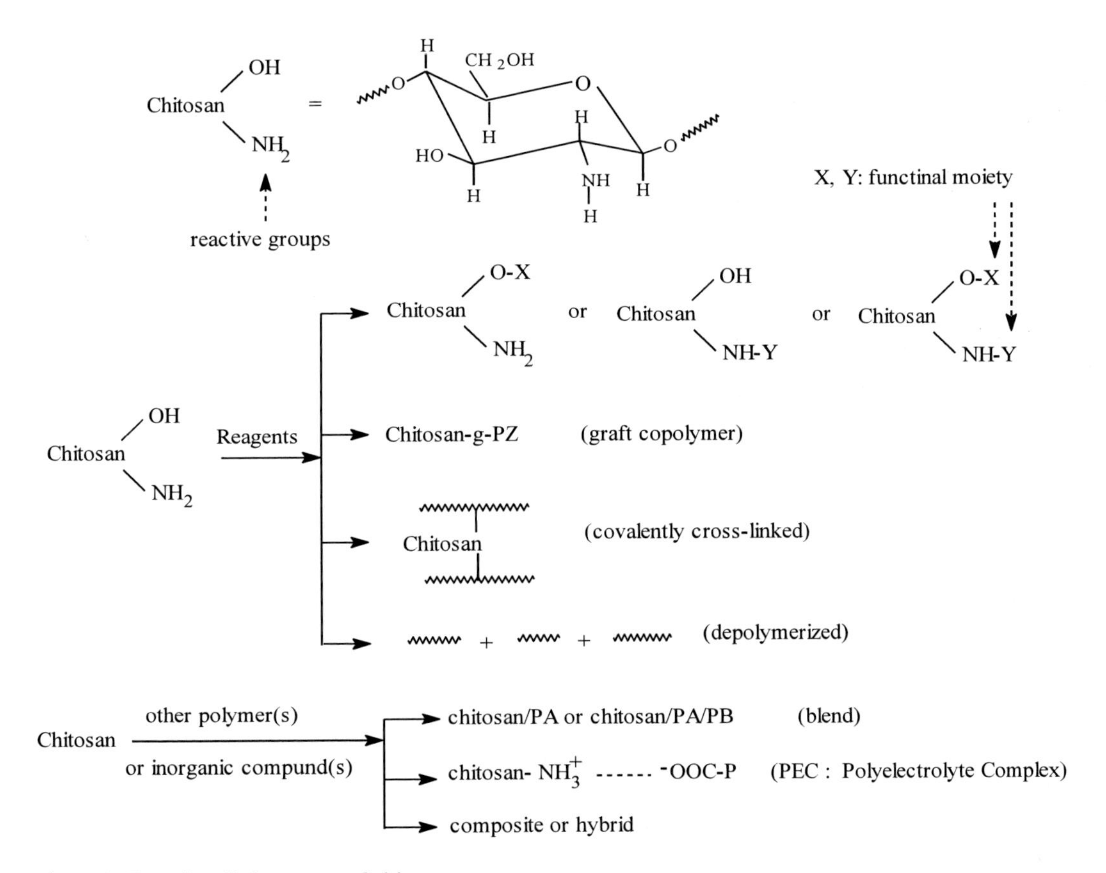

Figure 2. Functionalizing ways of chitosan.

2.2. Functionalized or Properties-Enhanced Chitosan

2.2.1. Water-Soluble Chitosan

Chemical, physical or enzymatic degradation of chitosan generate its oligomers. The chemical method shows several defects such as harsh condition and low yield. Low-molecular-weight chitosan (LMWC) can also be prepared by enzymic hydrolysis [6-7], enzymolysis and salts-removal [8], and ozone treatment [9]. LWMC possesses excellent solubility and keep its inherent characteristics. For instance, LMWC can inhibit the growth of S180 tumor cells in the mice [7], and water-soluble OCHT (chitosan-oligosaccharides) that prepared via enzymatic hydrolysis of chitosan exhibits strong antibacterial activity against *Staphylococcus aureus* isolated from bovin mastitis [10]. LMWC is also useful as drug carriers. For example, retinol-encapsulated LMW water-soluble chitosan is prepared for cosmic and pharmaceutical applications [11].

Chemical approaches are effective strategies to introduce hydrophilic groups onto the backbone of chitosan and produce multifarious water-soluble derivatives. Here are some common adopted methods. By introducing a phosphonic acid group or by quaternization of existing primary ammonium groups, water-soluble chitosan derivatives (NMPC and QC) are obtained. NPMC, QC and PVA are mixed in aqueous media to prepare charged composite membranes for separating mono- and divalent electrolytes [12]. N-acylation of chitosan with succinic or maleic anhydride not only incorporates carboxyl groups, but also the crystallinity decreases compared to chitosan. As consequence, water-soluble N-succinyl-chitosan (NSC) or N-maleic acyl-chitosan (NMCS) is provided. NSC behaves pH-sensitive reversibly solubility and can be utilized as a drug carrier, wound dressing, cosmetic materials and enzyme immobilization matrix [13-14]. NMCS is found to be capable of forming hydrogel for potential applications in tissue engineering and drug delivery [15]. Water-soluble hydroxypropyl chitosan (HPCS) is synthesized from chitosan by reacting with propylene epoxide under basic conditions. It is presented that HPCS has specific immunomodulatory effects and offers a new therapeutic modality for allergic asthma [16]. N,O-acetylation of chitosan is also a feasible way to prepare water-soluble chitosan derivatives, and it is affected with the degree of deacetylation and the molecular weight [17]. The solubility of chitosan can be improved by carboxymethylation. Carboxymethylchitosan harvested by reacting chitosan with mono-chloroacetic acid in the presence of sodium hydroxide is soluble in a wide range of pH [18]. A simple procedure is put forward to produce water-soluble N-carboxylethylchitosan by Michael reaction of chitosan with acrylic acid in water [19].

Conjugating with water-soluble polymers is a promising means to increase the solubility of chitosan. Poly(ethylene glycol)(PEG) is an ideal polymer because it is water-soluble and biocompatible. PEGylation can be achieved by various routines. Reductive amination and acylation of chitosan are two of them. Both PEG-N-chitosan and PEG-N,O-chitosan obtained are water-soluble [20]. To possess chitosan backbone and incorporate PEG chains synchronously, etherification reaction of N-phthaloyl chitosan with poly(ethylene glycol) monomethyl ether iodide (MPEG-I) is conducted. The O-PEGylated derivatives, i.e. chitosan-O-PEG graft copolymers are soluble in water and the aqueous solutions of wide pH range [21]. In order to obtain N-PEGylated chitosan-g-PEG, N-substitution of triphenylmethyl chitosan with MPEG-I and subsequent removal of protecting groups are performed. The copolymers are soluble in water of wide pH range [22].

The successful preparation of water-soluble chitosan is essential for the further applications of chitosan in biomedical and other fields.

2.2.2. Smart Chitosan

The physical/chemical properties of stimuli-responsive polymers are sensitive to environmental fluctuations such as variations in pH, temperature, ionic strength, or magnetic field (Figure 3). Incorporation of such a polymer will enable chitosan intelligent and suitable for some applications. Radical graft copolymerization of N-isopropylacrylamide (NIPAAm) with water-soluble chitosan generates a thermosensitive chitosan-g-PNIPAAm. It will exhibit reversible water soluble-insoluble transition around 32°C and is applied to differentiate from human mesenchymal stell cells to chondrocytes [23]. The low critical solution temperature (LCST) of chitosan-g-PNIAAm is close to human body surface temperature. Therefore, its thermosensitive in situ gel-forming property is considered in utilization for ocular drugs delivery [24]. In addition, the structure of chitosan-g-PNIPAAm can be controlled by reversible addition fragmentation chain transfer (RAFT) polymerization of NIPAAm in the presence of chitosan-RAFT agent [25]. Thermo-sensitive chitosan-based polymer may be chitosan-PEG diblock copolymer. It is prepared by radical copolymerization of PEG-macromer with chitosan and behaves thermoreversible transition at body temperature [26]. If the copolymerization of chitosan is carried out with pH- and temperature-sensitive monomers such as maleic anhydride or acrylic acid and NIPAAm, dual responsive copolymers are obtained. They are potential candidates for drug delivery [27-28].

Actually, chitosan itself is a pH-sensitive polymer owing to the presence of amine groups. On the other hand, poly(vinyl alcohol) (PVA) hydrogels can be prepared via freezing/thawing cycles. These characteristics are took advantage to form pH-sensitive physical chitosan-g-PVA hydrogel [29]. Chitosan-based pH-sensitive hydrogel can also be a semi-interpenetrating polymer network. It is synthesized by photopolymerizing water-soluble N-carboxyethyl chitosan, 2-hydroxyethyl methacrylate and tetra(ethylene glycol) dimethylacrylate and may be used as a drug sustained release matrix [30]. The environmental pH is an effect stimulus to trigger in situ forming hydrogel of N-palmitoyl chitosan (NPCS). Aqueous NPCS is an injectable hydrogel and may be useful for drug/cell delivery systems [31]. Enzyme-catalyzed reaction is another feasible routine for in situ forming chitosan-based hydrogel. Chitosan with phenolic hydroxyl groups is soluble at neutral pH and gellable in the presence of peroxidase within seconds. Such a chitosan derivative is recommended as injectable tissue engineering scaffold or drug carrier [32]. In addition, chitosan and its derivative carboxymethychitosan together can be cross-linked with glutaraldehyde to obtain electroactive hydrogel. The electromechanical response of the hydrogel is good and tunable by simply changing the pH of the buffer saline, which suggests its potential for microsensor and actuator applications [33].

Temperature and pH responsive chitosan-based polymers are used in development of drug delivery systems, bioseparation devices, tissue engineering scaffolds, cell culture supports, sensors or actuatoras systems [34]. Magnetic response is also applied in biomedical fields such as bioseparation and drug target delivery. To avoid the agglomeration of magnetic nanoparticles in biological fluids, the derivatives of chitosan are covalently bound onto Fe_3O_4 nanoparticles. These nanoparticles are superparamagnetic and stable in water [35-36].

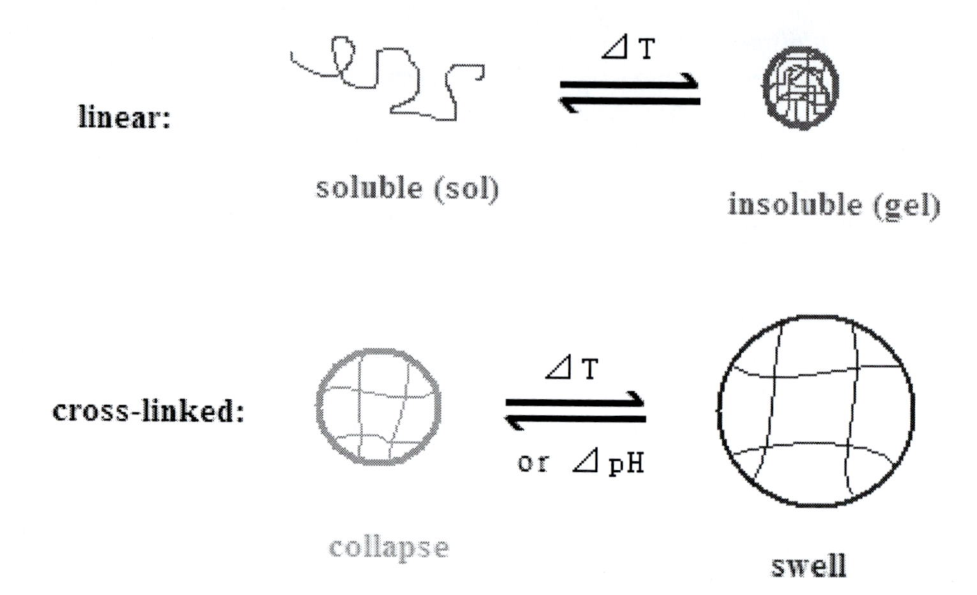

Figure 3. Outline of linear or cross-linked polymers' stimuli-sensitivity.

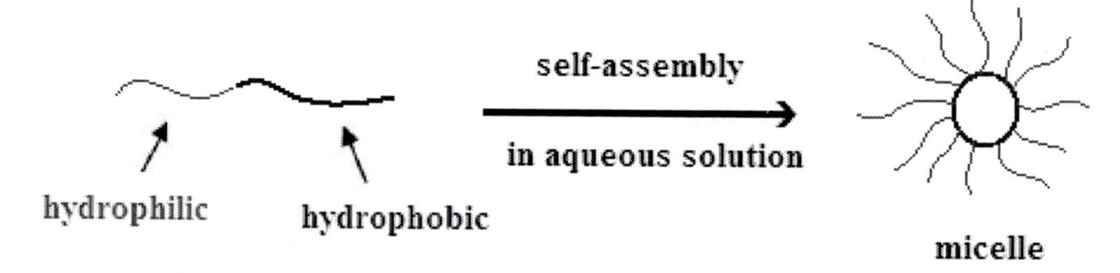

Figure 4. Formation of polymeric micelle.

2.2.3. Amphiphilic Chitosan

In aqueous solution, amphiphilic block or graft copolymer consisting of a hydrophobic and a hydrophilic segment will self-assemble into micelles. The hydrophobic moiety aggregates as a core and is stabilized with the hydrophilic shell (Figure 4). Chitosan-derived micelles are attractive for drug delivery owing to the unique properties of chitosan. Amphiphilic chitosan derivatives such as N-ackyl-O-sulfate chitosan, N-succinyl-chitosan, N-carboxyethylchitosan, chitosan-g-poly(L-lactic acid), phthaloylchitosan-MPEG, chitosan-O-poly(ε-caprolactone), and carboxymethyl-hexanoyl chitosan can self-assemble spontaneously in aqueous medium to form micelle or nanospheres [37-43]. The predominant driving forces are intermolecular hydrogen bonding and hydrophobic interaction among the hydrophobic moieties. Several factors such as the ratio of hydrophobic/hydrophilic chain length and pH of the medium will affect the self-aggregation of chitosan derivatives [44].

2.2.4. Complexible Chitosan

Polyelectrolyte complex (PEC) results from strong electrostatic interactions between two oppositely charged polyelectrolytes. Chitosan is a cationic polyelectrolyte as its amino groups will be protonated in acidic medium. Mixing the solutions of chitosan or its derivative consisting of free amine groups with negatively charged polymer leads to spontaneous aggregation, i.e. forming PEC by ionic self-assembly (ISA) (Figure 5). ISA technology has the advantages of low cost, reliability, mild preparation conditions and the simplicity of synthesis with high purity. It is a facile but powerful tool to create materials with interesting and versatile functions [45].

A polysaccharide-based polyelectrolyte complex of maleic starch half-ester acid (a carboxylic derivative of starch) with chitosan is formed via ISA from aqueous solution [46]. Such a kind of PEC is biodegradable and is potential for drug delivery and other biomedical applications. Chitosan and alginate is another pair of oppositely charged polysaccharides, their PEC beads with entrapped α-amylose is prepared by dropwise addition. The shelf-life of the enzyme is considerably improved [47]. One more example of negatively charged polysaccharide to self-assemble with chitosan is hyaluronic acid (HA). Chitosan/HA PEC film is obtained by drying their mixed solution [48].

As aforementioned, LMWC is water-soluble and inonizable. It is chosen to form PEC with biodegradable and water-soluble poly(L-aspartic acid) in the presence of the third component PEG for protein drug delivery. Here, PEG is used as protein-protective agent [49]. N-carboxyethylchitosan is water-soluble in the whole pH range and is a polyampholyte. It is readily form PEC with strong polyacid poly(2-acryloylamido-2-methylpropanesulfonic acid), weak polyacid poly(acrylic acid), or weak polybase poly(ethylene imine) [50]. Most of PEC is water-insoluble, yet water-soluble PEC can be created. Water-soluble chitosan-g-MPEG and HA at different weight ratios are used to prepare such a kind of PEC for various biomedical applications [51].

The electrostatic self-assembly of chitosan and its derivatives with the oppositely charged natural or synthetic ployelectrolyte is of interest due to its facile preparation and responsibility to suitable environmental stimuli.

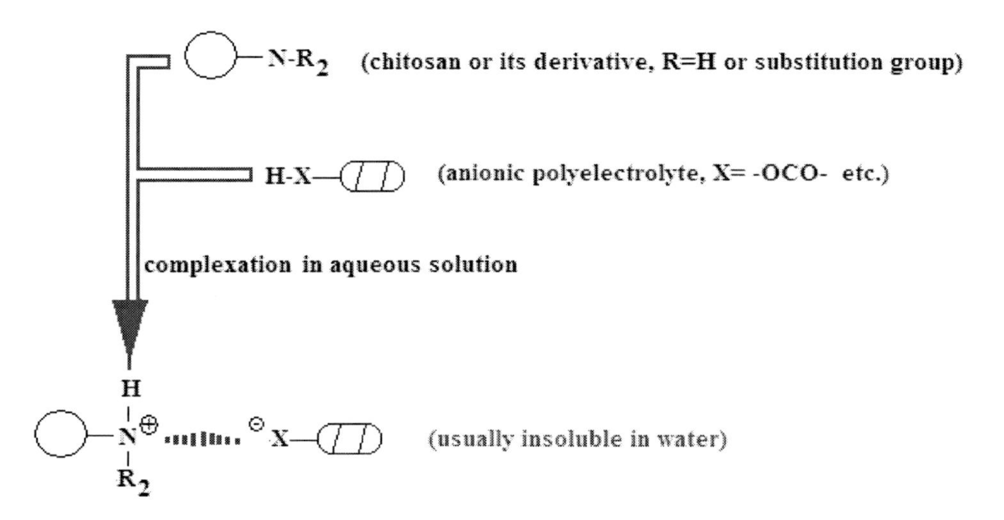

Figure 5. Ionic self-assembly of chitosan-based polyelectrolyte complex.

2.2.5. Miscible Chitosan

Blending is one of the effective modification approaches for chitosan to provide desirable properties for specific purposes. Chitosan is blended with synthetic hydrophilic polymers such as PVA, PEG and poly(vinyl pyrrolidone) (PVP). Compared to chitosan, the blends offer superior properties including improved comfort and reduced irritation, ease of processing, improved film quality and enhanced dissolution. They are promising candidates for formulation in oral drug delivery systems [52]. Water-soluble chitosan derivatives such as carboxyethyl chitosan and quaterized chitosan are blended with PVA via electrospining to give chitosan-based nanofiber. They may be used as wound dressing or show good antibacterial activity [53-54]. There are polar groups in both chitosan and hydrophilic polymers, which result in forming hydrogen bonds and exhibit good miscibility of the components. PVA is less miscible with chitosan [52].To enhance the compatibility of chitosan/PVA, the graft copolymer of chitosan and PVA is blended with PVA. Chitosan-g-PVA/PVA blends offer superior blood compatibility [55]. Ternary chitosan/PVA/pectin blend combines the characteristics of each component. PEC is formed between chitosan and anionic polysaccharide pectin, and the formation of hydrogen bonds among PVA and the other two polymer chains will reduce the phase separation. In viewing of this, the blend can be regarded as a physical semi-interpenetrating polymer network. The ternary film is edible and exhibits positive antimicrobial activity, which indicates such a film is potential as a food-packaging material [56].

Except for synthetic polymers, natural polymer is also often chosen as the hydrophilic polymers to be blended with chitosan. Water-soluble methylcellulose (MC) is more flexible than chitosan. It has excellent film-making property, efficient oxygen and lipid barrier, and compatible structures with chitosan. The mechanical and physical properties of chitosan/MC blend film are improved compared to individual component. This kind of biodegradable film is probably applied in food, pharmaceutical and cosmetic fields [57]. The other cellulose derivative hydroxypropylmethyl cellulose (HPMC), a flexible and nonionic polymer, is also misicible with chitosan [58]. Gelatin is a high molecular weight polypeptide derived from collagen and has good film forming capability. Blending of chitosan and gelatin overcomes the undesirable properties of both components. They are miscible due to the electrostatic interaction, i.e. PEC is formed probably, and hydrogen binding. The blend is stable in water and its electric response is significantly improved. Thus, this blend system is clarified as an actuator [59]. In order to enhance the flexibility and degradation rate of chitosan-based membrane, chitosan is blended with a softener PEG and gelatin. It is suggested that chitosan/PEG/gelatin blend membrane is satisfactory for fabricating guided tissue regeneration membrane [60].

3. Summary

Chitosan is an attractive natural polysaccharide and has received increasing attention. It possesses both unique properties and limitations. Hydroxyl and amino groups on the molecular chain enable chitosan to be easily funcationalized by various chemical and physical means.

Chitosan soluble in wide pH range is convenient to processing and further modifications such as forming PEC, blending and covalently cross-linking to generate hydrogels.

Chitosan or water-soluble chitosan are probably sensitive to environmental variations and capable of reacting with appropriate reagents to obtain stimuli-responsive polymers.

Especially, water-soluble chitosan consisting of some functional groups has ample room to introduce more specific properties.

The prospect of functionalized chitosan derivatives is bright. They are alleged to be potential for food, cosmetics, biomedical, and pharmaceutical applications. In fact, this conclusion is easily to be deduced from the context. The readers can also extend more about functionalizing chitosan such as surface/interface modifications, chitosan-based catalyst or imprinting template and so on.

REFERENCES

[1] Kurita, K. Controlled functionalization of the polysaccharides chitin, *Progress in Polymer Science*, 2001, 26: 1921-1971.

[2] Pillai, CKS; Paul, W; Sharma, CP. Chitin and Chitosan polymers: Chemistry, solubility and fiber formation, *Progress in Polymer Science*, 2009, 34: 641-678.

[3] Prashanth, KVH; Tharanathan, RN. Chitin/chitosan: modifications and their unlimited application potential-an overview, *Trend in Food Science and Technology*, 2007, 18: 117-131.

[4] Berger, J; Reist, M; Mayer, JM; Felt, O; Peppas, NA; Gurny, R. Structure and interactions in covalently and ionically crosslinked chitosan hydrogels for biomedical applications, *European Journal of Pharmaceutics Biopharmaceutics*, 2004, 57, 19-34.

[5] Mourya, VK; Inamdar, NN. Chitosan-modifications and applications: Opportunities and galore, Reactive and Functional Polymers, 2008, 68: 1013-1051.

[6] Kim, S-K; Rajapake, N. Enzymatic production and biological activities of chitosan oligosaccharides (COS): a review, *Carbohydrate Polymers*, 2005, 62: 357-368.

[7] Qin, CQ; Du, YM; Xiao, L; Li, Z; Gao, XH. Enzymic preparation of water-soluble chitosan and their antitumor activity, *International Journal of Biological Macromolecules*, 2002, 31: 111-117.

[8] Nah, J-W; Jang, M-K. Spectroscopic characterization and preparation of low molecular, water-soluble chitosan with free-amine group by novel method, *Journal of Polymer Science Part A: Polymer Chemistry*, 2002, 40: 3796-3803.

[9] Yue, W; Yao, PJ; Wei, YN; Li, SQ; Lai, F; Liu, XM. An innovative method for preparation of acid-free-water-soluble low-molecular-weight chitosan (AFWSLMWC), *Food Chemistry*, 2008, 108: 1082-1087.

[10] Moon, J-S; Kim, H-K; Koo, HC; Joo, Y-S; Nam, H-M; Park, Y-H; Kang, M-I. The antibacterial and immunostimulative effect of chitosan-oligosaccharides against infection by *Staphylococcus aureus* isolated from bovin mastitis, *Applied Microbiology and Biotechnology*, 2007, 75: 989-998.

[11] Kim, D-G; Jeong, Y-I; Choi, C; Roh, S-H; Kang, S-K; Jang, M-K; Nah, J-W; Retinol-encapsulated low molecular water-soluble chitosan nanoparticles, *International Journal of Pharmaceutics*, 2006, 319: 130-138.

[12] Saxena, A; Kumar, A; Shahi, VK. Preparation and characterization of N-methylene phosphonic and quaternized chitosan composite membranes for electrolyte separations, *Journal of Colloid and Interface Science*, 2006, 303: 484-493.

[13] Kto, Y; Onishi, H; Machida, Y. N-succinyl-chitosan as a drug carrier: water-insoluble and water-soluble conjugates, *Biomaterials*, 2004, 25: 907-915.

[14] Zhou, JQ; Wang, JW. Immobilization of alliinase with a water soluble-insoluble reversible N-succinyl-chitosan for allicin production, *Enzyme and Microbial Technology*, 2009, 45: 299-304.

[15] Zhu, AP; Pan, YN; Liao, TQ; Zhao, F; Chen T. The synthesis and characterization of polymerizable N-maleic acyl-chitosan, *Journal of Biomedical Materials Research Part B: Applied Biomaterials*, 2008, 85B: 489-495.

[16] Chen, C-L; Wang, Y-M; Liu, C-F; Wang, J-Y. The effect of water-soluble chitosan on manrophge activation and the attenuation of mite allergen-induced airway inflammation, *Biomaterials*, 2008, 29: 2173-2182.

[17] Yamamoto, N; Aiba, S-I. Chemical modification of chitosan. 14: Synthesis of water-soluble chitosan derivtives by simple acetylation, *Biomacromolecules*, 2002, 3: 1126-1128.

[18] de Abreu, FR; Campana-Filho, SP. Characteristics and properties of carboxymethylchitosan, *Carbohydrate Polymers*, 2009, 75: 214-221.

[19] Sashiwa, H; Yamamori, N; Ichinose, Y; Sunamoto J; Aiba, S-I. Chemical modification of chitosan. 17: Michael reaction of chitosan with acrylic acid in water, *Macromolecular Bioscience*, 2003, 3: 231-233.

[20] Du, J; Hsieh, Y-L. PEGylation of chitosan for improved solubility and fiber formation via electrospinning, *Cellulose*, 2007, 14: 543-552.

[21] Gorochovceva, N; Makuška, R. Synthesis and study of water-soluble chitosan-O-poly(ethylene glycol) graft copolymers, *European Polymer Journal*, 2004, 40: 685-691.

[22] Hu, YQ; Jiang, HL; Xu, CN; Wang, YJ; Zhu, KJ. Preparation and characterization of poly(ethylene glycol)-g-chitosan with water- and organosolubility, *Carbohydrate Polymers*, 2005, 61: 472-479.

[23] Cho, JH; Kim, S-H; Park, KD; Jung, MC; Yang, WI; Han, SW; Noh, JY; Lee, JW. Chondrogenic differentiation of human mesenchymal stell cells using a thermosensitive poly(N-isopropylacrylamide) and water-soluble chitosan copolymer, *Biomaterials*, 2004, 25: 5743-5751.

[24] Cao, YX; Zhang, C; Shen, WB; Cheng, ZH; Yu, L; Ping, QE. Poly (N-isopropylacrylamide)-chitosan as thermosensitive in situ gel-forming system for ocular drugs delivery, *Journal of Controlled Release*, 2007, 120: 186-194.

[25] Tang, J; Hua, DB; Cheng, JX; Jiang, J; Zhu, XL. Synthesis and properties of temperature-responsive chitosan by controlled free radical polymerization with chitosan-RAFT agent, *International Journal of Biological Macromolecules*, 2008, 43: 383-389.

[26] Ganji, F; Abdekhodaie MJ. Synthesis and characterization of a new thermosensitive chitosan-PEG diblock copolymer, *Carbohydrate Polymers*, 2008, 74: 435-441.

[27] Gou, B-L; Yuan, J-F; Gao, QY. Preparation and release behavior of temperature- and pH-responsive chitosan material, *Polymer International*, 2008, 57: 463-468.

[28] Chuang, C-Y; Don, T-M; Chiu, W-Y. Synthesis and properties of chitosan-based thermo- and pH-responsive nanoparticles and application in drug release, *Journal Polymer Science Part A: Polymer Chemistry*, 2009, 47: 2798-2810.

[29] Xiao, CM; Gao, F; Gao YK. Controlled preparation of physically cross-linked chitosan-g-poly(vinyl alcohol) hydrogel, *Journal of Applied Polymer Science*, 2010, 117: 2946-2950.

[30] Zhou, YS; Yang, DZ; Ma, GP; Tan, HL; Jin, Y; Nie, J. A pH-sensitive water-soluble N-carboxyethyl chitosan/poly(hydroxyethyl methacrylate) hydrogel as a potential drug sustained release matrix prepared by photopolymerization techniques, *Polymers for Advanced Technologies*, 2008, 19: 1133-1141.

[31] Chiu, YL; Chen, S-C; Su, C-J; Hsiao, C-W; Chen, Y-M; Chen, H-L; Sung, H-W. pH-triggered injectable hydrogels prepared from aqueous N-palmitol chitosan: In vitro characteristics and in vivo biocompatibility, *Biomaterials*, 2009, 30: 4877-4888.

[32] Sakai, S; Yamada, Y; Zenke, T; Kawakami, K. Novel chitosan derivative soluble at neutral pH and in-situ gellable via peroxidase-catalyzed enzymatic reaction, *Journal of Materials Chemistry*, 2009, 19: 230-235.

[33] Shang, J; Shao, ZZ; Chen, X. Chitosan-based electroactive hydrogel, *Polymer*, 2008, 49: 5520-5525.

[34] Carreira, AS; et al. Temperature and pH responsive polymers based on chitosan: Applications and new graft copolymerization strategies based on living radical polymerization, *Carbohydrate polymers* (2010), doi: 10.1016/j.carbpol.2009.12.047.

[35] Li, GY; Huang, K-L; Jiang, Y-R; Ding, P; Yang, D-L. Preparation and characterization of carboxyl functionalization of chitosan derivative magnetic nanoparticles, *Biochemical Engineering Journal*, 2008, 40: 408-414.

[36] López-Cruz, A; Barrera, C; Calero-DdelC, VL; Carlos, R. Water dispersible iron oxide nanoparticles coated with covalently linked chitosan, *Journal of Materials Chemistry*, 2009, 19: 6870-6876.

[37] Zhang, C; Ping, QE; Zhang, HJ; Shen, J. Preparation of N-ackyl-O-sulfate chitosan derivatives and micellar solubilization of taxol, *Carbohydrate Polymers*, 2003, 54: 137-141.

[38] Zhu, AP; Chen, T; Yuan, LH; Wu, H; Lu, P. Synthesis and characterization of N-succinyl-chitosan and its self-assembly of nanospheres, *Carbohydrate Polymers*, 2006, 66: 274-279.

[39] Mincheva, R; Bougard, F; Paneva, D; Vachaudez, M; Manolova, N; Dubois, P; Rashkov, I. Self-assembly of N-carboxyethylchitosan near the isoelectric point, *Journal of Polymer Science Part A: Polymer Chemistry*, 2008, 46: 6712-6721.

[40] Li, G; Zhuang, YL; Mu, Q; Wang, MZ; Fang, YE. Preparation, characterization behavior of amphiphilic chitosan derivative having poly(L-lactic acid) side chains, *Carbohydrate Polymers*, 2008, 72: 60-66.

[41] Yoksan, R; Akashi, M; Hiwatari, K-I; Chirachanchai, S. Controlled hydrophobic/hydrophilicity of chitosan for spheres without specific processing technique, *Biopolymers*, 2003, 69: 386-390.

[42] Cai, GQ; Jiang, HL; Chen, ZJ; Tu, KH; Wang, LQ; Zhu, KJ. Synthesis, characterization, and self-assemble behavior of chitosan-O-poly(ε-caprolactone), *European Polymer Journal*, 2009, 45: 1674-1680.

[43] Liu, K-H; Chen, S-Y; Liu, D-M; Liu, T-Y. Self-assembled hollow nanocapsule from amphiphatic carboxymethyl-hexanoyl chitosan as drug carrier, *Macromolecules*, 2008, 41: 6511-6516.

[44] Choochottios, C; Yoksan, R; Chirachanchai, S. Amphiphilic chitosan nanospheres: Factors to control nanosphere formation and its consequent pH responsive performance, *Polymers*, 2009, 1877-1886.

[45] Faul, CFJ; Antonietti, M. Ionic self-assembly: facile synthesis of supramolecular materials, *Advanced Materials*, 2003, 15: 673-683.

[46] Xiao, CM; Fang, F. Ionic self-assembly and characterization of a polysaccharide-based polyelectrolyte complex of maleic starch half-ester acid with chitosan, *Journal of Applied Polymer Science*, 2009, 112: 2255-2260.

[47] Sankalia, MG; Mashru, RC; Sankalia, JM; Sutariya, VB. Reversed chitosan-alginate polyelectrolyte complex for stability improvement of alpha-amylose: Optimization and physicochemical characterization, *European Journal of Pharmaceutics and Biopharmaceutics*, 2007, 65: 215-232.

[48] Kim, SJ; Lee, KJ; Kim, SI. Swelling behavior of polyelectrolyte complex hydrogels composed of chitosan and hyaluronic acid, *Journal of Applied Polymer*, 2004, 93: 1097-1101.

[49] Shu, SJ; Zhang, XG; Teng, DY; Wang, Z; Li, CX; Polyelectrolyte nanoparticles based on water-soluble chitosan-poly(L-aspartic acid)-polyethylene glycol for controlled protein release, *Carbohydrate Research*, 2009, 344: 1197-1204.

[50] Mincheva, R; Manolova, N; paneva, D; Rashkov, I. Novel polyelectrolyte complexes between N-carboxyethylchitosan and synthetic polyelectrolyte, *European Polymer Journal*, 2006, 42: 858-868.

[51] Wu, J; Wang, XF; Keum, JK; Zhou, HW; Gelfer, M; Avila-Orta, C-A; Pan, H; Chen, WL; Chiao, S-M; siao, BS; Chu, B. Water-soluble complexes of chitosan-g-MPEG and hyaluronic acid, *Journal of Biomedical Materials Research Part A*, 2007, 80A: 800-812.

[52] Khoo, CGL; Frantzich, S; Rosinski, A; Sjŏstrŏm, M; Hoogstraate, J. Oral gingival delivery systems from chitosan blends with hydrophilic polymers, *European Journal of Pharmaceutics and Biopharmaceutics*, 2003, 55: 47-56.

[53] Zhou, YS; Yang, DZ; Chen, XM; Xu, Q; Lu, FM; Nie, J; Electrospun water-soluble carboxyethyl chitosan/poly(vinyl alcohol) nanofibrous membrane as potential wound dressing for skin regeneration, *Biomacromolecules*, 2008, 9: 349-354.

[54] Alipour, SM; Nonri, M; Mokhtari, J; Bahrami, SH. Electrospining of water-soluble quaternized chitosan derivative blend, *Carbohydrate Research*, 2009, 344: 2496-2501.

[55] Don, T-M; King, C-F; Chiu, W-Y; Peng, C-A. Preparation and characterization of chitosan-g-poly(vinyl alcohol) blends used for the evaluation of blood-contacting compatibility, *Carbohydrate Polymers*, 2006, 63: 331-339.

[56] Tripathi, S; Mehrotra, GK; Dutta, PK. Preparation and physicochemical evaluation of citosan/poly(vinyl alcohol)/pectin ternary film for food-packaging application, *Carbohydrate Polymers*, 2010, 79: 711-716.

[57] Pinotti, A; García, MA; Martino, MN; Zartzky, NE. Study on microstructure and physical properties of composite films based on chitosan and methylcellulose, *Food Hydrocolloids*, 2007, 21: 66-72.

[58] Jayaraju, J; Raviprakash, SV; Keshavayya, J; Rai, SK. Miscibility studies on chitosan/hydroxypropylmethyl cellulose blend in solution by viscosity, ultrasonic velocity, density, and refractive index methods, *Journal o f Applied Polymer Science*, 2006, 102: 2738-2742.

[59] Haide S; Park, S-Y; Lee, S-H. Preparation, swelling and electro-mechano-chemical behaviors of a gelatin-chitosan blend membrane, *Soft Matter*, 2008, 4: 485-492.

[60] Hong, H; Liu, CS; Wu, WJ. Preparation and characterization of chitosan/PEG/gelatin composites for tissue engineering, *Journal of Applied Polymer Science*, 2009, 114: 1220-1225.

In: Handbook of Chitosan Research and Applications
Editors: Richard G. Mackay and Jennifer M. Tait
ISBN 978-1-61324-455-5

Chapter 17

The Unusual Bell-Like Dependence of the Activity of Chitosan against *Penicillium Vermoesenii* on Chitosan Molecular Weight

Vladimir Tikhonov*[1,1*]*, Evgeniya Stepnova*[1]*,
Sergey Lopatin*[2,2]*, Valery Varlamov*[2]*, Alla Ilyina*[2]
***and Igor Yamskov*[1]**

[1]. A.N. Nesmeyanov Institute of Organoelement Compounds,
Russian Academy of Sciences, Moscow, Russia
[2]. Centre "Bioengineering",
Russian Academy of Sciences, Moscow, Russia

Abstract

Chitosan has been reported to be a non-toxic, biocompatible antibacterial, antifungal and insecticidal agent. In many publications it has been shown that the biocidal activity of chitosan can grow or reduce with the reduction of chitosan molecular weight (MW). In some experiments it has been shown that the antimicrobial activity of chitosan does not depend on MW or even controversially differs for Gram-positive and Gram-negative bacteria. In this work the fungistatic efficacy of a set of low polydispersity chitosans varying in molecular weight from 1.2 to 90.0 kDa towards the filamentous palm parasitic fungus *Penicillium vermoesenii* has been studied *in vitro*. It has been shown for the first time that the activity of chitosan against the fungus has unusual bell-like dependence on chitosan molecular weight so that chitosans having the molecular weight between 5÷10

[1] A.N. Nesmeyanov Institute of Organoelement Compounds, Russian Academy of Sciences, Vavilov st. 28, 119991 Moscow, Russia.

* Corresponding author: Vladimir Tikhonov. Tel.: 007-499-1359375; fax: 007-499-1355085. E-mail address: tikhon@ineos.ac.ru.

[2] Centre "Bioengineering", Russian Academy of Sciences, Prospekt 60-Let Oktyabrya 7/1, 117312 Moscow, Russia.

kDa possess the highest fungistatic activity while chitosans with both lower and higher molecular weights are significantly less active.

Keywords: *Chitosan, fungistatic activity, Penicillium vermoesenii, Washingtonia filifera pink bud rot*

INTRODUCTION

Chitin, poly[β-(1→4)-2-acetamido-2-deoxy-D-glucose], is the most abundant among the natural amino polysaccharides and represents a constituent of the cell walls of most fungi, algae, insects and crustaceans. Chitosan is a collective name accepted for a group of partially and fully deacetylated chitins soluble in an acidic aqueous media. This poly(aminosaccharide) was found to be non-toxic to higher organisms, biocompatible and biodegradable, and has found several applications in food and pharmaceutical industries, mainly due to its high biodegradability and antimicrobial properties [1-6]. Chitooligosaccharides derived from chitin and chitosan by plant enzymes are important biological signal molecules because they play a role in plant defense response against plant pathogens in healthy and infected plants [7-10]. It should be mentioned that chitosan and its residues have been recognized to be environmentally friendly by US EPA since 1995 and they are used as bactericides and fungicides for commercial crops, ornamental plants and turf. Chitosan affects different stages of fungal development: mycelium growth, sporulation and spore germination [11-15]. It was also shown that chitosan suppressed growth of date-palm (*Phoenix dactylifera*) parasitic fungus *F.oxysporum f.sp. albedinis* and elicited defensive reactions in date-palm roots. When it was injected in roots of *Phoenix dactylifera*, chitosan elicited peroxidase and polyphenol oxidase activity, and increased the accumulation of nonconstitutive hydroxycinnamic acid derivatives known to be of great importance in date-palm resistance towards fungal infections [13].

Although numerous studies and reviews on microbiological activity of chitosan against many bacteria, filamentous fungi and yeasts have been published [11-25], the controversial evidences for a correlation between biocidal activity and molecular weight (MW) of chitosan have been found. The variation in chitosan molecular weight leads to two different mechanisms of chitosan and target microorganism interaction: the first – adsorption of chitosan onto cell walls leading to cell walls covering, membrane weakening, disruption and cell leakage - is mainly connected with a high molecular weight (HMW) chitosan; the second – penetration of chitosan into living cells leading to the inhibition of various enzymes and interfacing with the synthesis of mRNA and proteins - is mainly connected with a low molecular weight (LMW) chitosan [20, 21, 26-31]. It was found in some studies that the increase in chitosan molecular weight led to the decrease in biocidal activity of chitosan [11, 15, 18, 26, 32-34]. In the others an increased activity for high molecular weight chitosans in comparison with low molecular weight chitosans was found [35-43]. Despite the higher activity of HMW chitosan (MW 1671 ÷28 kDa) in comparison with LMW chitosan (MW 22÷1 kDa), no significant difference in activity was found between the LMW chitosans [44]. Although all chitosans (MW 0.5, 37, 57, and 290 kDa) at concentration of 0.1% completely suppressed Gray mold of tomato fruit caused by *Botrytis cinerea (Pers.)*, 57 kDa chitosan was found a bit more active than the others at lower concentrations [45]. On the other hand, when the antifungal activity of chitosans (MW 1, 3, 5, and 10 kDa) was examined against various human and plant pathogenic fungi, the activity of 10 kDa chitosan was found higher than that of the other LMW chitosans against all the tested fungal cells [46]. A set of LMW chitosans, which were used in the experiments against some Gram-negative and Gram-positive bacteria and yeast-like fungi *C. albicans* and *C. kruisei*, mainly revealed the insignificant variations in antibacterial activity between 5, 6, 10, 12, 15.7, and 27 kDa chitosans, although the activity was found to be specie-specific [17]. In general, HMW chitosan showed stronger bactericidal activity against Gram-positive bacteria than Gram-negative bacteria while LMW chitosan demonstrated stronger effect for Gram-negative bacteria [17, 30, 44, 47].

In our opinion, the controversial results in a correlation between biocidal activity and chitosan molecular weight have been found mainly because so far most investigators either have been using only few chitosan samples, or haven't been taking into account their polydispersity. No evidence of an extreme bell-like dependence of the biocidal activity of chitosan on chitosan molecular weight has been found so far.

In this paper we describe for the first time the *in vitro* fungistatic efficacy of a set of chitosan samples differing in molecular weights towards the filamentous fungus *Penicillium vermoesenii*, which represents one of the most lethal palm pathogenic fungi causing the Pink Bud Rot of *Washingtonia filifera*. The disease is widespread, affecting palms in the landscapes wherever *Washingtonia filifera* grows: in Spain, the Canary Islands, California and other tropical and subtropical areas [48-51].

EXPERIMENTAL

Materials

Low molecular weight chitosan (LMWChitosan) was purchased from ALDRICH. Antifungal assay agar (AA-agar) and glucosamine hydrochloride were from SIGMA. Chitosans having weight-average molecular masses 22.1, 29.2 and 40.5 kDa were prepared by enzymatic hydrolysis of LMWChitosan using the chitinolytic enzyme complex *Streptomyces kursanovii* following the described protocol [52]. The set of chitosan hydrochlorides was produced by means of acidic hydrolysis of LMWChitosan following the modified method [53] in which 6M hydrochloric acid/ethanol (1:1 v/v) mixture was used as a solvent and the hydrolysis was carried out at 50^0C for 0.2÷3h. After the hydrolysis the mixture was diluted with ethanol, and chitosan hydrochloride was collected by filtration, dried over P_2O_5 under vacuum at 20^0C and kept in a dark-glass tube. *P. vermoesenii* from the culture collection of the Laboratory for Plant Pathology, Department of Marine Sciences and Applied Biology (University of Alicante, Spain) was supplied by Dr. Luis V. Lopez-Llorca.

Methods

All chitosan samples were analysed by the High Performance Gel Permeation Chromatography (HPGPC) method in aqueous salt solution (0.15M ammonium acetate/

0.05M acetic acid, pH 5.2) on Ultrahydrogel 500 column 7.8 × 300mm «Waters» at 30^0C and the flow rate 0.5 ml/min. Chromatograms of the chitosan samples were recorded using RI detector Knauer K-2301. The weight-average molecular mass (Mw) the number-average molecular mass (Mn), and the index of polydispersity IP = M_w/M_n were calculated by MultiChrom software ("Ampersand", Russia). The column was calibrated with dextran molecular weight standards (1080, 4440, 9890, 43500, 66700, 123600 and 196300 Da) purchased from SIGMA.

^{1}H-Nuclear Magnetic Resonance was used to determine a degree of deacetylation of chitosan (DA) in accordance with the method described [54].

Fungistatic Assessments

Fungistatic activity of the samples was determined by a radial hyphal growth bioassay. A sample to be tested was dissolved in 50 ml water (in 0.05M hydrochloric acid in the case of LMWChitosan) and pH of the solution was adjusted to 5.0-5.5 with 1M $NaHCO_3$ solution. 7.5% (w/v) AA-agar media containing a sample of chitosan was sterilized at 121^0C for 10 min and then poured onto sterile 9-mm diameter Petri dishes (special experiments demonstrated the equality in fungistatic activities of chitosan samples sterilized at these conditions and activities of the samples sterilized by sterile filtration method). The dishes were inoculated with 6-mm-diameter plugs taken from the margins of 3-5 days old colonies of *P. vermoesenii* grown on AA-agar. Five replicas for each sample concentrations (0.1, 1.0, and 10 mg/ml) were used. Control plates (AA-agar only) were also inoculated with the fungus. All the dishes were incubated at 20^0C in the dark. The radial colony growth was measured after 5, 10, and 15 days. Control experiments were carried out under the same conditions without chitosan. All the experiments were carried out twice.

Results and Discussion

Samples Preparation and Characterization

In order to prepare the set of chitosan samples differing in molecular weights two methods were used for depolymerization of LMWChitosan and preparation of chitosans having molecular weights varying from 1.1 kDa to 90.0 kDa. Depolymerization of LMWChitosan with 6M hydrochloric acid was carried out at 50^0C for 0.5÷3 hours. This process led to the cleavage of glycosidic bonds and formation of chitosan of much lower molecular weight depending on the duration of the LMWChitosan hydrolysis (Tabl.1).

The partial deacetylation was also observed, so that a degree of acetylation of chitosan obtained after the acidic hydrolysis (samples 3-12) was lower than that of the parent chitosan in accordance with previously published data [55, 56]. Partial depolymerization of LMWChitosan was also carried out to reduce its molecular weight by the chitinolytic enzyme mixture produced by *S. kursanovii*. This enzyme mixture consisted of three endochitinases and N-acetylglucosaminidase, which were able to hydrolyze not only a colloidal chitin but also a soluble chitosan, produced chitosan (samples 13-15) with a reduced molecular weight compared with that of the parent chitosan [52, 57, 58].

Table 1. Chitosan samples characterization

Sample number	M_w, kDa	IP	DA, mol.%
1 (control)	-	-	-
2 (glucosamine)	0.161	1	0
3	1.2	1.2	10
4	1.3	1.2	12
5	1.5	1.4	6
6	1.6	1.2	6
7	2.8	1.2	4
8	4.1	1.4	5
9	5.1	1.6	4
10	7.5	1.6	7
11	9.9	1.4	8
12	15.0	1.9	11
13	24.1	2.2	12
14	29.2	2.8	10
15	40.5	3.2	10
16	90.0	5.5	15

After acidic or enzymatic hydrolysis, chitosan samples were analyzed by HPGPC to determine their molecular weights and polydispersity. These characteristics of the chitosan samples used in the investigations are listed in Table 1 together with the data of parent chitosan and glucosamine. Ten chitosan samples having Mw 1.2÷15.0 kDa and low polydispersity index PI 1.2÷1.9 obtained by acidic hydrolysis (samples 3-12) and 3 chitosan samples having Mw 24.1÷40.5 kDa and polydispersity PI 2.2-3.2 (samples 13-15) prepared by enzymatic hydrolysis were used in the bioassay together with glucosamine (sample 2) and the parent chitosan (sample 16). The degrees of deacetylation of chitosan samples were also determined, and these data represented additional characteristics which may be also useful for further investigations.

Molecular Weight Dependent Fungistatic Activity

In spite of the controversial results on the dependence of chitosan's biocidal activity versus its molecular weight published earlier, it is now commonly accepted that the biocidal activity of chitosan does depend on its molecular weight. Only few chitosan samples of different molecular weights have been used in earlier studies, and, furthermore, there is a lack of the data on biocidal activity of chitosans of widely different molecular weights and low polydispersities. Therefore, the basic question which we ought to clarify first of all was what molecular weight should chitosan have to possess the highest biocidal activity?

In our study the fungistatic activity of the chitosans having the molecular weights from 1.2 kDa to 90.0 kDA against *P. vermoesenii* was determined by measuring the colony diameter at several concentrations of chitosan: 0.1 mg/ml (0.01% w/v), 1 mg/ml (0.1% w/v), and 10 mg/ml (1% w/v).

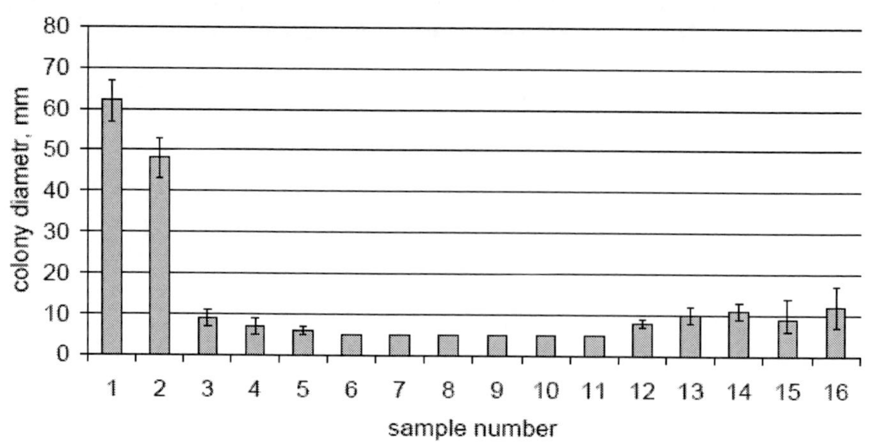

A.

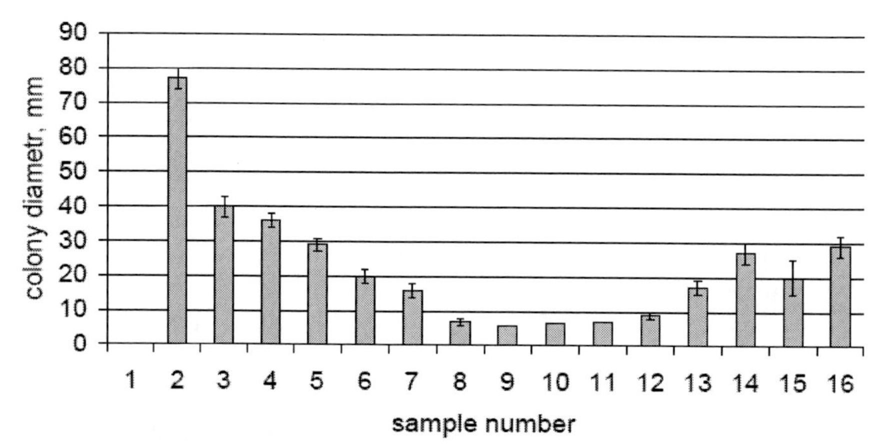

B.

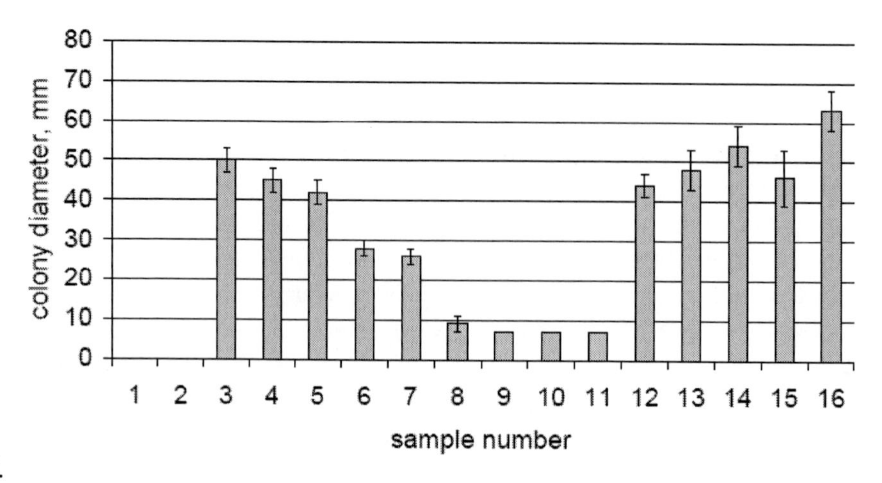

C.

Figure 1. Fungal colony diameters after 5 days (A), 10 days (B) and 15 days (C). Chitosan concentration: 1 mg/ml. Bars show standard errors.

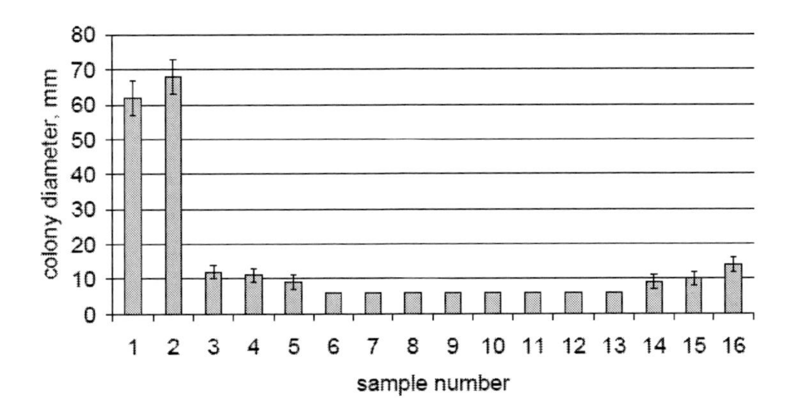

A.

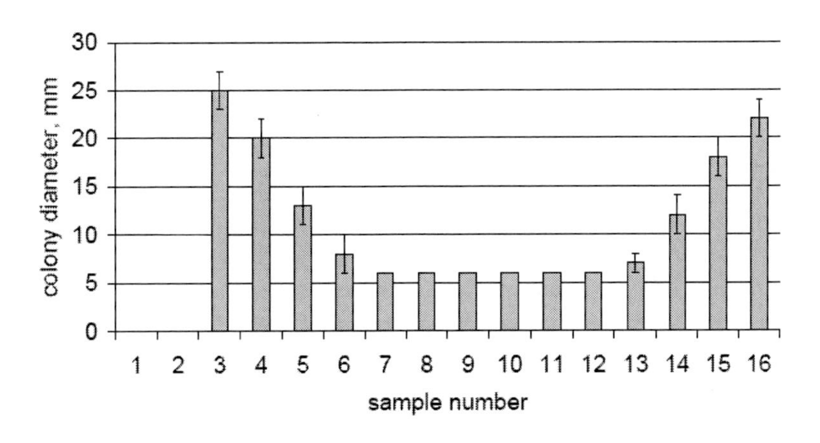

B.

Figure 2. Fungal colony diameters after 5 (A) and 15 (B) days after inoculation (chitosan concentration: 10 mg/ml). Bars show standard errors.

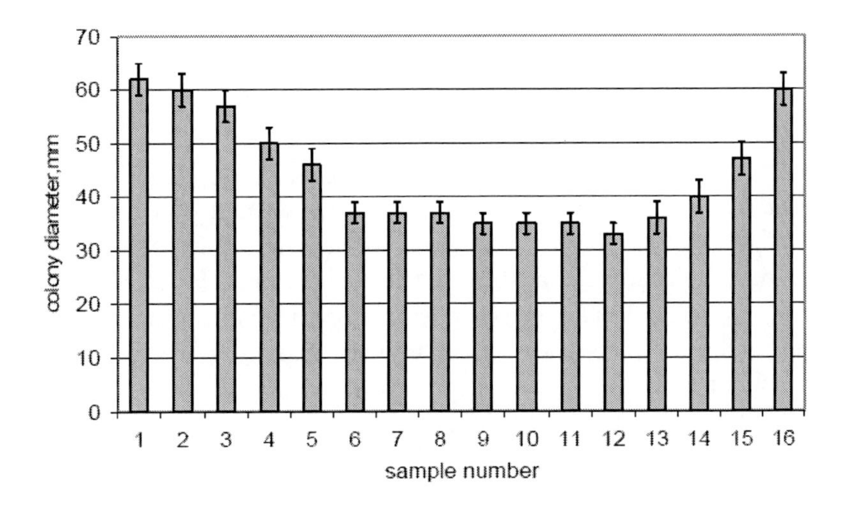

Figure 3. Fungal colony diameter after 5 days (chitosan concentration: 0.1 mg/ml). Bars show standard errors.

Antifungal assay agar used in these experiments represents a substrate rich in dextrose, vitamins, salts and minerals, and it has been recommended for assaying antifungal activity of fungicides at conditions similar to natural ones. Comparative data on fungistatic activities of 1 mg/ml (0.1% w/v) concentration of chitosans in 5, 10, and 15 days after the inoculation are shown in Figure1a-c. After 5 days the *P.vermoesenii* colony growth was slightly suppressed by glucosamine and significantly suppressed by chitosan (Figure 1a). Seven days after inoculation, the fungus grown on the control media (without a chitosan sample) reached the edges of the Petri dishes and therefore these dishes were excluded from the further measurements. As shown in Figure1b, glucosamine slightly and chitosans markedly inhibited the fungal growth, and the inhibitory effect varied with regards to the molecular weight of chitosan. Chitosan samples 3-8 suppressed the colony development in accordance with the increase in their molecular weights, while no fungal growth was observed on the Petri dishes containing chitosan samples 9-11. Further increase in chitosan molecular weight (samples 12-16) led to the gradual increase of fungal growth. Figure1c shows the colony diameter after 15 days of observation. Even after this time of inoculation chitosan samples 9-11 (concentration 1 mg/ml) completely inhibited the growth of the fungus. The inhibitory effect of the other chitosan samples shown in the left and right parts of the graph followed the above mentioned tendencies. The molecular weight dependence of antifungal activity of chitosan at higher concentration (10 mg/ml) was not clearly expressed since samples 3-15 practically completely suppressed *P. vermoesenii* growth after 5 days of incubation (Figure2a). Sample 16, which could not be completely dissolved at our experimental conditions and partially precipitated in the AA-agar medium, was found to be less effective than other samples. Meanwhile, *P. vermoesenii* grew in the presence of 10 mg/ml concentration of glucosamine even slightly better than at 1 mg/ml concentration (Figure2a). This effect may be related with the significant enrichment of the medium with a carbon/nitrogen source. The fungistatic effect of samples 7-12, which were more soluble at 10 g/l concentration in AA-agar media in comparison with samples 13-16 after 15 days of growth, is clearly shown in Figure2b. All chitosan samples at lower concentration (0.1 mg/ml) only slightly inhibited the fungal colony growth (Figure3).

CONCLUSION

The fungistatic activity of different molecular weight and low polydispersity chitosan set towards *Penicillium vermoesenii*, which causes the Pink Bud Rot of California fan palms (*Washingtonia filifera*), has been investigated for the first time. The experiments have demonstrated that the inhibitory activity of chitosan against the fungus has a bell-like dependence on chitosan molecular weight so that chitosans having the molecular weight 5÷10 kDa possess the highest fungistatic activity and the activity grows with the increase in chitosan concentration.

Due to the environmental compatibility of chitosan further studies focusing chitosan effect at different stages of fungal life cycle may help to create an ecologically safer substitutive fungicide based on 5÷10 kDa chitosan against palm and other plant parasitic fungi instead of synthetic chemical one.

REFERENCES

[1] Muzzarelli, RAA; Jeuniaux, C; Gooday, GW. Chitin in Nature and Technology. Plenum Press. New York and London; 1986.

[2] Bartnika-Garcia, S. Cell wall chemistry. Annual Rev. Microbiol, 1968, 22, 87-108.

[3] Muzzarelli, RAA. Chitin: Pergamon Press. Oxford; 1977.

[4] Knorr, D. Use of chitinous polymer in food – a challenge for food research and development. Food Technol., 1984, 38, 85-97.

[5] Knorr D. Nutritional quality food processing and biotechnology aspects of chitin and chitosan: a review. Proc. Biochem., 1986, 6, 90-92.

[6] Qin C; Gao J; Wang L; Zeng L; Lin Y. Safety evaluation of short-term exposure to chitooligomers from enzymic preparation. Food and Chemical Toxicology, 2006, 44, 855-861.

[7] Fry, SC; Hetherington, PR. Oligosaccharides as Signals and Substrates in the Plant Cell Wall. J. Plant Physiol., 1993, 103 1-5.

[8] Côté, F; Hahn, MG. Oligosaccharins: structure and signal transduction. Plant Mol. Biol., 1994, 26, 1397-1411.

[9] Shibuya, N; Minami, E. Oligosaccharide signaling for defense responses in plant. Physiol. Mol. Plant Pathol., 2001, 59, 223-233.

[10] Zhao, X; She X.; Du Y.; Liang X.; Induction of antiviral resistance and stimulatory effect by oligochitosan in tobacco. Pesticide Biochem. Physiol., 2007, 87, 78-84.

[11] Xu, J; Zhao, X; Han, X; Du Y. Antifungal activity of oligochitosan against Phytophthora capsici and other plant pathogenic fungi in vitro. Pesticide Biochem.Physiol., 2007, 87, 220-228.

[12] Muzzarelli, RAA; Muzzarelli, C; Tarsi, R; Miliani, M; Gabbanelli, F; Gartolari, M. Fungistatic activity of modified chitosan against *Saprolegnia parasitica.* Biomacromolecules, 2001, 2, 165-169.

[13] Barka, EA; El-Hadrani, A; El-Hadrani, I; El-Hassni, M; Daayf, F. Chitosan - antifungal product against *Fusarium oxysporum f. sp. albedinis* and elicitor of defence reactions in data palm roots. Phytopathologia Mediterranea, 2004, 43, 195-204.

[14] Palma-Guerrero, J; Jansson, H.-B; Salinas, J; Lopez-Llorca, LV. Effect of chitosan on hyphal growth and spore germination of plant pathogenic and biocontrol fungi. J. Appl. Microbiol., 2008, 104, 541-553.

[15] Hernández-Lauzardo, AN; Bautista-Baños, S; Velázquez-del Valle, MG; Méndez-Montealvo, MG; Sánchez-Rivera, MM; Bello-Perez, LA. Antifungal effects of chitosan with different molecular weight on in vitro development of *Rhizopuz stolonifer* (Ehrenb.:Fr) Vuill. Carbohydr. Polym., 2008, 73, 541-547.

[16] Barka, EA; Eullaffroy, P; Clément, P; Vernet, G. Chitosan improves development and protects *Vitis vinifera L.* against *Botrytis cinerea.* Plant Cell Rep., 2004, 22, 608-614.

[17] Gerasimenko, D.V; Avdienko, ID; Bannikova, G.E; Zueva, OYu; Varlamov, VP. Antibacterial effects of water-soluble low-molecular-weight chitosans on different microorganisms. Appl. Biochem.Microbiol., 2004, 40(3), 253-257.

[18] Tikhonov, VE; Stepnova, EA; Babak, VG; Yamskov, IA; Palma-Guerrero, J; Jansson, H-B; Lopez-Llorca, LV; Salinas, J; Gerasimenko, DV; Avdienko, ID; Varlamov, VP.

Bactericidal and antifungal activities of a low molecular weight chitosan and its N-/2(3)-(dodec-2-enyl)succinoyl/-derivatives. Carbohydrate Polymers, 2006, 64, 66-72.

[19] Bautista-Baños, S; Hernández-Lauzardo, AN; Velázquez-del Valle, MG; Hernández-López M; Barka, EA; Bosquez-Molina, E. Chitosan as a potential natural compound to control pre and post harvest diseases of horticultural commodities: review. Crop Protection, 2006, 25, 108-118.

[20] Rabea, EA; Badawy, MET; Stevens, CV; Smagghe, G; Steurbaut, W. Chitosan as antimicrobial agent: applications and mode of action. Biomacromolecules, 2003, 4, 1457-1465.

[21] Chirkov, SN. The antiviral activity of chitosan (review), Applied Biochem. Microbiol., 2002, 38, 1-8.

[22] Pospieszny, H. Antiviroid activity of chitosan. Crop Protection, 1997, 16, 105-106.

[23] Kim, S.-K; Rajapakse, N. Enzymatic production and biological activities of chitosan oligosaccharides (COS): A review. Carbohydrate Polymers, 2005, 62, 357-368.

[24] Singh, T; Vesentini, D; Singh, AP; Daniel, G. Effect of chitosan on physiological, morphological, and ultrastructural characteristics of wood-degrading fungi. Int. Biodeterior. Biodegrad., 2008, 62, 116-124.

[25] Palma-Guerrero, J; Jansson, H-B; Salinas, J; Lopez-Llorca, LV. Effect of chitosan on hyphal growth and spore germination of plant pathogenic and biocontrol fungi. J.Appl.Microbiol., 2008, 104, 541-553.

[26] Zheng, L-Y; Zhu, J-F. Study on antimicrobial activity of chitosan with different molecular weights. Carbohydrate Polymers, 2003, 54, 527-530.

[27] Liu, XF; Song, L; Li, L; Li, S; Yao, K. Antibacterial effects of chitosan and its water-soluble derivatives on E.coli, plasmids DNA, and mRNA. J. Applied Polym. Sci., 2007, 103(6), 3521-3528.

[28] Kong, M; Chen, XG; Liu; CS; Liu, CG; Meng, XH; Yu, LJ. Antibacterial mechanism of chitosan microspheres in a solid dispersing system against *E.coli.* Colloids & Surface B: Biointerfaces, 2008, 65, 197-202.

[29] Chung, YC; Chen, CY. Antibacterial characteristics and activity of acid-soluble chitosan. Bioresources Technol., 2008, 99, 2806-2814.

[30] Eaton, P; Fernandes, JC; Pereira, E; Pintado, ME; Malcata, FX. Atomic force microscopy study of the antibacterial effects od chitosans on Escherichia coli and Staphylococcus aureus. Ultramicroscopy, 2008, 108, 1128-1134.

[31] Palma-Guerrero, J; Huang, I-C; Jansson, H-B; Salinas, J; Lopez-Llorca, LV; Read, ND. Chitosan permeabilizes the plasma membrane and kills cells of *Neurospopra crassa* in an energy dependent manner. Fungal Genetic and Biology, 2009, 46 (8), 585-594.

[32] Yun, YS; Kim, KS; Lee, YN. Antibacterial and antifungal effects of chitosan. J.Chitin Chitosan, 1999, 4, 8-14.

[33] Jung, BO; Chung, SJ; Lee, GW. Effects of molecular weight of chitosan on its antimicrobial activity. J. Chitin Chitosan, 2002, 7, 231-236,

[34] Guerra-Sánchez, MG; Vega-Pérez, J; Velázquez-del Valle, MG; Nernandez-Lauzardo, AN. Antifungal activity and release of compounds *on Rhizopus stolonifer* (Ehrenb.:Fr.) *Vuill* by effect of chitosan with different molecular weights. Pesticide Biochem. Physiol., 2009, 93(1), 18-22.

[35] Liu, XF; Guan, YI; Yang, DZ; Li, Z; Yao, KD. Antibacterial action of chitosan and carboxylated chitosan. J. Appl. Polym. Sci., 2001, 79 (7), 1324-1335.

[36] Zhang, M; Tan, T; Yuan, H; Rui, C. Insecticidal and fungicidal activities of chitosan and oligochitosan. J. Bioact.Compat.Pol., 2003, 18(5), 391-400.
[37] Lin, S-B; Lin, Y-C; Chen, H-H. Low molecular weight chitosan prepared with the aid of cellulose, lysozyme and chitinase: characterization and antibacterial activity. Food Chemistry, 2009, 116(1), 47-53.
[38] Li, X-F; Feng, X-Q; Yang, S; Wang, T-P; Su, Z-X. Effects of molecular weight and concentration of chitosan on antifungal activity against *Aspergillus niger*. Iranian Polym. J, 2008, 17(11), 843-852.
[39] Kyung, WK; Thomas, RL; Chan, L; Park, H-J. J. Food Protection, 2003, 66, 1495-1498.
[40] Jeong, Y-J; Park, P-J: Kim, S-K. Antimicrobial effect of chitooligosaccharides produced by bioreactor. Carbohydrate Polymers, 2001, 44 (1), 71-76.
[41] Hirano, S; Nagao, N. Effects of chitosan, pectic acid, lysozyme and chitinase on the growth of several phytopathogens. Agricultural & Biol. Chem., 1989, 53, 3065-3066.
[42] Shahidi, F; Arachchi, JKV; Jeon, YJ. Food application of chitin and chitosans. Trends in Food Sciences and Technology, 1999, 10, 37-51.
[43] Qin, C; Li, H; Xiao, Q; Liu, Y; Zhu, J; Du, Y. Water-solubility of chitosan and its antimicrobial activity. Carbohydr. Polym. 2006, 63, 367-374.
[44] No, HK; Park, NY; Lee, SH; Meyers, SP. Antibacterial activity of chitosans and chitosan oligomers with different molecular weights. Int. J. Food Microbiol, 2002, 74, 65-72.
[45] Badawy, MEI; Rabea, EI, Potencial of the biopolymer chitosan with different molecular weights to control postharvest gray mold of tomato fruits. Postharvest Biol. Technol., 2009, 51 (1), 110-117.
[46] Yoonkyung, P; Kim, M-H; Park, S-C; Cheong, H; Jang, M-K; Nah, J-W; Hahm, K-S. Investigation of the antifungal activity and mechanism of action of LMWS-chitosan. J.Microbiol.Biotechnol., 2008, 18(10), 1729-1734.
[47] Fernandes, JC; Tavaria, FK; Soares, JC; Ramos, ÓC; Monteiro, MJ; Pintado, ME; Malcata, FX. Antimicrobial effects of chitosan and chitooligosaccharides, upon *Staphylococcus aureus* and *Escherichia coli*, in food model systems. Food Microbiol, 2008, 25, 922-928.
[48] Elliott, ML; Brodchat, TK; Uchida, JY; Simone, GW. Compendium of Ornamental Palm Diseases and Disorders. APS Press, MN, USA, 2004.
[49] Aragaki, A; Broschat, TK; Chase, AR; Ohr, HD; Simone, GW; Uchida, J. Diseases and Disorders of Ornamental Palms. APS Press, St.Paul, USA, 1991.
[50] Gallego, E; Garcia, E; Ortega, A. *Penicillium vermoesenii* Biourge (*Gliocladium vermoesenii* (Biourge) Thom) hongo causante de la 'seca' de las palmeras en el S.E. español. Agrícola vergel, 1991, 119, 725-729.
[51] Lopez-Llorca, LV; Orts, S. Histopathology of infection of the palm *Washingtonia filifera* with the pink bud rot fungus *Penicillium vermoesenii.* Mycol. Res., 1994, 98, 1195-1199.
[52] Ilyina AV, Tikhonov VE, Albulov AI, Varlamov VP. Enzymatic preparation of acid-free-water soluble chitosan. Process Biochemistry, 2000, 35, 563-568.
[53] Domard A, Carter N. Preparation, Separation and Characterization of the D-glucosamine oligomer series.Chitin and Chitosan, Elsevier Applied Science, London and New York, 1989, 383-388.

[54] Hirai A, Odani H, Nakajima A. Determination of Degree of Deacetylation of Chitosan by ^{1}H NMR Spectroscopy. Polym. Bull., 1991, 26, 87-94.

[55] Vårum KW, Ottøy MH, Smodsrød O. Acid hydrolysis of chitosan. Carbohydrate Polymers, 2001, 46, 89-98.

[56] Einbu A, Grasdalen H, Vårum KW. Kinetics of hydrolysis of chitin/chitosan oligomers in concentrated hydrochloric acid. Carbohydr. Res., 2007, 342, 1055-1062.

[57] Tikhonov VE, Radigina LA, Yamskov IA, Ilyina AV, Anisimova MV, Varlamov VP, Tatarinova Yu N. Affinity purification of major chitinases produced by *Streptomyces kurssanovii.* Enzyme & Microbial Technol., 1998, 22, 52-55.

[58] Ilyina AV, Tatarinova YuN, Tikhonov VE, Varlamov VP. Extracellular proteinase and chitinase produced by a culture of *Streptomyces kursanovii* . Appl. Biochem. Microbiol., 2000, 36, 146-149.

In: Handbook of Chitosan Research and Applications ISBN 978-1-61324-455-5
Editors: Richard G. Mackay and Jennifer M. Tait

Chapter 18

CHITOSAN-SILICATE HYBRIDS VIA SOL-GEL METHOD FOR SCAFFOLD APPLICATIONS

Yuki Shirosaki[1], Akiyoshi Osaka[1], Satoshi Hayakawa[1] and Kanji Tsuru[2]

[1]. Graduate School of Natural Science and Technology, Okayama University, Tsushima, Kita-ku, Okayama, Japan

[2]. Faculty of Dental Science, Kyushu University, Maidashi, Higashi-ku, Fukuoka, Japan

ABSTRACT

Novel strategies for regenerating or reconstructing damaged bone tissues are of urgent necessity, because of limitations in conventional therapies for trauma, congenital defects, cancer, and other bone diseases. The tissue engineering approach to repair and regeneration is founded upon the use of biodegradable polymer scaffolds, which may manipulate bone cell functions, encourage the migration of bone cells from border areas to the defect site, and provide a source of inductive factors to support bone cell differentiation. During the past decades, scaffolds from natural biodegradable polymers such as collagen, gelatin, fibrin, and alginates, or synthetic biodegradable polymers such as polyglycolide, polylactides, and copolymers of glycolide with lactides have been extensively explored. On the other hand, it has been confirmed that the bonelike apatite layer, deposited spontaneously on the biomaterial surfaces, can enhance osteoconductivity. The presence of such bone-like apatite layers is also believed to be a prerequisite to conduction of osteogenic cells into various porous scaffolds or onto the surface of bioactive glasses. The formation of such bone-like apatite is favored by the cooperative behavior of a hydrated silica or titania gel surface with many Si-OH or Ti-OH groups, and calcium ions to be released into the body fluid when implanted. Thus, hybrid materials derived from the integration of biodegradable polymers with bioactive inorganic species may construct a new group of scaffolds appropriate for tissue engineering. Moreover, tissue engineering approach depends on the use of porous scaffolds that serve to support and reinforce the regenerating tissue. Controlled porous structures of these scaffolds allow cell attachment and migration, tissue generation, or vascularization. We have been studying the synthesis of novel chitosan-silicate hybrids derived from the integration of chitosan and γ-glycidoxypropyltrimethoxysilane (GPTMS). We introduce some of the results on this hybrid for biomedical application.

INTRODUCTION

Tissue engineering approach to repairing, complementing and regenerating damaged tissue depends on polymer scaffolds that serve to support and reinforce the regenerating tissue. A number of natural and synthetic polymers are currently employed as tissue scaffolds (Lee, 2001; Park, 2002; Kuo, 2001; Widmer, 1998; Wang, 2003). Most of them are biodegradable and eliminated from the body due to bioresorption after implantation. Thus, their biodegradation is an important factor because these artificial supports are only temporarily required and may be taken out or should disappear after they accomplish their required and given roles. Chitosan and some of its complexes have been studied for a number of biomedical applications, including wound dressings, drug delivery systems, and space-filling implants (Chandy, 1990). For most medical applications, the polysaccharide network of chitosan should be crosslinked in order to improve its mechanical properties and to control its biodegradability. Various reagents have been used such as epoxy compounds, formaldehyde, and glutaraldehyde (Kawamura, 1993; Jammela, 1995), but these cross-linking agents are all highly cytotoxic and may impair the biocompatibility of the cross-linked biomaterials (Speer, 1980; Nishi, 1995). Therefore, non-toxic cross-linking agents are required. γ-glycidoxypropyltrimethoxysilane (GPTMS) is one of the silane-coupling agents, which has an epoxy group and methoxysilane groups. According to Ren et al. (Ren, 2001, 2002), the epoxy group is interacted with the amino groups of chitosan molecules, while the methoxysilane groups are hydrolyzed and form silanol groups, and the silanol groups are subject to the construction of a siloxane network due to the condensation. It is empirically accepted that bioactivity of silicate ceramics is due to the presence of Si-OH groups on the material surfaces (Hench, 1973; Hayakawa, 1996). Therefore, it is expected that Si-OH-involved molecules onto organic polymer surfaces by grafting or hybridizing Si-OH groups into the polymers should induce bioactivity and enlarge their applications in the medical field. Here, introduced were novel chitosan-silicate hybrids derived from integration of chitosan and γ-glycidoxypropyltrimethoxysilane (GPTMS) for biomedical application.

CHITOSAN-SILICATE HYBRID MEMBRANES

Shirosaki et al. (Shirosaki, 2005, 2009) prepared chitosan-silicate hybrid membranes derived from chitosan and GPTMS with 1/0.1 through 1/2.0 in chitosan/GPTMS monomer molar ratios. Figure 1 represents transparent and flexible chitosan-GPTMS hybrid membranes. The glycidoxy group at an end of a GPTMS molecule contains an epoxy ring as an active group (Innocenzi, 2000), while the methoxy group at the other end is so active to be hydrolyzed to yield a silanol group -Si-OH. Figure 2 illustrates that the epoxy group of GPTMS is reacted easily with the amino groups in the chitosan matrix, and that the silanols are condensed together under proper pH conditions, so that a cross-linked chitosan matrix is formed: (chitosan-NH-CH-CH(OH)-···Si-O-Si···-CH(OH)-NH-chitosan). The density of crosslinking estimated due to a ninhydrin method was more than 80%. When TEOS (tetraethoxyorthosilicate: without epoxy groups) was used instead of GPTMS, the density of crosslinking decreased to 60%. This means that silanol groups also were interacted with the amino groups in the chitosan chains. After Rashidva et al. (Rashidva, 2004), the silanol groups with negative charges and the amino groups with positive charges ($-NH_2 + H^+ \rightarrow -NH_3^+$) favor the crosslinking reactions. In contrast, chitosan-GMA (glycidilmethacrylate; with only epoxy groups) hybrids showed only 20% of the amino groups of chitosan were consumed in the crosslinking reactions. That is, the epoxy groups and silanol groups attacked 20 and 60% of the amino-groups, respectively, to form cross-links. According to Zhang et al. (Zhang, 1994), solid-state ^{13}C NMR spectroscopy is the most useful tool for structurally analyzing insoluble complex biopolymers without destroying their conformation. Figure 3 shows the ^{13}C NMR spectra for GPTMS homopolymer, the chitosan-GPTMS hybrid membrane and chitosan. Here, the peaks from the epoxy ring carbons C_f and C_e, found for GPTMS homopolymer, are absent in the hybrid spectrum, while a new peak appears for C_a at 7.8 ppm that is not detected in the spectrum of chitosan. Moreover, after Davis et al. (Davis, 2003), the change in the ^{13}C profile for chitosan at C_6 (61.0 ppm) and C_2 (57.8 ppm) peaks is attributed to Si-O-C bonds formed in the chitosan-GPTMS reactions. Therefore, Figure 3 confirms that most of the epoxy rings open to be grafted to the chitosan chain at the amino groups, and GPTMS and chitosan complex is formed.

Silicon atoms with one organic modifying group can form structure units T^n: $R\text{-}Si(\text{-}OSi)_n(X)_{3-n}$, where R and n stand for the organic group and the number of Si-O-Si bridging bonds, respectively, and X can be either hydroxyl or alkoxy groups when alkoxysilanes like GPTMS are concerned. When GPTMS is dissolved in the solution of chitosan/acetic acid, hydrolysis of the trimethoxysilane groups produces silanol (Si-OH) groups, and their condensation subsequently takes place to yield Si-O-Si bridging bonds or the siloxane units, i.e., unit T^0 is now changed to either of T^1, T^2 and T^3, after Ren et al. (Ren, 2001). From their FT-IR and ^{29}Si CP-MAS NMR spectra, the chitosan-GPTMS hybrids involve Si-OH groups and Si-O-Si bridging bonds in their matrix, and each Si atom has 2.6 bridging oxygen atoms in average: such a bridging oxygen fraction corresponds to the value for disilicate ions ($Si_2O_5^{2-}$), which crystallize into a 2D-sheet structure with one Si-O-non-bridging bond and three Si-O-Si bridging ones. It is thus strongly suggested that all of the silanol groups are involved in the hybrids with configurations similar to such disilicate structure: one (organic skeleton)-Si bond and three Si-O-Si bridging bonds. As the Si atoms in the hybrid have only 3 O atoms (in units T^n), a high fraction of O atoms (~90 %) take part in the siloxane network. Moreover, each Si atom in a GPTMS molecule is already connected to the organic group, and if the -C-Si bond is counted as a bridging one, about 60% of the Si atoms in the hybrids are in four-fold bridging like in pure silica SiO_2.

Many silanol groups remain still in the chitosan-GPTMS hybrids, as the fraction of T^1 and T^2 are assumed to be about 40% from the ^{29}Si NMR analysis. Wettability of material surfaces is one of the key factors for protein adsorption, cell attachment, and migration (Hench, 1982). Hydrophilic surfaces favor cell attachment and proliferation. Indeed, water contact angle for the hybrid membrane is 60° in contrast to 80° for chitosan. The decrease in contact angle due to the chitosan-GPTMS hybridization is then ascribed to the presence of the silanol groups derived from GPTMS that were orientated to the hybrid surface.

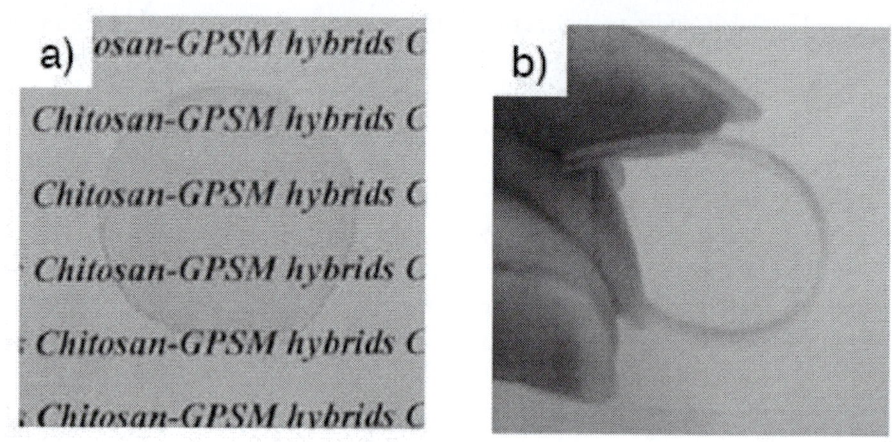

Shirosaki, 2009.

Figure 1. Photographs of the chitosa-GPTMS hybrid. The hybrid were transparent (a) and flexible (b).

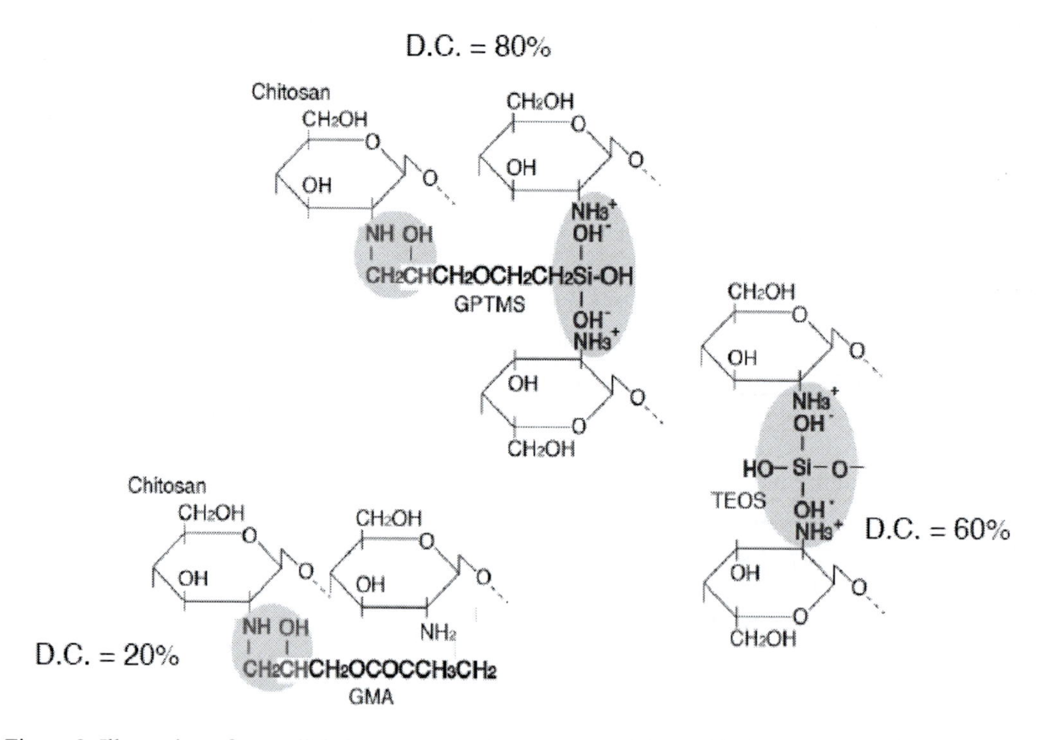

Figure 2. Illustration of cross-link formations in the chitosan hybrids, with the epoxy and silanol groups. D.C. = Degree of crosslinking estimated by ninhydrin method.

GPTMS increases hydrophilicity of the hybrids, and it reduces the water uptake. It is indicates that the chitosan-GPTMS hybrid membranes are significantly stiffened due to hybridization with GPTMS. In particular, the extensibility of the hybrids is reduced in comparison to the uncrosslinked chitosan. Therefore, the addition of GPTMS could lead to the variations of the mechanical properties.

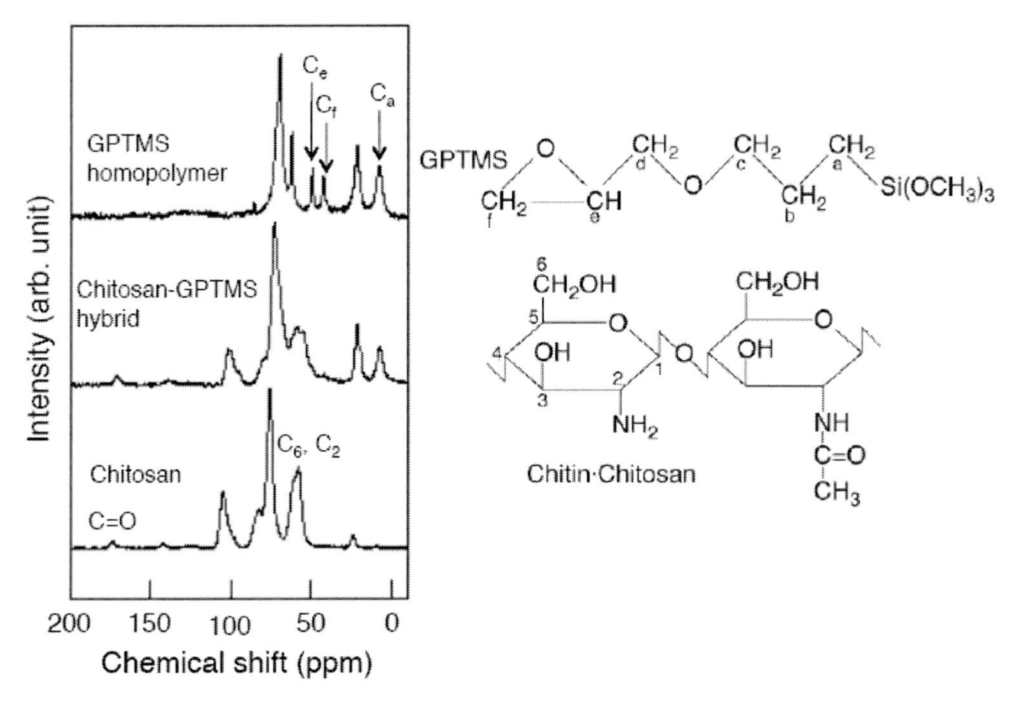

Figure 3. ^{13}C CP-MAS NMR spectra of chitosan, the chitosan-GPTMS hybrid, and GPTMS homopolymer. Each C is presented in the spectra and in the molecular structures.

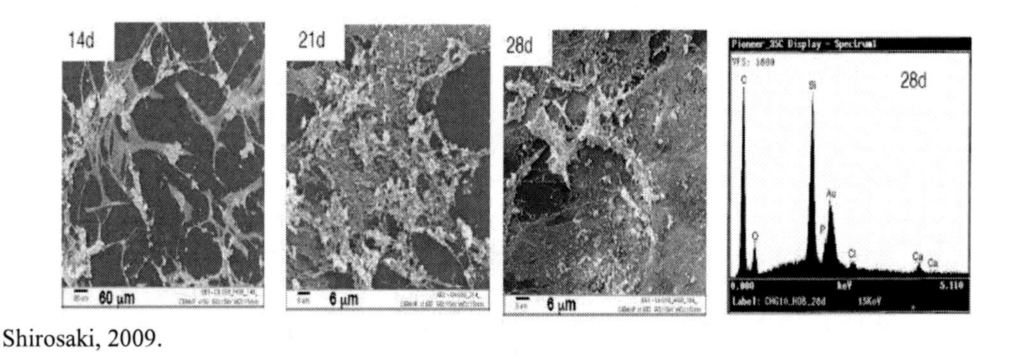

Shirosaki, 2009.

Figure 4. SEM photographs of HOB cells cultured on ChG10 in the presence of AA+β-GP at 14 (Magnification: x100), 21 (Magnification: x1000), and 28 days (Magnification: x1000).

Silicate ions influence the activity of bone cells, though their chemical states or oligomeric structures are unknown. Anderson et al. (Anderson, 1998) investigated the osteoblast response to a silica gel and observed an earlier formation of bone nodules in the cultures grown on the silica surfaces as compared to control by the uptake of sililic acid from the gels. The chitosan-GPTMS hybrids, as modified with silicon-containing species T^n, are expected to induce high osteocompatibility. Shirosaki et al. (Shirosaki, 2005, 2009) investigated the cytotoxicity of the chitosan-GPTMS hybrids as they examined human osteoblastic cell (MG63) and human bone marrow cell responses to their surfaces. Cell proliferation was evaluated by MTT assay and the total protein content of the cell layers; cell function was monitored by measuring ALP (Alkaline phosphatase) activity, since this enzyme is a common indicator of the expression of the osteoblastic phenotype (Martin, 1993, Aubin, 1993). The attachment, proliferation and function of MG63 cells are greatly improved in the hybrids as compared to chitosan membrane. Moreover, Figure 4 shows that human bone marrow cells, cultured on the hybrids show abundant cell growth and matrix mineralization, regardless of the absence or in the presence of dexamethasone (Dex). This is an exceptional event because Dex is added commonly and frequently to cell culture media to improve the proliferation and differentiation of osteoblastic cells (Coelho, 2000). Rhee et al. (Rhee, 2003) also confirmed the importance of Si(IV), as they reported that poly(m-ethylmethacrylate)(PMMA)/silica composites showed better responses to osteoblasts than PMMA due to the release of Si(IV) and Ca(II). It is fair for Gough et al. (Gough, 2004) to point out that a certain range (2.1 mM to 5.0 mM) of dissolution product concentrations is highly beneficial and favorable effects, but that excessively higher concentrations (more than 8.2 mM), on the other hand, appear to induce programmed cell death. Actually, the chitosan-GPTMS hybrids might release Si(IV) (0.5 mM) less than the range above, into the media after chemical analysis of the culture media soaked with the hybrids. In consequence, the surface structure involving Si-OH and Si-O-Si, and a little amount of Si(IV) species released from the hybrids can exert an important influence on cell proliferation and activity.

Porous Chitosan-Silicate Hybrid Scaffold

Tissue engineering approach to repairing, complementing and regenerating damaged tissue depends heavily on scaffolds that support and reinforce the regenerating tissue. Thus, porous scaffolds play key roles in manipulating cell functions. The scaffolds should not only promote cell adhesion, cell proliferation, and cell differentiation but also should be biocompatible, biodegradable, highly porous with a large surface to volume ratio, mechanically strong enough for handling, and capable of being formed into desired shapes (Tateishi, 2002). The high porosity is required because they have to provide sufficient spaces for cell adhesion/proliferation and to supply oxygen and nourishment. In addition, adequate interconnection of pores and pore-to-pore window opening to favor easy cell migration and blood vessel generation are important factors, too. Shirosaki et al (Shirosaki, 2008) formed porous chitosan-GPTMS hybrids in many shapes, such as sheet, pellet, column, and beads, employing a freeze-drying technique. Flexibility in shaping materials is one the very important aspects of the present hybrids because those tissue engineering scaffolds should fit themselves in any site. Moreover, the constitution of the porous hybrids involving GPTMS as the crosslinking agent is superior to that of the hybrids with other reagent in terms of maintaining the hybrid shape and sufficient mechanical strength for handling. The microstructure of the hybrids can be modified by the freezing rate, as the size and structure of the pores were found strongly dependent on the freezing temperature (Ren, 2001). The higher the freezing temperature is, the smaller is the number of ice nuclei that are initially formed in an aqueous system leading to an increased final size of ice crystals (Kang, 1999). When phase separation takes place due to immiscibility among the components in a homogeneous polymer solution system, another complex porous structures may be generated. After removal of the solvent in the solution, the space used to be occupied by the solvent (or water) remains as pores appropriate for scaffold applications. After Ho et al. (Ho, 2004), during the solvent removal, the polymer molecules surrounding the solvent should be rigid enough to prevent pore collapse and thus to retain the porous structure. The pore size of the chitosan-GPTMS hybrids is approximately 100 μm and 50 μm frozen at -20 °C and -85 °C, respectively regardless of the content of GPTMS. The water uptake of the porous hybrids reaches maximum values, ~2000 weight% within 2 hours, regardless of the composition, when they are soaked in phosphate buffered saline solution (PBS). Yet, the volume of the hybrids remains constant as before. It is no doubt that the water uptake depends on the porosity, but the fact that the chitosan-GPTMS porous hybrids adsorb as much water as their volume indicates that the 3-D structure of the chitosan-GPTMS porous hybrids is not deteriorated and retains much water in their matrix as well as in the pores. Such large water uptake is important as to stimulate plasma circulation inside pores, and subsequently to induce infiltration and migration of cells or tissues. The porous chitosan-GPTMS hybrids also have excellent biocompatibility in terms of MG63 osteoblastic cell culture (Shirosaki, 2008). The cells adhered and proliferated well on the pore walls. The cells extend many pseudopodia and connect with each other. Figure 5 indicates that the cells are infiltrated and migrated into the pores. Nutrition elements or cell dispersion deep into the matrix would not always be achieved for all scaffolds but only for those with appropriate porous structures. In this regard, too, the porous chitosan-GPTMS hybrids are suitable for scaffold applications.

Porous scaffolds may be sometimes obliged to incorporate some additives such as drugs or proteins like growth factors that have certain effects on cell growth, cell differentiation, and anti-inflammatory. Since those drugs are mostly unstable and diffused quickly into human plasma or blood when they are medicated to the patient, such scaffold materials are much favorable that release those impregnated drugs for short or long periods in a designed way so as to control their concentration profile without reaching toxic levels. Sinha et al. (Sinha, 2004) studied the use of chitosan as carrier for various classes of drugs such as antihypertensive agents, anticancer agents, proteins, peptide drugs, and vaccines. Two approaches are common for growth factors to be loaded in the materials; (1) after the material preparation or (2) during the formation of the material. In the former case, the growth factors will stay only on the surface or in the area just below the surface as diffusion is only the way for them, and so they are free to be detached, hence the stability of growth factor can be decreased (Ning, 1999). In the latter case (2), the growth factors will be more stable, surrounded by the matrix, and likely be easier for controlled release. Shirosaki et al. (Shirosaki, 2008) followed the latter procedure as they examined in vitro drug release characteristics for the porous chitosan-GPTMS hybrid, taking cytochrome C as a model protein for growth factors. A chitosan-GPTMS hybrid releases cytochrome C rapidly up to 5 days, then the release is slowed. Two models of release mechanisms work; diffusion of drug itself, and degradation of the scaffold. Considering that cytochrome C molecules form complexes with active groups on the hybrids surface, the degradation mechanism may be predominant: that is, cytochrome C is released as the hybrids are degraded to generate chitosan-GPTMS fragments. In contradiction to simple expectation, the cytochrome C release is decreased with increasing amount in the porous hybrid. Increase in interaction between cytochrome C molecules and the hybrid is probably responsible, and the results strongly suggest the presence of an appropriate range of holding each protein or drug in each scaffold material. With all considered, the porous chitosan-GPTMS hybrids have a potential for application in drug delivery system.

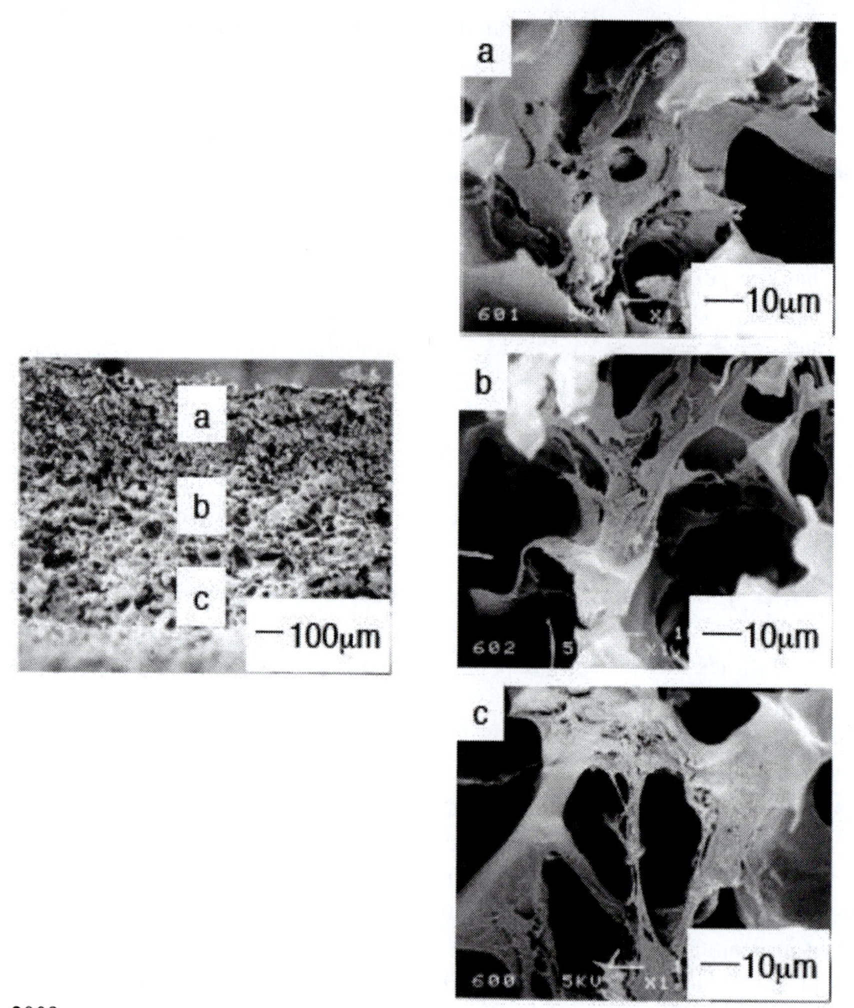

Shirosaki, 2008.

Figure 5. SEM photographs of cross-section of the porous chitosan-GPTMS hybrid after MG63 cells cultured for 7 days. (right) The larger magnification photographs for cross-section (a-c), indicating MG63 cells infiltrate into the hybrid.

Conclusion

To prepare the coming era with aged people and to ensue quality of life for all people, tissue engineering is an issue of keen interest. To respond to such a requirement from the human society, one should have excellent materials for scaffolds applicable to any status of medical care. A scheme of organic-inorganic hybridization is proposed, and the chitosan-GPTMS hybrids have been proposed as candidates of biocompatible and biodegradable organic-inorganic hybrids applicable to such scaffolds.

REFERENCES

Anredson S.I. Downes S. Perry C.C. Caballero A.M. Evaluation of the osteoblast response to a silica gel in vitro. *J. Mater. Sci. Mater. Med.* 1998; 9: 731-735.

Aubin J.E. Turksen K. Heersch J.N.M. Osteoblastic cell lineage. In Noda M. editor. Cellular and molecular biology of bone. Tokyo Academic Press Inc. 1993; 1-45.

Chandy T. Charma P.C. Chitosan- as a biomaterial. *Biomat. Art. Cells Art. Org.* 1990; 18: 1-24.

Coelho M.J. Fernandes M.H. Human bone cell cultures in biocompatibility testing. Part II: effect of ascorbic acid, β-glycerophosphate and dexamethasone on osteoblastic differentiation. *Biomaterials* 2000; 21: 1095-1102.

Davis R.S. Brough A.R. Atkinson A. Formation of silica/epoxy hybrid network polymers *J. Non-Cryst. Solid.* 2003; 315: 97-205.

Hayakawa S. Tsuru K. Iida H. Ohtsuki C. Osaka A. Studies of apatite formation on $50CaO \cdot 50SiO_2$ glass in a simulated body fluid. *Phys. Chem. Glasses* 1996; 37: 188-192.

Hench L.L. Ethridge E.C. Biomaterials An Interfacial Approach. Academic Press NewYork 1082, p.10.

Hench L.L. Splinter R.J. Allen W.C. Bonding mechanisms at the interface of ceramic prosthetic materials. *J. Biomed. Mater. Res.* 1973; 2: 117-141.

Ho M.H. Kuo P.Y. Hsieh H.J. Hsien T.Y. Hou L.T. Lai J.Y. Wang D.M. Preparation of porous scaffolds by using freeze-extraction and freeze-gelation methods. *Biomaterials* 2004; 25: 129-138.

Innocenzi P. Brustain G. Guglielmi M. Signorni R. Bozio R. Maggini M. 3-(Glycidoxypropyl)-trimethoxysilane-TiO_2 hybrid organic-inorganic materials for optical limiting. *J. Non-Cryst. Solids* 2000; 265: 68-74.

Jameela S.R. Jayakrishnan A. Glutalaldehyde crosslinked chitosan as a long acting biodegradable drug delivery vehicle: studies on the in vitro release of mitoxantrone and in vivo degradation of microspheres in rat muscle. *Bioamterials* 1995; 16: 769-775.

Kang H.W. Tabata Y. Ikada Y. Fabrication of porous gelatin scaffolds for tissue engineering. *Biomaterials* 1999; 20: 1339-1344.

Kawamura Y. Mitsuhashi M. Tanibe H. Yoshida H. Adsorption of metal ions on polyaminated highly porous chelating resin. *Ing. Eng. Chem. Res.* 1993; 32: 386-391.

Kuo C.K. Ma P.X. Ionically crosslinked alginate hydrogels as scaffolds for tissue engineering: Part 1 Structure, gelation rate and mechanical properties. *Biomaterials* 2001; 25: 511-521.

Lee Y.K. Mooney J.D. Hydrogels for tissue engineering. *Chemical Reviews* 2001; 101: 1869-1879.

Martin T.J. Findlay D.M. Heath J.K. Ng K.W. Osteoblasts: differentiation and function. In Mundy J.R., Martin T.J. editors. *Handbook of experimental pharmacology*. Berlin: Springer; 1993; 149-183.

Ning S. Yu N. Brown D.M. Kanekal S. Knox S.J. Radiosensitization by intratumoral administration of cisplatin in a sustained-release drug delivery system. *Radiother. Oncol.* 1999; 50: 215-223.

Nishi C. Nakajima N. Ikada Y. In vitro evaluation of cytotoxicity of diepoxy compounds used for biomaterial modification. *J. Biomed. Mater. Res.* 1995; 29: 829-834.

Park S.N. Park J.C. Kim H.O. Song M.J. Suh H. Characterization of porous collagen/hyaluronic acid scaffold modified by 1-ethyl-3-(3-dimethylaminopropyl) carbodiimide cross-linking. *Biomaterials* 2002; 23: 1205-1212.

Rashidva S.Sh. Shakarova D. Sh. Ruzimuradov O.N. Satabaldieva D.T. Zalyaliva S.V. Shpigun O.A. Varlamov V.P. Kabulov B.D. Bionanocompositional chitosan-silica sorbent for liquid chromatography. *J. Chromat.* B 2004; 800: 49-53.

Ren L. Tsuru K. Hayakawa S. Osaka A. Novel approach to fabricate porous gelatin-siloxane hybrids for bone tissue engineering. *Biomaterials* 2002; 23: 4765-4773.

Ren L. Tsuru K. Hayakawa S. Osaka A. Sol-gel preparation and in vitro deposition of apatite on porous gelatin-siloxane hybrids. *J. Non-Cryst. Solid* 2001; 285:116-122.

Ren L. Tsuru K. Hayakawa S. Osaka A. Sol-gel preparation and in vitro deposition of apatite on porous gelatin-siloxane hybrids. *J. Non-Cryst. Solid* 2001; 285: 116-122.

Shirosaki Y. Okayama T. Tsuru K. Hayakawa S. Osaka A. Synthesis and cytocompatibility of porous chitosan-silicate hybrids for tissue engineering scaffold application. *Chem. Eng. J.* 2008; 137: 122-128.

Shirosaki Y. Tsuru K. Hayakawa S. Osaka A. Biodegradable chitosan-silicate porous hybrids for drug delivery. *Key Eng. Mat.* 2008; 361-363: 1219-1222.

Shirosaki Y. Tsuru K. Hayakawa S. Osaka A. Lopes M.A. Santos J. D. Fernandes M.H. In vitro cytocompatibility of MG63 cells on chitosan-organosiloxane hybrid membranes *Biomaterials* 2005; 26: 485-493.

Shirosaki Y. Tsuru K. Hayakawa S. Osaka A. Lopes M.A. Santos J.D. Costa M.A. Fernandes M.H. Physical, chemical and in vitro biological profile of chitosan hybrid membrane as a function of organosiloxane concentration *Acta Biomater*. 2009; 5: 346-355.

Sinha V.R. Singla A.K. Wadhawan S. Kaushik R. Kumria R. Bansal K. Dhawan S. Chitosan microspheres as a potential carried for drugs. *Inter. J. Pharmace*. 2004; 274: 1-33.

Speer D.P. Chvapil M. Eskelson C.D. Ulreich J. Biological effects of residual glutaraldehyde in glutaraldehyde-tanned collagen biomaterials. *J. Biomed. Mater. Res.* 1980; 14: 753-764.

Wang Y.C. Lin M.C. Wang D.M. Hsieh H.J. Fabrication of a novel porous PGA-chitosan hybrid matrix for tissue engineering. *Biomaterials* 2003; 24: 1047-1057.

Widmer M.S. Gupta P.K. Lu L. Meszlenyi R.K. Evans G.R.D. Brandt K. Savel T. Gurlek A. Patrick Jr.C.W. Mikos A.G. Manufacture of porous biodegradable polymer conduits by an extrusion process for guided tissue regeneration. *Biomaterials* 1998; 19: 1945-1955.

Zhang M. Hisamori H. Yamata T. Hirano S. ^{13}C CP/MAS NMR spectral-analysis of 6-*o*-tosyl, 6-deoxy-6-iodo, and 6-deoxy derivatives of *N*-acetylchitosan in solid-state *Biosci. Biotech. Biochem.* 1994; 58: 1906-1908.

In: Handbook of Chitosan Research and Applications
Editors: Richard G. Mackay and Jennifer M. Tait
ISBN 978-1-61324-455-5

Chapter 19

APPLICATION OF CHITOSAN IN ENVIRONMENTAL ENGINEERING

Ashura A. Dulazi and Hui Liu
Key Laboratory of Biogeology and
Environmental Geology of Ministry of Education,
School of Environmental Studies,
China University of Geosciences,
Wuhan,China

ABSTRACT

Chitosan, the deacetylated form of chitin, has become a material of high potential in different fields such as environment, health, agriculture and medicine. The present review provides an insight of different applications of this biopolymer in the environment engineering especially in wastewater treatment, support for enzyme and microorganisms immobilization. Past and recent investigations are discussed in this paper.

Keywords: *Chitosan, application, adsorption, carrier, environment engineering*

Chitosan is a natural and abundant polysaccharide derived from deacetylation of chitin. Chitin is a major constituent of the exoskeleton and occurs in animals, mostly crustaceans, molluscs, insects and certain fungi cell wall [1]. The structure of chitosan is shown in figure 1 below.

Chitosan has unique properties such as biocompatibility, biodegradability, nontoxicity, bioactivity, polycationic plus antibacterial and adsorption properties [2]. It has three types of reactive functional groups; an amino group, primary and secondary hydroxyl groups at the C-2, C-3 and C-6 positions for each monomer [3]. Presence of these groups together with its biological and physicochemical properties have made it to be widely used in the field of environment particularly in water treatment plants for water purification, adsorption of organic compounds, metal ions, adsorption of dyes and protein, besides that it can also be used as carrier for enzymes and bacteria. In this paper most important applications of chitosan in environmental field are discussed.

1. Removal of Organic Compounds from Waste Water

Numerous studies have demonstrated that chitosan, utilizing its reactive amino group and hydroxyl groups can remove organic compounds such as chlorophenol, nitrophenol [4, 5], phthalate esters and other compounds from wastewater as shown by Chen and Chung, [6] who investigated the removal of phthalate esters from aqueous solutions using chitosan beads by batch and column experiments. Phthalates of different length such as Diethyl hexyl phthalate (DEHP), Dimethyl phthalate (DMP), Diethyl phthalate DEP), Dibutyl phthalate (DBP), Dipropyl phthalate (DPP) and Diheptyl phthalate (DHpP) were removed from aqueous solutions. Based on their findings the order of adsorption efficiency was DHpP> DBP> DEHP> DPP> DEP> DMP and the order of adsorption capacity was DHpP (1.52 mg/g)> DBP (0.54 mg/g)>DEHP (0.49 mg/g)> DPP (0.38 mg/g)> DEP (0.19 mg/g)> DMP (0.16 mg/g). The maximal adsorbed ratio for DHpP was 46%. Then, it was concluded that chitosan beads have the best adsorption capacity at 15°C and pH 8.0 and favored the adsorption of DHpP and not DMP.

Chitosan, despite having good adsorption capability, has weak mechanical properties thus studies on chemical and physical modifications have been done to improve its selectivity and the capacity for adsorption of metal ions [7] and organic compounds. Physical modifications increase the sorption properties and allow the expansion of the porous network due to the gel formation, which ultimately decreases the crystallinity of the adsorbent. Chemical modifications also increase the sorption properties while preventing the dissolution of the chitosan in strong acids and improving its mechanical strength. Some of the chemical modifications include cross-linking using cross-linking agents, grafting of a new functional group and acetylation [8, 9]. Chemical cross-linking can change the crystalline nature of chitosan and enhance its resistance against acid, alkali and chemicals. The most commonly used cross-linking agents are glutaraldehyde (GLA), epichlohydrine (ECH) and ethylene glycol diglycidyl ether (EGDE) [10, 11].

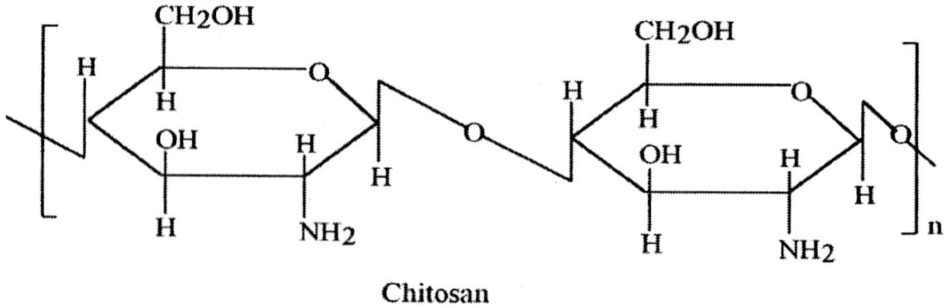

Chitosan

Kumar, 2000.

Figure 1. Structure of chitosan.

Modified chitosan beads have also been used to remove pollutants from water and proved to be more efficient than pure beads as it was observed by Cheng et al. [12] who studied the removal of the same phthalate esters using ά-cyclodextrin-linked chitosan beads in batch and column experiments. Same order of adsorption efficiency were observed and the order of adsorption capacity was (mg/g) DHpP (3.21)> DBP (3.16)> DEHP (3.09)> DPP (2.87)> DEP (2.82)> DMP (2.76) higher than when pure beads were used [6]. The maximal adsorbed ratio for DHpP was 96%. The adsorption capacity of ά-cyclodextrin-linked chitosan bead was higher than pure chitosan bead [6].

Almost similar results were obtained when the six phthalates were removed from aqueous solutions using molybdate impregnated chitosan beads [13]. The adsorption capacity of PAEs was DHpP (3.01mg/g) > DBP (2.96 mg/g) > DEHP (2.86 mg/g) > DPP (2.73 mg/g) > DEP (2.64 mg/g) > DMP (2.32 mg/g). The adsorption capacity of molybdate impregnated chitosan beads for various phthalates was found to be 2-to 14- fold higher than that of pure chitosan beads [6].

Adsorption of *p*-nitrophenol in aqueous solutions by chitosan flakes and chitosan-GLA beads was carried out by Ngah and Fatinathan, [14]. Chitosan-GLA beads was found to have higher adsorption capacity for *p*-nitrophenol of about 2.48mg/g while chitosan flakes had 0.63mg/g, chitosan-GLA showed a higher adsorption capacity due to enhanced resistance and mechanical strength attributed to the crosslinking process. Based on their adsorption capacity it was concluded that both chitosan flakes and chitosan-GLA can be used to treat wastewater containing *p*-nitrophenol.

Li et al, [15] studied the adsorption of phenol, *p*-nitrophenol and *p*-chlorophenol onto three adsorbents, chitosan chemically modified by salicylaldehyde (CS-SA), β-cyclodextrin (CS-CD) and a cross-linked β-cyclodextrin polymer (EPI-CD). Experimental results showed that the three adsorbents exhibited different adsorption abilities, and the adsorption of phenols onto EPI-CD, CS-CD, and CS-SA was predominated by hydrophobic interaction, hydrogen bonding and Π- Π- interaction. CS-CD was considered as a potential adsorption material due to its special adsorption ability for *p*-chlorophenol.

2. Removal of Heavy Metal from Waste Water

Chitosan has been regarded as a useful material for removal inorganic substances from waste water [16]. It combines with metal ions by three forms; ion exchange, sorption and chelation [17].

Jeon and Höll [18] investigated the removal of mercury ion using modified chitosan beads. Several chemical modification methods such as aminated chitosan beads, xanthated chitosan beads, phosphorylated chitosan beads, carboxylated chitosan beads and cross-linked chitosan beads (control) were tried to increase the uptake capacity of chitosan beads. Among them, aminated chitosan beads prepared through chemical reaction using ethylenediamine and carbodiimide showed the highest uptake capacity for mercury ions of about 2.3 mmol/g dry mass at pH 7, followed by xanthated chitosan (1.84mmol/g dry mass), phosphorylated beads (1.55mmol/g dry mass), carboxylated chitosan beads (1.51mmol/g dry mass) and crosslinked chitosan beads (1.47mmol/g dry mass). The uptake of mercury ions by aminated chitosan beads increased by about 47% compared to the control. Mercury(II) is classified as a soft ion according to HSAB theory, and soft ions form very strong bonds with groups containing nitrogen and sulphur atoms such as NH_2^-, SH^- [19], therefore the highest mercury ions uptake capacity for aminated chitosan beads was due to high affinity of mercury ions to the amine group in comparison with other functional group.

Adsorption of Cu(II) ions in aqueous solution using pure chitosan beads, chitosan-GLA 1:1, 2:1 ratio beads and chitosan-alginate beads was studied by Ngah and Fatinathan, [20]. The results indicated that the maximum monolayer adsorption capacity for pure chitosan beads was 64.62 mg/g, chitosan-GLA 1:1, 2:1 ratio beads were 31.20 mg/g, 19.51mg/g and 67.66 mg/g by chitosan-alginate beads and this was based on non-linear Langmuir isotherms. Chitosan-GLA beads showed a lower adsorption capacity due to the cross-linking which reduces the number of amino groups available for the uptake of Cu(II) ions, although the crosslinked chitosan beads with glutaraldehyde have lower percentage of swelling and several enhanced physical properties. Finally it was concluded that all the four adsorbents can be used to treat waste water containing Cu(II) ions.

The capacity of chitosan to adsorb Cr(III) and Cr(VI) ions from aqueous solutions was studied by Ngah et al [21]. Based on their findings the adsorption capacity of Cr(VI) was found to be two times higher than that of Cr(III), the optimum pH for Cr(III) adsorption was 5 while that of Cr(VI) was 3 and the maximum adsorption capacities of Cr(III) and Cr(VI) ions onto chitosan beads were 30.03 and 76.92 mg g^{-1}, respectively. The adsorption of Cr(III) on chitosan beads is a physical adsorption, while the adsorption of Cr(VI) is a chemical adsorption.

Gamage and Shahidi [22] used chitosan to remove Hg(II), Fe(II), Ni(II), Pb(II), Cu(II) and ZN(II) and protein from industrial waste water, the removal of the mentioned metal ions was most effective at pH of 7, this could be due to the greater availability of amino groups at higher pH values [23]. In the protein flocculation study the percentage of flocculation decreased with increasing protein concentration above 4mg/L and the highest protein flocculation was between 2 and 4mg/mL.

3. Removal of Dyes from Wastewater Effluents

Chitosan, due to its unique molecular structure, has an extremely high affinity for many classes of dyes, including dispersed, direct, reactive, acid, sulfur and naphthol dyes. At low pH values, chitosans free amine group are protonated and they adsorb the free anionic dyes by electrostatic attraction, higher pH value above 7 a decline in dye binding capacity is reported [24]. Acid effluents can severely limit the use of chitosan as an adsorbent in removing dyes and other metal ions due to the tendency of chitosan to dissolve in acid, it forms gels in pH below 5.5. Stabilization of chitosan molecule in acid solutions has been done using some cross-linking agents in order to overcome this problem, the cross-linked chitosan beads is reported to be insoluble in acid solution of pH 1-7 and with improved mechanical properties [25]. Chiou and Li [26] studied the adsorption of reactive dye RR 189 on cross-linked chitosan beads (epichlorohydrin (ECH) was used as a cross-linking agent) and found that the cross-linked chitosan beads had higher adsorption capacity to remove dye RR 189 than non crosslinked beads, a decrease in pH of solutions leads to a large increase in the adsorption capacity of dye on the cross-linked chitosan beads while non-cross-linked chitosan dissolves.

4. WATER PURIFICATION

For about three decades chitosan has been used in water purification process. When chitosan is spread over oil spills, it holds the oil mass together making it easier to clean up the spill. All over the world water purification plants use chitosan to remove oils, grease, heavy metals, and fine particulate matter that cause turbidity in wastewater streams [27].

Studies have shown that chitosan has high efficiency to coagulate the suspended particles in wastewater [28]. Due to the easy availability of free amino groups in chitosan, it carries a positive charge and thus reacts with many negatively charged surfaces.

Chi and Cheng, [28] used chitosan to coagulate fat and protein in wastewater from the system of cleaning in place (CIP) which contain high content of fat and protein. The results showed that the optimal efficiency was reached at of pH 7 with the coagulant dosage of 25 mg/L. Since, chitosan is non toxic and has antibacterial property the recovered fat and protein can be recycled.

5. CARRIER FOR ENZYME

Chitosan has been used as a matrix for enzyme immobilization [29, 30] and the immobilized enzymes have been used for hydrolysis of oil or fatty acid esterification and degradation of some organic compounds. Generally enzymes are often immobilized onto solid supports to increase their thermal and operational stability and recoverability [31].

Different ways of enzymes immobilisation are present and much more effort are still being devoted to the search for new novel techniques, so far the most common method is by crosslinking to polymeric material like chitosan using crosslinking reagents such as glutaraldehyde, which establishes intermolecular cross-links with the amino groups of the enzyme and those of the polymer. Another method is by activation of the hydroxyl group of chitosan using 1-Ethyl-3(3-dimethyl-aminopropyl) carbodiimide hydrochloride (EDC), and binary method utilizing both its amino and the hydroxyl groups [32].

In a study conducted by Ting et al., [33] soybean oil were hydrolyzed by Candida Rugosa immobilized on chitosan beads by binary method, based on their findings the immobilized lipase showed higher catalytic ability in soybean oil hydrolysis compared to the soluble lipase, 88% of the oil taken initially was hydrolyzed after 5 h under optimum conditions by immobilised lipase.

Zhang et al. [34] reported the removal of 2, 4-dichlorophenol by chitosan-immobilized laccase from *coriolus versilor*. In this study laccase was immobilized on chitosan by glutaraldehyde (as cross-linking agent). Results indicated that the immobilized enzyme has greater stability at room temperature over long time and retained higher removal efficiency above 50% for up to six usages.

6. CARRIER FOR BACTERIA

Chitosan has excellent properties as microbial delivery [35], it has high moisture retention to maintain continued microbial growth. Immobilization of viable cells is a versatile tool that serves to increase the stability of a microbial system, allowing its application under extreme environmental conditions, its reuse and the development of continuous bioprocesses. Applying immobilized microorganisms to water bioremediation enhances the rate of contaminant removal from water and avoids contaminant release into the environment during the biodegradation process. Studies have shown that bacteria can be immobilised on chitosan beads with some other modification where necessary, and applied in treatment plants to remove chemical compounds.

Gentili et al., [36] studied the bioremediation of crude oil polluted seawater using a hydrocarbon degrading bacterial strain(QBTo) of *Rhodococcus corynebacterioides* immobilized on chitin and chitosan flakes [37]. Results showed that the strain (QBTo) immobilized on chitin and chitosan flakes significantly enhanced the crude oil biodegradation while the free strain (QBTo) inoculated without carrier showed low hydrocarbon removal. The survival of the immobilised strain was possibly due to the protective effect of the carrier material and the biofilm structure formation by the cells.

Nitrate and phosphate uptake of a chitosan immobilised *Scenedesmus* spp. was investigated by [38]. It was pointed out that chitosan bead immobilized with *Scenedesmus sp.* (strain 1) was more efficient in removing phosphate and nitrate from water than the conventional free cell system. Immobilized *Scenedesmus sp.* (strain 1) had a higher growth rate and high viability than its free living counterpart.

In a study conducted by Cuero et al. [39] soil contaminated with heavy metals (e.g., Pb, Cu, and Zn) was treated with chitosan alone, bacterium *Bacillus subtilis* alone, and *Bacillus subtilis*-chitosan mixture to determine their effects on metal ion accumulation. The combined bacterial-chitosan treatment showed greater accumulation of metal ions compared to single treatments. Higher bacteria population and enzymes production (phosphatase and chitosanase) was observed in soils treated with *B. subtilis*-chitosan mixture as compared to soils treated with *B. subtilis* or chitosan alone. Perhaps, the polysaccharide chitosan stimulated the growth of *B. subtilis* in the soil, hence, increasing the microbial area for metal adsorption.

CONCLUSION

Chitosan, being natural, abundant, cheap, biodegradable and non toxic has proved to be a most appropriate alternative material in replacement of synthetic compounds. It improves the effectiveness of water treatment by removing many pollutants such as metal ions, some organic compounds, dyes, oils and fine particulate matters. When used as a carrier for bacteria, it provide favourable conditions for survival as well as functioning of the inoculants cells, resulting in a sufficiently long shelf life as well as improved survival and activity [40] and when used as a carrier for enzymes it increase the thermal and operational stability and recoverability of the immobilised enzymes. Interest of using chitosan in environmental field is growing rapidly; chitosan can now be employed to solve numerous problems in the environment especially in water treatment plants where "greener" methods have become an ecological necessity.

REFERENCES

[1] Muzzarelli RRA. 1977. Chitin. Pergamon Press, Oxford

[2] Juang RS and Shao HJ. 2002. Effect of pH on competitive adsorption of Cu(II), Ni(II), and Zn(II) from water onto chitosan beads. *Adsorption*. 8: 71.

[3] Fereidoon S, Kamil J, Arachichi V and You-Jin J. 1999. *Trends food Science Technology*. 1999. 10: 37-51.

[4] Zheng S, Yang Z, Jo DH, Park YH. 2004. Removal of chlorophenols from groundwater by chitosan sorption. *Water Resource*. 38: 2315-2322.

[5] Crini G. 2006. Non-conventional low-cost adsorbents for dye removal: a review. *Bioresource Technology*. 97: 1061-1085.

[6] Chen CY and Chung YC. 2006. Removal of Phthalate Esters from Aqueous Solutions by Chitosan Bead. *Journal of Environmental Science and Health Part A*. 41: 235-248.

[7] Qi L and Xu Z. 2004. Lead sorption from aqueous solutions on chitosan nanoparticles. *Colloids and Surfaces A: Physicochemical and Engineering Aspects*. 251: 183-190.

[8] Sun S and Wang A. 2006. Adsorption properties of N-succinyl-chitosan and cross-linked N-succinyl-chitosan resin with Pb(II) as template ions. *Separation Purification Technology*. 51: 409.

[9] Sun S and Wang A. 2006. Adsorption properties of carboxymethyl-chitosan and cross-linked carboxymethyl-chitosan resin with Cu(II) as template ions. *Separation Purification Technology*. 49: 197.

[10] Lee ST, Mi FL, Shen YJ and Shyu SS. 2001. Equilibrium and kinetic studies of copper (II) ion uptake by chitosan-tripolyphosphate chelating resin. *Polymer*. 42: 1879.

[11] Ngah WSW, Endud CS and Mayanar R. 2002. Removal of copper (II) ions from aqueous solution onto chitosan and cross-linked chitosan beads. *Reactive and Functional Polymers*. 50: 625.

[12] Chen CY, Chen CC and Chung YC. 2007. Removal of phthalate esters by ά-cyclodextrin-linked chitosan bead. *Bioresource Technology*. 98: 2578-2583.

[13] Chen CY and Chung YC. 2007. Removal of Phthalate Esters from Aqueous Solution by Molybdate Impregnated Chitosan Beads. *Environmental engineering science*. 24: 6.

[14] Ngah WSW and Fatinathan S. 2006. Chitosan flakes and chitosan-GLA beads for adsorption of *p*-nitrophenol in aqueous solution. Colloids and surfaces A: Physicochemical and Engineering Aspects. 277: 214-222.

[15] Li JM, Meng XG, Hu CW and Du J. 2009. Adsorption of phenols, p-chlorophenols and p-nitrophenols onto functional chitosan. *Bioresource Technology*. 100: 1168-1173.

[16] Ngah WSW, Ghani SA and Kamari A. 2005. Adsorption behavior of Fe(II) and Fe(III) ions in aqueous solution on chitosan and crosslinked chitosan beads. *Bioresource Technology*. 96: 443–450.

[17] Muzzarelli RAA. 1986. Pergamon Press, Oxford.

[18] Jeon C and Holl WH. 2003. Chemical modification of chitosan and equilibrium study for mercury ion removal. *Water research*. 37: 4770-4780.

[19] Volesky B. 1990. Biosorption of heavy metals. Boca Raton, FL: CRC Press. 253-275.

[20] Ngah WSW and Fatinathan S. 2008. Adsorption of Cu(II) ions in aqueous solution using chitosan beads, chitosan-GLA beads and chitosan-alginate beads. *Chemical engineering journal*. 143: 62-72.

[21] Ngah WSW, Kamari A, Fatinathan S, Ng PW. 2006. Adsorption of chromium from aqueous solution using chitosan beads. *Adsorption*. 12: 249-257.

[22] Gamage A and Shahidi F. 2007. Use of chitosan for the removal of metal ion contaminants and proteins from water. *Food Chemistry*. 104: 989-996.

[23] Bassi R, Prasher SO and Simpson BK. 2000. Removal of selected metal ions from aqueous solutions using chitosan flakes. *Separation Science and Technology*. 35: 547-560.

[24] Ravi Kumar MNV. 2000. A review of chitin and chitosan applications. Reactive and Functional Polymers. 46:1–27.

[25] Honarkar H and Barikani M. 2009. Applications of biopolymers I: Chitosan. *Journal of Monatsh chemical.* 1:18.

[26] Chiou MS and Li HY. 2002. Equilibrium and Kinetic modeling of adsorption of reactive dye on cross-linked chitosan beads. *Journal of Hazardous Materials B*. 93: 233-248.

[27] Hennen WJ. 1996. Chitosan. Woodland Publishing Inc.

[28] Chi FW and Cheng WP. 2006. Use of Chitosan as Coagulant to Treat Wastewater from Milk Processing Plant. Journal of Polymers and Environment. 14: 411-417.

[29] Betigeri SS and Neau SH. 2002. Immobilization of lipase using hydrophilic polymers in the form of hydrogel beads. *Biomaterials*. 23: 3627-3636.

[30] Alsarra IA, Betigeri SS, Zhang H, Erans BA and Neau SH. 2002. Molecular weight and degree of deacetylation effects on lipase- loaded chitosan bead characteristics. *Biomaterials*. 23:3627-3644.

[31] Chiou SH and Wu WT. 2004. Immobilization of Candida Rugosa lipase on chitosan with activation of the hydroxyl groups. *Biomaterials.* 25: 197-204.

[32] Hung TC, Giridhar R, Chiou SH, Wu WT. 2003. Binary immobilisation of Candida Rugosa lipase on chitosan. *Journal of Molecular Catalysis* B: Enzyme. 26: 69-78.

[33] Ting WJ, Tung KY, Giridhar R and Wu WT. 2006. Application of binary mobilized *Candida Rugosa* lipase for hydrolysis of soybean oil. *Journal of Molecular Catalysis* B: Enzyme. 42: 32-38.

[34] Zhang J, Xu Z, Chen H and Zong Y. Removal of 2, 4-dichlorophenol by chitosan-immobilised laccase from *Coriolus Versicolor. Biochemical engineering journal*. 45: 54-59.

[35] Dean JR. 1992. Advances in Chitin and Chitosan. Proc.: 5th International Conference.

[36] Gentili AR, Cubitto MA, Ferrero M and Rodriguez MS. 2006. Bioremediation of crude oil polluted seawater by a hydrocarbon degrading bacterial strain immobilized on chitin and chitosan flakes. *International Biodeterioration and Biodegradation*. 57: 222-228.

[37] Barengo NG, Cubitto MA, Sineriz F and Chiarello NM. 2002. Rhodococcus corynebacterioides strain QBTo isolated from soils polluted with crude oil refinery sludge. NCBI Genbank.

[38] Fierro S, Sanchez-Saavedra MP, Copalcua C. 2007. Nitrate and phosphate removal by chitosan immobilized Scenedesmus. *Bioresource Technology* 99: 1274–1279.

[39] Cuero RG. 1996. Enhanced heavy metal immobilization by a bacterial-chitosan complex in soil. *Biotechnology letters.* 18: 51 1-514.

[40] Van-Veen JA, Van-Overbeek LS and Van Elsas JD. 1997. Fate and activity of microorganisms introduced into soil. *Microbiology and Molecular Review.* 61: 121-135.

In: Handbook of Chitosan Research and Applications
Editors: Richard G. Mackay and Jennifer M. Tait

ISBN 978-1-61324-455-5

Chapter 20

LIPASE IMMOBILIZATION IN CHITOSAN FOR USE IN NON-AQUEOUS MEDIA CATALYSIS

Carlos E. Orrego* and Natalia Salgado

Plantas piloto de Biotecnología y Agroindustria,
Departamento de Física y Química,
Universidad Nacional de Colombia-Sede Manizales,
Bloque T, Campus La Nubia,
Manizales, Colombia

ABSTRACT

The most recent efforts in the development of cleaner sustainable chemistry are being driven by the production of raw materials from biofeedstocks using white biotechnology or industrial biotechnology instead of the conventional petrochemical technology.

Lipases are versatile biocatalysts and find applications in organic chemical, agrochemical, food, fine chemicals and pharmaceutical processing. Chitosan is a biopolymer prepared by deacetylation of chitin— a renewable resource obtainable from the wastes of the seafood industry or fungal sources— that has been used as a matrix for immobilization of enzymes for use in aqueous and non aqueous media because it has suitable properties such as insolubility, mechanical stability and rigidity in organic solvents, high affinity to water and proteins, availability of reactive functional groups, biodegradability, regenerability, and ease of preparation in different geometrical configurations. In this chapter, a comprehensive survey is made of different aspects of lipase activation and immobilization in chitosan hydrogels for catalysis in non-aqueous media. Topics on water activity, polarity of solvents, engineering support characteristics by physical-chemical modification of chitosan and, with special emphasis, surface characterization of chitosan supports and chitosan-lipase systems with differents techniques, are also discussed.

Keywords: *Biocatalysis; Lipase; Chitosan; Immobilization; Surface characterization*

* Corresponding author. Tel.: +57 68810000Ext 411; fax: E-mail address: corregoa@unal.edu.co.

1. INTRODUCTION

Enzymatic catalysis in nonaqueous solvents has gained considerable interest for the preparation of natural products, pharmaceuticals, fine chemicals and food ingredients [1-4]. Improved thermostability [5], synthesis favored over hydrolysis [6], tunable enzyme selectivity via medium engineering [7], and the ease of product recovery owing to the more easily evapourated reaction solvent [8], are reasons why nonaqueous bioprocessing is so attractive.

2. LIPASES

Lipases (EC 3.1.1.3) are triacylglycerolester hydrolases that catalyze the hydrolysis and interesterification of triglycerides at the interface between the insoluble substrate and water. Lipases from different organisms vary greatly in size, the smallest have molecular masses of 20–25 kDa (cutinase) while the largest are 60–65 kDa (*Candida rugosa* lipase) [9]. Apart from their natural substrates, such as water-insoluble esters and triglycerides, lipases catalyze the enantio- and regioselective hydrolysis and synthesis of a broad range of natural and synthetic esters [10-12]. Industrial applications of lipases have been thoroughly reviewed by various researchers [13-16]. Today lipases find applications in organic chemical processing, detergent formulations, synthesis of biosurfactants, the oleochemical industry, the dairy industry, the agrochemical industry, paper manufacture, nutrition, cosmetics, synthesis of fine chemicals and pharmaceutical processing [17]. One of the most dynamic areas of study is the use of enzymes for biodiesel production. Several studies have reported biodiesel production from different raw materials using lipases as catalyst, by enzymatic transesterification in non-aqueous media using both primary and secondary alcohols. A high yield of alkyl esters could be reached in the presence of organic solvents, but a low ester yield in the absence of organic solvents [18]. However, lipase is inactivated by excess amounts of lower linear alcohols, such as methanol and ethanol, conventionally used in the biodiesel process [19-21]. The inhibition is attributed to the poor miscibility of triacylglycerol with those alcohols which exist as droplets in the oil inactive lipases by contact or because alcohols, being polar in nature, get adsorbed to the immobilization support thereby blocking the entry of triacylglycerol, causing the reaction to stop [22, 23]. It is believed also that when a lower linear alcohol amount exceeds its solubility limits, lipase is deactivated by the insoluble alcohol that exists as drops in the oil, as proteins in general are unstable in short-chain alcohols [24].

3. LIPASE IMMOBILIZATION

Enzyme immobilization means the deliberate restriction of the mobility of the enzyme, which can also affect the mobility of the solutes [25]. After immobilization, enzymes are localized in a defined region of space, which is enclosed by an imaginary (or material) barrier that allows for physical separation of the enzyme from the bulk reaction medium and that is, at the same time, permeable to reactant and product molecules [26]. Immobilization may serve two objectives, first, to improve enzyme stability, and second, to facilitate a decrease in enzyme consumption as the enzyme can be recovered and reused for many repeated cycles of reactions. The extent of stabilization depends on the enzyme's structure, the immobilization method and the type of support [27].

Immobilization processes can be achieved by engineering: either the *microenvironment* of the enzyme, as is the case of immobilization by attachment to a carrier (e.g., covalent attachment, hydrophobic and ion exchange adsorption and cross-linking), and immobilization by containment in a barrier (e.g., microencapsulation using lipid vesicles, containment in reversed micelles, entrapment in polymeric matrices, and confinement in ultrafiltration hollow fibers); or, alternatively, its *macroenvironment* (as modification of the reaction medium, which is achieved, for example, via precipitation in an organic solvent) [12].

The immobilization of lipases has two important aspects: First, since most lipases appear to show some form of interfacial activation, immobilization should promote this process to the maximum possible extent and prevent the activated enzyme from reverting to the closed form, while minimizing any detrimental conformational changes. Studies have shown that matrices which provide both nonpolar/hydrophobic and polar/hydrophilic functionality are critical for achieving optimal lipase activation and stabilization. Second, since commercial lipases are often crude preparations with lipase contents as low as 0.1–5% w/w, the immobilization protocol should enable the selective binding of the enzyme from a complex protein mix [28]. Literature reports about the 3D structure of lipases indicate that lipases are fairly hydrophilic as well as hydrophobic proteins and their main hydrophobic area is the extremely hydrophobic region surrounding the catalytic site. According to this hypothesis, affinity between supports and enzyme is a key factor for the efficient immobilization of enzyme [29].

Methods of lipase-enzyme immobilization can be differentiated in two general categories: physical retention and chemical binding [30].

4. Chitosan

Chitosan is a copolymer amine polysaccharide of linked β(1-4),2-amino-2-deoxy-D-glucan and 2- acetamidodeoxy-D-glucan, obtained from alkaline D-acetylation of chitin, an unelastic and nitrogenated polysaccharide, which is found on the walls of the fungi and outer skeleton of arthropodes such as insects, crustaceans and scarabs. The C-2 amino groups of chitin are ideally acetylated. Chitin and chitosan are, consequently, copolymers of *N*-acetyl-D-glucosamine and D-glucosamine. The processing of crustacean shells mainly involves the removal of proteins and the dissolution of calcium carbonate which is present in crab shells in elevated concentration. In generally, 5–15% deacetylation takes place by strong alkaline treatment in the production process of chitin. Chitin is insoluble in water due to its rigid crystalline structure and intra- and intermolecular hydrogen-bonding network [31]. The resulting chitin is deacetylated to chitosan in high concentrated sodium hydroxide solutions [32].

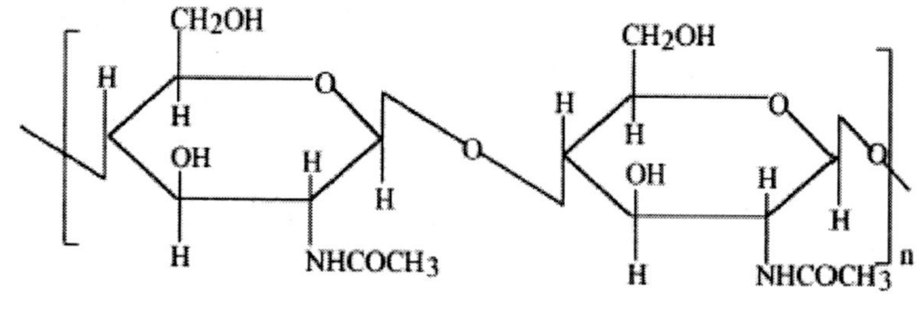

(a)

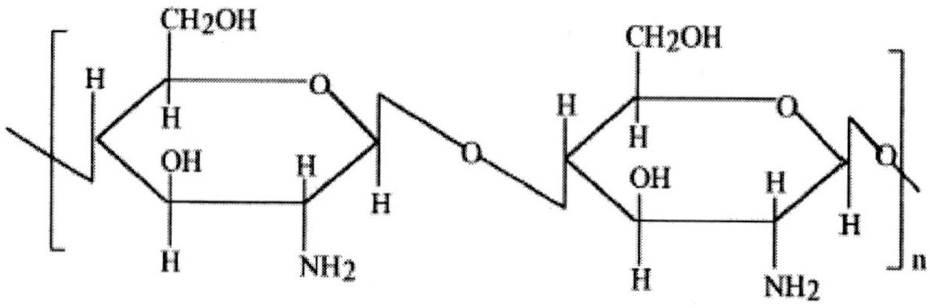

(b)

Figure 1. Idealized structures of chitin (a) and chitin 100% deacetylated (b). Chitosan is a copolymer intermediate between (a) and (b).

Chitin has the structure (a) shown in Figure 1, while chitosan is the *N*-deacetylated derivative with a composition between structures (a) and (b). Chitosan is characterized by the degree of deacetylation –DDA- and can be defined as chitin sufficiently deacetylated to form dilute acid soluble salts (DDA 80–85% or higher). The DDA is a measure of glucosamine moieties in chitosan (that carry free amine groups) , the fraction of free amine groups, or the positive charge distribution in the chitosan molecule that will be available for interactions with negatively charged surfaces or compounds [33]. However, some studies have shown that it is not really the total number of free amine groups that must be taken into account but the number of available free amine groups [34]. Therefore, amino groups make chitosan a cationic polyelectrolyte (p*Ka* ≈ 6.5), one of the few found in nature [35]. This basicity gives chitosan singular properties [36-38]:

1. It is insoluble in water, alkali and organic solvents.
2. Chitosan readily dissolves in aqueous acidic media, forming viscous solutions that can be used to produce gels.
3. It forms water – insoluble complexes with anionic polyelectrolytes that can also be used for the preparation of gel beads and membranes.
4. It chelates heavy metal ions.
5. The amino and hidroxyl groups allow easy chemical modification.

Chitosan is known for its biocompatibility allowing its use in various medical applications [39, 40], the production of value-added food products [41, 42] and can be considered as biodegradable [43]. Moreover, the concurrent presence of hydroxyl and amino groups, which are excellent functional groups for the anchoring of a large variety of organometallic complexes, makes chitosan a good precursor for heterogeneous molecular catalysts [44-47]. Engineering chitosan as catalytic carrier supports involves the understanding of the types and behavior of its hydrogels.

4.1. Chitosan Hydrogels

The term physical hydrogel describe non-covalently crosslinked network to distinguish it from chemical hydrogels formed by covalent bonds. The reversible bonds involved are generally non-covalent bonds such as hydrophobic interactions, micellar packing, hydrogen bonds, ionic bonds, crystallizing segments and combinations of these interactions [48]. Numerous papers can be found in the literature explaining how to produce physical hydrogels of chitosan. Hirano et al. [49] reported the formation of this kind of gels during the acylation of chitosan with various anhydrides in a hydro-alcoholic medium; Bernkop-Schnurch and Hopf [50] studied the reaction of chitosan with thioglycolic acid and Montembault et al. [51] produced a gel only containing water and chitosan.

Physical hydrogels are distinguished as ionically crosslinked hydrogels and polyelectrolyte complexes (PEC), or formed by secondary interactions as in chitosan/ poly(vinyl alcohol) complexed hydrogels, grafted hydrogels and entangled hydrogels [43].

4.1.1. Polyelectrolyte Complexed Hydrogels (PEC)

Polyelectrolyte complexes (PECs) are obtained by mixing aqueous solutions of two polymers carrying opposite charges [52]. Because of its cationic characteristics chitosan has been used for the preparation of various polyelectrolyte-complex products with natural (carboxymethylcellulose, alginic acid, dextran sulfate, carboxymethyldextran, heparin, carrageenan [53], pectin [54], xanthan [55], gelatin [56], alginic acid [57, 58]) or synthetic polyanions [53, 59]. The formation of protein–PEC complexes has been demonstrated for numerous enzymes. These complexes have the ability to improve the operational stability of the enzyme activity during catalysis because create an ionic microsystem which favours the stabilization of a protein polymer by interacting with the free acid and base functions. Furthermore, this type of hydrogel has a good porous structure that facilitates diffusion of both the substrate and the product of an enzymatic reaction [55].

4.1.2. Ionically Cross-Linked Chitosan Hydrogels

Compared to polyanions, the use of low molecular weight anions like phosphates to cross-link chitosan was found to be much simpler. Polyphosphate binds on the surface of chitosan, while tripolyphosphate can diffuse into chitosan droplets or films freely to form ionically cross-linked chitosan beads or films [60]. These gels have satisfactory enzyme-holding properties. Alsarra *et al.* [61, 62] used an entrapment technique of *Candida rugosa* lipase by gelling chitosan with tripolyphosphate while Szczesna- Antczac *et al.* used sodium hexametapolyphosphate to entrap *Mucor* lipase [63]. Ionotropicaly cross-linked lipase-chitosan beads are chemically and mechanically durable and do not shrink in organic solvents, usually used for ester synthesis. This high stability suggests that their network was formed via the reaction of polyphosphate groups with amine groups of both chitosan and proteins. This tight association with matrix protects the lipase from leakage [64, 65].

4.1.3. Chemical Hydrogels

Chemical hydrogels are formed by irreversible covalent links. Cross-linking procedure may be performed by reaction of chitosan with different bi-functional reagents such as glutaraldehyde [29, 66], carbondiimide [67], crown ether [68], di-carboxylic acids [69], poly (ethylene glycol) dialdehydes [70] and glyoxal [71]. It is also possible to use mono-functional reagents. Gümüşderelioğlu and Agi [72] prepared chitosan membranes cross-linked under alkaline conditions, using epichlorohydrin as the cross-linking agent. The most common crosslinker used with chitosan is glutaraldehyde. The aldehyde groups form covalent imine bonds with the amino groups of chitosan, due to the resonance established with adjacent double ethylenic bonds via a Schiff reaction [73]. However, links with hydroxyl groups of chitosan cannot be excluded [43]. Kinetics of chitosan cross-linked by glutaraldehyde is zero order with respect to the chitosan concentration but non-zero order with respect to the glutaraldehyde concentration [74].

4.1.4. Cryogels

Yolk exposed to temperatures bellow -6ºC will undergo gelation and upon thawing will have a consistency ranging from semifluid to pasty, depending on the freezing conditions [75]. Similarly aqueous solutions of many synthetic and natural polymers such as polyvinyl alcohol (PVA), chitosan, xanthan, locust bean gum, amylopectin, amylose and maltodextrins undergo physical changes as a result of their freezing, frozen storage, and subsequent thawing of their aqueous dispersions. Such cryogenic treatment of polymer solutions gives rise to the formation of physical crosslinked networks under a variety of conditions. This phenomenon is termed as cryostructuration or cryotropic gelation or cryogelation and the resultant gel-like polymer systems called cryogels or cryostructurates or cryotexturates. Cryogels are gel matrices that are formed in moderately frozen solutions of monomeric or polymeric precursors. The structure of cryogels, in combination with their osmotic, chemical and mechanical stability, makes them attractive matrices for chromatography of biological nanoparticles, whole cells [52, 76, 77], and enzymes [78-80] including lipases [81, 82]. When gelation is made by repeated freeze–thaw cycles, this enhances the functional properties of the gelled materials [83-85]. There are some studies that report the use of chitosan cryogels as porous scaffolds used in cell culture immobilization or as tissue regeneration templates [86, 87]. In these cases, chitosan acetic salt exclusion promoted by ice formation in freezing or cryogelling , produce a controlled average pore size and pore orientation material according to the rate of freezing which defines the ice crystal size [84, 88, 89].

4.2. Chitosan as Biocatalytic Support

Chitosan has been used as a matrix for immobilization of lipases [61, 64] and many other enzymes [90, 91]. Coupling polymers with enzymes or polymer conjugation has been reported to produce biocatalysts that respond to changes in pH, ionic strength, temperature, redox potential, and light [92]. Moreover, enzymes bound to sugars, or sugar-based polymers like chitosan are stabilized during lyophilization and in nonaqueous environments. This may be due to a reduction of autolysis, multipoint attachment that may limit enzyme distortions, or microenvironmental effects [93, 94].

A number of desirable characteristics should be common to any material considered for immobilizing enzymes. These include: high affinity to proteins, availability of reactive functional groups for direct reactions with enzymes and for chemical modifications, hydrophilicity, mechanical stability and rigidity, regenerability, and ease of preparation in different geometrical configurations that provide the system with permeability and surface area suitable for a chosen biotransformation. Chitosan have most of the above characteristics [90]. However, there are limiting factors for the catalytic activity of supported catalysts based on chitosan. They are the accessibility to internal sites, the poor porosity of the raw material, and the limitations due to diffusion. For improving diffusion properties one strategy consists of changing the conditioning of the polymer (dissolved state, fiber, flakes, beads, hollow fiber, supported on inorganic membrane, etc.). These modifications can offer technical solutions to minimize the effect of these limiting factors but they contribute to increase the cost of supported materials [95].

4.2.1. Control of Support Characteristics by Physical-Chemical Modification of Chitosan

Lipo- or hydrophilicity, porosity and permeability of enzyme-bound material need to be tailored with respect to reaction conditions and reactor configuration [96]. When using chitosan as enzyme support its properties may be affected in a variety of ways. The availability of amine groups of chitosan, related with its sorption ability, may be controlled by two parameters: (a) the crystallinity of the polymer and (b) the diffusion properties. The reduced polymer crystallinity can be maintained through freeze-drying of the chitosan solution, while an air-drying or oven-drying procedure partially reestablishes polymer crystallinity [95]. Although the larger crystalline regions in membranes will enhance the mechanical strength of the membrane, however, they may increase its fragility [97]. The protein encapsulation efficiency [65] and crystallinity degree are increased with increasing chitosan DDA [97]. Physical-chemical modifications including substitution and cross-linking on chitosan chains or physical modifications by means of successive freezing and thawing affect its sorption performance [98, 99], prevent chitosan solids from dissolution in aqueous acid solutions($pH<2$) [100], and control its structure [87, 89].

Chemical modification of chitosan can be used to attach functional groups and to control its hydrophobic, cationic, and anionic properties [101]. An useful method for this purpose is graft copolymerization of chitosan [102]. Tangpasuthadol *et al.* studied the amount of protein adsorbed related to the types of attached molecules on chitosan surface (stearoyl group, succinic anhydride and phthalic anhydride), after chemical modification with heptafluorobutyryl chloride and *m*-iodobenzoyl chloride. Their study demonstrated that it is conceivable to fine-tune surface properties which influence its response to bio-macromolecules by heterogeneous chemical modification [103]. Prabaharan and Mano discussed different methods of cyclodextrins grafting onto chitosan [104]. Crosslinking promotes the formation uniform porous network-like structure of chitosan, change its crystal structure from α- to β-type , and may also smooth the surface of chitosan [71, 105, 106].

Porosity, pore size distributions, internal surface area and size of surface pores could also be tailored with cryogelling, freezing and thawing treatments and drying procedure. The pore size and pore morphology of cryogels are controlled by the concentration of the monomers used and the freezing temperature [107]. Properties of ice crystals such as shape, size, and distribution play an important role in determining textural and physical properties of many frozen products. It is well-known that rapid freezing rather than slow freezing gives smaller size ice crystals in frozen foods [108, 109]. In chitosan-based materials obtained by freezing and lyophilization of chitosan solutions and gels, the ice crystal size and hence their mean pore size and pore orientation have been controlled by varying the freezing temperature/rate, by thermal gradients [110-112].

Figure 2 shows three-dimensional AFM photographs of membrane samples obtained by freezing and thawing procedures [89]. The surfaces of all membranes are not smooth and possess nodule-like structure and nodule aggregates. The nodules are seen as bright high peaks, whereas the pores are seen as dark depressions. It is evident the effect of thermal treatments on the surface structure.

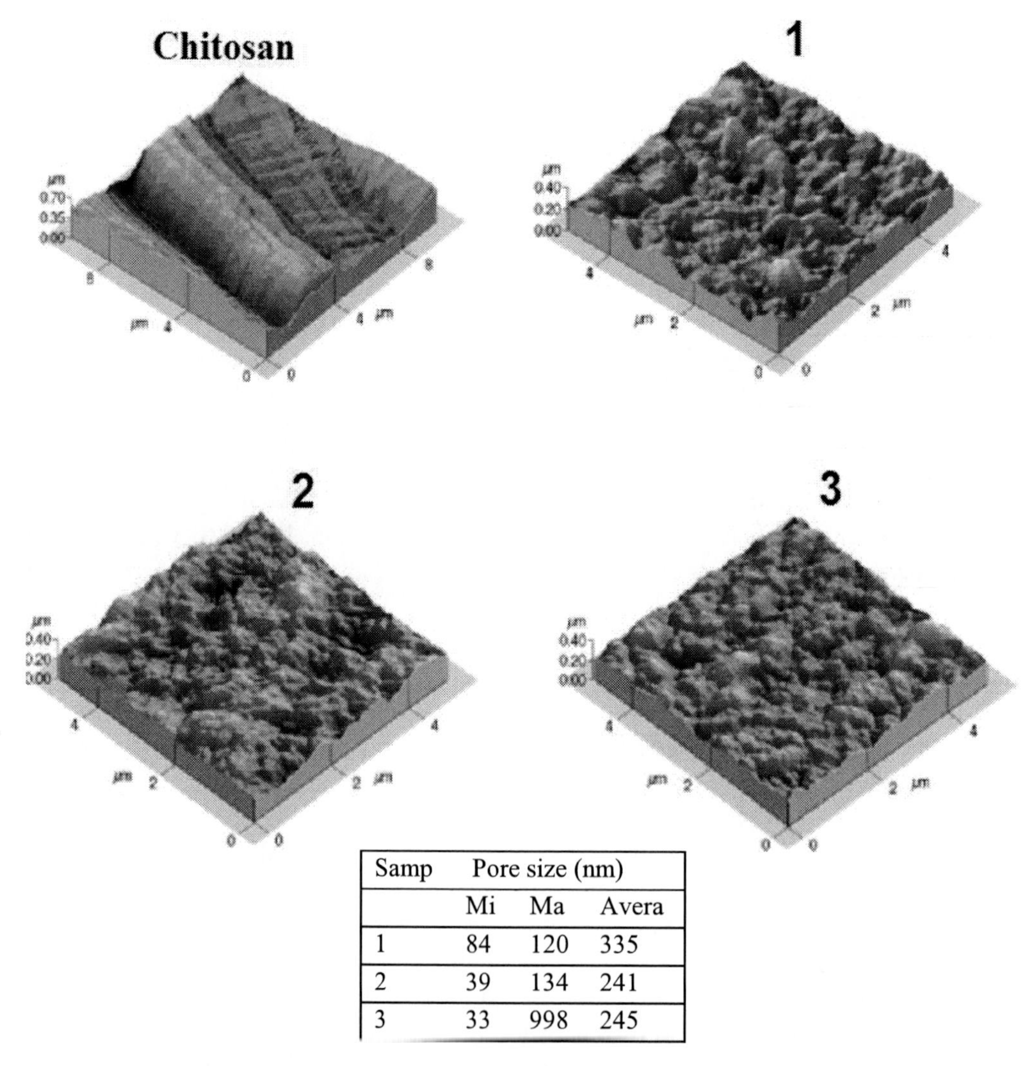

Samp	Pore size (nm)		
	Mi	Ma	Avera
1	84	120	335
2	39	134	241
3	33	998	245

Figure 2. AFM 3D images of chitosan membranes produced by freeze-thaw (F/T) treatments. 1. Chitosan without treatment, 2. Two F/T cycles, 3. Three F/T cycles and 4. Six of F/T cycles.

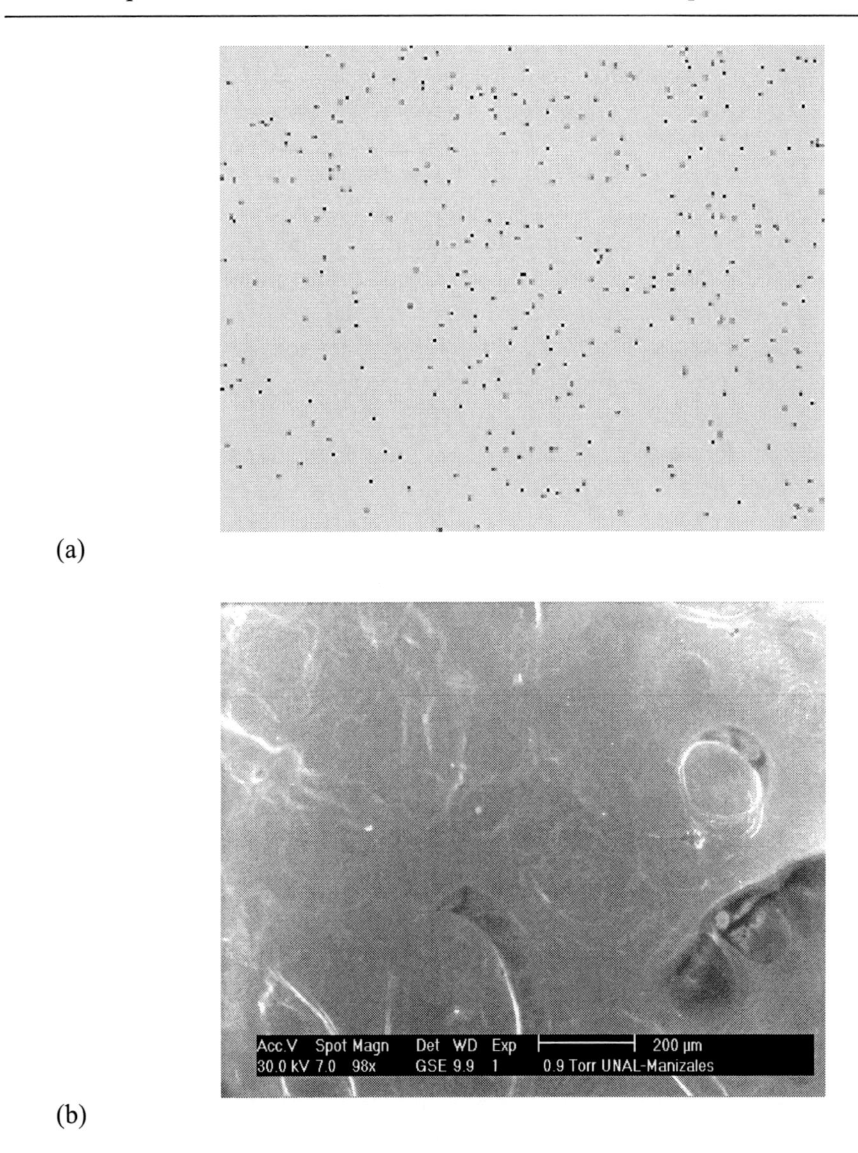

Figure 3. Nitrogen EDX mapping of glutaraldehyde crosslinked chitosan (a) The black points correspond to the higher nitrogen concentration (b) ESEM micrograph of the same area.

4.3. Surface Characterization of Chitosan Supports

Surface analysis methods applied to biomaterials include measurements such as adsorption isotherms, surface areas, pore size distribution, surface roughness, thickness and surface topography via microscopy. Spectroscopic methods, in contrast, provide information such as elemental composition, oxidation states, and functional groups, quantitative analysis and distribution of species. Some of these two groups of methods, that have been used to characterize chitosan supports and chitosan-enzyme biocatalytic systems, are described next.

4.3.1. XPS and EDX

X-ray photoelectron spectroscopy (XPS) is a quantitative spectroscopic technique that provides a direct chemical analysis of solid surfaces on a depth of approximately 5 nm. An X-ray beam induces the ejection of electrons from the atoms. The kinetic energy of the photoelectrons is analyzed and their binding energy is determined. Since the binding energy of electrons in the atom of origin is characteristic of the element and affected by its chemical environment, the method provides an elemental analysis and further information on functional groups [113]. XPS has been used to characterize biomaterial surfaces [114, 115], for detecting biomolecules adsorbed at surfaces [116, 117], verify biopolymer coatings in surface engineer metal films [118], the presence of functional groups onto the bipolopymer surfaces [119, 120] including enzymes [121] and analysis of whole cells [122, 123].

For chitosan based materials this technique brings information about crosslinking [124], evidence of covalent attachment [125, 126] and reactivity of chitosan surface [103].

Another close relative of XPS that can also be useful for qualitative and semi-quantitative analysis of chitosan surface is the SEM (Scanning Electron Microscopy) and EDX (Energy dispersive X-ray) spectroscopy. This method allows a fast and non-destructive chemical analysis with a spatial resolution in the micrometer regime. It is based on the spectral analysis of the characteristic X-ray radiation emitted from the sample atoms upon irradiation by the focussed electron beam of a SEM. [127, 128].

4.3.2. Elemental Analysis

Elemental analysis using CHN analyzers has been used for estimation of the nitrogen, hydrogen and carbon contents as well as N/C ratio in chitin and chitosan characterization [129, 130] and for the quantification of the total quantity of amino groups of chitosan support in enzyme immobilization [131]. X-ray energy dispersion spectroscopy (EDX) measurements are useful for confirm, from different elemental ratios, chitosan surface coatings, modification or the presence of some molecules inside chitosan fibers [132, 133]. Kjeldahl technique allows to measure nitrogen in organic compounds, proteins and food. It has been used for nitrogen determination in chitin and chitosan [134] and for the quantification of chitosan in fibers and composites [135, 136].

4.3.3. FTIR and Raman Spectroscopy

Samuels [137] and Urbanczyck [138] have shown that the FT-IR chitosan spectra may be related to the crystalline structure. They have identified two principal forms of chitosan: an "amorphous" (α) form and a crystalline (β) form. The β-form showed two absorption bands at 1350 and 1380 cm^{-1} and a third characteristic band at 760 cm^{-1}. The α-form, on the other hand, had a peak at 1380 cm^{-1}. At lower frequency the absorption was shifted to 800 cm^{-1}. In this manner, FT-IR allowed to follow the gradual transformation of the α into the β form [139].

The figure 4 show the IR spectra of chitosan and immobilized lipase on chitosan. The absorption bands at 1652.5 and 1588.68 cm^{-1} in this spectras are, respectively, attributed to the amide I and N–H bending mode of $-NH_2$ [140].

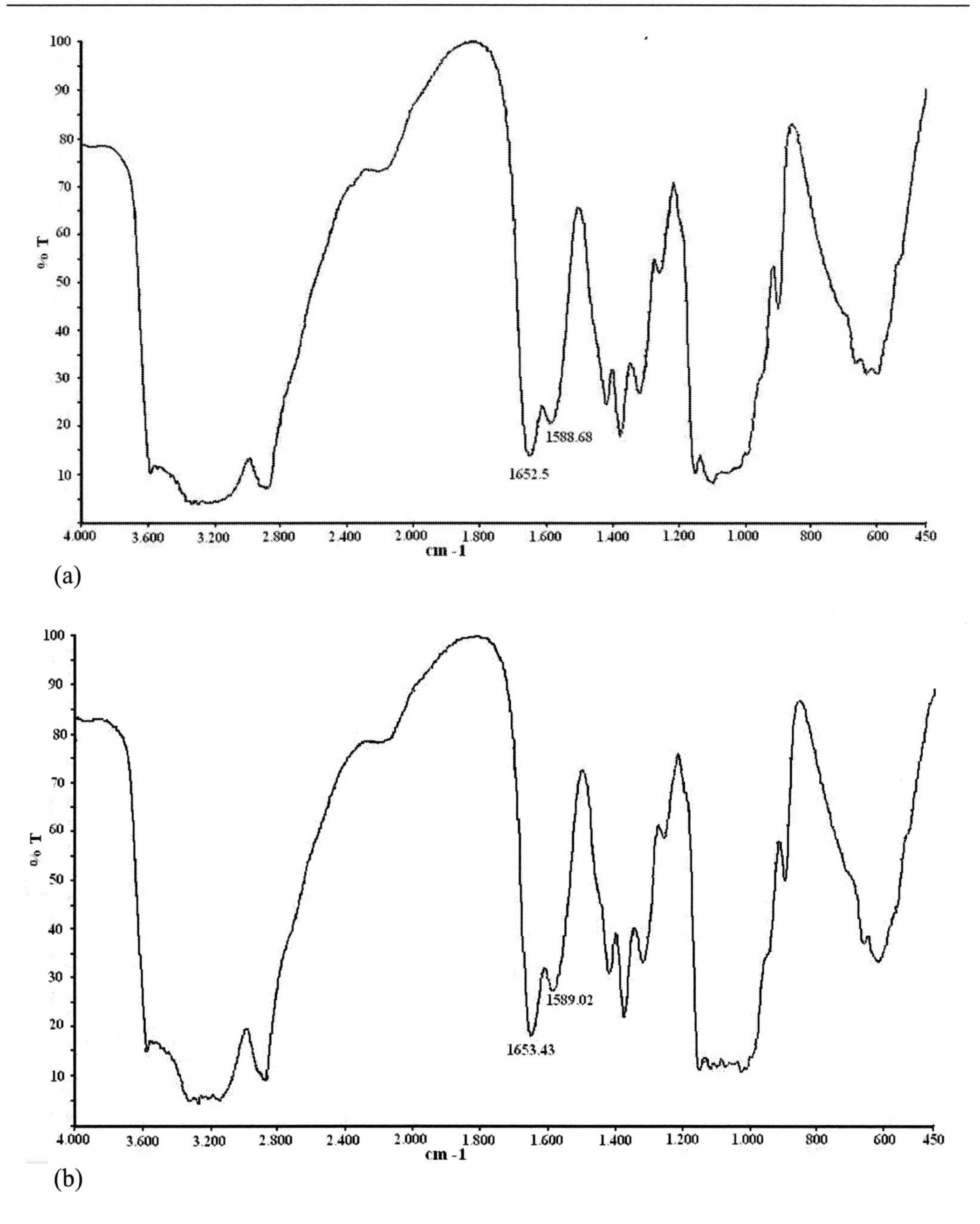

Figure 4. (a) FTIR spectra of chitosan support and (b) FTIR spectra of immobilized lipase on chitosan.

A recent study demonstrated the presence of immobilized lipase on chitosan supports by using Raman spectroscopy and elemental analysis. From Raman spectroscopy, when the spectra for supports and the immobilized lipase system were compared, an increasing on the signal from the carbonyl group in the amide I band was observed. The presence of amide II band and an increment on the CH_3 signal of the amide group (1380 cm^{-1}) was also observed in all immobilized lipase supports. The presence of lipases on the supports was verified with elemental analysis according to an increase in the nitrogen content (Kjeldahl analysis) as well as an increase on the C/N ratio (combustion and Kjeldahl analysis)[141].

4.3.4. X-Ray Diffraction

The X-Ray diffraction method is used to to characterize the crystallographic structure of solid samples.

The XRD results have shown that combined mild crosslinking and freezing and thawing (F/T) treatments can increase crystallinity of chitosan membranes, which, in turn, promotes the stability of the material and the availability of amine groups on surface support [89]. Some studies have found that chitosan crystallinity decreased by crosslinking [142, 143]. Apparently, freezing and thawing treatment of membranes has a superior effect than mild crosslinking on the biopolymer surface structuration.

4.3.5. Atomic Force Microscopy (AFM)

This technique has made possible to characterize surface structures of membranes – surface roughness, mean pore diameter, membrane pore density [144-148] – and study the presence of proteins on membranes [127]. This technique is able to image the surface of microfiltration membranes in air with single pore resolution, and permit to determine pore size distributions, internal surface area and size of surface pores showing similar results with the N_2 adsorption-desorption and a liquid displacement technique [149]. The formation uniform porous network-like structure of chitosan membrane surface by crosslinking has been studied with AFM [105].

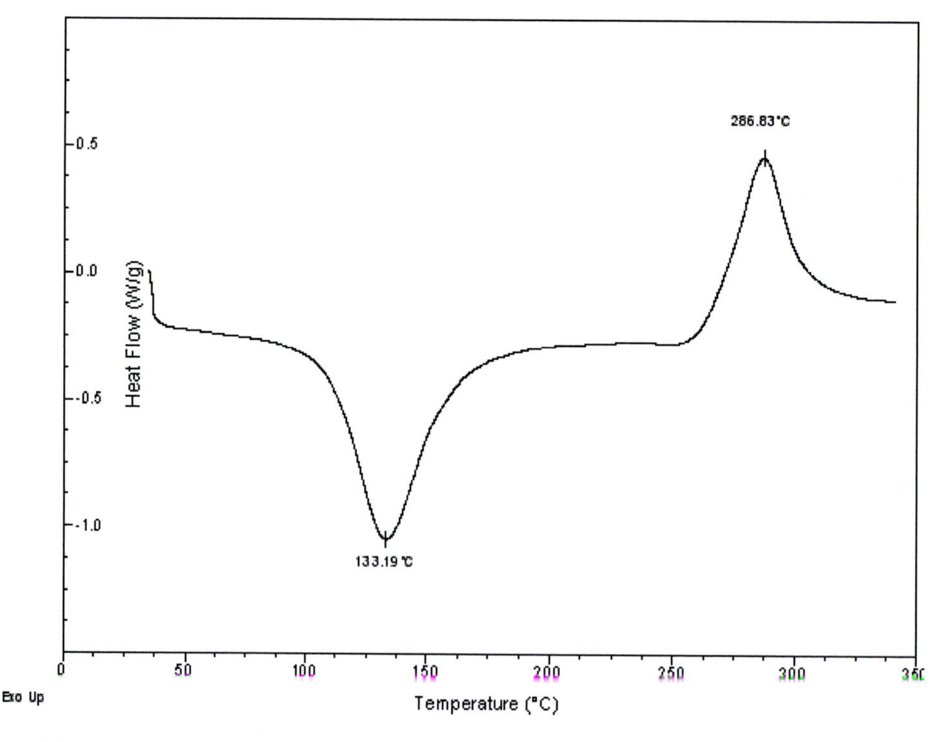

Figure 5. Thermal data plot of a chitosan membrane under nitrogen atmosphere.

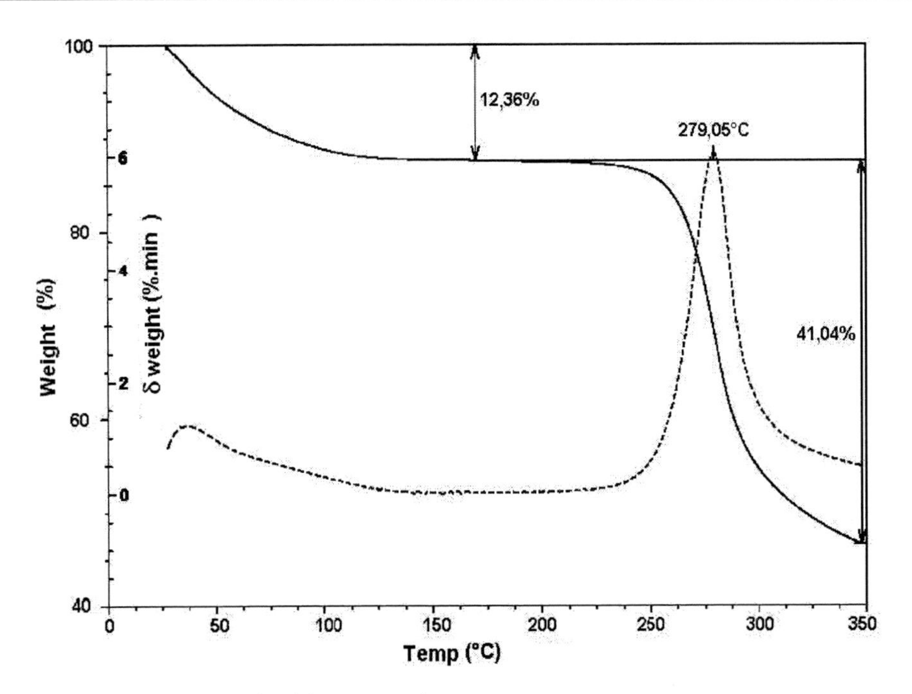

Figure 6. TGA y DTG curves of a chitosan membrane.

4.3.6. Thermal Analysis

Chitosan membranes prepared by using a combined freeze gelation and mild crosslinking method showed, in a DSC plot (Figure 5), a broad endothermic peak, related to dehydration phenomenon, was initially observed at about 100°C. The position of the endotherm peak was centered at about 119 °C. The DSC exothermic peak, at around 281 ° C, can be attributed to the thermal decomposition of the membrane. Figure 6 presents typical chitosan sample thermograms obtained with differential scanning calorimetry (DSC) and TGA. The first derivative of mass-change with respect to time [dm/dt, called derivative thermogravimetry (DTG)] was also plotted as a function of temperature in the same figure. In TGA curve, the evaporation of water, corresponding to the first weight loss, occurred above 100 °C, increased to 12.36% and remained in this level until approximately 200 °C. The second weight loss began at 260 °C. The DTG curve has a transition peak, also associated with the onset of the second stage of decomposition in nitrogen environment, at 279.05 °C [89].

The glass transition temperature (*Tg*) of chitosan has been reported by many authors. Sakurai *et al.* [150] found the *Tg* of chitosan to be 203 °C while Cheung *et al.* [151] observed the *Tg* of chitosan at around 103 °C and Shantha and Harding [152] and Suyatma *et al.* [153] found a sharp glass transition of chitosan around 195 °C. Mucha and Pawlak [154] reported *Tg* values between 156–170 °C and found that *Tg* decreases with increasing DDA of chitosan. Higher values of the *Tg* (> 200°C) has been reported for cross-linked chitosan membranes [155]. Thus, there are various values of the *Tg* of chitosan.

Illanes [156] have observed increase in thermal stability by immobilizing different enzymes to glutaraldehyde-activated chitin matrices, where multiple Schiff-base linkages are established between free amino groups in the protein and the aldehyde group in the glutaraldehyde linker. With the same crosslinker, covalently bound lipases on chitosan showed better thermal stability than free lipases in a solvent media alcoholysis of salicornia oil at all temperatures and time periods [157]. Besides the conventional use of different temperature activity assays, thermal stability measurements such as thermal stability or thermal decomposition of enzyme, chitosan and chitosan loaded enzyme membranes can also be made combining more than one thermal method , namely, thermogravimetric analysis (TGA) and differential scanning calorimetry (DSC) [155, 158]. Between a number of soluble additives (polyhydric alcohols and salts) on the thermal stability of the lipase of *R. miehei,* the most effective additive against thermal deactivation was the sorbitol. The study was developed by two methods: thermal denaturation at 50ºC as a function of time (kinetics of denaturation) and microcalorimetry with differential scanning. These two methods led to the same results for the classification of the protective effect of these additives [159]. Da Silva Crespo *et al.*[160] used the results of thermogravimetric analysis for thermal stability comparisons of commercial *Candida rugosa* and *Pseudomonas* sp. lipases. In the same work, the presence of the enzymes in the supports was confirmed by differential scanning calorimetry. Turner *et al.,*[5] studied the loss of water from protein samples by TGA, DSC and Karl Fisher titrations and concluded that temperatures in excess of 200°C are required to remove tightly bound water from various lipases. They hypothesised that those enzymes that require very little water for their catalytic activity should remain active at such elevated temperatures provided that they can be stabilised against thermodenaturation. This conclusion was verified by the observation that immobilised *Candida antarctica* lipase catalysed transesterification of octadecanol with palmityl stearate at 130°C for a considerable period of time [161].

4.3.7. Moisture Sorption Isotherms

In a recent work Orrego *et al.* analyzed the effect of the support water activity on the catalytic behavior of an immobilized lipase in a non-aqueous media.

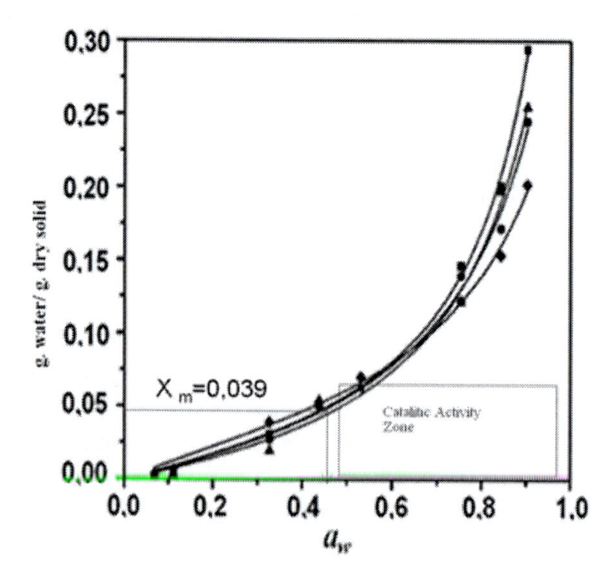

Figure 7. Chitosan supports adsorption isotherms. Symbols: □ 20°C ;• 30°C; ▲ 40°C; ◊ 50°C.

Figure 7 shows moisture sorption isotherms of chitosan membranes that were intended to find the Guggenheim-Anderson-de Boer model (GAB) parameters. They found that supports with water content above the–GAB- monolayer value, *X*m, will assure enough water availability for the lipase attached to chitosan to exert its catalytic function in organic media. The parameter *X*m can be considered as a suitable model boundary between bound and free water of the support. That is the reason because is a critical moisture content below of which water is not enough to allow reagent diffusion and, consequently, chemical and enzymatic reactions [146].

5. Immobilization of Lipases on Chitosan

Japanese researchers reported the use of chitosan as enzyme carrier since the 1970's. Kasumi *et al.* [162] observed that the amino groups of chitosan facilitate the immobilization of enzymes, either by adsorption or chemical reaction. Lipase from *Candida cylindracea* was immobilized on chitosan by a covalent binding method for the hydrolysis of beef tallow by Sakakibara *et al.* [163]. Itoyama *et al.* [164] immobilized cristallized lipoprotein lipase from *P. fluorescens* onto the surface of porous chitosan beads without or with spacers of different lengths. The relative activity of lipase immobilized without spacer decreased gradually with the decreasing surface concentration of the immobilized enzyme. On the other hand, lipase immobilized with spacer gave an almost constant activity for p-nitrophenyl laurate hydrolysis. The pH, thermal, and storage stabilities of the immobilized lipase were higher than those of the free one.

Dumitriu and Chornet [53] immobilized lipases on chitosan using matrix entrapment of the enzyme solution in polyelectrolyte complexes fluids. They used xanthan–chitosan hydrogels in order to determine the immobilization capacity of time and percentage of his type of gel and attempt to obtain a biocatalyst capable of hydrolyzing lipids and hemicelluloses present in food and industrial wastewaters. The immobilized lipase was quite stable in this type of hydrogel. The optimal conditions for lipase activity in aqueous medium were: incubation time =10 min, incubation temperature=37ºC, pH=7.5. Lipase activity in isooctane was also studied as a function of the concentration of olive oil, incubation time and percentage of water in the system. The optimal time and temperature of incubation were different than those derived from the aqueous solution of olive oil. Optimal time was 24 h and temperature was 34ºC [53]. In a further study [165] the lipase was immobilized in the same polyionic hydrogel. In the aqueous medium, the activity was twice as high for immobilized lipases as for free lipases. Immobilized lipases were also able to hydrolyze triacylglycerols in three distinct organic solvent media.

A chitosan/poly (vinyl alcohol) membrane has been prepared by Tan *et al.*[166] and used in the synthesis of a monoglyceride from palm oil triglyceride. The membrane was prepared mixing together a solution of chitosan, poly (vinyl alcohol), lipase, and cross-linking agent (glutaraldehyde or epichlorohydrin). The activity of the lipase was reduced by increasing the concentration of glutaraldehyde but not by increasing the concentration of epichlorohydrin. The presence of Ca^{2+} or Mg^{2+} was found to stabilize the immobilized lipase.

Betigeri and Neau [64] immobilized *Candida rugosa* lipase using agarose, alginate and chitosan beads that were prepared by ionic gelation using calcium chloride or sodium tripolyphosphate, respectively, as the cross-linking agent in the gelling solution.

Szczesna-Antczak *et al.*[63] geled chitosan by several methods including the cross-linking and ionotropic gelation with glutaraldehyde, NaOH, sodium tripolyphosphate and sodium hexametapolyphosphate and entrapped in the gel beads a *Mucor circinelloides* mycelium-bound lipase preparation. The immobilized preparations were used for *p*-NPA hydrolysis. The best biocatalysts were formed from 4.5% chitosan solutions, containing approximately the same weight of membrane-bound lipase, and enriched with polyvinylpyrrolidone which enhanced the hydrophilic character and porosity of the gel. Application of this biocatalyst for sucrose caprylate synthesis provided 70% conversion of the acid after more than 10 h of reaction.

Alsarra *et al.* [61] used an entrapment technique to produce *Candida rugosa* lipase loaded chitosan beads. A high molecular weight and degree of deacetylation can improve lipase loading and lessen the release of the entrapped enzyme [61]. In a further study, they observed the influences of the pH, tripolyphosphate concentration, and ionic strength of the gelling medium on the entrapment efficiency, release, and activity of *Candida rugosa* lipase in chitosan hydrogel beads [61]. The pH and the tripolyphosphate concentration each had an effect on the entrapment, retention, and activity of lipase. Activity was measured with the reaction for the conversion of triglycerides to diglycerides and fatty acids. Entrapped lipase retained a high degree of activity in five consecutive runs.

Pereira *et al.* [167] reported the immobilization of *Candida rugosa* and porcine pancreas lipases on a two types of chitosan (flake and porous) support by physical adsorption. The best results for recovery of total activity after immobilization were obtained for microbial lipase and porous chitosan beads. In organic medium, the direct synthesis of n-butyl butyrate in organic solvent was chosen as a model reaction. The use of this immobilized preparation was extended to the direct esterification of a large range of carboxylic acids with a variety of alcohols (from C_2 to C_{10}).

Lipase was also immobilized on modified chitosan by the introduction of hydrophobic groups into porous cross-linked chitosan beads which were further used for the transesterification of racemic 1-phenyl ethyl alcohol and vinyl acetate in organic solvents to investigate the effects of water and hydrophobic groups on the enzyme activities [168].

Da-Cunha *et al.* [169] used glutaraldehyde activated chitosan membranes prepared by evaporation of a 2.5% (w/v) solution in 0.4 M acetic acid. Lipolytic activity was found to be 245 μmol s^{-1} m^{-2}. Krajewska [36] prepared gel chitosan membranes supported with glass fabric by solvent evaporation technique followed by crosslinking of different concentrations of glutaraldehyde and coating with bovine serum albumina. While glutaraldehyde crosslinking (0.01-0.1%) was found to reduce the water content of the membrane and the pore size, protein immobilization was found to bring about the opposite effects.

The precursor of chitosan, chitin had also been used in lipase immobilization. Chitin, funcionalized with hexamethylenediamine followed by glutaraldehyde activation, was binded with *Candida rugosa* lipase. This methodology for immobilizing lipase on chitin by covalent binding gave high protein retention (80%) and activity yield of 27% at lipase loading of 250 U g^{-1} dry support. The thermal stability of the immobilized lipase was higher than that of the free one. The lipase immobilized on chitin was also effective for the application in the synthesis of butyl esters [170].

Two methods for immobilization of *Candida rugosa* lipase on chitosan beads by activating the hydroxyl groups of chitosan using carbodiimide (1-ethyl-3-(3-dimethylaminopropyl) carbodiimide hydrochloride –EDC-) coupling agent was studied by Wu [67, 171]. In these works, chitosan was pretreated with EDC to activate the glucosamine -OH groups, rather than the -NH_2 groups. Activated chitosan was then reacted with the lipase in either a dry or wet bead form. Wu also employed a dual approach to *Candida rugosa* lipase immobilization on chitosan. First, the lipase was attached to chitosan via -OH groups activated by EDC; second, further lipase was attached to the chitosan via the -NH_2 group using glutaraldehide-GA-. The method resulted in a higher loading of enzyme on chitosan (287.2μg/g) compared to EDC activation (86.4 μg/g) or GA activation alone (162.1 μg/g). The performance of the enzyme in catalyzing the release of *p*-nitrophenol from *p*-nitrophenyl palmitate was followed and found to be high (13.8 EU/g of chitosan). With these procedures the immobilized enzyme was found to be more stable with respect to pH, temperature, storage, and reuse. A third work also used this binary method of lipase immobilization on chitosan to study the hydrolysis of soybean oil. The effects of temperature, pH and oil to water ratio on the conversion, pH and thermal stability, reusability, storage stability and the kinetic properties were investigated. Under optimal conditions, 88% of the oil taken initially was hydrolyzed after 5 h. Better thermal stability was exhibited by the immobilized lipase and the pH stability was comparable to that of soluble lipase. Storage for 30 days at 4 °C, showed that the immobilized enzyme did not lose its activity. The relative activity upon six repeated uses was 80% [172].

Ye *et al.* [172] prepared a dual-layer biomimetic membrane by tethering chitosan on a polyacrylonitrile-based copolymer membrane surface (poly (acrylonitrile-co-maleic acid) –PANCMA-) and used for lipase immobilization. Lipase from Candida rugosa was immobilized on PANCMA using glutaraldehyde, and on the nascent PANCMA membrane using EDC/ N-hydroxylsuccin-imide as coupling agent. They found that both the activity retention (also measured in catalyzing the release of *p*-nitrophenol from *p*-nitrophenyl palmitate) of the immobilized lipase and the amount of bound protein on the dual-layer biomimetic membrane were higher than those on the nascent PANCMA membrane. After immobilization, the pH, thermal and reuse stabilities of the immobilized enzyme increased.

Desai *et al.* [157] immobilized Porcine pancreatic lipase on chitosan by covalent binding. They used various concentrations (0.01–5.0%) of gluteraldehyde for the activation of free amino groups. The activities of free and immobilized lipase were determined using olive oil as substrate. The free and immobilized enzymes showed pH 9 as optimum and retained 50% of activity after five cycles. Alcoholysis of salicornia oil, mediated by free and immobilized lipase, was carried out at 25ºC using methanol in hexane and acetone. Free and immobilized enzyme in hexane produced, respectively, 45% and 55% of fatty acid methyl ester after 12 h.

Kılınç et al. [173] compared the properties of free and chitin - chitosan immobilized porcine pancreatic lipase either by adsorption and subsequent crosslinking with glutaraldehyde, which was added before (conjugation) or after (crosslinking) washing unbound proteins. Conjugation proved to be the better method for both supports. Their results showed that the pH optimum was shifted from 8.5 to 9.0 for both the immobilized enzymes. Also, the optimum temperature was shifted from 30 to 40°C for chitin-enzyme and to 45°C for chitosan-enzyme conjugates. The immobilization efficiency is low, but the immobilized enzymes have good reusability and stability (storage and operational). Besides these properties, the immobilized lipases were also suitable for catalyzing esterification reactions of fatty acids and fatty alcohols, both with a medium chain length. Esterification activities of immobilized lipases were two- and four-fold higher for chitosan- and chitin-enzyme, than for the free enzyme, respectively.

Neutral lipase was inmmobilized on chitosan nano-particles (particle sizes smaller than 100 nm) prepared by the ionization gelation method. The optimal temperature and maximal activity of the immobilized enzyme for the hydrolysis of an olive oil-PVA emulsion were found at 45 °C and pH 7.5. The immobilized enzyme exhibited remarkably improved stability properties to various parameters, such as temperature, reuse, and storage time [174].

Table 1. Range of lipases immobilized on chitosan

Lipase	Method	Catalyzed reactions	Reference
C.cylindracea	Covalent binding	Hydrolysis of beef tallow	[163]
P. fluorescens	Adsorption Covalent binding	p-nitrophenyl laurate hydrolysis	[164]
C. rugosa	Entrapment on chitosan-xanthan	Hydrolysis of lipids in aqueous and organic media	[53, 165]
C. rugosa	Entrapment on chitosan/poly (vinyl alcohol) membrane	Synthesis of a monoglyceride from palm oil triglyceride	[166]
C. rugosa	Entrapment on chitosan microcapsules	Hydrolysis of lipid in aqueous media	[64]
M. circinelloides	Polyvinyl pyrrolidone- chitosan beads solidified with hexametapolyphosphate	Synthesis of sucrose, glucose, butyl and propyl oleates and caprylates, carried out in petroleum and di-*n*-pentyl ethers	[63]
C. rugosa	Entrapment in chitosan beads	Hydrolysis of lipid in aqueous media	[61, 62]
C. rugosa and porcine pancreas lipases	Adsorption in flakes and porous chitosan	Esterification of carboxylic acids with alcohols (from C_2 to C_{10}). Synthesis of n-butyl butyrate in organic solvent	[167]
	Porous cross-linked chitosan beads	Transesterification of racemic 1-phenyl ethyl alcohol and vinyl acetate in organic solvents	[172]
C. rugosa	Entrapment in glutaraldehyde activated chitosan membranes	Hydrolysis of triolein	[169]
C. rugosa	Binary covalent binding on chitosan beads	Release of *p*-nitrophenol in the enzymatic hydrolysis of *p*-NPP with etanol as substrate	[67, 171, 172]
C. rugosa	Covalent binding of the enzyme on a chitosan-polyacrylonitrile-based copolymer membrane	Release of *p*-nitrophenol in the enzymatic hydrolysis of *p*-NPP in aqueous and organic media	[172]
Porcine pancreatic lipase	Covalent binding	Hydrolysis of olive oil and alcoholysis of salicornia oil, using methanol in hexane and acetone	[157]
Porcine pancreatic lipase	Adsorption and subsequent crosslinking with glutaraldehyde	Esterification reactions of fatty acids and fatty alcohols	[173]
Neutral lipase	Entrapment in chitosan nano-particles	Hydrolysis of an olive oil-PVA emulsion	[174]
C. *rugosa*	Entrapment into photo-crosslinked chitosan-ACHCA	Hydrolytic activities by the olive oil emulsion method	[175]
C. *rugosa*	Magnetic Fe_3O_4-chitosan nanoparticles	Hydrolytic activities by the olive oil emulsion method	[176]
C. *rugosa*	Amino acid modified chitosan beads	Hydrolysis of *p*-nitrophenyl palmitate (*p*NPP)	[177]
C. *rugosa*	High crystallinity chitosan	Butyl oleate synthesis in isooctane	[99, 141]

Table 1 shows some of the range of methods and catalized reactions of lipases supported on chitosan.

6. Medium Effects and Kinetics for Esterification Reactions Catalyzed by Lipases in Non Aqueous Media

In the simplest approach, a one-substrate lipase catalized reaction follows Michaelis-Menten kinetics. This corresponds to the following sequence of reactions between the enzyme E and a single substrate S, in which ES designates the enzyme-substrate complex and P the product:

$$E + S \underset{k_{-1}}{\overset{k_1}{\rightleftarrows}} ES \overset{k_{cat}}{\rightarrow} E + P \quad (1)$$

In the Michaelis-Menten model, the initial reaction rate depends on the initial substrate concentration $[S_0]$ according to the law

$$v_0 = V_{max} \frac{[S_0]}{K_M + [S_0]} \quad (2)$$

where

$$K_M = \frac{k_{-1} + k_{cat}}{k_1} \quad (3)$$

K_M is known as the Michaelis constant. k_{cat} is also known as the turnover number as it represents the maximum number of substrate molecules that the enzyme can 'turn over' to product in a set time. The ratio k_{cat}/K_M determines the relative rate of reaction at low substrate concentrations, and is known as the specificity constant. A number of publications report data on k_{cat}, K_M, V_{max}, or on k_{cat}/K_M. However, many transformations (e.g; transesterifications) involve reactions with 2 substrates, hence having 2 different K_M values as well as possible inhibition constants. In most cases, the kinetics mechanism is not changed by enzyme immobilization. However, the magnitudes of the kinetics constants are modified.

In nonaqueous solvents, lipases carry out synthetic reactions and can also exhibit altered selectivities, pH memory, increased activity and stability at elevated temperatures, regio-, enantio- and stereoselectivity [5]; [178, 179].

Until date there is no general model to predict the effects of the medium on kinetics, on enzyme stability, on enantioselectivity and on reactor stability [7, 8, 179-183]. Attempts have been made to establish correlations between activity of the enzyme and different physicochemical properties of the solvents, such as accommodation of charge (dielectric constant) [183] and hydrophobicity (water/octanol partition coefficient, log P) [184, 185]

A theory was developed by van Tol *et al.* [186] according to the transition state theory. The transition state complex is assumed to be in equilibrium with the ground state of the substrate and enzyme, the corresponding thermodynamic equilibrium constant expressed in activity-based kinetic parameters that are the same in all solvents and also the resulting specificity constant (equal to V_{max} /[Enzyme]. K_M in the Michaelis equation) based on activity is independent of the solvent. Kinetic constants based on activity are called ''intrinsic parameters'' because the solvent does not affect them. Sandoval *et al.* [182] improved and verified this model using the esterification of oleic acid with ethanol catalyzed by an immobilized lipase to predict the kinetics in various organic solvents and in solvent-free systems.

Hazarika *et al.* [185] studied the effect of solvent on esterification of oleic acid with ethanol in a dispersed system of *Porcine pancreatic* lipase. The solvents were selected on the basis of their hydrophobicity (log *P*) values. The log *P* value has been proposed as a quantitative measure of solvent polarity and the enzyme activity for lipase catalysed reactions in general, increases with increasing hydrophobicity of the solvent. The relation of initial reaction rate with log *P* values of the solvent shown that the enzyme activity increases almost linearly with an increase of log *P*. Similar results were reported with lipase from *Candida rugosa* was coated with glutamic acid didodecyl ester ribitol amide as catalyst in organic solvents [185].

Desai *et al.*[157] used chitosan, in the form of beads, for covalent coupling of *Porcine pancreatic* lipase. The Michaelis constant (K_M) and the maximum reaction velocity (V_M) of the free and immobilized enzymes were almost the same, indicating no conformational change taking place during immobilization. On the same support, Hung *et al.*[67] studied the Michaelis–Menten kinetics of the hydrolytic activity of the free and the binary immobilized *Candida rugosa* lipase investigated using varying initial concentrations of *p*-nitrophenyl palmitate in ethanol as substrate. The kinetic constants V_M and K_M and the energy of activation were determined to be 96.1 U/mg-protein, 18.01mM and 15.3 kcal/gmol, respectively. The energy activation was lower than that of free lipase.

Hydrolytic activities in aqueous media of free lipase assayed by the olive oil emulsion method free *Candida rugosa* and porcine pancreas lipases obeys also the Michaelis-Menten equation with no inhibition by reaction products. For the case of the chitosan immobilized lipase detected a possible substrate inhibition or diffusion resistance [167].

Irrespective of the type of reaction catalyzed (i.e. hydrolysis, esterification, or interesterification), the most general and accepted description of the catalytic action of lipases is a Ping–Pong mechanism for two-substrate two-product (bi–bi) kinetic model [99, 187-190]. In the esterification of oleic acid with n-butanol, the first step is the binding of fatty acid to the enzyme to form a non-covalent lipase-oleic acid complex which then is transformed to by a unimolecular isomerization reaction to an acyl–enzyme intermediate and water is released. The second reactant, the n-butanol binds to the acyl–enzyme complex to form a modified enzyme–butanol complex which isomerizes unimolecularly to an enzyme oleic acid–butyl ester complex which then yields the butyl–oleate ester and the free enzyme. After the release of the ester, another molecule of oleic acid can bind to the enzyme. Nevertheless, above certain concentration of substrates, competitive inhibition is reported. The lipase may react with n-butanol to yield a dead end enzyme n-butanol complex or it may react with oleic acid to yield the effective lipase oleic acid complex [99].

The resultant equation for the Ping–Pong bi–bi reaction is expressed by Eq. (4)

$$v_o = \frac{V_{\max}[A][B]}{\frac{K_A}{K_{iB}}[B]^2 + K_A[B] + K_B[A] + [A][B]} \quad (4)$$

where v_o is the initial reaction rate, $V_{\max}$ the maximum reaction rate, [A] the oleic acid concentration, [B] the butyl alcohol concentration, K_A and K_B the Michaelis constants of oleic acid and butyl alcohol and K_{iB} is the butyl alcohol inhibition constant.

Kinetic parameters of Ping – Pong model of lipase catalyzed esterifications of oleic acid in non aqueous media are shown in Table 2. In these cases the systems are considered to operate under irreversible conditions and the data were obtained using the initial rate method focusing in the very beginning of the reaction since equilibrium of lipase catalized esterification reactions is normally reached after six or more hour [99]. In different esterifications in non aqueous media the lipase from *C. Rugosa* synthesizes more efficiently the short-chain fatty acids than for the long-chain fatty acids [188]. For this lipase, the higher values of *K*m for long chain fatty acids (Table 2) apparently corroborate this observation.

In the study of Orrego *et al.* (2009) the value of K_M for n-butanol (K_B = 0.149 mol L-1) is smaller of K_M value for oleic acid (K_A = 0.599 mol L-1), implying that the lipase in its immobilized form has higher affinity for the alcohol substrate. This could be partially explained by a slower diffusion rate of the long-chain oleic acid into the support. It was also observed, from the relatively large value of inhibition constant (K_{iB} = 1.933 mol L-1) that alcohol inhibition occurred above 0.25 M butanol (Figure 4) and is much less significant bellow this concentration. At superior concentration levels the alcohol, being polar in nature, is adsorbed by the immobilization support thereby blocks the entry of oleic acid and/or may be accumulated in the aqueous microenvironment that surrounds the enzyme, reaching a concentration level sufficient to denature the protein and causing the reaction to stop.

Table 2. Kinetic parameters for lipase catalysed esterifications of oleic acid

Substrate/ Solvent/ Lipase	*V*max (mmol.min^{-1}. g^{-1})	K_A (mol.L^{-1})	K_B (mol.L^{-1})	*K*i (mol.L^{-1})	Reference
Methanol/ Hexane/ *C. Antarctica*	4.9	0.013	0.016	0.003	[188]
Ethanol/ Hexane/ *R. Miehei*	1336.2	1.2	1.16×10^{-8}	9.46×10^{7}	[191]
Ethanol/ Hexane/ free *P. Pancreatic*	4.0	0.066	0.103	0.020	[192]
Ethanol/ Hexane/ *M. Miehei*	23.0	0.45	0.60	0.06	[193]
Ethanol/ $SCCO_2$/ *M. Miehei*	14.0	0.16	1.60	0.065	[193]
Butanol/ Hexane/ *C. Rugosa*	3.2	0.38	0.19	0.780	[190]
Butanol/ iso-octane/ *C. Rugosa*	18.2	0.599	0.149	1.933	[99]

Conclusion

Lipases are frequently used to catalyze synthetic reactions occurring in organic media, when the water content is sufficiently limited, because they have an inherent affinity for hydrophobic environment.

In ester production immobilization in water adsorbent supports like chitosan can allow high esterification activity without associated hydrolysis, permit to overcome the problem that enzymatic reaction rates in nearly anhydrous organic solvents are low and provide the local interface necessary for lipase activation. Chitosan has also a high affinity to proteins, availability of amino and hidroxyl groups for direct reactions with enzymes and for chemical modifications, hydrophilicity, mechanical stability and rigidity, regenerability, and ease of preparation in different geometrical configurations. To improve diffusional properties chitosan matrix porosity, pore size distributions, internal surface area and size of surface pores could be tailored with crosslinking, cryogelling, freezing and thawing treatments and drying procedures. For these reasons chitosan has proven promising as specialty material for the application in lipase catalyzed processes.

Surface analysis and spectroscopic methods applied to biocatalytic enzyme-chitosan systems provide information such as adsorption isotherms, surface areas, pore size distribution, surface roughness, surface topography, elemental composition, oxidation states, and functional groups, quantitative analysis and distribution of species.

The most accepted description of immobilized lipase esterification reaction kinetics of the catalytic action of lipases is a Ping Pong Bi Bi mechanism. The reactions approximately obey the Michaelis-Menten equation with alcohol inhibition effect.

References

[1] Carrea, G. ,Riva, S. (2000). Properties and synthetic applications of enzymes in organic solvents. *Angew.Chem. Int. Ed*, 39, 2226-2254.

[2] Faber, K. ,Franssen, M.C.R. (1993). Prospects for the increased application of biocatalysts in organic transformations. *Trends Biotechnol*, 11, 461-469.

[3] Ru, M.T., Hirokane, S.Y., Lo, A.S., Dordick, H., Reimer, J.A. ,Clark, D.S. (2000). On the Salt-Induced Activation of Lyophilized Enzymes in Organic Solvents: Effect of Salt Kosmotropicity on Enzyme Activity. *Journal of American Chemical Society*, 122, 1565-1571.

[4] Margolin, A.L. (1993). Enzymes in the synthesis of chiral drugs. *Enzyme and Microbial Technology*, 15, 266-280.

[5] Turner , N.A., Vulfson, E.N. (2000). At what temperature can enzymes maintain their catalytic activity? *Enzyme and Microbial Technology*, 27, 108-113.

[6] Van Unen, D.J., Engbersen, J.F.J. ,Reinhoudt, D.N. (2001). Studies on the mechanism of crown-ether-induced activation of enzymes in non-aqueous media. *Journal of Molecular Catalysis B: Enzymatic*, 11, 877-882.

[7] Vermuë, M.H., Tramper, J. (1995). Biocatalysis in non-conventional media: medium engineering aspects. *Pure and Applied Chemistry*, 67, 345-373.

[8] Lee, M.Y., Dordick, J.S. (2002). Enzyme activation for nonaqueous media. *Current Opinion in Biotechnology* 13, 379-384.

[9] Cygler, M., Schrag, J.D. (1999). Structure and conformational flexibility of Candida rugosa lipase. *Biochimica et Biophysica Acta* 1441, 205-214.

[10] Catoni, E., *Overexpression and protein engineering of lipase a and b from Geotrichum candidum cmicc335426, Fakultät Chemie*. 1999, Stuttgart.

[11] Domínguez de María , P., Carboni-Oerlemans, C., Tuin, B., Bargeman, G., van der Meer, A. ,van Gemert , R. (2005). Biotechnological applications of Candida Antarctica lipase A: State-of-the-art. *Journal of Molecular Catalysis B: Enzymatic*, 37, 36-46.

[12] Paiva, A.N., Balcão, V.M. ,Malcata, J. (2000). Kinetics and mechanisms of reactions catalyzed by immobilized lipases. *Enzyme and Microbial Technology* 27, 187-204.

[13] Vulfson, E.N. (1994). *Lipases, their structure, biochemistry and applications*, in *Industrial applications of lipases. Cambridge University Press*, P.S.E. Woolley P, Editor: Cambridge U.K. p. 271-288.

[14] Jaeger, K.E., Dijkstra, B.W. ,Reetz, M.T. (1999). Bacterial biocatalist: molecular biology, three dimensional structures and biotechnological applications of lipases. *Annual Review of Microbiology*, 53, 315-351.

[15] Sharma, R., Chisti, Y., Banerjee, U.C. (2001). Production, purification, characterization, and applications of lipases. *Biotechnology Advances*, 19, 627-662.

[16] Pandey, A., S.Benjamin, Soccol, C.R., Nigam., P., Krieger, N. ,Soccol, V.T. (1999). The realm of microbial lipases in biotechnology. *Biotechnology and Applied Biochemistry*, 29, 119-131.

[17] Sharma, R., Chisti, Y. ,Banerjee, U.C.P., purification, characterization, and applications of lipases (2001). Production, purification, characterization, and applications of lipases. *Biotechnology Advances*, 19, 627-662.

[18] Orrego, C.E. ,Valencia, J. (2008). *Lipase Catalised Biodiesel production. An overview*, in *Renewable Fuels Developments in Bioethanol and Biodiesel Production*, C.A. Cardona y J.S. Lee, Editores. Universidad Nacional de Colombia sede Manizales: Manizales. p. 228.

[19] Lee, K.-T., Foglia, T.A. ,Chang, K.-S. (2002). Production of alkyl ester as biodiesel from fractionated lard and restaurant grease *Journal of the American Oil Chemists' Society*, 79, 4.

[20] Noureddini, H., Gao, X. ,Philkana, R.S. (2005). Immobilized Pseudomonas cepacia lipase for biodiesel fuel production from soybean oil. *Bioresource Technology*, 96, 769-777.

[21] Shimada, Y., Watanabe, Y., Samukawa, T., Sugihara, A., Noda, H., Fukuda, H. ,Tominaga, Y. (2007). Conversion of vegetable oil to biodiesel using immobilized Candida antarctica lipase *Journal of the American Oil Chemists' Society*, 76, 4.

[22] Chen, J.-W. ,Wu, W.-T. (2003). Regeneration of immobilized Candida antarctica lipase for transesterification. *Journal of Bioscience and Bioengineering*, 95, 466-469.

[23] Shimada, Y., Watanabe, Y., Sugihara, A. ,Tominaga, Y. (2002). Enzymatic alcoholysis for biodiesel fuel production and application of the reaction to oil processing. *Journal of Molecular Catalysis B: Enzymatic*, 17, 133-142.

[24] Al-Zuhair, S., Ling, F.W. ,Jun, L.S. (2007). Proposed kinetic mechanism of the production of biodiesel from palm oil using lipase. *Process Biochemistry*, 42, 951-960.

[25] Tischer, W., Kasche, V. (1999). Immobilized enzymes: crystals or carriers? . *Tibtech*, 17, 326-335.

[26] Balcão, V.M., Paiva A.L. ,F.X., M. (1996). Bioreactors with immobilized lipases: state of the art. *Enzyme and Microbial Technology*, 18, 392-416.

[27] Dosanjh, N.S. ,Kaur, J. (2002). Immobilization, stability and esterification studies of a lipase from a Bacillus sp. *Biotechnol. Appl. Biochem.* , 36, 7-12.

[28] Gill, I., Pastor, E. ,Ballesteros, A. (1999). Lipase-Silicone Biocomposites: Efficient and Versatile Immobilized Biocatalysts. *J. Am. Chem. Soc.*, 121,

[29] Chiou, S.H. ,Wu, W.T. (2004). Immobilization of Candida rugosa lipase on chitosan with activation of the hydroxyl groups. *Biomaterials*, 25, 197-204.

[30] Arroyo, M. (1998). Inmovilización de enzimas. Fundamentos, métodos y aplicaciones. *Ars Pharmaceutica*, 39, 23-29.

[31] Morimoto, M., Saimoto, H. ,Shigemasa, Y. (2002). *Trends in Glycoscience and Glycotechnology* 14, 205-222.

[32] Kumar, M.N.V.R. (2000). A review of chitin and chitosan applications. *Reactive and Functional Polymers*, 46, 1-27.

[33] Guibal, E. (2004). Interactions of metal ions with chitosan-based sorbents: a review. *Sep. Purif. Technol.*, 48, 43.

[34] Piron, E. ,Domard, A. (1997). Interaction between chitosan and uranyl ions Part 1. Role of physicochemical parameters. *International Journal of Biological Macromolecules*, 21, 327-335.

[35] Sashiwa, H. ,Aiba, S. (2004). Chemically modified chitin and chitosan as biomaterials. *Prog. Polym. Sci.*, 29, 887-908.

[36] Krajewsja, B. (2001). Diffusional properties of chitosan hydrogel membranes. . *Journal of Chemical Technology and Biotechnology.*, 76, 636-642.

[37] Krajewska, B. (2004). *Enzyme and Microbial Technology* 35, 126-139.

[38] Krajewska, B. ,A.Olech (1996). Pore structure of gel chitosan membranes. I. Solute diffusion measurements. *Polymer Gels and Nerworks*, 4, 33-43.

[39] Chandy, T. ,Sharma, C. (1990). Chitosan as a biomaterial. *Biomater Artif Cell*, 18, 1-24.

[40] Okamoto, Y., Minami, S., Matsuhashi, A., Sashiwa, H., Saimoto, H., Shigemasa, Y., Tanigawa, T., Tanaka, Y. ,Tokura, S. (1993). Application of polymeric n-acetyl-d-glucosamine (chitin) to veterinary practice. *J. Vet. Med. Sci.*, 55, 743-747.

[41] Muzzarelli, R.A.A. (1996). Chitosan-based dietary foods. *Carbohydrate Polymers*, 29, 309-316.

[42] Shahidi, F., Arachchi, J.K.V. ,Jeon, Y.J. (1999). Food applications of chitin and chitosans. *Trends in Food Science and Technology*, 10, 37-51.

[43] Berger, J., Reist, M., Mayer, J.M., Felt, O., Peppas, N.A. ,Gurny, R. (2004). Structure and interactions in covalently and ionically crosslinked chitosan hydrogels for biomedical applications. . *European Journal of Pharmaceutics and Biopharmaceutics.*, 57, 19-34.

[44] Choplin, A. ,Quignard, F. (1998). *Coord. Chem. Rev.* , 178-180, 1679.

[45] Kasumi, T., Tsuji, M., Hayashi, K. ,Tsumura, N. (1977). Amino group in chitosan are useful for the immobilization of enzyme. *Agricultural Biological Chemistry*, 41, 1865.

[46] Quignard, F., Choplin, A. ,Domard, A. (2000). Chitosan: A Natural Polymeric Support of Catalysts for the Synthesis of Fine Chemicals. *Langmuir*, 16, 9106-9108.

[47] Ruiz, M., A., S. ,Guibal, E. (2002). Pd and Pt recovery using chitosan gel beads . I. Influence of the drying process on diffusion properties. *Sep. Sci. Technol.* , 37, 2385.
[48] Nishinari, K., Zhang, H. ,Ikeda, S. (2000). Hydrocolloid gels of polysaccharides and proteins. *Current Opinion in Colloid and Interface Science*, 5, 195-201.
[49] Hirano, S., Kondo, S. ,Ohe, Y. (1975). Chitosan gel: a novel polysaccharide gel. *Polymer*, 16, 622.
[50] Bernkop-Schnurch, A. ,Hopf, T.E. (2001). Synthesis and in-vitro evaluation of chitosan-thioglycolic acid conjugates. *Sci Pharm*, 69, 109-118.
[51] Montembault, A., C.Viton ,Domard, A. (2004). Physico-chemical studies of the gelation of chitosan in a hydroalcoholic medium. *Biomaterials*,
[52] Berger, J., Reist, M., Mayer, J.M., Felt, O. ,Gurny, R. (2004). Structure and interactions in chitosan hydrogels formed by complexation or aggregation for biomedical applications. *European Journal of Pharmaceutics and Biopharmaceutics* 57 b, 35-52.
[53] Dumitriu, S. ,Chornet, E. (1998). Inclusion and release of proteins from polysaccharide-based polyion complexes. *Advanced Drug Delivery Reviews*, 31, 223-246.
[54] Yao, K.D., Liu, J., Cheng, G.X., Lu, X.D. ,Tu., H.L. (1996). Swelling behavior of pectin/chitosan complex films. *J. Appl. Polym. Sci.* , 60, 279-283.
[55] Chu, C.-H., Sakiyama, T. ,Yano, T. (1995). pH-sensitive swelling of a polyelectrolyte complex gel prepared from xanthan and chitosan. *Biosci. Biotech. Biochem.*, 59, 717-719.
[56] Mao, J.S., Zhao, L.G., Yin, Y.J. ,Yao, K.D. (2003). Structure and properties of bilayer chitosan–gelatin scaffolds. *Biomaterials*, 24, 1067-1074.
[57] Bercherán-Marón, L., Peniche, C. ,Arguelles-Monal, W. (2004). Study of the interpolyelectrolyte reaction between chitosan and alginate: influence of alginate composition and chitosan molecular weight. . *International Journal of Biological Macromolecules*, 34, 127-133.
[58] Lai, H.L., Khalil, A.A., Duncan, Q.M. ,Craig, D.Q.M. (2003). The preparation and characterisation of drug-loaded alginate and chitosan sponges. *International Journal of Pharmaceutics*, 251, 175-181.
[59] Cascone, M.G. (1999). Effect of chitosan and dextran on the properties of poly(vinyl alcohol) hydrogels. *Journal of Materials Science: Materials in Medicine* 10, 431-435.
[60] Shu, X.Z. ,Zhu, K.J. (2002). The influence of multivalent phosphate structure on the properties of ionically cross-linked chitosan films for controlled drug release. *European Journal of Pharmaceutics and Biopharmaceutics*, 54, 235-243.
[61] Alsarra, I.A., Betigeri, S.S., Zhang, H., Erans, B.A. ,Neau, S.H. (2002). Molecular weight and degree of deacetylation effects on lipase-loaded chitosan beadcharacteristics. . *Biomaterials*, 23, 3637-3644.
[62] Alsarra, I.A., Neaub, S.H. ,Howard, M.A. (2004). Effects of preparative parameters on the properties of chitosan hydrogel beads containing Candida rugosa lipase . . *Biomaterials*, 25, 2645-2655.
[63] Szczesna-Antczak, M., Antczak, T., Rzyska, M., Modrzejewska, Z., Patura, J., Kalinowska, H. ,Bielecki, S. (2004). Stabilization of an intracellular Mucor circinelloides lipase for application in non-aqueous media. *Journal of Molecular Catalysis B: Enzymatic* 29, 163-171.
[64] Betigeri, S.S. ,Neau, S.H. (2002). Immobilization of lipase using hydrophilic polymers in the form hydrogel beads. *Biomaterials*, 23, 3627-3636.

[65] Xu, Y. ,Du, Y. (2003). Effect of molecular structure of chitosan on protein delivery properties of chitosan nanoparticles. *International Journal of Pharmaceutics*, 250, 215-226.

[66] Nakatsuka, S. ,Andrady, A.L. (1992). Permeability of vitamin B12 in chitosan membranes—effect of cross-linking and blending with poly(vinyl alcohol) on permeability. *J. Appl. Polym. Sci.*, 44, 17-28.

[67] Hung, T.C., Giridhar, R., Chiou, S.H. ,Wu, W.T. (2003). Binary immobilization of Candida rugosa lipase on chitosan. *Journal of Molecular Catalysis B: Enzymatic*, 26, 69-78.

[68] Yi, Y., Wang, Y. ,Liu, H. (2003). Preparation of new crosslinked chitosan with crown ether and their adsorption for silver ion for antibacterial activities. *Carbohydrate Polymers*, 53, 425-430.

[69] Bodnar, M., Hartmann, J.F. ,Borbely, J. (2005). Preparation and Characterization of Chitosan-Based Nanoparticles. *Biomacromolecules*, 6, 2521-2527.

[70] Pozzo, A.D., Vanini, L., Fagnoni, M., Guerrini, M., Benedittis, A.D. ,Muzzarelli, R.A.A. (2000). Preparation and characterization of poly(ethylene glycol)-crosslinked reacetylated chitosans. *Carbohydrate Polymers*, 42, 201-206.

[71] Yang, Q., Dou, F., Liang, B. ,Shen, Q. (2005). Investigations of the effects of glyoxal cross-linking on the structure and properties of chitosan fiber. *Carbohydrate Polymers*, 61, 393-398.

[72] Gümüşderelioğlu, M. ,Agi, P. (2004). Adsorption of concanavalin A on the well-characterized macroporous chitosan and chitin membranas. *Reactive and Functional Polymers*, 61, 211-220.

[73] Macquarrie, D.J. ,Hardy, J.J.E. (2005). Applications of Functionalized Chitosan in Catalysis. *Ind. Eng. Chem. Res.*, 44, 8499-8520

[74] Mi, F.L., Kuan, C.Y., Shyu, S.S., Lee, S.T. ,Chang, S.F. (2000). The study of gelation kinetics and chain-relaxation properties of glutaraldehyde-cross-linked chitosan gel and their effects on microspheres preparation and drug release. . *Carbohydrate Polymers*, 41, 389-396.

[75] Fennema, O.R., Powrie, W.D. ,Marth, E.H. (1973). *Low temperature preservation of foods and living matter*, ed. M.D. Inc. New York.

[76] Lozinsky, V.I., Galaev, I.Y., Plieva, F.M., I.N.Savina, H.Jungvid ,B.Mattiasson (2003). Polymeric cryogels as promising materials of biotechnological interest. *Trends in Biotechnology*, 21, 445.

[77] Lozinsky, V.I. ,Plieva, F.M. (1998). Poly(vinyl alcohol) cryogels employed as matrices for cell immobilization. 3. Overview of recent research and developments. *Enzyme and Microbial Technology*, 23, 227-242.

[78] Bacheva, A.V., Plieva, F.M., Lysogorskaya, E.N., Filippova, I.Y. ,Lozinsky, V. (2001). Peptide Synthesis in Organic Media with Subtilisin 72 Immobilized on Poly(vinyl alcohol)-Cryogel Carrier. *Bioorganic and Medicinal Chemistry Letters*, 11, 1005.

[79] Filippova, I.Y., Bacheva, A.V., Baivak, O.B., Plieva, F.M., Lysogorskaya, E.N., Oksenoit., E.S. ,Lozinsky, V. (2001). Proteinases immobilized on poly(vinil alcohol) cryogel: a novel biocatalysts for peptide synthesis in organic media. *Russian Chemical Bulletin*, 50, 1986.

[80] Kokufuta, E. ,Jinbo, E. (1992). A Hydrogel Capable of Facilitating Polymer Diffusion through the Gel Porosity and Its Application in Enzyme Immobilization. *Macromolecules*, 25, 3549.

[81] Plieva, F.M., Kotchetkov, K.A., Singh, I., Parmar, V.S., Belokon, Y.N. ,Lozinsky, V.I. (2000). Immobilization of hog pancreas lipase in macroporous poly(vinyl alcohol)-cryogel carrier for the biocatalysis in water-poor media. . *Biotechnology Letters*, 22, 551-554.

[82] Szczesna-Antczak, M., T.Antczak, Rzyska, M. ,S.Bielecki (2002). Catalytic properties of membrane-bound Mucor lipase immobilized in a hydrophilic carrier. *Journal of Molecular Catalysis B: Enzymatic*, 19-20, 261.

[83] Giannouli, P., Morris, E.R. (2003). Cryogelation of xanthan. *Food Hydrocolloids*, 17, 495–501.

[84] Lazaridou, A. ,Biliaderis, C.G. (2004). Cryogelation of cereal β-glucans: structure and molecular size effects. *Food Hydrocolloids* 18, 933-947.

[85] Xue, W., Champ , S., Huglin, M.B. ,Jones, T.G.J. (2004). Rapid swelling and deswelling in cryogels of crosslinked poly(N-isopropylacrylamide-co-acrylic) *European Polymer Journal*, 40, 467-476.

[86] Di Martino, A., Sittinger, M. ,Risbud, M.V. (2005). Chitosan: A versatile biopolymer for orthopaedic tissue-engineering. *Biomaterials*, 26, 5983–5990.

[87] Ho, M.H., Kuo, P.Y., Hsieh, H.J., Hsien, T.Y., Hou, L.T., Lai, J.Y. ,Wang, D.M. (2004). Preparation of porous scaffolds by using freeze-extraction and freeze-gelation methods. *Biomaterials* 25, 129-138.

[88] Madihally, S.S. ,Matthew, H.W. (1999). Porous chitosan scaffolds for tissue engineering. *Biomaterials*, 20, 1133-1142.

[89] Orrego, C.E. ,Valencia, J.S. (2008). Preparation and characterization of chitosan membranes by using a combined freeze gelation and mild crosslinking method. *Bioprocess and Biosystems Engineering*, 32, 197-206.

[90] Krajewska, B. (2004). Application of chitin- and chitosan-based materials for enzyme immobilizations: a review. *Enzyme and Microbial Technology*, 35, 126-139.

[91] Spagna, G., Barbagallo, R.N., Casarini, D. ,Pifferi, P.G. (2001). A novel chitosan derivative to immobilize α-1- rhamnopyranosidase from Aspergillus niger for applicationin beverage technologies. *Enzyme and Microbial Technology* 28, 427-438.

[92] Vazquez-Duhalt, R., Tinoco, R., D'Antonio, P., Topoleski, L.D.T. ,Payne, G.F. (2001). Enzyme Conjugation to the Polysaccharide Chitosan: Smart Biocatalysts and Biocatalytic Hydrogels. *Bioconjugate Chem.*, 12, 301-306.

[93] Fernandez-Lafuente, R., Armisén, P., Sabuquillo, P., Fernández-Lorente, G. ,Guisán, J.M. (1998). Immobilization of lipases by selective adsorption on hydrophobic supports. *Chemistry and Physics of Lipids*, 93, 185-197.

[94] Wang, P., Hill, T.G., Wartchow, C.A., Huston, M.E., Oehler, L.M., Smith, M.B., Bednarski, M.D. ,Callstrom, M.R. (1992). New carbohydrate-based materials for the stabilization of proteins. *Journal of American Chemical Society*, 114, 378-380.

[95] Guibal., E. (2005). Heterogeneous catalysis on chitosan-based materials: a review. *Progress Polymer Science*, 30 71–109. .

[96] Hilal, N., Al-ZoubP, H., Darwish , N.A., Mohammad , A.W. ,Abu Arabi, M. (2004). A comprehensive review of nanofiltration membranes: Treatment, pretreatment, modelling, and atomic force microscopy. *Desalination*, 170, 281-308.

[97] Wan, Y., Creber, K.A.M., Peppley, B. ,Bui, V.T. (2003). Ionic conductivity of chitosan membranes. *Polymer*, 44, 1057-1065.

[98] Guibal, E. (2004). Interactions of metal ions with chitosan-based sorbents: a review. *Separation and Purification Technology*, 38, 43-74.

[99] Orrego, C.E., Valencia, J.S. ,Zapata, C. (2009). Candida rugosa Lipase Supported on High Crystallinity Chitosan as Biocatalyst for the Synthesis of 1-Butyl Oleate *Catalysis Letters*, 129, 312-322.

[100] Juang, R.S., Wu, F.C. ,Tseng, R.L. (2002). Use of chemically modified chitosan beads for sorption and enzyme immobilization. *Advances in Environmental Research*, 6, 171-177.

[101] Sashiwa, H. ,Aiba, S. (2004). Chemically modified chitin and chitosan as biomaterials. *Progress in Polymer Science* 29, 887-908.

[102] Jayakumar, R., Prabaharan, M., Reis, R.L. ,Mano, J.F. (2005). Graft copolymerized chitosan—present status and applications. *Carbohydrate Polymers*, 62, 142-158.

[103] Tangpasuthadol, V., Pongchaisirikul, N. ,Hoven, V.P. (2003). Surface modification of chitosan films. Effects of hydrophobicity on protein adsorption. *Carbohydrate Research*, 338, 937-942.

[104] Prabaharan, M. ,Mano, J.F. (2006). Chitosan derivatives bearing cyclodextrin cavities as novel adsorbent matrices. *Carbohydrate Polymers*, 63, 153-166.

[105] Lin, J., Yan, F., Hu, X. ,Ju, H. (2004). Chemiluminescent immunosensor for CA19-9 based on antigen immobilization on a cross-linked chitosan membrane. *Journal of Immunological Methods* 291 165– 174.

[106] Peng, X. ,Zhang, L. (2005). Surface Fabrication of Hollow Microspheres from N-Methylated Chitosan Cross-Linked with Gultaraldehyde. *Langmuir*, 21, 1091-1095.

[107] Plieva , F.M., Savina , I.N., Deraz, S., Andersson, J., Galaev , I.Y. ,Mattiasson, B. (2004). Characterization of supermacroporous monolithic polyacrylamide based matrices designed for chromatography of bioparticles. *Journal of Chromatography B*, 807, 129-137.

[108] Hagiwara, T., Wang, H., Suzuki, T. ,Takai, R. (2002). Fractal Analysis of Ice Crystals in Frozen Food. *Journal of Agricultural and Food Chemistry*, 50, 3085-3089.

[109] O'Brien, F.J., Harley, B.A., Yannas, I.V. ,Gibson, L. (2004). Influence of freezing rate on pore structure in freeze-dried collagen-GAG scaffolds. *Biomaterials* 25, 1077–1086.

[110] Krajewska, B. (2005). Membrane-based processes performed with use of chitin/chitosan materials. *Separation and Purification Technology.*, 41, 305-312.

[111] Lai, H.L., Abu'Khalil, A. ,Craig, D.Q.M. (2003). The preparation and characterisation of drug-loaded alginate and chitosan sponges. *International Journal of Pharmaceutics*, 251, 175-181.

[112] Suh, J.K.F. ,Matthew, H.W.T. (2000). Application of chitosan-based polysaccharide biomaterials in cartilage tissue engineering: a review. *Biomaterials*, 21, 2589-2598.

[113] Bosquillon, C., Rouxhet, P.G., Ahimou, F., Simon, D., Culot, C., Préat, V. ,Vanbever, R. (2004). Aerosolization properties, surface composition and physical state of spray-dried protein powders. *Journal of Controlled Release*, 99, 357-367.

[114] Lausmaa, J. ,Kasemo, B. (1990). *Appl. Surf. Sci.*, 44, 133-146.

[115] Santerre, J.P., Labow, R.S. ,Adams, G.A. (1993). *J. Biomed. Mater. Res.*, 27, 97-109.

[116] Brevig, T., Holst, B., Ademovic, Z., Rozlosnik, N., Røhrmann, J.H., Larsen, N.B., Hansen, O.C. ,Kingshott, P. (2005). The recognition of adsorbed and denatured proteins

of different topographies by β2 integrins and effects on leukocyte adhesion and Activation. *Biomaterials*, 26, 3039–3053.

[117] Jacquemart, I., Pamuła, E., Cupere, V.M.D., Rouxhet, P.G. ,Dupont-Gillain, C.C. (2004). Nanostructured collagen layers obtained by adsorption and drying. *Journal of Colloid and Interface Science*, 278, 63-70.

[118] Cai, K., Rechtenbach, A., Hao, J., Bossert, J. ,Jandt, K.D. (2005). Polysaccharide-protein surface modification of titanium via a layer-by-layer technique: Characterization and cell behaviour aspects. *Biomaterials*, 26, 5960–5971.

[119] Bora, U., Kannan, K. ,Nahar, P. (2005). A simple method for functionalization of cellulose membrane for covalent immobilization of biomolecules. *Journal of Membrane Science*, 250, 215-222.

[120] Mao, J.S., liu, H.F., Yin, Y.J. ,Yao, K.D. (2003). The properties of chitosan–gelatin membranes and scaffolds modified with hyaluronic acid by different methods. *Biomaterials* 24, 1621–1629.

[121] Li, Z.F., Kang, E.T., Neoh, K.G. ,Tan, K.L. (1998). Covalent immobilization of glucose oxidase on the surface of polyaniline films graft copolymerized with acrylic acid. *Biomaterials*, 19, 45-53.

[122] Dufrêne, Y., WaL, A.v.d., Norde, W. ,Rouxhet, P. (1997). X-Ray Photoelectron Spectroscopy Analysis of Whole Cells and Isolated Cell Walls of Gram-Positive Bacteria: Comparison with Biochemical Analysis. *Journal of bacteriology*, 179, 1023–1028.

[123] Van der Mei, H.C., de Vries, J. ,Busscher, H.J. (2000). X-ray photoelectron spectroscopy for the study of microbial cell surfaces. *Surface Science Reports*, 39, 1-24.

[124] Musale, D.A. ,A.Kumar (2000). Effects of surface crosslinking on sieving characteristics of chitosan:poly(acrylonitrile) composite nanofiltration membranas. *Separation and Purification Technology.*, 21, 27-38.

[125] Wang, X.H., Li, D.P., Wang, W.J., Feng, Q.L., Cui, F.Z., Xu, Y.X. ,Song, X.H. (2003). Covalent immobilization of chitosan and heparin on PLGA surface. *International Journal of Biological Macromolecules*, 33, 95-100.

[126] Zhang, X. ,Bai, R. (2003). Immobilization of Chitosan on Nylon 6,6 and PET Granules Through Hydrolysis Pretreatment. *Journal of Applied Polymer Science*, 90, 3973–3979.

[127] Chan, R. ,Chen, V. (2004). Characterization of protein fouling on membranes: opportunities and challenges. *Journal of Membrane Science*, 242, 169-188.

[128] Heinemann, C., Heinemann, S., Lode, A., Bernhardt, A., Worch, H. ,Hanke, T. (2009). In Vitro Evaluation of Textile Chitosan Scaffolds for Tissue Engineering using Human Bone Marrow Stromal Cells. *Biomacromolecules*, 10, 1305–1310.

[129] Yen, M.T., Yang, J.H. ,Mau, J.L. (2008). Physicochemical characterization of chitin and chitosan from crab shells. *Carbohydrate Polymers*, doi:10.1016/j.carbpol.2008.06.006,

[130] Julkapli, N.M. ,Akil, H.M. (2008). Degradability of kenaf dust-filled chitosan biocomposites. *Materials Science and Engineering C*, 28, 1100-1111.

[131] Yi, S.S., Lee, C.W., Kim, J., Kyung, D., Kim, B.G. ,Lee, Y.S. (2007). Covalent immobilization of w-transaminase from Vibrio fluvialis JS17 on chitosan beads. *Process Biochemistry*, 42, 895-898.

[132] Xi, F., Liu, L., Wu, Q. ,Lin, X. (2008). One-step construction of biosensor based on chitosan–ionic liquid–horseradishperoxidase biocomposite formed by electrodeposition *Biosensors and Bioelectronics*, 24, 29-34.

[133] Ismail, Y.A., Shin, S.R., Shin, K.M., Yoon, S.G., Shon, K., Kim, S.I. ,Kim, S.J. (2007). Electrochemical actuation in chitosan/polyaniline microfibers for artificial muscles fabricated using an in situ polymerization. *Sensors and Actuators B*, 129, 834-840.

[134] Chandumpai, A., Singhpibulporn, N., Faroongsarng, D. ,Sornprasita, P. (2004). Preparation and physico-chemical characterization of chitin and chitosan from the pens of the squid species, Loligo lessoniana and Loligo formosana. *Carbohydrate Polymers*, 58, 467-474.

[135] Hou, Q., Liu, W., Liu, Z., Duan, B. ,Bai, L. (2008). Characteristics of antimicrobial fibers prepared with wood periodate oxycellulose. *Carbohydrate Polymers*, 74, 235-240.

[136] Liua, X.D., Nishi, N., Tokurab, S. ,Sakairi, N. (2001). Chitosan coated cotton fiber: preparation and physical properties. *Carbohydrate Polymers*, 44, 233-238.

[137] Samuels, R.J. (1981). Solid state characterization of the structure of chitosan films. *Journal of Polymer Science Part B: Polymer Physics* 19, 1081-1105.

[138] Urbanczyk, G.W. ,Lipp-Symonowicz, B. (1994). The influence of processing terms of chitosan membranes made of differently deacetylated chitin on the crystalline structure of membranes. *Journal of Applied Polymer Science* 51, 2191-2194.

[139] Fuentes, S., Retuert, P.J., Ubilla, A., Fernandez, J. ,Gonzalez, G. (2000). Relationship between Composition and Structure in Chitosan-based Hybrid Films. *Biomacromolecules*, 1, 239-243.

[140] Qina, C., Zhoub, B., Zenga, L., Zhanga, Z., Liub, Y., Dub, Y. ,Xiaob, L. (2004). The physicochemical properties and antitumor activity of cellulase-treated chitosan. *Food Chemistry*, 84, 107-115.

[141] Orrego, C.E., Salgado, N., Valencia, J.S., Giraldo, G.I., Giraldo, O.H. ,Cardona, C.A. (2010). Novel chitosan membranes as support for lipases immobilization: Characterization aspects *Carbohydrate Polymers*, 79, 9-16.

[142] Chen, J.P. ,Chiu, S.H. (1999). Preparation and characterization of urease immobilized onto porous chitosan beads for urea hydrolysis. *Bioprocess Engineering*, 21, 323-330.

[143] Qu, X., Wirsén, A. ,Albertsson, A.C. (2000). Novel pH-sensitive chitosan hydrogels: swelling behavior and states of water. *Polymer*, 41, 4589-4598.

[144] Hilal, N., Nigmatullin, R. ,Alpatova, A. (2004). Immobilization of cross-linked lipase aggregates within microporous polymeric membranes. *Journal of Membrane Science*, 238, 131-141.

[145] Bowen, W., Hilal, N., Lovitt, R., P. ,Williams, P. (1996). Atomic force microscope studies of membranes: Surface pore structures of Cyelopore and Anopore membranes. *Journal of Membrane Science*, 110, 233-238.

[146] Fritzsche, A.K., Arevalo, A.R., Connolly, A.F., Moore, M.D., Elings, V. ,Wu, C.M. (1992). The structure and morphology of the skin layer of polyethersulfone ultrafiltration membranes: a comparative atomic force microscope and scanning electron microscope study. *Journal of Applied Polymer Science*, 45, 1945-1956.

[147] Khayet, M. (2004). Membrane surface modification and characterization by X-ray photoelectron spectroscopy, atomic force microscopy and contact angle measurements. *Applied Surface Science*, 238, 269-272.

[148] Paredes, J.I., Martínez-Alonso, A. ,Tascón, J.M.D. (2003). Application of scanning tunneling and atomic force microscopies to the characterization of microporous and mesoporous materials. *Microporous and Mesoporous Materials*, 65, 93-126.
[149] Calvo, J.I., Pradanos, P., Hermindez, A., Bowen, R., Hilal, N., Lovitt, R.W. ,Williams, P.M. (1997). Bulk and surface characterization of composite UF membranes atomic force microscopy, gas adsorptiondesorption and liquid displacement techniques. *Journal of Membrane Science*, 128, 7-21.
[150] Sakurai, K., Maegawa, T. ,Takahashi, T. (2000). Glass transition temperature of chitosan and miscibility of chitosan/poly(N-vinyl pyrrolidone) blends. *Polymer*, 41, 7051-7056.
[151] Cheung, M.K., Wan, K.P.Y. ,Yu, P.H. (2002). Miscibility and morphology of chiral semicrystalline poly-(R)-(3-hydroxybutyrate)/chitosane and poly-(R)-(3-hydroxybutyrate-co-hydroxyvalerate)/ chitosan blends studied with DSC, 1H T1 and T1ρ CRAMPS. . *Journal of Applied Polymer Science*, 86, 1253-1258.
[152] Shantha, K.L. ,Harding, D.R.K. (2002). Synthesis and characterization of chemically modified chitosan microspheres. *Carbohydr.Polym.*, 48, 247-253.
[153] Suyatma, N.E., Tighzert, L., Copinet, A. ,V.Coma (2005). Effects of Hydrophilic plasticizers on Mechanical, Thermal, and Surface Properties of Chitosan Films. *J. Agric. Food Chem.*, 53, 3950-3957.
[154] Mucha, M. ,Pawlak, A. (2005). Thermal analysis of chitosan and its blends. *Thermochimica Acta*, 427, 69-76.
[155] Mukoma, P., Jooste, B.R. ,Vosloo, H.C.M. (2004). Synthesis and characterization of cross-linked chitosan membranes for application as alternative proton exchange membrane materials in fuel cells. *Journal of Power Sources* 136, 16–23.
[156] Illanes, A. (1999). Stability of biocatalysts. *EJB Electronic Journal of Biotechnology.*, 2, 1-9.
[157] Desai, P.D., Dave, A.M. ,Devi, S. (2006). Alcoholysis of salicornia oil using free and covalently bound lipase onto chitosan beads. *Food Chemistry*, 95, 193-199.
[158] Luckachan, G.E. ,Pillai, C.K.S. (2006). Chitosan/oligo L-lactide graft copolymers: Effect of hydrophobic side chains on the physico-chemical properties and biodegradability. *Carbohydrate Polymers*, 64,
[159] Noel, M. ,Combes, D. (2003). Rhizomucor miehei lipase: differential scanning calorimetry and pressure/temperature stability studies in presence of soluble additives. *Enzyme and Microbial Technology*, 33, 299-308.
[160] Crespo, J.d.S., Queiroz, N., Nascimento, M.d.G. ,Soldi, V. (2005). The use of lipases immobilized on poly(ethylene oxide) for the preparation of alkyl esters. *Process Biochemistry*, 40, 401-409.
[161] Persson, M., Wehtje, E. ,Adlercreutz, P. (2000). Immobilization of lipases by adsorption and deposition: high protein loading gives lower water activity optimum. *Biotechnol. Lett.*, 22, 1571.
[162] Kasumi, T., Tsuji, M., Hayashi, K. ,Tsumura, N. (1977). Amino group in chitosan are useful for the immobilization of enzyme. *Agricultural and Biological Chemistry Tokyo*, 41, 1865-1872.
[163] Sakakibara, M., Okada, F., Takahashi, K. ,Tokiwa, T. (1993). Hydrolysis of beef tallow by immobilized lipase in a biphasic organic-aqueous system. . *Nippon Kagaku Kaishi (Japan)*, 11, 1292–1294.

[164] Itoyama, K., Tokura, S. ,Hayashi, T. (1994). *Biotechnol. Prog.*, 10, 225.
[165] Magnin, D., Dumitriu, S., Magny, P. ,Chornet, E. (2001). Lipase Immobilization into Porous Chitoxan Beads: Activities in Aqueous and Organic Media and Lipase Localization. *Biotechnol. Prog.*, 17, 734-737.
[166] Tan, T., Wang, F. ,Zhang, H. (2002). Preparation of PVA/chitosan lipase membrane reactor and its application in synthesis of monoglyceride. *J. Mol. Catal., B: Enzym.*, 18, 325.
[167] Pereira, E.B., Zanin, J.M. ,Castro, H.F. (2003). Immobilization and catalytic properties of lipase on chitosan for hydrolysis and esterification reactions. *Brazilian Journal of Chemical Engineering*, 20, 343-355.
[168] Kawamura, Y., Okano, A. ,Tanibe, H. (1998). Transesterification catalyzed by lipase immobilized on porous chitosan beads carrier. . *Kichin Kitosan Kenkyu (Japan)*, 4, 126-127.
[169] Carneiro-da-Cunha, M.G., Rocha, J.M.S., Garcia, F.A.P. ,Gil, M.H. (1999). Lipase immobilisation on to polymeric membranes. *Biotechnology Techniques*, 13, 403-409.
[170] Gomes, F.M., Pereira, E.B. ,de Castro, H.F. (2004). Immobilization of Lipase on Chitin and Its Use in Nonconventional Biocatalysis. *Biomacromolecules*, 5, 17-23.
[171] Chiou, S.H. ,Wu, W.T. (2004). Immobilization of Candida rugosa lipase on chitosan with activation of the hydroxyl groups. *Biomaterials*, 25, 197–204.
[172] Ting, W.J., Tung, K.Y., Giridhar, R. ,Wu, W.T. (2006). *J. Mol. Cat. B: Enzym.*, 42, 32.
[173] Kılınç, A., Teke, M., Önal, S. ,Telefoncu, A. (2006). *Prep. Biochem.Biotechnol.* , 36, 153.
[174] Tang, Z.X., J.Q.Qian, Shi, L.E. ,Tang, Z.X. (2007). Characterizations of immobilized neutral lipase on chitosan nano-particles. *Materials Letters*, 61, 37-40.
[175] Monier, M., Weia, Y. ,Sarhanb, A.A. (2010). Evaluation of the potential of polymeric carriers based on photo-crosslinkable chitosan in the formulation of lipase from Candida rugosa immobilization. *Journal of Molecular Catalysis B: Enzymatic*, 63,
[176] Wu, Y., Wang, Y., Luo, G. ,Dai, Y. (2009). In situ preparation of magnetic Fe3O4-chitosan nanoparticles for lipase immobilization by cross-linking and oxidation in aqueous solution. *Bioresource Technology*, 100, 3459-3464.
[177] Yi, S.-S., Noh, J.-M. ,Lee, Y.-S. (2009). Amino acid modified chitosan beads: Improved polymer supports for immobilization of lipase from Candida rugosa. *Journal of Molecular Catalysis B: Enzymatic* 57 123–129.
[178] Ye, P., Xu, Z.K., Wang, Z.G., Wu, J., Deng, H.T. ,Seta, P. (2005). Comparison of hydrolytic activities in aqueous and organic media for lipases immobilized on poly(acrylonitrile-co-maleic acid) ultrafiltration hollow fiber membrana. *Journal of Molecular Catalysis B: Enzymatic*, 32, 115121.
[179] Zaks, A. ,Klibanov, A.M. (1988). Enzymatic catalysis in nonaqueous solvents. . *J. Biol. Chem.* , 263, 3194-3201.
[180] Colombie, S., Russel, J.T., Condoret, J.S. ,Marty, A. (1998). *Biotechnol. Bioeng.*, 60, 362.
[181] Qi, Z.H. ,Romberger, M.L. (1998). *Polysaccharide association structures in food*, in *Cyclodextrins*, W.R. H., Editor. Marcel Dekker Inc: New York. p. 207-225.
[182] Sandoval, G.C., A.Marty ,Condoret, J.S. (2001). *A.I.Ch.E J.*, 47, 718.

[183] Valivety, R.H., Halling, P.J., Peilow, A.D. ,Macrae, A.R. (1992). Lipases from different sources vary widely in dependence of catalytic activity on water activity. . *Biochim Biophys Acta*, 1122, 143-146.

[184] Bommarius, A.S. ,Riebel, B.R. (2004). *Biocatalysis. Fundamentals and applications*, Wiley-VC, Editor: Darmstadt. p. 349.

[185] Hazarika, S., Goswami, P., Dutta, N.N. ,Hazarika, A.K. (2002). Ethyl oleate synthesis by Porcine pancreatic lipase in organic solvents. *Chemical Engineering Journal*, 85, 61-68.

[186] Van Tol, J.B.A., Stevens, R.M.M., Veldhuizen, W.J., Jongejan, J.A. ,Duine, J.A. (1995). Do organic solvents affect the catalytic properties of lipase? Intrinsic kinetic parameters of lipases in ester hydrolysis and formation in various organic solvents. *Biotechnology and Bioengineering*, 47, 71-81.

[187] Chulalaksananukul, W., Condoret, J.S., Delorme, P. ,Willemot, R.M. (1990). Kinetic study of esterification by immobilized lipase in n-hexane. *FEBS Letters*, 276, 181-4.

[188] Janssen, A.E.M., Vaidya, A.M. ,Hailing, P.J. (1996). Substrate specificity and kinetics of Candida rugosa lipase in organic media. *Enzyme and Microbial Technology*, 18, 340-346.

[189] Rizzi, M., Stylos, P., Rick , A. ,Reuss, M. (1992). A kinetic study of immobilized lipase catalyzing the synthesis of isoamyl acetate by transesterification in n-hexane. *Enzyme and Microbial Technology*, 14, 709-713.

[190] Zaidi, A., Gainer, J.L., Carta, G., Mrani, A., Kadiri, T., Belarbi, Y. ,Mir, A. (2002). Esterification of fatty acids using nylon-immobilized lipase in n-hexane: kinetic parameters and chain-length effects. *Journal of Biotechnology*, 93, 209-216.

[191] Krishna, S.H. ,Karanth, N.G. (2001). Lipase-catalyzed synthesis of isoamyl butyrate A kinetic study. *Biochimica et Biophysica Acta*, 1547, 262-267.

[192] Garcia, T., Coteron, A., Martinez, M. ,Aracil, J. (2000). Kinetic model for the esterification of oleic acid and cetyl alcohol using an immobilized lipase as catalyst. *Chemical Engineering Science*, 55, 1411-1423.

[193] Kamiya, N. ,Goto, M. (1997). How Is Enzymatic Selectivity of Menthol Esterification Catalyzed by Surfactant-Coated Lipase Determined in Organic Media? *Biotechnology Progress*, 13, 488-492.

In: Handbook of Chitosan Research and Applications
Editors: Richard G. Mackay and Jennifer M. Tait
ISBN 978-1-61324-455-5

Chapter 21

EFFECT OF CHITOSAN ON THE CELLS OF PEA LEAF EPIDERMIS

Vitaly D. Samuilov*, Dmitry B. Kiselevsky, Elena V. Dzyubinskaya, Julia E. Kuznetsova and Lev A. Vasil'ev
Department of Immunology,
Faculty of Biology, Lomonosov Moscow State University,
Moscow, Russia

ABSTRACT

The effect of chitosan on epidermal and guard cells in the epidermis isolated from pea leaves was investigated. Chitosan induced the condensation and marginalization of chromatin and the subsequent destruction of nuclei in epidermal, but not in guard cells. The addition of H_2O_2 produced a similar effect on the cells. Nitroblue tetrazolium, pyruvate and mannitol, possessing antioxidant properties, inhibited the destruction of nuclei caused by chitosan or H_2O_2. The chitosan-induced destruction of epidermal cell nuclei was prevented under anaerobic conditions. Diphenylene iodonium and quinacrine, inhibitors of the NADPH oxidase of the plasma membrane, suppressed the destruction of nuclei induced by chitosan. Chitosan caused the generation of reactive oxygen species and, similar to H_2O_2, impaired the permeability barrier of the plasma membrane in guard cells. The data obtained demonstrate that chitosan and H_2O_2 produce similar effects on plant cells. Chitosan initiates the production of reactive oxygen species which induce programmed cell death, presumably by activating the NADPH oxidase of the plasma membrane.

* e-mail: vdsamuilov@mail.ru.

INTRODUCTION

Chitosan (poly(β-1,4)-N-acetylglucosamine) is a product of incomplete deacetylation of chitin, a component of the fungal cell wall. Chitosan is an elicitor, a signal molecule, which is recognized by plant cell. The interaction between chitosan and specific plant cell receptor molecules initiates a reaction cascade including the generation of reactive oxygen species (ROS) and the death of plant cells (Lin et al., 2005). Plant cell death provides a protection against pathogens.

Chitosan at low concentrations initiated the closure of stomata in plants. This effect probably suppresses the intrusion of pathogenic fungi into a plant tissue. Abscisic acid caused a similar effect on guard cells. The closure of stomata was accompanied with a rise in ROS generation in guard cells (Srivastava et al., 2009).

ROS are products of one-, two- and three-electron reduction of oxygen: superoxide anion-radical ($O_2^{\cdot-}$), hydrogen peroxide (H_2O_2) and hydroxyl radical ($OH^{\cdot}$). ROS stimulate programmed cell death (PCD) in plants. H_2O_2 significantly strengthened the cyanide-induced PCD of epidermal and guard cells in the epidermis of pea leaves (Samuilov et al., 2006; 2008) that was determined from the destruction of cell nuclei. The cyanide-induced destruction of cell nuclei was prevented by anaerobiosis and by antioxidants (Samuilov et al., 2002; 2003). H_2O_2 per se did not induce the destruction of guard cell nuclei, but caused the destruction of nuclei in epidermal cells (Samuilov et al., 2006). Like H_2O_2, chitosan at a concentration of 100 μg/ml caused the destruction of the nuclei of epidermal but not guard cells (Vasil'ev et al., 2009).

The goal of this work was to investigate studying the effect of chitosan on the state of epidermal and guard cells of the pea leaf epidermis.

EXPERIMENTAL PROCEDURES

Low-viscous chitosan (Fluka, Germany) used in experiments was a product of the alkaline deacetylation of chitin from crab shells. N-Deacetylation is accompanied by the cleavage of the glycoside bonds of polymer. Therefore, chitosan is a N-acetylglucoseamine and glucosamine heteropolymer that is inhomogeneous in terms of molecular weight.

The experiments were conducted with the peels of the leaf epidermis of pea (*Pisum sativum* L. cv Alpha) seedlings grown for 7–15 days at 20–24°C with 16 h light/8 h dark photoperiods. The light intensity of the luminescent lamps used in experiments was ~100 $\mu E \cdot m^{-2} \cdot s^{-1}$. The epidermis was separated with tweezers and placed into distilled water. A vacuum-infiltration method was used to cause a rapid influx of the added reagents into the cells. The epidermis was incubated at a pressure of 1–2 mm of Hg for 1–2 min. As experiments with pea seedlings demonstrated, this treatment was not inhibitory to their further growth. The samples were placed into polystyrene plates and incubated in distilled water with additives (the composition is given in the figure legends) at room temperature either in the dark or under illumination.

After incubation, the samples were treated with Battaglia fixative (a mixture of chloroform, 96% ethanol, glacial acetic acid, and 40% formaldehyde, 5:5:1:1) for 5 min. The samples were washed in ethanol for 10 min to remove the fixing mixture, incubated for 5 min in water, and stained with Carazzi's hematoxylin for 20 min. The stained epidermal peels were washed with tap water for light microscopy. The number of cells with destroyed and lacking nuclei was determined by examining 300–500 cells in a light microscope.

Anaerobic conditions was created by the incubation of epidermal peels for 30 min in distilled water containing 50 mM glucose, 0.1 μg ml^{-1} glucose oxidase, and catalase (2 activity units per ml) with a 3–5 mm thick layer of sunflower oil on the upper surface of the mixture.

Fluorescence microscopy was performed with epidermis that was fixed as described above and stained for 15 min with 0.2 μM 4′,6-diamidino-2-phenylindole dihydrochloride (DAPI) for monitoring the state of the cell nuclei. Confocal microscopy of non-fixed epidermal peels stained with 2 μM propidium iodide (PI) for 20 min was used for the determining of the state of the cell plasma membrane. The observations were performed with a Carl Zeiss Axiovert 200M microscope equipped with a confocal module LSM 510Meta (Germany). DAPI and PI fluorescence was excited at 300–390 or 543 nm and monitored at $\lambda > 420$ or 565–615 nm, respectively.

Intracellular content of H_2O_2 was estimated from the fluorescence of 2′,7′-dichlorofluorescein (DCF; LeBel et al., 1992) measured using a confocal microscopy with a Carl Zeiss Axiovert 200M microscope equipped with a confocal module LSM 510Meta (Germany). Isolated epidermis was stained with 20 μM 2′,7′-dichlorofluorescin (DCFH) diacetate. DCFH diacetate diffuses across the cell plasma membrane. It is hydrolyzed by intracellular esterases to non-fluorescent DCFH which is oxidized to DCF by H_2O_2 and peroxidase (LeBel et al., 1992). Fluorescence of DCF was excited at 488 nm and determined at 500–530 nm.

RESULTS

DAPI is used as a fluorescent dye binding to DNA. Chitosan caused chromatin condensation, margination (Figure 1), subsequent destruction and the disappearance of epidermal cell nuclei in the epidermis isolated from pea leaves. It had no visually detectable effect on guard cell nuclei. The destruction of epidermal cells nuclei was observed after chitosan treatment for 2 h. Guard cells retain their nuclei even after its treatment with chitosan for 3 days (data not shown).

The chitosan-induced destruction of epidermal cell nuclei was prevented by nitroblue tetrazolium (NBT), a $O_2^{\bar{\cdot}}$ quencher, and by pyruvate (Figure 2 A). Pyruvate is converted to acetate via non-enzymatic oxidative decarboxylation by H_2O_2 (Metzler, 1977). The effect of chitosan was abolished with mannitol, a trap for $OH^{\cdot}$. H_2O_2 and chitosan combined with H_2O_2 caused the destruction of epidermal cell nuclei (Figure 2 A). H_2O_2 did not strengthen the chitosan-induced nucleus destruction. NBT and mannitol suppressed the effects of H_2O_2 and chitosan with H_2O_2 (Figure 2 A). Anaerobiosis prevented the chitosan-induced destruction of epidermal cell nuclei (Figure 2 B).

Diphenylene iodonium (DPI) and quinacrine are inhibitors of the NADPH oxidase in the plasma membrane of plant cells. These compounds can also affect other flavin-containing enzymes. However, DPI and quinacrine at concentrations up to 100 μM did not influence respiration and photosynthetic O_2 evolution of pea leaf pieces (Samuilov et al., 2006). The chitosan-induced destruction of epidermal cell nuclei was suppressed by DPI and quinacrine (Figure 2 C).

In spite of the lack of a significant effect of chitosan on the state of guard cell nuclei, the incubation of the epidermis with chitosan increased the H_2O_2 amount in guard cells (Figure 3). A two-fold rise in DCF fluorescence occurred in cells treated with chitosan, compared to chitosan-untreated cells.

PI is a fluorescent dye binding to DNA, but it does not penetrate across an intact cell plasma membrane. The penetration of PI into cells and the staining of nuclear DNA (Figure 4 A) indicate the disruption of the permeability barrier of the cell plasma membrane. Chitosan impaired the guard cell plasma membrane permeability barrier (Figure 4 A, B). The effect of chitosan was manifested at concentrations above 0.1 and 0.5 mg/ml after the incubation of epidermal peels with chitosan for 3 and 23 h, respectively. H_2O_2 showed a similar effect on the plasma membrane of guard cells at concentrations above 1 mM (Figure 4 C, D).

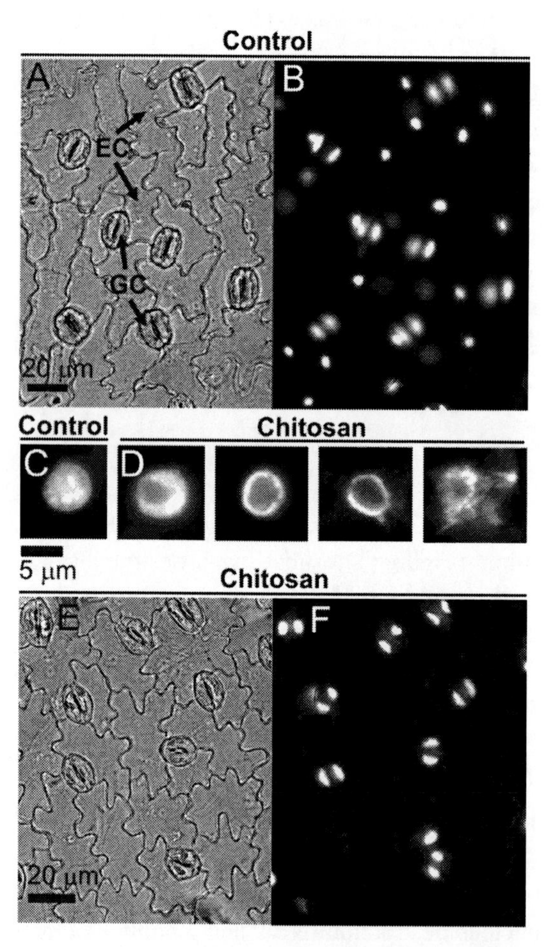

Figure 1. Effect of chitosan on the state of epidermal (EC) and guard cell (GC) nuclei. Data of the light (A, E – images in transmitted light) and fluorescence microscopy of epidermal and guard cells (B, F) and epidermal cell nuclei (C, D) in pea leaf epidermis. Epidermal peels without additives (control, A, B, C) or supplemented with 100 μg/ml chitosan (D, E, F) were mixed with a magnetic stirrer for 0.5 h and then incubated for 2 (C, D) or 5 (A, B, E, F) h without stirring in the dark. Epidermis was fixed, washed and stained with DAPI.

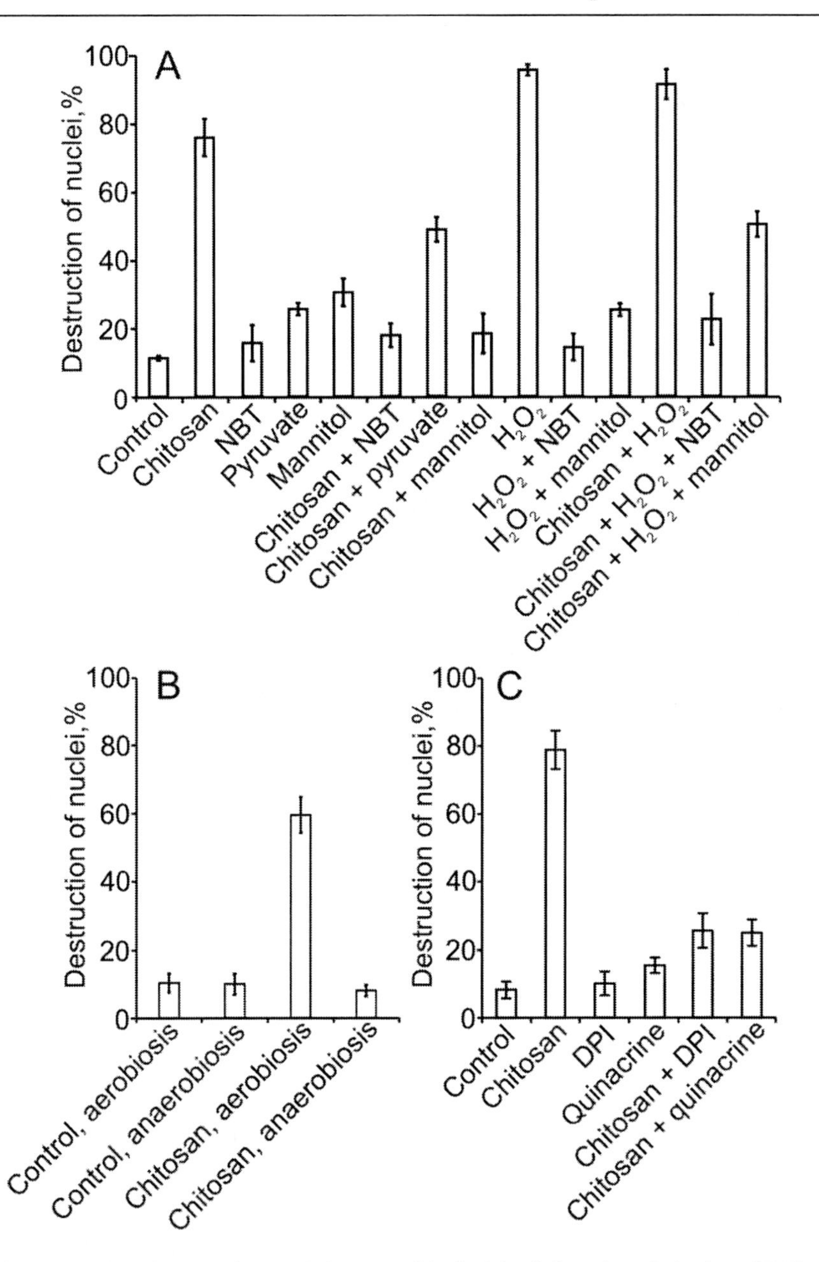

Figure 2. Effects of antioxidants and H_2O_2 (A), anaerobiosis (B), diphenylene iodonium (DPI) and quinacrine (C) on the chitosan-induced destruction of epidermal cell nuclei in pea leaf epidermis. (A) Epidermal peels were supplemented with 0.2 mM nutroblue tetrazolium (NBT), 5 mM potassium pyruvate, 125 mM mannitol and 100 μM H_2O_2, incubated for 20 min, supplemented with 100 μg/ml chitosan, incubated for 30 min with a magnetic stirrer and then incubated for 3 h in the dark without stirring. (B) Epidermal peels were incubated in aerobic or anaerobic condition, supplemented with 100 μg/ml chitosan, incubated for 30 min with a magnetic stirrer and then incubated for 3 h in the light without stirring. (C) Epidermal peels were supplemented with 100 μg/ml chitosan, incubated for 30 min with a magnetic stirrer and then supplemented with 25 μM DPI and 50 μM quinacrine and incubated for 4 h in the dark without stirring. Thereupon, the epidermal peels were fixed, washed and stained with Carazzi's hematoxylin. The number of cells with impaired nuclei was calculated.

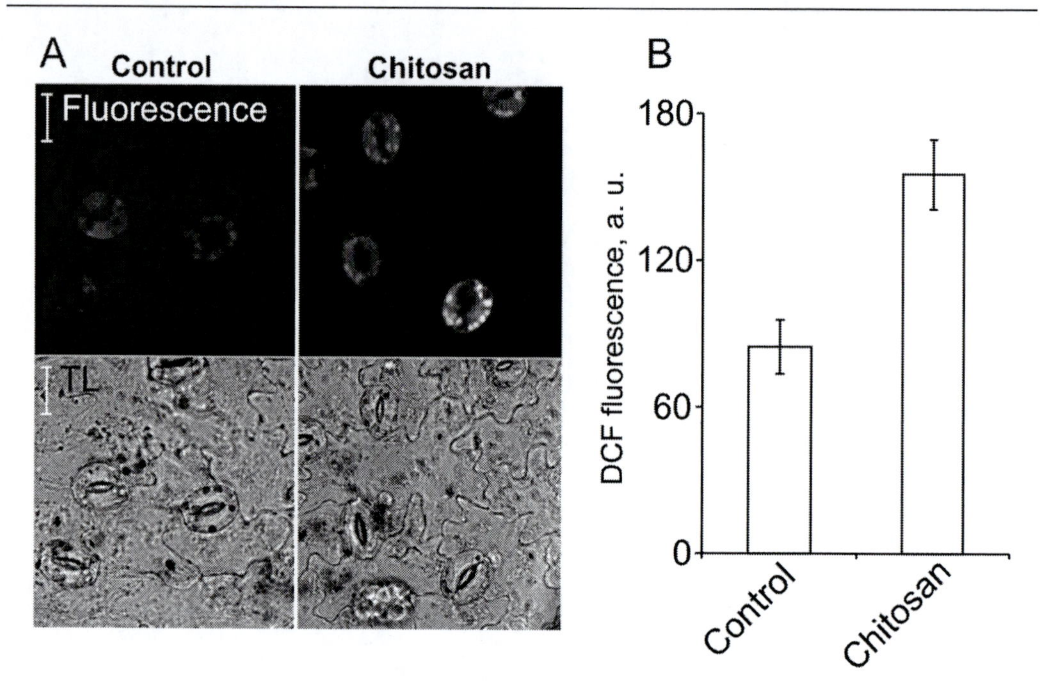

Figure 3. Effect of chitosan on the accumulation of H_2O_2 in guard cells of pea leaf epidermis. The intracellular content of H_2O_2 was determined from DCF fluorescence. Epidermal peels were incubated in distilled water for 3 h in the light, stained with DCFH diacetate and washed with distilled water in the dark. Thereupon, the samples were supplemented with 100 μg/ml chitosan and incubated with a magnetic stirrer for 20 min in the dark. Images with DCF fluorescence and transmitted light (TL) images are shown (A). Scale bars, 20 μm. The diagram (B) shows the DCF fluorescence in guard cells. The DCF fluorescence of 100 cells was calculated.

Discussion

Cyanide, an inhibitor of catalase and peroxidases, ribulose-1,5-bisphosphate carboxylase in cloroplasts and cytochrom *c* oxidase in mitochondria, caused the epidermal and guard cell nucleus destruction in pea leaf epidermis (Samuilov et al., 2002; 2003). The cyanide-induced destruction of guard cell nuclei was stimulated by H_2O_2 and suppressed under conditions preventing ROS generation (Samuilov et al., 2006; 2008). Data on the cell ultrastructure, internucleosomal DNA fragmentation and the effect of energy metabolism inhibitors demonstrate that cyanide induced PCD in pea guard cells with apoptosis-characteristic symptoms (Bakeeva et al., 2005; Dzyubinskaya et al., 2006).

The condensation and marginalization of nuclear chromatin with subsequent destruction of nuclei caused by chitosan (Figure 1) are markers of apoptosis, a form of PCD. Chitosan induces an immune reaction (a hypersensitive response) in plant cells accompanied by the generation of ROS and PCD (Iriti et al., 2006), which prevents the spreading of the pathogen in the plant tissue. The effects of anaerobiosis and antioxidants (Figure 2 A, B) demonstrate that the chitosan-induced PCD depends on ROS. The effect of H_2O_2 in our experiments was similar to the chitosan effect. The data obtained suggest that the effect of chitosan involves H_2O_2 accumulation in plant cells.

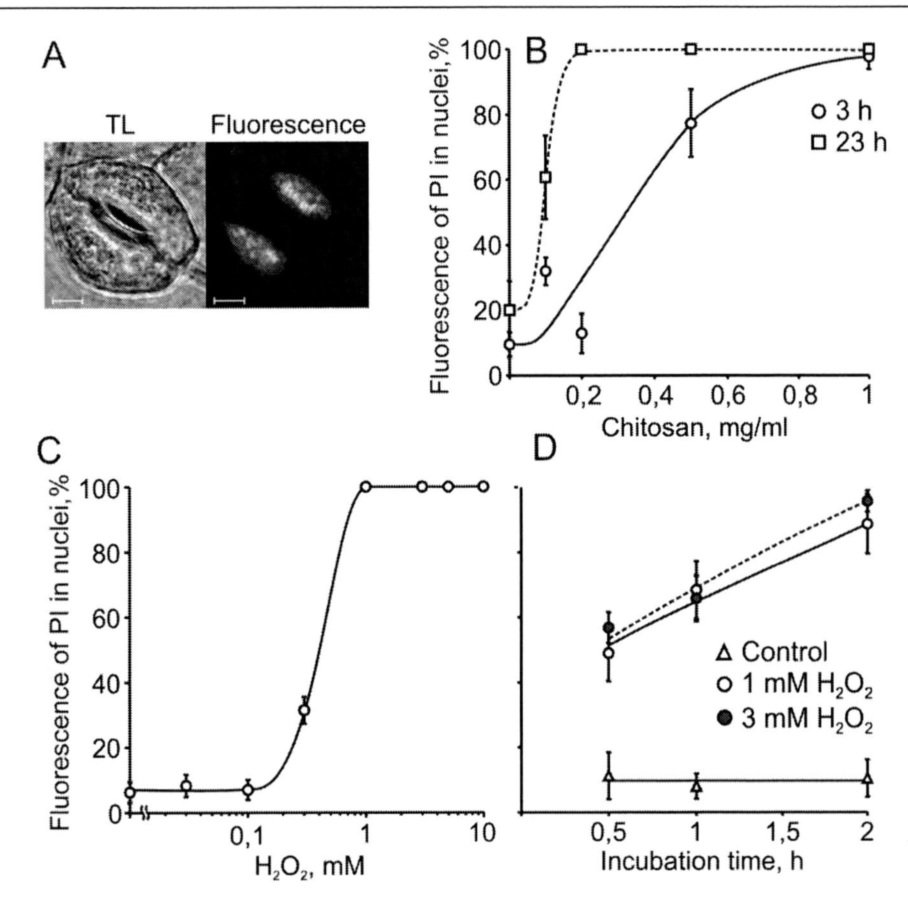

Figure 4. Effects of chitosan and H_2O_2 on the permeability of guard cell plasma membrane barrier. The permeability of the plasma membrane was estimated from the fluorescence of the cell nuclei that was stained with PI. (A, B) Epidermal peels were supplemented with chitosan and incubated with a magnetic stirrer in the light. The chitosan concentration was 1 mg/ml (A). The times of incubation of samples with chitosan were 3 h (A) or 3 and 23 h (B). (C, D) Epidermal peels were supplemented with H_2O_2 and incubated for 23 h (C) or 0.5–2 h (D) in the light. The number of cells with fluorescence in the cell nuclei as shown in picture (A) was determined. Scale bars, 5 μm.

The susceptibility of the epidermal cell nucleus destruction to inhibitors of the plasma membrane NADPH oxidase (Figure 2 C) demonstrates that the generation of ROS, in particular of H_2O_2, which stimulate PCD, occurs via the activation of the NADPH oxidase. DPI and quinacrine exerted no influence on respiration and photosynthetic O_2 evolution at concentrations used in experiments.

Chitosan caused the destruction of nuclei in epidermal cells, but not in guard cell. The fact that chitosan induced H_2O_2 accumulation in guard cells was in accordance with the DCF fluorescence data (Figure 3). H_2O_2 at concentrations up to 10–50 mM did not cause the destruction of guard cell nuclei (Samuilov et al., 2006).

The treatment of epidermal peels with chitosan stimulated the penetration of PI through plasma membrane (Figure 4 A, B). H_2O_2 at concentrations above 1 mM possessed a similar effect (Figure 4 C, D). These data show that the effect of chitosan on plant cells involves H_2O_2 generation.

The disruption of the barrier of the cell plasma membrane is characteristic of non-programmed cell death (necrosis). PI was used as a marker of cells undergoing necrosis (Koyama et al., 2001). However, different treatments of plant cells including freezing in liquid nitrogen, mechanical homogenization or addition of detergents induced necrotic-like lesions of the cells and passing of PI through the cell plasma membrane resulted in the internucleosomal DNA fragmentation, an apoptotic feature (Kuthanova et al., 2008). It is not clear whether the disruption of the barrier function of the plant cell plasma membrane is a criterion of non-programmed cell death only.

H_2O_2 induced changes in the permeability of the cell plasma membrane (Folmer et al., 2008). These changes can result in PI crossing the plasma membrane. Thus, the staining of guard cell nuclei with PI in the presence of chitosan may be due to the accumulation of H_2O_2 and the subsequent H_2O_2-stimulated changes in the cell plasma membrane properties.

Conclusion

The data obtained demonstrate that chitosan induces destructive changes in plant cells such as PCD or disrupting of the cell plasma membrane. The death of plant cells is an immune response of plants to pathogens. Chitosan, an activator of the plant immune system, protects plants against the infection induced by viruses (Pospieszny et al., 1991; Iriti et al., 2006). Chitosan initiates the production of H_2O_2 which induces destructive changes in cell structure, presumably associated with activating the NADPH oxidase of the plasma membrane.

References

Bakeeva, L. E., Dzyubinskaya, E. V., and Samuilov, V. D. (2005). Programmed cell death in plants: ultrastructural changes in pea guard cells. *Biochemistry (Moscow)*, *70*, 972–979.

Dzyubinskaya, E. V., Kiselevsky, D. B., Lobysheva, N. V., Shestak, A. A., and Samuilov, V. D. (2006). Death of stoma guard cells in leaf epidermis under disturbance of energy provision. *Biochemistry (Moscow)*, *71*, 1120–1127.

Folmer, V., Pedroso, N., Matias, A. C., Lopes, S. C. D. N., Antunes, F., Cyrne, L., and Marinho, H. S. (2008). H_2O_2 induces rapid biophysical and permeability changes in the plasma membrane of *Saccharomyces cerevisiae*. *Biochim. Biophys. Acta*, *1778*, 1141–1147.

Iriti, M., Sironi, M., Gomarasca, S., Casazza, A. P., Soave, C., and Faoro, F. (2006). Cell death-mediated antiviral effect of chitosan in tobacco. *Plant Physiol. Biochem.*, *44*, 893–900.

Koyama, H., Toda, T., and Hara, T. (2001). Brief exposure to low-pH stress causes irreversible damage to the growing root in *Arabidopsis thaliana*: pectin-Ca interaction may play an important role in proton rhizotoxicity. *J. Exp. Bot.*, *52*, 361–368.

Kuthanova, A., Opatrny, Z., and Fischer, L. (2008). Is internucleosomal DNA fragmentation an indicator of programmed death in plant cells? *J. Exp. Bot.*, *59*, 2233–2240.

LeBel, C. P., Ischiropoulos, H., and Bondy, S. C. (1992). Evaluation of the probe 2',7'-dichlorofluorescin as an indicator of reactive oxygen species formation and oxidative stress. *Chem. Res. Toxicol.*, *5*, 227–231.

Lin, W., Hu, X., Zhang, W., Rogers, W. J., and Cai, W. (2005). Hydrogen peroxide mediates defence responses induced by chitosans of different molecular weights in rice. *J. Plant Physiol.*, *162*, 937–944.

Metzler, D. E. (1977). *Biochemistry: The Chemical Reactions of Living Cell*, Chap. 8: K4, Academic Press, New York.

Pospieszny, H., Chirkov, S. N., and Atabekov, I. G. (1991). Induction of antiviral resistance in plants by chitosan. *Plant Sci.*, *79*, 63–68.

Samuilov, V. D., Lagunova, E. M., Dzyubinskaya, E. V., Izyumov, D. S., Kiselevsky, D. B., and Makarova, Ya. V. (2002). Involvement of chloroplasts in the programmed death of plant cells. *Biochemistry (Moscow)*, *67*, 627–634.

Samuilov, V. D., Lagunova, E. M., Kiselevsky, D. B., Dzyubinskaya, E. V., Makarova, Ya. V., and Gusev, M. V. (2003). Participation of chloroplasts in plant apoptosis. *Biosci. Rep.*, *23*, 103–117.

Samuilov, V. D., Kiselevsky, D. B., Sinitsyn, S. V., Shestak, A. A., Lagunova, E. M., and Nesov, A. V. (2006). H_2O_2 intensifies CN^--induced apoptosis in pea leaves. *Biochemistry (Moscow)*, *71*, 384–394.

Samuilov, V. D., Kiselevsky, D. B., Shestak, A. A., Nesov, A. V., & Vasil'ev, L. A. (2008). Reactive oxygen species in programmed death of pea guard cells. *Biochemistry (Moscow)*, *73*, 1076–1084.

Srivastava, N., Gonugunta, V. K., Puli, M. R., and Raghavendra, A. S. (2009). Nitric oxide production occurs downstream of reactive oxygen species in guard cells during stomatal closure induced by chitosan in abaxial epidermis of *Pisum sativum*. *Planta*, *229*, 757–765.

Vasil'ev, L. A., Dzyubinskaya, E. V., Zinovkin, R. A., Kiselevsky, D. B., Lobysheva, N. V., and Samuilov, V. D. (2009). Chitosan-induced programmed cell death in plants. *Biochemistry (Moscow)*, *74*, 1035–1043.

Reviewers:

Agepati S. Raghavendra,

Department of Plant Sciences, School of Life Sciences, University of Hyderabad, Hyderabad 500046, India

e-mail: asrsl@uohyd.ernet.in

Lev S. Yaguzhinsky,

A.N.Belozersky institute of physic-chemical biology, Lomonosov Moscow State University, Moscow 119991, Russia.

e-mail: yag@genebee.msu.su

In: Handbook of Chitosan Research and Applications
Editors: Richard G. Mackay and Jennifer M. Tait
ISBN 978-1-61324-455-5

Chapter 22

CHITOSAN AS ELICITOR OF HEALTH BENEFICIAL SECONDARY METABOLITES IN *IN VITRO* PLANT CELL CULTURES

Maura Ferri and Annalisa Tassoni*
Department of Experimental Evolutionary Biology,
University of Bologna, Bologna, Italy

ABSTRACT

Chitosan is an important structural component of the cell wall of several plant pathogen fungi. In plants, it can be applied *in vivo* as antimicrobial agent against fungi, bacteria and viruses, and *in vitro* can be used as elicitor of defense mechanisms. Its degrees of polymerization and of acetylation strongly influence the elicitation activity. Chitosan seems to be particularly effective in increasing the content of a large spectrum of plant polyphenols, such as stilbenes (i.e. resveratrol) and flavonoids (i.e. anthocyanins), very important antioxidants for plants, animals and humans. All these compounds, can be ingested by humans through daily diet and are absorbed in the small intestine, providing health benefits, such as increase of the antioxidant capacity of blood and potential prevention of cancer and cardiovascular diseases.

In this chapter we present a comprehensive overview of the applications and effects of chitosan in plant cell cultures. Although the exact mechanism of chitosan action in plants is still unknown, different hypotheses have been proposed. Plant cell cultures have been viewed as a promising alternative to whole plant extraction for obtaining valuable metabolites, with the elicitation of the interesting compounds and the scale-up process as key points. In particular, in grape cell suspensions, the synthesis and release of several metabolites, were up-regulated by chitosan treatment. Concomitantly, the expression levels of some enzymes of the phenylpropanoid pathway as well as those of pathogen related proteins, together with the *de-novo* synthesis and/or accumulation of different proteins, were promoted. With the aim to further ameliorate the polyphenol yield, batch and fed batch grape cell fermentations were optimized also with the addition of chitosan, which proved to be highly effective in improving metabolite production in bioreactors.

* Corresponding author Annalisa Tassoni: e-mail: annalisa.tassoni2@unibo.it

INTRODUCTION

Chitin and Chitosan: Discovery and Chemical Characteristics

Chitin and chitosan are important animal and fungi biopolymers whose history have been closely related to the plant world since their discovery in a botanical garden two centuries ago. In 1811 Henry Braconnot, director of the Botanical Garden and Professor of Natural History in Nancy University, discovered the compound then named "chitin". Those years were crucial for the connections between botany, chemistry and medicine. Braconnot was also interested in the definition of the nutritional value of mushrooms and the discovery of chitin was essentially based on some reactions carried out on raw material isolated from *Agaricus, Hydnum* and *Boletus* mushrooms (Muzzarelli and Muzzarelli, 2003). About 20 years later, a compound with similar structure was found in insect exoskeleton and it was termed "chitin", which means "envelope" or "tunic" in the Greek glossary. Chitosan was discovered by C. Rouget in 1859 by accident while he was boiling chitin in water plus concentrate soda to produce natural soap: boiling in alkaline solution led to chitin deacetylation. The chemical structure of chitosan, however, was identified only in 1950.

From an ecological point of view, chitin (polymer of N-acetyl-D-glucosamine) is the most naturally abundant nitrogen compound widely distributed among invertebrates and ubiquitous in the fungi. At least ten gigatons ($1x10^{13}$ kg) of chitin are synthesized and degraded each year in the biosphere. It is therefore important for making nitrogen available to countless living organisms. Chitins isolated from various sources differ from degree of acetylation (typically close to 0.90), elemental analysis (N typically close to 7%, N/C ratio 0.146 for fully acetylated chitin), molecular size and polydispersity. The average molecular weight of chitin *in vivo* is probably in the order of the MDa. Unmodified chitin is practically insoluble in water and organic solvent, but it is soluble in concentrated acids. Generally chitin is processed to chitosan (Ravi Kumar et al., 2004).

The term "chitosan" indicates a family of biopolymers naturally and industrially obtained by *N*-deacetylation of chitin, although the degree of the *N*-deacetylation is almost never complete. Chitin deacetylase is an enzyme that catalyzes the hydrolysis of acetamine groups of N-acetyl-D-glucosamine in chitin, converting it to chitosan. Chitosan itself occurs naturally as the major structural component of the cell wall of particular groups of fungi belonging to Zygomycetes class and in particular to Mucorales order (*Absidia* and *Mucor* species), but it can easily recovered also from other fungi and yeasts, such as *Rhizopus oryzae* (Pochanavanich and Suntornsuk, 2002). However, to date, chitosans have been commercially produced by alkaline deacetylation of crustacean chitins, obtained from the abundant waste of crustacean canning industry. Chitosan is therefore a linear D-(1,4)-glucosamine polymer, and the only largely available cationic polysaccharide, whilst most polysaccharides are anionic, such as alginates and pectins. In general, chitosans have a nitrogen content higher than 7% and a degree of acetylation lower than 0.40. Because of the positive charge on the C-2 of the glucosamine monomer below pH 6, chitosan is more soluble than chitin, but it is insoluble in most solvents. Chitosan is soluble in dilute organic acids such as acetic acid, formic acid, succinic acid, lactic acid and malic acid (Ravi Kumar et al., 2004).

A number of derivatization/functionalization reactions useful to enhance chitosan properties and to impart new ones, are reported in the literature (Ravi Kumar et al., 2004). For instance, the chemical modifications of chitosan lead to more soluble polymers (i.e. certain chitosan salts are water soluble), to have higher biodegradability in animal bodies or to acquire physical properties interesting for applications in the solid state or in solution.

Chitosan Applications

Since chitosan is relatively inexpensive, biodegradable, non-toxic, non-immunogenic and biocompatibile in animal tissues, much research was directed toward its use in medical applications such as drug delivery, artificial skin and blood anticoagulants (Prasitsilp et al., 2000; Ravi Kumar et al., 2004). Other suggested uses were: flocculating agent in wastewater treatment, additive in the food industry (i.e. as food thickener), hydratant in cosmetics, paper and textile adhesive, component of packaging membrane, chelator of metals, etc.

Chitosan and its derivatives were also considered versatile biopolymers in agriculture application as antimicrobial agents against a wide variety of microorganisms including algae, bacteria, virus and mainly fungi, due to their ability in eliciting natural plant defense mechanisms. Chitosan has a better antimicrobial activity than chitin, due to its positive charges and its higher solubility. Most physiological activities and functional properties of chitosan and its derivatives clearly depend on their molecular weight, even if their mode of action needs to be further investigated. The *in vivo* and *in vitro* antimicrobial activity of chitosan is influenced by intrinsic factors such as the average degree of chitosan polymerization, the degree of *N*-deacetylation, the positive charge value, the type of host organism, the natural nutrient constituency, the chemical and/or nutrient composition of the substrates and the environmental conditions (i.e. substrate water activity and/or moisture). Chitosan was utilized *in vivo* in soil amendment, in seed coating and as a foliar treatment to control fungus infections. For instance, high protection against fungi in tomato seeds occurred with chitosan concentrations between 0.5 and 1 mg/mL, while at 0.1 mg/mL it induced a delay in disease development (Benhamou et al., 1994). Wheat seeds, instead, required higher chitosan concentration (2-8 mg/mL) to control seed-borne *F. graminearum* infection and to significantly improve germination and vigor (Reddy et al., 1999). Chitosan reduced the *in vitro* growth of numerous fungi with the exception of *Zygomycetes*, that contain chitosan as a major component of their cell walls (Rabea et al., 2003).

Although both native chitosan and its derivatives are effective as antimicrobial agents, a clear difference was evidence. Chitosan shows its antimicrobial activity only in acidic medium because of its poor solubility above pH 6.5. Thus, water-soluble chitosan derivatives (soluble in both acidic and basic physiological circumstances) may be good candidates as a polycationic biocide. The effect of the polymer molecular weight on some antibacterial and antifungal activities was explored. Chitosan with a molecular weight ranging from 10 to 100 kDa might be helpful in restraining the growth of bacteria. Moreover, the antibacterial activity of chitosan is influenced by its degree of deacetylation, its concentration in solution, and the pH of the medium (Rabea et al., 2003).

In addition, chitosan has potential as an edible antifungal coating material for postharvest produce. *In vivo* studies reported a delay in infection and decay in chitosan coated fruits after storage, while they ripened normally and did not show any apparent sign of phytotoxicity. The antifungal properties of chitosan are of interest in the food industry especially because it was demonstrated to be a safe biopolymer suitable for oral administration and was also successfully used in the production of food wraps (Rabea et al., 2003).

Chitosan Possible Mechanisms of Action in Plants

As previously reported chitosan is a cell wall structural component found in many pathogenic fungi. Upon pathogen infection, chitosan molecules (oligosaccharins) are released from the fungi by cell-wall-degrading enzymes (i.e. chitinases, glucanases) secreted by the plants. The interaction between oligosaccharins and receptors located in the plant membranes, results in the final accumulation of many plant defense secondary metabolites, such as phytoalexins. Chitosan receptors were identified (Chen and Xu, 2005), but the signaling pathway from the point at which the oligosaccharin contacts the cell surface to that at which the final response is realized, is not yet well clarified. Recently, Amborabè and coworkers (2008) have investigated the early events induced by chitosan in plant cells. It was found that chitosan triggers, in a dose-dependent manner, the rapid transient depolarization of cell membrane with consequent modification of the proton fluxes and alteration of the cell permeability, with the plasma membrane H^+-ATPase as primary site of action. However, the oligosaccharin/receptor interaction seems to be followed by an unknown step that is thought to activate more than one defense response, among which specific gene transcription activities that lead to the synthesis of phenylpropanoid enzymes. Possible proposed second messengers are calcium and cAMP but also the variation of potassium concentration and pH level (Rabea et al., 2003). In poplar it was demonstrated that the chitosan-induced signal transduction involves a rapid and transient activation of at least two myelin basic protein kinases (MBPKs), which have characteristics of mitogen activated protein-kinases (MAPKs) and may function as convergence points in defense signaling cascades activated both by biotic and abiotic elicitors (Hamel et al., 2005).

The exact mechanism of the antimicrobial action of chitosan, chitin and their derivatives is still unknown, but different hypotheses have been proposed. In whole plants, chitosan induces the reduction of stomatal aperture (Klüsener et al., 2002), limiting pathogen access to the inner leaf tissues. At cellular level, interaction between positively charged chitosan molecules and negatively charged microbial cell surfaces, leads to the leakage of proteinaceous and other intracellular constituents and causes bacterial agglutination (Savard et al., 2002). Chitosan acts as a chelating agent that selectively binds trace metals and thereby inhibits the production of toxins and microbial growth. It can also bind water and inhibit various enzyme activities (Rabea et al., 2003). It has been proposed that chitosan liberated from the cell wall of fungal pathogens by plant host hydrolytic enzymes, penetrates to the nuclei of fungi where it interacts with DNA (Hadwiger and Beckman, 1980) interfering with mRNA and protein synthesis (Rabea et al., 2003; Yin et al., 2006). When exogenously supplied, chitosan is also able to reach the plant nucleus (Rabea et al., 2003), where it breaks and complexes with DNA strands in competition with positively charged nuclear proteins (Klosterman et al., 2003). Anyway, by imitating the contact of the plant with a phytopathogen, chitosan induces a wide spectrum of protective reactions in the plant, triggering early cellular responses and leading to the development of the systemic acquired resistance. For example, its ability to suppress viral infections does not depend on the virus type but to chitosan itself that affects the plant inducing resistance to the viral infection (Rabea et al., 2003; Iriti et al., 2006).

In summary, chitosan behaves like a general elicitor, inducing a non-host resistance and priming a systemic acquired immunity. The responses elicited by chitosan include rising of cytosolic H^+ and Ca^{2+}, activation of MAP-kinases, callose apposition, oxidative burst, hypersensitive response (HR), synthesis of abscisic acid (ABA), of jasmonate, of phytoalexins and of pathogenesis related (PR) proteins. Nevertheless, it must be pointed out that biological activity of chitosan, besides the plant model, strictly depends on its physicochemical properties (deacetylation degree, molecular weight and viscosity), and that there is a threshold for chitosan concentration able to switch the induction of a cell death program into necrotic cell death (cytotoxicity) (Iriti and Faoro, 2009).

Chitosan as Elicitor in Plants

Elicitation is a process of induced or enhanced synthesis of secondary metabolites by the plant cells to ensure their survival, persistence and competitiveness. The eliciting molecules could be abiotic (such as ozone, heavy metals, UV light, etc) or biotic in which case they are usually macromolecules originating from host plant (endogenous elicitors) or plant pathogen (exogenous elicitors). Plant cell and organ cultures grown usually exhibit changes in physiological and biochemical responses upon exposure to biotic and abiotic elicitors.

Chitosan has been extensively studied to determine its ability to elicit natural plant defense responses in *in vitro* cultures. Several investigators reported its effectiveness in inducing callose and lignin formation and deposition (Hadwiger 1999; Bézier et al., 2002; Ndimba et al., 2003; Aziz et al., 2003; Kim et al., 2005). The synthesis of lignin precursors such as *p*-coumaric, ferulic, sinapic and phenolic acids, that show antimicrobial activity, is also stimulated proportionally with the elicitor concentration (Rabea et al., 2003; Chakraborty et al., 2009). Chitosan induces the synthesis of pathogenesis-related (PR) proteins and of several defense enzymes (such as phenylalanine ammonia lyase and peroxidase) (Hadwiger 1999; Bézier et al., 2002; Ndimba et al., 2003; Aziz et al., 2003; Kim et al., 2005; Ferri et al., 2009). The transcription of several gene families (resistance genes) is rapidly activated in the presence of pathogens as consequence of the interaction chitosan/membrane receptors. Chitosan promotes the production of phytoalexins, like stilbenes (Hadwiger 1999; Righetti et al., 2007; Ferri et al., 2009), and of many phenolic and terpenic compounds (Rabea et al., 2003; Kim et al., 2005). The plant defense compounds synthesized after chitosan elicitation, have usually relevant physiological activity, mainly as antioxidants (i.e. polyphenols).

Chitosan Elicitation of Plant Secondary Metabolites in *In Vitro* Cell Cultures

Plant cultures can often serve as model systems to study the cell changes in relation to plant defense responses against pathogens both at biochemical and molecular levels (Ferri et al., 2009; Chakraborty et al., 2009). Cell cultures not only have a higher metabolism rate than differentiated plants, but also have shorter biosynthetic cycles (Zenk, 1991). As *in vitro* cell systems are relatively easy to manipulate, they can provide a better control of external factors that can interfere with the metabolic activities. Therefore, they have been viewed as a promising alternative to whole plant extraction, to obtain valuable secondary metabolites. In fact, *in vitro* cultures provide a source of homogeneous highly active cells that allow to overcome some plant limits such as slow growth, seasonal and environmental variations and disease susceptibility.

Most of the literature published about chitosan supplementation to *in vitro* cell plant cultures reports data on plant defense mechanisms. However, several authors have tested chitosan as elicitor to improve the production of compounds with pharmaceutical, medical and general industrial applications (see examples in Table 1). In particular, cell suspensions of *Plumbago rosea* treated with chitosan were utilized for the production of plumbagin (an anticancer, antimutagenic, antimicrobial, antifertility and insecticidal compound) (Komaraiah et al., 2002). The same elicitated cells were immobilized in calcium alginate: immobilization enhanced plumbagin synthesis and more than 70% of the metabolite was released into the medium, result that is highly desirable for easy recovery of the product in an hypothetical industrial production (Komaraiah et al., 2003). The use of chitosan to improve in *Vitis vinifera* cell suspensions (Ferri et al., 2009) the production of stilbenes (antioxidant metabolites used as cardiovascular protectors, chemopreventive and chemotherapeutic agents) in bioreactor scale, will be discussed later in this chapter.

Chitosan was used to improve the yield of many useful compounds also in *in vitro* culture of tissues and whole plants. For example, in basil chitosan increased the total amount of phenolic and terpenic compounds, especially of rosmarinic acid and eugenol that have strong antioxidant, antiviral, antibacterial, anti-inflammatory, antiseptic and anesthetic properties, and show application in perfumery and flavoring (Kim et al., 2005). Tissue cultures treated with chitosan were mostly represented by hairy root systems (obtained after *Agrobacterium rhizogenes* infection), such as *Brugmansia candida* roots for scopolamine and hyosciamine synthesis (anticholinergic tropane alkaloids employed as antispasmodics drugs) (Pitta-Alvarez and Giulietti, 1999), *Armoracia lapathifolia* roots for peroxidases production (applied in industrial processes to produce food colorants, dyes for perfume industries, in biobleaching and in the removal of toxic and carcinogenic pollutants from industrial effluents) (Flocco and Giulietti, 2003) and *Artemisia annua* roots for artemisinin synthesis (antimalarian) (Putalun et al., 2007).

The effectiveness of polysaccharides, among which chitosan, as active elicitors is dependent on their molecular structures. In general, polysaccharides with a high degree of branching are more active than those with a lower number of side chains (Dörnenburg and Knorr, 1994). In the case of chitosan, the degrees of acetylation and of polymerization are important in inducing defense metabolisms in plant cell cultures. In fact, in cell suspensions and protoplasts of *Catharanthus roseus* it was demonstrated that callose synthesis increased with the degree of polymerization of chitosan and decreased when the homopolymer was often interrupted by N-acetyl groups (Kauss et al., 1989). Also in *Morinda citrifolia* suspensions, 14 days after chitosan treatment, the amount of the produced anthraquinone (a phytoalexin) increased in relation to the chitosan degree of acetylation (Dörnenburg and Knorr, 1994). In addition, the rapid and transient production of reactive oxygen species (oxidative burst) occurring as one of the earliest reactions of plant cells against pathogen attack, was strongly dependent on the degree of acetylation of chitosan oligomers and polymers, with highly acetylated chitosans being most active (dos Santos et al., 2008).

Table 1. Use of chitosan as elicitor in different plant cell cultures to improve the production of compounds with useful applications in pharmacy, medicine and various other industries

Plant	Compounds	Applications	References
Catharanthus roseus	terpenoid indole alkaloids from strictosidine	chemotherapeutic agents	Pasquali et al., 1992
Cistanche deserticola	phenylethanoid glycosides	free radicals scavengers, sedatives, improver male sexual function	Cheng et al., 2006
Cocoa nucifera	p-hydroxybenzoic acid, p-coumaric acid, ferulic acid; total phenolic content	antioxidant	Chakraborty et al., 2009
Hypericum perforatum	xanthone,euxanthone, cadensin G	antioxidant used in case of allergies, infections (microbial, fungus, viral), cholesterol levels, inflammation, skin disorders, gastro-intestinal disorders, fatigue; vasorelaxant, immune system stimulator; insecticide, larvicide	Tocci et al., 2010
Morinda citrifolia	anthraquinone	laxatives, antimalarials, antineoplastics; bird repellants on seeds, gas generator in satellite balloons, building block of many dyes, bleaching pulp for papermaking	Dörnenburg and Knorr, 1994
Oryza sativa	momilactone A, monilactone B	chemotherapeutic agent	Ren and West, 1992
Panthoea agglomerans	peroxidase	food colorants, dyes for perfume industries, biobleaching, removal of pollutants from industrial effluents	Ortmann and Moerschbacher, 2006
Petroselinum crispum	coumarins	antilymphedema, precursor of anticoagulant; sweet scented perfume	Conrath et al., 1989
Physcomitrella patens	peroxidase	food colorants, dyes for perfume industries, biobleaching, removal of toxic and carcinogenic pollutants from industrial effluents	Lehtonen et al., 2009
Plumbago rosea	plumbagin	anticancer, antimutagenic, antimicrobial, antifertility; insecticidal	Komaraiah et al., 2002; Komaraiah et al., 2003
Rubia tinctorum	anthraquinone	anthracenediones, laxatives, antimalarials, antineoplastics; bird repellant on seeds, gas generator in satellite balloons, building block of many dyes, bleaching pulp for papermaking	Vasconsuelo et al., 2004
Silybum marianum	silymarin	antihepatotoxic, hepatoprotective	Sanchez-Sampedro et al., 2005
Taxus baccata	dolichols and polyprenols (polyisoprenoid alcohols)	hepatoprotector, neuroprotector, diseases, anticancer, immune system stimulator, antiviral, membranes protector from peroxidation	Skorupinska-Tudek et al., 2007
Taxus canadensis	paclitaxel	chemotherapeutic agent, mitotic inhibitor	Linden and Phisalaphong, 2000
Taxus yunnanaensis	paclitaxel	chemotherapeutic agent, mitotic inhibitor	Zhang et al., 2002
Vitis vinifera	resveratrol, piceid, resveratroloside, anthocyanins	cardiovascular protection, chemopreventive and chemotherapeutic agents	Ferri et al., 2009

Other studies have been focused on the chitosan polymerization degree required for biological activity. For instance, it was determined that the octameric oligomer optimally induced pisatin accumulation in pea tissue and inhibited fungal growth in comparison to other oligomer sizes (Hadwiger et al., 1994). The effectiveness of chitosans with different molecular weights (MW 6-753 kDa) in the elicitation of callose synthesis in *Phaseolus vulgaris* and in inducing resistance to tobacco necrosis virus, was evaluated by Faoro and Iriti (2007) that found the intermediate polymer sizes (MW 76, 120, 139 kDa) to be the most effective.

Chitosan Treatment of Grapevine Cell Cultures: Polyphenol Production

Plants are an abundant source for numerous useful compounds including pharmaceuticals, nutraceuticals and food additives. In particular, polyphenols, synthesized through the phenylpropanoid pathway, are one of the most widespread groups of plant secondary metabolites present in many crop plants and have largely proven to be health-beneficial, due to their known antioxidant properties (Winkel-Shirley, 2001a). Through the diet, humans ingest significant quantities of plant polyphenols due to their widespread distribution and variety, and heat stability. These metabolites are, for example, responsible for the brightly colored pigments of many fruits and vegetables and for plant protection from diseases and UV light (Winkel-Shirley, 2001a; Winkel-Shirley, 2001b). Several epidemiologic studies have indicated that regular consumption of foods rich in polyphenols (fruits, vegetables, nuts, whole grain cereals, red wine, green tea, cocoa, etc.) is associated with reduced risk of developing cardiovascular disease and several types of cancer (Espín et al., 2007). The "French paradox" classically associates the regular moderate consumption of red wine (rich of polyphenols) with low rates of coronary artery disease (Kopp, 1998). Numerous *in vitro* studies have shown that plant polyphenolic compounds (i.e. stilbenes) are powerful antioxidants that can protect cell membranes, cellular DNA and low density lipoproteins (LDL) from the damaging effects of free radicals induced oxidative damage (Lin and Tsai, 1999; Das et al., 1999).

The importance of grapevine as a crop makes *V. vinifera* a good model system for studies on the improvement of the nutraceutical properties of food products. The most relevant grape polyphenols families, such as stilbenes (i.e. resveratrol) and flavonoids (i.e. anthocyanins and catechins), have in common a strong antioxidant activity in plants, animals and humans. Flavonoids are ubiquitous in plants (Winkel-Shirley, 2001a) while stilbenes, the group of metabolites derived from resveratrol, are present at a basal level in grape and can be elicited by biotic and abiotic stimuli. These compounds are very important in plant defense mechanisms (Jeandet et al., 2002). Both stilbenes and flavonoids include free and glycosylated forms: glucosylation increases their water solubility, protects the aglycones from enzymic oxidation, extending their half-life and bioavailability, often not influencing their reactivity (Regev-Shoshani et al., 2003). After ingestion by humans through daily diet of table grape, wine and other derived products, both glycosides and aglycones are absorbed in the small intestine, providing health benefits, such as increase of the antioxidant capacity of blood and potential prevention of cancer and cardiovascular diseases (Espín et al., 2007).

Grape cell cultures synthesize the major polyphenols found in table grapes and in wines and some authors have undertaken studies on *V. vinifera* cell suspensions to investigate the factors able to induce and/or modify their biosynthesis and metabolism (Waffo-Teguo et al., 1998; Krisa et al., 1999a; Decendit et al., 2002; Aziz et al., 2003; Tassoni et al., 2005; Ferri et al., 2007; Righetti et al., 2007; Ferri et al., 2009). Chitosan, is commonly used to stimulate grape defense mechanisms and to promote the production of health-beneficial secondary metabolites (Righetti et al., 2007; Ferri et al., 2009). Moreover, *in vivo* and *in vitro* treatments of grape plantlets inhibit the development of the pathogenic fungi *Botrytis cinerea* and *Plasmopara viticola* (Bézier et al., 2002; Ait Barka et al., 2004; Aziz et al., 2006; Trotel-Aziz et al., 2006).

Vitis vinifera cell suspensions were obtained from different grape cultivars (Waffo-Teguo et al., 1996; Larronde et al., 1998; Krisa et al., 1999b; Larronde et al., 2005). Several studies were performed on cv. Barbera cell suspensions with the aim to optimize the eliciting conditions of health-beneficial secondary metabolites, mainly polyphenols such as stilbenes (resveratrol and its glucosides), catechins, anthocyanins and other flavonoids. The effect of several elicitors was investigated, among which salicylic acid, Na-orthovanadate, jasmonates, chitosan and derivatives, and the antibiotics ampicillin and rifampicin (Tassoni et al., 2005; Ferri et al., 2007; Righetti et al., 2007; Ferri et al., 2009). Most of the tested elicitors were able to promote resveratrol accumulation in the cells and some of them also stimulated its released in the culture medium without significantly affecting cell growth and viability (Righetti et al., 2007). Resveratrol biosynthesis was elicited by 10 μM methyl-jasmonate, that caused an increased accumulation of *trans*- and *cis*-isomers and of the stilbene synthase protein expression (Tassoni et al., 2005). However the best results were obtained after treatment with chitosan, as reported in the following part of this chapter (Ferri et al., 2009).

Elicitation with Chitin and Chitosan Monomers

The effect on resveratrol synthesis of 50 μg/mL N-acetyl-D-glucosamine and 50 μg/mL D-glucosamine, respectively chitin and chitosan monomers, was evaluated. *Cis*- and *trans*-resveratrol, intracellular and released in the cell culture medium, were quantified by HPLC (Tassoni et al., 2005) during a 14-day time-course after elicitation. The two compounds did not affect cell growth, but influenced resveratrol production. In the control culture, maximum accumulation (1.8 μmol/gDW) of resveratrol was observed at day 4, with 96% of the compound released in the media (Fig. 1). Both treatments anticipated the maximum of resveratrol production at day 2. With respect to the control, the intracellular accumulation was improved of 10.3-fold and 2.6-fold with D-glucosamine and N-acetyl-D-glucosamine respectively, but the total amount and the exogenous release were strongly reduced (Fig. 1). Therefore chitin and chitosan monomers did not seem to be effective in improving the total level of resveratrol, confirming the importance of the polymerization degree of chitosan in relation its efficacy as elicitor.

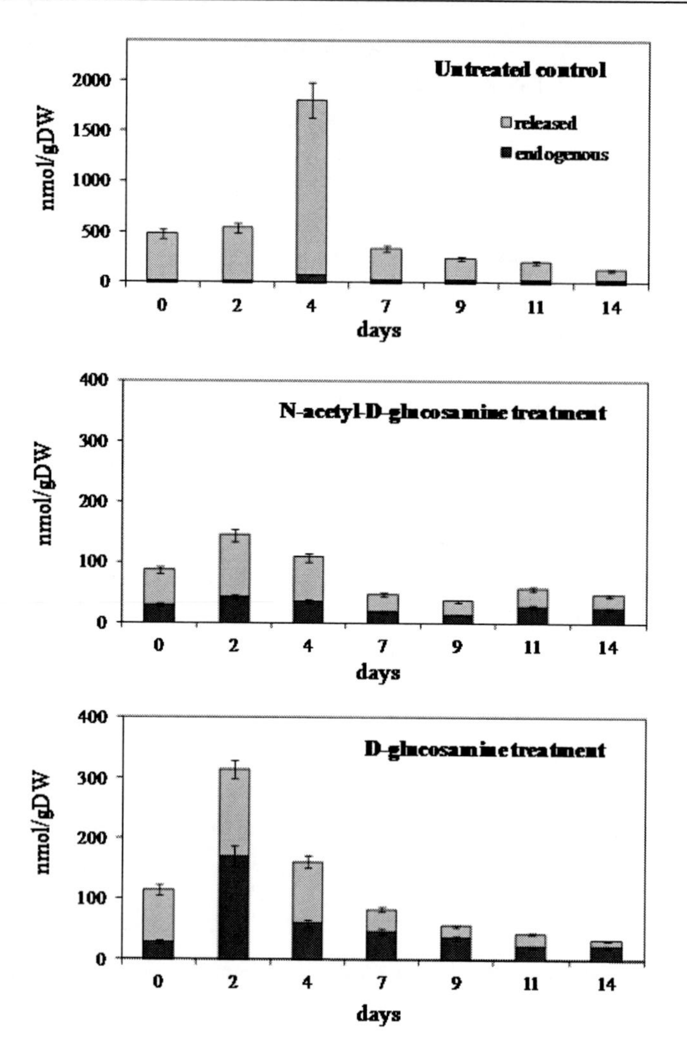

Figure 1. Resveratrol levels in Vitis vinifera cv. Barbera cell suspensions after N-acetyl-D-glucosamine and D-glucosamine (50 μg/mL) treatments and in untreated control. Data are the mean ± SE (n = 3).

Elicitation with Chitosan: Metabolomic and Proteomic Analyses

Low viscous chitosan (76-139 kDa, 50 μg/mL dissolved in 0.1% v/v acetic acid) was used to promote plant defense responses in cv. Barbera cells with the aim to provide a comprehensive picture of the activation and regulation of stilbenes and related compounds metabolism and to analyze the occurring proteomic changes. The results demonstrated that chitosan increases the production of stilbenes and anthocyanins in grape cell suspensions in comparison to untreated and acetic acid (chitosan solvent) controls (Ferri et al., 2009). The addition of chitosan and acetic acid did not significantly affect cell growth and slightly reduced cell viability (around 95% of living cells in untreated control and 80-85% in acetic acid and chitosan treated cultures). A wide range of health beneficial secondary metabolites, representative of the major grape polyphenolic families, were quantified by HPLC-DAD and spectrophotometric analyses both in the cells and in the culture media: free and monoglucosilated stilbenes (*cis*- and *trans*-resveratrol, *cis*- and *trans*-piceid and *cis*- and *trans*-resveratroloside); catechins (catechin, epicatechin, epigallocatechin-gallate); other flavonoids (naringenin, quercetin, rutin); gallic acid; hydroxycinnamic acids (*trans*-cinnamic, *p*-coumaric, caffeic, chlorogenic, ferulic and sinapic acids); total anthocyanins.

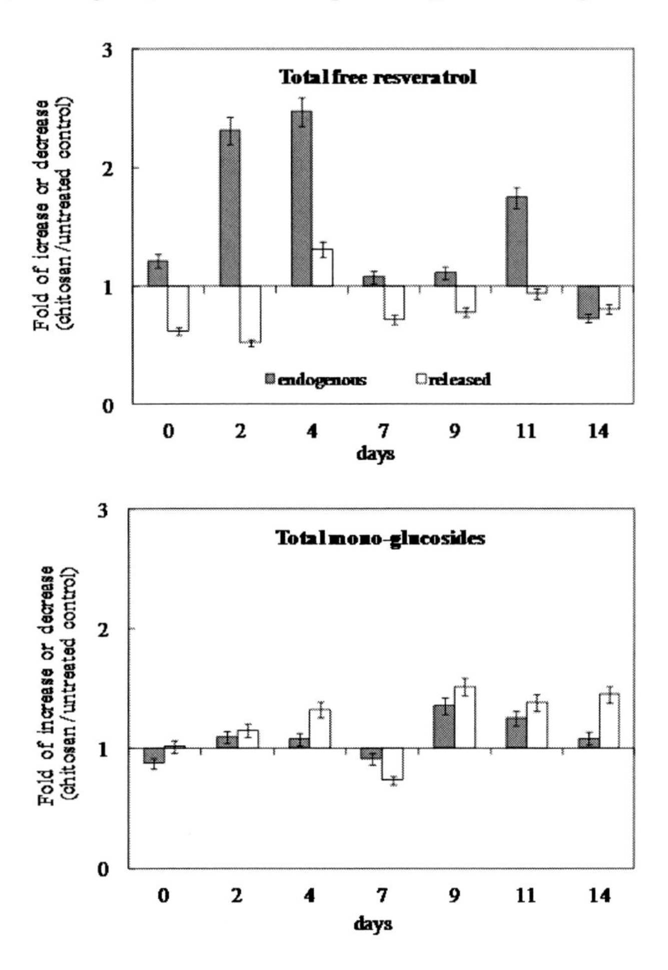

Figure 2. Fold of increase or decrease of total (trans plus cis isomers) resveratrol and total mono-glucoside levels (piceid plus resveratroloside, trans- plus cis-isomers) calculated as ratio between the amount of chitosan-treated cells to that of the untreated control cells. Modified from Ferri et al. (2009).

Ferri and colleagues (2009) showed that, in cv. Barbera cell suspensions, *trans*-resveratrol reached the maximum intracellular level 4 days after chitosan treatment, exceeding acetic acid and untreated controls by 36% and 63% respectively. Conversely, chitosan decreased the release in the media by 62% at day 2, with respect to acetic acid control. The *cis*-isomer and the mono-glycosylated forms (resveratroloside and piceid) were only slightly influenced by chitosan treatment. Interestingly, the resveratrol mono-glucosides were 100-200-fold higher than the aglycone, with the exception of the free resveratrol peak days. This results seemed of particular interest, due to the fact that glucosylation improves resveratrol half-life and bioavailability without reducing its antioxidant activity. These mono-glucosylated stilbenes could be much more functional and valuable than free resveratrol for future medical and nutraceutical applications. Figure 2 shows the fold of increase or decrease, inside and outside the chitosan-treated cells, of stilbene total amounts (free resveratrol *trans*- plus *cis*-isomers; resveratroloside plus piceid, *trans*- plus *cis*-forms) during the 14-day observation period, in comparison to the untreated control. The ratio chitosan/untreated control total levels allows to take into account the sum of the effects of both acetic acid and chitosan. Using this type of data elaboration, the maximum of resveratrol accumulation was confirmed at day 4 and another peak was detected at day 11 (Fig. 2). In the case of mono-glucosides, it was demonstrated an increase of both accumulation and release in the chitosan-treated cells throughout the culture period and particularly during the second week (Fig. 2). It is worth noting that in terms of absolute value the mono-glucosylated forms were 10-fold higher in concentration respect to the free resveratrol.

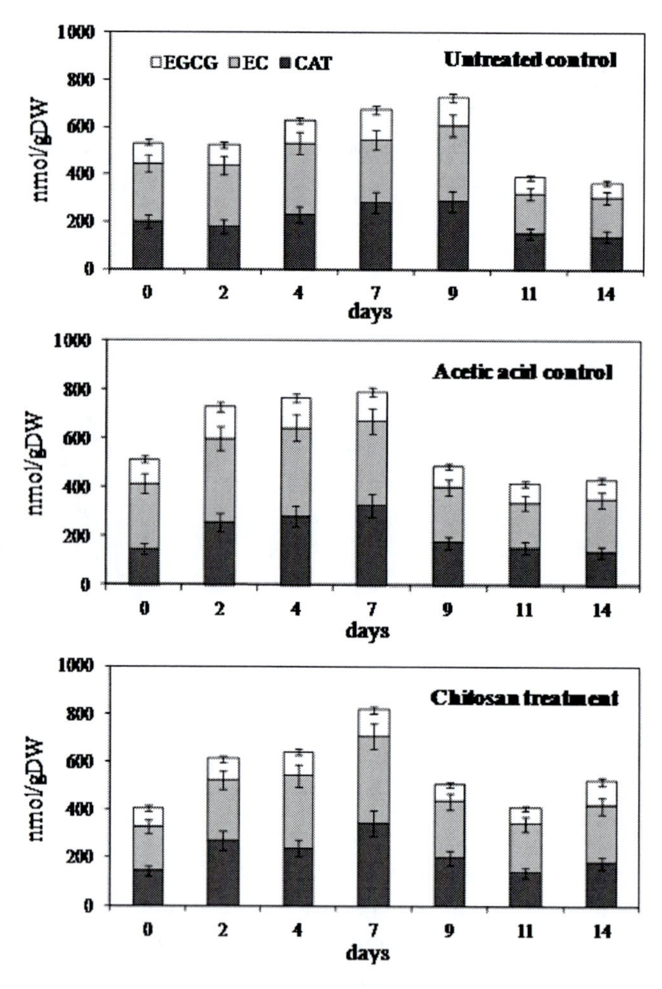

Figure 3. Levels of total endogenous and released catechins in Vitis vinifera cv. Barbera cell suspensions after treatment with chitosan (50 µg/mL) in comparison with acetic acid and untreated controls. EGCG, epigallocatechin-gallate; EP, epicatechin, CAT, catechin. Data are the mean ± SE (n = 3).

The accumulation of catechins was poorly influenced by chitosan treatment. Catechin (CAT), epicatechin (EP) and epigallocatechin-gallate (EGCG) were detected both inside the cells and in the culture media at constant levels, 38%, 46% and 16% respectively of the total amount. The treatment with chitosan did not significantly stimulate their synthesis and/or endogenous accumulation when compared to the acetic acid control. Conversely, with respect to the untreated control, a slightly higher amount of catechins was detected in the acetic acid culture from day 2 to 7 (Fig. 3). Some other flavonoids (naringenin, quercetin, rutin), gallic acid and several hydroxycinnamic acids were either not present or detectable in trace amounts throughout the observation period (Ferri et al., 2009). Total anthocyanin level was about 2-fold higher in chitosan-supplemented cells than in both controls, confirming the macroscopic observations in which the treated suspensions showed an intense brown-red color, from day 3 onwards (Fig. 4) (Ferri et al., 2009). The elicitation of anthocyanins and not of catechins, suggests that chitosan selectively up-regulates specific target branches of the phenylpropanoid pathway. A summary of the metabolomic effects of chitosan treatment on polyphenols synthesized from flavonoid pathway in *Vitis vinifera* cv. Barbera cell suspensions is presented in Fig. 5 (grey arrows).

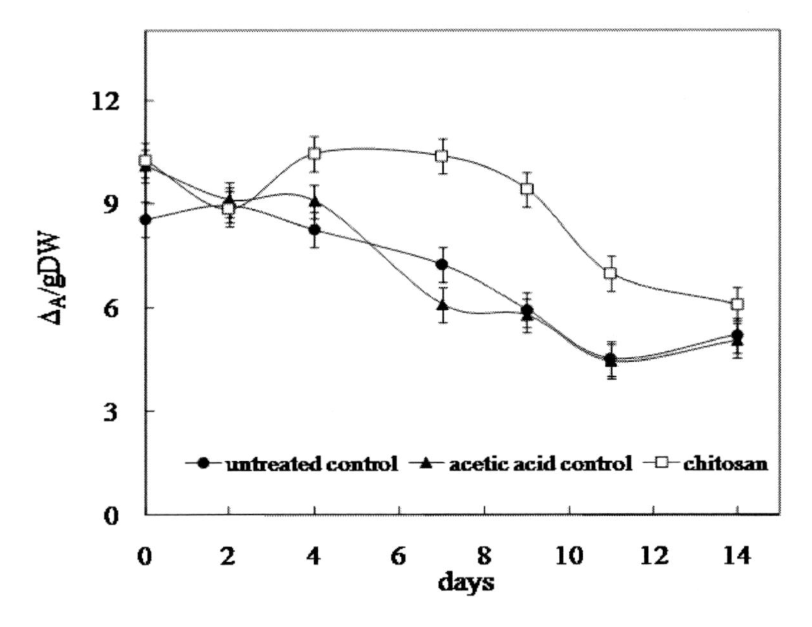

Figura 4. Total anthocyanin levels in Vitis vinifera cv. Barbera cell suspensions after chitosan (50 μg/mL) treatment. Data are expressed as the variation of absorbance units for grams of dry weight (□A/gDW). Data are the mean ± SE (n = 3). Modified from Ferri et al. (2009).

In the same cell system, Ferri and coworkers (2009) studied the gene expression of some key enzymes (Fig. 5) involved in the phenylpropanoid pathway as well as the variation of total protein expression profile after chitosan treatment. Cell samples were collected at day 4 of culture, corresponding to the maximum *trans*-resveratrol production, from cultures supplied with 50 μg/mL chitosan and from acetic acid and untreated controls. Northern blot analyses showed an increase of the expression of phenylalanine ammonia lyase (PAL, the first enzyme of the pathway), chalcone synthase (CHS, that produces chalcone, the first flavonoid) and chalcone-flavanone isomerase (CHI, which converts chalcone into naringenin switching the pathway from chalcones to flavones and anthocyanins), while the mRNA levels of stilbene synthase (STS, the resveratrol biosynthetic enzyme), were not particularly influenced (Fig. 5, blue arrows).

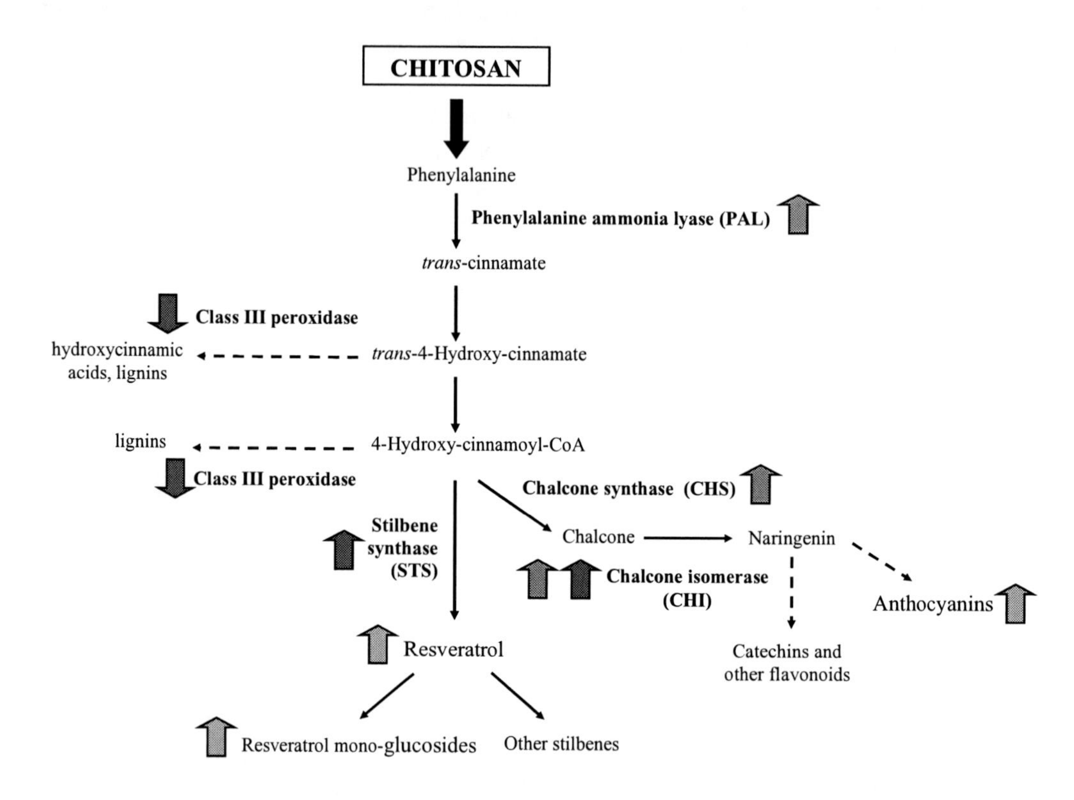

Figure 5. Effect of chitosan treatment (50 μg/mL) in Vitis vinifera cv. Barbera cell suspensions, on the expression of some of the genes and enzymatic proteins of phenylpropanoid pathway and on the accumulation of the related metabolites. Blue arrows, gene expression; red arrows, protein expression; grey arrows, metabolite accumulation. The figure was constructed from data obtained from Ferri et al. (2009).

Total protein extracts were analyzed on pH 3-10 non-linear 2-D gels (Fig. 6) and the intensity of 73 spots was consistently changed (at least 2-fold higher or lower) in the treated samples compared with the controls, or between the two controls. These spots were analyzed by different mass spectrometry techniques (MALDI-ToF, Q-Tof, MALDI-MS/MS and *de novo* sequencing) and 56 proteins were identified (Table 2). Nine proteins were involved in the biosynthetic pathway for stilbene and related compounds (Table 2 and 3), of which six were identified as *V. vinifera* stilbene synthase 1 (STS Vst1), two as *V. vinifera* chalcone-flavanone isomerase (CHI) and one was a class III peroxidase involved in the lignin biosynthetic pathway, a lateral branch of the phenylpropanoid pathway. All these proteins were up-regulated by chitosan but not influenced by acetic acid, with the exception of the class III peroxidase which was down-regulated (Table 3, Fig. 5 red arrows). In particular, stilbene synthase proteins were highly induced by treatment with chitosan with a 3.2 to 10.1-fold increase (Fig. 6D-F, spots 43-48). The 6 STS proteins, all encoded by the gene *STS Vst1*, appeared as a train of six distinct spots in the 2-D gels (Fig. 6D-F, spots 43-48), due to different degrees of phosphorylation of the same protein. In fact, the addition of one or more phosphate groups changes the protein pI but not the molecular weight, shifting the spot pI toward acidic values (Ferri et al., 2009). Furthermore, as northern blot analyses did not evidence a significant increase of STS transcript in response to chitosan treatment, it was confirmed that the STS up-regulation was only at post-translational level. This chitosan-dependent regulation was probably due to a higher protein stability or to a reduction of protein degradation. These hypotheses were partially supported by STS phosphorylation degree, and by proteomic results that showed a decrease in some proteolytic enzymes, such as a mitochondrial processing peptidase β subunit and two 20S proteasome β subunit (Ferri et al., 2009). Furthermore, the proteomic analyses suggested that one mechanism to increase resveratrol production could have been a *de-novo* biosynthetic activity due to stilbene pathway enzyme proteins whose accumulation was specifically promoted by the elicitor (Fig. 6, Table 3), rather than to a remobilization of stored or previously present resveratrol mono-glucosides.

With regard to specific flavonoid biosynthesis, proteomic analyses (Fig. 6, spots 22 and 23) showed an up-regulation of CHI protein after chitosan treatment (2.4 to 3.3 fold-increase), while CHS did not significantly change (Fig. 5, red arrows). HPLC analyses detected only trace amounts of naringenin and quercetin, leading to the hypothesis of a rapid utilization of these compounds as precursors for the synthesis of anthocyanins, as demonstrated by spectrophotometric determinations that showed about a 30% (Fig. 4, day 7) higher anthocyanin level in chitosan-supplemented cells in comparison to acetic acid control.

Interestingly, fourteen of the identified proteins seem to be involved in other defense mechanisms (pathogenesis-related or oxidative stress responses) (Table 2) of which eleven were *Vitis* pathogenesis-related protein 10 (PR-10) (Fig. 6, spots 29-39). PR-10s are a large group of proteins, encoded by a multigene family, constitutively expressed in several plant species, whose expression could be further induced by several stresses, suggesting a general defense role. In general, the expression of these proteins was up-regulated by chitosan and not by acetic acid only, both at proteomic and transcriptional levels (Ferri et al., 2009).

The other 29 identified proteins belonged to primary metabolisms and generally decreased after chitosan treatment, in particular proteins involved in respiratory chain and energetic metabolisms (Table 2). The induction of defense mechanisms leads inevitably to the consumption of considerable amounts of available cellular resources, including substrates and energy. It was suggested that reduction or even repression of some cellular functions occurs to ensure a metabolic balance during response to the stress caused by pathogen attack (Lo and Nicholson, 1998), in this case mimicked by chitosan.

In other plant species, such as wheat seeds (Reddy et al., 1999) and *Catharanthus roseus* cell suspensions (Kauss et al., 1989), the treatment with chitosan induced a set of defense responses among which there were the synthesis of lignin precursors, the formation of callose and cell wall lignification to create a physical barrier to prevent pathogen penetration and spread. Conversely, in Barbera cells HPLC analyses did not indicate any increase of lignin precursors such as *trans*-cinnamic, *p*-coumaric, caffeic, ferulic, sinapic and acids. In addition, proteomic analysis showed a down-regulation, in chitosan treated cells, of class III peroxidase (that catalyses the last step of the lignin synthesis) (Fig. 5 red arrows) and of catalase proteins (Ferri et al., 2009).

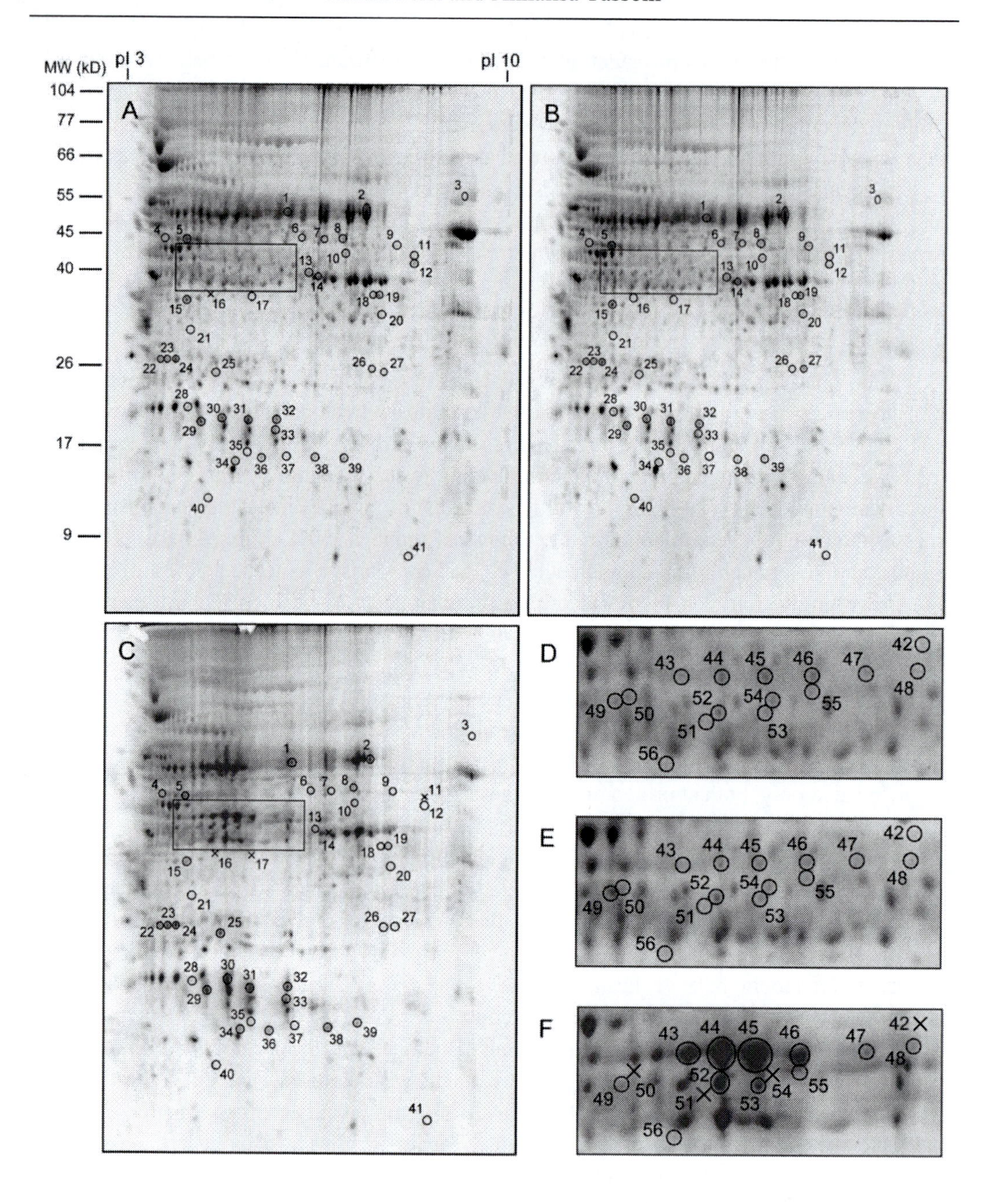

Figure 6. Proteomic analysis of chitosan-treated Vitis vinifera cv. Barbera cell suspensions. Total proteins were extracted from cells at day 4 of culture and separated on 2-D gels (first dimension pI 3-10 non linear, second dimension 10% Duracryl gels). (A) Untreated control. (B) Acetic acid (0.01% v/v) control cells. (C) Chitosan-treated (50 μg/mL) cells. (D, E, F) Enlargements of boxed areas in A, B and C from top to bottom respectively. For a detailed description of the spots see Ferri et al. (2009). Non-identified proteins are not shown.

The studies on grape cell suspensions cv. Barbera therefore confirmed the ability of chitosan to induce or improve the capacity of plants to resist to pathogen attack that seems to be directly related with its effectiveness in eliciting a complex set of plant defense responses such as phytoalexin accumulation and the expression of defense-related genes and proteins (Rabea et al., 2003; Ferri et al., 2009).

Table 2. Summary of proteins identified by mass spectrometry (MALDI-ToF, Q-T of, MALDI-MS/MS and *de novo* sequencing) after 2-D gel separation of total proteins extracted from untreated control, 0.01% (v/v) acetic acid control and 50 µg/mL chitosan-treated *V. vinifera* (cv. Barbera) cell suspensions harvested at day 4 of culture.Modified from Ferri et al. (2009)

Metabolism	Number of identified proteins	Acetic acid effect	Chitosan effect
Stilbene and related compounds biosynthetic pathway	9	≈	+
Other defense mechanisms	14	≈	+
DNA transcription, mRNA and protein processing	13	+/-	≈/-
Energetic metabolisms	11	+/≈/-	+/≈/-
Aminoacid metabolisms	3	+/≈/-	-
Other proteins	2	-	-
Unknown function proteins	4	+/≈/-	+/≈/-
Unidentified proteins	17	+/≈/-	+/≈/-

+, improved; ≈, not significantly changed; -, decreased

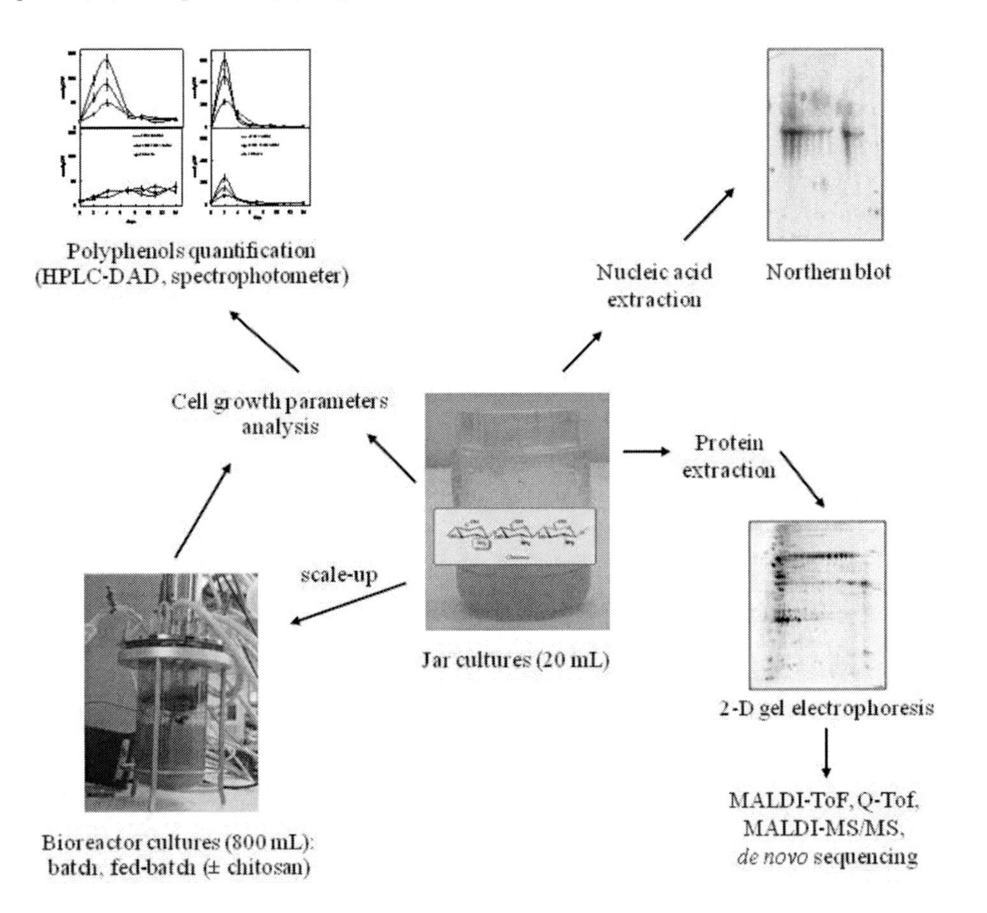

Figure 7. Scheme of the different experimental procedures adopted on Vitis vinifera cv. Barbera cell suspensions with the aim of improving polyphenols yields.

Table 3. Proteins of the stilbene and related compounds biosynthetic pathway identified by mass spectrometry after 2-D gel separation of untreated control, 0.01% (v/v) acetic acid control and 50 µg/mL chitosan-treated *V. vinifera* (cv. Barbera) cell suspensions harvested at day 4 of culture. Modified from Ferri et al. (2009). FC, fold change of protein amount

Protein	FC acetic acid/untreated	FC chitosan/ untreated	FC chitosan/ acetic acid
Stilbene synthase 1 (EC 2.3.1.95)	-	3.7 ± 0.1	3.8 ± 0.1
Stilbene synthase 1 (EC 2.3.1.95)	-	4.9 ± 0.1	4.6 ± 0.1
Stilbene synthase 1 (EC 2.3.1.95)	-	6.0 ± 0.1	10.1 ± 0.1
Stilbene synthase 1 (EC 2.3.1.95)	-	5.5 ± 0.2	8.3 ± 0.2
Stilbene synthase 1 (EC 2.3.1.95)	-	4.0 ± 0.1	3.2 ± 0.1
Stilbene synthase 1 (EC 2.3.1.95)	-	6.2 ± 0.1	4.6 ± 0.1
Chalcone-flavonone isomerase (EC 5.5.1.6)	-	-	2.4 ± 0.1
Chalcone-flavonone isomerase (EC 5.5.1.6)	-2.6 ± 0.1	-	3.3 ± 0.1
Class III peroxidase (EC 1.11.1.7)	-	-3.7 ± 0.1	-2.9 ± 0.1

Bioreactor Scale-Up

Cell cultures can be scaled-up to large volumes. The major problems hindering the development of large-scale culture of plant cells include low productivity, cell line instability and difficulty in the process scale-up. Media, elicitors, culture conditions, bioreactor design and other critical parameters influence the behavior of plant cell culture, allowing to improve the production of one or few compounds, which would not normally accumulate in the intact plant. At present only few plant cell culture processes are commercially conducted for the synthesis of bioactive compounds. An example is the production of the anticancer drug Taxol® (registered trademark of Bristol-Myers Squibb, the generic name is paclitaxel). Recently this process was switched from traditional plant cell cultures to the latest Plant Cell Fermentation (PCF) technology, which improved the sustainability of the supply of paclitaxel while reducing the amount of waste and pollutants. Through the application of the PCF technology, paclitaxel is being made using only plant cell cultures (www.bms.com).

Previously described grape cv. Barbera cell suspensions were also utilized to study chitosan elicitation of useful compounds in a scale-up bioreactor system with a productive point of view (Ferri et al., sent for publication). Firstly, the cultures were transferred from 20 mL jars (Tassoni et al., 2005; Ferri et al., 2009) to 1 L bioreactor scale (800 mL culture volume) (Fig. 7), without changing medium composition and inoculum size and age, and biomass yield and productivity were compared. The biomass growth rate of fermented cultures was faster than in the jars, with an average duplication time of 10.5 and 13.5 days respectively. The cell viability of fermented cultures reached 90% of living cells at day 14. The simple scale-up step, besides improving culture growth and viability, increased also secondary metabolite production. In fact, total resveratrol level (*trans*- plus *cis*-isomers, endogenous plus released amounts) was improved up to 13.9-times at day 13. These data support a higher metabolic activity of the cells when cultured in bioreactor with respected to the jars (Ferri et al, sent for publication). Previous studies on other grape cultivars (i.e. Gamay Fréaux) reported that polyphenols biosynthesis in bioreactor cultures were lower or approximately the same as in shake flasks (Decendit et al., 1996), on the contrary to what obtained in Barbera cell cultures. Taking into consideration the obtained results, a further optimization of the Barbera cell fermentative parameters was performed to improve the production of stilbenes and catechins.

Table 4. Levels (µmol/gDW) of stilbenes and catechins in the tested culture conditions. The data are reported at the day of maximum production

	Jars (20 mL)		Bioreactor (800 mL)		
	untreated	chitosan	batch	fed-batch	fed-batch + chitosan
Total stilbenes	1.57	1.62	14.02	11.23	32.73
Total catechins	0.73	0.79	5.12	4.33	1.78

Several culture conditions were tested and trial fermentations were carried out to set up the best operational conditions such as impeller type, configuration, rotational speed, air flow rate, time and amount of fresh medium feeding, length of culture. Batch and fed batch fermentation processes were carried out to assess the effects of operative parameters, such as sucrose concentration and cell inoculum size, on the biomass yield and stilbene and catechin productions. Good results were obtained in the process with 50% (v/v) of 40 g/L cell inoculum fresh weight and medium added with 30 g/L sucrose and 50 mg/L of rifampicin, an antibiotic that not only reduces the risk of microbial contamination, but also elicitates the synthesis of polyphenols (Ferri et al., 2007). In that batch fermentation, catechin and stilbene yields were improved with respect to the jar scale both with and without chitosan (Table 4). The fresh medium feeding at day 14 (fed-batch process) led to a strong reduction of the average duplication times (8.7 days), but also to a slight decrease in catechin and stilbene productions. When the fed fresh medium was added with 50 mg/L chitosan, stilbene yield rose, on the contrary to catechin one (Table 4, fed-batch+chitosan), confirming that this elicitor specifically induces the synthesis only of some polyphenols families. This selectivity could be advantageous for the production of specific polyphenols, such as stilbenes. At day 28 (maximum production), the total stilbene amount accumulated in chitosan-treated fed batch cell cultures was equivalent to 32.73 µmol/gDW (about 48 mg/L) (Table 4) of which 12.19 µmol/gDW were represent by free resveratrol and 20.54 µmol/gDW by mono-glucosides. This total stilbene level was 11-fold higher than in the untreated fed batch culture at the same day (data not shown) and 3-fold higher than in the untreated fed batch culture at the day of maximum (11.23 µmol/gDW, day 15) (Table 4). With respect to the jar cultures treated with chitosan at the same concentration, the final improvement of stilbene production was of about 20-fold (Table 4). In general in bioreactor cultures, over 74% of the produced stilbenes was represented by resveratrol mono-glucosides (73% resveratroloside, 1% piceid), which are physiologically as active as free resveratrol but more stable and bioavailable. Due to these characteristics these stilbenes could be much more functional and valuable than free resveratrol for nutraceutical and medical applications. Moreover, in the described culture conditions, over 85% of the stilbenes produced were released in the culture medium. This result seems therefore to be advantageous for a quick and less expensive procedure of extraction and purification of the compounds from the medium, if compared to the cells. Chitosan-treated fed batch Barbera cultures could be suitable for a possible future industrial production of stilbenes for medical, nutraceutical or pharmaceutical purposes. The specificity of the elicitor, the improved yield, the large release and the prevalence of mono-glycosylated forms, are clear promising advantages. Additional improvement of the optimized method could be reached by utilizing different kinds of bioreactors, i.e. air-lift system that seem to be particularly suitable for plant cells from other species. The possibility of setting up a continuous production by utilizing a process with biomass recycle in fresh media, and contemporary using the exhausted media for extraction, could also be explored.

Conclusion

In conclusion chitosan has proven to be a suitable elicitor for plant cell culture in order to increase the production of several types of secondary metabolites. Its use could be extended to scale-up biotechnological processes which will be, in the near future, the preferred method for industrial production of numerous useful compounds including pharmaceuticals, nutraceuticals and food additives. The use of chitosan as elicitor in bioreactor processes will also allow to conduct industrial processes in a more sustainable way both from an energetic and ecological point of view, with the use of lower initial biomass amount and the consequent preservation of the plant biodiversity and of many natural resources.

Acknowledgments

The work on grape cell suspensions was financed by the project "Changes in the grape berry polyphenolic profiles: molecular and ecophysiological aspects during ripening and post harvest", PRIN 2005 funds from Ministry of University and Scientific Research of Italy to N. Bagni and by the project "Plant polyphenols and neurodegeneration: production, identification and functional analysis in cell models of optical mitochondrial neuropathy" financed by the CARISBO Foundation, (Bologna, Italy) to A. Tassoni.

References

Ait Barka, E.; Eullaffroy, P.; Clément, C.; Vernet, G. Chitosan Improves Development, And Protects *Vitis Vinifera* L. Against *Botrytis Cinerea*. *Plant Cell Rep.,* 2004, 22, 608-614.

Amborabè, B. E.; Bonmort, J.; Fleurat-Lessard, P.; Roblin, G. Early Events Induced By Chitosan On Plant Cells. *J. Exp. Bot.,* 2008, 59, 2317-2324.

Aziz, A.; Poinssot, B.; Daire, X.; Adrian, M.; Bézier, A.; Lambert, B.; Joubert, J. M.; Pugin, A. Laminarin Elicits Defense Responses In Grapevine And Induces Protection Against *Botrytis Cinerea* And *Plasmopara Viticola*. *Mol. Plant-Microb. Interact.,* 2003, 16, 1118-1128.

Aziz, A.; Trotel-Aziz, P.; Dhuicq, L.; Jeandet, P.; Couderchet, M.; Vernet, G. Chitosan Oligomers And Copper Sulfate Induce Grapevine Defense Reactions And Resistance To Gray Mold And Downy Mildew. *Phytopathology,* 2006, 96, 1188-1194.

Benhamou, N.; Lafontaine, P. J.; Nicole, M. Induction Of Systemic Resistance To *Fusarium* Crown And Root Rot In Tomato Plants By Seed Treatment. *Phytopathology,* 1994, 84, 1432-1444.

Bézier, A.; Lambert, B.; Baillieul, F. Study Of Defense-Related Gene Expression In Grapevine Leaves And Berries Infected With *Botrytis Cinerea. Eur. J. Plant Pathol.,* 2002, 108, 111-120.

Chakraborty, M.; Karun, A.; Mitra, A. Accumulation Of Phenylpropanoid Derivatives In Chitosan-Induced Cell Suspension Culture Of *Cocos Nucifera. J. Plant Physiol.,* 2009, 166, 63-71.

Chen, H. P.; Xu, L. L. Isolation And Characterization Of A Novel Chitosan-Binding Protein From Non-Heading Chinese Cabbage Leaves. *J. Integrat. Plant Biol.,* 2005, 47, 452-456.

Cheng, X. Y.; Zhou, H. Y.; Cui, X.; Ni, W.; Liu, C. Z. Improvement Of Phenylethanoid Glycosides Biosynthesis In *Cistanche Deserticola* Cell Suspension Cultures By Chitosan Elicitor. *J. Biotechnol.,* 2006, 121, 253-260.

Conrath, U.; Domard, A.; Kauss, H. Chitosan-Elicited Syntheis Of Callose And Of Coumarin Derivatives In Parsley Cell Suspension Cultures. *Plant Cell Rep.,* 1989, 8, 152-155.

Das, D. K.; Sato, M.; Ray, P. S.; Maulik, G.; Engelman, R. M.; Bertelli, A. A.; Bertelli, A. Cardioprotection Of Red Wine: Role Of Polyphenolic Antioxidants. *Drugs Exp. Clin. Res.,* 1999, 25, 115-120.

Decendit, A.; Ramawat, K. G.; Waffo-Teguo, P.; Deffieux, G.; Badoc, A.; Mérillion, J. M. Anthocyanins, Catechins, Condensed Tannins And Piceid Production In *Vitis Vinifera* Cell Bioreactor Cultures. *Biotech. Lett.,* 1996, 18, 659-662.

Decendit, A.; Waffo-Teguo, P.; Richard, T.; Krisa, S.; Vercauteren, J.; Monti, J. P.; Deffieux, G.; Merillon, J. M. Galloylated Catechins And Stilbene Diglucosides In *Vitis Vinifera* Cell Suspension Cultures. *Phytochemistry,* 2002, 60, 795-798.

Dörnenburg, H.; Knorr, D. Effectiveness Of Plant-Derived And Microbial Polysaccharides As Elicitors For Anthraquinone Synthesis In *Morinda Citrifolia* Cultures. *J. Agric. Food Chem.,* 1994, 42, 1048-1052.

Dos Santos, A. L. W.; El Gueddari, N. E.; Trobotto, S.; Moerschbacher, B. M. Partially Acetylated Chitosan Oligo- And Polymers Induce An Oxidative Burst In Suspension Cultured Cells Of The Gymnosperm *Araucaria Angustifolia. Biomacromolecules,* 2008, 9, 3411-3415.

Espín, J. C.; García-Conesa, M. T.; Tomás-Barberán, F. A. Nutraceuticals: Facts And Fiction. *Phytochemistry,* 2007, 68, 2986-3008.

Faoro, F.; Iriti, M. Callose Synthesis As A Tool To Screen Chitosan Efficacy In Inducing Plant Resistance To Pathogens. *Caryologia,* 2007, 60, 121-124.

Ferri, M.; D'Errico, A. L.; Vacchi Suzzi, C.; Franceschetti, M.; Righetti, L.; Bagni, N.; Tassoni, A. Use Of Rifampicin For Resveratrol Production In *Vitis Vinifera* Cell Suspensions. In: Jeandet P., Clément C., Conreux A. (Eds.), *Macromolecules And Secondary Metabolites Of Grapevine And Wine*, Paris; Lavoisier; 2007; Pp 95-98.

Ferri, M.; Tassoni, A.; Franceschetti, M.; Righetti, L.; Naldrett, M. J.; Bagni, N. Chitosan Treatment Induces Changes Of Protein Expression Profile And Stilbene Distribution In *Vitis Vinifera* Cell Suspensions. *Proteomics,* 2009, 9, 610-624.

Flocco, C. G.; Giulietti, M. Effect Of Chitosan On Peroxidase Activity And Isoenzyme Profile In Hairy Root Cultures Of *Armoracia Lapathifolia. Appl. Biochem. Biotechnol.,* 2003, 110, 175-183.

Hadwiger, L. A. Host-Parasite Interactions: Elicitation Of Defense Responses In Plants With Chitosan. In: Jolles P., Muzzarelli R. A. A. (Eds.), *Chitin And Chitosanes*, Basel; Birkhauser Verlag; 1999; Pp 185-200.

Hadwiger, L. A.; Beckman, J. M. Chitosan As A Component Of Pea-*Fusarium Solani* Interactions. *Plant Physiol.,* 1980, 66, 205-211.

Hadwiger, L. A.; Ogawa, T.; Kuyama, H. Chitosan Polymer Sizes Effective In Inducing Phytoalexin Accumulation And Fungal Suppression Are Verified With Synthesized Oligomers. *Mol. Plant Microbe Interact.,* 1994, 7, 531-533.

Hamel, L. P.; Miles, G. P.; Samuel, M. A.; Ellis, B. E.; Seguin, A.; Beaudoin, N. Activation Of Stress-Responsive Mitogen-Activated Protein Kinase Pathways In Hybrid Poplar (*Populus Trichocarpa X Populus Deltoides*). *Tree Physiol.,* 2005, 25, 277-288.

Iriti, M.; Faoro, F. Chitosan As A MAMP, Searching For A PRR. *Plant Signaling & Behavior,* 2009, 4, 66-68.

Iriti, M.; Sironi, M.; Gomarasca, S.; Casazza, A. P.; Soave, C.; Faoro, F. Cell Death-Mediated Antiviral Effect Of Chitosan In Tobacco. *Plant Physiol. Biochem.,* 2006, 44, 893-900.

Jeandet, P.; Douillet-Breuil, A. C.; Bessis, R.; Debord, S.; Sbaghi, M.; Adrian, M. Phytoalexins From The Vitaceae: Biosynthesis, Phytoalexin Gene Expression In Transgenic Plants, Antifungal Activity, And Metabolism. *J. Agric. Food Chem.,* 2002, 50, 2731-2741.

Kauss, H.; Jeblick, W.; Domard, A. The Degree Of Polymerization And N-Acetylation Of Chitosan Determine Its Ability To Elicit Callose Formation In Suspension Cells And Protoplasts Of *Catharanthus Roseus*. *Planta,* 1989, 178, 385-392.

Kim, H. J.; Chen, F.; Wang, X.; Rajapakse, N. C. Effect Of Chitosan On The Biological Properties Of Sweet Basil (*Ocinum Basilicum* L.). *J. Agric. Food Chem.,* 2005, 53, 3696-3701.

Klosterman, S. J.; Choi, J. J.; Hadwiger, L. A. Analysis Of Pea HMG-I/Y Expression Suggests A Role In Defence Gene Regulation. *Mol. Plant Pathol.,* 2003, 4, 249-258.

Klüsener, B.; Young, J. J.; Murata, Y.; Allen, G. J.; Mori, I. C.; Hugouvieux, V.; Schroeder, J. I. Convergence Of Calcium Signaling Pathways Of Pathogenic Elicitors And Abscisic Acid In Arabidopsis Guard Cells. *Plant Physiol.,* 2002, 130, 2152-2163.

Komaraiah, P.; Naga Amrutha, R. N.; Kavi Kishor, P. B.; Ramakrishna, S. V. Elicitor Enhanced Production Of Plumbagin In Suspension Cultures Of *Plumbago Rosea* L. *Enz. Microb. Technol.,* 2002, 31, 634-639.

Komaraiah, P.; Ramakrishna, S. V.; Reddanna, P.; Kavi Kishor, P. B. Enhanced Production Of Plumbagin In Immobilized Cells Of *Plumbago Rosea* By Elicitation And *In Situ* Adsorption. *J. Biotechnol.,* 2003, 101, 181-187.

Kopp, P. Resveratrol, A Phytoestrogen Found In Red Wine. A Possible Explanation For The Conundrum Of The 'French Paradox'? *Eur. J. Endocrinol.,* 1998, 138, 619-620.

Krisa, S.; Larronde, F.; Budzinski, H.; Decendit, A.; Deffieux, G.; Merillon, J. Stilbene Production By *Vitis Vinifera* Cell Suspension Cultures: Methyljasmonate Induction And ^{13}C Biolabeling. *J. Nat. Prod.,* 1999a, 62, 1688-1690.

Krisa, S.; Waffo-Teguo, P.; Decendit, A.; Deffieux, G.; Vercauteren, J.; Mérillion, J. M. Production Of ^{13}C-Labelled Anthocyanins By *Vitis Vinifera* Cell Suspension Cultures. *Phytochemistry,* 1999b, 51, 651-656.

Larronde, F.; Krisa, S.; Decendit, A.; Chèze, C.; Deffieux, G.; Mérillion, J. M. Regulation Of Polyphenol Production In *Vitis Vinifera* Cell Suspension Cultures By Sugars. *Plant Cell Rep.,* 1998, 17, 946-950.

Larronde, F.; Richard, T.; Delaunay, J. C.; Decendit, A.; Monti, J. P.; Krisa, S.; Merillon, J. M. New Stilbenoid Glucosides Isolated From *Vitis Vinifera* Cell Suspension Cultures (Cv. Cabernet Sauvignon). *Planta Med.,* 2005, 71, 888-890.

Lehtonen, M. T.; Akita, M.; Kalkkinen, N.; Hola-Iivarinen, E.; Ronnholm, G.; Somervuo, P.; Thelander, M.; Valkonen, J. P. Quickly-Released Peroxidase Of Moss In Defense Against Fungal Invaders. *New Phytol.,* 2009, 183, 432-443.

Lin, J. K.; Tsai, S. H. Chemoprevention Of Cancer And Cardiovascular Disease By Resveratrol. *Proc. Natl. Sci. Counc. Repub. China,* 1999, 23, 99-106.

Linden, J. C.; Phisalaphong, M. Oligosaccharides Potentiate Methyl Jasmonate-Induced Production Of Paclitaxel In *Taxus Canadensis*. *Plant Sci.,* 2000, 158, 41-51.

Lo, S. C.; Nicholson, R. L. Reduction Of Light-Induced Anthocyanin Accumulation In Inoculated Sorghum Mesocotyls. Implications For A Compensatory Role In The Defense Response. *Plant Physiol.,* 1998, 116, 979-989.

Muzzarelli, C.; Muzzarelli, R. A. A. Chitin Related Food Science Today (And Two Centuries Ago). *Agrofood Industry Hi-Tech,* 2003, Sept/Oct, 39-42.

Ndimba, B. K.; Chivasa, S.; Hamilton, J. M.; Simon, W. J.; Slabas, A. R. Proteomic Analysis Of Changes In The Extracellular Matrix Of *Arabidopsis* Cell Suspension Cultures Induced By Fungal Elicitors. *Proteomics,* 2003, 3, 1047-1059.

Ortmann, I.; Moerschbacher, B. M. Spent Growth Medium Of *Pantoea Agglomerans* Primes Wheat Suspension Cells For Augmented Accumulation Of Hydrogen Peroxide And Enhanced Peroxidase Activity Upon Elicitation. *Planta.,* 2006, 224, 963-970.

Pasquali, G.; Goddijn, O. J.; De, W. A.; Verpoorte, R.; Schilperoort, R. A.; Hoge, J. H.; Memelink, J. Coordinated Regulation Of Two Indole Alkaloid Biosynthetic Genes From *Catharanthus Roseus* By Auxin And Elicitors. *Plant Mol. Biol.,* 1992, 18, 1121-1131.

Pitta-Alvarez, S. I.; Giulietti, A. M. Influence Of Chitosan, Acetic Acid And Citric Acid On Growth And Tropane Alkaloid Production In Transformed Roots Of *Brugmansia Candida*. Effect Of Medium Ph And Growth Phase. *Plant Cell Tiss. Org. Cult.,* 1999, 59, 31-38.

Pochanavanich, P.; Suntornsuk, W. Fungal Chitosan Production And Its Chracetrization. *Lett. App. Microbiol.,* 2002, 35, 17-21.

Prasitsilp, M.; Jenwithisuk, R.; Kongsuwan, K.; Damrongchai, N.; Watts, P. Cellular Responses To Chitosan *In Vitro*: The Importance Of Deacetylation. *J. Mat. Sci. : Materials In Medicine,* 2000, 11, 773-778.

Putalun, W.; Luealon, W.; De-Eknamkul, W.; Tanaka, H.; Shoyama, Y. Improvement Of Artemisinin Production By Chitosan In Hairy Root Cultures Of *Artemisia Annua* L. *Biotechnol. Lett.,* 2007, 29, 1143-1146.

Rabea, E. I.; Badway, M. E. T.; Stevens, C.; Smagghe, G.; Steurbaunt, W. Chitosan As Antimicrobial Agent: Applications And Mode Of Action. *Biomacromolecules,* 2003, 4, 1457-1465.

Ravi Kumar, M. N. V.; Muzzarelli, R. A. A.; Muzzarelli, C.; Sashiwa, H.; Domb, A. J. Chitosan Chemistry And Pharmaceutical Perspectives. *Chemical Reviews,* 2004, 104, 6017-6084.

Reddy, M. V. B.; Arul, J.; Angers, P.; Couture, L. Chitosan Treatment Of Wheat Seeds Induces Resistance To *Fusarium Graminearum* And Improves Seed Quality. *J. Agric. Food Chem.,* 1999, 47, 1208-1216.

Regev-Shoshani, G.; Shoseyov, O.; Bilkis, I.; Kerem, Z. Glycosylation Of Resveratrol Protects It From Enzymic Oxidation. *Biochem. J.,* 2003, 374, 157-163.

Ren, Y. Y.; West, C. A. Elicitation Of Diterpene Biosynthesis In Rice (*Oryza Sativa* L.) By Chitin. *Plant Physiol.,* 1992, 99, 1169-1178.

Righetti, L.; Franceschetti, M.; Ferri, M.; Tassoni, A.; Bagni, N. Resveratrol Production In *Vitis Vinifera* Cell Suspensions Treated With Several Elicitors. *Caryologia,* 2007, 60, 169-171.

Sanchez-Sampedro, M. A.; Fernandez-Tarrago, J.; Corchete, P. Yeast Extract And Methyl Jasmonate-Induced Silymarin Production In Cell Cultures Of *Silybum Marianum* (L.) Gaertn. *J. Biotechnol.,* 2005, 119, 60-69.

Savard, T.; Beaulieu, C.; Boucher, I.; Champagne, C. P. Antimicrobial Action Of Hydrolyzed Chitosan Against Spoilage Yeasts And Lactic Acid Bacteria Of Fermented Vegetables. *J. Food Prot.,* 2002, 65, 828-833.

Skorupinska-Tudek, K.; Pytelewska, A.; Zelman-Femiak, M.; Mikoszewski, J.; Olszowska, O.; Gajdzis-Kuls, D.; Urbanska, N.; Syklowska-Baranek, K.; Hertel, J.; Chojnacki, T.; Swiezewska, E. *In Vitro* Plant Tissue Cultures Accumulate Polyisoprenoid Alcohols. *Acta Biochim. Pol.,* 2007, 54, 847-852.

Tassoni, A.; Fornalè, S.; Franceschetti, M.; Musiani, F.; Michael, A. J.; Perry, B.; Bagni, N. Jasmonates And Na-Orthovanadate Promote Resveratrol Production In *Vitis Vinifera* Cv. Barbera Cell Cultures. *New Phytol.,* 2005, 166, 895-905.

Tocci, N.; Ferrari, F.; Santamaria, A. R.; Valletta, A.; Rovardi, I.; Pasqua, G. Chitosan Enhances Xanthone Production In *Hypericum Perforatum Subsp. Angustifolium* Cell Cultures. *Nat. Prod. Res.,* 2010, 24, 286-293.

Trotel-Aziz, P.; Couderchet, M.; Vernet, G.; Aziz, A. Chitosan Stimulates Defence Reactions In Grapevine Leaves And Inhibits Development Of *Botrytis Cinerea*. *Eur. J. Plant Pathol.,* 2006, 114, 405-413.

Vasconsuelo, A.; Giulietti, A. M.; Boland, R. Signal Transduction Events Mediating Chitosan Stimulation Of Anthraquinone Synthesis In *Rubia Tinctorum*. *Plant Science,* 2004, 166, 405-413.

Waffo-Teguo, P.; Decendit, A.; Krisa, S.; Deffieux, G.; Vercauteren, J.; Mérillion, J. M. The Accumulation Of Stilbene Glycosides In *Vitis Vinifera* Cell Suspensions. *J. Nat. Prod.,* 1996, 59, 1189-1191.

Waffo-Teguo, P.; Fauconneau, B.; Deffieux, G.; Huguet, F.; Vercauteren, J.; Mérillon, J. M. Isolation, Identification, And Antioxidant Activity Of Three Stilbene Glucosides Newly Extracted From *Vitis Vinifera* Cell Cultures. *J. Nat. Prod.,* 1998, 61, 655-657.

Winkel-Shirley, B. Flavonoid Biosynthesis. A Colorful Model For Genetics, Biochemistry, Cell Biology, And Biotechnology. *Plant Physiol.,* 2001a, 126, 485-493.

Winkel-Shirley, B. It Takes A Garden. How Work On Diverse Plant Species Has Contributed To An Understanding Of Flavonoid Metabolism. *Plant Physiol.,* 2001b, 127, 1399-1404.

Yin, H.; Li, S.; Zhao, X.; Du, Y.; Ma, X. Cdna Microarray Analysis Of Gene Expression In *Brassica Napus* Treated With Oligochitosan Elicitor. *Plant Physiol Biochem.,* 2006, 44, 910-916.

Zenk, M. H. Chasing The Enzymes Of Secondary Metabolism: Plant Cell Cultures As A Pot Of Gold. *Phytochemistry,* 1991, 30, 3861-3863.

Zhang, C. H.; Wu, J. Y.; He, G. Y. Effects Of Inoculum Size And Age On Biomass Growth And Paclitaxel Production Of Elicitor-Treated *Taxus Yunnanensis* Cell Cultures. *Appl. Microbiol. Biotechnol.,* 2002, 60, 396-402.

In: Handbook of Chitosan Research and Applications
Editors: Richard G. Mackay and Jennifer M. Tait
ISBN 978-1-61324-455-5

Chapter 23

EFFECT OF CHITOSAN ON THE PRODUCTION OF SECONDARY METABOLITES IN EMBRYOGENIC CULTURES OF NUTMEG, *MYRISTICA FRAGRANS* HOUTT.

R. Indira Iyer*[*], *G. Jayaraman and A. Ramesh
Department of Genetics,
Dr. A.L. Mudaliar Post Graduate Institute of Basic Medical Sciences,
University of Madras (Taramani Campus),
Chennai, India

ABSTRACT

Nutmeg (*Myristica fragrans* Houtt.) is a tropical evergreen dioecious tree which yields the spices, nutmeg (endosperm) and mace and has a long history of medicinal usage. It has a rich diversity of phytochemicals of commercial and therapeutic value. However it is of restricted distribution. Though nutmeg tissues are extremely recalcitrant to tissue culture technology owing to the high content of phenolics, embryogenic cultures of nutmeg were established from zygotic embryos by using media with activated charcoal. The present study investigated the effect of chitosan, a biomolecular tool with tremendous potential in biotechnology, on the production of phytochemicals by the embryogenic cultures of nutmeg using GCMS analysis. The induction of the biosynthesis of several novel metabolites by chitosan including methyl salicylate, farnesol and anthraquinones as compared with unelicited cultures is being reported. The results indicate the enhancement of the biosynthetic potential of embryogenic cultures of nutmeg by chitosan and point to its application in strategies for large scale *in vitro* production of valuable secondary metabolites from nutmeg.

Keywords : *chitosan, nutmeg, methyl salicylate, anthraquinones*

[*] E-mail: riiyer@yahoo.co.in.

INTRODUCTION

Nutmeg (*Myristica fragrans* Houtt.), a tropical evergreen dioecious tree is the source of the valuable spices, nutmeg (endosperm) and mace (the reddish aril).These spices are renowned since ages for their flavor, fragrance and medicinal properties. Nutmeg with its diverse array of phytochemicals including several essential oils, lignans and malabaricones and others [1] is reported to possess a wide range of pharmacological activities - anti-cancer, anti-viral, anti-microbial, hepatoprotective, anti-inflammatory, anti-ulcer, anti-diabetic and hypolipidaemic [2]. The rich diversity of biomolecules in nutmeg reflects the immense biosynthetic potential of its tissues. The use of natural molecules derived from plant extracts for healthcare is on the rise especially due to the absence of side-effects. There is an escalation of interest in the medicinal importance of nutmeg as witnessed by several recent reports indicating its tumoricidal, antibacterial, cardioprotective, cytotoxic and antioxidant [3-6] activities. Therefore there is a need to explore further the chemosyntheic potential of nutmeg in order to develop bioactive molecules with therapeutic value and lead molecules for designing new drugs. However nutmeg is of limited distribution since it is a dioecious tree with single-seeded fruits and a long generation time. The seeds of nutmeg are recalcitrant and are not suitable for preservation in gene banks [7]. Plant tissue culture technology is of inherent importance for the conservation of relatively scarce medicinal plants threatened by heightened demand for herbal medicinal formulations. Cultured plant tissues and organs represent stable and readily renewable potential alternative sources of valuable secondary metabolites [8]. Productivity of useful metabolites in plant tissue cultures can be greatly enhanced and modified by elicitation [9] though the response to different elicitors varies with different plant species. Though nutmeg tissues were highly recalcitrant to tissue culture technology owing to their high content of phenolics, embryogenic cultures could be established in media with activated charcoal and the embryogenic cultures possessed phenylalanine ammonia-lyase activity [10].

Chitosan is a biomolecule of great potential [11] and has been reported to elicit the production of paclitaxel, anthracene derivatives, azadirachtin and plumbagin [12-15] in different plant species. There is a need for application of *in vitro* technology coupled with the use of elictors in *Myristica fragrans* not only to facilitate its conservation and sustainable utilization but also for manipulation and enhancement of its biosynthetic potential.

In the present investigation attempts have been made to study the effect of elicitors including chitosan to manipulate the biosynthetic capacity of the tissues of *Myristica fragrans* using *in vitro* technology and to achieve directed synthesis of metabolites. Our report reviews the induction of biosynthesis of several novel metabolites by chitosan including methyl salicylate and anthraquinones as compared with unelicited cultures and those treated with methyl jasmonate and salicylic acid.

MATERIALS AND METHODS

Establishment of Long Term Embryogenic Cultures of *M. fragrans* from Zygotic Embryos

Establishment of embryogenic cultures in *M. fragrans* from intact and fragmented zygotic embryos was described earlier [10, 16-17]. Zygotic embryos from freshly harvested, fully ripe split fruits (about 5 cm diameter) of *Myristica fragrans* were used as explants. The fruits were sourced from elite trees in State Horticultural Farm, Kallar, Tamil Nadu, washed thoroughly with liquid detergent and surface-sterilized using 0.1% mercuric chloride ($HgCl_2$) for 10 min, and rinsed several times with sterilized distilled water in a laminar flow hood. After detachment of the aril and removal of the shell of the seed the zygotic embryos surrounded by massive endosperm were excised with a sharp scalpel and a pair of forceps. The intact zygotic embryos and the pieces of zygotic embryos fragmented during excision were used as explants and cultured in MS [18] medium with different combinations of growth regulators, supplements and with or without activated charcoal (Table 1). The pH of the medium was adjusted to 5.6-5.8 prior to addition of 0.8% agar followed by sterilization in an autoclave at 121°C and 1.1 kg cm^{-2}. All cultures were incubated at 25 ± 2°C in continuous light with 35 μmol m^{-2} s^{-1} illumination provided by white, cool fluorescent tubes.

Analysis of the Embryogenic Mass by Headspace GCMS

The embryogenic mass from the cultured zygotic embryos were harvested at 1, 3 and 5 weeks after culture initiation. After recording the fresh weights they were directly analyzed by headspace GCMS (gas chromatography-mass spectrometry) using an Agilent 5973-6890 GCMS system with a DB-5 capillary column (30m × 0.25 mm ; film thickness 0.25 μm). The operating conditions were as follows - Helium carrier gas flow: 1mL/min; Temperature programming: 50ºC (1 min), ramp of 10°C per min to 250°C, hold for 10 min, ramp of 10°C per min to 280°C, hold for 2 min; Injector temperature : 250°C; E.I.: 70 eV . The Chemstation software with the Wiley GCMS library was used to analyze the eluted volatiles and identify the compounds.

Elicitor Treatment of Zygotic Embryos and Metabolite Profiling by Headspace GCMS

The elicitors used were chitosan (100 mg/L), salicylic acid (100 μM) or jasmonic acid (40 μM) and were added to the medium before sterilization. Chitosan solution was prepared by dissolving 1g of chitosan (Sigma-Aldrich Co.) in 2 mL glacial acetic acid added dropwise at 60°C for 15 min and the final volume was made up to100 mL. The pH was adjusted to 5.7 with NaOH before autoclaving and used as an elicitor as described by Komariah et al. [15]. Chitosan and salicylic acid were directly added to the medium before sterilization. Jasmonic acid (Sigma-Aldrich Co.) was dissolved in absolute ethanol and was filter-sterilized before addition to the autoclaved medium. Headspace GC-MS analysis of the embryogenic mass was carried out 4 weeks after culture of the zygotic embryos.

RESULTS AND DISCUSSION

Establishment of Long Term Embryogenic Cultures of *M. fragrans* from Zygotic Embryos

Direct formation of green somatic embryos (12-14 per explant) without any intervening callus stage was obtained from intact as well as broken zygotic embryos in media with various hormonal combinations and with activated charcoal in 3-5 weeks. All combinations of plant growth regulators tried could elicit embryogenic response provided activated charcoal was added to the medium. The best responses were obtained with combinations of naphthaleneacetic acid (NAA) and benzyl adenine (BA). For all further experiments, MS media with NAA at 5 mg/L and BA at 1 mg/L was used along with 0.3% activated charcoal. High frequency somatic embryogenesis was obtained with fragmented zygotic embryos (Figure1). There was no response in explants cultured in the dark indicating that light was absolutely essential for induction of somatic embryogenesis in *M. fragrans*. The embryogenic competence could be maintained by repeated subculture for about ten months and more and establishment of long term embryogenic cultures, which could serve as stable sources of the various secondary metabolites of nutmeg was achieved.

Figure 1. Embryogenic cultures of nutmeg , *Myristica fragrans* Houtt.(bar= 1 mm).

Table1. Abundance of compounds detected by headspace GCMS in unelicited 3 week old and 5 week old cultures of nutmeg

	Compounds detected	Abundance(%) in 3 wk old cultures	Abundance (%) 5 wk old cultures
1	α-pinene	23	16.615
2	sabinene	5	4.46
3	*β*–pinene	15	10.84
4	*β*- myrcene	2	-
5	limonene	1	-
6	*cis β*–ocimene	7	-
7	γ- terpinene	3	-
8	p-methoxyamphetamine	-	0.8
9	terpinolene	3.6	-
10	terpinen-4-ol	15	-
11	myrcenol	8.4	-
12	myristicin	-	8.95

Metabolite Profiling by Headspace GCMS of Unelicited Cultures

Headspace GC-MS analysis (Table 1) of unelicited 3 week old and 5 week old embryogenic cultures revealed the presence of several known phenylpropenes and terpenes in the cultured tissues indicating the biosynthetic potential of the cultured tissues. The diversity of monoterpenes was higher in 3 week old cultures. In 5 week old embryogenic cultures (Table 1) myristicin was also detected in addition to α – pinene, sabinene and β–pinene which were present at a lower concentration than in 3 week old cultures. The presence of p-methoxyamphetamine, which has not been reported *in vivo* in nutmeg, has been detected in the embryogenic cultures.

Metabolite Profiling by Headspace GCMS after Elicitor Treatment

Headspace GC-MS profile after treatment with various elicitors indicated that chitosan was the most efficient among the elicitors tested in nutmeg embryogenic cultures (Table 2).

In contrast treatment with jasmonic acid and salicylic acid (Table 3) did not result in the biosynthesis of even the constitutive terpenes or phenylpropenes and other important medicinal compounds at all.

Elicitation with chitosan not only resulted in the production of constitutive secondary metabolites of nutmeg i.e., α–pinene, sabinene and β–pinene but also elicited the production of farnesol and the phenolic compounds - methyl salicylate, anthraquinone compounds and compounds similar to anisaldehyde (Table 2). The detection of high levels of anthraquinones on elicitation with chitosan can be correlated with the development of red pigmentation in the cultured tissues (Figure 2). The stimulation of anthraquinone production by chitosan has been also observed in madder, *Rubia akane,* cell cultures [19]. The effect of chitosan on elicitation of secondary metabolite production in nutmeg can be correlated with its reported ability to induce various enzymes of phenylpropanoid metabolism [20-21] and accumulation of phenylpropanoid derivatives [22] in other species. Induction of *de novo* production of metabolites by chitosan has also been reported in *Hyoscyamus muticus* [23].

Table 2. Compounds detected by headspace GC-MS of the embryogenic mass after treatment with chitosan (100 mg L^{-1})

Sl no.	Retention time (min)	Compounds Identified	Abundance (%)
1	5.41	α-pinene	3.15
2	6.03	Sabinene	3.52
3	6.15	β-pinene	3.53
4	9.62	methyl salicylate	6.89
5	14.81	benzenedicarboxylic acid ,diethyl ester	8.04
6	16.57	Guanosine	0.91
7	18.65	palmitic acid	8.38
8	20.36	oleic acid	6.01
9	20.55	stearic acid	2.83
10	27.35	2-amino-N-(4-nitrophenyl)-propanamide(L Ala p-nitroanilide)	8.37
11	29.02	farnesol	2.42
12.	30.32	eicosane	14.86
13	31.53	1-amino-2- phenoxy-4-hydroxy anthraquinone	15.97
14	31.71	1-amino-2- phenoxy-4-hydroxy anthraquinone	11.37
15	31.89	(2,4-dinitrophenyl)-[(4-methoxybenzylidene)amino]amine	3.77

Table 3. Compounds detected by headspace GC-MS of the embryogenic mass after treatment with salicylic acid (100 μM)

Sl no.	Retention time (min)	Compounds Identified	Abundance (%)
1	14.53	lauric acid	22.36
2	16.79	myristic acid	27.94
3	18.67	2-palmitoleic acid	7.67
4	18.85	palmitic acid	24.70
5	21.35	biphenol A	7.86
6.	27.43	Squalene	9.47

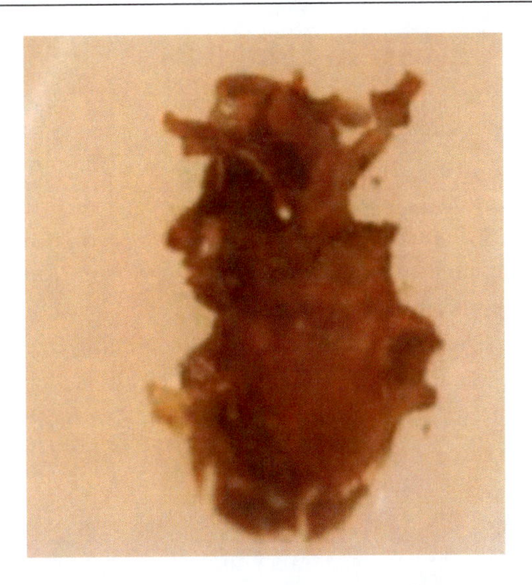

Figure 2. Development of red pigmentation after treatment with chitosan.

The results are significant since the metabolites produced by nutmeg cultures in response to elicitation with chitosan including methyl salicylate and others have considerable medicinal potential. Farnesol is reported to have anti-cancer and anti-bacterial effects **[24-25].** Anthraquinones have a wide range of biological activities attributed to them - anti-HIV, anti-fungal and anti-cancer [26-28].

However the growth response in chitosan elicited cultures was slightly diminished. Decrease in growth during elicitation with chitosan has been observed in *Panax ginseng* hairy root cultures [8].

Conclusion

The results reveal that chitosan is a powerful elicitor of medicinal compounds in nutmeg embryogenic cultures and elicited the production of several new metabolites not present in the unelicited cultures of nutmeg or those treated with salicylic acid or jasmonic acid. Future strategies employing varying concentrations of chitosan in conjunction with use of suitable precursors and other elicitors and manipulation of hormonal levels in the culture medium can lead to optimization of production of useful secondary metabolites in nutmeg embryogenic cultures. Elicitation by chitosan has therefore opened up new possibilities for exploiting the biosynthetic potential of nutmeg embryogenic cultures and can help in conservation and sustainable utilization of this valuable bioresource.

Acknowledgment

R.Indira Iyer thanks Department of Science and Technology (DST), Govt. of India for financial support in the form of a project (SR/WOS-A/LS-400/2003) under the DST Women Scientists' Scheme.

References

[1] Olaleye MT, Akinmoladun AC, Akindahunsi AA. Antioxidant properties of *Myristica fragrans* (Houtt.) and its effect on selected organs of albino rats. *Afr. J. Biotechnol.* . 2006 3 Jul;5(13):1274-8.

[2] Somani R, Karve S, Jain D, Jain K, Singhai AK (2008). Phytochemical and pharmacological potential of *Myristica fragrans* . *Phcog. Rev.* [serial online] 2008;2(4 Suppl.):68-77.Available from: URL:http:// www.phcogrev.com

[3] Mazzio EA, Karam FA, Soliman KFA. *In vitro* screening for the tumoricidal properties of international medicinal herbs. *Phytother. Res. 2009;23:385–98.*

[4] Yanti, Rukayadi Y, Kim K-H, Hwang J-K. *In vitro* anti-biofilm activity of macelignan isolated from Myristica fragrans Houtt. against oral primary colonizer bacteria. Phytother. Res. 2008;22:308-12.

[5] Kareem MA, Krushna GS, Hussain SA, Devi KL. Effect of aqueous extract of nutmeg on hyperglycaemia, hyperlipidaemia and cardiac histology associated with isoproterenol-induced myocardial infarction in rats. *Trop. J. Pharm. Res*. 2009; 8:337-44.

[6] Duan L, Tao, HW, Hao X, Gu Q-Q, Zhu W-M. Cytotoxic and antioxidative phenolic compounds from the traditional Chinese medicinal plant, *Myristica fragrans*. *Planta Med*. 2009; 75:1241-5.

[7] Gepts P. Plant genetic resources conservation and utilization.The accomplishments and future of a societal insurance policy. *Crop. Sci.* 2006; 46:2278-92.

[8] Jeong GA, Park DH. Enhanced secondary metabolite biosynthesis by elicitation in transformed plant root system. *Appl. Biochem. Biotechnol*. 2006;130:436-46.

[9] Namdeo AG. Plant cell elicitation for production of secondary metabolites. Phcog Rev. [serial online] 2007;1:69-79. Available from: URL:http:// www.phcogrev.com

[10] Iyer RI, Jayaraman G, Gopinath PM, Lakshmi Sita G. Direct somatic embryogenesis. *Trop. Agric.* (Trinidad). 2000;77: 98-105.

[11] Tharanathan RN, Kittur FS. Chitin--the undisputed biomolecule of great potential. *Crit. Rev. Food Sci. Nutr.* 2003;43:61-87.

[12] Luo J, He GY. Optimization of elicitors and precursors for paclitaxel production in cell suspension culture of *Taxus chinensis* in the presence of nutrient feeding. *Process Biochem*. 2004; 39:1073-9.

[13] Kasparová M, Siatka T. Effect of chitosan on the production of anthracene derivatives in tissue culture of *Rheum palmatum* L. *Ceska Slov. Farm.* 2001;50:249-53.

[14] Prakash G, Srivastava AK. Statistical elicitor optimization studies for the enhancement of azadirachtin production in bioreactor *Azadirachta indica* cell cultivation. *Biochem. Eng. J.* 2008; 40:218-22 .

[15] Komaraiah P, Naga Amrutha R, Kavi Kishor PB, Ramakrishna SV. Elicitor enhanced production of plumbagin in suspension cultures of *Plumbago rosea* L. *Enzyme Microb. Technol.* 2002; 31:634-9.

[16] Iyer RI. In vitro propagation of nutmeg, *Myristica fragrans* Houtt. In:Jain SM, Haggman H editors. Protocols for micropropagation of woody trees and fruits.The Netherlands: Springer; 2007. p. 335-44.

[17] Iyer RI, Jayaraman G, Ramesh A. *In vitro* responses and production of phytochemicals of potential medicinal value in nutmeg, *Myristica fragrans* Houtt. *Indian J. Sci. Technol.* 2009 Apr. 2(4):65-70.

[18] Murashige T, Skoog F. A revised medium for rapid growth and bioassays with tobacco tissue cultures. *Physiol. Plant.* 1962;15:473-97.

[19] Jin JH, Shin JH, Kim JH, Chung IS, Lee HJ. Effect of chitosan elicitation and media components on the production of anthraquinone colorants in Madder(*Rubia akane* Nakai) cell culture. *Biotechnol. Bioprocess Eng* . 1999;4:300-4.

[20] Funk C, Brodelius P. Influence of growth regulators and an elicitor on phenylpropanoid metabolism in suspension cultures of *Vanilla planifolia. Phytochemistry* 1990; 29:845-8.

[21] Khan W, Prithiviraj B, Smith DL. Chitosan and chitin oligomers increase phenylalanine ammonia-lyase and tyrosine ammonia-lyase activities in soybean leaves. *J. Plant Physiol.* 2003;160:859–63.

[22] Chakraborty M, Karun A, Mitra A. Accumulation of phenylpropanoid derivatives in chitosan-induced cell suspension culture of *Cocos nucifera. J. Plant Physiol.* 2009; http://www.sciencedirect.com/science/journal/01761617166 :63-71.

[23] Singh G. Manipulating and enhancing secondary metabolite production. In: Fu TJ, Singh G, Curtis WR. Editors . Plant cell and tissue culture for the production of food ingredients. New York: Kluwer Academic / Plenum Publishers; 1999. p. 101-120.

[24] Unnanuntana A, Bonsignore L, Shirtliff ME, Greenfield EM. The effects of farmesol on *Staphylococcus aureus* biofilms and osteoblasts. *J. Bone Joint Surg. Am.* 2009;91:2683-92.

[25] Chaudhary SC, Alam MS, Siddiqui MS, Athar M. Chemopreventive effect of farnesol on DMBA/TPA-induced skin tumorigenesis: Involvement of inflammation, Ras-ERK pathway and apoptosis. *Life Sci.* 2009; 85:196-205.

[26] Schinazi RF, Chu CK, Babu JR, Oswald BJ, Saalmann V, Cannon DL, Eriksson BF, Nasr M. Anthraquinones as a new class of antiviral agents against human immunodeficiency virus. *Antiviral-Res.* 1990;13:265-72.

[27] Agarwal SK, Singh SS, Verma S, Kumar S. Antifungal activity of anthraquinone derivatives from *Rheum emodi* . *J. Ethnopharmacol.* 2000;72:43-6.

[28] Huang Q, Lu G, Shen H-M, Chung MC, Ong CN. Anti-cancer properties of anthraquinones from rhubarb. *Med. Res. Rev.* 2007;27:609-30.

In: Handbook of Chitosan Research and Applications
Editors: Richard G. Mackay and Jennifer M. Tait
ISBN 978-1-61324-455-5

Chapter 24

WHOLE-CELL BIOCATALYST FOR UTILIZATION OF CHITOSAN BY YEAST CELL SURFACE ENGINEERING OF CHITOSANASE

Danya Isogawa, Kouichi Kuroda and Mitsuyoshi Ueda*
Division of Applied Life Sciences,
Graduate School of Agriculture, Kyoto University,
Sakyo-ku, Kyoto 606-8502, Japan

ABSTRACT

Chitosan is mainly degraded by chemical methods and the degree of degradation cannot be easily controlled. On the other hand, the degree of degradation of chitosan can be mildly controlled with enzymatic degradation properties of chitosanase. Yeast cell surface engineering is an attractive strategy for molecular preparation of novel whole-cell biocatalysts that can degrade chitosan polymers. Using this technique, chitosanase from *Paenibacillus fukuinensis* could be displayed on the yeast cell surface with the retention of the enzymatic activity and its application has been widely developed. Interestingly, chitosanase from *P. fukuinensis* is a unique enzyme that belongs to glucanase family 8 and has dual hydrolysis activities of chitosanase and β-1, 4 glucanase. A mutant library of chitosanase in which the putative catalytic proton acceptors were comprehensively substituted by the other amino acids was constructed using yeast cell surface engineering. As the results of this research, we demonstrated that the amino acid residues of Glu302 and Asn312 were catalytic proton acceptors and essential for having both chitosanase and β-1, 4 glucanase activities. Utilization of chitosanase for the production of useful chitosan oligosaccharides has been made further attractive. Promotion of efficient utilization of chitosan by enzymatic degradation may contribute to the reduction of the amount of waste crustacean shell and to the remaking of functional materials with bioactivities for the construction of a sustainable society with recycling of natural resources.

* E-mail: miueda@kais.kyoto-u.ac.jp.

INTRODUCTION

Chitin is a natural polysaccharide obtained mainly from crustacean shells and is composed of β-1, 4 linkages of *N*-acetyl-D-glucosamine. The amount of chitin in existence is relatively abundant and is next to that of cellulose on earth. However, chitin is an insoluble polymer and has a rigid structure for utilization. Chitosan is also a natural polysaccharide that is a partly deacetylated form of chitin and composed of β-1, 4 linkages of *N*-acetyl-D-glucosamine and D-glucosamine. It is a soluble polymer, thus, chitosan can be utilized more easily than chitin. Chitosan has various bioactivities such as antibacterial activity, antitumor activity, and anticancer activity, and a cholesterol lowering effect [1, 2]. Therefore, chitosan is utilized in various fields, such as food, cosmetics, and agriculture owing to its bioactivities. Recently, it has been reported that chitosan oligosaccharides obtained after the hydrolysis of chitosan polymers also have such bioactivities and that the activities were higher than those of chitosan. The strength of activity differs with the degree of polymerization of glucosamine, a monomer of chitosan.

Currently, chitosan is mainly degraded by chemical methods. Using such methods, the degree of degradation cannot be easily controlled and there is the risk of environmental pollution by the chemical agent used. If an enzymatic method is employed for the degradation of chitosan, the degree of degradation can be mildly controlled using enzymatic degradation properties without consuming energy (Figure 1).

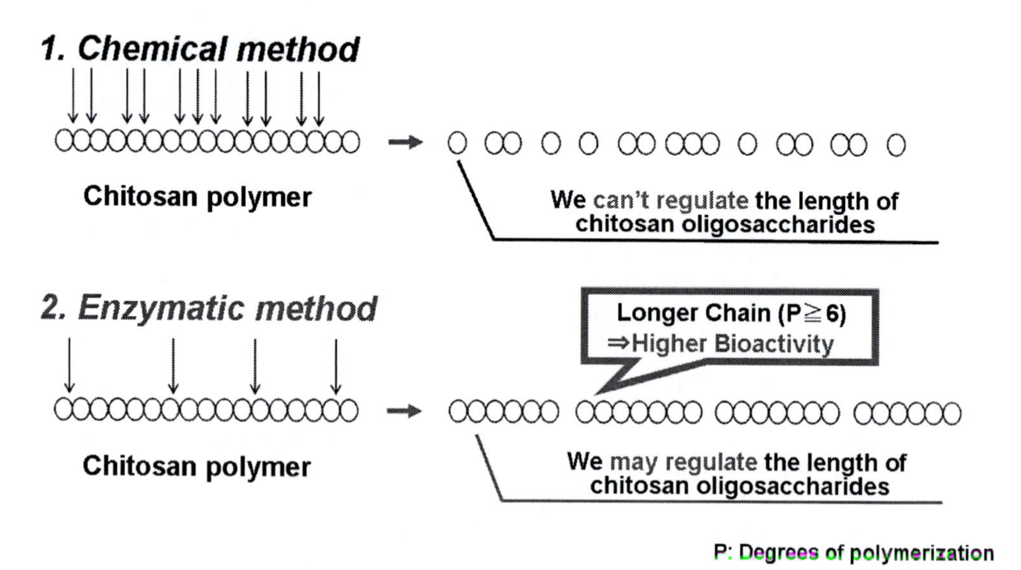

Figure 1. Methods of chitosan degradation.

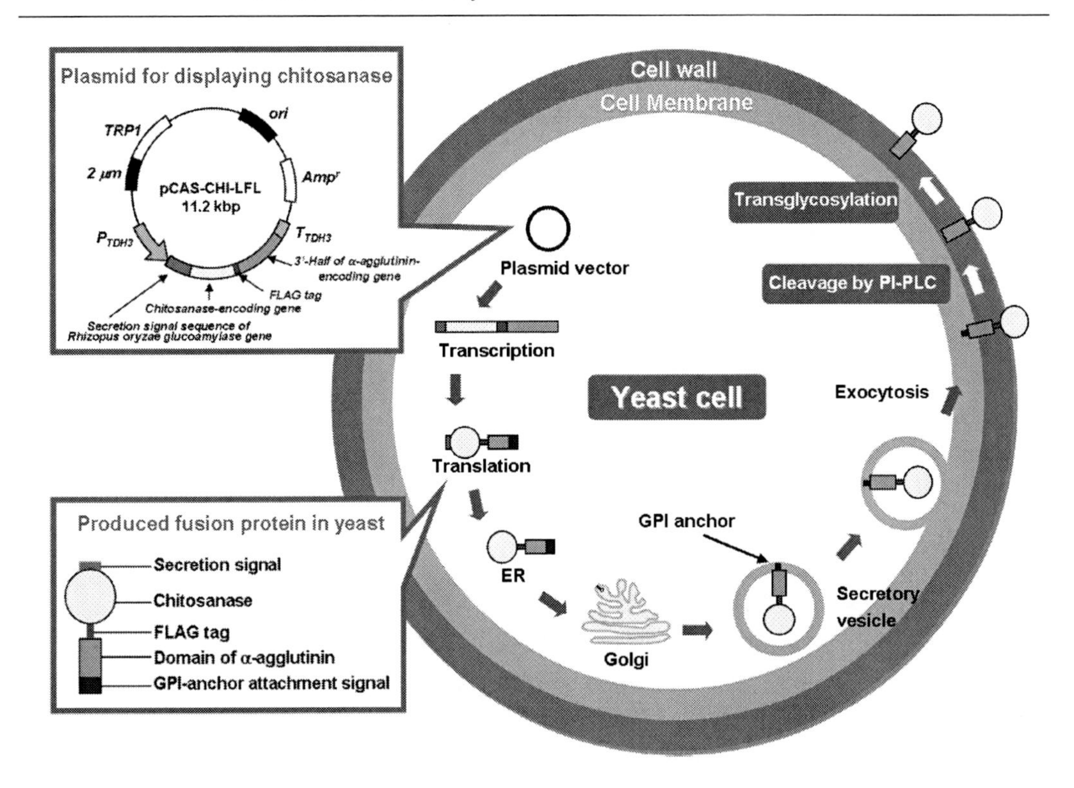

Figure 2. Yeast cell surface display of chitosanase by yeast cell surface engineering.

Paenibacillus fukuinensis D2 was isolated in Fukui prefecture and secretes a unique chitosanase that can hydrolyze chitosan [3]. In the classification of glycosyl hydrolases by sequence-based determination, chitosanase is generally categorized into family 46 (GH-46) as the chitosanase group and has only chitosanase activity. However, chitosanase from *P. fukuinensis* D2 belongs to family 8 (GH-8) of glucanase groups and has dual hydrolysis activities of both chitosanase activity and β-1, 4 glucanase activity. Natural chitosanases degrade chitosan to short oligomers such as monomers, dimers, and trimers and cannot produce longer oligomers with higher bioactivities. In this chapter, we constructed the chitosanase-displaying yeast as a whole-cell biocatalyst using yeast cell surface engineering and demonstrated the catalytic amino acid residues of the chitosanase by utilizing them to contribute to the reduction of the amount of the waste crustacean shell by the promotion of efficient utilization of chitosan.

1. Construction of Chitosanase-Displaying Yeast as a Whole-Cell Biocatalyst Using Yeast Cell Surface Engineering

Yeast cell surface engineering is an excellent gene-expression method that allows the display of proteins and peptides on the cell surface of the yeast *Saccharomyces cerevisiae* linked to the C-terminal half of α-agglutinin (Figure 2) [4, 5] . The yeast cells, which display proteins on the cell surface, can be used as a protein cluster [6]. Generally, the gene expression and protein production have been achieved by an intracellular or extracellular expression system using *E. coli* etc. Such conventional methods take a long time and much labor to collect proteins by the many complicated purification steps and the yield is insufficient. On the other hand, this molecular displaying technique has several advantages. For instance, the enzyme molecules are rapidly prepared after activation of the promoter and are anchored to the cell surface without many time-consuming purification and separation steps (Figure 3). In addition, a polymeric substrate such as chitosan is easily hydrolyzed by the cell-surface locating enzyme. This is because it is not necessary for the polymeric substrates to be diffused into the cell. We constructed the yeast displaying chitosanase from *P. fukuinensis* (Figure 2). The chitosanase-displaying yeast acted as a unique and excellent whole-cell biocatalyst with hydrolysis activity and degraded chitosan to chitosan oligosaccharides [7].

2. Demonstration of Catalytic Amino Acid Residues of Chitosanase by Constructed Chitosanase-Displaying Yeast

To utilize this unique chitosanase for the production of useful chitosan oligosaccharides, we have to clarify the hydrolysis mechanism of chitosan. The catalytic proton donor and substrate recognition site have been already clarified as Glu115 and Asp176, respectively [3]. However, the catalytic proton acceptor has not yet been clarified.

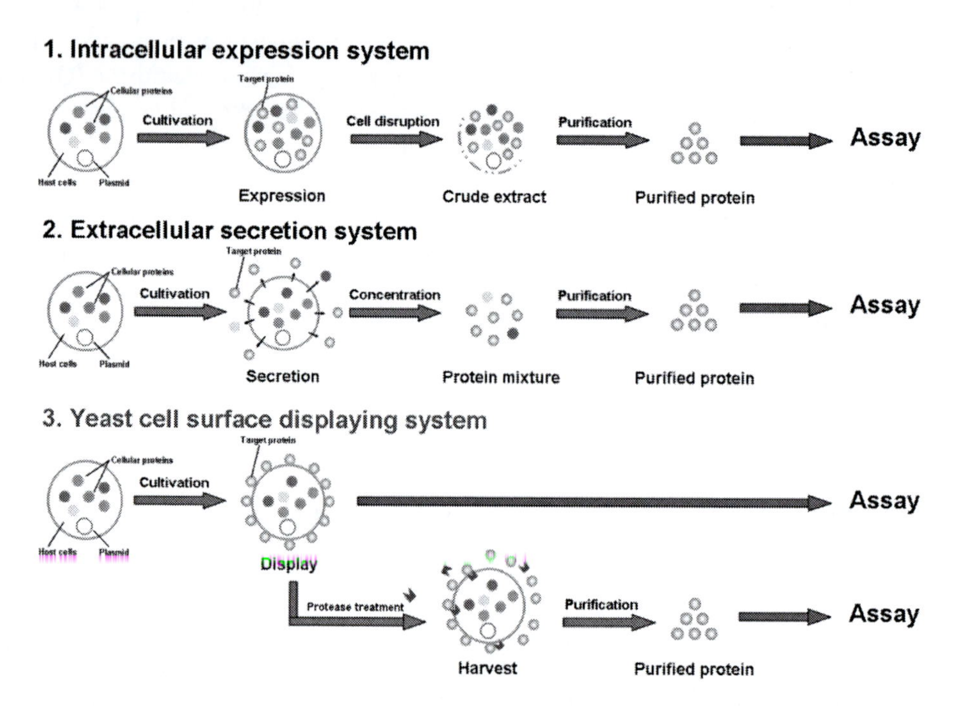

Figure 3. Advantage of yeast cell surface displaying system.

Table 1. Comparison of catalytic amino acid residues in family GH-8 enzymes

Function	Family GH-8a	Family GH-8b	In case of Chitosanase from *Paenibacillus fukuinensis*
Catalytic Acid	Glu	Glu	(Glu115)[a]
Substrate Recognition Site	Asp	Asp	(Asp176)[a]
Catalytic Base	Asp	Glu	(Glu302?)[b]
Assistance of Catalytic Base	No corresponding amino acid residue	Asn?	(Asn312?)[b]

a. Kimoto et al. (2002), b. This study.

The catalytic proton acceptor of this chitosanase was demonstrated using the constructed chitosanase-displaying yeast. A large mutant library was rapidly prepared by this engineered enzyme and the mutual correlation among residues of interest was comprehensively analyzed. The enzymes in the family GH-8 glucanase group to which chitosanase from *P. fukuinensis* D2 belongs can be further classified into three subfamilies - subfamily GH-8a, subfamily GH-8b, and subfamily GH-8c - on the basis of these proton acceptors in the enzymatic reaction (Table 1). In the subfamily GH-8a, Asp comprising the active site is conserved as a proton acceptor, and in the subfamilies GH-8b and GH-8c, Glu in the additional inserted loop domain at the active site is conserved [8]. Interestingly, Asp, which acts as a proton acceptor in the family GH-8a, might be changed to Glu and/or Asn in subfamilies GH-8b and GH-8c. On the chitosanase from *P. fukuinensis* D2 classified into subfamily GH-8b, Glu302, which is the putative proton acceptor in the inserted loop domain, and Asn312, which is conserved at the point of Asp acting as a proton acceptor in family GH-8a, are present. However, Glu302 in the inserted loop domain as a proton acceptor for exhibiting chitosanase activity and Asn312 for mediating the hydrolysis of chitosan have not yet been clarified in terms of their roles in the enzymatic reaction of chitosanase from *P. fukuinensis*. The comprehensive mutant library in which Glu302 and Asn312 were substituted by the other amino acid residues was prepared and evaluated by immunofluorescence staining using anti-FLAG antibody (Figure 4) and determination of enzymatic activities (Figure 5).

Chitosanase activity in nearly all of the Glu302 or Asn312 chitosanase mutants remained approximately 70% of that of the parent strain. In the comparison of chitosanase activity among the constructed mutants, we observed that the mutants with a basic amino acid and aromatic amino acid at residue 302, such as E302F, E302K, E302P, E302R, E302W, and E302Y, exhibited a markedly lower activity than the parent strain with the original Glu. These results are probably due to the repulsion of positive poles among the basic amino acids, protons in the active water, and amino group of the substrate, chitosan, and to the steric hindrance of the side chain of the amino acid in the active center. Moreover, the degradation rate obtained by the substitution of Asn312 also decreased similarly to that obtained by the substitution of Glu302. Asn312 in the enzyme certainly participated in the hydrolysis of chitosan. As a result of analysis of the correlation between Glu302 and Asn312 in the Glu302/Asn312 double mutants, Asn312 probably participated in the hydrolysis of the substrate and was an important amino acid in addition to Glu302 [9].

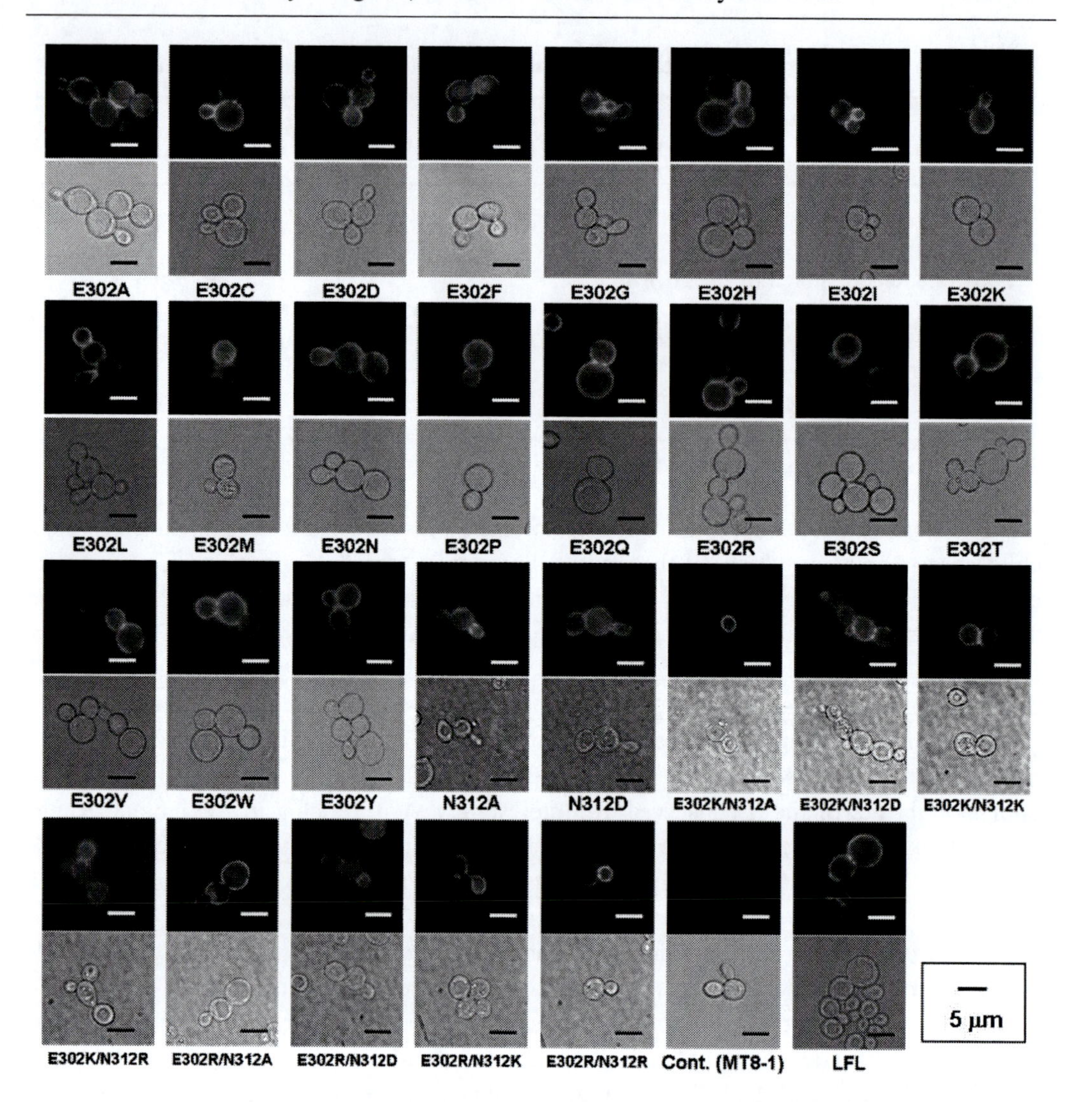

Figure 4. Immunofluorescence staining of chitosanase- and its mutants-displaying cells using an anti-FLAG antibody and an Alexafluor488-conjugated goat anti-mouse IgG antibody. Phase contrast micrographs and immunofluorescence micrographs of parent (pCAS-CHI-LFL) LFL, single and double mutants, and MT8-1 (negative cont.); Bars, 5 μm.

Similarly to the fact that the conserved Asn312 contributed to not only β-1, 4 glucanase activity but also chitosanase activity, to demonstrate whether Glu302 also contributed to β-1, 4 glucanase activity, β-1, 4 glucanase activity among the constructed mutants was evaluated. The mutants with a basic amino acid at residue 302, such as E302K and E302R, exhibited a lower activity than the parent strain with the original Glu. In comparison with the activities of N312A and N312D, the activity of N312D was higher than that of N312A, owing to substitution by acidic amino acid residues. The β-1, 4 glucanase activity of the prepared E302K/N312K, E302K/N312R, E302R/N312A, E302R/N312K, and E302R/N312R markedly decreased as similarly observed for chitosanase activity assay. However, the β-1, 4 glucanase activity of E302K/N312A, E302K/N312D, and E302R/N312D was higher than that of other double mutants. Except for E302K/N312A, these tendencies were the same as those observed in mutants of substitution of Asn312 only (Figures 5A and 5B). Therefore, the substitution of asparagine by acidic amino acid residues made Asn312 slightly effective to act as an assistance of proton acceptor, compared with substitution by nonpolar amino acid residues. Consequently, Glu302 and Asn312 probably contribute to both chitosanase activity and β-1, 4 glucanase activity and are essential amino acid residues for having those activities.

A

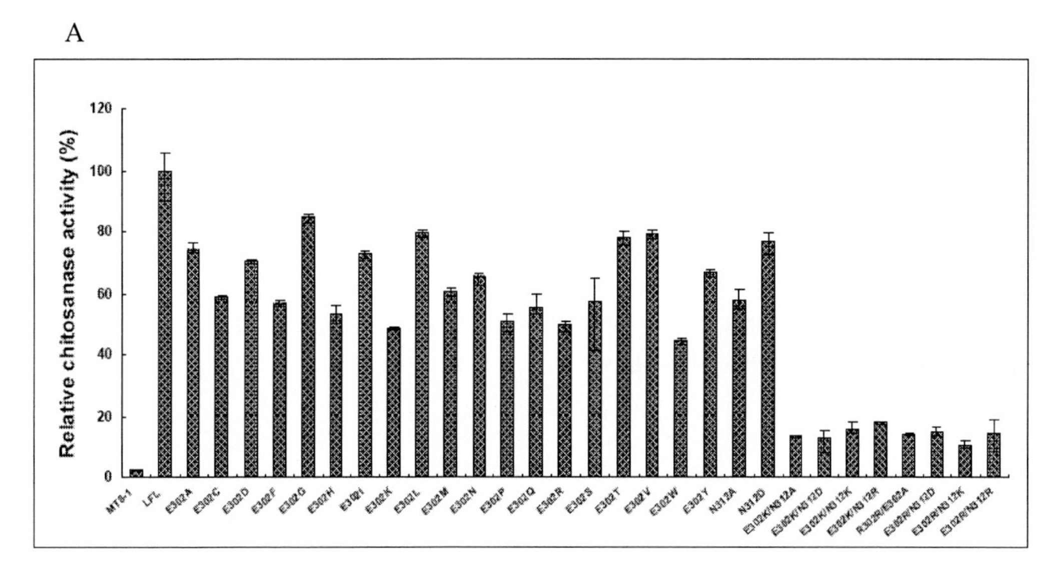

B

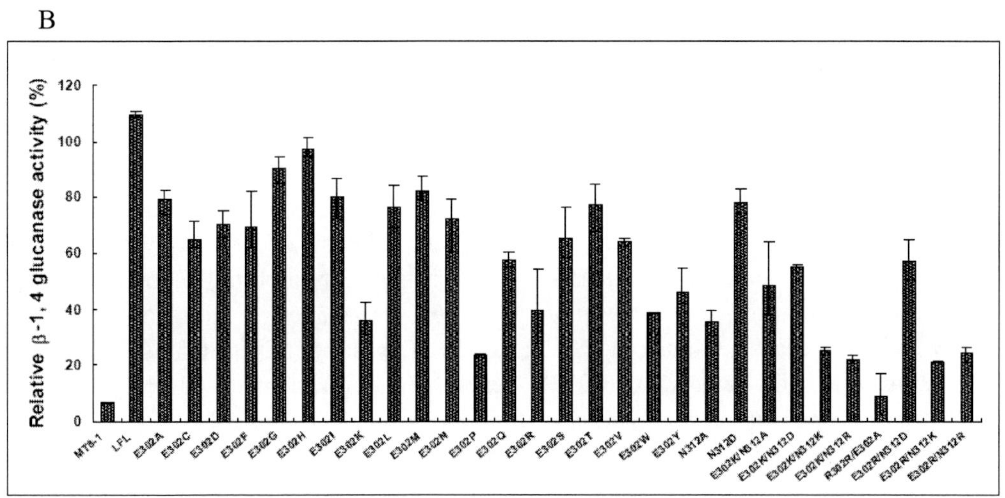

Figure 5. Measurement of chitosanase activity (A) and β-1, 4 glucanase activity (B); The value of the parent was regulated as 100%, and relative ratios of values of mutants to that of the parent (pCAS-CHI-LFL) were calculated. MT8-1 was used as the negative control; Values represent the means ± standard deviations of the results from three independent experiments.

3. Prospectives

For the reduction of the amount of waste crustacean shell by the promotion of efficient utilization of chitosan, chitosanase- and its mutant-displaying yeasts were constructed by yeast cell surface engineering. By comprehensive analysis among the mutant library, Glu302 and Asn312 in the chitosanase from *P. fukuinensis* were determined to be essential as the catalytic proton acceptors, and engaged in the dual hydrolysis activities of chitosanase and β-1, 4 glucanase. On the diversity in family GH-8b, the property of enzymes with bifunctions will be derived from the difference in effect of Glu302 and Asn312. Using these results that the catalytic amino acid residues of chitosanase from *P. fukuinensis* were essential for the hydrolysis of chitosan and cellulose, we may construct mutants that produce longer chitosan oligosaccharides with higher bioactivities by modification of the substrate binding site. In the biomass degradation, we can promote efficient utilization of chitosan by enzymatic degradation and may contribute to the reduction of the amount of waste crustacean shell and the remaking of functional materials with bioactivities for the construction of a sustainable society with recycling of natural resources.

References

[1] Qin, C., Du, Y., Xiao, L., Li, Z. and Gao, X. (2002). Enzymic preparation of water-soluble chitosan and their antitumor activity. *Int J Biol Macromol*, 31, 111-117.

[2] Huang, R., Mendis, E., Rajapakse, N. and Kim SK. (2006). Strong electronic charge as an important factor for anticancer activity of chitooligosaccharides (COS). *Life Sci,* 78, 2399-2408.

[3] Kimoto, H., Kusaoke, H., Yamamoto, I., Fujii, Y., Onodera, T. and Taketo, A. (2002). Biochemical and genetic properties of *Paenibacillus* glycosyl hydrolase having chitosanase activity and discoidin domain. *J Biol Chem,* 277, 14695-14702.

[4] Murai, T., Ueda, M., Yamamura, M., Atomi, H., Shibasaki, Y., Kamasawa, N., Osumi, M., Amachi, T. and Tanaka, A. (1997). Construction of a starch-utilization yeast by cell surface engineering. *Appl Environ Microbiol*, 63, 1362-1366.

[5] Murai, T., Ueda, M., Atomi, H., Shibasaki, Y., Kamasawa, N., Osumi, M., Kawaguchi, T., Arai, M. and Tanaka, A. (1997). Genetic immobilization of cellulose on the cell surface of *Saccharomyces cerevisiae*. *Appl Microbiol Biotechnol*, 48, 499-503.

[6] Shibasaki, S., Ueda, M., Iizuka, T., Hirayama, M., Ikeda, Y., Kamasawa, N., Osumi, M. and Tanaka A. (2001). Quantitative evaluation of the enhanced green fluorescent protein displayed on the cell surface of *Saccharomyces cerevisiae* by fluorometric and confocal laser scanning microscopic analysis. *Appl Microbiol Biotechnol*, 55, 471-475.

[7] Fukuda, T., Isogawa, D., Takagi, M., Kato-Murai, M., Kimoto, H., Kusaoke, H., Ueda, M. and Suye, S. (2007). Yeast cell-surface expression of chitosanase from *Paenibacillus fukuinensis*. *Biosci Biotechnol Biochem,* 71, 2845-2847.

[8] Adachi, W., Sakihama, Y., Shimizu, S., Sunami, T., Fukazawa, T., Suzuki, M., Yatsunami, R., Nakamura, S. and Takenaka, A. (2004). Crystal structure of family GH-8 chitosanase with subclass II specificity from *Bacillus* sp. K17. *J Mol Biol,* 343, 785-795.

[9] Isogawa, D., Fukuda, T., Kuroda, K., Kusaoke, H., Kimoto, H., Suye, S. and Ueda, M. (2009). Demonstration of catalytic proton acceptor of chitosanase from *Paenibacillus fukuinensis* by comprehensive analysis of mutant library. *Appl Microbiol Biotechnol,* 85, 95-104.

In: Handbook of Chitosan Research and Applications
Editors: Richard G. Mackay and Jennifer M. Tait
ISBN 978-1-61324-455-5

Chapter 25

AMIDE BOND FORMATION DURING OXIDATIVE DESTRUCTION OF CHITOSAN

Yu. I. Murinov*, A. R. Kuramshina, R. A. Hisamutdinov and N. N. Kabal'nova
Institute of Organic Chemistry, Ufa Scientific Centre of the Russian Academy of Sciences, Russian Federation, , Prospekt Oktyabrya

ABSTRACT

The interest to investigation of the properties of nature polysaccharide chitosan is caused by this wide prevalence in the nature, low toxicology and high biological activity. The unique physicochemical properties of chitosan, its derivatives and water-soluble oligomers allow to use them as supports, prolongating the action of medicine forms. We found that in the reactions of the destruction and oxidation of chitosan by the different reagents (ozone, hydrogen peroxide, NaClO, $NaClO_2$, ClO_2) can be occur amide bond formation. Formation of the amide bond was shown by the methods of IR and ^{13}C NMR, potentiometric titration. The inverse gas chromatography (IGC) is a useful method to observe the interaction between polymer and solutes. In the present work the given method is applied for study of adsorption properties of chitosan and its modified analogue in relation to some *n*-alkanes and *n*-alcohols. It is shown, that ability of adsorbents to dispersive interactions with test probes is higher on the modified chitosan. The amide bond formation were observed when the chitosan oxidation occur which rupture of glycoside bond and formation of the carboxyl groups at C(6).

INTRODUCTION

Chitosan is a polymer of β-1,4-linked 2-amino-2-deoxy-D-glucopyranose derived by *N*-deacetylation of chitin in aqueous alkaline medium.

* Institute of Organic Chemistry, Ufa Scientific Centre of the Russian Academy of Sciences, Russian Federation, 450054, Ufa, Prospekt Oktyabrya, 71; Phone/Fax: (347)235-60-66, e-mail: oxboss@anrb.ru

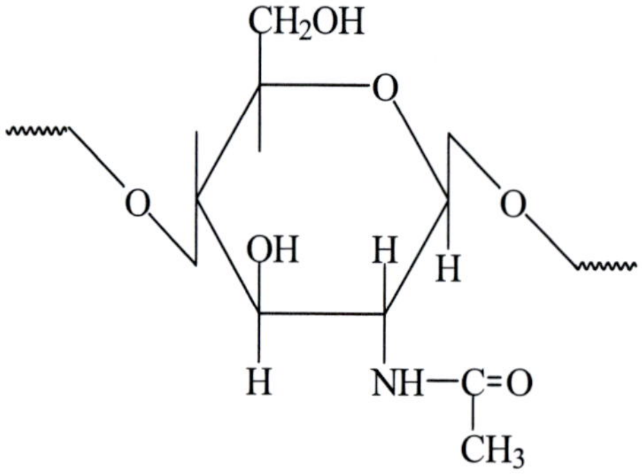

chitin 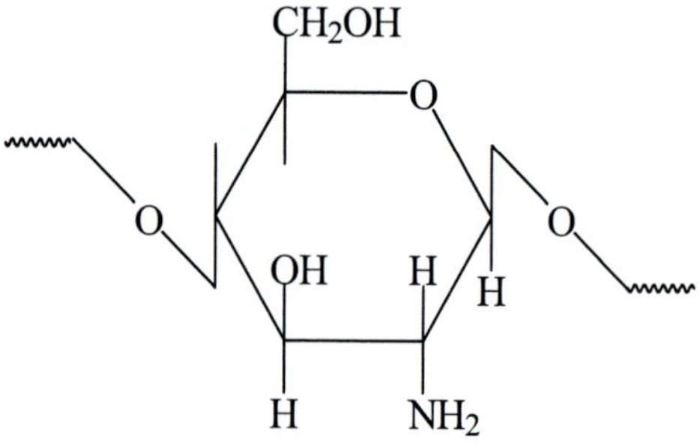 chitosan

The indefatigable interest of a science and the industry to search of renewable biopolymers of a natural origin, such as chitosan and its derivatives is connected, first of all, with uniqueness of its properties: high biological activity, biocompatibility with fabrics of the human, animals and plants, non-polluting in relation to environment, application possibility to carrying out of nature protection actions. The presence of functional groups in the modified chitosan derivatives increases the physical and chemical properties of the polymer allow significantly expand the range of their application.

We found that the oxidative degradation of chitosan in some cases leads to the formation of amide bonds. The formation of amide bond is an important reaction in organic synthesis and occurs either in hard conditions or catalytically and with the participation of activating reagents. We observed the formation of amide bonds under mild conditions by oxidative degradation of chitosan or separation of the reaction product.

Results and Discussion

Oxidative degradation of chitosan under the effect of hydrogen peroxide, chlorine dioxide, ozone, hypochlorite and sodium chlorite, oxygen in the presence of catalytic amount of Co(II)-Mn(II) nitrates was conducted. When sodium hypochlorite and sodium chlorite, oxygen in the presence catalytic amount of Co(II)- Mn(II) nitrates, chlorine dioxide were used the oxidation of chitosan C(6) was observed together with the degradation of macromolecules.

In those cases we observed the formation of amide bond, presumably between the carboxyl and amino groups of neighboring molecules, for example:

Oxidation of chitosan with sodium chlorite. In the IR spectrum of chitosan (Fig.1) , has oxidized at pH 4.5 the following changes have been noted: the reduction of intensity of absorption band in the field of 1530 cm^{-1}, an increase of the absorption bands intensities at 1570 and 1640 cm^{-1}, the appearance of intensive bands of absorption at 1420 and 1740 cm^{-1}.

CH_2OH OH H H NH_3^+ n

$\xrightarrow[\text{pH 4.5}]{NaClO_2}$

COO^- OH H H NH CO OH H H NH_3^+ m

n > m

The authors of works [1, 2] shown that at deacetylation of chitin in the IR spectrum observed a decrease of the intensity of the absorption bands at 1650 cm^{-1} (amide I). The occurrence of the absorption bands of stretching vibrations of C=O bond at 1640 cm^{-1} (amide I) and the absorption bands at 1570 cm^{-1} (amide II) were attributed to the amide group [3-5].

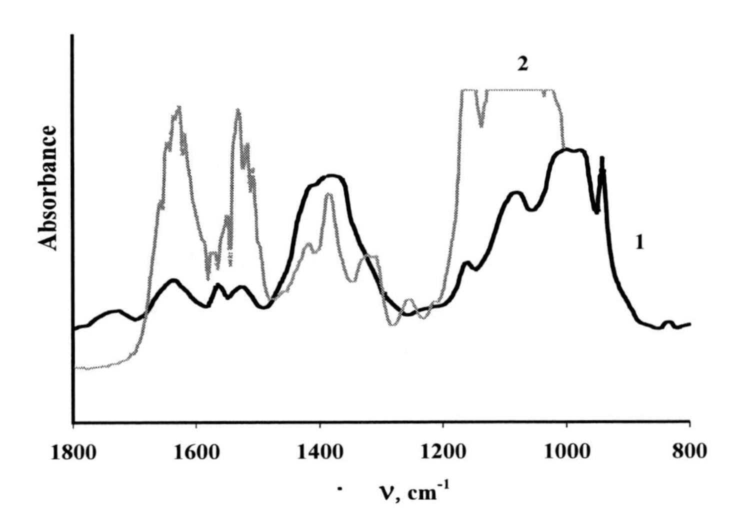

Figure 1. IR spectra of oxidized chitosan with sodium chlorite at pH 4.5 (1) and original hydrochloric acid solution of chitosan (2).

The presence of intensive absorption bands in the 1375 - 1420 cm^{-1}, we attributed to the interaction of stretching vibrations of C-O and planar deformation vibrations of O-H groups [3, 4], which appeared in the spectrum of carboxylic acids. The absorption band at 1740 cm^{-1} was related to the stretching vibrations of carbonyl groups. Thus the interaction of chitosan with sodium chlorite at pH 4.5 has occurred the oxidation of primary alcohol group at C(6) to the carboxyl group. Kinematic viscosity of the received sample had decreased in 3-4 times.

Fig. 2 illustrated the IR spectrum of chitosan oxidized in a heterogeneous aqueous medium at pH 6.4, an increase of the intensity absorption band at 1530 cm^{-1} and a decrease of the intensity absorption band at 1640 cm^{-1} was observed.

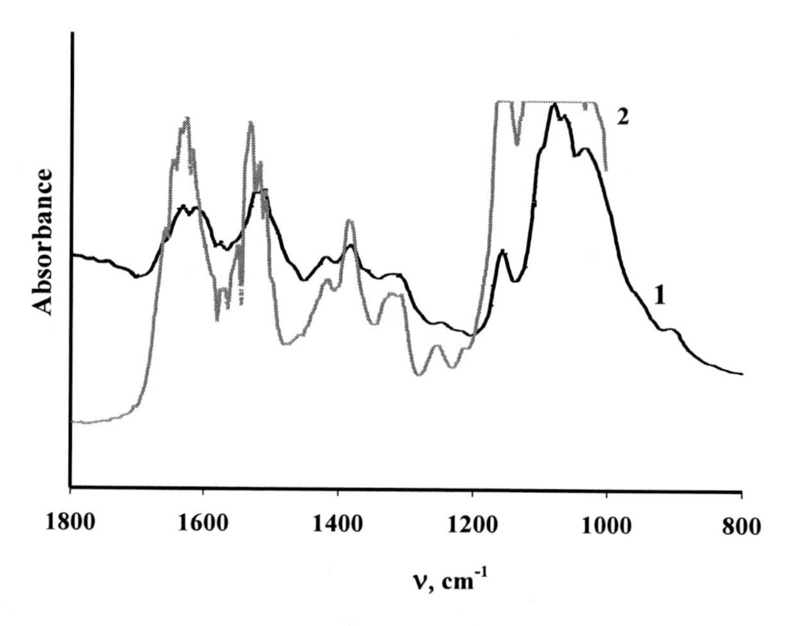

Figure 2. IR spectra of oxidized chitosan by sodium chlorite in a heterogeneous medium at pH 6.4 and original hydrochloric acid solution of chitosan (2).

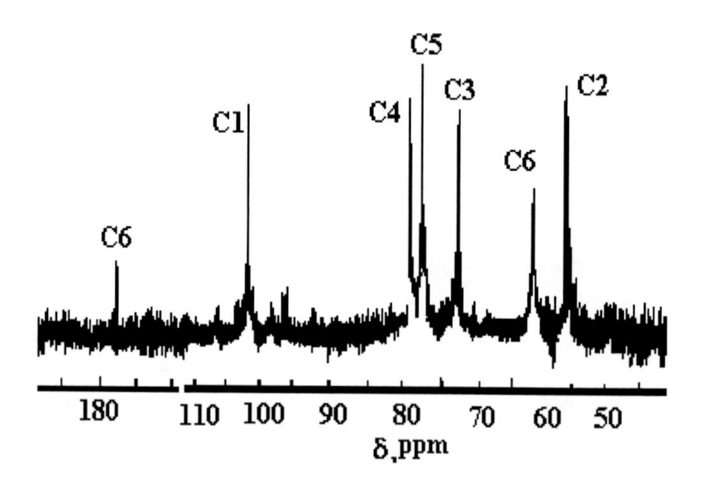

Figure3. ^{13}C NMR - spectrum of oxidized chitosan by sodium chlorite at pH 6.4.

The 13C NMR spectrum (Fig. 3) demonstrated the presence of a signal of group C = O in the field 179 - 180 ppm. Probably, the changes to the ratio of intensity of the absorption bands at 1640 (amide I) and 1570 cm-1 (amide II), has indicated by the oxidation of chitosan sodium chlorite is happened by the formation of a covalent bond between the nitrogen of the amino group and the carbon of the carboxyl group of chitosan.

Kinematic viscosity of solutions of oxidized chitosan at pH 6.4 decreased in 2 times. Introduction of 2,2,5,5-tetramethyl-4-phenyl-3-imidazoline-3-oxide-1-oxyl (R) to the reaction medium has practically no influence on the process of oxidation.

Oxidation of chitosan of sodium chlorite with the addition of catalytic amounts of sodium hypochlorite and R, pH 6.4. The IR spectrum of the oxidized sample was an increase of the intensity of the absorption bands at 1530 (amide II) and 1650 cm^{-1} (amide I). The change of intensity occurs due to formation of a covalent bond between carbon atoms of the carboxyl group and nitrogen atom of the amino group of chitosan (Fig. 4).

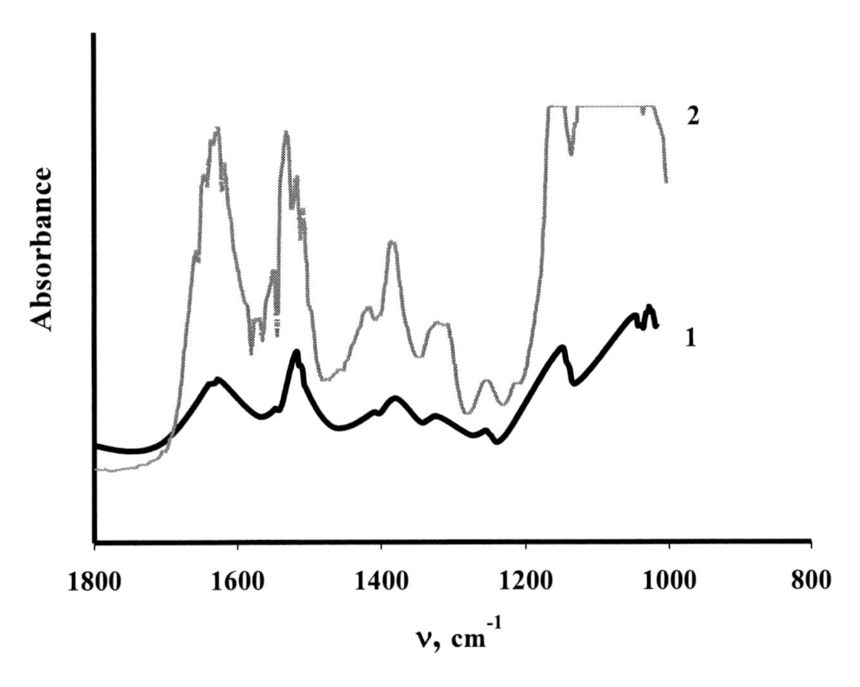

Figure 4. IR spectrum of oxidized chitosan by $NaClO_2$, in the presence of catalytic amounts of NaClO and 2,2,5,5-tetramethyl-4-phenyl-3-imidazoline-3-oxide-1-oxyl, KBr and original hydrochloric acid solution of chitosan (2).

Measurement of kinematic viscosity of the samples showed that after the first hour of reaction the viscosity decreases by 5 times. In the future, there is an increase of kinematic viscosity of the oxidized samples after 6 h the viscosity almost reaches the original value. It can be assumed that such a change of viscosity in the process of chitosan oxidation in aqueous medium under the action of sodium chlorite, catalytic amounts of R, KBr, and sodium hypochlorite associated with the formation of interchain cross-linking in oxidized chitosan. The decrease of the kinematic viscosity at the first hour of reaction was resulted due to of oxidative degradation (Fig. 5).

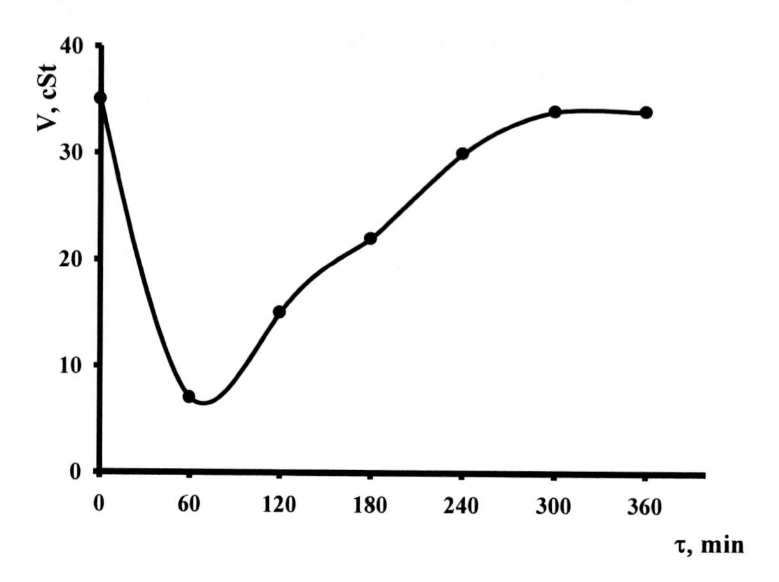

Figure 5. Change of kinematic viscosity in the system chitosan - $NaClO_2$, in the presence of catalytic amounts of NaClO and 2,2,5,5-tetramethyl-4-phenyl-3-imidazoline-3-oxide-1-oxyl and KBr.

Figure 6 showed the titration curves of the original chitosan and oxidized by sodium chlorite with the addition of sodium hypochlorite and R. By the method described in [1], the number of amino groups is calculated in the original - 74.50% and the oxidized chitosan - 34.24%. The number of free carboxyl groups in the oxidized sample is 31%. Given the number of carboxyl groups involved in covalent bonding with amino groups, it was calculated that the total degree of oxidation is 70%.

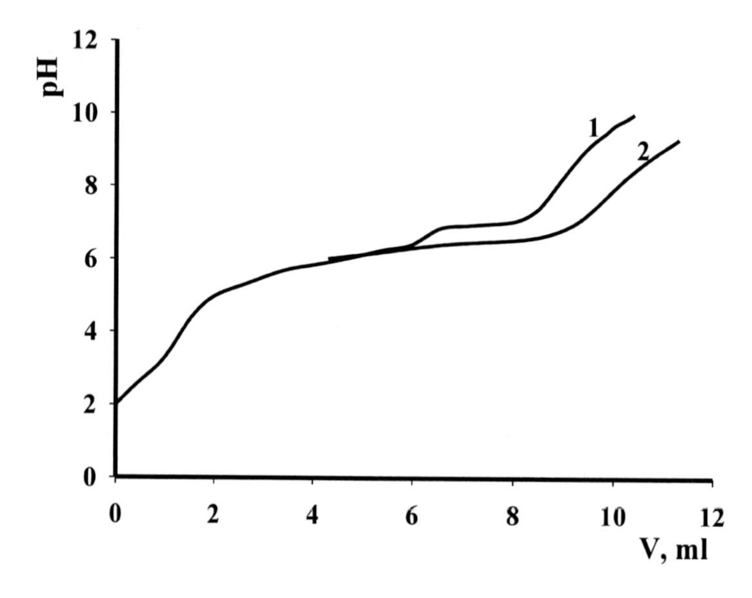

Figure 6. Curves of titration of oxidized chitosan by $NaClO_2$, in the presence of catalytic amounts of NaClO and 2,2,5,5-tetramethyl-4-phenyl-3-imidazoline-3-oxide-1-oxyl (1) and original hydrochloric acid solution of chitosan (2), titrant 0.1 N NaOH.

Thus, the interaction of chitosan and sodium chlorite in a homogeneous medium at pH 4.5 at the same time proceed oxidative degradation of chitosan and the oxidation of primary alcohol group at C6 to the carboxyl group. The interaction of chitosan and sodium chlorite in a heterogeneous medium at pH 6.4 passes the oxidation of chitosan with the formation of a covalent bond between amino nitrogen and carbon the carboxyl group. The method of potentiometric titration of the oxidized chitosan determined the number of carboxyl groups, which is 30%.

Destruction of chitosan under the effect of hydrogen peroxide it was spent in the homogeneous medium at 90°C. A certain volume of hydrochloric acid solution of chitosan (pH 1) was placed into a constant temperature reactor then necessary volume of hydrogen peroxide of the set concentration ([Chitosan]:[H_2O_2] = 1:1.7) was added.

The time of reaction was 3.5 h. The iodometric analysis of reaction mixture has shown that as a result of oxidative destruction no appreciable presence of H_2O_2 is observed. The reaction mixture was brought to pH 7 by addition of concentrated solution of NH_3. Water-soluble oligomers chitosan which remained in a filtrate, landed acetone, filtered and dried on air to constant weight. The yield was 38 %.

The amount of amino groups in original chitosan is 74.50 % and oxidized chitosan – 53.00 %. The quantity of free carboxylic groups is 17 %. IR spectrum of the product the insignificant increase of strips intensity of absorption amide groups in absence of characteristic strips of absorption carboxylic groups is observed. Kinematic viscosity of chitosan oligomers is close to the viscosity of the solvent.

The analysis of data obtained by IR spectroscopy methods, viscosimetry, potentiometric titration showed that the interaction of chitosan with hydrogen peroxide is processed oxidative destruction of the polymer with a small sewing together disrupt the sample at the expense of formation amide bonds.

Interaction of oligomers chitosan which received as a result of the destruction under the effect of hydrogen peroxide with sodium chlorite by addition catalytic amount of sodium hypochlorite and stable nitroxyl radical was studied in similar conditions. Ratio of reagents in the reaction mixture was [Chitosan]:[$NaClO_2$]: [NaClO]: [R]: [KBr] = 1:2:0.1:0.07:1.

By the technique resulted in work [1], the quantity of amino groups in original oligomers – 53 % and oxidized chitosan – 42 % are calculated.

IR spectrum of the product (Fig. 7) indicates an increase of the intensity of the absorption band at 1650 cm^{-1} (amide I) and the absorption band at 1570 cm^{-1} (amide II) - amide groups.

Changes in the ratio of the intensity of the absorption bands at 1650 cm^{-1} (amide I) and 1570 cm^{-1} (amide II), probably indicate that the oxidation of chitosan oligomers by sodium chlorite in the presence of catalytic amounts of sodium hypochlorite and 2,2,5,5-tetramethyl-4-phenyl-3-imidazoline-3-oxide-1-oxyl is formed covalent bond between the amino group nitrogen and carbon the carboxyl group of chitosan.

Oxidation of oligomers chitosan (obtained under the effect of hydrogen peroxide) and chlorine dioxide was carried out in an aqueous solution of hydrochloric acid at pH 4.5, T = 30 ° C, with the ratio [Chitosan]: [ClO_2] = 1:2 ([Chitosan] = $5.84 \cdot 10^{-3}$ mole/l). Response time is 6 hours. The yield of the product was 78%.The reaction products were studied by ^{13}C NMR and IR spectroscopy.

The number of amino groups was calculated by the method described in [1] in the oxidized oligomers chitosan does not exceed 48%.

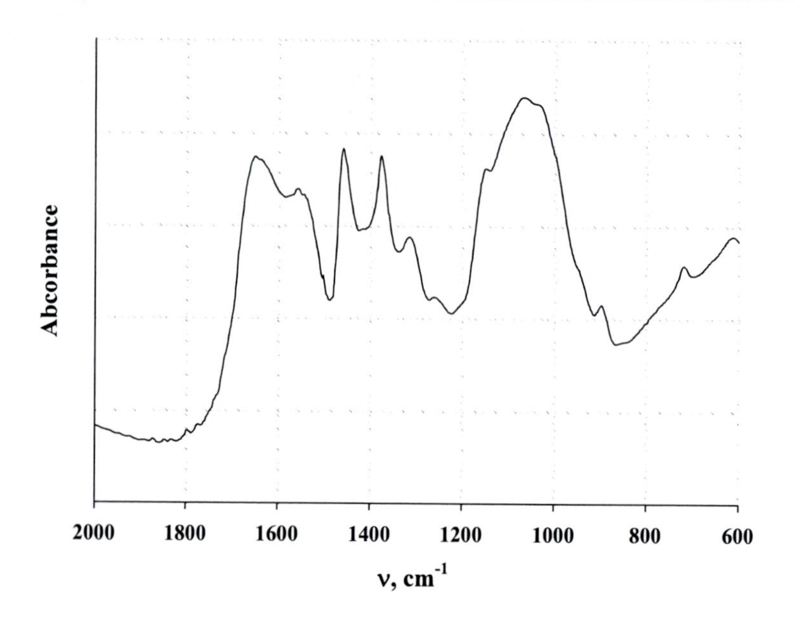

Figure 7. IR spectrum of oligomers chitosan oxidized $NaClO_2$, in the presence of catalytic amounts of NaClO and 2,2,5,5-tetramethyl-4-phenyl-3-imidazoline-3-oxide-1-oxyl and hydrochloric acid solution of chitosan.

In the IR spectrum of the product marked change: the emergence of the absorption band of 1740 cm^{-1} - absorption band of carboxylic acids and an increase of the intensity of the absorption band of 1420 cm^{-1} - carboxylate ion absorption band, indicating that the oxidation of primary alcohol group at C(6) and the oxidative destruction of 1,4-β-glycosidic bond. A small change in the ratio of the intensity of the absorption bands at 1645 cm^{-1} (amide I) and at 1570 cm^{-1} (amide II) associated with the possibility of matching a small percentage of the final product.

Kinematic viscosity of the samples obtained close to the viscosity of the solvent. Based on these data we can conclude that the interaction of chitosan oligomers with chlorine dioxide the oxidation of primary alcohol group at C (6) and the oxidative destruction of 1,4-β-glycosidic bond, the formation of amide bonds.

Oxidation of chitosan in the heterogeneous medium by oxygen under mild conditions in the presence of catalytic amount of Co(II)-Mn(II) nitrates. Oxidation of primary and secondary alcohols to aldehydes and ketones under the influence of atmospheric oxygen under mild conditions in the presence of Mn(II) - Co(II) or Mn(II) - Cu(II) nitrates, in combination with 2,2,6,6-tetrametilpiperidin-1-oxyl (TEMPO) has been investigated earlier [6].

Oxidation of chitosan (MW = 300 kDa) by atmospheric oxygen in the presence of catalytic amount of Co(II)-Mn(II) nitrates with the addition of TEMPO was spent in ice acetic acid at the room temperature. Reaction time is 2 hours. The yield was 93%. Potentiometric titration calculates the degree of deacetylation of the received product is 43 %.

In the IR spectrum oxidized chitosan observed the following changes: an increase of the intensity of the absorption bands at 1420 cm^{-1} – a carboxylate ion and a significant increase of the intensity of the absorption bands at 1650 cm^{-1} and at 1530 cm^{-1}. Changing of intensity of the absorption bands of amide groups NH-C=O at 1530 cm^{-1} and 1650 cm^{-1} occurs through the formation of bonds between carbon atoms of the carboxyl group and nitrogen atom of the amino group of chitosan

Kinematic viscosity of the received sample chitosan is close to the viscosity of solvent.

Thus, the interaction of chitosan with atmospheric oxygen in ice acetic acid in the presence of catalytic amount of Co(II)-Mn(II) nitrates with the addition of TEMPO proceeds both oxidizing destruction chitosan, and oxidation of polymer with the formation of amide bonds.

Sorption Properties of Chitosan Derivatives, Received by Molecular Self-Assembly as a Result of Oxidizing and Complexation

For the purpose of change the physical and chemical properties of chitosan (for example, the size of a time, chemical stability) works on its modification under the effect of various reagents are conducted now. Such updating can change the structure of the surface of chitosan and its adsorption [7, 8, 9].

By method of the reversed-phase gas chromatography (RPGC) in the Henry region adsorption properties original chitosan (Chit) and its modified analogue (Chit-1) received by oxidation with sodium chlorite in the presence of catalytic amount of sodium hypochlorite and nitroxyl radical [10] in relation to n-alkanes ($C_6 - C_9$) and n- alcohols ($C_1 - C_3$) in the range of temperatures 41 – 97 °C are studied.

Adsorption Properties Chitosan Modified by Oxidation with Sodium Chlorite in the Presence of Catalytic Amount of Sodium Hypochlorite and 2,2,5,5-Tetramethyl-4-Phenyl-3-Imidazoline-3-Oxide-1-Oxyl

Schematically the structure of the original and the modified chitosan can be represented as follows:

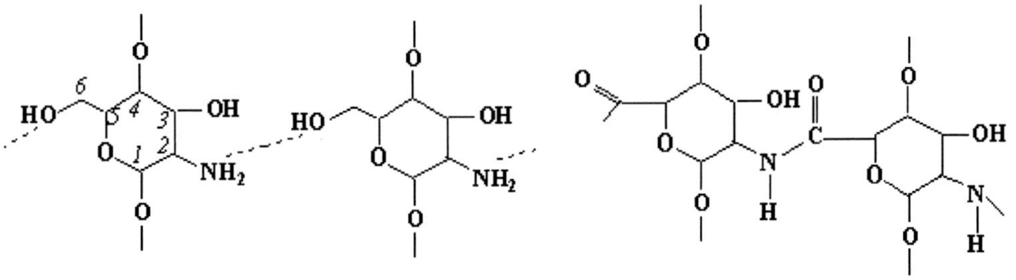

Chitosan Chitosan -1

Adsorption isotherms (Fig. 8), calculated under known formulas [11], are linear for n-alkanes (C_6H_{14} - C_9H_{20}).

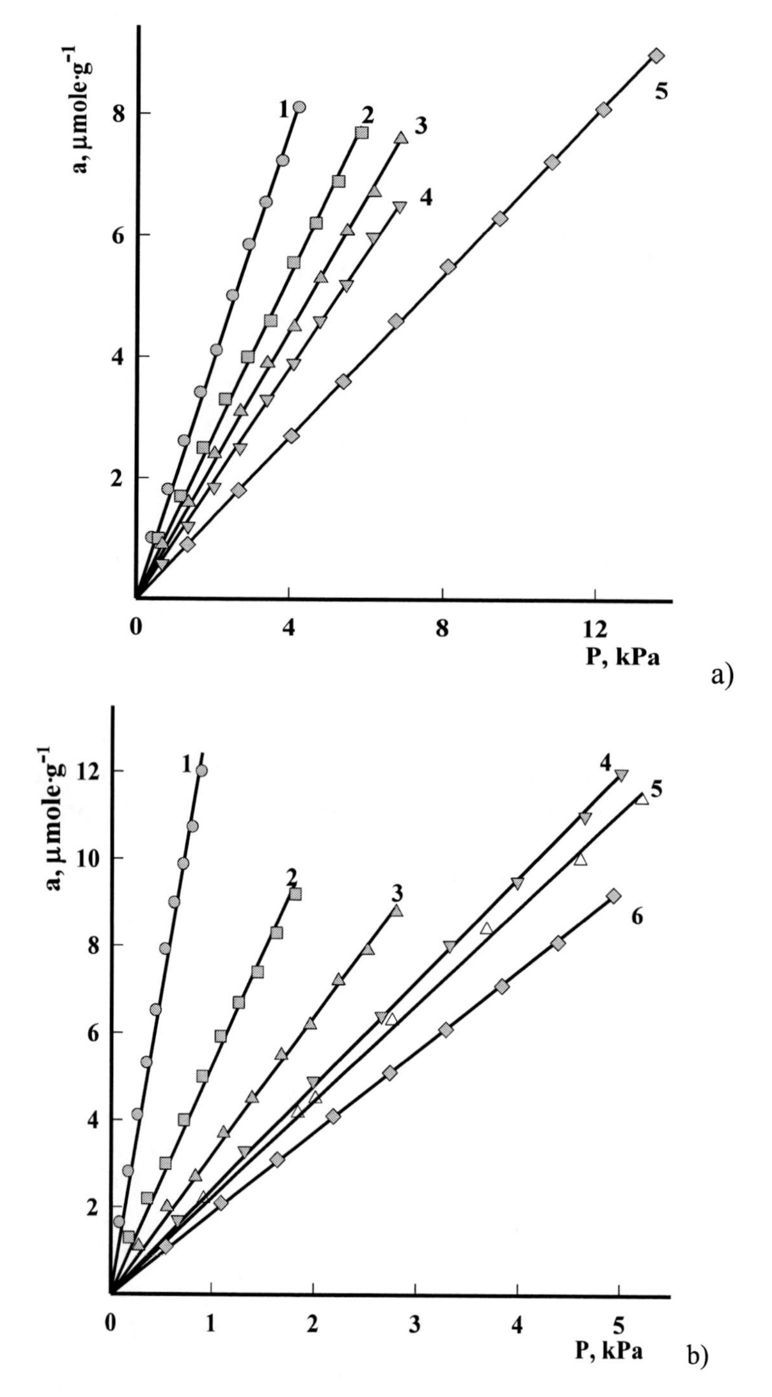

Figure 8. Adsorption isotherms of n-heptane on original chitosan (a): 1 – 41 °C; 2-58 °C; 3 – 72 °C; 4 – 87 °C; 5 – 92 °C; chitosan-1 (b):1 – 41 °C; 2 – 58 °C; 3 – 72 °C; 4-80 °C; 5 – 87 °C; 6 – 90 °C.

The linearity of adsorption isotherm specifies that the value of retention volume does not depend on volume of adsorbate probe injected, and the equilibrium constant represents the Henry adsorbate – adsorbent equilibrium constant.

As n-alkanes are capable only to nonspecific interaction the equation becomes simpler and looks like ΔG_A^D = -RT ln H for them, where H – the Henry adsorbate – adsorbent equilibrium constant. Dispersive interactions possess property of additivity, therefore increment ΔG_A on CH_2 groups is defined by a tangent of angle of inclination ΔG_A from number of carbon atoms in a molecule n-alkanes (Fig. 9).

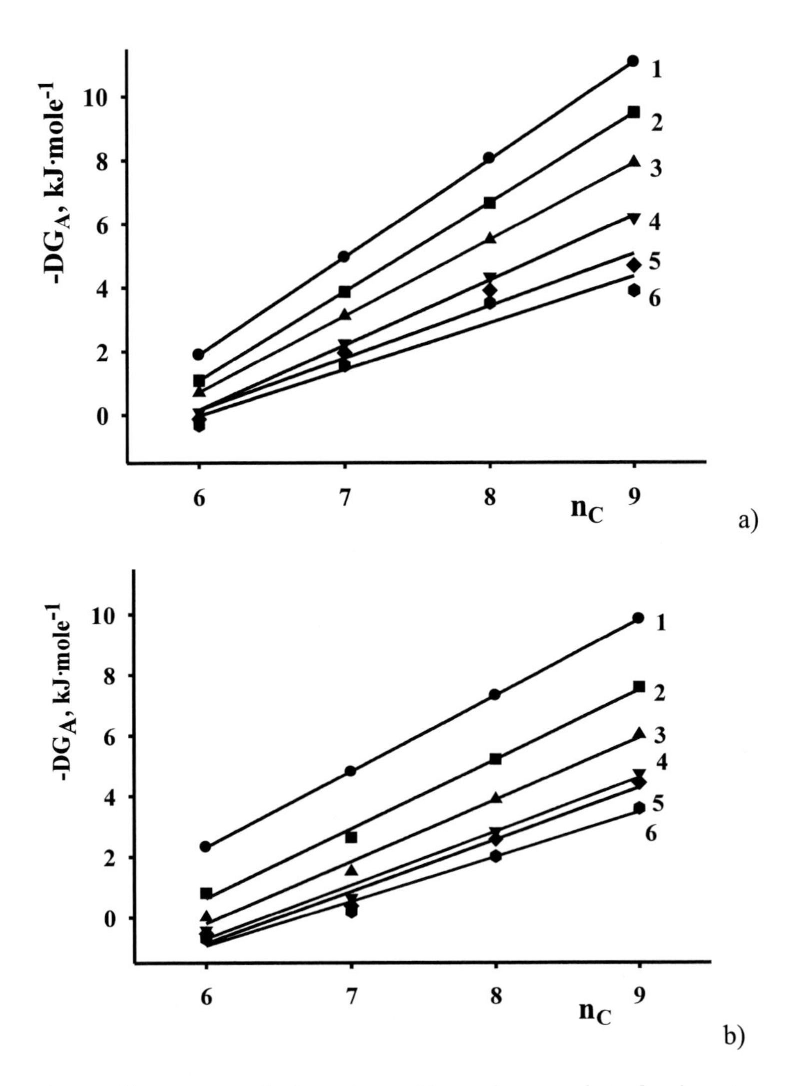

Figure 9. Dependence of free energy of adsorption n-alkanes from number of carbon atoms: a) original chitosan: 1 – 41 °C; 2 – 58 °C; 3 – 72 °C; 4 – 87 °C; 5 – 92 °C; 6 – 97 °C; b) chitosan-1: 1 – 41 °C; 2 – 58 °C; 3 – 72 °C; 4 – 80 °C; 5 – 87 °C; 6 – 90 °C.

Figure 9 illustrate the difference of the adsorption of n-alkanes is insignificant, with an increase in temperature decreases the adsorption, and the increasing number of carbon atoms - is increasing as a result of greater dispersion interaction increases the CH_2 groups. At the same time Chitosan-1 adsorption is more when increasing temperature, and the energy dispersive interaction is reduced, reflecting a decrease of adsorption. Table 1 shows the contribution of the methylene group free energy to the total energy of adsorption, determined from the slope of lines shown in Figure 9.

Table 1. Group CH_2 contribution ($-\Delta G_{CH_2}$, kJ·mole^{-1}) in free energy of adsorption n-alkanes on chitosan

Temperature, °C	41	58	72	80	87	90	92	97
Chitosan	2.51	2.15	1.74		1.37		1.06	0.88
Chitosan-1	2.50	2.29	1.94	1.77	1.61	1.57		

The insignificant distinction of the adsorption of n-alkanes passes due to the fact that the ability of Chitosan-1 to the dispersion interaction is some more, as follows from the values calculated by equation $\gamma_S{}^D = (\Delta G_{CH2})^2 / (4N^2 \cdot a^2{}_{CH2} \cdot \gamma_{CH2})$ calculated on the equation dispersive making $\gamma_S{}^D$ than free energy of adsorbents surface (Fig. 10) .

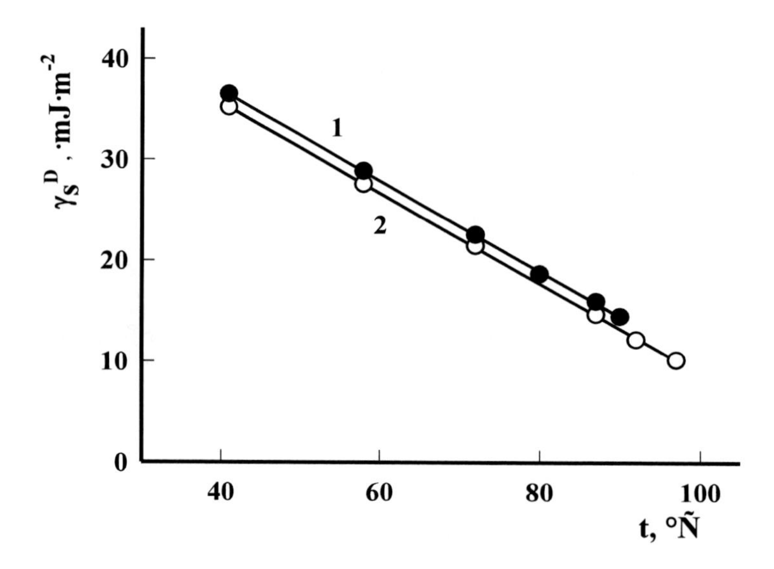

Figure 10. Temperature change of the dispersion properties of surface: 1 - original chitosan, 2 - chitosan-1.

Generally, reduction $\gamma_S{}^D$ at temperature increase is explained by entropic the contribution to free energy of a surface [12]. Linearity of the dependence on Figure 10 shows that in the studied temperature range the structure of both samples chitosan does not undergo changes, and a little higher values $\gamma_S{}^D$ Chitosan-1, probably, is explained by partial shielding of the nitrogen atom.

N-alcohols adsorption isotherms on all samples are nonlinear, therefore the calculation is carried out on initial parts in the Henry region. A dispersive component of adsorption $\Delta G_A{}^D$ of alcohols is calculated by a method of the least squares on equation $\Delta G_A{}^D = b + tg\alpha \cdot T_{b\cdot alc}$ (°C), intersection and slope of straight of dependence ΔG_A - $T_{b(Hydrocarbon)}$; energy of specific interaction ΔG_A^S is determined on a difference of total adsorption ΔG_A energy of alcohol and its dispersive part ΔG_A^D (Tabl.1).

Values of adsorption energy of alcohols and its components are resulted in Table 2. As can be seen from the table, at a constant in the range of alcohols in both samples of chitosan on the rise of the dispersion component of the free energy of adsorption as a result of increased hydrophobic effect and the decrease of the energy of the hydrogen bond (Fig. 11).

The quantitative characteristic of alcohols adsorption is the ratio between hydrophilic and hydrophobic parts, or hydrophilic-lipophilic balance (HLB). It can be seen, change ΔG_A^S in series of alcohols are well correlated with their HLB- coefficients.

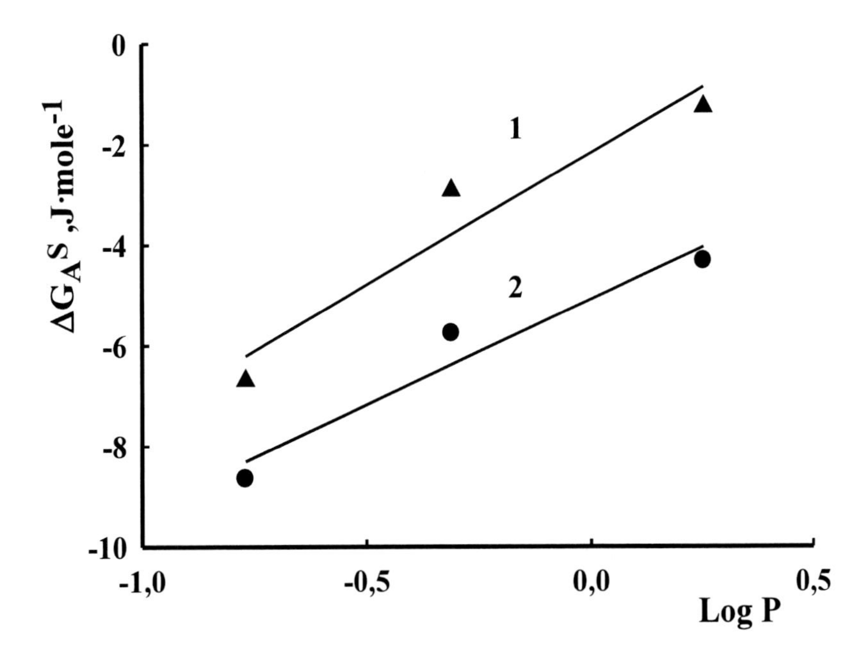

Figure 11. Change of specific making adsorption of n-alcohols, 41°C, depending on their factors GLB: 1 – original chitosan; 2 - chiosant-1 (P - factor of distribution of alcohol in system octanol – water).

As a result of oxidation free amino group is replaced to amide group N-C=O with 40.26 % in the structure of Chitosan-1 additional oxygen donor atom appears that is reflected of the increase of adsorption of n-alcohols (Tabl.2). The data in Table 2 show that on these samples with an increase in quantity CH_2 groups in a molecule of alcohol the energy of specific interaction decreases (shielding OH groups), respectively, the dispersion component increases.

In all cases the numerical values of the parameters of adsorption above in case of Chitosan-1. A possible explanation is that as a result of the oxidation and conversion of amino groups in the amide in the Chitosan-1 formed more ordered supramolecular structure which increases its adsorption capacity with respect to the n-alcohols through the formation of stronger hydrogen bond compared with the original chitosan.

Table 2. Dispersive (ΔG_A^D) and specific(ΔG_A^S) components of adsorption energy of n- alcohols

The parameters of adsorption of n- alcohols	T,°C	Methanol		Ethanol		*n*-Propanol	
		1	2	1	2	1	2
ΔG_A^D, kJ·mole^{-1}	41	-1.14	-1.86	-2.40	-3.12	-4.13	-4.84
	58	-0.38	-0.40	-1.49	-1.64	-3.04	-3.18
	72	-0.09	0.41	-1.01	-0.63	-2.28	-1.98
	80		1,01				-1.24
	87	0.41	1.25	-0.34	0.01	-1.36	-0.72
	90		1.46				-0.44
	92	0.34		-0.09	0.42	-1.07	
	97	0.49		0.04	0.65	-058	
ΔG_A^S, kJ·mole^{-1}	41	-6.69	-8.65	-2.92	-5.76	-1.26	-4.32
	58	-7.86	-9.66	-3.31	-6.95	-2.29	-5.52
	72	-8.11	-10.20	-3.74	-7.66	-2.79	-6.33
	80		-10.40		-8.08		-6.85
	87	-8.99	-10.20	-3.50	-8.39	-1.88	-7.20
	90		-10.30		-8.40		-7.36
	92	-8.52		-2.39		-1.31	
	97	-8.01		-2.70		-1.04	

EXPERIMENTAL

In work we used chitosan a crab Far East (poly [(1→4)-2-amino-2-desoxy-β-D-glucose]. Crushed chitosan (the size of particles < 0.6 mm) were treated 4 times at room temperature and constant mixing during 3 h in triple volume of 1N HCl before colourless of the mother solution. The precipitate was kept for 1.5 h in 2 N NaOH, washed by distilled water to neutral reaction and dried at a room temperature to constant weight. Molecular weight chitosan (1.5–2.0)•10^5, humidity – 11%, degree deacetylation – 74.5 % were identified as described in [1, 13]. Element structure: C 44.0 %; H 7.5 %; N 7.9 %; O 42.6 %; ashes – traces.

Oxidation of chitosan in the homogeneous medium at pH 4.5 sodium chlorite and with the addition of a stable nitroxyl radical 2,2,5,5-tetramethyl-4-phenyl-3-imidazoline-3-oxide-1-oxyl (R) in the presence of KBr was carried out in thermostatically controlled reactor and at constant mixing at 35°C, [Chitosan]:[$NaClO_2$] = 1:10 mole/mole. A sample of chitosan (1.15g taking into account humidity) was dissolved in 100 ml 0.1 N HCl and slowly added dropwise 2 N KOH to pH 4.5.

The control for pH carried out potentiometrically. Then through each 30 minutes was added 0.2 ml of solution $NaClO_2$ in water. The concentration of an oxidant in the reaction medium was determined iodometric titration [14]. Stable nitroxyl radical R and KBr (in separate experiments) were paid after loading chitosan ([Chitosan]:[$NaClO_2$]:[R]:[KBr] = 1:10:0.01:1). After 6 h a full expenditure of oxidant the samples were analyzed by IR spectroscopy in 4000 - 400 cm^{-1}. For this oxidized chitosan precipitated with acetone from the reaction mixture, washed by distilled water, dried the sample at a room temperature to constant weight. The yield was 75%.

For the ^{13}C NMR sample was dissolved in a mixture of D_2O and DCl. The ^{13}C NMR spectrum were recorded on a device "Bruker AM = 300" (operating frequency of 75 MHz for ^{13}C, the internal standard tetrametilsilan).

Oxidation of chitosan in the homogeneous medium at pH 4.5 sodium chlorite and with the addition of a stable nitroxyl radical 2,2,5,5-tetramethyl-4-phenyl-3-imidazoline-3-oxide-1-oxyl (R) in the presence of KBr was carried out in thermostatically controlled reactor and at constant mixing at 35°C, [Chitosan]:[$NaClO_2$] = 1:10 mole/mole. A sample of chitosan (1.15g taking into account humidity) was dissolved in 100 ml 0.1 N HCl and slowly added dropwise 2 N KOH to pH 4.5.

Oxidation of chitosan in the heterogeneous medium at pH 6.4 sodium chlorite ([Chitosan]:[$NaClO_2$] = 1:10 mole/mole) and sodium chlorite in the presence of a stable nitroxyl radical 2,2,5,5-tetramethyl-4-phenyl-3-imidazoline-3-oxide-1-oxyl ([Chitosa-n]:[$NaClO_2$]:[R]= 1:10:0.01) was conducted in thermostatically controlled reactor and at constant mixing at 35°C. In experiences with the additive R then each half hour added water solution of $NaClO_2$ during 4 h. The time of reaction was 6 h. The product yield was 87 %. A reaction product analyzed by IR spectroscopy and NMR.

Oxidation of chitosan sodium chlorite in the presence of catalytic amount of sodium hypochlorite, R and KBr was conducted in thermostatically controlled reactor and at constant mixing by pH 6.4 at 35°C [15]. Ratio of reagents in the reaction mixture was [Chitosan]:[$NaClO_2$]:[NaClO]:[R]:[KBr] = 1:10:0.1:0.01:1. For 1 h sequentially added aqueous solutions of sodium hypochlorite and sodium chlorite. Six experiences have realized. The time of reaction was 1, 2, 3, 4, 5, 6 h, respectively. The product yield was 84 %. A reaction product analyzed by IR spectroscopy and NMR.

Oxidative destruction of chitosan by hydrogen peroxide at pH 1 was carried out in thermostatically controlled reactor and at constant mixing at 90°C. A sample of chitosan (0.11 mole/l) was dissolved in 50 ml of hydrochloric acid water solution (pH 1) and hydrogen peroxide (0.19 mole/l) was added. During reaction sampling for expense monitoring of hydrogen peroxide by iodometric method every 30 minutes was conducted. The time of reaction was 3.5 h. The product yield was 84 %. A reaction product analyzed by IR spectroscopy and NMR. Element structure: C 39.33 %; H 7.27 %; N 6.57 %; O 42.64 %; ashes – traces.

Oxidation of oligomers chitosan received as a result destruction under action hydrogen peroxide ([Chitosan]:[H_2O_2]=1:1.7) and sodium chlorite with addition of catalytic quantities of sodium hypochlorite and stable nitroxyl radical R (2,2,5,5-tetramethyl-4-phenyl-3-imidazoline-3-oxide-1-oxyl) and KBr in the heterogeneous medium at constant mixing by pH 6.4 at 35°C was conducted. The molar ratio equal to [Chitosan]:[$NaClO_2$]:[NaClO]:[R]:[KBr] = 1:2:0.1:0.07:1. Then consistently added water solutions of sodium hypochlorite and sodium chloride on 1.25 ml of everyone during 3.5 h. The time of reaction was 5 h. Further reaction mixing was cooled to a room temperature, solution of Na_2S_2O3 (0.36 mole/l) was added, cooled to 0°C. For allocation of a reaction product a dialysis was realized. The product yield was 34 %.

Oxidation of oligomers chitosan received under effect of hydrogen peroxide by chlorine dioxide in a water solution of hydrochloric acid at pH 4.5and 30 °C was conducted. Chlorine dioxide was obtained by the method described in [16]. The molar ratio equal to [Chitosan]: [ClO_2] = 1:2 ([Chitosan] = $5.84 \cdot 10^{-3}$ mole/l). The received product landed with acetone, filtered, washed by the distilled water, dried on air to constant weight. The product yield was 84 %.

For the reception of water-soluble oligomers chitosan the oxidative destruction of hydrochloric acid water solution chitosan (pH 4.5) under the action of ozone at 20 °C was realized. To the reactor placed in a ceramic glass, flowed 50 ml of hydrochloric acid water solution chitosan ([Chitosan = $5.84 \cdot 10^{-2}$ mole/l]). Then the ozone-oxygen mixture (speed of giving of oxygen was 134.8 ml/mines) was submitted. A maximum of ozone absorption band λ_{max} = 300 nm. The time of reaction was 6 h. The molar ratio equal was [Chitosan]: $[O_3]$ = 18:1. The received product landed with acetone, washed by distilled water, dried on air to constant weight.

The quantity of carboxyl groups and the degree of deacetylation of the original and the oxidized chitosan by potentiometric titration (OP-112 / 1 "Radzelkis") with a modified electrode was determined. The sample was dissolved in 0.1 N HCl, then titrated with 0.1 N NaOH. The degree of deacetylation was calculated by the methods described in [14, 17, 18].

Kinematic viscosity of chitosan of samples aqueous solutions of chitosan samples in 0.33 M acetic acid in capillary viscosimeter with a hanging level at 25±0.1 °C were determined.

Chromatographic measurements using a Chrom-5 gas chromatograph equipped with a thermal conductivity detector were made. Helium was used as a carrier gas. Flow-rates were determined with a soap-bubble flow meter.

Chitosan with the size of particles 0.1 - 0.2 mm was packed in a stainless steel column (1000x3 mm) as the adsorbent phase. The packed column had been aged for 28 h in flowing highly pure helium at 150 °C before determination of retention times. All sorbates have chromatographic grade and their purity was checked by gas chromatography prior to use.

Calculation of parameters of adsorption. Intermolecular interactions in the adsorbate-adsorbent system may be dispersive and specific which corresponds to the dispersive, γ_S^D, and specific, γ_S^S, component of free surface energy, γ_S, of the adsorbent [19]:

$$\gamma_S = \gamma_S^D + \gamma_S^S \tag{1}$$

Similarly, the free energy of adsorption can be split into dispersion and specific components:

$$\Delta G_A = \Delta G_A^D + \Delta G_A^S \tag{2}$$

Energy of dispersive interaction of a chitosan with n-алканами expected on the equation:

$$\gamma_S^D = (\Delta G_{CH2})^2 / (4N^2 * a^2_{CH2} * \gamma_{CH2}), \tag{3}$$

where $a_{CH2\,2}$ - the area of one methylen-group surface (0.06 nm^2)

$\gamma_{CH2} = 35.6 + 0.058(293 - T)$ - – a part of adsorption energy, corresponding to adsorption of CH2-group, N – Avogadro's number, T – temperature, K

The relative error of definition of thermodynamic parametres of adsorption did not exceed 3 %.

Conclusions

As a result of the done researches various samples of chitosan are received and identified. The possibility of amide bond formation by interaction of chitosan with different oxidation systems. The mechanism of formation of the covalent bond between the amino group nitrogen and carbon of the carboxyl group of chitosan

Whether the question is formed amide bond during a reaction or separation of the product from the reaction mixture remains opened.

References

[1] Domard A., Rinaudo M. // *Int. J. Biol. Macromol*. 1983. Vol. 5. N 1. P. 49.

[2] Brugnerotto J., Lizardi J., Goycoolea F.M., Arguelles–Monal W., Desbrieres J., Rinaudo M. // *Polymer*. 2001. Vol. 42. N 8. P. 3569.

[3] Silverstein, R., Bassler, G., Morrill T. *Identification of Organic Compounds*. New York: Wiley. 1974. 590 p.

[4] [4] Comprehensive Analytical Chemistry. Ed. C.L. Wilson, D.W. Wilson. Amsterdam: *Elsevier*. 1976. Vol. 7. 550 p.

[5] Kramareva N.V., Stakheev A.Yu, Tkachenko, O.P.; Klementiev K.V., Grünert W., Finashina E.D., Kustov L.M. // *J. Mol. Catal*. A: Chem. 2004. Vol. 209. N 1 - 2. P. 97.

[6] Cecchetto A., Fontana F., Minisci F., Recupero F. //*Tetrahedron Lett*. 2001. Vol. 42. P. 6651.

[7] Suder B.J., Wightman J.R. // *J. Appl. Polym. Sci*. 1982. Vol. 27. N 21. P. 4827.

[8] Inoue K., Baba Y., Yoshiguza K. // *Bull. Chem. Soc. Jpn*. 1993. Vol. 66. P. 2915.

[9] Zeng X., Ruckenstein E. // *J. Membr. Sci*. 1998. Vol. 148. P. 195.

[10] Kolyadina O.A., Murinov K.Yu., Murinov Yu.I. // *Russ. J. Phys. Chem*. 2002. Vol. 76. N 5. P. 905.

[11] Vigdergaus M.S. *Chromatographia. Moscow: Science*. 1978. 184 p.

[12] Shen W., Parker I.H., Sheng Y.J. // *J. Adhesion Sci. Technol*. 1998. Vol. 12. P. 161.

[13] Chitin and Chitosan: Sources, Chemistry, Biochemistry, Physical Properties, and Application. Ed. T. Anthonsen. London: Elsevier Applied Science. 1990. 830 p.

[14] Muzarelli R.A.A. Chitin. Oxford: Pergamon Press. 1970. 309 p.

[15] Zhao M., Li J., Mano E,. Song Z., Tschaen D.M., Grabowski E.J.J., Reider P.J. // *J. Org. Chem*. 1999. Vol. 64. N 7. P. 2564.

[16] Ganieva E.S., Ganiev I.M., Grabovskiy S.A., Kabal'nova N.N. // *Russ. Chem. Bull*. 2008. Vol. 57. N 11. P. 2328

[17] Masschelein W.J. *Chlorine Dioxide. Chemistry and Environmental Impact of Oxychlorine Compounds*. Ann Arbor Science Publishers. 1979. 190 p.

[18] Park J.W., Choi K.H., Park K.K. // *Bull. Korean Chem. Soc*. 1983. Vol. 4. N 2. P. 68.

[19] Dorris G.M., Gray D.G. // *J. Colloid. Interface Sci*. 1979. Vol. 71. N 2. P. 931

In: Handbook of Chitosan Research and Applications
Editors: Richard G. Mackay and Jennifer M. Tait
ISBN 978-1-61324-455-5

Chapter 26

CHITOSAN MICROSPHERES

Sonia Touriño[1], Tzanko Tzanov[*1], N. Skirtenko[2], U. Angel[2] and Aharon Gedanken[2]
[1] Department of Chemical Engineering; Universitat Politècnica de Catalunya; Escola Universitària d'Enginyeria Tècnica Industrial; Terrassa; Catalonia, Spain
[2] Department of Chemistry and Kanbar Laboratory for Nanomaterials at Bar-Ilan University, Center for Advanced Materials and Nanotechnology, Bar-Ilan University, Ramat-Gan, Israel.

1. INTRODUCTION

Over the last few decades, researchers have tried to develop systems capable of delivering active compounds to specific target sites. Microparticulate carriers, e.g. microspheres, are considered good potential oral delivery systems to provide a constant therapeutic level of the drug and, in this way, to avoid frequent administration of the solid oral dosage form. Microspheres are generally defined as small spheres made of any material and sized from one micron (one one-thousandth of a millimetre) to a few mm. Similarly, smaller spheres, sized from 10 to 500 nm are called nanospheres.

Microspheres are in strict sense, spherical empty particles whereas the term "microcapsule" is defined, as a spherical particle with the size varying in between 50 nm to few millimeters in diameter containing a core substance. However, the terms microcapsules and microspheres are often used synonymously. In addition, some related terms are used as well. For example, "microbeads" and "beads" are used alternatively (Kumar, 2000). In the current review the latter terms will be considered synonymous.

The ideal microspheres are completely spherical and homogeneous in size, though less perfect particles are often termed microspheres as well. Microspheres have many applications in medicine, with the main uses being for encapsulation of drugs and proteins. The medical

* Corresponding author: Tzanko Tzanov, Department of Chemical Engineering, Universitat Politècnica de Catalunya, Colom 1, 08222 Terrassa, Spain, Tel (off.): +34 937398761, Fax: +34 937398225, email : tzanko.tzanov@upc.edu

uses of particulate drug delivery systems cover all areas of medicine such as cardiology, endocrinology, gynaecology, immunology, pain management, and oncology among others. The drug loaded microspheres are delivered to the target area by passive means (e.g. trapping by size) or active means (e.g. magnetic targeting) and slowly release the encapsulated drug over a desired time period, the length of which is determined mainly by the drug's biological half-life and the release kinetics of the microsphere matrix. The biodistribution and final fate of the microspheres is highly dependent on their size and surface charge. Microspheres not only have potential as drug delivery carriers for non-invasive routes of administration such as oral, nasal and ocular, but also show to be good adjuvant for vaccines. Despite these advantages, there is no ideal microsphere system available.

Many materials have been proposed for preparing biodegradable microspheres including synthetic (polylactic acid, copolymers of lactic and glycolic acids) and natural polymers. Polymeric carriers are selected according to different criteria: (i) ease to synthesize and characterize; (ii) inexpensive; (iii) biocompatible; (iv) biodegradable (v) non-immunogenic; (vi) non toxic and (vii) water soluble. In terms of the mechanism, it must be decided to which is given priority among safe delivery, prolonged release, and targeting. Mechanistic studies involve: (a) strategy set-up (choice of the system, mechanism and kinetics); (b) microsphere preparation (choice of materials and preparative methods; (c) drug incorporation (in some cases simultaneously with step b); (d) *in vitro/in vivo* testing (for release rate and/or distribution). The drug is released either via diffusion through the microsphere walls, or via microsphere degradation or corrosion, depending on the quality of the carrier microsphere and the combination of drug and carrier. The release mechanism may determine the kinetics of drug release that might change from a zero to second order (Kawaguchi, 2000).

Biopolymer microspheres were first designed using albumin (Scheffel et al., 1972) and non-biodegradable synthetic polymers such as polyacrylamide and poly(methylacrylate) (Birrenbach and Speiser, 1976; Kreuter and Speiser, 1976). However, the risks of chronic toxicity due to the intracellular and/or tissue overloading of non-degradable polymers were soon considered as a main limitation for the systemic administration in humans. As a consequence, the development of micro and nanospheres from synthetic biodegradable polymers including polyalkylcyanoacrylate, poly (lactic-co-glycolic acid) and polyanhydride received much attention (Iwata and McGinity, 1991); (Allémann et al., 1993; Bhagat et al., 1994). Despite the interesting results reported in the literature, these systems may also cause toxicological problems (Fernández-urrusuno et al., 1995) and therefore the natural polymers began to gain relevance. In particular, the excellent characteristics of chitosan makes it a suitable biomaterial candidate for production of microcapsules/microspheres used in medicine and pharmacy (Chandy and Sharma, 1990). Chitosan was used for its hypobilirubinaemic and hypocholesterolemic effects (Furda, 1980; Nagyvary, 1982), antacid and antiulcer activities (Ito et al., 2000), wound and burn healing properties (Tachihara et al., 1997), as a platform for immobilization of enzymes and living cells and in ophthalmology (Felt et al., 1999). Chitosan has been shown to possess mucoadhesive properties (He et al., 1998; Kockisch et al., 2003) due to the electrostatic interaction between the positively charged chitosan macromolecules and the negatively charged mucosal surfaces. These properties may be attributed to the presence of groups forming hydrogen bonding like –OH and –COOH (Schipper et al., 1997), strong charges (Dodane et al., 1999), high molecular weight (Kotzé et al., 1998; Schipper et al., 1996), chain flexibility (He et al., 1998) and surface energy properties favouring the rapid spread of chitosan into the mucus (Lueßen et al., 1994).

As a result of chitosan bioactivity these formulations can present special therapeutic effects. Therefore, new formulations and a greater increase in the applicability of chitosan microspheres/microcapsules are expected in the future.

2. Preparation of Chitosan Microspheres

A great variety of methods have been designed for preparing chitosan microparticulate systems. The properties of these microspheres, e.g. size, porosity, burst strength, swelling capacity, degradability and their composition could be changed by selecting the most appropriate technique in order to meet the end-use requirements.

2.1. Interaction with Anions

The amino groups of the polysaccharide are responsible for its pH dependent solubility. Reacting chitosan with controlled amounts of multivalent anion results in crosslinking between chitosan macromolecules. The crosslinking may be achieved in acidic, neutral or basic environments depending on the method applied. Ionic crosslinking has been extensively used for preparation of chitosan microspheres and the main experimental procedures are briefly described below.

2.1.1. Ionotropic Gelation

Ionotropic gelation is based on the ability of polyelectrolytes to crosslink in the presence of counter ions to form hydrogels. The natural polyelectrolytes contain certain anions on their chemical structure. These anions form meshwork structure by combining with the polyvalent cations as chitosan and induce gelation by binding mainly to the anionic blocks. Hydrogel beads have been produced by dropping a drug-containing polymeric solution into the aqueous solution of polyvalent cations. The cations diffuse into the drug-loaded polymeric drops, forming a three dimensional ionically crosslinked lattice. Biomolecules can also be loaded into these hydrogel beads under mild conditions to retain their three dimensional structure (Patil et al., 2010). The complex systems formed by macromolecules (interpolymer complexes) can be classified according to the dominant type of intermolecular interaction (the van der Waals interactions, hydrogen bonding and coordination, covalent and polyelectrolyte complexes stabilised by intermolecular ionic bonds) (Krayukhina and et al., 2008). The formation and stability of these polyelectrolyte complexes depend on many factors such as the degree of ionization of each of the oppositely charged polyelectrolytes, the density of the charges on the polyelectrolytes, the charge distribution over the polymeric chains, the concentration of the polyelectrolytes, their mixing ratio, the mixing order, the duration of the interaction, the nature of the ionic groups, the position of the ionic groups on the polymeric chains, the molecular weight of the polyelectrolytes, the polymer chain flexibility as well as the temperature, ionic strength, and the pH of the reaction medium (Etrych et al., 2005; Il'ina and Varlamov, 2005; Park et al.).

The cationic amino groups on the C2 position of the repeating glucopyranose units of chitosan can interact electrostatically with the anionic groups (usually carboxylic acid groups) of other polyions to form polyelectrolyte complexes. Many different polyanions from natural origin (e.g. pectin, alginate, carrageenan, xanthan gum, carboxymethyl cellulose, chondroitin sulphate, dextran sulphate, hyaluronic acid) or synthetic origin (e.g., poly (acrylic acid)), polyphosphoric acid, poly (L-lactide) have been used to form polyelectrolyte complexes with chitosan in order to provide the required physicochemical properties for the design of specific drug delivery systems (Berger et al., 2004).

Microspheres of chitosan-alginate are among the most frequently developed microcapsules. They are formed by means of anionic complexation. The polyelectrolyte complex formed between these two polymers is still biodegradable and biocompatible, but mechanically stable, at low pHs where chitosan dissolves. The chitosan-alginate complex formation is due to the strong electrostatic interaction of amine groups of chitosan with the carboxyl groups of alginate. Chitosan-alginate gel beads composed from a chitosan core and a chitosan-alginate skin are prepared by dropping a solution of alginate into chitosan solution. It was found that the macromolecular chitosan rapidly bind onto the alginate droplet surface, but is limited to diffuse into the inner core (Gåserød et al., 1998). In order to increase the stability of chitosan-alginate complexes, calcium chloride containing chitosan solution has been used for alginate gelation. The presence of divalent calcium ions greatly enhanced the gel formation. The rate and extent of chitosan binding process increased linearly with the increase of calcium chloride concentration. Several authors (Hari et al., 1996; Huguet et al., 1994) developed insulin-containing calcium-alginate-chitosan microspheres. The shrinkage of calcium alginate capsules at low pH enables the encapsulated drug to be released into the gastrointestinal tract. In general, they provided an excellent gastrointestinal delivery system for insulin. Chitosan–alginate microcapsules can also be an effective method of protecting egg yolk immunoglobulin (IgY) from gastric inactivation, enabling its use for the widespread prevention and control of enteric diseases (Li et al., 2009).

Dambies et al.(Dambies et al., 2001) prepared chitosan gel beads using molybdate as a gelling agent. It was observed that this new gelation technique led to a structure different from the one produced during alkaline coagulation of chitosan solutions. Large open pores are characteristic for the morphology of the gel beads produced in a molybdate solution under optimum conditions (pH 6, molybdate concentration, 7 g/l). These conditions correspond to the predominance of molybdate polyoxoanions (polynuclear hydrolyzed species), identified as the most favourable molybdate species for both adsorption on chitosan and chitosan coagulation. Different interaction mechanisms are proposed for these reactions since (a) the molar ratio between Mo and $-NH_3^+$ differs with the process; (b) the maximum molybdate uptake was lower for the coagulation process than for the sorption technique (except when a very high molybdate concentration is used (20 g/L^{-1})); (c) the fraction of molybdate released from the gel beads when treated with phosphoric acid is significantly higher for beads saturated by sorption than for beads directly coagulated in a molybdate solution. A double layer structure corresponding to a very compact 100 μm thick external layer and an internal structure of small pores was generated.

2.1.2. Emulsification and Ionotropic Gelation

In this method the dispersed phase, which consists of an aqueous solution of chitosan, is added to a non-aqueous continuous phase to form water in oil emulsion. The emulsion cross-linking method utilizes the reactive functional groups of chitosan to cross-link with a cross-linking agent. Then, the microspheres thus formed are removed by filtration, washed and dried. Acidic chitosan solution was emulsified in an oil phase by mechanical stirring, followed by cross-linking to stabilize the droplets. As emulsion medium, most authors use petroleum ether–paraffin oil mixture (Lim et al., 1997), but also *n*-hexadecane and sunflower oil mixture (Biró et al., 2008) have been reported.

Shu and collaborators (Shu and Zhu, 2001) investigated the capacity of three kinds of anions (tripolyphosphate, citrate and sulphate) to interact with chitosan by turbidimetric titration. Tripolyphosphate, citrate and sulphate are multivalent anions, which may interact with positive charged chitosan to form complexes. However, these electrostatic interactions between anions and chitosan are mainly controlled by the charge density of both anions and chitosan, which is determined by the solution pH. Out of the pH region where anions interacted with chitosan, no microspheres were formed. In the case of the conventional emulsification and ionotropic gelation method the anions interact with chitosan, forming only irregular microparticles, while spherical microspheres with diameters in the range of tens of microns have been obtained in the modified process.

2.1.3. Coacervation and Phase Separation.

Coacervation is the separation of colloidal systems into two liquid phases. The more concentrated in colloidal component phase is the coacervate and the other phase is the equilibrium solution. This method consists of blowing with compressed air an acetic or formic acid chitosan solution into a precipitating medium, such as aqueous or methanolic NaOH. The size of the droplets depends on the nozzle diameter and the extrusion rate of the polymer solution. The strength and porosity of the beads formed is influenced by the concentration and deacetylation degree of chitosan, and the type and concentration of the coacervating agent used. The beads are hardened by successive washings with hot and cold water (Figure 1).

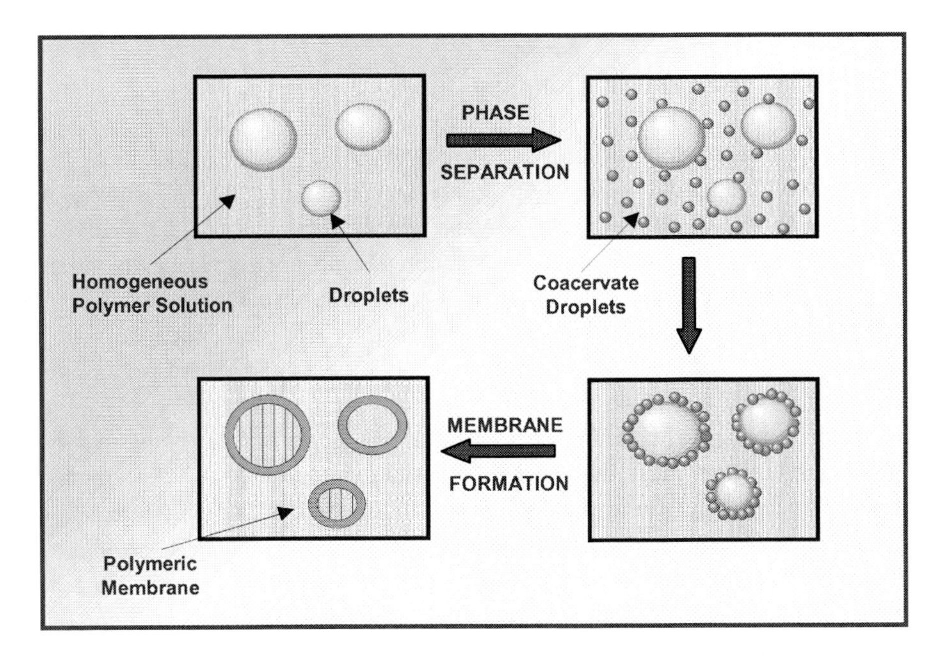

Figure 1. Schematic procedure for preparation of chitosan microspheres by coacervation/phase separation method.

Complex coacervation is a process of spontaneous phase separation that occurs when two oppositely charged polyelectrolytes are mixed in an aqueous solution. It is based on the ability of cationic and anionic water-soluble polymers to interact in water to form a liquid, polymer-rich phase called a complex coacervate. This method is based on the separation of an aqueous polymeric solution into two miscible liquid phases - a dense coacervate phase and a dilute equilibrium phase. The dense coacervate phase wraps as a uniform layer around suspended core materials. Complex coacervation can result spontaneously upon mixing of oppositely charged polyelectrolytes in aqueous media. The charges must be large enough to induce interaction, but not too large to cause precipitation. The core material such as an oily phase is dispersed in an aqueous solution of the two polymers. A change is made in the aqueous phase (pH) to induce the formation of a polymer rich phase that becomes the wall material. The coacervates are usually stabilized by thermal treatment, crosslinking or desolvation techniques. Microcapsules developed by this method are usually collected by filtration or centrifugation, washed with an appropriate solvent and subsequently dried (air-dried or by standard techniques such as by fluid bed or spray drying). This method has been extended for the preparation of microspheres loaded with small-molecule substrates, DNA and proteins (Sinha et al., 2004). Efficient chitosan based delivery systems have been designed employing the coacervation technique for sustained release of appropriately modified brassinosteroids in agricultural applications (Quiñones et al., 2010).

Sodium alginate, sodium carboxymethylcellulose, κ-carregeenan and sodium polyacrylic acid can be used for complex coacervation with chitosan to form microspheres. The microcapsules are formed by dropwise addition of a 3 % chitosan (polycation) solution into a 0.75 % alignate (polyanion) solution. The resultant capsules consist of a liquid chitosan core with a hard alginate coating. The addition of $CaCl_2$ to chitosan and glucose to alginate increased the capsule strength (Mary and Dietrich, 1988).

2.2. Crosslinking with Other Chemicals

To increase the time frame for drug delivery, hydrophilic polymers such as chitosan need to be cross-linked. Several cross-linking reagents have been used for cross-linking chitosan such as glutaraldehyde, tripolyphosphate, ethylene glycol, formaldehyde, diglycidyl ether and diisocyanate. The most common chitosan crosslinking agents are glyoxal, glutaraldehyde and the naturally occurring compound genipin. Glutaraldehyde is a highly reactive molecule, being able to undergo polymerization in aqueous media. The glutaraldehyde reaction with compounds which have primary amino groups is well known and has been extensively used. However, reports (Monteiro Jr and Airoldi, 1999; Thacharodi and Rao, 1993) indicate that variables like chitosan and glutaraldehyde concentration, degree of deacetylation, solvent, pH, and reaction temperature must be carefully considered, because they could change the physical and chemical properties of the final product (Neto et al., 2005). Different studies (Mi et al., 2002) have demonstrated that when the crosslinking agent is genipin, chitosan microspheres had a superior biocompatibility and a slower degradation rate than the glutaraldehyde-crosslinked chitosan microspheres. Moreover, there are concerns about the toxicity of the most studied cross-linking agents, especially glutaraldehyde, which may impair the biocompatibility of the chitosan delivery system, and therefore, there is a need to provide a crosslinking agent with low cytotoxicity. It was concluded that the genipin-crosslinked chitosan microspheres may be the most suitable polymeric carrier for long-acting injectable drug delivery.

The general procedure to obtain the cross-linked chitosan microspheres consists in dissolving purified chitosan in acid solution (HOAc) under vigorous stirring for about 3h at room temperature, followed by subsequent blowing of the chitosan solution through a nozzle into a methanolic solution of NaOH (1.0 M) to form the microspheres. The microspheres are coacervated and settled in the container with methanolic solution of alkali. Then, the microspheres obtained by centrifugation and washing with deionized water are transferred to a solution of glutaraldehyde, formaldehyde or genipin for cross-linking. After 6 h, the microspheres are removed and washed with hot and cold water and vacuum dried at 30 °C.

Genipin-chitosan delivery systems have shown to be useful to control the release of a number of drugs, such as clarithromycin, tramadol hydrochloride, and low molecular weight heparin loaded in spray-dried microspheres. Nevertheless, *in vivo* studies are needed to support the use of genipin for controlled drug release(Harris et al., 2010). In particular, for pharmaceutical applications, specific crosslinking conditions should be studied in each case depending on the nature of the drug, dose, release location in human body, and pharmacokinetic factors, among others.

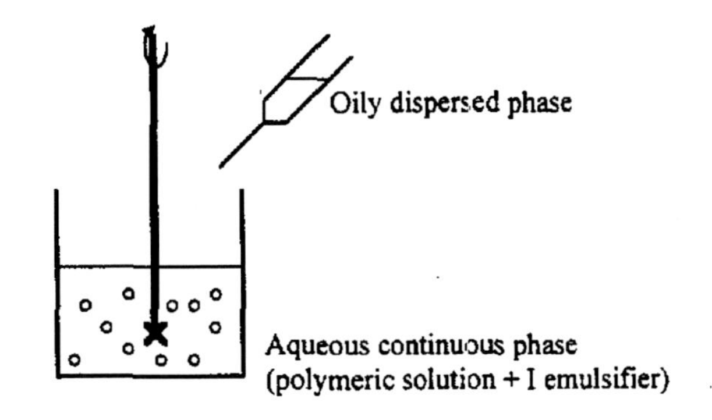

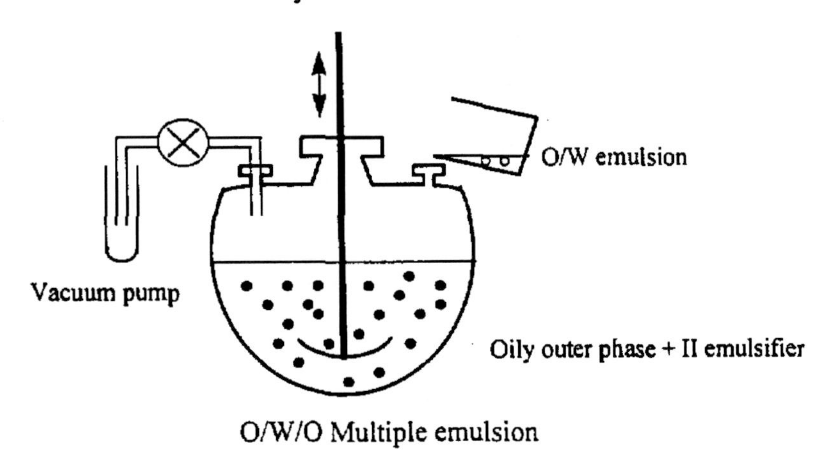

Figure 2. Schematic procedure for preparation of chitosan microspheres by the “dry-in-oil multiple emulsion” method. Reproduced from Genta et al. 1997.

Encapsulation of lipophylic drugs in chitosan has been achieved by multiple emulsion method proposed by Genta et al. (Genta et al., 1997) and recently used by Wu and Li (Wu and Li 2002) to form genistein-chitosan microspheres. This method consists of an emulsification/solvent evaporation process performed under mild conditions (Figure2). Furthermore, the technique allows the combination with physical and chemical cross-linking in order to modulate the drug release.

Other techniques to prepare porous chitosan microspheres use a wet phase-inversion method (Mi et al., 1999). The pore structure of the chitosan microsphere could be modified altering the pH of the coagulation medium by introducing reagents with different functional groups; carboxyl, hydrophobic acyl and quaternary ammonium groups. This kind of microspheres were suitable for the delivery of antigen (controlled-release vaccine) and showed higher adsorption efficiency and slower release rate of the antigen when modified chemically with 3-chloro-2 hydroypropyltrimethylamonium chloride.

2.3. Other Techniques

Other approaches for the generation of microspheres include: spray drying (He et al., 1996), layer-by-layer deposition (Ye et al., 2005), supercritical assisted atomization (Reverchon and Antonacci, 2006) and sonication (Skirtenko et al., 2010).

The layer-by-layer (LBL) self-assembly technique provides microparticles with a narrow size distribution, however, the fabrication procedure is complicated and template cores are needed for the formation of microparticles. Nevertheless, some authors employed this method to encapsulate mainly low water-soluble drugs. For instance, Ye *et al.*(Ye et al., 2005) developed microcapsules of chitosan and alginate through LBL to encapsulate Indomethacin (IDM).

The supercritical assisted atomization (SAA) technique used by Reverchon (Reverchon and Antonacci, 2006) allowed different degrees of crystallinity which can be easily modulated as a function of the temperature in the precipitator. This approach could be particularly relevant, for example, in modifying drug release mechanisms in CS/drugs formulations.

Spray drying is a well-known process, which is used to produce dry powders, granules, or agglomerates from drug-excipient solutions and suspensions (Chawla et al., 1994). When the liquid is fed into the nozzle with a peristaltic pump, atomization occurs under the effect of the compressed air, disrupting the liquid into small droplets. The droplets, together with hot air, are blown into a chamber where the solvent in the droplets is evaporated. Eventually, the dry microspheres are collected in a collection bottle.

Lately, several studies have been published describing the preparation of microspheres by the combination of cross-linking agents and spray drying methods (Oliveira et al., 2005) (Alhalaweh et al., 2009; Desai and Park, 2005b). Other complementary technique that could be carried out by spray-drying equipment is the spray congealing where protective coating is to be applied as a melt. Briefly, core material is dispersed in a coating material melt rather than in a coating solution. Coating solidification is accomplished by spraying the hot mixture into cool air stream. Waxes, fatty acids and alcohols, polymers which are solid at room temperature but meltable at reasonable temperature, are applicable in spray congealing (Jyothi et al., 2010)

Skirtenko and collaborators (Skirtenko et al., 2010) developed a fast one-step process in solutions with a wide range of pHs (4–9) for the preparation of stable chitosan microspheres. Chitosan is dissolved in double-distilled water (pH 5.5) in a cylindrical vessel and a layer of dodecane or soybean oil was placed over the aqueous solution. The microspherization was carried out using ultrasonic waves and was completed in only three minutes. After the synthesis, the microspheres were separated from the unreacted chitosan by leaving the reaction mixture at 4 °C for 24 – 36 h, and washed with distilled water by centrifugation at 800 rpm for 15 min. Free amino groups which are available on the external surface of the microsphere have been further conjugated, under mild conditions, with different fluorescent dyes, thereby suggesting possible modification of the surface of the microsphere with targeting vectors such as peptides, proteins, or small molecules for therapeutic or diagnostic purposes. Additionally, the particle size, surface morphology and drug-entrapment efficiency of the microspheres have been characterized (Figure 3). The stability of chitosan microspheres was demonstrated either when the organic overlayer was dodecane or soybean oil by shaking the microspheres under different acidic and basic conditions, including formic acid buffer (100 mm, pH 4.0), PBS buffer (100 mm, pH 7.4), and Tris/HCl buffer (100 mm, pH 9).

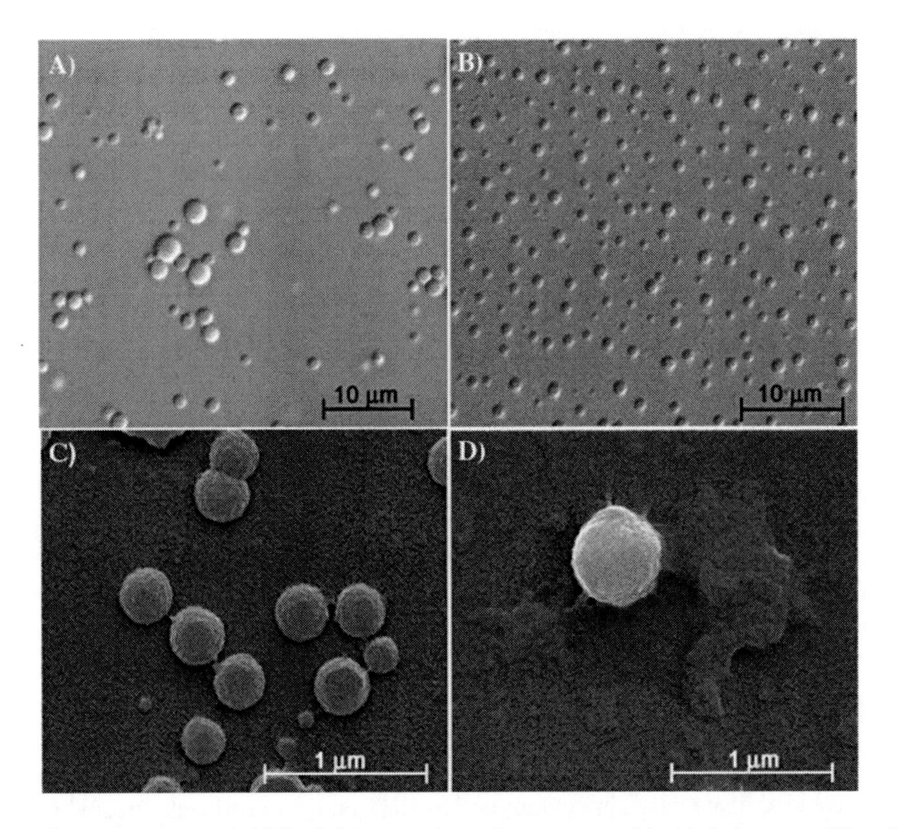

Figure 3. Microscope images (x100) of chitosan microspheres prepared in A) dodecane or B) soybean oil, and corresponding SEM pictures prepared in dodecane (C, D). Reproduced from Skirtenko et al. with permission.

3. Applications of Chitosan Microspheres.

3.1. Food Industry

The current trend for healthier way of living includes a growing consumers' awareness about what they eat and what health benefits certain ingredients may have. Preventing illness by diet is an opportunity for innovative so called "functional foods", many of which are augmented with health-promoting ingredients. However, simply adding functional ingredients to food products to improve their nutritional value could compromise the taste, colour, texture and aroma, due to the slow degradation, oxidation and lost of activity. Ingredients can also react with components present in the food system, which may limit their bioavailability. Microencapsulation is used to overcome these drawbacks by providing viable texture blending, appealing aroma release and taste, odour and colour masking. The encapsulation technology allows the incorporation in the food of minerals, vitamins, flavours and essential oils. In addition, microencapsulation can simplify the food manufacturing process by converting liquids to solid powder in powder handling equipment, thereby decreasing the production costs. Microencapsulation also helps unstable materials to withstand the processing and packaging conditions and extends the shelf life of the active ingredient. (Schrooyen et al., 2001).

Several antioxidants such vitamins and fatty acids were encapsulated into chitosan microspheres. Desai and Park encapsulate vitamin C in chitosan microspheres using tripolyphosphate (TPP) as a cross-linking agent for the first time by spray drying method (Desai and Park, 2006; Desai and Park, 2005a). Unlike formaldehyde or gluteraldehyde, TPP is a non-toxic cross-linking agent. The volume of TPP and the molecular weight of the chitosan had a noticeable influence on the particle size, loading and encapsulation efficiency range.

α-lipoic acid is an *antioxidant* able to scavenge directly free radicals and protect cells from oxidative damage. Weerakody and Kosaraju (Weerakody et al., 2008) encapsulated α-lipoic acid in chitosan microspheres by a spray-drying process achieving extended over the time release of antioxidant activity. Urucum pigment, similarly to vitamins, is widely used in food industry. Urucum pigment, mainly composed by antioxidant compounds belonging to the class of carotenoids, has been successfully incorporated into chitosan by means of a spray-drying process obtaining a dry, water soluble and colourful powders (Souza et al., 2005). However, the spray-drying encapsulation presents the disadvantage of the high temperature of the drying process, potentially accelerating the oxidation of fatty acids. To avoid this, Klaypradit and Huang (Klaypradit and Huang, 2008) studied the encapsulation of fish oil into chitosan microspheres using an ultrasonic atomizer.

Krasaekoopt, Bhandari, and Deeth (Krasaekoopt et al., 2006) evaluated the survival of probiotics, such as L. acidophilus 547, L. casei 01 and B. bifidum 1994, encapsulated in chitosan-coated alginate beads in yoghurt and in UHT and conventionally treated milk during storage. The survival of the encapsulated bacteria was higher than that of the free cells by approximately 1 log. The number of probiotic bacteria was maintained above the recommended therapeutic minimum (107 cfu/g) throughout storage for the lactobacilli but not for the bifidobacteria. Lee et al. (Lee et al., 2004) carried out a similar study and compared different molecular weights chitosans for coating conventional alginate beads. They investigated the effects of chitosan alginate microparticles on the survival of L. bulgaricus KFRI763 in simulated gastric juices and intestinal fluid and on their storage stability at 4 and 22 °C. The probiotic loaded in alginate microparticles were prepared by spraying a mixture of sodium alginate and cell culture into a $CaCl_2$ chitosan solution using an air-atomizing device. None of the microorganism exposed to the gastric fluid (pH 2.0) for 1 h survived. In contrast, an impressively high survival rate was observed when the sprayed particles were coated with chitosan. The authors concluded that the microencapsulation of lactic acid bacteria (LAB) in alginate and the chitosan coating offers an effective means of delivering viable bacterial cells to the colon and maintaining their survival during refrigerated storage

3.2. Agriculture

Nowadays insect pheromones are becoming a biorational alternative to the conventional hard pesticides. Specifically, sex attractant pheromones can reduce insect populations by disrupting their mating process. Small amounts of species-specific pheromones are dispersed during the mating season, raising the background level of pheromone to the point where it hides the pheromone plume released by its female mate (Dubey et al., 2009) Brassinosteroids (BS) are one group of steroid plant hormones with important regulatory functions in the physiological factors of plant growing. These hormones have potential uses as agrochemicals promoting the vegetable growth i.e. the productivity of the crops, due to their specific ability to stimulate cell enlargement and division. Recently, Quiñones and collaborators (Quiñones et al., 2010) encapsulated two synthetic analogues of brassinosteroids (BA) and diosgenin derivatives into chitosan microspheres (CS) by simple coacervation–cross-linking in the presence of sodium tripolyphosphate.

3.3. Pharmaceutics

One of the major application areas of the encapsulation technique is the pharmaceutical/ biomedical one for controlled/sustained drug delivery (Langer, 1990). Potential applications of such drug delivery system are the replacement of therapeutic agents, gene therapy and vaccines. Microparticles prepared using chitosan are being further investigated for delivery of several classes of drugs. The different categories of microencapsulated in chitosan drugs are reviewed by Sinha (Sinha et al., 2004). In this work more than nineteen different classes of substances were described. Anticancer, anti-inflammatory drugs, cardiac and antithrombotic agents, antibiotics among other therapeutic agents have been entrapped in chitosan microspheres. The efficiency of drugs in the chitosan microspheres is dependent upon the chitosan concentration. The entrapment efficiency increases with increase in chitosan concentration. The drugs can also be loaded via a passive absorption method by adding microspheres to a drug solution, relying on the swelling properties of the microspheres in the drug solution. The drug release from the chitosan microspheres is dependent upon the molecular weight and concentration of chitosan, drug content and density of crosslinking

3.4. Textile Industry

Examples of microencapsulation technology applied in textiles include fabrics coloration and finishing, with i.e. durable fragrances and skin softeners (Nelson, 1991). Other potential applications include insect repellents, cosmetics, dyes, antimicrobials, phase change materials, fire retardants, polychromic and thermochromic effects etc. (Monllor et al., 2007).

Very recently our group developed "antibacterial fabrics" by sonochemical coating of cotton bandages with chitosan microspheres (Angel and Gedanken, *in progress*). An organic solvent, such as dodecane, has been used as a co-solvent, and the sonochemical process has been carried out for only 3 min. The creation and the attachment of the chitosan microspheres were achieved in a one-step process. The sonication cell contained the aqueous solution of chitosan, the overlayering organic solution (dodecane or cotton oil), the antibiotic (tetracycline) and a 5 x 5 cm sample of cotton or polyester. At the end of the reaction chitosan microspheres containing encapsulated tetracycline were embedded on the fabric. Light microscopy observations revealed the morphology of the cotton bandages coated with chitosan microspheres (Figure 4). The antibacterial effect of the tetracycline loaded microspheres was demonstrated against two strains of bacteria as shown in Fig. 5.

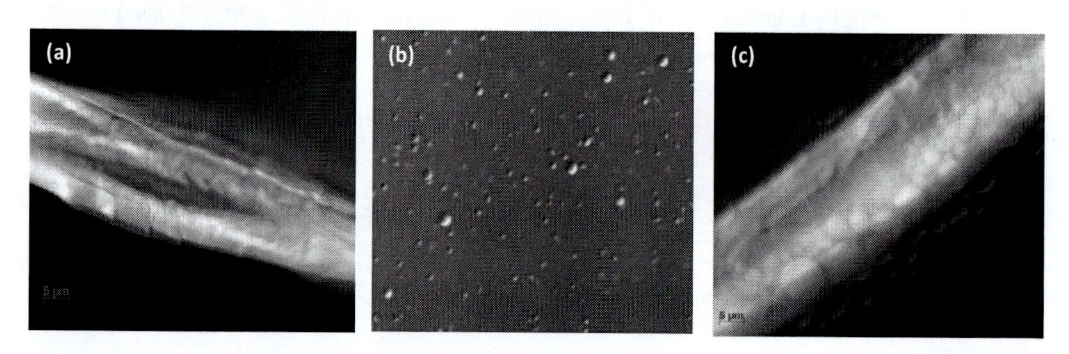

Figure 4. Morphology of the cotton bandages coated with chitosan microspheres determined by light microscopy (a) Apo-Tome image of pristine cotton fibre; (b) Apo-Tome image of chitosan microspheres (average size ~1 μm); (c)Apo-Tome image of cotton fiber coated with chitosan microspheres.

4. General Conclusions

Microspheres of chitosan, also called indistinctively microparticles or microbeads, can be developed by several methods based on the chemical structure of the biopolymer. The cationic amino groups of the repeating glucopyranose units of chitosan interact electrostatically with anionic groups (usually carboxylic acid groups) of other polyions to form polyelectrolyte complexes. So far, this crosslinking approach has been the most extensively used for the preparation of chitosan microspheres. Examples of techniques using ionic crosslinking are emulsification and ionotropic gelation and coacervation.

Other techniques rely on particles formation through chemical cross-linking of chitosan amino groups with reagents such as glutaraldehyde. However, due to the toxicity of the conventional chemical crosslinkers the tendency is to progressively replace them with natural compounds, i.e. genipin in order to assure the biocompatibility and immune response especially in drug delivery applications. Nevertheless, further *in vivo* studies are still needed to confirm the suitability of genipin for medical application.

Recently, an innovative sonochemical method revealed to be suitable for a one step simple microspheres generation in a considerably shorter reaction time (3 min) compared to the time necessary for preparation of the microspheres by means of emulsification and ionotropic gelation (about 3h). Moreover, the sonochemical method was also suitable for the encapsulation of several functional molecules or nanoparticles. In general, the manufacturing approach is always selected in order to ensure the most suitable for specific application properties (size, porosity, swelling capacity, burst strength and degradability) to meet the end-use requirements.

The biological properties of chitosan as a material for microspheres preparation have shown to improve the dissolution rate of poorly soluble drugs and thus could improve the bioavailability and enhance the therapeutic effects of the active agent incorporated.

Among the therapeutics successfully incorporated in chitosan microparticles are anticancer, anti-inflammatory, antidiabetic, antithrombotic and diuretic drugs, antibiotics, steroids, proteins and amino acids (Sinha et al., 2004). Moreover, the intrinsic antimicrobial activity of chitosan could synergistically enhance the effect of the therapeutic agents.

Microparticular delivery systems based on biocompatible carriers, such as chitosan are expected to even further expand into the food, textile, agricultural, medical and cosmetic fields. This will certainly stimulate the work to improve their production technologies and the search for novel, versatile approaches in the future.

REFERENCES

Alhalaweh, A., Andersson, S., Velaga, S.P., 2009. Preparation of zolmitriptan-chitosan microparticles by spray drying for nasal delivery. *European Journal of Pharmaceutical* Sciences 38, 206-214.

Allémann, E., Leroux, J.-C., Gurny, R., Doelker, E., 1993. *In vitro* extended-release properties of drug-loaded poly(DL-lactic acid) nanoparticles produced by a salting-out procedure. *Pharmaceutical Research* 10, 1732-1737.

Angel, U., Gedanken, A., *in progress. Chitosan microspheres attached to cotton fabric by a sonochemical method.*

Berger, J., Reist, M., Mayer, J.M., Felt, O., Gurny, R., 2004. Structure and interactions in chitosan hydrogels formed by complexation or aggregation for biomedical applications. *European Journal of Pharmaceutics and Biopharmaceutics* 57, 35-52.

Bhagat, H.R., Hollenbeck, R.G., Pande, P.G., Bogdansky, S., D., F.C., Rock, M., 1994. Preparation and evaluation of methotrexate-loaded biodegradable polyanhydride microspheres. *Drug Development and Industrial Pharmacy* 10, 1725-1737.

Biró, E., Németh, Á.S., Sisak, C., Feczkó, T., Gyenis, J., 2008. Preparation of chitosan particles suitable for enzyme immobilization. *Journal of Biochemical and Biophysical* Methods 70, 1240-1246.

Birrenbach, G., Speiser, P.P., 1976. Polymerized micelles and their use as adjuvants in immunology. *Journal of Pharmaceutical Sciences* 65, 1763-1766.

Chandy, T., Sharma, C.P., 1990. Chitosan-as a biomaterial. *Artificial Cells, Blood Substitutes and Biotechnology* 18, 1-24.

Chawla, A., Taylor, K.M.G., Newton, J.M., Johnson, M.C.R., 1994. Production of spray dried salbutamol sulphate for use in dry powder aerosol formulation. *International Journal of Pharmaceutics* 108, 233-240.

Dambies, L., Vincent, T., Domard, A., Guibal, E., 2001. Preparation of Chitosan Gel Beads by Ionotropic Molybdate Gelation. *Biomacromolecules* 2, 1198-1205.

Desai, K.G., Park, H.J., 2006. Effect of manufacturing parameters on the characteristics of vitamin C encapsulated tripolyphosphate-chitosan microspheres prepared by spray-drying. *Journal of Microencapsulation* 23, 91-103.

Desai, K.G.H., Park, H.J., 2005a. Encapsulation of vitamin C in tripolyphosphate cross-linked chitosan microspheres by spray drying. *Journal of Microencapsulation* 22, 179-192.

Desai, K.G.H., Park, H.J., 2005b. Preparation of cross-linked chitosan microspheres by spray drying: Effect of cross-linking agent on the properties of spray dried microspheres. Journal of Microencapsulation: *Micro and Nano Carriers* 22, 377 - 395.

Dodane, V., Amin Khan, M., Merwin, J.R., 1999. Effect of chitosan on epithelial permeability and structure. *International Journal of Pharmaceutics* 182, 21-32.

Dubey, R., Shami, T.C., Bhasker, R.K.U., 2009. Microencapsulation technology and applications. *Defence Science Journal* 59.

Etrych, T., Leclercq, L., Boustta, M., Vert, M., 2005. Polyelectrolyte complex formation and stability when mixing polyanions and polycations in salted media: A model study related to the case of body fluids. *European Journal of Pharmaceutical Sciences* 25, 281-288.

Felt, O., Furrer, P., Mayer, J.M., Plazonnet, B., Buri, P., Gurny, R., 1999. Topical use of chitosan in ophthalmology: tolerance assessment and evaluation of precorneal retention. *International Journal of Pharmaceutics* 180, 185-193.

Fernández-urrusuno, R., Fattal, E., Porquet, D., Feger, J., Couvreur, P., 1995. Evaluation of liver toxicological effects induced by polyalkylcyanoacrylate nanoparticles. *Toxicology and Applied Pharmacology* 130, 272-279.

Furda, I., 1980. *Nonabsorbable lipid binder*, in: (16664 Meadowbrook La., W., MN, 55391) (Ed.).

Gåserød, O., Jolliffe, I.G., Hampson, F.C., Dettmar, P.W., Skjåk-Bræk, G., 1998. The enhancement of the bioadhesive properties of calcium alginate gel beads by coating with chitosan. *International Journal of Pharmaceutics* 175, 237-246.

Genta, I., Perugini, P., Conti, B., Pavanetto, F., 1997. A multiple emulsion method to entrap a lipophilic compound into chitosan microspheres. *International Journal of Pharmaceutics* 152, 237-246.

Hari, P.R., Thomas, C., Chandra, P.S., 1996. Chitosan/calcium-alginate beads for oral delivery of insulin. *Journal of Applied Polymer Science* 59, 1795-1801.

Harris, R., Lecumberri, E., Heras, A., 2010. Chitosan-genipin microspheres for the controlled release of drugs: clarithromycin, tramadol and heparin. *Marine Drugs* 8, 1750-1762.

He, P., Davis, S.S., Illum, L., 1996. Chitosan microspheres prepared by spray dying method. *European Journal of Pharmaceutical Sciences* 4, S173-S173.

He, P., Davis, S.S., Illum, L., 1998. In vitro evaluation of the mucoadhesive properties of chitosan microspheres. *International Journal of Pharmaceutics* 166, 75-88.

Huguet, M.L., Groboillot, A., Neufeld, R.J., Poncelet, D., Dellacherie, E., 1994. Hemolglobin encapsulation in chitosan/calcium alginate beads. *Journal of Applied Polymer Science* 51, 1427-1432.

Il'ina, A.V., Varlamov, V.P., 2005. Chitosan-based polyelectrolyte complexes: A review. *Applied Biochemistry and Microbiology* 41, 5-11.

Ito, M., Ban, A., Ishihara, M., 2000. Anti-ulcer effects of chitin and chitosan, healthy foods, in rats. *The Japanese Journal of Pharmacology* 82, 218-225.

Iwata, M., McGinity, J.W., 1991. Preparation of multi-phase microspheres of poly(D,L-lactic acid) and poly(D,L-lactic-co-glycolic acid) containing a W/O emulsion by a multiple emulsion solvent evaporation technique. *Journal of Microencapsulation* 9, 201-214.

Jyothi, N.V.N., Prasanna, P.M., Sakarkar, S.N., Prabha, K.S., Ramaiah, P.S., Srawan, G.Y., 2010. Microencapsulation techniques, factors influencing encapsulation efficiency. *Journal of Microencapsulation* 27, 187-197.

Kawaguchi, H., 2000. Functional polymer microspheres. *Progress in Polymer Science* 25, 1171-1210.

Klaypradit, W., Huang, Y.-W., 2008. Fish oil encapsulation with chitosan using ultrasonic atomizer. *LWT - Food Science and Technology* 41, 1133-1139.

Kockisch, S., Rees, G.D., Young, S., A, Tsibouklis, J., Smart, J., D., 2003. Polymeric microspheres for drug delivery to the oral cavity: An *in vitro* evaluation of mucoadhesive potential. *Journal of Pharmaceutical Sciences* 92, 1614-1623.

Kotzé, A.F., Lueßen, H.L., de Leeuw, B.J., de Boer, B.G., Coos Verhoef, J., Junginger, H.E., 1998. Comparison of the effect of different chitosan salts and N-trimethyl chitosan chloride on the permeability of intestinal epithelial cells (Caco-2). *Journal of Controlled Release* 51, 35-46.

Krasaekoopt, W., Bhandari, B., Deeth, H.C., 2006. Survival of probiotics encapsulated in chitosan-coated alginate beads in yoghurt from UHT- and conventionally treated milk during storage. *LWT - Food Science and Technology* 39, 177-183.

Krayukhina, M.A., et al., 2008. Polyelectrolyte complexes of chitosan: formation, properties and applications. *Russian Chemical Reviews* 77, 799.

Kreuter, J., Speiser, P.P., 1976. New adjuvants on a polymethylmethacrylate base. *Infect. Immun.* 13, 204-210.

Kumar, M.N.V.R., 2000. Nano and microparticles as controlled drug delivery devices. *J Pharm Pharmaceut Sci* 3, 234-258.

Langer, R., 1990. New methods of drug delivery. *Science* 249, 1527-1533.

Lee, J.S., Cha, D.S., Park, H.J., 2004. Survival of Freeze-Dried Lactobacillus bulgaricus KFRI 673 in Chitosan-Coated Calcium Alginate Microparticles. *Journal of Agricultural and Food Chemistry* 52, 7300-7305.

Li, X.-Y., Jin, L.-J., Uzonna, J.E., Li, S.-Y., Liu, J.-J., Li, H.-Q., Lu, Y.-N., Zhen, Y.-H., Xu, Y.-P., 2009. Chitosan-alginate microcapsules for oral delivery of egg yolk immunoglobulin (IgY): In vivo evaluation in a pig model of enteric colibacillosis. *Veterinary Immunology and Immunopathology* 129, 132-136.

Lim, L.Y., Wan, L.S.C., Thai, P.Y., 1997. Chitosan microspheres prepared by emulsification and ionotropic gelation. *Drug Development and Industrial Pharmacy* 23, 981 - 985.

Lueßen, H.L., Lehr, C.M., Rentel, C.O., Noach, A.B.J., de Boer, A.G., Verhoef, J.C., Junginger, H.E., 1994. Bioadhesive polymers for the peroral delivery of peptide drugs. *Journal of Controlled Release* 29, 329-338.

Mary, M.D., Dietrich, K., 1988. Chitosan-Alginate Complex Coacervate Capsules: Effects of Calcium Chloride, Plasticizers, and Polyelectrolytes on Mechanical Stability. *Biotechnology Progress* 4, 76-81.

Mi, F.-L., Shyu, S.-S., Chen, C.-T., Schoung, J.-Y., 1999. Porous chitosan microsphere for controlling the antigen release of Newcastle disease vaccine: preparation of antigen-adsorbed microsphere and in vitro release. *Biomaterials* 20, 1603-1612.

Mi, F.-L., Tan, Y.-C., Liang, H.-F., Sung, H.-W., 2002. In vivo biocompatibility and degradability of a novel injectable-chitosan-based implant. *Biomaterials* 23, 181-191.

Monllor, P., Bonet, M.A., Cases, F., 2007. Characterization of the behaviour of flavour microcapsules in cotton fabrics. *European Polymer Journal* 43, 2481-2490.

Monteiro Jr, O.A.C., Airoldi, C., 1999. Some studies of crosslinking chitosan-glutaraldehyde interaction in a homogeneous system. *International Journal of Biological Macromolecules* 26, 119-128.

Nagyvary, J.J.B., TX) 1982. *Method for treating hyperbilirubinemia*

Nelson, G., 1991. Microencapsulates in textile coloration and finishing. *Review of Progress in Coloration and Related Topics* 21, 72-85.

Neto, C.G.T., Dantas, T.N.C., Fonseca, J.L.C., Pereira, M.R., 2005. Permeability studies in chitosan membranes. Effects of crosslinking and poly(ethylene oxide) addition. *Carbohydrate Research* 340, 2630-2636.

Oliveira, B.F., Santana, M.H.A., Ré, M.I., 2005. Spray-dried chitosan microspheres cross-linked with d, l-glyceraldehyde as a potential drug delivery system: preparation and characterization. *Brazilian Journal of Chemical Engineering* 22, 353-360.

Park, J.H., Saravanakumar, G., Kim, K., Kwon, I.C., Targeted delivery of low molecular drugs using chitosan and its derivatives. *Advanced Drug Delivery Reviews* 62, 28-41.

Patil, J.S., Kamalapur, S.C., Marapur, D.V., Kadam, D.V., 2010. Ionotropic gelation and polyectrolyte complexation:the novel tecnhiques to design hydrogel particulate sustained, modulated drug delivery system: A review. *Digest Journal of Nanomaterials and Biostructures* 5, 241-248.

Quiñones, J.P., García, Y.C., Curiel, H., Covas, C.P., 2010. Microspheres of chitosan for controlled delivery of brassinosteroids with biological activity as agrochemicals. *Carbohydrate Polymers* 80, 915-921.

Reverchon, E., Antonacci, A., 2006. Chitosan Microparticles Production by Supercritical Fluid Processing. *Industrial & Engineering Chemistry Research* 45, 5722-5728.

Scheffel, U., Rhodes, B.A., Natarajan, T.K., Wagner, H.N., Jr., 1972. Albumin microspheres for study of the reticuloendothelial system. *J Nucl Med* 13, 498-503.

Schipper, N.G.M., Olsson, S., Hoogstraate, J.A., deBoer, A.G., Vårum, K.M., Artursson, P., 1997. Chitosans as Absorption Enhancers for Poorly Absorbable Drugs 2: Mechanism of Absorption Enhancement. *Pharmaceutical Research* 14, 923-929.

Schipper, N.G.M., Vårum, K.M., Artursson, P., 1996. Chitosans as absorption enhancers for poorly absorbable drugs. 1: influence of molecular weight and degree of acetylation on drug transport across human intestinal epithelial (Caco-2) cells. *Pharmaceutical Research* 13, 1686-1692.

Schrooyen, P.M.M., Meer, R.v.d., Kruif, C.G.D., 2001. *Microencapsulation: its application in nutrition.* Proceedings of the Nutrition Society 60, 475-479.

Shu, X.Z., Zhu, K.J., 2001. Chitosan/gelatin microspheres prepared by modified emulsification and ionotropic gelation. *Journal of Microencapsulation* 18, 237-245.

Sinha, V.R., Singla, A.K., Wadhawan, S., Kaushik, R., Kumria, R., Bansal, K., Dhawan, S., 2004. Chitosan microspheres as a potential carrier for drugs. *International Journal of Pharmaceutics* 274, 1-33.

Skirtenko, N., Tzanov, T., Gedanken, A., Rahimipour, S., 2010. One-step preparation of multifunctional chitosan microspheres by a simple sonochemical method. *Chemistry - A European Journal* 16, 562-567.

Souza, T.C.R., Parize, A.L., Brighente, I.M.C., FÃ¡vere, V.T., Laranjeira, M.C.M., 2005. Chitosan microspheres containing the natural urucum pigment. *Journal of Microencapsulation: Micro and Nano Carriers* 22, 511 - 520.

Tachihara, K., Onishi, H., Machida, Y., 1997. Preparation of silver sulfadiazine-containing spongy membranes of chitosan and chitin–chitosan mixture and their evaluation as burn wound dressings. *Arch. Practical Pharm.* 57, 159-167.

Thacharodi, D., Rao, K.P., 1993. Propranolol hydrochloride release behaviour of crosslinked chitosan membranes. *Journal of Chemical Technology & Biotechnology* 58, 177-181.

Weerakody, R., Fagan, P., Kosaraju, S.L., 2008. Chitosan microspheres for encapsulation of [a]-lipoic acid. *International Journal of Pharmaceutics* 357, 213-218.

Wu, W.Y., Li , Y.G., 2002. *Preparation of genistein-loaded chitosan microspheres.* Zhongguo Yaoxuehui, Beijing, CHINE.

Ye, S., Wang, C., Liu, X., Tong, Z., 2005. Deposition temperature effect on release rate of indomethacin microcrystals from microcapsules of layer-by-layer assembled chitosan and alginate multilayer films. *Journal of Controlled Release* 106, 319-328.

INDEX

2

21st century, 202

3

3D images, 350

A

access, x, 27, 41, 49, 390
accessibility, 349
accommodation, 361
acetic acid, 31, 43, 63, 105, 110, 111, 165, 172, 183, 184, 195, 196, 213, 215, 216, 222, 238, 241, 244, 264, 271, 284, 316, 327, 358, 378, 388, 396, 397, 398, 399, 400, 401, 403, 404, 415, 440, 441, 448
acetone, 105, 284, 359, 360, 439, 446, 447, 448
acetylation, xvii, 2, 19, 28, 43, 51, 58, 87, 92, 100, 170, 172, 174, 232, 282, 283, 302, 308, 316, 336, 345, 387, 388, 392, 466
acidic, 11, 37, 43, 52, 57, 60, 61, 62, 63, 67, 69, 73, 74, 77, 95, 141, 180, 183, 184, 194, 197, 199, 218, 234, 305, 314, 315, 316, 317, 346, 389, 401, 428, 453, 459
acidity, 59, 74, 166, 171
acquired immunity, 391
acrylic acid, 41, 48, 71, 227, 231, 236, 302, 303, 305, 308, 371, 454
acrylonitrile, 216, 223, 359, 371, 374
active compound, xviii, 451
active site, 115, 181, 427
actuation, 372
actuators, xi, 103, 104, 105
acute infection, 292
acylation, 90, 93, 301, 302, 347
additives, 282, 331, 356, 373, 378, 380
adenine, 33, 416
adhesion, xiv, 75, 181, 183, 231, 232, 237, 257, 258, 259, 260, 261, 262, 263, 264, 265, 266, 267, 268, 330, 371
adhesions, xiv, 257, 258, 259, 260, 261, 262, 263, 264, 265, 266, 267, 268
adjustment, 11, 64
ADP, 97
adrenal gland, 97
adrenal glands, 97
adsorption, xviii, 6, 53, 54, 59, 61, 63, 64, 68, 69, 70, 74, 75, 77, 91, 140, 160, 164, 183, 210, 212, 213, 218, 221, 233, 234, 235, 236, 238, 314, 327, 335, 336, 337, 338, 340, 341, 342, 345, 351, 354, 356, 357, 358, 359, 364, 368, 369, 370, 371, 373, 433, 441, 443, 444, 445, 446, 448, 454, 458
adsorption isotherms, 351, 356, 364, 444
adverse effects, 288
AF350, 354
agar, 241, 242, 315, 316, 320, 415
age, 333, 404
agglutination, 95, 170, 171, 390
aggregation, 52, 53, 54, 97, 99, 186, 194, 210, 213, 221, 237, 304, 305, 367, 463
agriculture, xvi, 61, 68, 177, 281, 335, 389, 424
Agrobacterium, 392
airway inflammation, 308
alcohols, xviii, 284, 344, 356, 358, 360, 393, 433, 440, 441, 444, 445, 446, 458
aldehydes, 77, 440
alfalfa, 21
algae, 73, 180, 314, 389
alkalinity, 74
alkaloids, 392, 393
alkylation, 34, 278, 301
amine, 2, 6, 13, 15, 30, 63, 141, 170, 181, 210, 303, 305, 307, 338, 345, 346, 347, 349, 354, 418, 454
amine group, 2, 6, 13, 15, 30, 63, 141, 170, 181, 210, 303, 305, 307, 338, 346, 347, 349, 354, 454
amines, 28, 34, 36, 40, 75, 142, 161

amino acid, xviii, 165, 171, 173, 284, 423, 425, 427, 428, 430, 463
amino acids, xviii, 173, 284, 423, 427, 463
amino-groups, 327
ammonia, xii, 163, 174, 175, 213, 228, 255, 256, 391, 399, 414, 421
ammonium, 28, 36, 43, 60, 67, 68, 69, 122, 184, 229, 230, 232, 302, 315
amphiphilic compounds, 92, 93
amylase, 4, 76
anastomosis, 263
anchoring, 346
anger, 461
aniline, 105, 113, 114, 117, 122, 131, 132, 133, 136
ANOVA, 244
anthocyanin, 255, 399, 401
antibiotic, 292, 293, 405, 462
antibody, 20, 101, 427, 428
anticancer activity, 424, 430
anticancer drug, 404
anticholinergic, 165, 392
anticoagulant, 229, 232, 393
antigen, 86, 370, 458, 466
antihypertensive agents, 331
anti-inflammatory drugs, 461
antimalarials, 393
antioxidant, xvii, 4, 5, 6, 171, 174, 195, 196, 197, 199, 201, 207, 377, 387, 392, 393, 394, 397, 414, 460
antispasmodics, 392
antitumor, ix, 1, 2, 24, 47, 231, 307, 372, 424, 430
antiviral agents, 421
apoptosis, 45, 382, 385, 421
apples, 167, 174, 175, 182, 186, 240, 253
aqueous solutions, 3, 35, 75, 80, 99, 144, 195, 223, 229, 231, 233, 234, 302, 336, 337, 338, 341, 342, 347, 348, 447, 448
Arabidopsis thaliana, 171, 384
aromatic rings, 115
arthritis, 37, 47
ascorbic acid, xii, 163, 171, 176, 183, 204, 333
aspartate, 220
aspartic acid, 305, 310
assessment, 37, 188, 464
asymmetry, 273, 281
atmosphere, 167, 172, 174, 175, 354
atomic force, 185, 369, 372, 373
atomic force microscope, 372
atoms, 62, 68, 69, 111, 112, 184, 327, 338, 352
attachment, xvi, 39, 96, 112, 160, 210, 289, 326, 327, 330, 345, 348, 352, 462
attractant, 461
autolysis, 348
avoidance, 260
awareness, x, 28, 460
azadirachtin, 414, 420

B

Bacillus subtilis, 340
bacteria, xv, xvii, 4, 42, 68, 86, 94, 95, 100, 101, 180, 181, 184, 189, 190, 196, 197, 204, 292, 293, 313, 314, 336, 340, 387, 389, 420, 460, 462
bacterial infection, 86, 293
bacterial strains, 86
bactericides, 314
bacteriocins, 196
bacteriostatic, 180, 293
bacterium, 231, 292, 340
barriers, 13, 261
basal lamina, xv, 287, 289, 291
base, ix, xii, 2, 8, 9, 12, 13, 53, 59, 61, 67, 68, 74, 77, 139, 143, 144, 148, 149, 159, 347, 355, 465
basic research, xv, 17, 287, 294
batteries, xi, 104, 105, 137
beef, 187, 191, 357, 360, 373
behaviors, 219, 231, 233, 234, 311
Beijing, 81, 235, 236, 467
bending, 110, 115, 352
beneficial effect, 165, 196, 197
benefits, xi, xvii, 139, 159, 194, 196, 199, 267, 295, 387, 394, 460
beta particles, 72
bile, 197, 203, 214, 288
bile acids, 288
binding energy, 94, 132, 352
bioactive materials, 261
bioassay, 316, 317
bioavailability, xiii, 193, 194, 198, 199, 394, 397, 460, 463
biocatalysts, xvi, xviii, 161, 219, 343, 348, 358, 364, 368, 373, 423
biochemistry, 25, 73, 173, 365
biocompatibility, xv, 28, 30, 31, 34, 35, 37, 38, 42, 44, 46, 194, 211, 219, 226, 231, 261, 265, 270, 299, 300, 309, 326, 331, 333, 335, 346, 434, 456, 462, 466
Biocompatibility, 30, 31, 43, 45
biocompatibility test, 333
biodegradability, ix, xii, xv, xvi, 38, 77, 104, 179, 194, 219, 227, 265, 299, 300, 314, 326, 335, 343, 373, 389
biodegradable materials, 70, 186
biodegradation, 210, 326, 340
biodiesel, 344, 365
biodiversity, 406

biofuel, 210
biological activities, ix, xi, 2, 5, 22, 85, 97, 98, 170, 220, 307, 322, 419
biological activity, ix, xi, xv, xviii, 85, 86, 96, 101, 231, 275, 299, 300, 391, 394, 433, 434, 466
biological fluids, 303
biological systems, 47
biomass, 404, 405, 406, 430
biomass growth, 404
biomaterials, xii, xiv, 28, 179, 180, 202, 257, 260, 261, 264, 265, 268, 326, 334, 351, 366, 370
biomedical applications, 28, 29, 30, 162, 262, 305, 307, 326, 366, 367, 463
biomolecules, 167, 352, 371, 414
biopolymer, xii, xiv, xvi, 2, 68, 104, 139, 140, 141, 179, 187, 190, 255, 257, 290, 291, 299, 323, 335, 343, 352, 354, 369, 390, 462
biopolymers, xi, 20, 122, 139, 159, 180, 183, 185, 186, 195, 212, 214, 327, 342, 388, 389, 434
bioremediation, 340
biosensors, xi, 103, 210, 213
bioseparation, 21, 303
biosphere, 388
biosynthesis, xii, xvii, 96, 163, 166, 167, 169, 170, 172, 177, 253, 395, 401, 404, 413, 414, 417, 420
biosynthetic pathways, 168
biotechnological applications, 164, 365
biotechnology, xiii, xvi, xvii, 23, 160, 194, 204, 225, 226, 321, 343, 365, 413
biotic, 165, 170, 178, 390, 391, 394
bleaching, 393
bleeding, 77, 260
blend films, 222
blends, 71, 76, 189, 191, 192, 306, 310, 373
blood, xvii, 3, 5, 6, 23, 28, 31, 97, 140, 199, 232, 258, 263, 264, 289, 306, 310, 330, 331, 387, 389, 394
blood pressure, 5, 6, 199
blood vessels, 258, 263
body fluid, xvi, 265, 325, 333, 464
body weight, 29
bonding, 112, 199, 229, 300, 304, 337, 345, 452, 453
bonds, 36, 59, 67, 75, 76, 93, 112, 281, 282, 283, 306, 316, 327, 338, 347, 348, 378, 434, 439, 440, 441, 453
bone, x, xv, 51, 110, 112, 127, 231, 325, 329, 333, 334
bone cells, xvi, 325, 329
bone marrow, 329
brain tumor, 45
branching, xiv, 91, 239, 246, 247, 252, 392
Brazil, 179, 257
breakdown, 214
breast cancer, 45, 105
Brownian motion, 53
burn, ix, 1, 452, 467
burn dressings, ix, 1
by-products, x, 51, 210

C

Ca^{2+}, 166, 168, 176, 178, 196, 231, 357, 391
cabbage, 173
calcium, xii, xvi, 76, 163, 166, 168, 169, 171, 174, 177, 178, 189, 191, 195, 197, 203, 204, 205, 229, 231, 294, 325, 345, 358, 390, 392, 454, 464, 465
calcium carbonate, 345
cancer, xv, xvii, 199, 220, 325, 387, 394, 419, 421
candida, 165, 392
candidates, 219, 300, 303, 306, 332, 389
capillary, 238, 415, 448
capsule, 196, 204, 456
carbohydrate, 86, 88, 93, 94, 369
carbohydrates, 23
carbon, 5, 61, 70, 72, 76, 105, 112, 172, 185, 219, 226, 227, 253, 262, 320, 352, 437, 439, 441, 443, 449
carbon atoms, 437, 441, 443
carbon dioxide, 172, 185, 226, 253, 262
carbon materials, 219
carbon nanotubes, 70, 72
carbon-centered radicals, 5
carbonyl groups, 93, 282, 436
carboxyl, xviii, 61, 74, 77, 94, 218, 227, 234, 281, 302, 309, 433, 434, 436, 437, 438, 439, 441, 448, 449, 454, 458
carboxylic acid, 36, 167, 178, 283, 284, 348, 358, 360, 435, 440, 453, 462
carboxylic acids, 283, 284, 348, 358, 360, 435, 440
carboxylic groups, 282, 439
carboxymethyl cellulose, 162, 278, 279, 454
carcinogenicity, 260
cardiac surgery, 268
cardiovascular disease, xvii, 200, 387, 394
carotenoids, 200, 460
cartilage, 268, 370
casting, 233
catalysis, xvi, 213, 343, 344, 347, 369, 374
catalyst, 6, 307, 344, 362, 375
catalytic activity, xiii, 11, 209, 220, 349, 356, 364, 375
catalytic properties, 374, 375
Catharanthus roseus, 392, 393, 401
cation, xiii, 7, 8, 9, 12, 14, 59, 60, 61, 62, 64, 66, 67, 73, 74, 76, 184, 218, 225, 226, 229, 230, 231, 233
C-C, 112

cDNA, 178
CEC, 62, 67, 70
cell culture, xvii, 32, 40, 166, 170, 303, 330, 331, 333, 348, 387, 392, 393, 395, 404, 405, 406, 417, 421, 461
cell death, xvii, 168, 170, 171, 172, 178, 181, 330, 377, 378, 384, 385, 391
cell differentiation, xvi, 290, 325, 330, 331
cell line, 31, 35, 38, 40, 42, 44, 45, 237, 288, 295, 333, 404
cell lines, 35, 38, 40, 42, 44, 45, 237
cell membranes, 30, 31, 33, 38, 94, 167, 288, 394
cell signaling, 172, 292
cell surface, x, xviii, 27, 42, 94, 181, 252, 292, 295, 371, 390, 423, 425, 426, 430
cellulose, xii, 2, 14, 17, 55, 75, 104, 140, 162, 179, 180, 185, 222, 262, 281, 282, 306, 311, 323, 371, 424, 430
cellulose derivatives, 281
ceramic, 160, 333, 448
CGL, 310
chain scission, 110
chain transfer, 303
challenges, 176, 203, 371
charge density, 30, 35, 37, 54, 92, 100, 140, 160, 180, 181, 184, 185, 212, 234, 263, 455
chelates, 288, 346
chemical characteristics, 262, 263
chemical degradation, 201
chemical interaction, 55, 59
chemical properties, xiii, 99, 193, 201, 209, 225, 226, 276, 303, 373, 434, 441, 456
chemical reactions, 210, 270, 301
chemical stability, 104, 200, 441
chemical structures, 90, 91, 214, 228
chemicals, xvi, 19, 33, 70, 336, 343, 344
chemotherapeutic agent, 392, 393
chicken, 35, 46
children, 266
China, xiii, 193, 209, 225, 239, 240, 241, 253, 256, 263, 299, 335, 409
chitinase, 4, 19, 21, 165, 171, 240, 323, 324
chlorine, 105, 111, 434, 439, 440, 447
chloroform, 284, 378
cholesterol, 5, 20, 74, 196, 199, 289, 393, 424
chromatography, xviii, 348, 370, 415, 433, 441, 448
chromium, 161, 342
chymotrypsin, 210
circadian rhythm, 290
classes, 2, 169, 331, 338, 461
classification, 356, 425
clay minerals, 51, 62, 63, 64, 69, 76
cleaning, 339
cleavage, 177, 316, 378
clinical application, 42, 262, 291, 294
closure, 174, 175, 266, 378, 385
clusters, 53, 54, 59, 76, 77
CMC, 104, 227, 230, 231, 232, 233, 234, 235, 279, 280, 281
CO2, 31, 167, 175, 255
coagulation process, 454
coatings, xii, 140, 165, 167, 171, 173, 174, 175, 177, 179, 180, 182, 183, 189, 190, 191, 192, 207, 352
collagen, xvi, 259, 262, 306, 325, 334, 370, 371
collisions, 53, 56, 57
Colombia, 343, 365
colon, 5, 23, 197, 461
colon carcinogenesis, 5, 23
color, xii, 31, 139, 147, 149, 178, 180, 184, 235, 399
combined effect, 73
combustion, 272, 354
commercial, xi, xvii, 22, 33, 42, 104, 139, 140, 141, 198, 203, 204, 215, 255, 256, 314, 345, 356, 413
commercial crop, 314
communities, 292
community, 30, 194, 199, 292
compaction, 63
compatibility, 28, 31, 72, 171, 262, 306, 310, 320
competition, 52, 96, 196, 390
competitiveness, 391
complement, 263
complex interactions, 73, 172
complications, xiv, 257, 260
composites, x, 69, 70, 71, 72, 76, 77, 85, 86, 185, 187, 210, 220, 311, 330, 352
composition, 61, 63, 86, 198, 200, 206, 263, 264, 265, 267, 279, 331, 346, 351, 364, 367, 370, 378, 389, 453
compounds, xii, xiv, xvii, 2, 11, 19, 20, 21, 51, 69, 73, 77, 86, 90, 92, 93, 97, 100, 111, 163, 165, 166, 177, 180, 181, 182, 192, 194, 195, 197, 200, 201, 234, 239, 244, 253, 256, 261, 262, 265, 282, 283, 322, 326, 333, 336, 340, 346, 379, 387, 391, 392, 393, 394, 395, 396, 400, 401, 403, 404, 405, 406, 415, 417, 419, 456, 460
compression, 53, 54
computer, 94
condensation, xvi, 228, 326, 327, 377, 379, 382
conduction, xvi, 11, 126, 127, 132, 136, 325
conductivity, xi, 11, 72, 103, 104, 105, 118, 125, 126, 127, 129, 130, 131, 132, 133, 136, 370, 448
configuration, 8, 11, 12, 14, 52, 182, 349, 405
conjugation, 212, 213, 220, 348, 359
conjunctiva, 38
connective tissue, 31, 258
conservation, 414, 419, 420

constituents, 88, 94, 180, 252, 390
construction, xviii, 93, 326, 372, 423, 430
consumers, 194, 199, 203, 460
consumption, xiv, 62, 142, 194, 200, 269, 272, 277, 278, 345, 394, 401
contaminant, 340
contaminated water, 61
contamination, 183, 191, 192, 405
contradiction, 331
controversial, 264, 314, 315, 317
convergence, 206, 390
coordination, 59, 231, 453
copolymer, 47, 180, 194, 220, 288, 299, 303, 304, 306, 308, 345, 346, 359, 360
copolymerization, xiii, 122, 123, 225, 227, 301, 303, 309, 349
copolymers, xvi, 46, 262, 302, 303, 308, 325, 345, 373, 452
coronary artery disease, 394
correlation, 62, 121, 212, 262, 314, 315, 427
correlations, 166, 361
corrosion, 452
cosmetic, 37, 278, 282, 302, 306, 463
cosmetics, ix, xv, 47, 68, 104, 281, 299, 307, 344, 389, 424, 462
cost, 13, 140, 159, 233, 305, 341, 349
cotton, 372, 462, 463, 466
counterbalance, 294
covalent bond, 75, 142, 199, 283, 347, 437, 438, 439, 449
covalent bonding, 75, 199, 438
covering, 197, 314
crabs, 140, 200, 240
crops, 165, 461
cross-linked polymers, 304
crown, 254, 348, 364, 368
crude oil, 340, 342
crystal structure, 52, 349
crystalline, 72, 106, 112, 117, 127, 232, 281, 300, 336, 345, 349, 352, 372
crystallinity, 35, 37, 52, 61, 117, 280, 302, 336, 349, 354, 360, 458
crystals, 31, 32, 220, 330, 349, 366
CTA, 230, 234
cultivars, 395, 404
cultivation, 420
culture, x, xvii, 27, 31, 32, 33, 41, 69, 165, 173, 174, 175, 184, 241, 242, 315, 324, 330, 392, 395, 397, 398, 399, 402, 403, 404, 405, 413, 414, 415, 416, 419, 420, 421
culture conditions, 404, 405
culture media, 33, 184, 330, 397, 399
culture medium, 31, 32, 165, 395, 405, 419
cure, 142
cyanide, 378, 382
cycles, 177, 290, 295, 303, 345, 348, 350, 359, 391
cyclodextrins, 294, 349
cyclooxygenase, 101
cyclophosphamide, 290, 296
cysteine, 36
cystitis, xv, 287, 292, 293, 295
cytochrome, 331
cytocompatibility, x, 27, 334
cytokines, 86, 96, 258, 262
cytometry, 41
cytoplasm, xiv, 169, 239, 247, 248, 252, 289, 292, 295
cytoplasmic tail, 292
cytoskeleton, 288, 290
cytotoxicity, x, 27, 28, 30, 31, 33, 34, 35, 36, 37, 38, 39, 40, 41, 43, 45, 46, 48, 220, 261, 263, 264, 329, 333, 391, 456

D

dairy industry, 73, 344
DBP, 336, 337
deacetylation, xii, xiii, xiv, xvi, 2, 4, 5, 6, 37, 45, 51, 52, 53, 63, 140, 164, 170, 179, 180, 181, 188, 190, 192, 193, 194, 209, 210, 212, 216, 223, 226, 263, 269, 270, 271, 272, 275, 276, 277, 288, 299, 302, 316, 317, 335, 342, 343, 345, 346, 358, 367, 378, 388, 389, 391, 433, 435, 440, 446, 448, 455, 456
decay, xii, 164, 165, 167, 171, 178, 189, 192, 254, 255, 389
decomposition, 71, 112, 119, 160, 355
decomposition temperature, 112
decontamination, 61
defect site, xvi, 132, 325
defects, xv, 214, 226, 260, 267, 302, 325
defence, 173, 175, 176, 240, 255, 256, 321, 385
defense mechanisms, xvii, 42, 387, 389, 392, 394, 395, 401, 403
deformation, xiv, 269, 270, 435
degradation, xviii, 3, 38, 39, 43, 70, 104, 105, 107, 110, 121, 136, 151, 153, 160, 166, 172, 174, 182, 200, 201, 261, 262, 263, 264, 272, 273, 276, 302, 306, 331, 333, 339, 401, 423, 424, 427, 430, 434, 437, 439, 452, 456, 460
degradation mechanism, 331
degradation rate, 263, 264, 306, 427, 456
degree of crystallinity, 227, 232, 280
dehydration, 71, 112, 166, 175, 355
demonstrations, 132
denaturation, 11, 155, 356

depolarization, 390
depolymerization, 20, 22, 23, 316
deposition, 166, 205, 215, 253, 259, 261, 263, 334, 373, 391, 458
derivative thermogravimetry, 355
desorption, 70, 74, 76, 233, 354
destruction, xvi, xviii, 11, 274, 290, 377, 378, 379, 380, 381, 382, 383, 433, 439, 441, 447
detachment, xv, 287, 290, 291, 415
detectable, 31, 379, 399
detection, 70, 417
detergents, 219, 384
detoxification, 98
developed countries, 164, 292
developing countries, 142
diacylglycerol, 168
dialysis, 447
diaphragm, 258
dielectric constant, 132, 133, 134, 135, 136, 361
dielectric permittivity, 133
diet, xvii, 20, 29, 140, 387, 394, 460
differential scanning, xi, 103, 145, 355, 356, 373
differential scanning calorimetry, 145, 355, 356, 373
diffraction, 280, 354
diffusion, 15, 35, 59, 61, 72, 76, 155, 158, 183, 210, 233, 283, 289, 331, 347, 349, 357, 362, 363, 366, 367, 452
diffusivity, 172
digestibility, 198, 205, 255
dimethylformamide, 229
diseases, xiii, xv, 176, 189, 190, 194, 239, 240, 252, 254, 255, 256, 292, 322, 325, 393, 394, 454
disorder, 87, 127
dispersion, 32, 52, 54, 55, 56, 57, 58, 59, 62, 63, 65, 66, 69, 72, 74, 79, 133, 136, 185, 200, 214, 291, 293, 331, 352, 443, 444, 445, 448
dispersity, 277
displacement, 91, 109, 354, 373
dissociation, ix, 2, 8, 30, 87, 88, 93, 112
distilled water, 143, 144, 145, 148, 200, 243, 378, 379, 382, 415, 446, 448, 459
distortions, 348
distribution, xiv, xvii, 44, 54, 88, 90, 91, 92, 122, 212, 213, 214, 216, 233, 239, 252, 267, 272, 273, 274, 275, 276, 346, 349, 351, 364, 394, 413, 414, 445, 452, 453, 458
diuretic, 463
diversity, xii, xvii, 179, 413, 414, 417, 430
DMF, 229, 232
DNA, 28, 30, 33, 34, 35, 40, 41, 42, 44, 46, 48, 49, 252, 322, 379, 380, 382, 384, 390, 394, 403, 456
doping, xi, 103, 114, 115, 118, 121, 127, 129, 133, 136
dosage, xviii, 55, 339, 451
double bonds, 227
double helix, 33
down-regulation, 401
dream, 140
dressing material, 37
dressings, 42, 326, 467
drug carriers, 49, 302
drug delivery, ix, x, xiii, 1, 27, 29, 37, 42, 46, 47, 49, 75, 78, 193, 202, 209, 211, 219, 220, 221, 288, 289, 294, 295, 302, 303, 304, 305, 306, 326, 331, 333, 334, 389, 452, 454, 456, 461, 463, 465, 466
drug design, 93
drug interaction, 75
drug release, 37, 47, 48, 71, 75, 202, 223, 231, 288, 309, 331, 367, 368, 452, 457, 458, 461
drugs, x, xv, 27, 28, 30, 36, 48, 77, 86, 165, 194, 211, 231, 232, 275, 287, 288, 289, 294, 303, 308, 331, 334, 364, 392, 414, 451, 457, 458, 461, 463, 464, 465, 466, 467
drying, 32, 63, 195, 203, 206, 244, 264, 265, 305, 330, 349, 364, 367, 371, 456, 458, 460, 463, 464
DSC, xi, 103, 107, 108, 120, 121, 143, 145, 150, 151, 355, 356, 373
dura mater, xiv, 257
dyeing, 238
dyes, 41, 73, 74, 160, 235, 335, 338, 340, 392, 393, 459, 462
dynamic viscosity, 36

E

E.coli, xv, 287, 296, 322
economic efficiency, 180
editors, 99, 173, 204, 333, 420, 421
effluent, 104
effluents, 12, 73, 76, 234, 338, 392, 393
Egypt, 27, 139
elaboration, 86, 398
electric charge, 13, 53, 54, 62, 66, 233
electric current, 17
electric field, 7, 8, 13, 14, 15, 17, 18, 127, 185, 191, 214
electrical conductivity, 104, 126, 131
electrical resistance, 35, 288
electrodeposition, 372
electrodes, xi, 7, 70, 104, 137
electrolyte, 54, 65, 67, 166, 233, 308
electromigration, 14, 15, 16, 17, 18, 20, 233
electro-migration, 14
electron, xi, xiv, 24, 39, 48, 68, 91, 94, 103, 105, 110, 127, 132, 136, 199, 217, 239, 266, 352, 372, 378

electron microscopy, xiv, 39, 48, 91, 94, 199, 239
electrons, 54, 352
electrospinning, 71, 214, 215, 216, 219, 222, 223, 308
elongation, 70, 89, 94, 182, 183, 184, 185, 252
elucidation, 97
e-mail, 103, 377, 385, 387, 433
embryogenesis, 416, 420
emulsions, 56, 197, 198, 201, 205, 207, 211, 284
encapsulation, 194, 195, 196, 198, 199, 200, 205, 206, 210, 211, 214, 349, 451, 460, 461, 463, 465, 467
endocrinology, 452
endosperm, xvii, 166, 413, 414, 415
endothelial cells, 5, 23, 43, 48, 99, 259
endothermic, 107, 121, 355
endotoxemia, 97, 98, 101
endotoxins, 86, 90, 96, 97, 98, 99
energy, 3, 7, 9, 11, 12, 53, 72, 93, 105, 111, 112, 270, 322, 352, 362, 382, 384, 401, 424, 444, 445, 446, 448
energy consumption, 12
energy efficiency, 9, 11
engineering, ix, xiii, xv, xvi, xviii, 29, 44, 68, 78, 82, 194, 209, 212, 219, 261, 266, 267, 268, 302, 303, 311, 325, 326, 330, 332, 333, 334, 335, 341, 342, 343, 344, 345, 364, 369, 370, 423, 425, 430
England, 21
enlargement, 461
entrapment, 140, 141, 142, 204, 220, 231, 345, 347, 357, 358, 459, 461
environment, xvi, 12, 19, 34, 38, 54, 62, 67, 212, 226, 240, 335, 340, 352, 355, 364, 434
environmental aspects, 202
environmental awareness, 211
environmental conditions, 205, 340, 389
environmental factors, 179
environmental stimuli, 305
enzymatic activity, xviii, 147, 155, 423
enzyme immobilization, xi, xiii, 76, 139, 140, 141, 142, 144, 158, 160, 161, 164, 193, 209, 210, 211, 219, 220, 302, 339, 345, 352, 361, 369, 370, 463
enzyme-linked immunosorbent assay, 34
enzymes, ix, xi, xii, xvi, xvii, 1, 2, 4, 76, 139, 140, 142, 143, 146, 152, 153, 155, 160, 162, 165, 166, 167, 170, 171, 194, 199, 202, 210, 211, 216, 219, 220, 252, 284, 314, 336, 339, 340, 343, 344, 347, 348, 349, 352, 355, 357, 359, 364, 366, 379, 387, 390, 391, 399, 418, 427, 430, 452
EPA, 198, 314
epicardium, 260
epidemiologic studies, 394
epidemiology, 267
epidermis, xvi, 377, 378, 379, 380, 381, 382, 384, 385
epithelia, 45, 46, 288, 290
epithelial cells, xv, 38, 46, 287, 288, 295, 296, 465
epithelium, xv, 48, 287, 289, 295, 296
epoxy groups, 327
equality, 316
equilibrium, 52, 56, 60, 66, 76, 94, 341, 362, 363, 443, 455, 456
equipment, 13, 141, 143, 144, 160, 271, 272, 458, 460
erosion, 75
erythrocyte membranes, 31
ESR, 22
ESR spectroscopy, 22
ester, 14, 17, 33, 87, 305, 310, 344, 347, 359, 362, 364, 365, 375, 418
ethanol, 71, 72, 160, 171, 198, 230, 243, 244, 315, 344, 362, 378, 415
etherification, xiii, 225, 227, 229, 302
ethers, 277, 282, 360
ethyl alcohol, 358, 360
ethylene, xii, 37, 46, 47, 71, 163, 165, 167, 168, 170, 175, 177, 178, 220, 221, 264, 302, 303, 305, 308, 336, 348, 368, 373, 456, 466
ethylene glycol, 37, 46, 220, 302, 303, 308, 336, 348, 368, 456
ethylene oxide, 47, 221, 264, 373, 466
etiology, 142
EU, 287, 359
Europe, 203
evaporation, 35, 196, 200, 215, 355, 358, 458, 465
evidence, 61, 74, 88, 97, 194, 258, 315, 352, 389, 401
evolution, 379, 383
excision, 415
excitation, 33
exclusion, 33, 90, 348
exocytosis, 48, 291
exoskeleton, 140, 164, 335, 388
exothermic peaks, 151
experimental condition, 320
exploitation, 13
exposure, 71, 95, 167, 264, 291, 321, 384, 391
expulsion, 214
external environment, 97
extracellular matrix, 262, 263
extraction, xvii, 71, 243, 333, 369, 387, 392, 405
extracts, 207, 254, 400, 414
extrusion, 200, 206, 271, 272, 273, 274, 277, 279, 334, 455
exudate, 263

F

fabrication, 206, 214, 222, 458
Fabrication, 187, 222, 333, 334, 370
families, xi, 139, 140, 391, 394, 396, 405
fat, 56, 73, 74, 171, 173, 260, 288, 294, 339
fatty acids, 87, 182, 191, 197, 198, 201, 205, 288, 358, 360, 363, 375, 458, 460
fermentation, 204, 405
fiber, 3, 53, 54, 162, 214, 215, 216, 222, 258, 278, 307, 308, 349, 368, 372, 374, 462
fiber bundles, 258
fibers, 51, 53, 54, 55, 59, 63, 69, 74, 75, 76, 140, 172, 202, 214, 215, 216, 222, 262, 345, 352, 372
fibrin, xvi, 261, 263, 325
fibrinolysis, 258, 267
fibrinolytic, 258, 260
fibroblasts, 38, 259
fibrosis, xiv, 257, 259, 260
fibrous cap, 261, 264
fibrous tissue, 261
field crops, xii, 163, 165
filler particles, 185
fillers, 172, 185
film thickness, 200, 415
films, xii, xiv, 70, 71, 75, 76, 105, 107, 110, 113, 114, 118, 167, 172, 173, 175, 179, 180, 182, 183, 184, 185, 186, 187, 188, 189, 190, 191, 192, 199, 206, 216, 257, 262, 263, 264, 268, 277, 295, 297, 310, 347, 352, 367, 370, 371, 372
filtration, ix, 2, 8, 13, 14, 140, 232, 264, 315, 316, 454, 456
financial, 419
financial support, 419
fire retardants, 462
first dimension, 402
fish, 25, 198, 199, 200, 460
fish oil, 198, 199, 460
flavonoids, xvii, 166, 250, 251, 387, 394, 395, 397, 399
flavor, 200, 201, 206
flavour, 206, 414, 466
flexibility, 30, 90, 92, 121, 182, 306, 365, 452, 453
flight, 214
flocculation, x, 51, 53, 55, 56, 57, 58, 59, 62, 63, 64, 66, 69, 73, 74, 238, 338
flowers, xii, 163, 164, 167
fluctuations, 303
fluid, 3, 33, 291, 456, 461
fluorescence, x, 28, 33, 213, 379, 380, 382, 383
foams, 284
folate, 42, 49
food additives, 394, 406
food industry, 186, 195, 198, 389, 460
food products, xii, xiii, 171, 179, 193, 200, 346, 394, 460
food safety, 180, 240
forage crops, 255
force, 7, 13, 15, 17, 53, 54, 67, 74, 92, 215, 218, 322, 372
formaldehyde, 326, 378, 456, 457, 460
fouling, 4, 9, 19, 23, 233, 371
fractional composition, xiv, 269, 275
fragility, 349
fragments, 88, 175, 281, 331
France, 79, 80
free energy, 93, 443, 444, 445, 448
free radical copolymerization, 122
free radicals, 6, 105, 110, 112, 393, 394, 460
free surface energy, 448
freezing, 303, 330, 348, 349, 350, 354, 364, 370, 384
friction, 17, 258
fruits, xii, 163, 165, 167, 171, 173, 174, 175, 177, 179, 184, 187, 188, 192, 201, 252, 253, 254, 323, 389, 394, 414, 415, 420
FTIR, xi, xii, 35, 99, 103, 110, 111, 114, 115, 116, 136, 139, 143, 145, 148, 149, 150, 229, 352, 353
FTIR spectroscopy, 99
FTIR technique, xii, 139
fuel cell, 72, 373
full capacity, 9
functional analysis, 406
functional food, 194, 195, 198, 201, 460
functionalization, 202, 299, 307, 309, 371, 388
funds, 406
fungal infection, 293, 314
fungi, ix, xvii, 1, 2, 6, 25, 42, 140, 164, 180, 181, 185, 252, 254, 255, 256, 288, 314, 315, 320, 321, 322, 335, 345, 378, 387, 388, 389, 390, 395
fungus, xv, 241, 242, 313, 314, 315, 316, 320, 323, 389, 393

G

gadolinium, 220
gamma rays, 105, 136
gastrointestinal tract, 194, 454
GCE, 76
gel, xi, xii, xvi, 36, 39, 52, 56, 75, 76, 105, 139, 141, 142, 143, 144, 145, 146, 148, 151, 152, 153, 154, 155, 158, 159, 160, 161, 162, 182, 184, 185, 191, 194, 195, 197, 204, 211, 219, 220, 262, 263, 264, 265, 266, 267, 303, 308, 325, 329, 333, 334, 336, 346, 347, 348, 357, 358, 366, 367, 368, 403, 404, 454, 464
gel formation, 56, 336, 454

gelation, 76, 195, 196, 198, 212, 333, 348, 355, 358, 360, 367, 368, 369, 453, 454, 455, 462, 463, 465, 466
gene expression, 40, 41, 48, 176, 178, 399, 400, 426
gene therapy, 40, 46, 47, 461
gene transfer, 48
genes, xii, 30, 40, 47, 163, 167, 168, 169, 170, 173, 255, 391, 400, 403
genomics, 201
Germany, 73, 162, 378, 379
germination, xiv, 165, 174, 239, 241, 245, 246, 252, 314, 321, 322, 389
gingival, 310
ginseng, 170, 176, 419
glass transition, 182, 186, 190, 355
glass transition temperature, 182, 186, 355
glasses, xvi, 325
glucoamylase, 76
glucose, 71, 110, 142, 146, 147, 148, 157, 164, 180, 226, 227, 229, 240, 243, 314, 360, 371, 379, 446, 456
glucose oxidase, 146, 371, 379
glucoside, 397
glutamic acid, 46, 166, 181, 221, 362
glutamine, 31
glycerol, 87, 182, 186, 188, 189, 191, 192
glycidilmethacrylate, 327
glycine, 144
glycol, 42, 47, 48, 49, 71, 182, 302, 310
glycolysis, 97, 175
glycoproteins, 36, 262
glycosaminoglycans, 289, 295
glycoside, xviii, 378, 433
granules, 223, 458
Great Britain, 265
groundwater, 341
growth, xiv, 5, 6, 21, 32, 42, 45, 86, 95, 165, 169, 170, 171, 174, 176, 177, 180, 183, 184, 185, 187, 191, 195, 199, 231, 239, 240, 241, 242, 244, 245, 246, 252, 253, 262, 264, 266, 302, 314, 316, 320, 321, 322, 323, 330, 331, 339, 340, 378, 389, 390, 392, 394,뱀395, 396, 404, 415, 419, 421, 461
growth factor, 42, 262, 266, 331
growth rate, 340
guard cell, xvi, 377, 378, 379, 380, 382, 383, 384, 385

H

half-life, 394, 397, 452
hardness, xi, 103, 109, 110, 121, 122, 223
harmful effects, 254
HE, 44, 45, 46
healing, 265, 452
health, xvi, xvii, 194, 195, 197, 199, 203, 335, 387, 394, 395, 396, 460
heating rate, 145
heavy metals, 61, 64, 77, 80, 234, 235, 339, 340, 341, 391
height, 132, 145, 165
helical conformation, 262
helium, 448
hemocompatibility, 262
hemoglobin, 31, 75, 87
hemorrhage, 258
heptane, 442
hernia, xiv, 257
heterogeneity, 272
hexane, 213, 244, 284, 359, 360, 375
histidine, 48
histology, 420
history, xvii, 388, 413
HIV, 419
homeostasis, 165, 265
homogeneity, 63, 69, 215
hormone, 169
hormones, 97, 168, 290, 461
horses, 258
horticultural commodities, xii, 164, 165, 170, 240, 254, 322
host, 86, 97, 98, 196, 219, 252, 267, 292, 296, 389, 390, 391
human, x, 5, 23, 30, 38, 43, 45, 48, 51, 77, 89, 96, 97, 99, 101, 142, 192, 198, 240, 261, 266, 267, 288, 292, 295, 303, 308, 315, 329, 331, 332, 421, 434, 457, 466
human body, 303, 457
human health, x, 51, 77, 240, 292
human immunodeficiency virus, 421
humidity, 31, 70, 215, 446, 447
humoral immunity, 96
Hunter, 267
hybrid, ix, xvi, 2, 8, 13, 45, 69, 71, 72, 77, 159, 325, 326, 327, 328, 329, 330, 331, 332, 333, 334
hybridization, 327, 328, 332
hydrogels, xvi, 151, 161, 162, 194, 231, 236, 303, 307, 309, 310, 333, 343, 347, 348, 357, 366, 367, 372, 453, 463
hydrogen, xviii, 5, 20, 23, 36, 52, 54, 56, 58, 67, 68, 72, 88, 93, 112, 115, 146, 165, 182, 183, 184, 196, 199, 204, 218, 229, 230, 262, 300, 304, 306, 337, 345, 347, 352, 378, 433, 434, 439, 445, 447, 452, 453
hydrogen atoms, 54
hydrogen bonds, 56, 58, 67, 72, 88, 182, 183, 184, 230, 262, 306, 347

hydrogen fluoride, 20
hydrogen peroxide, xviii, 5, 23, 36, 146, 165, 196, 204, 378, 433, 434, 439, 447
hydrogenation, 228
hydrolysis, ix, xviii, 1, 2, 3, 4, 9, 10, 11, 19, 22, 110, 111, 112, 142, 147, 158, 159, 160, 161, 180, 252, 302, 315, 316, 317, 324, 327, 339, 342, 344, 357, 358, 359, 360, 362, 364, 372, 374, 375, 388, 423, 424, 425, 426, 427, 430
hydrophilicity, 37, 211, 309, 328, 349, 364
hydrophobicity, 89, 92, 95, 172, 361, 362, 370
hydrothermal synthesis, 72
hydroxyethyl methacrylate, 303, 309
hydroxyl, ix, 2, 5, 8, 35, 43, 59, 76, 93, 95, 105, 164, 194, 210, 218, 227, 277, 279, 281, 282, 300, 303, 327, 335, 336, 339, 342, 346, 348, 359, 366, 374, 378
hydroxyl groups, 35, 59, 76, 95, 105, 164, 194, 279, 281, 282, 303, 335, 336, 339, 342, 348, 359, 366, 374
hyperbilirubinemia, 466
hyperglycaemia, 420
hypersensitivity, 96
hypertrophy, 97
hypotension, 86
hypothesis, 88, 95, 345, 356, 401

I

ICAM, 81
ideal, 231, 261, 302, 451
identification, 41, 169, 203, 255, 406
IL-8, 96
illumination, 290, 295, 378, 415
image, 41, 109, 117, 259, 354, 462
images, 39, 107, 109, 199, 380, 382, 459
imbalances, 180
immobilization, xi, xii, xiii, xvi, 36, 76, 139, 140, 142, 143, 144, 146, 147, 151, 154, 155, 157, 159, 160, 161, 162, 178, 202, 204, 209, 210, 211, 216, 218, 219, 221, 223, 335, 342, 343, 344, 345, 348, 357, 358, 359, 362, 363, 364, 366, 368, 370, 371, 372, 373, 374, 392, 430, 452
immobilized enzymes, 140, 142, 339, 359, 362
immune reaction, 382
immune response, 45, 384, 462
immune system, 86, 196, 199, 292, 384, 393
immunity, 5, 231, 293
immunofluorescence, 427, 428
immunogenicity, 262
immunoglobulin, 454, 465
immunomodulatory, 302
immunostimulatory, 47
immunotherapy, 223
implants, 326
impregnation, 172
imprinting, 307
improvements, xii, 155, 163, 184, 185, 186
impurities, 53, 74, 226
in vitro, xi, xv, xvii, 25, 28, 31, 35, 36, 40, 42, 44, 45, 46, 48, 49, 85, 197, 199, 240, 253, 256, 262, 264, 288, 292, 294, 313, 315, 321, 331, 333, 334, 387, 389, 391, 392, 394, 395, 413, 414, 452, 465, 466
in vivo, xi, xvii, 31, 40, 42, 44, 45, 48, 85, 97, 178, 198, 205, 207, 220, 240, 262, 263, 264, 265, 289, 290, 294, 295, 309, 333, 387, 388, 389, 395, 417, 452, 457, 463
incidence, xiii, 169, 176, 194, 239, 252, 261, 266, 296
incubation period, 154
incubation time, 147, 154, 155, 243, 251, 252, 357
India, 103, 163, 385, 413, 419
inducer, xii, 97, 163, 290
induction, xvii, 101, 167, 168, 169, 173, 174, 189, 253, 254, 255, 290, 293, 391, 401, 413, 414, 416
industries, 24, 69, 77, 140, 159, 278, 314, 392, 393
industry, ix, xvi, 1, 2, 200, 343, 344, 388, 434
infection, xiv, 22, 196, 240, 241, 255, 256, 257, 260, 292, 296, 307, 323, 384, 389, 390, 391, 392
inflammation, 229, 232, 293, 393, 421
inflammatory cells, 258, 263
influenza, 5
infrared spectroscopy, xi, 38, 103, 145
ingest, 394
ingestion, 195, 394
ingredients, xiii, 74, 77, 193, 194, 195, 198, 201, 212, 344, 421, 460
inhibition, 6, 11, 32, 96, 157, 165, 168, 170, 171, 187, 198, 201, 232, 240, 244, 254, 294, 314, 344, 361, 362, 363, 364
inhibitor, 42, 77, 382, 393
inhomogeneity, 275
initiation, 256, 415
inoculation, 241, 242, 250, 251, 252, 319, 320
inoculum, 404, 405
inositol, 168
insecticide, 393
insects, ix, 1, 288, 314, 335, 345
insertion, 70, 71, 72, 76, 94
insulation, 185
insulin, 5, 21, 30, 47, 454, 464
integration, xvi, 325, 326
integrins, 371
integrity, 20, 31, 33, 36, 38, 75, 258

interface, ix, 2, 8, 9, 197, 201, 214, 261, 301, 307, 333, 344, 364
interference, 48, 180, 199
internalization, 28, 38, 42, 48
intestinal tract, 196, 199
intestine, xiv, 3, 142, 257, 288
intoxication, xi, 85
intravesical instillation, 291
inversion, 458
invertebrates, 388
ion-exchange, 7, 8, 64, 65, 66, 67, 68
ionization, 64, 88, 97, 277, 360, 453
ionizing radiation, 104, 110
ions, ix, xvi, 2, 6, 8, 12, 21, 54, 55, 58, 63, 64, 74, 76, 172, 180, 181, 229, 233, 234, 262, 293, 325, 327, 329, 337, 338, 341, 366, 453, 454
IR spectra, 218, 282, 352, 435, 436
IR spectroscopy, xiv, 269, 439, 446, 447
Iran, 81
Ireland, 265
iron, 309
irradiation, xi, 103, 105, 106, 107, 109, 110, 112, 114, 117, 121, 136, 174, 352
ischemia, xiv, 257, 258, 260
islands, 133
isoflavonoids, 168
isolation, 278
isomerization, 362
isomers, 395, 397, 398, 404
isotherms, 338, 357, 442
Israel, 81, 172, 451
issues, x, 27
Italy, 387, 406

J

Japan, 37, 79, 145, 175, 177, 243, 263, 325, 373, 374, 423

K

K^+, 233
KBr, 437, 438, 439, 446, 447
ketones, 440
kidney, 96
kill, 47, 253
kinase activity, 177
kinetic constants, 157, 362
kinetic model, 147, 362
kinetic parameters, 200, 362, 375
kinetic studies, 341
kinetics, 20, 61, 76, 143, 197, 218, 356, 361, 362, 364, 368, 375, 452
KOH, 446, 447

L

lactate dehydrogenase, 43, 180
lactic acid, 69, 183, 192, 196, 204, 227, 237, 241, 242, 244, 304, 309, 388, 461, 463, 465
lactose, 73, 142, 146, 147, 148, 157, 158, 159, 162, 196
lactose intolerance, 142
lamella, 62, 65, 67, 68, 76
laminar, 261, 415
landscapes, 315
lanthanide, 61
laparoscopy, 267
laparotomy, 258
L-arginine, 37
larvae, 205
laxatives, 393
LDL, 394
lead, xiv, 58, 70, 87, 92, 94, 104, 132, 198, 257, 260, 282, 328, 389, 390, 414, 419
leakage, x, 27, 33, 94, 142, 166, 180, 248, 252, 292, 314, 347, 390
lecithin, 198, 205
lesions, 258, 262, 265, 384
leucine, 43
liberation, 31
life cycle, 320
life quality, 260, 265
lifetime, 76, 142, 292
ligand, x, 42, 85, 90, 93, 231
light, 33, 39, 90, 177, 194, 195, 199, 200, 241, 348, 378, 379, 380, 381, 382, 383, 415, 416, 462
light scattering, 39, 90
lignans, 414
lignin, xiv, 166, 168, 239, 241, 250, 251, 253, 255, 391, 400, 401
limestone, x, 51
linoleic acid, 171, 211, 221
lipases, 216, 219, 344, 345, 348, 353, 355, 357, 358, 359, 360, 361, 362, 364, 365, 366, 369, 372, 373, 374, 375
lipid oxidation, 182, 198
lipid peroxidation, 20, 97
lipids, 6, 20, 94, 182, 198, 205, 289, 357, 360
lipoproteins, 394
liposomes, 38
liquid chromatography, 334
liquid phase, 283, 455, 456
liquids, 37, 86, 191, 270, 277, 460

Listeria monocytogenes, 187, 191
liver, 97, 259, 260, 464
liver cells, 97
living radical polymerization, 309
localization, 41, 92, 129
longevity, 265
long-term retention, 37
lubricants, 182
luciferase, 35, 40
Luo, 159, 171, 176, 221, 238, 255, 374, 420
lymphocytes, 100
lysine, 28, 37, 40, 46, 89, 100, 197
lysis, 31, 33
lysozyme, 4, 21, 233, 262, 263, 323

M

machinery, 172
macromolecules, 30, 38, 51, 52, 53, 55, 58, 59, 61, 66, 68, 74, 76, 77, 86, 133, 212, 217, 252, 277, 281, 349, 391, 434, 452, 453
macrophages, 38, 39, 47, 96, 101, 262, 263, 264
magnetic field, 303
magnitude, 71, 133, 218
majority, xv, 4, 287
mammal, 158, 258
mammalian cells, 41
mammals, 6, 87
manipulation, 20, 414, 419
mannitol, xvii, 377, 379, 381
manufacturing, 195, 203, 460, 463, 464
mapping, 351
marginalization, xvi, 377, 382
MAS, 327, 329, 334
masking, 460
mass, xiii, 7, 169, 199, 200, 209, 212, 232, 234, 274, 276, 289, 316, 337, 339, 355, 400, 403, 404, 415, 416, 418
mass spectrometry, 400, 403, 404, 415
mastitis, 302, 307
material surface, 326, 327
materials, xi, xii, xiii, xvi, xviii, 28, 33, 37, 63, 68, 69, 70, 71, 75, 78, 79, 80, 88, 104, 119, 139, 164, 165, 179, 180, 186, 187, 188, 191, 194, 197, 198, 202, 204, 210, 211, 219, 220, 222, 225, 226, 227, 229, 230, 235, 261, 262, 268, 278, 302, 305, 310, 325, 330, 331, 332, 333, 348, 349, 350, 352, 368, 369, 370, 373, 423, 430, 452, 456, 460, 462
materials science, 261
matrix, xvi, 39, 70, 71, 72, 73, 75, 104, 114, 126, 136, 160, 172, 183, 185, 186, 188, 197, 198, 200, 204, 206, 211, 302, 303, 309, 326, 327, 330, 331, 334, 339, 343, 347, 348, 357, 364, 452
matrixes, 77
matter, iv, 73, 235, 261, 339, 368
maturation process, 293
measurement, 31, 136, 276
measurements, 90, 320, 351, 352, 356, 366, 372, 448
mechanical degradation, 276
mechanical properties, xi, 70, 71, 103, 109, 121, 136, 182, 184, 185, 186, 187, 188, 326, 328, 333, 336, 338
media, x, xvi, xvii, 19, 37, 51, 52, 53, 56, 57, 59, 60, 61, 64, 65, 66, 69, 71, 73, 75, 77, 92, 93, 104, 214, 215, 277, 281, 284, 302, 314, 316, 320, 330, 343, 344, 346, 356, 357, 360, 362, 363, 364, 365, 367, 368, 369, 374, 375, 395, 397, 406, 413, 414, 416, 421, 456, 464
medical, ix, 1, 71, 140, 142, 211, 232, 326, 332, 346, 389, 392, 398, 405, 451, 463
medical care, 332
medicine, xvi, xviii, 61, 77, 104, 194, 201, 220, 231, 262, 267, 268, 278, 335, 388, 393, 433, 451, 452
medium composition, 404
melt, 200, 206, 458
melting, 87
membrane permeability, 35, 178, 380
membranes, ix, 2, 7, 8, 9, 11, 12, 13, 14, 19, 20, 21, 22, 41, 71, 72, 76, 77, 86, 95, 140, 162, 172, 197, 198, 205, 223, 232, 233, 238, 255, 258, 261, 277, 290, 291, 302, 308, 326, 328, 334, 346, 348, 349, 350, 354, 355, 356, 357, 358, 360, 366, 368, 369, 370, 371, 372, 373, 390, 393, 466, 467
memory, 361
mercury, 337, 341
Mercury, 76, 337
mesoderm, 258
mesoporous materials, 373
mesothelium, 258
messengers, 168, 390
metabolism, 165, 172, 174, 382, 391, 395, 396, 418, 421
metabolites, xvii, 164, 165, 166, 167, 168, 169, 178, 387, 390, 391, 392, 394, 395, 396, 400, 406, 413, 414, 416, 417, 419, 420
metabolized, 31
metal complexes, 25, 181, 192
metal ion, 181, 226, 231, 234, 277, 288, 333, 335, 336, 337, 338, 340, 342, 346, 366, 370
metal ions, 181, 226, 231, 234, 277, 288, 333, 335, 336, 337, 338, 340, 342, 346, 366, 370
metal salts, 181
metals, 42, 61, 181, 201, 234, 238, 389, 390
meter, 448
methanol, 71, 72, 105, 106, 107, 111, 244, 344, 359, 360

methodology, xii, 140, 218, 358
methyl group, 284
methyl groups, 284
methylation, 35, 36, 184
methylcellulose, 186, 188, 189, 306, 310
methylene blue, 268
methylene chloride, 211
Mg^{2+}, 233, 357
mice, xi, 29, 45, 85, 95, 97, 302
microbial cells, 181
microbial community, 292
microcalorimetry, 356
microemulsion, 223
micrometer, 352
microorganism, 91, 180, 181, 314, 461
microorganisms, xii, xvi, 4, 21, 86, 90, 160, 179, 180, 181, 183, 196, 197, 226, 321, 335, 340, 342, 389
microscope, 33, 242, 372, 379
microscopy, 33, 41, 185, 322, 351, 369, 372, 373, 379, 380, 462
microspheres, xviii, 43, 45, 46, 72, 177, 195, 203, 217, 223, 322, 333, 334, 368, 373, 451, 452, 453, 454, 455, 456, 457, 458, 459, 460, 461, 462, 463, 464, 465, 466, 467
microstructure, 63, 69, 71, 190, 199, 206, 310, 330
microstructures, 200
migration, xvi, 15, 17, 59, 264, 325, 327, 330
Ministry of Education, 335
mitochondria, 31, 382
mitogen, 168, 390
mixing, 32, 87, 147, 198, 212, 230, 244, 262, 265, 270, 274, 347, 357, 446, 447, 453, 456, 464
model system, xv, 287, 296, 323, 391, 394
modelling, 369
models, x, 85, 93, 132, 258, 263, 288, 290, 292, 331, 406
modifications, xii, xiii, xv, 51, 70, 76, 77, 104, 143, 163, 172, 184, 194, 198, 211, 225, 242, 243, 244, 263, 299, 301, 307, 336, 349, 364, 389
modules, 93
modulus, xi, 103, 109, 110, 121, 122, 182, 184, 186
modus operandi, 38
moisture, xii, 107, 119, 179, 182, 187, 192, 198, 232, 339, 357, 389
moisture content, 198, 357
moisture sorption, 357
molar ratios, 326
mold, 169, 177, 187, 191, 254, 314, 323
mole, 87, 91, 439, 444, 446, 447, 448
molecular biology, 333, 365
molecular mass, 3, 87, 89, 92, 99, 199, 274, 275, 276, 277, 281, 315, 316, 344
molecular mobility, 185
molecular structure, 122, 172, 221, 232, 329, 338, 368, 392
molecular weight, ix, xiii, xv, 1, 2, 3, 5, 6, 11, 13, 14, 19, 22, 23, 28, 30, 34, 36, 43, 45, 46, 48, 52, 65, 87, 88, 93, 95, 96, 98, 153, 170, 171, 173, 176, 177, 179, 180, 181, 187, 188, 192, 194, 195, 197, 203, 214, 215, 216, 218, 226, 232, 235, 262, 272, 288, 300, 302, 306, 313, 314, 315, 316, 317, 320, 321, 322, 323, 347, 358, 367, 378, 385, 388, 389, 391, 394, 401, 452, 453, 457, 460, 461, 466
molecules, ix, xii, 1, 2, 8, 13, 14, 17, 19, 34, 36, 45, 47, 52, 54, 56, 65, 67, 68, 69, 71, 72, 74, 75, 77, 86, 87, 88, 89, 90, 92, 94, 98, 100, 154, 163, 167, 168, 169, 170, 181, 185, 188, 191, 194, 195, 199, 200, 215, 262, 263, 314, 326, 331, 344, 349, 352, 361,뱀378, 390, 391, 414, 426, 434, 459, 463
monolayer, 70, 258, 338, 357
monomers, 3, 4, 35, 51, 59, 60, 264, 303, 349, 395, 425
monosaccharide, 89, 93
Moon, 23, 44, 48, 160, 220, 307
morbidity, 292, 296
mordenite, 71, 72
morphology, x, 43, 85, 91, 195, 196, 214, 215, 216, 222, 246, 265, 349, 372, 373, 454, 459, 462
mortality, 86
Moscow, 100, 269, 313, 377, 384, 385, 449
mRNA, 252, 314, 322, 390, 400, 403
mRNAs, 177
MTS, 32, 38
multilayer films, 467
muscles, 105, 372
mutant, xviii, 100, 169, 423, 427, 430, 431
mycelium, 246, 247, 252, 314, 358
myelin, 390
myelin basic protein, 390
myocardial infarction, 420

N

Na^{+}, 63, 185, 233
NaCl, 74, 76, 88, 233
NAD, 33
nanocomposites, 69, 70, 71, 76, 81, 185, 186, 189, 190, 192
nanofibers, xiii, 185, 209, 210, 211, 215, 216, 219, 220, 223
nanoindentation, xi, 103, 109
nanomaterials, xiii, 31, 32, 209, 210
nanometer, 219
nanometric range, 185

nanoparticles, x, xiii, 27, 28, 32, 35, 37, 38, 39, 40, 41, 42, 45, 46, 47, 48, 49, 71, 72, 186, 187, 188, 194, 195, 199, 202, 206, 209, 210, 211, 212, 213, 214, 216, 217, 218, 219, 220, 221, 223, 293, 297, 303, 307, 309, 310, 341, 348, 360, 368, 374, 463, 464
nanostructured materials, 210
nanostructures, 69, 71, 222
nanosystems, 31
nanotechnology, 214
nanotube, 72
natural compound, x, 51, 253, 254, 322, 462
natural polymers, 55, 104, 182, 226, 348, 452
natural resources, xviii, 406, 423, 430
necrosis, 290, 291, 384, 394
nerve, 259, 267
nerve fibers, 259, 267
Netherlands, 204, 255, 420
network polymers, 333
neurodegeneration, 406
neuropathy, 406
neutral, 2, 14, 30, 34, 44, 61, 63, 69, 76, 99, 185, 218, 223, 271, 300, 303, 309, 374, 446, 453
neutrophils, 262
New Zealand, 187
NH2, xi, 51, 52, 59, 60, 61, 75, 110, 115, 136, 139, 141, 142, 148, 159, 170, 180, 181, 218, 226, 227, 229, 279, 327, 338, 352, 359
nicotinamide, 33
NIR, xi, 103, 114, 115, 127, 136
NIR spectra, 114, 115, 127
nitrates, 434, 440, 441
nitric oxide, xii, 163, 168
nitrogen, 184, 243, 283, 320, 338, 351, 352, 354, 355, 384, 388, 437, 439, 441, 444, 449
NMR, xviii, 35, 188, 229, 234, 324, 327, 329, 334, 433, 436, 439, 447
nodules, 259, 329, 350
nontoxicity, 194, 211, 231, 278, 335
Northern blot, 399
novel materials, 61
Nuclear Magnetic Resonance, 316
nuclei, xvi, 33, 330, 377, 378, 379, 380, 381, 382, 383, 384, 390
nucleic acid, 180
nucleus, 33, 41, 170, 379, 382, 383, 390
nutraceutical, ix, 1, 2, 29, 202, 394, 398, 405
nutrient, 389, 420
nutrients, 165, 171, 181, 194, 197, 199
nutrition, 68, 205, 344, 466

O

ofloxacin, 45
OH, xvi, 8, 60, 61, 64, 73, 110, 112, 115, 210, 212, 227, 229, 262, 325, 326, 327, 330, 359, 445, 452
oil, 166, 195, 196, 197, 198, 199, 200, 205, 206, 207, 211, 216, 339, 340, 342, 344, 356, 357, 359, 360, 365, 373, 379, 454, 457, 459, 462, 465
oil spill, 339
oleic acid, 166, 171, 177, 192, 362, 363, 375, 418
oligomeric structures, 329
oligomerization, 301
oligomers, ix, xviii, 1, 2, 3, 4, 5, 6, 8, 9, 10, 11, 12, 13, 14, 15, 17, 19, 20, 21, 22, 23, 25, 31, 35, 40, 44, 45, 48, 171, 172, 175, 237, 302, 323, 324, 392, 421, 425, 433, 439, 440, 447, 448
oligosaccharide, 5, 23, 71, 86, 88, 172, 177, 221
olive oil, 161, 357, 359, 360, 362
one dimension, 185
operations, 44
opportunities, x, 85, 86, 164, 176, 371
optimization, 221, 223, 405, 419, 420
oral cavity, 200, 465
orchid, 176
organ, 86, 295, 346, 391
organic compounds, 69, 73, 74, 335, 336, 339, 340, 352
organic matter, 73, 74
organic solvents, xvi, 38, 220, 232, 278, 284, 300, 343, 344, 346, 347, 358, 360, 362, 364, 375
organism, 97, 194, 261, 262, 389
organs, xiv, 257, 258, 260, 261, 414, 420
ornamental plants, 176, 314
osmosis, 13, 233
overproduction, 165
overweight, 288
oxidation, xviii, 6, 72, 167, 184, 195, 198, 199, 201, 212, 221, 227, 253, 351, 364, 374, 394, 433, 434, 436, 437, 438, 439, 440, 441, 445, 449, 460
oxidation products, 201
oxidative damage, 5, 394, 460
oxidative destruction, 439, 440, 448
oxidative stress, 256, 385, 401
oxide nanoparticles, 309
oxygen, xii, xvii, 62, 68, 73, 172, 178, 179, 180, 182, 183, 185, 194, 196, 199, 204, 218, 227, 306, 327, 330, 377, 378, 385, 434, 440, 441, 445, 448
ozone, xviii, 302, 391, 433, 434, 448
ozone-oxygen mixture, 448

P

PAA, 41, 238
Pacific, 85, 102
paclitaxel, 45, 221, 393, 404, 414, 420
pain, 260, 452
pain management, 452
palm oil, 357, 360, 365
pancreas, 358, 360, 362, 369
parallel, 10, 63, 69, 141, 143, 144, 148, 290
partition, 156, 361
pathogenesis, 97, 168, 296, 391, 401
pathogens, 168, 177, 183, 192, 196, 240, 252, 253, 254, 255, 314, 378, 384, 390, 391
pathophysiological, 97
pathways, x, xii, 27, 42, 48, 58, 59, 98, 163, 165, 169, 170, 296
peptidase, 401
peptide, 14, 21, 24, 30, 36, 87, 100, 221, 227, 237, 288, 289, 294, 295, 331, 368, 465
peptides, 20, 21, 24, 30, 42, 86, 94, 99, 100, 194, 212, 289, 425, 459
pericardium, xiv, 257, 258, 260
peritoneal cavity, 258, 260, 263, 264
peritoneum, xiv, 257, 258, 261, 263, 265, 266
peritonitis, 258
permeability, xvii, 44, 70, 72, 95, 165, 180, 181, 182, 183, 185, 186, 188, 191, 194, 196, 199, 204, 211, 288, 289, 292, 295, 349, 368, 377, 380, 383, 384, 390, 464, 465
permeation, 36, 45, 187, 191, 289, 294
permission, iv, 7, 8, 459
permit, 354, 364
permittivity, 133
peroxidation, 171, 393
peroxide, 37, 146, 172, 176, 385, 439, 447
petroleum, 226, 360, 455
phage, 22
phagocytosis, 38
pharmaceutical, ix, xi, xv, xvi, 2, 19, 44, 69, 98, 139, 140, 159, 194, 195, 202, 211, 220, 232, 278, 288, 294, 299, 302, 306, 307, 314, 343, 344, 392, 405, 457, 461
pharmaceuticals, ix, xv, 1, 68, 165, 287, 344, 394, 406
pharmaceutics, 28
pharmacology, 333
phase inversion, 75
phase transitions, 212
phenol, 74, 337
phenolic compounds, xii, 163, 168, 240, 417, 420
phenotype, 330
phenotypes, 297
phenylalanine, xii, xiv, 163, 174, 175, 240, 241, 244, 256, 391, 399, 414, 421
phonons, 127
phosphate, 88, 90, 94, 145, 146, 213, 229, 231, 243, 288, 331, 340, 342, 367, 401
phosphates, 347
phospholipids, 99, 258, 288
phosphorylation, 166, 168, 169, 176, 229, 237, 301, 401
photoelectron spectroscopy, 371, 372
photographs, 329, 332, 350
photopolymerization, 309
photosynthesis, 200
phthalates, 337
physical interaction, 301
physical properties, 25, 92, 104, 173, 188, 192, 306, 310, 338, 349, 372, 389
physical structure, 99
physical treatments, xiii, 179
physicochemical characteristics, 288
physicochemical properties, xviii, 31, 35, 70, 76, 174, 190, 335, 361, 372, 391, 433, 454
Physiological, 255
physiological factors, 461
physiological mechanisms, 261
physiology, 255
PI3K, 168
pigmentation, 417, 419
plant growth, 416
plants, xii, xvii, 4, 6, 24, 163, 165, 166, 169, 170, 171, 172, 173, 174, 177, 178, 226, 252, 254, 255, 314, 335, 339, 340, 378, 384, 385, 387, 390, 391, 392, 394, 402, 414, 434
plasma membrane, xvii, 33, 165, 167, 168, 169, 170, 175, 288, 291, 292, 322, 377, 379, 380, 383, 384, 390
plasmid, 35, 43, 44, 48
plasminogen, 258, 263
plasticity, 184
plasticization, 182
plasticizer, 187, 189
platelet aggregation, 97
platelets, 97, 101, 232
platform, 452
pleura, 258, 260
PM, 420
PMMA, 330
polar, 54, 69, 200, 281, 306, 344, 363
polar groups, 200, 306
polarity, xvi, 343, 362
polarization, 132, 133
policy, 420
pollutants, 61, 235, 337, 340, 392, 393, 404

pollution, 142, 424
poly(vinylpyrrolidone), 71
polyacrylamide, 235, 370, 452
polyamine, xii, 94, 140, 163
polyamines, xii, 139, 142, 151, 167, 255
polydispersity, xv, 313, 315, 316, 317, 320, 388
polyelectrolyte complex, ix, xii, 35, 44, 98, 139, 141, 142, 145, 148, 150, 151, 152, 153, 159, 160, 161, 223, 305, 310, 347, 357, 367, 453, 454, 462, 465
polymer chain, 52, 53, 69, 77, 92, 107, 132, 133, 182, 184, 185, 213, 216, 306, 453
polymer chains, 52, 53, 69, 77, 107, 133, 182, 184, 306
polymer films, 113
polymer materials, 226
polymer matrix, 71, 183
polymer molecule, 92, 183, 214, 331
polymer solubility, 274
polymer solutions, 261, 348
polymer structure, 36, 43, 52
polymer systems, 348
polymeric chains, 55, 59, 69, 453
polymeric materials, 172
polymeric matrices, 345
polymeric membranes, 372, 374
polymerization, xvii, 3, 19, 87, 88, 93, 94, 97, 99, 122, 171, 223, 272, 303, 372, 387, 389, 392, 394, 395, 424, 456
polymers, x, xvi, xviii, 27, 28, 31, 32, 35, 36, 37, 38, 41, 42, 44, 55, 61, 66, 67, 74, 77, 78, 87, 88, 104, 105, 109, 116, 117, 119, 121, 126, 127, 129, 131, 133, 136, 137, 180, 182, 185, 190, 197, 214, 215, 237, 255, 262, 272, 276, 288, 294, 295, 303, 306, 307,뫰309, 310, 325, 326, 342, 347, 348, 367, 389, 392, 423, 424, 452, 454, 456, 458, 465
polymethylmethacrylate, 465
polymorphism, 91
polypeptide, 306
polyphenols, xvii, 20, 253, 387, 391, 394, 395, 399, 403, 404, 405, 406
polypropylene, xiv, 257, 259, 261, 263, 265, 267, 268
polysaccharide, ix, xi, xii, xiii, xiv, xv, xviii, 1, 2, 47, 52, 87, 88, 89, 92, 98, 104, 139, 140, 164, 179, 180, 182, 193, 202, 209, 222, 225, 226, 235, 257, 263, 268, 269, 270, 272, 284, 287, 288, 299, 300, 305, 306, 310, 326, 335, 340, 345, 367, 370, 388, 424, 433, 453
polysaccharide chains, 89
Polysaccharides, 104, 137, 202, 237, 238, 255, 262, 407
polystyrene, 378
polyunsaturated fat, 198, 205
polyunsaturated fatty acids, 198, 205
polyvinyl alcohol, 75, 348
pools, xii, xv, 163, 287, 293
population, 142, 293, 340
porosity, 59, 62, 69, 261, 330, 349, 358, 364, 453, 455, 463
potassium, 63, 64, 69, 166, 171, 180, 381, 390
potassium persulfate, 63, 69
potato, xiii, 167, 239, 240, 241, 252, 253, 254, 255, 256
potential benefits, 203
precipitation, 9, 11, 58, 73, 74, 77, 202, 211, 212, 234, 238, 275, 281, 345, 456
preparation, iv, ix, xiii, xvi, xviii, 2, 3, 12, 22, 23, 24, 35, 38, 46, 72, 91, 154, 160, 161, 174, 186, 197, 198, 200, 209, 211, 212, 216, 219, 220, 221, 223, 225, 226, 227, 228, 229, 230, 235, 236, 243, 244, 264, 270, 278, 303, 305, 307, 309, 316, 321, 323, 331, 334, 343, 344, 346, 347, 349, 358, 364, 367, 368, 370, 372, 373, 374, 423, 430, 452, 453, 455, 456, 457, 458, 459, 462, 463, 466, 467
preservation, 25, 42, 164, 165, 173, 179, 190, 254, 261, 368, 406, 414
preservative, 192
prevention, xiv, xvii, 164, 201, 231, 257, 258, 260, 261, 262, 264, 265, 266, 267, 294, 387, 394, 454
priming, 174, 391
principles, 78
probe, 40, 95, 126, 385, 443
probiotic, 196, 197, 203, 204, 460
probiotics, 194, 196, 197, 201, 205, 460, 465
process duration, 17
production costs, 460
progenitor cells, 266
programming, 415
pro-inflammatory, 96
project, 406, 419
proliferation, 45, 219, 262, 327, 329, 330
proline, 165
promoter, 174, 426
propagation, 420
proposition, 181
propylene, 47, 302
prosthetic materials, 333
protection, 97, 100, 171, 173, 174, 175, 177, 197, 201, 227, 278, 378, 389, 393, 394, 434
protective coating, 458
protective mechanisms, 194
protein engineering, 365
protein immobilization, 358
protein synthesis, 252, 390
proteinase, 4, 76, 218, 223, 324

proteins, xvi, xvii, 34, 47, 55, 56, 86, 92, 100, 101, 168, 169, 172, 175, 176, 180, 194, 210, 214, 233, 288, 289, 291, 296, 314, 331, 342, 343, 344, 345, 347, 349, 352, 354, 359, 364, 367, 369, 370, 387, 390, 391, 400, 401, 402, 403, 425, 451, 456, 459, 463
proteoglycans, xv, 287, 288
proteolytic enzyme, 401
protons, ix, 2, 8, 12, 53, 56, 60, 68, 227, 229, 427
Pseudomonas aeruginosa, 95
pseudopodia, 331
public concern, 240
public health, 194, 211
pulp, 75, 192, 393
pure water, 152, 184
purification, 9, 13, 14, 19, 32, 71, 324, 335, 339, 365, 405, 426
purity, ix, 1, 2, 13, 180, 305, 448
PVA, 302, 303, 306, 348, 360, 374
PVP, 71, 306

Q

quality of life, 332
quantification, 41, 204, 352
quantum dot, 72
quantum dots, 72
quaternary ammonium, 22, 184, 190, 226, 229, 230, 231, 232, 237, 238, 293, 458
quercetin, 397, 399, 401
quinacrine, xvii, 377, 379, 381, 383

R

race, 358, 360
radiation, xi, 103, 104, 105, 110, 114, 136, 137, 162, 352
Radiation, v, 103, 104, 105, 113, 174
radical copolymerization, 303
radical polymerization, 227, 308
radicals, 110, 112, 122, 123, 171, 206
radius, 63, 64
Raman spectroscopy, 353
ramp, 415
rationalisation, 19, 22
raw materials, xvi, 343, 344
reactant, 344, 362
reactants, 32, 270, 283
reaction mechanism, 6
reaction medium, 146, 227, 344, 345, 437, 446, 453
reaction rate, 11, 155, 167, 361, 362, 363, 364
reaction temperature, 87, 283, 456
reaction time, 35, 197, 228, 230, 272, 463
reactions, xiv, xviii, 173, 201, 211, 240, 252, 253, 255, 257, 260, 269, 270, 271, 282, 283, 301, 314, 321, 327, 345, 349, 357, 359, 360, 361, 362, 363, 364, 365, 374, 388, 390, 392, 433, 454
reactive groups, 104, 164
reactive oxygen, xvii, 168, 176, 177, 200, 377, 378, 385, 392
reactive sites, 75
reactivity, 181, 283, 352, 394
reagents, xviii, 32, 217, 228, 270, 272, 277, 278, 283, 307, 326, 339, 348, 378, 433, 434, 439, 441, 447, 456, 458, 462
reception, 252, 448
receptors, 42, 48, 96, 101, 168, 169, 172, 390, 391
recognition, 42, 75, 98, 99, 167, 370, 426
recommendations, iv, 174
reconstruction, 260, 261, 262
recovery, 61, 140, 260, 291, 293, 344, 358, 367, 392
recrystallization, 121
recycling, xviii, 423, 430
red blood cells, 36
refractive index, 311
regeneration, xv, xvi, 38, 231, 261, 262, 266, 287, 290, 291, 293, 306, 310, 325, 334, 348
regenerative capacity, 258
regulations, xii, 163
reinforcement, 6, 151, 252
rejection, 233
relaxation, 132, 133, 368
relaxation properties, 368
relevance, 190, 452
reliability, 33, 305
repair, xv, 195, 258, 266, 267, 325
reparation, 160, 223
repression, 252, 401
reproduction, 22, 266
repulsion, 37, 53, 55, 65, 66, 89, 233, 427
requirements, 68, 174, 216, 260, 265, 453, 463
researchers, xviii, 105, 122, 210, 228, 277, 344, 357, 451
residuals, 3
residues, ix, xviii, 1, 2, 36, 37, 42, 87, 88, 93, 142, 252, 262, 314, 423, 425, 427, 428, 430
resistance, xii, xiii, 24, 72, 86, 97, 102, 164, 165, 168, 172, 175, 176, 182, 197, 200, 209, 239, 240, 241, 253, 254, 255, 256, 260, 289, 292, 314, 321, 336, 337, 362, 385, 390, 391, 394
resolution, 262, 352, 354
resources, 226, 401, 420
respiration, xii, 167, 171, 179, 182, 379, 383

response, 96, 98, 162, 168, 169, 177, 218, 252, 256, 258, 261, 267, 290, 291, 295, 303, 306, 314, 329, 333, 349, 382, 390, 391, 401, 414, 416, 419
responsiveness, 101, 236
restoration, 112, 291
restrictions, 93
resveratrol, xvii, 387, 393, 394, 395, 397, 399, 401, 404, 405
retardation, 39
retention volume, 443
retinol, 302
reusability, xi, 139, 140, 148, 158, 159, 359
rhenium, 74
rheology, 63, 69, 70
rings, 115, 182, 327
risk, xiv, 199, 257, 260, 394, 405, 424
risk factors, 199
risks, x, 51, 194, 267, 295, 452
RNA, 28, 48, 255
rodents, 258
rods, 144
room temperature, 107, 119, 132, 141, 147, 213, 229, 242, 264, 271, 273, 339, 378, 440, 446, 447, 457, 458
root, 165, 177, 254, 384, 392, 419, 420
root rot, 254
root system, 392, 420
roots, 165, 170, 174, 255, 314, 321, 392
roughness, 70, 261, 351, 354, 364
routes, 64, 229, 262, 452
routines, 302
Russia, 85, 269, 313, 316, 377, 385

S

safety, 24, 36, 44, 104, 289, 300
salmon, 200, 294
salt formation, 281
salts, ix, x, 1, 2, 12, 51, 62, 70, 172, 184, 187, 197, 203, 215, 277, 281, 302, 320, 346, 356, 389, 465
SAP, 166
saponin, 170
saturation, 87, 89, 115, 181
Saudi Arabia, 27
scanning calorimetry, 356
scanning electron microscopy, xi, xiv, 103, 195, 239
scavengers, 206, 393
science, 79, 80, 202, 206, 233, 341, 421, 434
seafood, xvi, 184, 343
secondary metabolism, 169, 172
secrete, 258
secretion, 96, 288
sedatives, 393
sedimentation, 53, 62, 87, 88
seed, 178, 195, 240, 293, 389, 415
seedlings, 177, 378
selectivity, 13, 15, 72, 76, 336, 344, 405
self-assembly, 213, 305, 309, 310, 458
SEM micrographs, 232
semiconductor, 131
semiconductors, 127
senescence, xii, 163, 167, 171
sensitivity, 227, 231, 232, 236, 237, 252, 304
sensors, xi, 70, 104, 105, 137, 169, 303
sepsis, 86
sequencing, 400, 403
serum, 31, 89, 97, 145, 196, 214, 221, 231, 244, 264, 358
serum albumin, 89, 145, 214, 221, 231, 244, 358
shape, 44, 195, 200, 232, 330, 349
shear, xiv, 269, 270, 272, 276, 277, 281, 282, 283, 284
shear deformation, xiv, 269, 270, 272, 276, 277, 281, 282, 283, 284
sheep, 30, 267
shelf life, xii, 165, 171, 173, 175, 177, 179, 182, 186, 187, 188, 190, 191, 340, 460
shock, 86, 97, 98
shoot, 165, 177
shoots, 165
showing, 39, 41, 110, 171, 262, 265, 354
shrimp, 25, 192, 296
side chain, 232, 237, 309, 373, 392, 427
side effects, xv, 287, 289, 293
SIGMA, 315, 316
signal transduction, 96, 168, 169, 172, 252, 321, 390
signaling pathway, 169, 170, 390
signals, 40, 169, 174, 175
signs, 290, 291
silane, 72, 326
silanol groups, 69, 326, 327, 328
silica, xvi, 71, 75, 165, 219, 220, 325, 327, 329, 333, 334
silicon, 62, 159, 329
silver, 111, 293, 297, 368, 467
Singapore, 222
SiO2, 327
siRNA, 48
skeleton, 301, 327, 345
skin, 183, 188, 263, 310, 372, 389, 393, 421, 454, 462
sludge, x, 51, 63, 74, 342
small intestine, xvii, 199, 289, 387, 394
SO42-, 233
society, xviii, 194, 332, 423, 430

sodium, xiv, 35, 36, 37, 47, 63, 74, 75, 162, 171, 195, 198, 200, 201, 206, 213, 216, 223, 229, 241, 243, 264, 269, 270, 278, 302, 345, 347, 358, 434, 435, 436, 437, 438, 439, 441, 446, 447, 456, 461
sodium dodecyl sulfate (SDS), 201
sodium hydroxide, xiv, 35, 171, 198, 200, 216, 229, 264, 269, 270, 302, 345
softener, 306
software, 93, 316, 415
sol-gel, 71, 162
solid phase, 277, 278, 283, 284
solid state, 110, 271, 276, 389
solid surfaces, 352
solidification, 458
solubility, xiii, xv, 2, 30, 35, 52, 70, 73, 75, 76, 93, 104, 142, 181, 184, 194, 199, 225, 226, 227, 228, 229, 230, 231, 232, 262, 271, 272, 274, 275, 276, 281, 284, 299, 302, 307, 308, 323, 344, 389, 394, 453
solvents, xvi, 100, 111, 199, 215, 230, 244, 245, 281, 284, 343, 344, 361, 362, 374, 375, 388
sorption, 21, 218, 277, 336, 337, 341, 349, 370, 454
Spain, 315, 451
speciation, 21
species, xvi, xvii, 7, 37, 51, 52, 53, 54, 56, 59, 61, 64, 69, 73, 74, 77, 156, 168, 169, 177, 192, 200, 258, 260, 262, 263, 325, 329, 351, 364, 372, 377, 378, 385, 388, 392, 401, 406, 414, 418, 454, 461
spectroscopy, xi, 103, 327, 352, 353
spin, 24
spleen, 260
sponge, 41, 265
spore, xiv, 239, 241, 245, 246, 252, 314, 321, 322
stability, xi, xii, xvi, 30, 40, 41, 44, 45, 56, 88, 103, 136, 139, 143, 151, 158, 159, 160, 171, 180, 195, 197, 198, 203, 204, 205, 206, 210, 211, 212, 213, 216, 218, 220, 262, 265, 290, 310, 331, 339, 340, 343, 344, 347, 348, 349, 354, 356, 359, 360, 361, 364, 366, 373, 394, 401, 453, 454, 459, 461, 464
stabilization, 54, 55, 56, 58, 65, 160, 192, 205, 345, 347, 369
stabilizers, 35, 284
stable complexes, x, 85, 97
standard deviation, 429
standard error, 245, 246, 249, 318, 319
starch, 55, 104, 186, 187, 190, 305, 310, 430
state, x, 51, 52, 53, 55, 56, 70, 73, 87, 101, 140, 190, 200, 270, 281, 290, 327, 334, 349, 362, 366, 370, 372, 378, 379, 380
states, 70, 329, 351, 364, 372
steel, 144, 243, 448
sterile, 32, 242, 243, 316
sternum, 260
steroids, 463
stimulant, 267
stimulus, 256, 303
stoichiometry, x, 55, 74, 85, 87, 89, 95, 97
stomata, 165, 378
storage, xiii, 12, 98, 167, 174, 175, 176, 177, 178, 186, 187, 189, 191, 197, 199, 200, 201, 203, 204, 205, 213, 218, 239, 240, 253, 254, 255, 256, 348, 357, 359, 360, 389, 460, 465
stress, 23, 165, 169, 174, 177, 290, 295, 384, 401
stretching, 110, 115, 435
strong interaction, 61, 62, 115, 136
structural barriers, 253
structural characteristics, 6, 262
structural defects, 62, 65
structural knowledge, 98
structure formation, 340
styrene, 61, 74
substitutes, 88, 232
substitution, 35, 40, 62, 93, 95, 188, 227, 230, 231, 232, 234, 270, 279, 280, 282, 283, 284, 302, 349, 427, 428
substrate, 54, 59, 74, 100, 148, 157, 158, 161, 167, 183, 189, 210, 243, 320, 344, 347, 359, 360, 361, 362, 363, 389, 426, 427, 430
substrates, 73, 76, 77, 210, 344, 361, 362, 389, 401, 426, 456
sucrose, 87, 255, 358, 360, 405
sulfate, 21, 37, 74, 75, 122, 123, 141, 195, 237, 295, 304, 309, 347
sulfur, 228, 338
sulfuric acid, 23, 71
sulphur, 338
Sun, 184, 190, 192, 206, 237, 240, 256, 296, 297, 341
supplementation, 392
supplier, 227, 231
suppliers, 13, 180
suppression, 96, 173, 174
surface area, xiii, 144, 158, 209, 210, 211, 349, 351, 354, 364
surface energy, 192, 452
surface modification, 371, 372
surface properties, x, 27, 187, 189, 192, 349
surface reactions, 184
surface structure, 330, 350, 354
surface tension, 183, 215
surfactant, xii, 75, 163, 183, 187, 196, 200, 211, 213, 216, 258
surfactants, 74, 226, 234, 281
survival, 45, 197, 203, 204, 205, 292, 340, 391, 460
survival rate, 197, 461
susceptibility, 187, 197, 383, 392

suspensions, xvii, 53, 62, 63, 68, 73, 74, 171, 172, 387, 392, 395, 396, 397, 398, 399, 400, 401, 402, 403, 404, 406, 458
sustainability, 13, 19, 404
suture, 263, 266
swelling, xi, xiv, 52, 61, 103, 111, 231, 236, 239, 252, 265, 270, 277, 311, 338, 367, 369, 372, 453, 461, 463
symptoms, 86, 382
syndrome, 86
synergistic effect, 62, 73, 232, 294
synthetic analogues, 461
synthetic polymers, xii, 55, 179, 180, 185, 214, 306, 326, 452

T

talc, 263, 264
tannins, 160
target, xviii, 28, 86, 95, 168, 181, 194, 199, 214, 303, 314, 399, 451, 452
techniques, x, xi, xv, xvi, 35, 41, 85, 139, 140, 143, 199, 210, 269, 309, 339, 343, 373, 400, 456, 458, 462, 465
technologies, 19, 38, 235, 463
technology, ix, xiii, xvi, xvii, 2, 8, 9, 13, 14, 19, 20, 23, 75, 79, 82, 160, 193, 198, 202, 205, 206, 214, 233, 235, 255, 305, 343, 404, 413, 414, 460, 462, 464
teflon, 162
TEM, xiv, 42, 212, 214, 239, 248
temperature, x, 71, 73, 76, 85, 87, 97, 98, 99, 127, 131, 132, 133, 135, 136, 143, 145, 147, 155, 159, 165, 174, 175, 190, 194, 195, 196, 199, 201, 213, 214, 218, 227, 232, 234, 262, 272, 273, 279, 283, 303, 308, 330, 348, 349, 355, 356, 357, 359, 360, 364, 368, 373,뱀415, 439, 443, 444, 446, 448, 453, 458, 460, 467
temperature dependence, 132
template polymerization technique, 161
tensile strength, 70, 183, 185, 186, 199, 264
tension, 183, 197
TEOS, 327
terminals, 76
terpenes, 417
testing, x, 27, 31, 32, 43, 201, 452
tetraethoxysilane, 71
tetrahydrofuran, 71, 222
textiles, 42, 68, 462
texture, 206, 260, 460
TGA, xi, 36, 103, 108, 119, 120, 355, 356
therapeutic agents, 42, 461, 463
therapeutic effects, 453, 463
therapeutic use, 289
therapeutics, 293, 463
therapy, 98, 220, 266, 293
thermal decomposition, 355, 356
thermal degradation, 108, 187
thermal properties, 222
thermal stability, xi, xii, 35, 70, 71, 73, 139, 141, 142, 143, 144, 148, 151, 155, 159, 161, 185, 186, 213, 218, 232, 355, 358, 359
thermal treatment, 206, 350, 456
thermodynamic equilibrium, 362
thermograms, 107, 108, 121, 151, 355
thermogravimetric analysis, 107, 356
thermostability, 344
thin films, 104
thinning, 56
thorax, 258
thrombin, 232
time frame, 456
time periods, 356
time use, xii, 139, 180
tissue, ix, xiv, xv, xvii, 29, 44, 47, 48, 78, 167, 191, 243, 244, 252, 255, 257, 258, 259, 260, 261, 262, 263, 266, 267, 268, 287, 288, 290, 295, 302, 303, 306, 311, 325, 326, 330, 332, 333, 334, 348, 369, 370, 378, 382, 394, 413, 414, 420, 421, 452
titania, xvi, 325
titanium, 371
TLR, 96
TLR4, 96
TMC, 30, 34, 35, 41, 42, 45
TNF, 96, 101
TNF-α, 96, 101
tobacco, 14, 20, 170, 174, 255, 256, 321, 384, 394, 421
tocopherols, 182
total energy, 444
toxic effect, 36, 86, 97, 291
toxic metals, 76
toxicity, xv, 28, 35, 36, 37, 38, 39, 40, 42, 46, 87, 95, 98, 100, 217, 219, 226, 262, 270, 275, 287, 291, 296, 300, 452, 456, 462
toxicology, xviii, 433
TPA, 421
trace elements, 231, 288
trachea, 35, 46
trafficking, x, 27, 41, 48
transcription, xii, 6, 163, 170, 390, 391, 403
transcription factors, 170
transcripts, 176
transduction, 96, 169, 177, 296
transesterification, 344, 356, 358, 365, 375
transfection, 31, 35, 39, 40, 41, 42, 43, 46, 48

transformation, 8, 12, 19, 22, 64, 278, 352
transformations, 277, 361, 364
transition metal, 181
transition metal ions, 181
transition temperature, 182, 191, 373
translation, 100
translocation, 288, 295
transmission, 13, 115, 182, 266
Transmission Electron Microscopy, 242
Transmission Electron Microscopy (TEM), 242
transparency, 199
transpiration, 169
transport, x, 17, 27, 30, 47, 48, 127, 194, 240, 288, 294, 466
trauma, xiv, xv, 257, 258, 260, 265, 267, 325
trial, 158, 405
trifluoroacetic acid, 215
triggers, 166, 261, 390
triglycerides, 288, 344, 358
Trinidad, 420
trypsin, 213, 221
tumor, 24, 38, 47, 86, 220, 266, 294, 302
tumor cells, 38, 47, 302
tumor necrosis factor, 86, 266
tumorigenesis, 421
tumors, xv, 287, 294, 297
tumours, 37
tunneling, 132, 373
turbulence, 7
turnover, 172, 200, 361
tyrosine, xii, 163, 175, 421
Tyrosine, 176

U

UK, 25, 78, 138, 191, 206
ulcer, 414, 465
ultrastructure, 382
uniform, 91, 141, 143, 160, 185, 200, 215, 242, 247, 248, 265, 272, 273, 349, 354, 456
updating, 441
uranium, 74
urea, 88, 219, 372
urinary bladder, xv, 287, 288, 289, 290, 291, 292, 294, 295, 296
urinary tract, 292, 296, 297
urinary tract infection, 292, 296, 297
urine, 289, 290, 293
urothelium, 289, 290, 291, 293, 294, 295, 296
USA, 78, 80, 137, 145, 242, 255, 296, 323
USSR, 285
UV, xi, 71, 77, 103, 114, 115, 127, 136, 243, 264, 267, 391, 394
UV light, 391, 394

V

vaccine, 20, 37, 458, 466
vacuum, 200, 315, 378, 457
valence, 63
Valencia, 365, 369, 370, 372
validation, 218
vapor, 172, 182, 186
variables, 198, 456
variations, 143, 263, 265, 303, 307, 315, 328, 392
varieties, 4
vascular bundle, 165
vascularization, xvi, 326
vector, x, 28, 46, 85, 86, 232
vegetable oil, 199, 365
vegetables, xii, xiv, 163, 164, 165, 167, 173, 174, 175, 177, 179, 184, 188, 192, 201, 240, 253, 254, 256, 394
vehicles, 197
vein, 5, 23, 43
velocity, 15, 87, 88, 148, 155, 157, 158, 311, 362
vinyl monomers, 122
viral gene, 30, 43
viral infection, 6, 390
viruses, xvii, 5, 42, 170, 384, 387
viscera, 258, 261
viscosity, 2, 3, 4, 20, 24, 87, 153, 154, 213, 214, 215, 216, 263, 272, 276, 277, 280, 281, 288, 311, 391, 436, 437, 438, 439, 440, 441, 448
visualization, 40
vitamin B1, 368
vitamin B12, 368
vitamin B2, 195, 203
vitamin C, 195, 203, 460, 464
Vitamin C, 195
vitamin E, 171
vitamins, 194, 288, 320, 460
volatility, 200

W

waste, xviii, 61, 70, 73, 74, 270, 337, 338, 388, 404, 423, 425, 430
waste water, 337, 338
wastewater, xiii, xiv, xvi, 68, 209, 225, 226, 230, 231, 233, 234, 235, 269, 335, 336, 337, 339, 389
water vapor, 172, 180, 182, 183, 184, 185, 186, 189, 277
water-soluble polymers, 302, 456
wave number, 115, 136

WAXS, 99
weight loss, 107, 119, 355
weight ratio, 305
wettability, 183
working conditions, 215
worldwide, 140, 180, 292
wound healing, 240, 262, 263, 267, 288, 294

X

xanthan gum, 454
xanthophyll, 175, 177
XPS, 237, 352
X-ray diffraction, xi, 35, 80, 103, 280
X-ray diffraction (XRD), xi, 35, 103
X-ray photoelectron spectroscopy (XPS), 352
XRD, xi, 35, 103, 106, 107, 112, 117, 127, 354

Y

yeast, xviii, 315, 423, 425, 426, 427, 430
Yeasts, 181, 410
yield, xvii, 3, 31, 32, 34, 218, 272, 274, 275, 302, 326, 327, 344, 358, 362, 387, 392, 404, 405, 426, 439, 440, 446, 447, 448
yolk, 454, 465

Z

zeolites, 51, 62, 72
zirconium, 162